# Modern Matrix
# Algebra

# Modern Matrix Algebra

**David R. Hill**
*Temple University*

**Bernard Kolman**
*Drexel University*

PRENTICE HALL, Upper Saddle River, New Jersey 07458

*Library of Congress Cataloging-in-Publication Data*

Hill, David R. (David Ross)
    Modern matrix algebra / David R. Hill, Bernard Kolman.
        p.    cm.
    Includes index.
    ISBN 0-13-948852-9
    1. Matrices.    2. Algebra.    I. Kolman, Bernard.
II. Title.
    QA188 .H55   2001
    512.9' 434--dc21                                        00-037489

**Acquisitions Editor:** George Lobell
**Production Editor:** Betsy A. Williams
**Senior Managing Editor:** Linda Mihatov Behrens
**Executive Managing Editor:** Kathleen Schiaparelli
**Assistant Vice President of Production and Manufacturing:** David W. Riccardi
**Marketing Manager:** Angela Battle
**Marketing Assistant:** Vince Jansen
**Manufacturing Buyer:** Alan Fischer
**Manufacturing Manager:** Trudy Pisciotti
**Supplements Editor/Editorial Assistant:** Gale Epps
**Art Director/Cover Designer:** Ann France/Jon Boylan
**Director of Creative Services:** Paul Belfanti
**Assistant to Art Director:** John Christiana
**Art Manager:** Gus Vibal
**Art Editor:** Grace Hazeldine
**Interior Designer:** Ann France
**Cover Designer:** Kiwi Design
**Cover Image:** Patrick Henry Bruce (United States, 1881–1936), "Composition II",
© 1916, oil on canvas, signed lower right: "Bruce", Yale University Art Gallery,
New Haven, Connecticut, Gift of Collection Societe Anonyme

 © 2001 by Prentice-Hall, Inc.
Upper Saddle River, New Jersey 07458

Printed in the United States of America
10   9   8   7   6   5   4   3   2   1

ISBN 0-13-948852-9

Prentice-Hall International (UK) Limited, *London*
Prentice-Hall of Australia Pty. Limited, *Sydney*
Prentice-Hall Canada Inc., *Toronto*
Prentice-Hall Hispanoamericana, S.A., *Mexico*
Prentice-Hall of India Private Limited, *New Delhi*
Prentice-Hall of Japan, Inc., *Tokyo*
Pearson Education Asia Pte. Ltd.
Editora Prentice-Hall do Brasil, Ltda., *Rio de Janeiro*

MATLAB is a registered
trademark of the MathWorks, Inc.

The MathWorks, Inc.
24 Prime Park Way
Natick, MA 01760-1520
Phone: 508-647-7000
FAX: 508-647-7001
info@mathworks.com
http://www.mathworks.com

*To Suzanne*

D. R. H.

*To Judith*

B. K.

# CONTENTS

# PREFACE

Linear algebra is an important course for an increasing number of students in diverse disciplines. It is the course in which the majority of students have the opportunity to learn abstract concepts and deal with a wide variety of applications that can be drawn from fields such as physics, chemistry, biology, geology, economics, engineering, computer science, psychology, and sociology.

This book provides an introduction to the basic algebraic, geometric, and computational tenets of linear algebra at the sophomore level. It also stresses the abstract foundations of the topics involved, using a spiral approach. To this end we introduce many of the foundations of linear algebra by first dealing with matrices, their properties, and the algebra associated with them, while showing the geometric foundations of the topics. It is important for students to be able to associate a visual perception with certain topics, and we try to emphasize that algebra and geometry are complementary to one another. We next use a linear system of equations to develop the basic abstract notions of linear algebra. Rather than use abstract vector spaces, we concentrate on the subspaces associated with a linear system and its coefficient matrix. Chapters 1 through 4 establish these foundations and provide the opportunity to explore a variety of applications that illustrate the underlying concepts.

The second level of the spiral involves Chapters 5 through 7, in which the major ideas of beginning linear algebra are developed from a more abstract point of view. At this point the student has worked with the notions and jargon in the special cases of matrices and the subspaces associated with a linear system. Hence the reformulation of the topics to the general setting of abstract vector spaces is not as intimidating from either a conceptual or algebraic standpoint. We feel that the experience obtained in Chapters 1 through 4 will make the assimilation of ideas in the more abstract setting easier since we have established conceptual hooks for all the major concepts.

*Throughout the text there is the opportunity to use* MATLAB® *to complement the topics.* **It is not required**, but in today's computer-oriented society and workplace our experience is that it adds a level of understanding through hands-on involvement with computations and concepts. This certainly is in line with the Linear Algebra Curriculum Study Group recommendations and is supported by the experiences of a wide segment of mathematical educators. The idea is to use software tools within MATLAB to enhance the presentation of topics and provide opportunities for students to explore topics without the algebraic overhead involved in solving linear systems. To this end there are tools available for both the instructor and the student.

Our approach is somewhat different from that developed elsewhere. However, *the major objective is to present the basic ideas of linear algebra in a manner that the student will understand by utilizing both algebraic and geometric constructs.*

## EXERCISES

The exercises are an integral part of this text. Many of them are numerical in nature, whereas others are more conceptually oriented. The conceptual exercises often call for an explanation, description, or supporting reasoning to establish credence of a statement. It is extremely important in our technological age to be able to communicate with precision and to express ideas clearly. Thus exercises of this type are designed to sharpen such skills. An instructor can choose whether to incorporate complex numbers and matrices as part of the course. While **the use of MATLAB is optional**, we highly encourage both instructors and students to incorporate it in some way within the instructional and study processes associated with this text. Hence, where appropriate we have included MATLAB exercises, denoted ML exercises, in many sections. These are grouped separately from the standard exercises. If an instructor desires to use MATLAB as part of the course, then naturally it can be used with the computationally oriented standard exercises.

## PRESENTATION

We have learned from experience that at the sophomore level, abstract ideas must be introduced gradually and must be supported by firm reasoning. Thus we have chosen to use a spiral approach so that basic linear algebra concepts are seen first in the context of matrices and solutions of linear systems, and are then extended to a more general setting involving abstract vector spaces. Hence there is a natural grouping of the material. Chapters 1 through 4 emphasize the structure and algebra of matrices and introduce the requisite ideas to determine and analyze the solution space of linear systems. We tie the algebra to geometric constructs when possible and emphasize the interplay in these two points of view. We also provide the opportunity to use the notions of linear algebra in a variety of applications. (A list of applications appears on pages xiv–xv.) Chapters 5 through 7 extend the concepts of linear algebra to abstract vector spaces. We feel that the experiences from Chapters 1 through 4 make the assimilation of ideas in this more abstract setting easier. Instead of encountering a dual battle with abstract vector space notions and the concepts for manipulating such information, there is only the generalization to objects that are not necessarily vectors or matrices. Since we have already established the major concepts of closure, span, linear independence/dependence, subspaces, transformations, and eigenvalues/eigenvectors, the accompanying language and major manipulation steps are familiar.

We have designed features throughout the text to aid students in focusing on the topics and for reviewing the concepts developed. At the end of each section is a set of True/False Review Questions on the topics introduced. In addition, there is a list of terminology used in the section together with a set of review/discussion questions. This material can be used by the student for self study or by the instructor for writing exercises. At the end of each chapter is a Chapter Test, which provides additional practice with the main topics.

Rather than include only answers to exercises in a section at the back of the book, we have also included complete solutions to selected exercises. A number of these provide guidance on the strategy for solving a particular type of exercise, together with comments on solution procedures. Thus our **Answers/Solutions to Selected Exercises** incorporates features of a student study guide. The answers to all of the True/False Review Questions and the Chapter Tests are included at the back of the book.

## MATERIAL COVERED

Chapters 1 through 4 cover the major topics of linear algebra using only matrices and linear systems. A goal is to establish early on the language and concepts of (general) linear algebra with these simple, pervasive, and concrete constructs. Hence the order of topics is not traditional, yet carefully builds the solution space of a linear system, basic notions of linear transformations, and the eigen concepts associated with matrices. There is ample flexibility in the topics, in Chapters 3 and 4 especially, to develop a variety of paths for instruction. By careful selection of materials an instructor can also incorporate selected sections of Chapters 5 through 7, instead of using the linear ordering given in the table of contents. Varying the amount of time spent on the theoretical material can readily change the level and pace of the course.

**Chapter 1** deals with matrices, their properties, and operations used to combine the information represented within this simple data structure. We use the simple versatile matrix to introduce many of the fundamental concepts of linear algebra that are later extended to more general settings. **Chapter 2** develops techniques for solving linear systems of equations and for studying the properties of the sets of solutions. In particular, we investigate ways to represent the entire set of solutions of a linear system with the smallest amount of information possible. This adds an important concept to those previously developed. **Chapter 3** introduces the basic properties of determinants. The concepts in this chapter can be viewed as extensions of previous material or can be treated more traditionally using recursive computation. The instructor has flexibility in choosing the extent to which the variety of material in this chapter is used. **Chapter 4** considers the concepts of eigenvalues and eigenvectors of a matrix. We initially provide a geometric motivation for the ideas and follow this with an algebraic development. In this chapter we do consider that a real matrix can have complex eigenvalues and eigenvectors. We also have presented material that uses eigen information for image compression, and hence efficient transmission of information. This approach serves to illustrate geometrically the magnitude of the information that is distilled into the eigenvalues and eigenvectors. Optionally the important notion of singular value decomposition can be discussed. **Chapter 5** introduces abstract vector spaces. Here we use the foundations developed for matrices and linear systems in previous chapters to extend the major concepts of linear algebra to more general settings. **Chapter 6** discusses inner product spaces. The material here is again an extension of ideas previously developed for vectors in $R^n$ (and $C^n$). We revisit topics in more general settings and expand a number of applications developed earlier. We investigate the important notion of function spaces. **Chapter 7** presents material on linear transformations between vector spaces that generalize the matrix transformations developed in Chapter 1. We further analyze properties and associated subspaces of linear transformations. We also extend the notion of eigenvalues and eigenvectors to this more general setting. As an application we present an introduction to fractals, which requires the use of MATLAB in order to provide a visualization component to this important area of mathematics. **Chapter 8** provides a quick introduction to MATLAB and to some of the basic commands and language syntax needed to effectively use the optional ML exercises included in this text. **Appendix 1** introduces in a brief but thorough manner complex numbers. **Appendix 2** reviews material on summation notation.

## MATLAB SOFTWARE

MATLAB is a versatile and powerful software package whose cornerstone is its linear algebra capabilities. MATLAB incorporates professionally developed computer routines for linear algebra computation. The code employed by MATLAB is written in the C language and is upgraded as new versions are released. MATLAB is available from The MathWorks, Inc., 24 Prime Park Way, Natick, MA 01760, (508-653-1415), e-mail info@mathworks.com and not distributed with this book or the instructional routines developed for the ML exercises.

The instructional M-files that have been developed to be used with the ML exercises of this book are available from the following Prentice Hall Web site: www.prenhall.com/hillkolman . (A list of these M-files appears in Section 8.2.) These M-files are designed to transform MATLAB's capabilities into instructional courseware. This is done by providing pedagogy that allows the student to interact with MATLAB, thereby letting the student think through the steps in the solution of a problem and relegating MATLAB to act as a powerful calculator to relieve the drudgery of tedious computation. Indeed, this is the ideal role for MATLAB in a beginning linear algebra course, for in this course, more than in many others, the tedium of lengthy computations makes it almost impossible to solve modest-sized problems. Thus by introducing pedagogy and reining in the power of MATLAB, these M-files provide a working partnership between the student and the computer. Moreover, the introduction to a powerful tool such as MATLAB early in the student's college career opens the way for other software support in higher-level courses, especially in science and engineering.

## SUPPLEMENTS

**Instructor's Solutions Manual (0-13-019725-4):**   Contains answers/solutions to all exercises and is available (to instructors only) from the publisher at no cost.

**Optional combination packages:**   Provide a MATLAB workbook at a reduced cost when packaged with this book. Any of the following three MATLAB manuals can be wrapped with this text for a small extra charge:

- Hill/Zitarelli, *Linear Algebra Labs with MATLAB*, 2/e (0-13-505439-7)
- Leon/Herman/Faukenberry, *ATLAST Computer Exercises for Linear Algebra* (0-13-270273-8)
- Smith, *MATLAB Project Book for Linear Algebra* (0-13-521337-1)

## ACKNOWLEDGMENTS

We benefited greatly from the class testing and useful suggestions of David E. Zitarelli of Temple University. We also benefited from the suggestions and review of the MATLAB material provided by Lila F. Roberts of Georgia Southern University.

We are pleased to express our thanks to the following reviewers: John F. Bukowski, Juniata College; Kurt Cogswell, South Dakota State University; Daniel Cunningham, Buffalo State College; Rad Dimitric, University of California, Berkeley; Daniel Kemp, South Dakota State University; Daniel King, Sarah Lawrence College; George Nakos, US Naval Academy; Brenda J. Latka, Lafayette College; Cathleen M. Zucco Teveloff, State University of New York, New Paltz; Jerome Wolbert, University of Michigan at Ann Arbor.

The numerous suggestions, comments, and criticisms of these people greatly improved this work.

We thank Dennis Kletzing, who typeset the entire manuscript.

We also thank Nina Edelman, Temple University, for critically reading page proofs. Thanks also to Blaise DeSesa for his aid in editing the exercises and solutions.

Finally, a sincere expression of thanks to Betsy Williams, George Lobell, Gale Epps, and to the entire staff at Prentice Hall for their enthusiasm, interest, and cooperation during the conception, design, production, and marketing phases of this book.

D.R.H.
B.K.

# LIST OF APPLICATIONS

# 1

# *MATRICES AND MATRIX ALGEBRA*

The revolution of the latter part of the twentieth century that continues into the twenty-first century is the information age. The collection, organization, transmission, and interpretation of information are the cornerstones of science, industry, business, and government. Each of these aspects of information management uses mathematics as a tool; sampling techniques are used to collect data, data structures and encoding procedures are used to organize information, transformation/compression schemes are employed to distill information for rapid transmission, and a wide variety of mathematical models can be constructed to provide the means to analyze and visualize information content. The role of mathematics is to make the invisible (information) visible.[1]

Mathematical models can be as simple as tables, charts, and graphs or as complex as DNA chains. In this chapter we investigate the contributions of the simple versatile structure known as a **matrix**. Basically a matrix is an arrangement of rows and columns that provides an organizational structure for data. We investigate how such structures can be manipulated. In later chapters we show how certain manipulations can be used to reveal intrinsic properties of the information within a matrix. The matrix is just one of the structures that "fits" a variety of situations and hence provides a means to unify the study of properties common to diverse applications.

```
»> A=[1 2;0 6;−5 7]
A=
    1    2
    0    6
   −5    7
»> row2=A(2,:)
row2=
    0    6
»> col1=A(:,1)
col1=
    1
    0
   −5
»> diagonal=diag(A)
diagonal=
    1
    6
```

Matrices in MATLAB

[1] Keith Devlin, *Life by the Numbers*, John Wiley and Sons, Inc., 1998.

Since matrices provide an important structure for organizing data, it is not surprising that computer programs take advantage of such a device. In this book we refer to the computing environment MATLAB®. **The MATLAB material is completely optional** but does provide an instructional opportunity for those wanting to incorporate the use of technology in linear algebra. Where appropriate we will display features that show how a topic can use MATLAB and its graphics. See Chapter 8 for a brief introduction to MATLAB.

## 1.1 ■ MATRICES

A matrix is an arrangement of rows and columns that organizes information. It is often called the mother of all data structures. Examples 1 through 5 illustrate a variety of matrix structures. We give a formal definition later in this section. We also introduce a notation for matrices and the data entries within a matrix that will be used throughout this book.

**EXAMPLE 1**   A table of values that is a sample of the function $y = x^2$ can be used to generate a graph that provides a visual representation of the curve described by the function. Such a table is a matrix.

| $x$ | $-2$ | $-1$ | 0 | 0.5 | 1 | 1.5 |
|---|---|---|---|---|---|---|
| $y$ | 4 | 1 | 0 | 0.25 | 1 | 2.25 |

■

**EXAMPLE 2**   United States postal rates for first class mailings can also be shown in a matrix.

| wt = weight in oz. | wt $\leq$ 1 | 1 < wt $\leq$ 2 | 2 < wt $\leq$ 3 | 3 < wt $\leq$ 4 |
|---|---|---|---|---|
| rate | $0.33 | $0.55 | $0.77 | $0.99 |

■

**EXAMPLE 3**   The wind chill table that follows is a matrix.

| mph | °F 15 | 10 | 5 | 0 | $-5$ | $-10$ |
|---|---|---|---|---|---|---|
| 5 | 12 | 7 | 0 | $-5$ | $-10$ | $-15$ |
| 10 | $-3$ | $-9$ | $-15$ | $-22$ | $-27$ | $-34$ |
| 15 | $-11$ | $-18$ | $-25$ | $-31$ | $-38$ | $-45$ |
| 20 | $-17$ | $-24$ | $-31$ | $-39$ | $-46$ | $-53$ |

A combination of air temperature and wind speed makes a body feel colder than the actual temperature. For example, when the temperature is 10°F and the wind is 15 miles per hour, this causes a body heat loss equal to that when the temperature is $-18$°F with no wind.

■

EXAMPLE 4   It is known that the three points $(1, -5)$, $(-1, 1)$, $(2, 7)$ lie on a parabola, $p(x) = ax^2 + bx + c$. In order to determine its $y$-intercept, we first want to find the coefficients $a$, $b$, and $c$. We note that each of the three points must satisfy the quadratic equation, so we have the following **system of linear equations**:

$$
\begin{aligned}
a(1)^2 + b(1) + c &= -5 \\
a(-1)^2 + b(-1) + c &= 1 \qquad \text{or} \\
a(2)^2 + b(2) + c &= 7
\end{aligned}
\qquad
\begin{aligned}
a + b + c &= -5 \\
a - b + c &= 1 \\
4a + 2b + c &= 7.
\end{aligned}
$$

Observe that like unknowns are aligned in a column form, so a matrix representing this system of equations will be taken as follows:

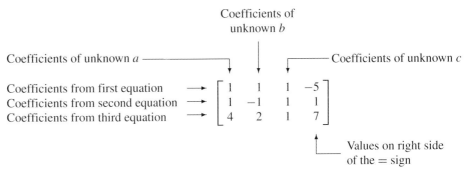

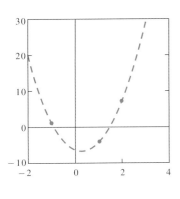

FIGURE 1

The solution of the linear system of equations will provide values for the coefficients $a$, $b$, and $c$. We can then compute the $y$-intercept. Figure 1 shows the three data points and the graph of the quadratic polynomial obtained from solving the system of linear equations.   ■

EXAMPLE 5   The director of a trust fund has \$100,000 to invest. The rules of the trust state that both a certificate of deposit (CD) and a long-term bond must be used. The director's goal is to have the trust yield \$7800 on its investments for the year. The CD chosen returns 5% per annum and the bond 9%. The director determines the amount $x$ to invest in the CD and the amount $y$ to invest in the bond as follows:

Total investment = \$100,000   gives equation   $x + y = 100,000$
Investment return = \$7800   gives equation   $0.05x + 0.09y = 7800$
                                                                                    (1)

The linear system of equations in (1) can be represented in a matrix like that used in Example 4:

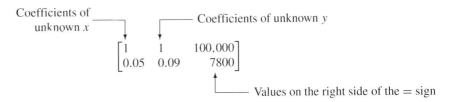

Later we will show how to use the matrix representing the linear system to determine a solution. But for now we can solve the linear system of equations in (1) using the familiar elimination technique outlined next.

Solve the first equation for $x$ :   $x = 100,000 - y$.   (2)

Substitute (2) into the second equation of the linear system in (1) to get

$$0.05(100,000 - y) + 0.09y = 7800. \tag{3}$$

Solving (3) for $y$, we get $y = 70,000$. Substitute this value for $y$ into (2) and solving for $x$, we get $x = 30,000$. Thus the trust director should invest \$30,000 in the CD and \$70,000 in the bond.  ■

We will learn how to solve linear systems of equations like those in Examples 4 and 5, and much larger systems containing more unknowns and more equations. We will show how to use the matrix structure that contains the coefficients and numbers to the right of the equal sign in the solution process. Our techniques will operate directly on the matrix structure rather than use algebraic elimination of variables, as done in Example 5. Thus we make the following definition.

---

**Definition**   A **matrix** is a rectangular array consisting of rows and columns of real numbers, complex numbers, expressions, etc. We say matrix $\mathbf{A}$ is of size $m \times n$ if $\mathbf{A}$ has $m$ rows and $n$ columns. The contents of a matrix are called **entries** or **elements** and are placed at the intersection of rows and columns. We use $a_{ij}$ to denote the entry of matrix $\mathbf{A}$ that lies at the $i$th row and $j$th column. We denote matrix $\mathbf{A}$ as $\mathbf{A} = \begin{bmatrix} a_{ij} \end{bmatrix}$ where

$$\mathbf{A} = \begin{bmatrix} a_{11} & a_{12} & \cdots & \cdots & \cdots & a_{1n} \\ a_{21} & a_{22} & \cdots & \cdots & \cdots & a_{2n} \\ \vdots & \vdots & \cdots & \cdots & \cdots & \vdots \\ \vdots & \vdots & \cdots & \cdots & a_{ij} & \vdots \\ \vdots & \vdots & \cdots & \cdots & \cdots & \vdots \\ a_{m1} & a_{m2} & \cdots & \cdots & \cdots & a_{mn} \end{bmatrix} \longleftarrow i\text{th row}$$

$\uparrow j\text{th column}$

Row $i$ of $\mathbf{A}$ is $\begin{bmatrix} a_{i1} & a_{i2} & \cdots & \cdots & a_{in} \end{bmatrix}$ and column $j$ of $\mathbf{A}$ is $\begin{bmatrix} a_{1j} \\ a_{2j} \\ \vdots \\ \vdots \\ a_{mj} \end{bmatrix}$.

An additional notation for the entries of a matrix is $\text{ent}_{ij}(\mathbf{A})$, which we read as entry $(i, j)$ of $\mathbf{A}$ or the $(i, j)$-entry of $\mathbf{A}$.

---

A matrix with exactly one row, that is, of size $1 \times n$, is called a **row matrix** or **row vector**. A matrix with exactly one column, that is, of size $m \times 1$, is called a **column matrix** or **column vector**. Often we will use the term *row* or *column* for brevity, or just the term *vector*. If a vector has $n$ entries we call it an ***n*-vector**. At times we will want to emphasize that we are working with a particular row or column of a matrix $\mathbf{A}$; then we use the notation

$$\text{row}_k(\mathbf{A}) \text{ is row } k \text{ of matrix } \mathbf{A}$$

$$\text{col}_k(\mathbf{A}) \text{ is column } k \text{ of matrix } \mathbf{A}$$

Linear algebra and geometry are interchangeable, as we will see throughout this book. We will often use geometric models to help us visualize algebraic objects and concepts. It is helpful to have a geometric model of an $n$-vector when its entries are real numbers. Here we give an intuitive introduction to such "geometric" vectors and provide a detailed development in Section 1.4.

If the entries of a 2-vector $\mathbf{v} = \begin{bmatrix} v_1 & v_2 \end{bmatrix}$ are real numbers, then there is a natural geometric model for $\mathbf{v}$ as a directed line segment in the plane connecting the origin $O(0, 0)$ with the point $P(v_1, v_2)$ as shown in Figure 2. Similarly, for a 3-vector $\mathbf{w} = \begin{bmatrix} w_1 & w_2 & w_3 \end{bmatrix}$ with real entries we have the 3-dimensional geometric model shown in Figure 3.

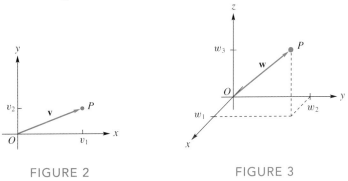

FIGURE 2                    FIGURE 3

It is customary to think of an $n$-vector, $n > 3$, with real entries in an analogous fashion, even though we cannot draw a corresponding picture.

An $m \times n$ matrix $\mathbf{A}$ with the same number of rows as columns, $m = n$, is called a **square matrix** of order $n$. The entries $a_{11}, a_{22}, \ldots,$ at the intersection of like numbered rows and columns are called the (main) **diagonal entries** of matrix $\mathbf{A}$. A square matrix $\mathbf{A}$ of order $n$ has $n$ diagonal entries $a_{11}, a_{22}, \ldots, a_{nn}$, while an $m \times n$ matrix $\mathbf{A}$ has diagonal entries $a_{11}, a_{22}, \ldots, a_{kk}$, where $k = \min\{m, n\}$.

EXAMPLE 6    Let

$$\mathbf{A} = \begin{bmatrix} 1.1 & -3.2 \\ 4.0 & 6.1 \\ -0.1 & 17.3 \\ 2.0 & p \end{bmatrix}.$$

Then we have $\text{ent}_{21}(\mathbf{A}) = 4.0$, $\text{ent}_{32}(\mathbf{A}) = 17.3$,

$$\text{row}_2(\mathbf{A}) = \begin{bmatrix} 4.0 & 6.1 \end{bmatrix}$$

a 2-vector,

$$\text{col}_1(\mathbf{A}) = \begin{bmatrix} 1.1 \\ 4.0 \\ -0.1 \\ 2.0 \end{bmatrix}$$

a 4-vector, and the diagonal entries of $\mathbf{A}$ are $a_{11} = 1.1$ and $a_{22} = 6.1$.    ■

**Definition**    A square matrix $\mathbf{A} = \begin{bmatrix} a_{ij} \end{bmatrix}$ for which $a_{ij} = 0$ when $i \neq j$, that is, for which every entry not on the main diagonal is zero, is called a **diagonal matrix**.

EXAMPLE 7    $\mathbf{A} = \begin{bmatrix} 5.5 & 0 \\ 0 & -8.0 \end{bmatrix}$ and $\mathbf{B} = \begin{bmatrix} 12 & 0 & 0 \\ 0 & -9 & 0 \\ 0 & 0 & -1 \end{bmatrix}$ are diagonal matrices. The diagonal entries of a diagonal matrix can be zero; for instance,

$$\mathbf{C} = \begin{bmatrix} -4 & 0 & 0 \\ 0 & 8 & 0 \\ 0 & 0 & 0 \end{bmatrix}$$

has diagonal entry $\text{ent}_{33}(\mathbf{C}) = 0$.                                                                     ■

A diagonal matrix with each diagonal entry equal to 1 is called an **identity matrix**. We designate an $n \times n$ identity matrix by $\mathbf{I}_n$. Thus,

$$\mathbf{I}_2 = \begin{bmatrix} 1 & 0 \\ 0 & 1 \end{bmatrix}, \quad \mathbf{I}_3 = \begin{bmatrix} 1 & 0 & 0 \\ 0 & 1 & 0 \\ 0 & 0 & 1 \end{bmatrix}, \quad \mathbf{I}_4 = \begin{bmatrix} 1 & 0 & 0 & 0 \\ 0 & 1 & 0 & 0 \\ 0 & 0 & 1 & 0 \\ 0 & 0 & 0 & 1 \end{bmatrix}.$$

The table of values in Example 1 can be written as the $2 \times 6$ matrix

$$\mathbf{A} = \begin{bmatrix} -2 & -1 & 0 & 0.5 & 1 & 1.5 \\ 4 & 1 & 0 & 0.25 & 1 & 2.25 \end{bmatrix}$$

or, equivalently, as the $6 \times 2$ matrix

$$\mathbf{B} = \begin{bmatrix} -2 & 4 \\ -1 & 1 \\ 0 & 0 \\ 0.5 & 0.25 \\ 1 & 1 \\ 1.5 & 2.25 \end{bmatrix}.$$

Matrix $\mathbf{B}$ is obtained from matrix $\mathbf{A}$ by writing $\text{row}_k(\mathbf{A})$ as $\text{col}_k(\mathbf{B})$ for $k = 1$, 2. This interchange of rows to columns (or equivalently columns to rows) will be convenient in upcoming topics, so we define the following operation on a matrix.

---

**Definition**   If $\mathbf{A} = \begin{bmatrix} a_{ij} \end{bmatrix}$ is an $m \times n$ matrix, then $\mathbf{A}^T = \begin{bmatrix} b_{ij} \end{bmatrix}$ is the $n \times m$ matrix where $b_{ij} = a_{ji}$, and matrix $\mathbf{A}^T$ is called the **transpose** of matrix $\mathbf{A}$. [That is, the $(i, j)$-entry of $\mathbf{A}^T$ is the $(j, i)$-entry of $\mathbf{A}$.] The transpose of a matrix $\mathbf{A}$ is obtained by interchanging its rows and columns.

---

It follows that the transpose of a row matrix is a column matrix and vice versa.

EXAMPLE 8   Let $\mathbf{C} = \begin{bmatrix} 1 & 2 & 4 & -6 \\ 0 & -3 & 5 & 7 \end{bmatrix}$ and $\mathbf{D} = \begin{bmatrix} 5 & -2 \\ 1 & 0 \end{bmatrix}$. Then we have

$$\mathbf{C}^T = \begin{bmatrix} 1 & 0 \\ 2 & -3 \\ 4 & 5 \\ -6 & 7 \end{bmatrix} \quad \text{and} \quad \mathbf{D}^T = \begin{bmatrix} 5 & 1 \\ -2 & 0 \end{bmatrix}.$$

For $\mathbf{E} = \begin{bmatrix} 1.8 & -6.3 & 12 \end{bmatrix}$, we have $\mathbf{E}^T = \begin{bmatrix} 1.8 \\ -6.3 \\ 12 \end{bmatrix}$.             ■

**Definition**  A square matrix $\mathbf{A}$ with real entries is called **symmetric** if $\mathbf{A}^T = \mathbf{A}$. That is, $\mathbf{A}$ is symmetric if $\text{ent}_{ij}(\mathbf{A}) = \text{ent}_{ji}(\mathbf{A})$.

EXAMPLE 9  Every diagonal matrix is symmetric and in particular each identity matrix is symmetric; $\mathbf{I}_n^T = \mathbf{I}_n$. Each of the following matrices is symmetric.

$$\mathbf{A} = \begin{bmatrix} 5 & -8 \\ -8 & 2 \end{bmatrix}, \quad \mathbf{B} = \begin{bmatrix} 2 & 1 & 1 \\ 1 & 2 & 1 \\ 1 & 1 & 2 \end{bmatrix}, \quad \mathbf{C} = \begin{bmatrix} 0 & p & -4.4 & 0 \\ p & -1 & 3.9 & -8 \\ -4.4 & 3.9 & 72 & 3 \\ 0 & -8 & 3 & 12 \end{bmatrix}$$

The arrangement of entries in a symmetric matrix is symmetric about the diagonal.

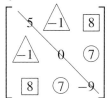

There are two other special types of matrices that have a particular pattern for the arrangement of zero entries that will be used in later sections. We define these next. Compare these types with diagonal matrices.

**Definition**  An $m \times n$ matrix $\mathbf{U}$ is called **upper triangular** if $\text{ent}_{ij}(\mathbf{U}) = 0$ for $i > j$; that is, all the entries below the diagonal entries are zero. A square upper triangular matrix is shown below:

$$\mathbf{U} = \begin{bmatrix} u_{11} & u_{12} & \cdots & \cdots & u_{1n} \\ 0 & u_{22} & \cdots & \cdots & u_{2n} \\ 0 & 0 & u_{33} & \cdots & u_{3n} \\ \vdots & \vdots & \cdots & \ddots & \vdots \\ 0 & 0 & \cdots & \cdots & u_{nn} \end{bmatrix}.$$

**Definition**  An $m \times n$ matrix $\mathbf{L}$ is called **lower triangular** if $\text{ent}_{ij}(\mathbf{L}) = 0$ for $i < j$; that is, all the entries above the diagonal entries are zero. A square lower triangular matrix is shown below:

$$\mathbf{U} = \begin{bmatrix} l_{11} & 0 & \cdots & \cdots & 0 \\ l_{21} & l_{22} & \cdots & \cdots & 0 \\ l_{31} & l_{32} & l_{33} & \cdots & 0 \\ \vdots & \vdots & \cdots & \ddots & \vdots \\ l_{n1} & l_{n2} & \cdots & \cdots & l_{nn} \end{bmatrix}.$$

We can represent diagonal matrices, square upper triangular matrices, and square lower triangular matrices schematically as follows:

Diagonal

Upper triangular

Lower triangular

The black shaded areas represent the regions in which nonzero entries can appear. Examples 10 and 11 show several nonsquare upper and lower triangular matrices.

EXAMPLE 10    Each of the following matrices is upper triangular.

$$\begin{bmatrix} -2 & 5 \\ 0 & 7 \end{bmatrix}, \quad \begin{bmatrix} 4 & 0 & 1 \\ 0 & 3 & -9 \\ 0 & 0 & 12 \end{bmatrix}, \quad \begin{bmatrix} 1 & 0 & 0 \\ 0 & 1 & 0 \\ 0 & 0 & 1 \end{bmatrix}, \quad \begin{bmatrix} 0 & 0 \\ 0 & 0 \end{bmatrix},$$

$$\begin{bmatrix} -2 & 0 & 0 & 0 \\ 0 & 5 & 0 & 0 \\ 0 & 0 & 8 & 0 \\ 0 & 0 & 0 & 6 \end{bmatrix}, \quad \begin{bmatrix} -2 & 2 & 1 & 3 \\ 0 & 5 & -3 & 15 \\ 0 & 0 & 8 & -7 \\ 0 & 0 & 0 & 6 \end{bmatrix}, \quad \begin{bmatrix} 5 & -3 & 1 & 2 \\ 0 & 4 & 3 & 6 \\ 0 & 0 & 8 & 1 \end{bmatrix}, \quad \begin{bmatrix} 7 & -5 \\ 0 & 2 \\ 0 & 0 \\ 0 & 0 \end{bmatrix}. \quad ■$$

EXAMPLE 11    Each of the following matrices is lower triangular.

$$\begin{bmatrix} 5 & 0 \\ 13 & -4 \end{bmatrix}, \quad \begin{bmatrix} -4 & 0 & 0 \\ 9 & 8 & 0 \\ -5 & 6 & 1 \end{bmatrix}, \quad \begin{bmatrix} 1 & 0 & 0 \\ 0 & 1 & 0 \\ 0 & 0 & 1 \end{bmatrix}, \quad \begin{bmatrix} 0 & 0 \\ 0 & 0 \end{bmatrix},$$

$$\begin{bmatrix} -2 & 0 & 0 & 0 \\ 0 & 5 & 0 & 0 \\ 0 & 0 & 8 & 0 \\ 0 & 0 & 0 & 6 \end{bmatrix}, \quad \begin{bmatrix} -2 & 0 & 0 & 0 \\ 2 & 5 & 0 & 0 \\ 1 & -3 & 8 & 0 \\ 3 & 15 & -7 & 6 \end{bmatrix}, \quad \begin{bmatrix} 4 & 0 & 0 & 0 \\ 2 & 1 & 0 & 0 \\ 3 & -1 & 6 & 0 \end{bmatrix}, \quad \begin{bmatrix} 2 & 0 \\ 4 & 5 \\ 1 & 6 \\ 0 & -2 \end{bmatrix}. \quad ■$$

The entries of our matrices will usually be real numbers, but they can be complex also; in fact, entries can be functions. Basic information about complex numbers appears in Appendix 1. When matrices with complex entries are used, the notion of a transpose is changed to **conjugate transpose**; that is, we take the conjugate of each entry and then the transpose (or take the transpose of the matrix and then the conjugate of each entry). The conjugate transpose of a matrix $\mathbf{Q}$ is denoted by $\mathbf{Q}^*$. For example,

$$\mathbf{Q}^* = \begin{bmatrix} 3 - 2i & 5 + i \\ 2i & 7 \end{bmatrix}^* = \begin{bmatrix} 3 + 2i & -2i \\ 5 - i & 7 \end{bmatrix}.$$

We conclude this section by showing two ways by which particular types of information can be represented using a matrix.

By a **graph** we mean a set of points called **nodes** or **vertices**, some of which are connected by **edges**. The nodes are usually labeled as $P_1, P_2, \ldots, P_k$ and for now we allow an edge to be traveled in either direction. One mathematical representation of a graph is constructed from a table. For example, the following table represents the graph shown in Figure 4.

|       | $P_1$ | $P_2$ | $P_3$ | $P_4$ |
|-------|-------|-------|-------|-------|
| $P_1$ | 0     | 1     | 0     | 0     |
| $P_2$ | 1     | 0     | 1     | 1     |
| $P_3$ | 0     | 1     | 0     | 1     |
| $P_4$ | 0     | 1     | 1     | 0     |

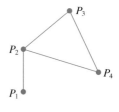

FIGURE 4

We have that the $(i, j)$-entry $= 1$ if there is an edge connecting vertex $P_i$ to vertex $P_j$, otherwise $(i, j)$-entry $= 0$. The **incidence matrix A** is the $k \times k$ matrix obtained by omitting the row and column labels from the preceding table. The incidence matrix for the graph in Figure 4 is

$$\mathbf{A} = \begin{bmatrix} 0 & 1 & 0 & 0 \\ 1 & 0 & 1 & 1 \\ 0 & 1 & 0 & 1 \\ 0 & 1 & 1 & 0 \end{bmatrix}.$$

Note that incidence matrix **A** is symmetric. (See Exercise 15 at the end of this section.)

A **Markov matrix** or **probability matrix** is a square matrix **A** whose entries satisfy $0 \leq a_{ij} \leq 1$ and the sum of the entries in each column is 1. Such matrices are used in mathematical models that predict changes in the behavior of a process. (See Section 1.3.) For example, $\mathbf{A} = \begin{bmatrix} 0.2 & 0.3 & 0 \\ 0.7 & 0.3 & 0.5 \\ 0.1 & 0.4 & 0.5 \end{bmatrix}$ is a $3 \times 3$ Markov matrix.

## EXERCISES 1.1

1. Determine a matrix that holds the information you would use to graph $y = x^3$ at equispaced values of $x$ starting at $x = -2$, with spacing of 0.5, and ending at $x = 1.5$.

2. Determine a matrix that holds the information you would use to start a sketch of the surface $z = 2x + y$ where $x$ takes values in the set $\{-1, 0, 1\}$ and $y$ takes values in the set $\{0, 1, 2\}$.

3. From Example 3, extract a matrix that gives the wind chill information when the temperature in $°F$ takes values in the set $\{5, 0, -5\}$ and the wind speed in miles per hour takes on values in the set $\{5, 15\}$.

4. Determine a matrix that represents the system of equations that arises when you want the quadratic polynomial $p(x) = ax^2 + bx + c$ to go through the points $\{(0, 0), (2, 3), (3, -1)\}$.

5. Water freezes at $32°F$ and $0°C$ and it boils at $212°F$ and $100°C$. Determine the linear equation $y = mx + b$, where $x$ is in $°F$ and $y$ is in $°C$, that relates temperatures in $°F$ to those in $°C$. Using this equation, construct a $2 \times 5$ matrix whose first row is $40°F$, $50°F$, $60°F$, $70°F$, $80°F$ and whose second row is the corresponding temperature in $°C$.

6. Many autos have speedometers that show both mph (miles per hour) and kph (kilometers per hour). Typically maximum auto speeds in a school zone are 15 mph or 24 kph and on limited access highways 65 mph or 105 kph. Determine the linear equation $y = mx + b$, where $x$ is in mph and $y$ is in kph, that relates speeds in mph to those in kph. Using this equation construct a $2 \times 6$ matrix whose first row is 15 mph, 25 mph, 35 mph, 45 mph, 55 mph, 65 mph and whose second row is the corresponding speed in kph. (The stated data in kph has been rounded to the nearest whole number. Do the same for the speeds in kph that you record in the matrix.)

7. Construct a linear system of three equations in three unknowns that provides a model for the following and then represent the linear system in a matrix as done in Examples 4 and 5.

   An inheritance of \$24,000 is divided into three parts, with the second part twice as much as the first. The three parts are invested and earn interest at the annual rates of 9%, 10%, and 6%, respectively, and return \$2210 at the end of the first year.

8. Construct a linear system of three equations in three unknowns that provides a model for the following and then represent the linear system in a matrix as done in Examples 4 and 5.

   A local health food store is running a special on a dried fruit mixture of apricots, apples, and raisins. The owner wants to make 12-pound bags of a 3:2:1 mixture with the weight of the apples three times the weight of the apricots and the weight of the raisins twice that of the apricots.

*In Exercises 9 and 10, determine the incidence matrix associated with each given graph.*

9.   10.

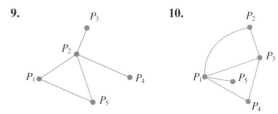

*In Exercises 11 and 12, determine the incidence matrix associated with each given graph.*

**11.**

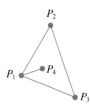

**12.**

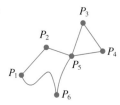

*In Exercises 13 and 14, construct a graph for each given matrix. Label the vertices $P_1, P_2, \ldots, P_5$.*

**13.** $\mathbf{A} = \begin{bmatrix} 0 & 1 & 0 & 1 & 1 \\ 1 & 0 & 0 & 0 & 1 \\ 0 & 0 & 0 & 1 & 1 \\ 1 & 0 & 1 & 0 & 1 \\ 1 & 1 & 1 & 1 & 0 \end{bmatrix}$.

**14.** $\mathbf{A} = \begin{bmatrix} 0 & 1 & 0 & 0 & 0 \\ 1 & 0 & 1 & 1 & 1 \\ 0 & 1 & 0 & 0 & 0 \\ 0 & 1 & 0 & 0 & 0 \\ 0 & 1 & 0 & 0 & 0 \end{bmatrix}$.

**15.** Explain why every incidence matrix associated with a graph is a symmetric matrix.

**16.** Let $\mathbf{A} = \begin{bmatrix} 5 & -3 & 1 & 2 \\ 4 & 9 & 6 & 7 \end{bmatrix}$.

    (a) Compute $\mathbf{A}^T$.

    (b) Compute $(\mathbf{A}^T)^T$.

    (c) Explain why the result in part (b) could have been anticipated.

    (d) Which entries in $\mathbf{A}$ and $\mathbf{A}^T$ did not change positions? Give a name for these entries.

*In Exercises 17–20, determine whether each given matrix is a Markov matrix.*

**17.** $\begin{bmatrix} 0.1 & 0.7 \\ 0.9 & 0.3 \end{bmatrix}$.

**18.** $\begin{bmatrix} 0.2 & 0.4 & 0.3 \\ 0.8 & 0.4 & 0.6 \\ 0.1 & 0.2 & 0.1 \end{bmatrix}$.

**19.** $\begin{bmatrix} 1 & 0 & 0 \\ 0 & 0.5 & 0.8 \\ 0 & 0.5 & 0.2 \end{bmatrix}$.

**20.** $\begin{bmatrix} 0.7 & 0.3 \\ 0.6 & 0.4 \end{bmatrix}$.

*In Exercises 21–23, determine a value for each missing entry, denoted by □, so that the matrix is a Markov matrix. (In some cases there can be more than one correct answer.)*

**21.** $\begin{bmatrix} □ & 0.5 & 0.4 \\ 0.3 & □ & 0.1 \\ 0.5 & 0.2 & □ \end{bmatrix}$.

**22.** $\begin{bmatrix} □ & 0.6 \\ □ & □ \end{bmatrix}$.

**23.** $\begin{bmatrix} □ & 0.2 & □ \\ 0.2 & □ & 0.2 \\ □ & 0.2 & □ \end{bmatrix}$.

An $n \times n$ matrix $\mathbf{A} = \begin{bmatrix} a_{ij} \end{bmatrix}$ is called **strictly diagonally dominant** if $|a_{ii}| >$ sum of the absolute values of the other entries in $\text{row}_i(\mathbf{A})$ for each value $i = 1, 2, \ldots, n$. Strictly diagonally dominant matrices appear in mathematical models involving approximations to data and in the solution of systems of equations that involve rates of change.

*In Exercises 24–27, determine whether each given matrix is strictly diagonally dominant.*

**24.** $\begin{bmatrix} 5 & -2 \\ 1 & -3 \end{bmatrix}$.

**25.** $\begin{bmatrix} 6 & -3 & 2 \\ 8 & 0 & 1 \\ 7 & 3 & 12 \end{bmatrix}$.

**26.** $\begin{bmatrix} 1 & 0 & 0 \\ 0 & 1 & 0 \\ 0 & 0 & 1 \end{bmatrix}$.

**27.** $\begin{bmatrix} 7 & 3 \\ 5 & 3 \end{bmatrix}$.

**28.** (a) Construct a $3 \times 3$ strictly diagonally dominant matrix that has no zero entries.

    (b) Construct a $4 \times 4$ strictly diagonally dominant matrix with integer entries where each diagonal entry is equal to its row number, the matrix is symmetric, and there are no zero entries.

    (c) Can zero ever be a diagonal entry for a strictly diagonally dominant matrix? Explain.

    (d) Is every diagonal matrix strictly diagonally dominant? Explain.

**29.** (a) If $\mathbf{U}$ is upper triangular, then describe $\mathbf{U}^T$.

    (b) If $\mathbf{L}$ is lower triangular, then describe $\mathbf{L}^T$.

    (c) Give an example of a matrix that is both upper and lower triangular.

    (d) A _____ matrix is both upper and lower triangular.

**30.** Suppose we had a system of four linear equations (see Example 4) in four unknowns, $x_1, x_2, x_3,$ and $x_4$. Construct such a system where equation 1 has only unknown $x_1$, equation 2 has only unknowns $x_1$ and $x_2$, and so on. As in Example 4, build the matrix representing this system of equations. Looking at just the part that corresponds to the coefficients, what type of matrix is it?

**31.** For the following system of linear equations, construct a matrix that represents the system. (See Example 4.)

$$5x_1 + 8x_2 - 4x_3 = 2$$
$$- 4x_2 + 6x_3 = 0$$
$$2x_3 = 10.$$

Looking at just the part that corresponds to the coefficients, what type of matrix is it?

*In Exercises* 32 *and* 33, *compute the conjugate transpose of each given matrix.*

**32.** $\mathbf{A} = \begin{bmatrix} 1 + 4i & -2 - 5i & 3 \\ -8i & 2 + i & 3 - 6i \end{bmatrix}$.

**33.** $\mathbf{B} = \begin{bmatrix} i & 0 & 2 - 4i \end{bmatrix}$.

**34.** What is the conjugate transpose of a matrix all of whose entries are real numbers?

**35.** A matrix $\mathbf{A}$ that is equal to its conjugate transpose, $\mathbf{A} = \mathbf{A}^*$, is called **Hermitian**. Note that a Hermitian matrix must be square.

(a) Is $\mathbf{A} = \begin{bmatrix} 3 & 2 - 5i \\ 2 + 5i & 8 \end{bmatrix}$ Hermitian?

(b) Is $\mathbf{B} = \begin{bmatrix} 0 & -2i & 1 + i \\ 2i & 5 + 3i & 3 + 2i \\ 1 - i & 8 & 12 \end{bmatrix}$ Hermitian?

(c) Is $\mathbf{C} = \begin{bmatrix} 5 & -2 \\ -2 & 0 \end{bmatrix}$ Hermitian?

(d) Construct a $4 \times 4$ Hermitian matrix.

(e) If a matrix is Hermitian, describe its diagonal entries.

(f) If $\mathbf{A}$ is Hermitian with all of its entries real numbers, then $\mathbf{A}$ is called _____.

**36.** Explain how you would use a mirror to help describe a symmetric matrix.

### In MATLAB

*In order to do the following* MATLAB *exercises you must have read Chapter* 8 *and used* MATLAB *to work the examples included in this brief introduction.*

**ML. 1.** Enter the matrix $\mathbf{A} = \begin{bmatrix} 1 & 2 \\ 0 & 6 \\ -5 & 7 \end{bmatrix}$ into MATLAB.

(a) Enter the command **size(A)** and explain the resulting output.

(b) Define matrix $\mathbf{B}$ to be the same as $\mathbf{A}$ using the command $\mathbf{B} = \mathbf{A}$. Then change the $(3, 2)$-entry of $\mathbf{B}$ to be 21.

(c) Define the column vector $\mathbf{x} = \begin{bmatrix} 3 \\ 9 \\ 15 \end{bmatrix}$ in MATLAB by using the command **x=[3;9;15]**. Now enter command **[A x]** and explain the result.

(d) Using vector $\mathbf{x}$ from part (c), enter command **[A;x]**. Explain the message that appears.

**ML. 2.** In Example 1 we had a sample of the function $y = x^2$. In MATLAB define the $x$-coordinates in a row vector $\mathbf{x}$ and the corresponding $y$ coordinates in a row vector $\mathbf{y}$. To plot these six ordered pairs and connect them with straight line segments, use the following commands:

**figure, plot(x,y,'\*b',x,y,'-k')**

Note that the points are designated by blue asterisks (the meaning of '\*b') and connected by black line segments (the meaning of '-k'). For more information about plotting in MATLAB, type **help plot**. (You may not understand the entire description shown.)

**ML. 3.** Enter each of the following commands and explain the output. (Remember, you can use **help** followed by a command name to get explanations.)

(a) **eye(3)**   (b) **eye(5)**   (c) **ones(4)**

(d) **zeros(3)**   (e) **8\*ones(3)**

**ML. 4.** In MATLAB we compute the transpose of a matrix using the single quote or prime symbol after the name of the matrix.

(a) If $\mathbf{A}$ is the matrix of Exercise ML. 1, then its transpose in MATLAB is computed by $\mathbf{A}'$. (Try it. Of course you must enter the matrix $\mathbf{A}$ if you have not already done so.) The transpose operator is useful when we want to enter a column with real entries. For instance the column vector $\mathbf{x}$ in ML. 1 could be entered using the command **x=[3 9 15]'**.

(b) Enter matrix $\mathbf{C} = \begin{bmatrix} 2 & 0 & 4 \\ 9 & 8 & -3 \end{bmatrix}$ into MATLAB. Use each of the following commands and explain the output shown.

**C'   (C')'   C(1,:)'   C(:,2)'**

(c) Enter command **[A C']** and describe the resulting matrix in terms of matrices $\mathbf{A}$ and $\mathbf{C}$.

(d) Enter command **[C [5 6]' [12 -7]']** and describe the resulting matrix in terms of matrix $\mathbf{C}$.

(e) Compare the matrices generated by the following commands: **eye(3)** and **eye(3)'**. Complete the following statement: Matrix **eye(3)** is _____.

(f) Compare the matrices generated by the following commands: **ones(4)** and **ones(4)'**. Complete the following statement: Matrix **ones(4)** is _____.

**ML. 5.** If a matrix has complex entries, then we can enter it into MATLAB with no special treatment. Here we use $i$ to denote $\sqrt{-1}$, a square root of $-1$. For

$$\mathbf{Q} = \begin{bmatrix} 3 - 2i & 5 + i \\ 2i & 7 \end{bmatrix}$$

enter the MATLAB command

**Q = [3-2i   5+i;2i   7]**,

where we were careful not to use any extra spaces between the real and imaginary parts of the complex entries. In MATLAB the transpose operator computes the conjugate transpose of a complex matrix. Enter the command $\mathbf{Q}'$ and explain the result.

**ML. 6.** In Exercises 13 and 14 you were asked to construct the graphs that corresponded to the incidence matrices listed. To check your results we can use MATLAB. Type **help igraph** for directions; then use the routine **igraph** to generate the graphs and compare the output with your hand-drawn solutions.

**ML. 7.** A graph has vertices $P_1$, $P_2$, $P_3$, $P_4$, and $P_5$.

(a) Construct an incidence matrix that corresponds to the situation in which there is an edge connecting $P_3$ to every other vertex, but no other connections. Verify that your matrix is correct by displaying and checking the picture generated by routine **igraph**.

(b) Construct an incidence matrix that corresponds to the situation in which every odd-numbered vertex is connected to every other odd-numbered vertex and the same is true for the even-numbered vertices. Verify that your matrix is correct by displaying and checking the picture generated by routine **igraph**.

(c) In terms of moving from vertex to vertex traversing any number of edges, describe a distinct difference between the pictures generated in parts (a) and (b).

## True/False Review Questions

*Determine whether each of the following statements is true or false.*

**1.** A diagonal matrix is symmetric.

**2.** An identity matrix is symmetric.

**3.** An identity matrix is lower triangular.

**4.** An identity matrix is upper triangular.

**5.** An identity matrix is strictly diagonally dominant.

**6.** If a matrix and its transpose are the same, then it is symmetric.

**7.** An incidence matrix is symmetric.

**8.** The incidence matrix of graph ⟨graph with $P_1$, $P_2$, $P_3$⟩ has every non-diagonal entry equal to 1.

**9.** $\begin{bmatrix} 0.7 & 0.3 & 0 \\ 0.2 & 0.5 & 0.5 \\ 0.1 & 0.2 & 0.5 \end{bmatrix}$ is a Markov matrix.

**10.** $\begin{bmatrix} 0.7 & 0.3 & 0 \\ 0.2 & 0.5 & 0.5 \\ 0.1 & 0.2 & 0.5 \end{bmatrix}$ is strictly diagonally dominant.

## Terminology

| | |
|---|---|
| System of linear equations | Matrix; entries; elements |
| Row matrix or vector; column matrix or vector | Square matrix |
| Diagonal entries of a matrix | Diagonal Matrix[(P)] |
| Identity matrix[(P)] | Transpose of a matrix |
| Symmetric matrix[(P)] | Upper and lower triangular matrices[(P)] |
| Conjugate transpose of a matrix | Graph; vertices; edges |
| Incidence matrix | Markov matrix[(P)] |
| Diagonally dominant matrix[(P)] | Hermitian matrix[(P)] |

You should be able to describe the *pattern* of the entries for each type of matrix denoted with a [(P)] and state several examples. Such matrices will appear frequently throughout this book. Other types of patterns will also play a prominent role in our study of linear algebra.

[(P)]There is a pattern involved in the entries of this type of matrix.

- What is a symmetric matrix symmetric about?
- What type of matrix is both upper and lower triangular?
- Describe how an incidence matrix of a graph is constructed.
- Describe a Markov matrix.
- Is the transpose of a Markov matrix another Markov matrix? Explain your answer.
- Is the transpose of a strictly diagonally dominant matrix another strictly diagonally dominant matrix? Explain your answer.

## 1.2 ■ MATRIX OPERATIONS

As shown in Section 1.1, a matrix can be used to hold and organize information from a wide variety of sources. But to be truly useful such a structure must permit the information contained therein to be easily manipulated, changed, and processed to derive new information. In this section we introduce operations that provide tools for further analysis of the information contained in a matrix and we see how matrices can be combined to yield new information on a topic.

| In the text. | In MATLAB. |
|---|---|
| $a_{ij}$ | $\mathbf{A}(i, j)$ |
| $\mathbf{A} + \mathbf{B}$ | $\mathbf{A} + \mathbf{B}$ |
| $\mathbf{A} - \mathbf{B}$ | $\mathbf{A} - \mathbf{B}$ |
| $k\mathbf{A}$ | $k * \mathbf{A}$ |
| $c_1\mathbf{A}_1 + c_2\mathbf{A}_2$ | $c1 * \mathbf{A1} + c2 * \mathbf{A2}$ |
| $\mathbf{A}^T$ | $\mathbf{A}'$ |

Matrix operations[1]

**Definition** Two $m \times n$ matrices $\mathbf{A} = \begin{bmatrix} a_{ij} \end{bmatrix}$ and $\mathbf{B} = \begin{bmatrix} b_{ij} \end{bmatrix}$ are said to be equal if $\text{ent}_{ij}(\mathbf{A}) = \text{ent}_{ij}(\mathbf{B})$ or, equivalently, $a_{ij} = b_{ij}$, for $1 \leq i \leq m$, $1 \leq j \leq n$; that is, if corresponding entries are equal.

One way to generate equal matrices is to identify the information contained in the matrix in different ways. If $\mathbf{A} = \begin{bmatrix} a_{ij} \end{bmatrix}$ is $m \times n$, then

$$\mathbf{A} = \begin{bmatrix} \text{row}_1(\mathbf{A}) \\ \text{row}_2(\mathbf{A}) \\ \vdots \\ \text{row}_m(\mathbf{A}) \end{bmatrix}$$

where $\text{row}_k(\mathbf{A})$ is an $n$-vector and

$$\mathbf{A} = \begin{bmatrix} \text{col}_1(\mathbf{A}) & \text{col}_2(\mathbf{A}) & \cdots & \text{col}_n(\mathbf{A}) \end{bmatrix}$$

where $\text{col}_k(\mathbf{A})$ is an $m$-vector are two ways to group the information in $\mathbf{A}$. For

[1]See Chapter 8 and Exercise ML. 1 at the end of this section.

matrix

$$\mathbf{A} = \begin{bmatrix} 2 & 4 & 7 \\ -3 & 0 & 1 \\ 9 & 5 & 6 \\ 10 & 3 & 8 \end{bmatrix},$$

identify

$$\mathbf{P}_1 = \begin{bmatrix} 2 & 4 & 7 \\ -3 & 0 & 1 \end{bmatrix}, \quad \mathbf{P}_2 = \text{row}_3(\mathbf{A}), \quad \text{and} \quad \mathbf{P}_3 = \text{row}_4(\mathbf{A});$$

then $\mathbf{A} = \begin{bmatrix} \mathbf{P}_1 \\ \mathbf{P}_2 \\ \mathbf{P}_3 \end{bmatrix}$. The rows, columns, or portions of $\mathbf{A}$, like $\mathbf{P}_1$, contain subsets of the information in $\mathbf{A}$. Accordingly, we make the following definition.

---

**Definition**    A **submatrix** of $\mathbf{A} = \begin{bmatrix} a_{ij} \end{bmatrix}$ is any matrix of entries obtained by omitting some, but not all, of its rows and/or columns.

---

EXAMPLE 1    Let

$$\mathbf{B} = \begin{bmatrix} 1 & 2 & 3 & 4 \\ 5 & 6 & 7 & 8 \\ 9 & 10 & 11 & 12 \\ 13 & 14 & 15 & 16 \end{bmatrix}.$$

Each of the following is a submatrix of $\mathbf{B}$. (Determine which rows and columns were omitted to obtain the submatrix.)

$$\mathbf{Q}_1 = \begin{bmatrix} 1 & 2 \\ 5 & 6 \end{bmatrix}, \quad \mathbf{Q}_2 = \begin{bmatrix} 3 & 4 \\ 7 & 8 \end{bmatrix}, \quad \mathbf{Q}_3 = \begin{bmatrix} 9 & 10 \\ 13 & 14 \end{bmatrix},$$

$$\mathbf{Q}_4 = \begin{bmatrix} 11 & 12 \\ 15 & 16 \end{bmatrix}, \quad \mathbf{T}_1 = \begin{bmatrix} 1 & 3 \\ 9 & 11 \\ 13 & 15 \end{bmatrix}, \quad \mathbf{T}_2 = \begin{bmatrix} 1 & 4 \\ 13 & 16 \end{bmatrix}.$$    ■

A matrix can be partitioned into submatrices or blocks by drawing horizontal lines between rows and vertical lines between columns. Each of the following is a partition of matrix $\mathbf{B}$ from Example 1, and we see that a partitioning can be carried out in many different ways.

$$\begin{bmatrix} 1 & 2 & 3 & 4 \\ 5 & 6 & 7 & 8 \\ 9 & 10 & 11 & 12 \\ 13 & 14 & 15 & 16 \end{bmatrix}, \quad \begin{bmatrix} 1 & 2 & 3 & 4 \\ 5 & 6 & 7 & 8 \\ 9 & 10 & 11 & 12 \\ 13 & 14 & 15 & 16 \end{bmatrix}, \quad \begin{bmatrix} 1 & 2 & 3 & 4 \\ 5 & 6 & 7 & 8 \\ 9 & 10 & 11 & 12 \\ 13 & 14 & 15 & 16 \end{bmatrix}$$

In addition,

$$\mathbf{A} = \begin{bmatrix} a_{11} & a_{12} & a_{13} & a_{14} & a_{15} \\ a_{21} & a_{22} & a_{23} & a_{24} & a_{25} \\ a_{31} & a_{32} & a_{33} & a_{34} & a_{35} \end{bmatrix} \quad \text{and} \quad \mathbf{A} = \begin{bmatrix} a_{11} & a_{12} & a_{13} & a_{14} & a_{15} \\ a_{21} & a_{22} & a_{23} & a_{24} & a_{25} \\ a_{31} & a_{32} & a_{33} & a_{34} & a_{35} \end{bmatrix}$$

are partitions of a $3 \times 5$ matrix $\mathbf{A}$. From Example 1, we see that $\mathbf{B} = \begin{bmatrix} \mathbf{Q}_1 & \mathbf{Q}_2 \\ \mathbf{Q}_3 & \mathbf{Q}_4 \end{bmatrix}$ is a **partition** or subdivision of $\mathbf{B}$ into $2 \times 2$ blocks.

We say that a matrix is **block diagonal** provided that there exists a partitioning such that each nondiagonal block consists of all zeros. See Example 2.

EXAMPLE 2   Let

$$A = \begin{bmatrix} 1 & 2 & 0 & 0 & 0 \\ 3 & 4 & 0 & 0 & 0 \\ 0 & 0 & 5 & 0 & 0 \\ 0 & 0 & 0 & 6 & 7 \\ 0 & 0 & 0 & 8 & 9 \end{bmatrix};$$

then $A$ can be considered a block diagonal matrix in several ways, as indicated by the following partitionings:

$$\begin{bmatrix} 1 & 2 & 0 & 0 & 0 \\ 3 & 4 & 0 & 0 & 0 \\ 0 & 0 & 5 & 0 & 0 \\ 0 & 0 & 0 & 6 & 7 \\ 0 & 0 & 0 & 8 & 9 \end{bmatrix}, \quad \begin{bmatrix} 1 & 2 & 0 & 0 & 0 \\ 3 & 4 & 0 & 0 & 0 \\ 0 & 0 & 5 & 0 & 0 \\ 0 & 0 & 0 & 6 & 7 \\ 0 & 0 & 0 & 8 & 9 \end{bmatrix}, \quad \begin{bmatrix} 1 & 2 & 0 & 0 & 0 \\ 3 & 4 & 0 & 0 & 0 \\ 0 & 0 & 5 & 0 & 0 \\ 0 & 0 & 0 & 6 & 7 \\ 0 & 0 & 0 & 8 & 9 \end{bmatrix}.$$  ■

In Example 4 in Section 1.1 we represented the linear system of equations

$$a + b + c = -5$$
$$a - b + c = 1$$
$$4a + 2b + c = 7$$

by the matrix

$$\begin{bmatrix} 1 & 1 & 1 & -5 \\ 1 & -1 & 1 & 1 \\ 4 & 2 & 1 & 7 \end{bmatrix}.$$

In Section 2.1 we will use the partitioning

$$\begin{bmatrix} 1 & 1 & 1 & -5 \\ 1 & -1 & 1 & 1 \\ 4 & 2 & 1 & 7 \end{bmatrix}$$

where submatrix $A = \begin{bmatrix} 1 & 1 & 1 \\ 1 & -1 & 1 \\ 4 & 2 & 1 \end{bmatrix}$ is called the **coefficient matrix** of the linear system and submatrix $b = \begin{bmatrix} -5 \\ 1 \\ 7 \end{bmatrix}$ is called the **right side** of the linear system.

Thus we can use the partitioned or block matrix $C = \begin{bmatrix} A & | & b \end{bmatrix}$ to represent the linear system.[2] Also in Section 2.1 we will find it convenient to use a row partition of $C$ and perform operations on the rows to refine the information in order to solve the linear system.

Partitioning matrices into submatrices or blocks can also be used to speed up the transmission of information. The $5 \times 5$ matrix

$$A = \begin{bmatrix} a & b & 0 & 0 & 0 \\ c & d & 0 & 0 & 0 \\ 0 & 0 & e & f & 0 \\ 0 & 0 & g & h & 0 \\ 0 & 0 & 0 & 0 & k \end{bmatrix}$$

[2]If the coefficient matrix $A$ and right side $b$ have been entered into MATLAB, then command [A b] generates a matrix that represents the linear system. In Section 2.1 we call this the **augmented matrix** of the linear system.

is easily partitioned into a block diagonal form:

$$\left[\begin{array}{cc|cc|c} a & b & 0 & 0 & 0 \\ c & d & 0 & 0 & 0 \\ \hline 0 & 0 & e & f & 0 \\ 0 & 0 & g & h & 0 \\ \hline 0 & 0 & 0 & 0 & k \end{array}\right].$$

Instead of transmitting the 25 entries of $\mathbf{A}$, we need only send the 9 entries of the diagonal blocks $\begin{bmatrix} a & b \\ c & d \end{bmatrix}$, $\begin{bmatrix} e & f \\ g & h \end{bmatrix}$, and $\begin{bmatrix} k \end{bmatrix}$. This type of partitioning is especially useful on large matrices in which a high percentage of the entries are zero; such matrices are said to be **sparse** and can be partitioned in a variety of ways to avoid transmitting blocks of zeros.

Partitioning implies a subdivision of the information into blocks or units. The reverse process is to consider individual matrices as blocks and adjoin them to form a partitioned matrix. The only requirement is that after joining the blocks, all rows have the same number of entries and all columns have the same number of entries.

EXAMPLE 3    Let $\mathbf{B} = \begin{bmatrix} 2 \\ 3 \end{bmatrix}$, $\mathbf{C} = \begin{bmatrix} 1 & -1 & 0 \end{bmatrix}$, and $\mathbf{D} = \begin{bmatrix} 9 & 8 & -4 \\ -1 & 7 & 5 \end{bmatrix}$. Then we have

$$\begin{bmatrix} \mathbf{B} & \mathbf{D} \end{bmatrix} = \left[\begin{array}{c|ccc} 2 & 9 & 8 & -4 \\ 3 & -1 & 7 & 5 \end{array}\right], \quad \begin{bmatrix} \mathbf{D} \\ \mathbf{C} \end{bmatrix} = \left[\begin{array}{ccc} 9 & 8 & -4 \\ -1 & 7 & 5 \\ \hline 1 & -1 & 0 \end{array}\right] \quad \text{and}$$

$$\left[\begin{bmatrix} \mathbf{D} \\ \mathbf{C} \end{bmatrix} \quad \mathbf{C}^T \right] = \left[\begin{array}{ccc|c} 9 & 8 & -4 & 1 \\ -1 & 7 & 5 & -1 \\ 1 & -1 & 0 & 0 \end{array}\right].$$

■

Adjoining matrix blocks to expand information structures is done regularly in a variety of applications. It is common to keep monthly sales data for a year in a $1 \times 12$ matrix and then adjoin such matrices to build a sales history matrix for a period of years. Similarly, results of new laboratory experiments are adjoined to existing data to update a database in a research facility.

EXAMPLE 4    In Example 5 in Section 2.1 we use the formation of block matrices as follows. Certain applications give rise to linear systems of equations where the coefficient matrix is the same but the right side of the linear system is different. If a pair of linear systems have the same coefficient matrix $\mathbf{A}$ but different right sides $\mathbf{b}$ and $\mathbf{c}$, then the block matrix $\begin{bmatrix} \mathbf{A} & | & \mathbf{b} & | & \mathbf{c} \end{bmatrix}$ can be manipulated to solve both systems at once.    ■

Next we define operations on matrices that produce new matrices. These operations allow us to combine information from individual matrices and derive new information. We begin with element-by-element operations; that is, operations that combine corresponding entries of the matrices.

---

**Definition**    If $\mathbf{A} = \begin{bmatrix} a_{ij} \end{bmatrix}$ and $\mathbf{B} = \begin{bmatrix} b_{ij} \end{bmatrix}$ are both $m \times n$ matrices, then their **sum**, $\mathbf{A} + \mathbf{B}$, is the $m \times n$ matrix whose $(i, j)$-entry is $a_{ij} + b_{ij}$; that is, we add corresponding entries. The **difference**, $\mathbf{A} - \mathbf{B}$, is the $m \times n$ matrix whose $(i, j)$-entry is $a_{ij} - b_{ij}$; that is, we subtract corresponding entries.

---

EXAMPLE 5   Let

$$\mathbf{A} = \begin{bmatrix} 2 & 4 & -6 \\ 1 & 0 & 3 \end{bmatrix}, \quad \mathbf{B} = \begin{bmatrix} 5 & 1 & 2 \\ -3 & 4 & 7 \end{bmatrix}, \quad \text{and} \quad \mathbf{C} = \begin{bmatrix} 1 & 3 \\ 4 & -5 \end{bmatrix}.$$

Then

$$\mathbf{A} + \mathbf{B} = \begin{bmatrix} 2+5 & 4+1 & -6+2 \\ 1+(-3) & 0+4 & 3+7 \end{bmatrix} = \begin{bmatrix} 7 & 5 & -4 \\ -2 & 4 & 10 \end{bmatrix}$$

and

$$\mathbf{A} - \mathbf{B} = \begin{bmatrix} 2-5 & 4-1 & -6-2 \\ 1-(-3) & 0-4 & 3-7 \end{bmatrix} = \begin{bmatrix} -3 & 3 & -8 \\ 4 & -4 & -4 \end{bmatrix}.$$

However, $\mathbf{A} + \mathbf{C}$ is not defined and neither is $\mathbf{B} - \mathbf{C}$, since the matrices involved in these expressions do not have the same size.   ■

The next operation involves numbers (possibly real or complex), which we call **scalars**, and a matrix. The entries of the matrix could be real numbers, complex numbers, or even functions; we have not restricted the entries of a matrix.

---

**Definition**   If $\mathbf{A} = \begin{bmatrix} a_{ij} \end{bmatrix}$ is an $m \times n$ matrix and $k$ is a scalar, then the **scalar multiple** of $\mathbf{A}$ by $k$, $k\mathbf{A}$, is the $m \times n$ matrix whose $(i, j)$-entry is $ka_{ij}$; that is, each entry of matrix $\mathbf{A}$ is multiplied by $k$.

---

EXAMPLE 6   Let $\mathbf{S}_1 = \begin{bmatrix} 18.95 & 14.75 & 8.98 \end{bmatrix}$ be a 3-vector that represents the current price of three items at store 1 and $\mathbf{S}_2 = \begin{bmatrix} 17.80 & 13.50 & 10.79 \end{bmatrix}$ be a 3-vector of prices of the same three items at store 2. Then

$$\mathbf{P} = \begin{bmatrix} \mathbf{S}_1 \\ \mathbf{S}_2 \end{bmatrix} = \begin{bmatrix} 18.95 & 14.75 & 8.98 \\ 17.80 & 13.50 & 10.79 \end{bmatrix}$$

represents the combined information about the prices of the items at the two stores. If each store announces a sale so that the price of each of the three items is reduced by 20%, then

$$0.80\mathbf{P} = \begin{bmatrix} 0.80\mathbf{S}_1 \\ 0.80\mathbf{S}_2 \end{bmatrix} = \begin{bmatrix} 0.80(18.95) & 0.80(14.75) & 0.80(8.98) \\ 0.80(17.80) & 0.80(13.50) & 0.80(10.79) \end{bmatrix}$$

$$= \begin{bmatrix} 15.16 & 11.80 & 7.18 \\ 14.24 & 10.8 & 8.63 \end{bmatrix}$$

represents the sale prices at the stores.   ■

The scalar multiple $(-1)\mathbf{A}$ is often written $-\mathbf{A}$ and is called the **negative** of $\mathbf{A}$; thus the difference $\mathbf{A} - \mathbf{B} = \mathbf{A} + (-1)\mathbf{B}$. The difference $\mathbf{A} - \mathbf{B}$ is also referred to as subtracting $\mathbf{B}$ from $\mathbf{A}$.

## Linear Combinations

We can combine scalar multiples of matrices with addition and subtraction into expressions of the form $3\mathbf{A} - 5\mathbf{B} + \mathbf{C}$, where it is assumed that matrices $\mathbf{A}$, $\mathbf{B}$, and $\mathbf{C}$ are of the same size. The scalars multiplying the matrices are called **coefficients** or **weights**.

---

**Definition**   A **linear combination** of two or more quantities is a sum of scalar multiples of the quantities. The linear combination in which each scalar is zero is called the **trivial linear combination**.

---

EXAMPLE 7

(a) Let

$$\mathbf{b} = \begin{bmatrix} 1 \\ 2 \end{bmatrix}, \quad \mathbf{c} = \begin{bmatrix} -2 \\ 3 \end{bmatrix}, \quad \text{and} \quad \mathbf{d} = \begin{bmatrix} 4 \\ 0 \end{bmatrix}.$$

Then $2\mathbf{b} - \mathbf{c} + 3\mathbf{d}$ is a linear combination of the 2-vectors $\mathbf{b}$, $\mathbf{c}$, and $\mathbf{d}$ with coefficients 2, $-1$, and 3, respectively. We have

$$2\mathbf{b} - \mathbf{c} + 3\mathbf{d} = 2\begin{bmatrix} 1 \\ 2 \end{bmatrix} - \begin{bmatrix} -2 \\ 3 \end{bmatrix} + 3\begin{bmatrix} 4 \\ 0 \end{bmatrix} = \begin{bmatrix} 16 \\ 1 \end{bmatrix}.$$

(b) Let

$$\mathbf{A}_1 = \begin{bmatrix} 1 & 0 \\ 0 & 0 \end{bmatrix}, \quad \mathbf{A}_2 = \begin{bmatrix} 0 & 1 \\ 0 & 0 \end{bmatrix}, \quad \mathbf{A}_3 = \begin{bmatrix} 0 & 0 \\ 1 & 0 \end{bmatrix}, \quad \text{and} \quad \mathbf{A}_4 = \begin{bmatrix} 0 & 0 \\ 0 & 1 \end{bmatrix}.$$

Then $c_1\mathbf{A}_1 + c_2\mathbf{A}_2 + c_3\mathbf{A}_3 + c_4\mathbf{A}_4$ is a linear combination of the $2 \times 2$ matrices $\mathbf{A}_i$, $i = 1, \ldots, 4$ with coefficients $c_1$, $c_2$, $c_3$, and $c_4$, respectively.

(c) Let $g_i$ be the score (out of 100) of the $i$th exam, $i = 1, 2, 3$, and $w_i$ be the weight of the $i$th exam (that is, the percent of the final grade assigned to the $i$th exam; $0 \le w_i \le 1$). Then the linear combination $w_1 g_1 + w_2 g_2 + w_3 g_3$ gives the numerical course grade (out of 100). It is assumed that the sum of the weights is 1. For example, a student with exam scores $g_1 = 81$, $g_2 = 97$, and $g_3 = 71$ for which the corresponding weights are $w_1 = w_2 = 0.25$ and $w_3 = 0.50$ has the numerical course grade

$$(0.25)(81) + (0.25)(97) + (0.50)(71) = 80.$$

(d) Polynomial $x^3 - 7x^2 + 8x + 4$ is a linear combination of functions $x^3$, $x^2$, $x$, and 1 with coefficients 2, $-7$, 8, and 4, respectively.

(e) Function $4 + 3\sin(x) + \sin(2x) - 2\cos(x) + 0.5\cos(2x)$ is a linear combination of the functions 1, $\sin(x)$, $\sin(2x)$, $\cos(x)$, and $\cos(2x)$ with weights 4, 3, 1, $-2$, and 0.5, respectively.   ■

## Span

Linear combinations play an important role in our analysis of the information contained in a matrix that represents a linear system of equations or other processes. The following two questions involving linear combinations are the focus of some topics in Chapter 5:

1. Given a quantity, can it be expressed as a linear combination of a particular set of quantities?
2. Given a particular set of quantities, what is the nature of the set of all possible linear combinations of these quantities?

We use the following terminology:

- The set of all possible linear combinations of a particular set $S$ of quantities is called the **span** of set $S$ and is denoted span($S$).

- If it is possible to find scalars so that a given quantity can be expressed as a linear combination of a particular set of quantities, we say that the given quantity is **in the span** of the set.

EXAMPLE 8    Given the set of 2-vectors

$$S = \left\{ \begin{bmatrix} 2 \\ -1 \end{bmatrix}, \begin{bmatrix} 1 \\ 3 \end{bmatrix} \right\}$$

and 2-vector $\mathbf{b} = \begin{bmatrix} 7 \\ 0 \end{bmatrix}$, is $\mathbf{b}$ in span($S$)? If it is, then we should be able to find scalars $c_1$ and $c_2$ so that the linear combination

$$c_1 \begin{bmatrix} 2 \\ -1 \end{bmatrix} + c_2 \begin{bmatrix} 1 \\ 3 \end{bmatrix} = \mathbf{b} = \begin{bmatrix} 7 \\ 0 \end{bmatrix}. \tag{1}$$

Perform the scalar multiplication and matrix addition on the left side of (1) and an equivalent expression is

$$\begin{bmatrix} 2c_1 + c_2 \\ -c_1 + 3c_2 \end{bmatrix} = \begin{bmatrix} 7 \\ 0 \end{bmatrix}.$$

Using equality of matrices (that is, corresponding entries are equal), we see that this is a system of linear equations

$$\begin{aligned} 2c_1 + \ c_2 &= 7 \\ -c_1 + 3c_2 &= 0. \end{aligned}$$

If there is a solution to this system, then $\mathbf{b}$ is in span($S$); otherwise it is not in span($S$). Note that this system of linear equations can be represented by the partitioned matrix

$$\left[ \begin{array}{cc|c} 2 & 1 & 7 \\ -1 & 3 & 0 \end{array} \right].$$

In Sections 2.1 and 2.2 we develop a systematic procedure for determining whether this system has a solution or not by working directly with this partitioned matrix. The procedure forms linear combinations of rows to aid in this analysis. For now, verify that $c_1 = 3$ and $c_2 = 1$ solves (1); hence $\mathbf{b}$ is in span($S$).    ■

EXAMPLE 9    Given the set of matrices

$$S = \left\{ \begin{bmatrix} 1 & 0 \\ 0 & 0 \end{bmatrix}, \begin{bmatrix} 0 & 1 \\ 0 & 0 \end{bmatrix}, \begin{bmatrix} 0 & 0 \\ 1 & 0 \end{bmatrix}, \begin{bmatrix} 0 & 0 \\ 0 & 1 \end{bmatrix} \right\},$$

describe the set of all linear combinations of the members of $S$; that is, describe span($S$). If $c_i$, $i = 1, 2, 3, 4$, are any scalars, then a linear combination of the members of $S$ is

$$c_1 \begin{bmatrix} 1 & 0 \\ 0 & 0 \end{bmatrix} + c_2 \begin{bmatrix} 0 & 1 \\ 0 & 0 \end{bmatrix} + c_3 \begin{bmatrix} 0 & 0 \\ 1 & 0 \end{bmatrix} + c_4 \begin{bmatrix} 0 & 0 \\ 0 & 1 \end{bmatrix}. \tag{2}$$

Performing the scalar multiplications and the matrix additions, we get

$$c_1 \begin{bmatrix} 1 & 0 \\ 0 & 0 \end{bmatrix} + c_2 \begin{bmatrix} 0 & 1 \\ 0 & 0 \end{bmatrix} + c_3 \begin{bmatrix} 0 & 0 \\ 1 & 0 \end{bmatrix} + c_4 \begin{bmatrix} 0 & 0 \\ 0 & 1 \end{bmatrix} = \begin{bmatrix} c_1 & c_2 \\ c_3 & c_4 \end{bmatrix}.$$

Since the scalars $c_i$ are arbitrary, the linear combination in (2) could produce any $2 \times 2$ matrix of numbers we desire. If we restrict the scalars $c_i$ to be real numbers, then span($S$) is the set of all $2 \times 2$ matrices with real entries. We say that $S$ **generates** the set of all $2 \times 2$ real matrices. If we permit the scalars $c_i$ to be complex numbers, then span($S$) is the set of all $2 \times 2$ matrices with complex entries. (Why does the set of all $2 \times 2$ matrices with complex entries include all of the $2 \times 2$ matrices with real entries?)  ■

## Properties of Matrix Operations

We have seen that linear combinations provide a way to combine information in matrices. We will show that we can use them to extract information that provides insight into the fundamental nature of matrix behavior. The following properties are basic to using and manipulating linear combinations. These properties constitute the algebraic rules that we will use repeatedly with linear combinations from here on.

> **Properties of Matrix Addition**
>
>   Let $\mathbf{A}$, $\mathbf{B}$, and $\mathbf{C}$ be matrices of the same size; then
>
>   (a) $\mathbf{A} + \mathbf{B} = \mathbf{B} + \mathbf{A}$.  (*commutativity of addition*)
>   (b) $\mathbf{A} + (\mathbf{B} + \mathbf{C}) = (\mathbf{A} + \mathbf{B}) + \mathbf{C}$.  (*associativity of addition*)

These properties are operations that we use instinctively because of experience with addition of real and complex numbers. However, we must verify that they are valid for the data structure of a matrix. Our primary tools for step-by-step verification are the definition for addition of matrices and the definition of equality of matrices. Exercise 52 asks you to show that these rules are valid, and we illustrate the process on a property of linear combinations that follows the next set of properties.

> **Properties of Linear Combinations**
>
>   Let $r$ and $s$ be scalars and $\mathbf{A}$ and $\mathbf{B}$ be matrices; then
>
>   (a) $r(s\mathbf{A}) = (rs)\mathbf{A}$.
>   (b) $(r + s)\mathbf{A} = r\mathbf{A} + s\mathbf{A}$.
>   (c) $r(\mathbf{A} + \mathbf{B}) = r\mathbf{A} + r\mathbf{B}$.

The verification (or proof) of these properties, which seem quite natural, requires that we use the definitions established earlier. We illustrate this by showing that $r(\mathbf{A} + \mathbf{B}) = r\mathbf{A} + r\mathbf{B}$. To show two matrices are equal, we show that corresponding entries are equal and we use the notation $\text{ent}_{ij}(\mathbf{C})$ to denote the $(i, j)$-entry of a matrix $\mathbf{C}$:

$$\text{ent}_{ij}(r(\mathbf{A} + \mathbf{B})) = r\,\text{ent}_{ij}(\mathbf{A} + \mathbf{B})$$

$$= r(\text{ent}_{ij}(\mathbf{A}) + \text{ent}_{ij}(\mathbf{B}))$$

(*This expression contains only scalars so we can use standard arithmetic properties.*)

$$= r\,\text{ent}_{ij}(\mathbf{A}) + r\,\text{ent}_{ij}(\mathbf{B})$$

$$= \text{ent}_{ij}(r\mathbf{A}) + \text{ent}_{ij}(r\mathbf{B})$$

$$= \text{ent}_{ij}(r\mathbf{A} + r\mathbf{B}).$$

Since corresponding entries are equal, we have shown that $r(\mathbf{A} + \mathbf{B}) = r\mathbf{A} + r\mathbf{B}$. This property is often described as the **distributivity** of scalar multiplication over matrix addition.

The operation of a transpose can be applied to linear combinations of matrices. The following properties are basic to using the transpose on linear combinations.

**Properties of the Transpose**

Let $r$ be a real scalar and **A** and **B** be matrices with real entries; then

(a) $(\mathbf{A}^T)^T = \mathbf{A}$.

(b) $(\mathbf{A} + \mathbf{B})^T = \mathbf{A}^T + \mathbf{B}^T$.    *(The transpose of a sum is the sum of the individual transposes.)*

(c) $(r\mathbf{A})^T = r(\mathbf{A}^T)$.

While these properties seem natural, they require proof. As before, the basic strategy is to show that corresponding entries are equal. We illustrate by showing that $(\mathbf{A}^T)^T = \mathbf{A}$. Using the previous notation, we have

$$\text{ent}_{ij}((\mathbf{A}^T)^T) = \text{ent}_{ji}(\mathbf{A}^T) = \text{ent}_{ij}(\mathbf{A}).$$

## Closure

Let $S$ be a set of quantities on which we have defined two operations: one called addition, which operates on a pair of members of $S$, and a second called scalar multiplication, which combines a scalar and a member of $S$. We say that $S$ is **closed** if every linear combination of members of $S$ belongs to the set $S$. This can also be expressed by saying that the sum of any two members of $S$ belongs to $S$ (**closed under addition**) and that the scalar multiple of any member of $S$ belongs to the set $S$ (**closed under scalar multiplication**). We sometimes refer to this as the **closure property** of a set and it implies closure with respect to both addition and scalar multiplication.

Here we deal exclusively with matrices and polynomials and their standard operations of addition and scalar multiplication. In Chapter 5 we consider other sets and operations, which can be quite different from the standard forms that we use in a rather natural way here.

*The key for determining whether a set $S$ is closed or not is to carefully determine the criteria for membership to the set $S$.* While this seems easy, it is at the heart of closure since we must add two members of $S$ and ask if the result is in $S$ and take a scalar times a member of $S$ and ask if the result is also in $S$. If the answer to either question is no, then the set is not closed. This essentially means that we are not guaranteed that linear combinations of members of set $S$ produce quantities that are also in $S$. The best training in this important concept is to do lots of exercises, since each set $S$ can be different in its criteria for membership. We illustrate this in Examples 10 through 12.

EXAMPLE 10    Let $S$ be the set of all $3 \times 1$ matrices of the form $\begin{bmatrix} a \\ b \\ 1 \end{bmatrix}$ where $a$ and $b$ are any real numbers. The criteria for membership in $S$ are that the third entry must be 1 and the first and second entries are any real numbers. To determine if $S$ is closed, we check sums and scalar multiples of members of $S$ to see if the result is in $S$.

Checking sums of members of $S$:

$$\begin{bmatrix} a_1 \\ b_1 \\ 1 \end{bmatrix} + \begin{bmatrix} a_2 \\ b_2 \\ 1 \end{bmatrix} = \begin{bmatrix} a_1 + a_2 \\ b_1 + b_2 \\ 2 \end{bmatrix}.$$

Since the third entry of the result is not a 1, it is not in $S$. Hence it follows that $S$ is not closed.

There is no need to check scalar multiples. However, it is not closed under scalar multiplication. (Verify.)    ■

**EXAMPLE 11**    Let $T$ be the set of all $2 \times 2$ matrices of the form $\begin{bmatrix} a & 0 \\ b & c \end{bmatrix}$ where $c = 2a$ and $a$ and $b$ are any real numbers. The criteria for membership in $T$ is that the $(1, 2)$-entry be zero, the $(2, 2)$-entry is twice as big as the $(1, 1)$-entry, and the $(2, 1)$-entry is any real number. To determine if $T$ is closed, we check sums and scalar multiples of members of $T$ to see if the result is in $T$.

Checking sums of members of $T$:

$$\begin{bmatrix} a_1 & 0 \\ b_1 & 2a_1 \end{bmatrix} + \begin{bmatrix} a_2 & 0 \\ b_2 & 2a_2 \end{bmatrix} = \begin{bmatrix} a_1 + a_2 & 0 \\ b_1 + b_2 & 2a_1 + 2a_2 \end{bmatrix}$$

$$= \begin{bmatrix} a_1 + a_2 & 0 \\ b_1 + b_2 & 2(a_1 + a_2) \end{bmatrix}.$$

We see that this result follows the pattern for membership in the set $T$. Hence $T$ is closed under addition.

Checking scalar multiples of members of $T$ (let $k$ be any real scalar):

$$k \begin{bmatrix} a_1 & 0 \\ b_1 & 2a_1 \end{bmatrix} = \begin{bmatrix} ka_1 & 0 \\ kb_1 & k(2a_1) \end{bmatrix} = \begin{bmatrix} ka_1 & 0 \\ kb_1 & 2(ka_1) \end{bmatrix}.$$

We see that this result follows the pattern for membership in the set $T$. Hence $T$ is closed under scalar multiplication.

Since $T$ is closed under both operations, we can say $T$ is a closed set.    ■

**EXAMPLE 12**    Let $V$ be a set of two $3 \times 1$ matrices. (You can choose them, but once you make your choices they are fixed.) Denote the vectors in $V$ as $\mathbf{u}$ and $\mathbf{w}$. Let $S = \text{span}(V)$. Determine if $S$ is a closed set. We proceed as in Examples 10 and 11.

Criterion for membership in $S$: a vector must be a linear combination of $\mathbf{u}$ and $\mathbf{w}$. Thus every vector in $S$ is of the form $k_1 \mathbf{u} + k_2 \mathbf{w}$ for some scalars $k_1$ and $k_2$.

Checking sums of members of $S$:

$$(k_1 \mathbf{u} + k_2 \mathbf{w}) + (c_1 \mathbf{u} + c_2 \mathbf{w}) = (k_1 + c_1) \mathbf{u} + (k_2 + c_2) \mathbf{w}.$$

The preceding expression was obtained using properties of matrix operations. Note that the result is a linear combination of vectors $\mathbf{u}$ and $\mathbf{w}$, thus satisfying the pattern for membership in $S$. So $S$ is closed under addition.

Checking scalar multiples of members of $S$ (let $t$ be any scalar):

$$t(k_1 \mathbf{u} + k_2 \mathbf{w}) = tk_1 \mathbf{u} + tk_2 \mathbf{w}.$$

The preceding expression was obtained using properties of matrix operations. Note that the result is a linear combination of vectors $\mathbf{u}$ and $\mathbf{w}$, thus satisfying the pattern for membership in $S$. So $S$ is closed under scalar multiplication.

Since $S$ is closed under both operations, we can say $S$ is a closed set.    ■

In Example 12 the set $V$ contained only two matrices; however, it can be shown that if $V$ is any finite set of matrices (of the same size), then $\text{span}(V)$ is a closed set. The steps in Example 12 are mimicked with more terms corresponding to the number of vectors in $V$. Hence *the span of a set is a closed set.* We meet this idea in a more general setting in Chapter 5.

Examples 10 through 12 are more instances of patterns in mathematics. In Example 10 it was the pattern of the entries in the 3-vector that defined membership in the set $S$. In Example 11 there was another pattern that determined the members of set $T$. The pattern used in Example 12 was a bit more general since we did not specify entries or relationships between entries, but it was still easy enough to recognize. In Exercises 32 through 40 various patterns are used to define sets of matrices. When working on such closure problems, it is helpful to write a verbal statement that describes the criteria for membership *before* you begin the algebra of addition and scalar multiplication. In this way you stay focused on how to recognize new members of the set.

As we accumulate more ideas on matrices and operations on them, patterns will play an increasing role in developing ways to think about systems of equations and applications. In order to describe prevalent patterns that arise, we will need more terminology and notation. We have already begun accumulating this language of linear algebra. (See the "Terminology" list at the end of Section 1.1 and at the end of this section.) The items in these lists provide verbal aids to the patterns that are associated with the matrix or concept. We need such language in order to communicate efficiently with one another. This further emphasizes that mathematics is the science of patterns and in many ways plays an invisible role in much of what we do and how we see things in life.

## Application:  RGB Color Displays

Many color displays use an RGB (red, green, blue) system to create the color of a dot or **pixel**. Associated with each pixel is a 3-vector $\begin{bmatrix} r & g & b \end{bmatrix}$ which tells the intensity of the red, green, and blue light sources that are blended to create the color of the pixel. The intensities r, g, and b are in the range [0, 1]. For instance, $\begin{bmatrix} 1 & 0 & 0 \end{bmatrix}$ is pure red, $\begin{bmatrix} 0 & 1 & 0 \end{bmatrix}$ is pure green, $\begin{bmatrix} 0 & 0 & 1 \end{bmatrix}$ is pure blue, $\begin{bmatrix} 0 & 0 & 0 \end{bmatrix}$ is black, and $\begin{bmatrix} 1 & 1 & 1 \end{bmatrix}$ is white. The color of a pixel is determined by a linear combination of the 3-vectors in

$$C = \left\{ \begin{bmatrix} 1 & 0 & 0 \end{bmatrix}, \begin{bmatrix} 0 & 1 & 0 \end{bmatrix}, \begin{bmatrix} 0 & 0 & 1 \end{bmatrix} \right\}$$

where the coefficients are restricted to have values between 0 and 1. Note that black is the trivial linear combination, the absence of all the red, green, and blue light (see Figure 1), while white consists of the full intensity of each contributing color. In

### RGB COLOR Example

| | 0 | 1 | VALUE |
|---|---|---|---|
| RED | ◀ ⦀ | ▶ | 0 |
| GREEN | ◀ ⦀ | ▶ | 0 |
| BLUE | ◀ ⦀ | ▶ | 0 |

Use your mouse to position the sliders.
As a linear combination:

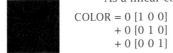

COLOR = 0 [1 0 0]
+ 0 [0 1 0]
+ 0 [0 0 1]

FIGURE 1  From MATLAB routine **rgbexamp**; see Exercise ML.4.

Exercise ML. 4, we can show that in MATLAB's color display the linear combination

$$0.5\begin{bmatrix} 1 & 0 & 0 \end{bmatrix} + 1\begin{bmatrix} 0 & 1 & 0 \end{bmatrix} + 0.83\begin{bmatrix} 0 & 0 & 1 \end{bmatrix} = \begin{bmatrix} 0.5 & 1 & 0.83 \end{bmatrix}$$

is aquamarine.

A display screen $S$ is really a matrix of pixels and each entry of matrix $S$ has a 3-vector of RGB intensities associated with it. Let $R$, $G$, and $B$ be matrices of the same size as $S$, where matrix $R$ contains the red intensities of the pixels of $S$, $G$ the green intensities, and $B$ the blue intensities. Thus the configuration of colors shown on screen $S$ depends upon the information in the three matrices $R$, $G$ and $B$. Since the color of each pixel is a blend of the three light sources, this situation amounts to saying that $S = R + G + B$. That is, the screen $S$ is generated by the intensity matrices $R$, $G$, and $B$.

## Application:  Polynomials

Let's agree to write a polynomial (see Example 7(d)) so that the highest-degree term is written first and the succeeding terms have decreasing exponents. For example $-8x^2 + 3x + 5$ and $2x^3 + 4x - 7$. With a polynomial we associate a coefficient vector:

$$-8x^2 + 3x + 5 \longrightarrow \begin{bmatrix} -8 & 3 & 5 \end{bmatrix},$$
$$2x^3 + 4x - 7 \longrightarrow \begin{bmatrix} 2 & 0 & 4 & -7 \end{bmatrix}.$$

In the latter example we used the convention that if a term is missing from the polynomial we assign a zero coefficient to its position. That is, $2x^3 + 4x - 7$ is considered to be $2x^3 + 0x^2 + 4x - 7$. Since we add and subtract polynomials term-by-term, we can use addition and subtraction of the corresponding coefficient vectors of polynomials of the same degree. It also follows that the multiplication of a polynomial by a constant is the same as scalar multiplication of its coefficient vector. Thus linear combinations of polynomials correspond to the same linear combination of their coefficient vectors. (The only requirement is that the polynomials have the same degree, so the coefficient vectors are the same size.) This association of polynomials with coefficient vectors gave early digital computers and programming languages a way to manipulate polynomials without the need for a symbolic representation in terms of an unknown like $x$.

Let $P_k$ be the set of all polynomials of degree $k$ or less with real coefficients. Any member of $P_k$ corresponds to a $(k + 1)$-vector as follows:

$$a_k x^k + a_{k-1} x^{k-1} + \cdots + a_1 x + a_0 \longrightarrow \begin{bmatrix} a_k & a_{k-1} & \cdots & a_1 & a_0 \end{bmatrix}.$$

For example, in $P_2$ we have

$$q(x) = x^2 + 3x \longrightarrow \begin{bmatrix} 1 & 3 & 0 \end{bmatrix},$$
$$r(x) = 2x + 7 \longrightarrow \begin{bmatrix} 0 & 2 & 7 \end{bmatrix},$$
$$2q(x) - 3r(x) \longrightarrow 2\begin{bmatrix} 1 & 3 & 0 \end{bmatrix} - 3\begin{bmatrix} 0 & 2 & 7 \end{bmatrix} = \begin{bmatrix} 2 & 0 & -21 \end{bmatrix},$$

which corresponds to $2x^2 - 21$. $P_k$ also includes the "zero polynomial" whose coefficients are all zero. The zero polynomial in $P_2$ corresponds to the 3-vector $\begin{bmatrix} 0 & 0 & 0 \end{bmatrix}$.

Application:    **Parametric Form of a Line in the Plane**

Let $Q(a, b)$ and $R(d, e)$ be a pair of points in the plane. Then the equation of the line $L$ through $Q$ and $R$ written in point-slope form is

$$y - b = \frac{e - b}{d - a}(x - a) \tag{3}$$

where we assume $d \neq a$. When we are interested in the line segment between $Q$ and $R$ we require that $x$ be restricted to values between $a$ and $d$. We call (3) the **algebraic equation** of $L$. For work in algebra and calculus an algebraic expression is quite useful, but for some geometric applications, like graphing, an equivalent equation in parametric[3] form is often more useful.

To construct the parametric form of (3), we proceed as follows. Rearrange the expression in (3) into the form

$$\frac{y - b}{e - b} = \frac{x - a}{d - a} \tag{4}$$

where we assume $d \neq a$ and $b \neq e$. Observe that each side of (4) is independent of the other side. Hence it follows that each side is equal to a common value $t$; thus,

$$t = \frac{x - a}{d - a} \quad \text{and} \quad t = \frac{y - b}{e - b}.$$

Solving these expressions for $x$ and $y$, respectively, gives

$$x = (1 - t)a + td \tag{5a}$$

and

$$y = (1 - t)b + te. \tag{5b}$$

The expressions in (5a) and (5b) give us equations for the $x$- and $y$-coordinates of points lying on line $L$. We call (5a) and (5b) the **parametric equations** of $L$. For $-\infty < t < \infty$, we can use (5a) and (5b) to generate points on $L$. Using (5a) and (5b), we can express the set of all points on $L$ as a linear combination of the coordinates of points $Q$ and $R$ as follows:

$$\begin{bmatrix} x \\ y \end{bmatrix} = (1 - t)\begin{bmatrix} a \\ b \end{bmatrix} + t\begin{bmatrix} d \\ e \end{bmatrix}. \tag{6}$$

The expression in (6) is called the **geometric equation** of line $L$.

If we require that $0 \leq t \leq 1$, then (6) represents the **line segment** between $Q$ and $R$. The convention is that the positive sense of direction along the line segment is in the direction determined by points along the segment as $t$ increases. For $0 \leq t \leq 1$, Equation (6) is also called a **convex linear combination** of vectors

$$\begin{bmatrix} a \\ b \end{bmatrix} \quad \text{and} \quad \begin{bmatrix} d \\ e \end{bmatrix}.$$

We investigate convex linear combinations graphically in Section 1.4.

---

[3]A parametric curve is one for which the defining equations are given in terms of a common independent variable called a parametric variable.

## EXERCISES 1.2

**1.** Let $\mathbf{A} = \begin{bmatrix} 2 & p+q \\ -1 & 4 \end{bmatrix}$ and $\mathbf{B} = \begin{bmatrix} p & p+q \\ -1 & q \end{bmatrix}$.
If $\mathbf{A} = \mathbf{B}$, then determine values for $p$ and $q$ and write out the matrix $\mathbf{B}$ with numerical entries.

**2.** Let $\mathbf{A} = \begin{bmatrix} 2 & p+q \\ -1 & 4 \end{bmatrix}$ and $\mathbf{B} = \begin{bmatrix} 2 & 3 \\ p-q & 4 \end{bmatrix}$.
If $\mathbf{A} = \mathbf{B}$, then determine values for $p$ and $q$.

**3.** Let $\mathbf{A} = \begin{bmatrix} 2 & p+q \\ -1 & 4 \end{bmatrix}$ and $\mathbf{B} = \begin{bmatrix} q & p+q \\ -1 & 4 \end{bmatrix}$.
If $\mathbf{A} = \mathbf{B}$, then determine all possible values for $p$ and $q$.

**4.** Let $\mathbf{A} = \begin{bmatrix} 2 & p^2+1 \\ -1 & 4 \end{bmatrix}$ and $\mathbf{B} = \begin{bmatrix} q^2 & 10 \\ -1 & 4 \end{bmatrix}$.
If $\mathbf{A} = \mathbf{B}$, then determine all possible values for $p$ and $q$.

**5.** Let $\mathbf{B} = \begin{bmatrix} 0 & 2 & 3 \\ 1 & -1 & 5 \\ 4 & 7 & 9 \\ -8 & 6 & 0 \end{bmatrix}$. Determine which of the following are submatrices of $\mathbf{B}$, and for those that are, list the row and column numbers that were omitted.

$$\mathbf{B}_1 = \begin{bmatrix} 0 & 3 \\ 1 & 5 \end{bmatrix}, \quad \mathbf{B}_2 = \begin{bmatrix} 2 & 5 \\ -1 & 9 \end{bmatrix},$$

$$\mathbf{B}_3 = \begin{bmatrix} 0 & 3 \\ -8 & 0 \end{bmatrix}, \quad \mathbf{B}_4 = \begin{bmatrix} 0 & 3 \\ 4 & 9 \\ -8 & 0 \end{bmatrix}.$$

**6.** Determine the coefficient matrix and right side associated with the linear system

$$\begin{aligned} 3x + 4y - 2z &= 3 \\ -x \quad\quad + z &= -1 \\ x - 3y + 2z &= 5 \end{aligned}$$

and display them in a partitioned matrix.

**7.** A linear system of equations is represented by the partitioned matrix

$$\begin{bmatrix} 4 & -2 & 1 & | & 3 \\ 2 & 0 & -5 & | & 7 \end{bmatrix}.$$

Construct the linear system of equations.

*In Exercises 8–11, construct the partitioned matrix that represents the linear system of equations that you develop to find the intersection point of lines $L_1$ and $L_2$.*

**8.** $L_1$: goes through point $(1, 1)$ with slope 3.
$L_2$: goes through point $(2, 1)$ with slope $-2$.

**9.** $L_1$: goes through points $(1, 2)$ and $(3, 0)$.
$L_2$: goes through point $(2, 3)$ with slope 0.

**10.** $L_1$: goes through points $(3, 1)$ and $(2, 2)$.
$L_2$: goes through point $(5, 0)$ with slope $-\frac{1}{2}$.

**11.** $L_1$: goes through point $(2, 2)$ with slope 1.
$L_2$: goes through points $(5, 5)$ and $(0, 0)$.

**12.** Let $\mathbf{Q}_1 = \begin{bmatrix} 2 & 1 \\ 3 & 8 \end{bmatrix}$, $\mathbf{Q}_2 = \begin{bmatrix} 4 & -2 \\ 5 & 1 \end{bmatrix}$, and $\mathbf{Q}_3 = \begin{bmatrix} 7 \end{bmatrix}$.

(a) What size is the block diagonal matrix formed from $\mathbf{Q}_1$ and $\mathbf{Q}_2$?

(b) If a block diagonal matrix is formed from $\mathbf{Q}_1$, $\mathbf{Q}_2$, and $\mathbf{Q}_3$, what percent of the entries are zero?

(c) How many different block diagonal matrices can be formed using $\mathbf{Q}_1$, $\mathbf{Q}_2$, and $\mathbf{Q}_3$?

**13.** If an $n \times n$ matrix $\mathbf{A}$ is quite large and sparse, then instead of transmitting all $n^2$ entries of $\mathbf{A}$, one way of reducing transmission time is to send just the numerical value of the nonzero entries and its (row, column) location. That is, we send a set of 3-vectors (triples) of the form $\begin{bmatrix} a_{ij} & i & j \end{bmatrix}$ whenever $a_{ij} \neq 0$.

(a) If $\mathbf{A}$ is $100 \times 100$ and upper triangular (see Section 1.1), is there any savings to using this "triple" strategy to transmit $\mathbf{A}$? Explain how you arrive at your answer.

(b) Let $\mathbf{A}$ be $100 \times 100$ and tridiagonal; that is, the only possible nonzero entries appear in the set of entries

$$\{a_{11}, a_{12}, a_{21}, a_{22}, a_{23}, \dots,$$
$$a_{ii}, a_{i\,i-1}, \dots, a_{100\,99}, a_{100\,100}\}$$

where $i = 2, 3, \dots, 99$. (Figure 2 shows a schematic diagram of a tridiagonal matrix.) Determine the percent of savings if we use the triple strategy to transmit $\mathbf{A}$.

FIGURE 2

**14.** (a) For the matrix in Exercise 13(a), suggest a more efficient way to transmit the data than using the triple strategy. (*Hint*: Use the pattern of upper triangular.)

(b) For the matrix in Exercise 13(b), suggest a more efficient way to transmit the data than using the triple strategy. (*Hint*: Use the pattern of tridiagonal.)

**15.** Let $\mathbf{A}$ be a $100 \times 100$ block diagonal matrix with diagonal blocks of sizes 8, 15, 10, 17, 20, 10, 10, and 10.

(a) Develop a transmission strategy different from the triples of Exercise 13 that will be efficient for block diagonal matrices.

(b) Compute the percent of savings if your strategy is used compared to sending the entire matrix.

*In Exercises 16–22, let*

$$\mathbf{A} = \begin{bmatrix} 2 & 3 \\ 1 & -4 \\ 5 & 0 \end{bmatrix}, \quad \mathbf{B} = \begin{bmatrix} 4 & 2 \\ -2 & 1 \end{bmatrix}, \quad \mathbf{C} = \begin{bmatrix} 5 & -6 \\ 2 & 8 \\ 0 & 1 \end{bmatrix},$$

$$\mathbf{D} = \begin{bmatrix} 3 & 5 \end{bmatrix}, \quad \mathbf{E} = \begin{bmatrix} 1 & 0 & -1 \\ 2 & 2 & 4 \end{bmatrix}, \quad \mathbf{F} = \begin{bmatrix} 2 \\ -1 \end{bmatrix}.$$

*Compute the indicated linear combination, if possible. When it is not possible, give a reason why it is not possible.*

**16.** $2\mathbf{A} - \mathbf{C}$.

**17.** $3\begin{bmatrix} \mathbf{B} \\ \mathbf{D} \end{bmatrix} + \mathbf{C} - 2\mathbf{E}^T$.

**18.** $4\mathbf{B} + \mathbf{F}$.

**19.** $2\begin{bmatrix} \mathbf{F} & \mathbf{D}^T \end{bmatrix} - 3\mathbf{B}$.

**20.** $5\mathbf{A}^T + 3\mathbf{E}$.

**21.** $\begin{bmatrix} \mathbf{B} & \mathbf{D}^T \end{bmatrix} - \mathbf{C}$.

**22.** $2\begin{bmatrix} \mathbf{B} & \mathbf{F} \end{bmatrix} - 3\begin{bmatrix} \mathbf{B}^T & \mathbf{D}^T \end{bmatrix}$.

*In Exercises 23–27, let*

$$\mathbf{A} = \begin{bmatrix} 2+3i & 1-i \\ 3-4i & 7 \end{bmatrix} \quad and \quad \mathbf{B} = \begin{bmatrix} 4 & 2+i \\ 3i & -1+4i \end{bmatrix}.$$

*Compute the indicated linear combination, if possible. When it is not possible, give a reason why it is not possible.*

**23.** $2\mathbf{A} + 3\mathbf{B}$.

**24.** $(1+i)\mathbf{A} - 2\mathbf{B}$.

**25.** $\mathbf{A} + \mathbf{A}^*$.

**26.** $\mathbf{B} - \mathbf{B}^*$.

**27.** $\begin{bmatrix} \mathbf{A} & \mathbf{B} \end{bmatrix} + \begin{bmatrix} \mathbf{B} & \mathbf{A} \end{bmatrix}^*$.

**28.** Let

$$\mathbf{C} = \begin{bmatrix} 2+i & -3i & 1-i \\ 4 & 5+6i & -2+3i \\ 1+i & 0 & 4+2i \end{bmatrix}.$$

Express $\mathbf{C}$ as a linear combination $p\mathbf{A} + q\mathbf{B}$ where $\mathbf{A}$ and $\mathbf{B}$ are $3 \times 3$ matrices with only real entries.

**29.** An $n \times n$ diagonal matrix $\mathbf{A}$ with each diagonal entry the same value $k$ is called a **scalar matrix**, since $\mathbf{A}$ is a scalar multiple of the identity matrix; that is, $\mathbf{A} = k\mathbf{I}_n$.

(a) If $\mathbf{A} = p\mathbf{I}_n$ and $\mathbf{B} = q\mathbf{I}_n$, then show that $\mathbf{A} + \mathbf{B}$ is a scalar matrix.

(b) If $\mathbf{A} = p\mathbf{I}_n$, then show that $r\mathbf{A}$ is a scalar matrix where $r$ is any scalar.

(c) Complete the following statement:
Any linear combination of scalar matrices (of the same size) is a _____.

**30.** Let $\mathbf{A}$ and $\mathbf{B}$ be $n \times n$ symmetric matrices.

(a) Show that $\mathbf{A} + \mathbf{B}$ is symmetric. (*Hint*: Show that $(i, j)$-entry $= (j, i)$-entry.)

(b) Show that $k\mathbf{A}$ is symmetric, where $k$ is any scalar.

(c) Complete the following statement:
Any linear combination of symmetric matrices (of the same size) is a _____.

**31.** A square matrix $\mathbf{A}$ with real entries is called **skew symmetric** provided that $\mathbf{A}^T = -\mathbf{A}$.

(a) What are the diagonal entries of a skew symmetric matrix?

(b) Construct a $3 \times 3$ skew symmetric matrix in which the entries are not all zeros.

*In Exercises 32–37, verify that each statement is true.*

**32.** The set of all 3-vectors is closed.

**33.** The set of all $2 \times 4$ matrices is closed.

**34.** The set of all $4 \times 4$ scalar matrices is closed. (See Exercise 29.)

**35.** The set of all $3 \times 3$ diagonal matrices is closed.

**36.** The set of all $n \times n$ symmetric matrices is closed. (See Exercise 30.)

**37.** The set of all $m \times n$ matrices is closed.

*In Exercises 38–40, verify that each statement is false.*

**38.** The set of all 2-vectors $\begin{bmatrix} a & b \end{bmatrix}$ where $a > 0$ and $b > 0$ is closed.

**39.** The set of all $3 \times 4$ matrices whose $(1, 1)$-entry is 7 is closed.

**40.** The set of all $2 \times 2$ diagonally dominant matrices is closed. (See Section 1.1.)

**41.** Each branch store of World Communications Marketing sells seven products. Each week, every branch sends a 7-vector containing weekly sales (in units sold) to company headquarters. The sales manager at headquarters uses a partitioned matrix strategy to build a sales history week by week for each store and uses a linear combination strategy to determine companywide weekly sales. Develop a proposal for the way the manager stores and processes this data.

**42.** A computer program uses a matrix to supply information regarding colored squares it will display. (The squares are all of the same size.) For an $m \times n$ matrix the program displays $m$ squares vertically and $n$ squares horizontally, while the value of an entry in the data matrix indicates the color of the corresponding square. Suppose that 0 corresponds to black and 1 corresponds to white. The matrix $\mathbf{M} = \begin{bmatrix} 0 & 1 \\ 1 & 0 \end{bmatrix}$ corresponds to the display

(a) Draw the figure corresponding to $\begin{bmatrix} \mathbf{M} & \mathbf{M} \\ \mathbf{M} & \mathbf{M} \end{bmatrix}$.

(b) Using $\mathbf{M}$, construct a data matrix $\mathbf{C}$ that corresponds to a game board that has 12 squares along each side.

(c) Let $\mathbf{P} = \begin{bmatrix} 1 & 1 \\ 1 & 1 \end{bmatrix}$. Draw the figure corresponding to

$$\begin{bmatrix} \mathbf{M} & \mathbf{P} & \mathbf{P} \\ \mathbf{P} & \mathbf{M} & \mathbf{P} \\ \mathbf{P} & \mathbf{P} & \mathbf{M} \end{bmatrix}.$$

**43.** Let $\mathbf{M} = \begin{bmatrix} 0 & 1 \\ 1 & 0 \end{bmatrix}$ and $\mathbf{N} = \begin{bmatrix} 1 & 0 \\ 0 & 1 \end{bmatrix}$. Using the ideas from Exercise 42, do the following.

(a) Express the data matrix for a white $2 \times 2$ square as a linear combination of $\mathbf{M}$ and $\mathbf{N}$.

(b) Express the data matrix for a black $2 \times 2$ square as a linear combination of $\mathbf{M}$ and $\mathbf{N}$.

(c) Construct a data matrix $\mathbf{X}$ that corresponds to the letter $x$, which is a $6 \times 6$ matrix.

**44.** (a) Using the correspondence between polynomials and their coefficient vectors, explain why we can say $P_2$ is closed. (*Hint*: See Exercise 32.)

(b) Using the correspondence between polynomials and their coefficient vectors, explain why we can say that $P_k$ is closed. (*Hint*: See Exercise 37.)

(c) What familiar set of quantities is $P_0$?

**45.** (a) Describe span($S$) where

$$S = \left\{ \begin{bmatrix} 1 \\ 0 \end{bmatrix}, \begin{bmatrix} 0 \\ 1 \end{bmatrix} \right\}.$$

(b) Describe span($S$) where

$$S = \left\{ \begin{bmatrix} 1 \\ 0 \\ 0 \end{bmatrix}, \begin{bmatrix} 0 \\ 0 \\ 1 \end{bmatrix} \right\}.$$

(c) Describe span($V$) where

$$V = \left\{ \begin{bmatrix} 1 & 0 \\ 0 & 0 \end{bmatrix}, \begin{bmatrix} 0 & 0 \\ 0 & 1 \end{bmatrix} \right\}.$$

(d) Describe span($T$) where

$$T = \left\{ \begin{bmatrix} 1 & 0 \\ 0 & 0 \end{bmatrix}, \begin{bmatrix} 0 & 0 \\ 0 & 1 \end{bmatrix}, \begin{bmatrix} 0 & 1 \\ 1 & 0 \end{bmatrix} \right\}.$$

(e) Explain why span($T$) contains span($V$).

**46.** Explain why $\begin{bmatrix} 7 \\ 8 \\ 9 \end{bmatrix}$ is not in span($S$) where

$$S = \left\{ \begin{bmatrix} 1 \\ 0 \\ 0 \end{bmatrix}, \begin{bmatrix} 0 \\ 0 \\ 1 \end{bmatrix} \right\}.$$

**47.** Explain why $\begin{bmatrix} 1 & 2 \\ 3 & 4 \end{bmatrix}$ is not in span($T$) where

$$T = \left\{ \begin{bmatrix} 1 & 0 \\ 0 & 0 \end{bmatrix}, \begin{bmatrix} 0 & 0 \\ 0 & 1 \end{bmatrix}, \begin{bmatrix} 0 & 1 \\ 1 & 0 \end{bmatrix} \right\}.$$

**48.** Explain why $x$ is not in span($V$) where $V = \{x^2, 1\}$.

**49.** Show that $\begin{bmatrix} 3 \\ 1 \end{bmatrix}$ is in span($S$) where $S = \left\{ \begin{bmatrix} 1 \\ 0 \end{bmatrix}, \begin{bmatrix} 0 \\ 1 \end{bmatrix} \right\}$. That is, find scalars $p$ and $q$ so that

$$p \begin{bmatrix} 1 \\ 0 \end{bmatrix} + q \begin{bmatrix} 0 \\ 1 \end{bmatrix} = \begin{bmatrix} 3 \\ 1 \end{bmatrix}.$$

**50.** Show that $\begin{bmatrix} 8 \\ -5 \\ 0 \end{bmatrix}$ is in span($T$) where

$$T = \left\{ \begin{bmatrix} 2 \\ 0 \\ 0 \end{bmatrix}, \begin{bmatrix} 1 \\ -1 \\ 0 \end{bmatrix} \right\}.$$

That is, find scalars $p$ and $q$ so that

$$p \begin{bmatrix} 2 \\ 0 \\ 0 \end{bmatrix} + q \begin{bmatrix} 1 \\ -1 \\ 0 \end{bmatrix} = \begin{bmatrix} 8 \\ -5 \\ 0 \end{bmatrix}.$$

**51.** Show that $5x^2 - 3x$ is in span($V$) where $V = \{x^2, x^2 - x\}$. That is, find scalars $p$ and $q$ so that $p(x^2) + q(x^2 - x) = 5x^2 - 3x$.

**52.** Show that the following properties of matrix addition hold by verifying that the corresponding entries are equal.

(a) $\mathbf{A} + \mathbf{B} = \mathbf{B} + \mathbf{A}$

(b) $\mathbf{A} + (\mathbf{B} + \mathbf{C}) = (\mathbf{A} + \mathbf{B}) + \mathbf{C}$

**53.** Show that the following properties of scalar multiplication hold by verifying that the corresponding entries are equal.

(a) $r(s\mathbf{A}) = (rs)\mathbf{A}$

(b) $(r + s)\mathbf{A} = r\mathbf{A} + s\mathbf{A}$

**54.** Show that the following properties of the transpose hold by verifying that the corresponding entries are equal.

(a) $(\mathbf{A} + \mathbf{B})^T = \mathbf{A}^T + \mathbf{B}^T$

(b) $(r\mathbf{A})^T = r(\mathbf{A}^T)$

**55.** In Section 1.1 we defined $\mathbf{A}$ to be a symmetric matrix provided $a_{ij} = a_{ji}$ or, equivalently, if $\mathbf{A}^T = \mathbf{A}$. Rather than show that "symmetrically" located entries are equal, we need only show that a matrix equals its transpose. In certain instances this approach is simpler. Let $\mathbf{B}$ be any square matrix with real entries. Show that matrix $\frac{1}{2}(\mathbf{B} + \mathbf{B}^T)$ is symmetric.

**56.** Following the approach from Exercise 55, for any square matrix $\mathbf{B}$ with real entries, show that matrix $\frac{1}{2}(\mathbf{B} - \mathbf{B}^T)$ is skew symmetric. (See Exercise 31.)

**57.** Using the results in Exercises 55 and 56, show that any square matrix $\mathbf{B}$ with real entries can be written as a linear combination of a symmetric matrix and a skew symmetric matrix. That is, every square matrix can be written in terms of symmetric and skew symmetric matrices.

**58.** Explain why any nonzero scalar $k$ could be used in place of $\frac{1}{2}$ in Exercises 55 and 56. If $k \neq \frac{1}{2}$ is used, can Exercise 57 be verified using these matrices? Explain.

**59.** A square matrix **A** with complex entries is called **skew Hermitian** provided that $\mathbf{A}^* = -\mathbf{A}$.

    (a) What are the diagonal entries of a skew Hermitian matrix?

    (b) Construct a $3 \times 3$ skew Hermitian matrix in which the entries are not all zeros.

**60.** Using the information in Exercises 55 and 56 as a guide, formulate a conjecture for a Hermitian matrix **P** and for a skew Hermitian matrix **Q** such that a square matrix **A** with complex entries can be written as a linear combination of **P** and **Q**. Verify your conjecture.

**61.** Let $Q(2, -3)$ and $R(1, 4)$ be points in the plane.

    (a) Determine the algebraic equation of a line $L$ through $Q$ and $R$.

    (b) Determine the geometric equation of a line $L$ through $Q$ and $R$.

    (c) Draw the line segment between $Q$ and $R$ and place an arrow head pointing in the positive direction.

    (d) Determine the points on $L$ corresponding to the parameter values $t = \frac{1}{2}$ and $t = \frac{2}{3}$.

**62.** Write out the expressions for the parametric equations of $L$ in (5a) and (5b) for points $Q(a, b)$ and $R(a, e)$.

**63.** Write out the expressions for the parametric equations of $L$ in (5a) and (5b) for points $Q(a, b)$ and $R(d, b)$.

**64.** If $Q(a, b, c)$ and $R(d, e, f)$ are points in 3-space, then it can be shown that a parametric representation of the line $L$ through $Q$ and $R$ is given by

$$x = (1 - t)a + td$$
$$y = (1 - t)b + te$$
$$z = (1 - t)c + tf.$$

That is, $(x, y, z)$ is a point on $L$ for any real value of the parameter $t$. Construct the geometric form of $L$; that is, express the set of all points on $L$ as a linear combination of the coordinates of $Q$ and $R$.

**65.** The parametric form of a cubic curve $g$ in 3-space is given by

$$x = a_1 t^3 + b_1 t^2 + c_1 t + d_1$$
$$y = a_2 t^3 + b_2 t^2 + c_2 t + d_2$$
$$z = a_3 t^3 + b_3 t^2 + c_3 t + d_3.$$

where $t$ is the parameter and $a_j, b_j, c_j, d_j, j = 1, 2, 3$ are real numbers. Express the set of points on curve $g$ as a linear combination of 3-vectors whose coefficients involve the parameter $t$. (In Section 2.1 we show how to obtain the geometric form of a cubic curve.)

## In MATLAB

*The notation for matrix operations in* MATLAB *is very close to the way the corresponding statement is written in this book. A major exception is scalar multiplication where* MATLAB *requires an* $*$, *as in* $k * \mathbf{A}$. *Another exception is that in the text we can use subscripts on scalars and matrices like* $c_1, c_2, \ldots, c_n$ *and* $\mathbf{A}_1, \mathbf{A}_2, \ldots, \mathbf{A}_n$, *but* MATLAB *does not display subscripts. However, the values of the scalars* $c_1, c_2, \ldots, c_n$ *can be stored in an n-vector* **c** *and then accessed by writing* **c(1), c(2), …, c(n)** *in* MATLAB.

**ML. 1.** Enter the following matrices into MATLAB.

$$\mathbf{A} = \begin{bmatrix} 2 & -1 \\ 0 & 3 \\ 4 & 7 \end{bmatrix}, \quad \mathbf{B} = \begin{bmatrix} 1 & 3 & 7 \\ -4 & 2 & -1 \end{bmatrix},$$

$$\mathbf{C} = \begin{bmatrix} 5 & 1 \\ 1 & -5 \\ 2 & 0 \end{bmatrix}, \quad \mathbf{x} = \begin{bmatrix} 4 \\ -1 \\ 2 \end{bmatrix},$$

$$\mathbf{y} = \begin{bmatrix} 6 \\ 0 \\ 1 \end{bmatrix}, \quad \mathbf{z} = \begin{bmatrix} 3 \\ 5 \end{bmatrix}.$$

Execute each of the following MATLAB commands. If any errors arise, explain why they occurred.

    (a) **A + B**         (b) **A + C**

    (c) **B – C**         (d) **[A x] + [C y]**

    (e) **3*A – 4*C**     (f) **z(1)*x + z(2)*y**

    (g) **x(1)*A + x(2)*B' + x(3)*C**

**ML. 2.** We want to form the linear combination

$$3\begin{bmatrix} 0 \\ -1 \\ 1 \end{bmatrix} + 5\begin{bmatrix} 1 \\ -2 \\ 1 \end{bmatrix} - 6\begin{bmatrix} 1 \\ 1 \\ 3 \end{bmatrix}$$

in MATLAB. Each of the following is a way to accomplish this. Execute each set of statements and show that the result is the same.

    (a) Enter vectors

$$\mathbf{c} = [3\ 5\ {-6}], \quad \mathbf{x} = [0\ {-1}\ 1]',$$
$$\mathbf{y} = [1\ {-2}\ 1]', \quad \mathbf{z} = [1\ 1\ 3]'.$$

Then enter the command

$$\mathbf{lincomb} = \mathbf{c}(1)*\mathbf{x} + \mathbf{c}(2)*\mathbf{y} + \mathbf{c}(3)*\mathbf{z}.$$

    (b) Enter **c = [3 5 –6], A = [0 –1 1; 1 –2 1; 1 1 3]'**. Then enter the command

$$\mathbf{lincomb} = \mathbf{c}(1)*\mathbf{A}(:,1) + \mathbf{c}(2)*\mathbf{A}(:,2)$$
$$+ \mathbf{c}(3)*\mathbf{A}(:,3)$$

(c) Enter **c** and **A** as in part (b). Then enter the set of commands

> **lincomb=0; for j=1:3,**
> **lincomb = lincomb + c(j)∗A(:,j), end**.

**ML. 3.** We can extract submatrices in MATLAB by specifying the rows and columns we want to include. Enter matrix **B** into MATLAB where

$$\mathbf{B} = \begin{bmatrix} 1 & 2 & 3 & 4 \\ 5 & 6 & 7 & 8 \\ 9 & 10 & 11 & 12 \\ 13 & 14 & 15 & 16 \end{bmatrix}.$$

Execute each of the following commands and then explain how the submatrix that is displayed was obtained.

(a) **B(:,[1 3])**          (b) **B([1 3],[1 3])**
(c) **B([2:4],3)**          (d) **B([1:3],[2,4])**

**ML. 4.** The application RGB Color Displays shows how to build colors for a video display as linear combinations of the colors Red, Green, and Blue. In MATLAB we can demonstrate this principle. Type **help rgbexamp** and read the display for a set of directions. Next execute the example by typing command **rgbexamp**. Experiment with the sliders to set the intensities and clicking the APPLY button to see the result. Determine, as best you can, the intensities that produce the following colors. (*Note*: Because we are relying on our visual perception, different individuals may determine a different set of intensities for a specified color. Cyan is blue-green in hue.)

(a) white       (b) yellow       (c) magenta
(d) cyan        (e) black        (f) pale purple

**ML. 5.** The application "Polynomials" discussed how to use vectors to represent a polynomial. MATLAB uses this scheme to compute the value of a polynomial in routine **polyval**. For example, with the polynomial $p(x) = -8x^2 + 3x + 5$, we associate the coefficient vector $\mathbf{v} = \begin{bmatrix} -8 & 3 & 5 \end{bmatrix}$. To evaluate this at $x = -1$ in MATLAB enter the following commands.

> **v = [-8 3 5]**
> **x = -1**
> **y = polyval(v,x)**

Check by hand that the value displayed for $y$ is $p(-1)$. The **polyval** command can be used to evaluate $p(x)$ at a set of $x$-values. To see this, enter the following commands into MATLAB:

> **x = [-2 -1 0 1 2]**
> **y = polyval(v,x)**

Check that $\mathbf{y(1)} = p(-2)$, $\mathbf{y(2)} = p(-1)$, $\mathbf{y(3)} = p(0)$, $\mathbf{y(4)} = p(1)$, and $\mathbf{y(5)} = p(2)$. To graph $p(x)$ over $[-2, 2]$ we define a set of $x$-values at equispaced points and then plot them. Enter the following commands.

> **x = -2:0.1:2;**
> **y = polyval(v,x);**
> **figure,plot(x,y)**

Use this procedure to plot the following polynomials over the interval specified using a spacing of 0.1.

(a) $p(x) = 2x^2 - x + 1$, $[2, -2]$.
(b) $p(x) = x^3 - 2x^2 + 2$, $[-1, 3]$.

## True/False Review Questions

*Determine whether each of the following statements is true or false.*

**1.** For matrices **A**, **B**, and **C** of the same size, if $\mathbf{A} = \mathbf{B}$ and $\mathbf{B} = \mathbf{C}$, then $\mathbf{A} = \mathbf{C}$.

**2.** $\begin{bmatrix} 1 & 2 \\ 3 & 4 \end{bmatrix}$ is a submatrix of $\begin{bmatrix} 1 & 2 & 3 \\ 4 & 5 & 6 \\ 7 & 8 & 9 \end{bmatrix}$.

**3.** $\mathbf{A} - \mathbf{B}$ is the same as $\mathbf{B} - \mathbf{A}$.

**4.** If $\mathbf{B} = 3\mathbf{A}$ and $\mathbf{C} = -2\mathbf{B}$, then $\mathbf{C} = -6\mathbf{A}$.

**5.** $\begin{bmatrix} 5 \\ 4 \end{bmatrix}$ is in span $\left\{ \begin{bmatrix} 1 \\ 0 \end{bmatrix}, \begin{bmatrix} 0 \\ \frac{1}{2} \end{bmatrix} \right\}$.

**6.** $\begin{bmatrix} 1 & 0 \\ 0 & 1 \end{bmatrix}$ will generate the set of all $2 \times 2$ diagonal matrices.

**7.** $\begin{bmatrix} 1 & 0 \\ 0 & 1 \end{bmatrix}$ and $\begin{bmatrix} 0 & 1 \\ 1 & 0 \end{bmatrix}$ will generate the set of all $2 \times 2$ matrices of the form $\begin{bmatrix} a & b \\ b & a \end{bmatrix}$ where $a$ and $b$ are any real numbers.

**8.** If $\mathbf{A}$ and $\mathbf{B}$ are $2 \times 2$ with $\mathbf{A}$ symmetric, then $(5\mathbf{A} - 6\mathbf{B})^T = -1(\mathbf{A} + \mathbf{B}^T)$.

**9.** The set of all $2 \times 2$ matrices of the form $\begin{bmatrix} a & 0 \\ b & c \end{bmatrix}$, $a, b, c$ any real numbers, is closed.

**10.** The set of all $1 \times 3$ matrices of the form $\begin{bmatrix} a & 5 & b \end{bmatrix}$ where $a$ and $b$ are any real numbers is closed.

## Terminology

| | |
|---|---|
| Equal matrices | Submatrix |
| Partition | Block diagonal matrix[(P)] |
| Coefficient matrix and right side | Sum and difference of matrices |
| Scalar multiple of a matrix | Negative of a matrix |
| Linear combination | Span of a set |
| Properties of matrix operations | Properties of linear combinations |
| Closure; closed sets | The set of polynomials $P_k$ |
| Scalar matrix[(P)] | Skew symmetric matrix[(P)] |

You should be able to describe the *pattern* of the entries for each type of matrix denoted with a [(P)]. In addition to patterns of entries in a matrix, there are patterns for operations and other concepts. For example,

- What pattern do we use to check to see if two matrices are the same?
- Two block matrices are to be added. What patterns must be present to accomplish this?
- Describe the pattern of the entries that form a column of a coefficient matrix of a linear system of equations.
- What pattern is used to form a linear combination?
- What is the pattern associated with span$\{\mathbf{v}_1, \mathbf{v}_2, \mathbf{v}_3\}$?
- What is the pattern we use to determine if a set is closed?
- What is the pattern that associates a polynomial with a matrix?
- What does it mean to say matrix addition is commutative?

Answer or complete each of the following using only words, no matrix equations.

- How is matrix subtraction performed?
- How is scalar multiplication of a matrix performed?
- The transpose of the transpose of a matrix is _____.
- The transpose of a sum of matrices is _____.
- The sum of two scalar matrices (of the same size) is a _____.
- The sum of two symmetric matrices (of the same size) is a _____.

## 1.3 ■ MATRIX PRODUCTS

The matrix operations developed in Section 1.2 involved element-by-element manipulations. Here we introduce operations that involve row-by-column computations. Such operations provide another algebraic tool for deriving additional information involving matrices.

| In the text. | | In MATLAB. |
|---|---|---|
| $\mathbf{x} \bullet \mathbf{y}$ | $\mathbf{x}$ and $\mathbf{y}$ real $n \times 1$ vectors. | $\mathbf{x'} * \mathbf{y}$ |
| $\bar{\mathbf{x}} \bullet \mathbf{y}$ | $\mathbf{x}$ and $\mathbf{y}$ complex $n \times 1$ vectors. | $\mathbf{x'} * \mathbf{y}$ |
| $\mathbf{AB}$ | Matrix Products. | $\mathbf{A} * \mathbf{B}$ |
| $\mathbf{A}^k$ | Matrix Powers. | $\mathbf{A} \wedge k$ |

Products[1]

As we saw in Section 1.2, a linear combination is a sum of scalars times quantities. Such expressions arise quite frequently, as we illustrated in Example 7 of Section 1.2. In fact, the type of linear combination that appears in Example 7(c) of Section 1.2 is of the form

$$(\text{weight}_1)(\text{grade}_1) + (\text{weight}_2)(\text{grade}_2) + (\text{weight}_3)(\text{grade}_3).$$

This pattern arises so naturally and frequently that we define the following operation.

---

**Definition**    The **dot product** or **inner product** of two $n$-vectors of scalars

$$\mathbf{a} = \begin{bmatrix} a_1 & a_2 & \cdots & a_n \end{bmatrix} \quad \text{and} \quad \mathbf{b} = \begin{bmatrix} b_1 & b_2 & \cdots & b_n \end{bmatrix}$$

is[2]

$$\mathbf{a} \cdot \mathbf{b} = a_1 b_1 + a_2 b_2 + \cdots + a_n b_n = \sum_{j=1}^{n} a_j b_j.$$

---

Thus we see that $\mathbf{a} \cdot \mathbf{b}$ follows the pattern of a linear combination of entries of $n$-vectors $\mathbf{a}$ and $\mathbf{b}$. The result of a dot product is a scalar. We further note that $\mathbf{a} \cdot \mathbf{b} = \mathbf{b} \cdot \mathbf{a}$. (See "The Angle between 2-Vectors," near the end of this section, for a geometric perspective on the dot product.)

In the definition of the dot product we wrote the $n$-vectors as rows for convenience. The dot product of two $n$-vectors is defined between two row vectors, two column vectors, and one row vector and one column vector as long as the vectors have the same number of entries. If $\mathbf{a}$ is a row vector with $n$ entries and $\mathbf{b}$ is a column vector with $n$ entries, then

$$\mathbf{a} \cdot \mathbf{b} = \begin{bmatrix} a_1 & a_2 & \cdots & a_n \end{bmatrix} \cdot \begin{bmatrix} b_1 \\ b_2 \\ \vdots \\ b_n \end{bmatrix} = \sum_{j=1}^{n} a_j b_j$$

is referred to as a **row-by-column product**.

EXAMPLE 1    Let $\mathbf{a} = \begin{bmatrix} 4 & -5 & 2 \end{bmatrix}$ and $\mathbf{b} = \begin{bmatrix} 2 & 1 & -3 \end{bmatrix}$. Then

$$\mathbf{a} \cdot \mathbf{b} = (4)(2) + (-5)(1) + (2)(-3) = -3.$$ ■

---

[1] See Chapter 8 for definitions and examples of these MATLAB operations.

[2] You may already be familiar with this useful notation, the summation notation. It is discussed in Appendix 2.

EXAMPLE 2   Let $\mathbf{u} = \begin{bmatrix} 1 & 1 & 3 & 5 \end{bmatrix}$ and $\mathbf{v} = \begin{bmatrix} -2 \\ 0 \\ 1 \\ 2 \end{bmatrix}$; then the row-by-column product is

$$\mathbf{u} \cdot \mathbf{v} = \begin{bmatrix} 1 & 1 & 3 & 5 \end{bmatrix} \cdot \begin{bmatrix} -2 \\ 0 \\ 1 \\ 2 \end{bmatrix} = (1)(-2) + (1)(0) + (3)(1) + (5)(2) = 11.$$

■

Both the dot product and the row-by-column product are computed as a sum of the products of the corresponding entries of the $n$-vectors. Naturally these operations are not defined between vectors that have a different number of entries.

From Example 8 of Section 1.2, we have the linear combination

$$c_1 \begin{bmatrix} 2 \\ -1 \end{bmatrix} + c_2 \begin{bmatrix} 1 \\ 3 \end{bmatrix} = \begin{bmatrix} 2c_1 + c_2 \\ -c_1 + 3c_2 \end{bmatrix}. \tag{1}$$

The entries on the right side of (1) are linear combinations. In fact, they are row-by-column products:

$$2c_1 + c_2 = \begin{bmatrix} 2 & 1 \end{bmatrix} \cdot \begin{bmatrix} c_1 \\ c_2 \end{bmatrix} \quad \text{and} \quad -c_1 + 3c_2 = \begin{bmatrix} -1 & 3 \end{bmatrix} \cdot \begin{bmatrix} c_1 \\ c_2 \end{bmatrix}.$$

Now we use matrices to organize these two dot products by letting

$$\begin{bmatrix} \begin{bmatrix} 2 & 1 \end{bmatrix} \\ \begin{bmatrix} -1 & 3 \end{bmatrix} \end{bmatrix} = \begin{bmatrix} 2 & 1 \\ -1 & 3 \end{bmatrix}$$

and then writing

$$\begin{bmatrix} 2 & 1 \\ -1 & 3 \end{bmatrix} * \begin{bmatrix} c_1 \\ c_2 \end{bmatrix} = \begin{bmatrix} 2c_1 + c_2 \\ -c_1 + 3c_2 \end{bmatrix}$$

where the $*$ means to use row-by-column multiplication with each row of the $2 \times 2$ matrix and the vector of coefficients. Let $\mathbf{A} = \begin{bmatrix} 2 & 1 \\ -1 & 3 \end{bmatrix}$ and $\mathbf{c} = \begin{bmatrix} c_1 \\ c_2 \end{bmatrix}$. Then

$$\mathbf{A} * \mathbf{c} = \begin{bmatrix} 2 & 1 \\ -1 & 3 \end{bmatrix} * \begin{bmatrix} c_1 \\ c_2 \end{bmatrix} = \begin{bmatrix} \text{row}_1(\mathbf{A}) \cdot \mathbf{c} \\ \text{row}_2(\mathbf{A}) \cdot \mathbf{c} \end{bmatrix} = \begin{bmatrix} 2c_1 + c_2 \\ -c_1 + 3c_2 \end{bmatrix}.$$

From (1) we have that the linear combination can be expressed as

$$c_1 \begin{bmatrix} 2 \\ -1 \end{bmatrix} + c_2 \begin{bmatrix} 1 \\ 3 \end{bmatrix} = \begin{bmatrix} 2 & 1 \\ -1 & 3 \end{bmatrix} * \begin{bmatrix} c_1 \\ c_2 \end{bmatrix} = \mathbf{A} * \mathbf{c}$$

thus $\mathbf{A} * \mathbf{c} =$ linear combination of the columns of $\mathbf{A}$. This is a compact way to indicate a linear combination of columns.

We call $\mathbf{A} * \mathbf{c}$ a **matrix-vector product**. In fact, we can view $\mathbf{A} * \mathbf{c}$ as a kind of dot product of matrix $\mathbf{A}$ partitioned into columns and $\mathbf{c}$ a column vector of scalars:

$$\mathbf{A} * \mathbf{c} = \begin{bmatrix} \text{col}_1(\mathbf{A}) & \text{col}_2(\mathbf{A}) \end{bmatrix} \cdot \begin{bmatrix} c_1 \\ c_2 \end{bmatrix} = c_1 \text{col}_1(\mathbf{A}) + c_2 \text{col}_2(\mathbf{A}).$$

It is easy to see that a matrix-vector product $\mathbf{A} * \mathbf{c}$ makes sense only if the number of columns of $\mathbf{A}$ equals the number of entries in vector $\mathbf{c}$. For simplicity we omit writing the $*$ for a matrix-vector product and use the notation $\mathbf{Ac}$. We make the following definition, which shows *two different but equivalent forms for a matrix-vector product: one in terms of linear combinations and another in terms of dot products.* Both forms will prove to be useful in upcoming work.

---

**Definition**   Let $\mathbf{A}$ be an $m \times n$ matrix and $\mathbf{c}$ a column vector with $n$ entries; then the **matrix-vector product $\mathbf{Ac}$** is the linear combination

$$c_1\text{col}_1(\mathbf{A}) + c_2\text{col}_2(\mathbf{A}) + \cdots + c_n\text{col}_n(\mathbf{A})$$

and the entries of $\mathbf{Ac}$ can be computed directly as dot products using

$$\mathbf{Ac} = \begin{bmatrix} \text{row}_1(\mathbf{A}){\cdot}\mathbf{c} \\ \text{row}_2(\mathbf{A}){\cdot}\mathbf{c} \\ \vdots \\ \text{row}_n(\mathbf{A}){\cdot}\mathbf{c} \end{bmatrix}.$$

---

**EXAMPLE 3**   Let $\mathbf{A} = \begin{bmatrix} 2 & 1 & 4 & 0 \\ -1 & 3 & 5 & 7 \\ 1 & 6 & -2 & -3 \end{bmatrix}$ and $\mathbf{c} = \begin{bmatrix} 1 \\ 2 \\ 0 \\ -1 \end{bmatrix}$. Then the matrix-vector product is the linear combination

$$\mathbf{b} = \mathbf{Ac} = 1\begin{bmatrix} 2 \\ -1 \\ 1 \end{bmatrix} + 2\begin{bmatrix} 1 \\ 3 \\ 6 \end{bmatrix} + 0\begin{bmatrix} 4 \\ 5 \\ -2 \end{bmatrix} + (-1)\begin{bmatrix} 0 \\ 7 \\ -3 \end{bmatrix}.$$

We can compute this linear combination by performing the indicated operations or compute the entries directly as dot products $b_i = \text{row}_i(\mathbf{A}){\cdot}\mathbf{c}$, $i = 1, 2, 3$. Either way we get $\mathbf{b} = \begin{bmatrix} 4 \\ -2 \\ 16 \end{bmatrix}$. ■

The two interpretations of a matrix-vector product $\mathbf{Ac}$ provide us with a versatile tool that we use frequently, especially in Sections 1.5 and 2.1. The next example illustrates an important connection between linear systems of equations and matrices. We will use the technique from this example repeatedly when we focus on solving linear systems of equations.

**EXAMPLE 4**   An important use for a matrix-vector product is to express a linear system of equations as a matrix equation. For the linear system of equations

$$\begin{aligned} 5x_1 - 3x_2 + \phantom{6}x_3 &= \phantom{-}1 \\ -4x_1 + 2x_2 + 6x_3 &= -8 \end{aligned}$$

we have from Section 1.2 that the coefficient matrix is $\mathbf{A}$ and the right side is $\mathbf{b}$ where

$$\mathbf{A} = \begin{bmatrix} 5 & -3 & 1 \\ -4 & 2 & 6 \end{bmatrix}, \quad \mathbf{b} = \begin{bmatrix} 1 \\ -8 \end{bmatrix}.$$

Define the vector of unknowns as $\mathbf{x} = \begin{bmatrix} x_1 \\ x_2 \\ x_3 \end{bmatrix}$. Then

$$\mathbf{Ax} = \begin{bmatrix} 5x_1 - 3x_2 + \phantom{6}x_3 \\ -4x_1 + 2x_2 + 6x_3 \end{bmatrix}$$

and it follows by using equality of matrices that we have **matrix equation Ax = b**. In fact, any linear system of equations can be written as a matrix equation in the form

$$(\textit{\textbf{coefficient matrix}}) * (\textit{\textbf{column vector of unknowns}}) = \textit{\textbf{right side}}. \quad ■$$

In Example 4 of Section 1.2 we showed how to use partitioned matrices to simultaneously represent a pair of linear systems of equations that had the same coefficient matrix but different right sides. (We will use this in Section 2.1.) We follow a similar route with a set of matrix-vector products **Ac**, **Ab**, and **Ax**. Let $\mathbf{B} = \begin{bmatrix} \mathbf{c} & \mathbf{b} & \mathbf{x} \end{bmatrix}$, a partitioned matrix; then by **AB** we mean the partitioned matrix $\begin{bmatrix} \mathbf{Ac} & \mathbf{Ab} & \mathbf{Ax} \end{bmatrix}$.

EXAMPLE 5   Let

$$\mathbf{A} = \begin{bmatrix} 3 & 4 & 2 \\ -1 & 1 & 0 \end{bmatrix}, \quad \mathbf{c} = \begin{bmatrix} 1 \\ 2 \\ 1 \end{bmatrix}, \quad \mathbf{b} = \begin{bmatrix} 0 \\ 1 \\ -1 \end{bmatrix}, \quad \text{and} \quad \mathbf{x} = \begin{bmatrix} -1 \\ 1 \\ 3 \end{bmatrix}.$$

Define

$$\mathbf{B} = \begin{bmatrix} \mathbf{c} & \mathbf{b} & \mathbf{x} \end{bmatrix} = \begin{bmatrix} 1 & 0 & -1 \\ 2 & 1 & 1 \\ 1 & -1 & 3 \end{bmatrix}.$$

Then

$$\mathbf{AB} = \begin{bmatrix} \mathbf{Ac} & \mathbf{Ab} & \mathbf{Ax} \end{bmatrix}$$
$$= \begin{bmatrix} \text{row}_1(\mathbf{A}){\cdot}\mathbf{c} & \text{row}_1(\mathbf{A}){\cdot}\mathbf{b} & \text{row}_1(\mathbf{A}){\cdot}\mathbf{x} \\ \text{row}_2(\mathbf{A}){\cdot}\mathbf{c} & \text{row}_2(\mathbf{A}){\cdot}\mathbf{b} & \text{row}_2(\mathbf{A}){\cdot}\mathbf{x} \end{bmatrix}$$
$$= \begin{bmatrix} 13 & 2 & 7 \\ 1 & 1 & 2 \end{bmatrix}$$

and note that we can also write

$$\mathbf{AB} = \begin{bmatrix} \text{row}_1(\mathbf{A}){\cdot}\text{col}_1(\mathbf{B}) & \text{row}_1(\mathbf{A}){\cdot}\text{col}_2(\mathbf{B}) & \text{row}_1(\mathbf{A}){\cdot}\text{col}_3(\mathbf{B}) \\ \text{row}_2(\mathbf{A}){\cdot}\text{col}_1(\mathbf{B}) & \text{row}_2(\mathbf{A}){\cdot}\text{col}_2(\mathbf{B}) & \text{row}_2(\mathbf{A}){\cdot}\text{col}_3(\mathbf{B}) \end{bmatrix}.$$

Thus it follows that the $(i, j)$-entry of **AB** is $\text{row}_i(\mathbf{A}){\cdot}\text{col}_j(\mathbf{B})$. ■

In Example 5 we showed that $\text{ent}_{ij}(\mathbf{AB}) = \text{row}_i(\mathbf{A}){\cdot}\text{col}_j(\mathbf{B})$ for $i = 1, 2$ and $j = 1, 2, 3$. The next definition generalizes Example 5 and makes precise the idea of a product of matrices. This concept unifies the ideas of dot product, row-by-column product, and matrix-vector product, which we developed previously.

---

**Definition**   Let $\mathbf{A} = \begin{bmatrix} a_{ij} \end{bmatrix}$ be an $m \times n$ matrix and $\mathbf{B} = \begin{bmatrix} b_{ij} \end{bmatrix}$ an $n \times p$ matrix. Then the **product** of matrices **A** and **B**, denoted **AB**, is the $m \times p$ matrix $\mathbf{C} = \begin{bmatrix} c_{ij} \end{bmatrix}$ where

$$c_{ij} = \text{row}_i(\mathbf{A}){\cdot}\text{col}_j(\mathbf{B}) = \begin{bmatrix} a_{i1} & a_{i2} \cdots & a_{in} \end{bmatrix} \begin{bmatrix} b_{1j} \\ b_{2j} \\ \vdots \\ b_{nj} \end{bmatrix} = \sum_{k=1}^{n} a_{ik}b_{kj},$$

$1 \le i \le m$ and $1 \le j \le p$. (The product **AB** is defined only when the number of columns of **A** equals the number of rows of **B**.)

---

Figure 1 illustrates a matrix product $\mathbf{AB}$, where $\mathbf{A}$ is $m \times n$ and $\mathbf{B}$ is $n \times p$.

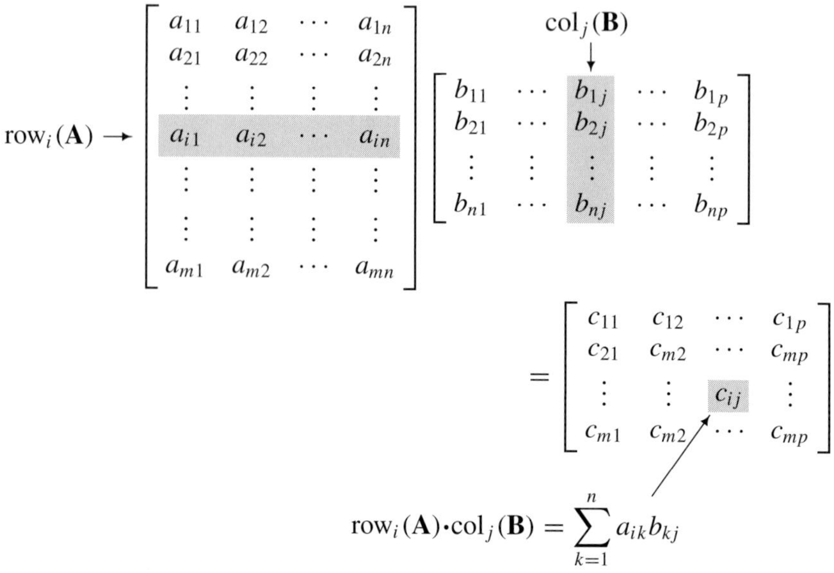

$$\text{row}_i(\mathbf{A}) \rightarrow \begin{bmatrix} a_{11} & a_{12} & \cdots & a_{1n} \\ a_{21} & a_{22} & \cdots & a_{2n} \\ \vdots & \vdots & \vdots & \vdots \\ a_{i1} & a_{i2} & \cdots & a_{in} \\ \vdots & \vdots & \vdots & \vdots \\ \vdots & \vdots & \vdots & \vdots \\ a_{m1} & a_{m2} & \cdots & a_{mn} \end{bmatrix} \begin{bmatrix} b_{11} & \cdots & b_{1j} & \cdots & b_{1p} \\ b_{21} & \cdots & b_{2j} & \cdots & b_{2p} \\ \vdots & \vdots & \vdots & \vdots & \vdots \\ b_{n1} & \cdots & b_{nj} & \cdots & b_{np} \end{bmatrix}$$

$$= \begin{bmatrix} c_{11} & c_{12} & \cdots & c_{1p} \\ c_{21} & c_{m2} & \cdots & c_{mp} \\ \vdots & \vdots & c_{ij} & \vdots \\ c_{m1} & c_{m2} & \cdots & c_{mp} \end{bmatrix}$$

$$\text{row}_i(\mathbf{A}) \cdot \text{col}_j(\mathbf{B}) = \sum_{k=1}^{n} a_{ik} b_{kj}$$

FIGURE 1

EXAMPLE 6

(a) For $\mathbf{A} = \begin{bmatrix} 3 & 4 & 2 \\ -1 & 1 & 0 \end{bmatrix}$ and $\mathbf{B} = \begin{bmatrix} 1 & 0 & -1 \\ 2 & 1 & 1 \\ 1 & -1 & 3 \end{bmatrix}$ we have

$$\mathbf{AB} = \begin{bmatrix} 13 & 2 & 7 \\ 1 & 1 & 2 \end{bmatrix}. \quad \text{(Verify.)}$$

(b) For $\mathbf{C} = \begin{bmatrix} 3 & 2 & -1 & 4 \end{bmatrix}$ and $\mathbf{D} = \begin{bmatrix} 1 & 3 \\ 0 & 5 \\ -1 & -2 \\ 0 & 1 \end{bmatrix}$ we have

$$\mathbf{CD} = \begin{bmatrix} 4 & 25 \end{bmatrix}. \quad \text{(Verify.)}$$

(c) For $\mathbf{E} = \begin{bmatrix} 2 & 1 \\ 2 & 0 \end{bmatrix}$ and $\mathbf{F} = \begin{bmatrix} 1 & -2 \\ 3 & 1 \\ -1 & 2 \end{bmatrix}$ we have that $\mathbf{EF}$ is undefined, but note that

$$\mathbf{FE} = \begin{bmatrix} -2 & 1 \\ 8 & 3 \\ 2 & -1 \end{bmatrix}. \quad \text{(Verify.)}$$

(d) For $\mathbf{G} = \begin{bmatrix} 1 & 3 & 8 \end{bmatrix}$ and $\mathbf{H} = \begin{bmatrix} 2 \\ -6 \\ 2 \end{bmatrix}$ we have $\mathbf{GH} = \mathbf{G} \cdot \mathbf{H} = 0$. (Verify.) We also have

$$\mathbf{HG} = \begin{bmatrix} 2 \\ -6 \\ 2 \end{bmatrix} \begin{bmatrix} 1 & 3 & 8 \end{bmatrix} = \begin{bmatrix} 2 & 6 & 16 \\ -6 & -18 & -48 \\ 2 & 6 & 16 \end{bmatrix}.$$

■

## Products of Matrices versus Products of Real Numbers

Example 6 illustrates several major distinctions between multiplication of matrices and multiplication of scalars.

1. Multiplication of matrices is not defined for all pairs of matrices. The number of columns of the first matrix must equal the number of rows of the second matrix. (See Example 6(c).)
2. Matrix multiplication is not commutative; that is, $\mathbf{AB}$ need not equal $\mathbf{BA}$. (See Examples 6(c) and (d).)
3. The product of two matrices can be zero when neither of the matrices consists of all zeros. (See Example 6(d).)

The definition of a matrix product includes row-by-column products as a $1 \times n$ matrix times an $n \times 1$ matrix and it also includes matrix-vector products as an $m \times n$ matrix times an $n \times 1$ matrix. Thus we will only refer to matrix products from here on, except when we want to emphasize the nature of a product as one of these special cases.

In addition, note that

$$\mathbf{C} = \mathbf{AB} = \mathbf{A} \begin{bmatrix} \text{col}_1(\mathbf{B}) & \text{col}_2(\mathbf{B}) & \cdots & \text{col}_p(\mathbf{B}) \end{bmatrix}$$
$$= \begin{bmatrix} \mathbf{A}\text{col}_1(\mathbf{B}) & \mathbf{A}\text{col}_2(\mathbf{B}) & \cdots & \mathbf{A}\text{col}_p(\mathbf{B}) \end{bmatrix}$$

which says that $\text{col}_j(\mathbf{AB})$ is a linear combination of the columns of $\mathbf{A}$ with scalars from $\text{col}_j(\mathbf{B})$ since

$$\mathbf{A}\text{col}_j(\mathbf{B}) = b_{1j}\text{col}_1(\mathbf{A}) + b_{2j}\text{col}_2(\mathbf{A}) + \cdots + b_{nj}\text{col}_n(\mathbf{A}).$$

Hence linear combinations are the fundamental building blocks of matrix products.

## Matrix Powers

If $\mathbf{A}$ is a square matrix, then products of $\mathbf{A}$ with itself, such as $\mathbf{AA}, \mathbf{AAA}, \ldots$, are defined and we denote them as **powers** of the matrix $\mathbf{A}$ in the form

$$\mathbf{AA} = \mathbf{A}^2, \quad \mathbf{AAA} = \mathbf{A}^3, \ldots.$$

If $p$ is a positive integer, then we define

$$\mathbf{A}^p = \underbrace{\mathbf{AA}\cdots\mathbf{A}}_{p \text{ factors}}.$$

For an $n \times n$ matrix $\mathbf{A}$ we define $\mathbf{A}^0 = \mathbf{I}_n$, the $n \times n$ identity matrix. For nonnegative integers $p$ and $q$, matrix powers obey familiar rules for exponents, namely

$$\mathbf{A}^p\mathbf{A}^q = \mathbf{A}^{p+q}$$

and

$$(\mathbf{A}^p)^q = \mathbf{A}^{pq}.$$

However, since matrix multiplication is not commutative, for general square matrices

$$(\mathbf{AB})^p \neq \mathbf{A}^p \mathbf{B}^p.$$

Matrix powers are useful in applications. In Example 7 we show how to use matrix powers to predict the future distribution of warehouse stock.

EXAMPLE 7    A government employee at an obscure warehouse keeps busy by shifting portions of stock between two rooms. Each week, two-thirds of the stock from room $R_1$ is shifted to room $R_2$ and two-fifths of the stock from $R_2$ to $R_1$. The following transition matrix $\mathbf{T}$ provides a model for this activity:

$$\text{From Room}$$
$$R_1 \quad R_2$$
$$\mathbf{T} = \begin{bmatrix} \frac{1}{3} & \frac{2}{5} \\ \frac{2}{3} & \frac{3}{5} \end{bmatrix} \begin{matrix} R1 \\ R2 \end{matrix} \quad \text{To Room}$$

Entry $t_{ij}$ is the fraction of stock that is shipped to $R_i$ from $R_j$ each week. If $R_1$ contains 3000 items and $R_2$ contains 9000 items, at the end of the week the stock in a room is the amount that remains in the room plus the amount that is shipped in from the other room. We have

$$\tfrac{1}{3}(\text{stock in } R_1) + \tfrac{2}{5}(\text{stock in } R_2) = \tfrac{1}{3}(3000) + \tfrac{2}{5}(9000) = 4600$$

$$\tfrac{2}{3}(\text{stock in } R_1) + \tfrac{3}{5}(\text{stock in } R_2) = \tfrac{2}{3}(3000) + \tfrac{3}{5}(9000) = 7400.$$

So the new stock in each room is a linear combination of the stocks at the start of the week and this can be computed as the product of the transition matrix and $\mathbf{S}_0 = \begin{bmatrix} 3000 \\ 9000 \end{bmatrix}$, the original stock vector. Thus the stock vector at the end of the first week is

$$\mathbf{TS}_0 = \mathbf{S}_1 = \begin{bmatrix} 4600 \\ 7400 \end{bmatrix}.$$

The stock vector at the end of the second week is computed as

$$\mathbf{TS}_1 = \mathbf{T}^2 \mathbf{S}_0 = \begin{bmatrix} 4493 \\ 7507 \end{bmatrix} = \mathbf{S}_2$$

and it follows that

$$\mathbf{S}_3 = \mathbf{TS}_2 = \mathbf{T}^3 \mathbf{S}_0 = \begin{bmatrix} 4500 \\ 7500 \end{bmatrix}.$$

(We have rounded the entries of the stock vectors to whole numbers.)

The following table is a portion of the records that the employee keeps. We see that in just a few weeks the employee can claim to do the shifting of the stock without actually performing the task since the number of items in each room does not change after the third week. We say the process has reached a **steady state**. This

| Original stock | Stock at end of week 1 | Stock at end of week 2 | Stock at end of week 3 | Stock at end of week 4 | Stock at end of week 5 |
|---|---|---|---|---|---|
| 3000 | 4600 | 4493 | 4500 | 4500 | 4500 |
| 9000 | 7400 | 7507 | 7500 | 7500 | 7500 |

type of iterative process is called a **Markov chain** or **Markov process**. The basic question that we pose for such a process is to determine the limiting behavior of the sequence of vectors $\mathbf{S}_0, \mathbf{S}_1, \mathbf{S}_2, \mathbf{S}_3, \ldots$. We show later that the ultimate behavior of the process depends upon the information contained in the transition matrix and we show how to determine the steady state, if it exists. ■

## Properties of Matrix Products

**Properties of Scalar Multiplication and Matrix Products**

Let $r$ be a scalar and $\mathbf{A}$ and $\mathbf{B}$ be matrices. Then

$$r(\mathbf{AB}) = (r\mathbf{A})\mathbf{B} = \mathbf{A}(r\mathbf{B}).$$

The verification (or proof) of this property, which seems quite natural, requires that we use the definitions established earlier. We show this property using the notion that two matrices are equal provided corresponding entries are equal. If we use the notation $\text{ent}_{ij}(\mathbf{C})$ to denote the $(i, j)$-entry of a matrix $\mathbf{C}$, then

$$\text{ent}_{ij}(r(\mathbf{AB})) = r\,\text{ent}_{ij}(\mathbf{AB})$$

$$= r\,\text{row}_i(\mathbf{A}){\cdot}\text{col}_j(\mathbf{B})$$

$$= r \sum_{k=1}^{n} a_{ik}b_{kj} \qquad \begin{array}{l}\textit{(This expression contains only scalars so we} \\ \textit{can use standard arithmetic properties. See} \\ \textit{also Appendix 2.)}\end{array}$$

$$= \sum_{k=1}^{n} (r a_{ik})b_{kj}$$

$$= \text{row}_i(r\mathbf{A}){\cdot}\text{col}_j(\mathbf{B}) = \text{ent}_{ij}((r\mathbf{A})\mathbf{B}).$$

**Properties of Matrix Products and Linear Combinations**

Let $\mathbf{A}$, $\mathbf{B}$, and $\mathbf{C}$ be matrices, $\mathbf{x}$ and $\mathbf{y}$ be vectors, and $r$ and $s$ be scalars. Then

$$\mathbf{A}(\mathbf{B} + \mathbf{C}) = \mathbf{AB} + \mathbf{AC} \quad \text{and} \quad \mathbf{A}(r\mathbf{x} + s\mathbf{y}) = r\mathbf{Ax} + s\mathbf{Ay}.$$

Each of these properties is called a **distributive rule** for matrix multiplication over matrix addition (or linear combinations). The verification of these expressions uses the same technique as previously: Show that corresponding entries are equal. (See Exercise 42.) It also follows that there can be any number of terms in the sums in either expression. For instance,

$$\mathbf{A} \sum_{k=1}^{n} c_k \mathbf{x}_k = \sum_{k=1}^{n} c_k \mathbf{Ax}_k,$$

where the $c$'s are scalars and the $\mathbf{x}$'s are vectors.

**Properties of Matrix Products and the Transpose**

If $\mathbf{A}$ and $\mathbf{B}$ are matrices, then

$$(\mathbf{AB})^T = \mathbf{B}^T\mathbf{A}^T.$$

This expression implies that the transpose of a product is the product of the transposes in reverse order. The verification (or proof) of this property requires that we use the definitions established earlier. If $\mathbf{A}$ is $m \times n$ and $\mathbf{B}$ is $n \times p$, then we show that corresponding entries of $(\mathbf{AB})^T$ and $\mathbf{B}^T\mathbf{A}^T$ are equal. We have

$$\text{ent}_{ij}((\mathbf{AB})^T) = \text{ent}_{ji}(\mathbf{AB}) = \text{row}_j(\mathbf{A})\cdot\text{col}_i(\mathbf{B})$$

$$= \sum_{k=1}^{n} a_{jk}b_{ki} = \sum_{k=1}^{n} b_{ki}a_{jk}$$

$$= \sum_{k=1}^{n} \text{ent}_{ik}(\mathbf{B}^T)\text{ent}_{kj}(\mathbf{A}^T)$$

$$= \text{row}_i(\mathbf{B}^T)\cdot\text{col}_j(\mathbf{A}^T) = \text{ent}_{ij}(\mathbf{B}^T\mathbf{A}^T).$$

EXAMPLE 8    Let

$$\mathbf{A} = \begin{bmatrix} 1 & 3 & 2 \\ 2 & -1 & 3 \end{bmatrix} \quad \text{and} \quad \mathbf{B} = \begin{bmatrix} 0 & 1 \\ 2 & 2 \\ 3 & -1 \end{bmatrix}.$$

Then the product $\mathbf{AB} = \begin{bmatrix} 12 & 5 \\ 7 & -3 \end{bmatrix}$ and therefore $(\mathbf{AB})^T = \begin{bmatrix} 12 & 7 \\ 5 & -3 \end{bmatrix}$ and we see that

$$\mathbf{B}^T\mathbf{A}^T = \begin{bmatrix} 0 & 2 & 3 \\ 1 & 2 & -1 \end{bmatrix}\begin{bmatrix} 1 & 2 \\ 3 & -1 \\ 2 & 3 \end{bmatrix} = \begin{bmatrix} 12 & 7 \\ 5 & -3 \end{bmatrix}.$$ ■

### The Angle between 2-vectors[3]

The dot product of a pair of $n$-vectors provided us with an algebraic tool for computing matrix-vector products and matrix products. However, the dot product is more than just an algebraic construction; it arises naturally when we determine the angle between "geometric" $n$-vectors; see Section 1.1. Here we show how $\mathbf{a}\cdot\mathbf{b}$ is related to the angle between the 2-vectors $\mathbf{a} = \begin{bmatrix} a_1 & a_2 \end{bmatrix}$ and $\mathbf{b} = \begin{bmatrix} b_1 & b_2 \end{bmatrix}$.

Using the geometric representation for 2-vectors given in Section 1.1, we can view vectors $\mathbf{a}$ and $\mathbf{b}$ as shown in Figure 2 where $\alpha$ is the angle between vector $\mathbf{a}$ and the horizontal axis and $\beta$ is the angle between vector $\mathbf{b}$ and the horizontal axis. Hence the angle between vectors $\mathbf{a}$ and $\mathbf{b}$ shown in Figure 2 is $\alpha - \beta$.

From basic trigonometric relationships we have

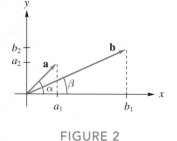

FIGURE 2

$$\cos(\alpha) = \frac{a_1}{\sqrt{a_1^2 + a_2^2}}, \qquad \cos(\beta) = \frac{b_1}{\sqrt{b_1^2 + b_2^2}},$$

$$\sin(\alpha) = \frac{a_2}{\sqrt{a_1^2 + a_2^2}}, \qquad \sin(\beta) = \frac{b_2}{\sqrt{b_1^2 + b_2^2}}.$$

From the trigonometric identity $\cos(\alpha - \beta) = \cos(\alpha)\cos(\beta) + \sin(\alpha)\sin(\beta)$, it

[3]This development was given in "Going Dotty with Vectors," by S. Taverner, in *The Mathematical Gazette*, vol. 81, No. 491, July 1997, pp. 293–295.

follows that

$$\cos(\alpha - \beta) = \frac{a_1 b_1}{\sqrt{a_1^2 + a_2^2} \sqrt{b_1^2 + b_2^2}} + \frac{a_2 b_2}{\sqrt{a_1^2 + a_2^2} \sqrt{b_1^2 + b_2^2}}$$

$$= \frac{a_1 b_1 + a_2 b_2}{\sqrt{a_1^2 + a_2^2} \sqrt{b_1^2 + b_2^2}}$$

$$= \frac{\mathbf{a} \cdot \mathbf{b}}{\sqrt{a_1^2 + a_2^2} \sqrt{b_1^2 + b_2^2}} = \frac{\mathbf{a} \cdot \mathbf{b}}{\sqrt{\mathbf{a} \cdot \mathbf{a}} \sqrt{\mathbf{b} \cdot \mathbf{b}}}.$$

Thus the cosine of the angle between 2-vectors $\mathbf{a}$ and $\mathbf{b}$ has been expressed in terms of dot products. Later we generalize this result to a pair of $n$-vectors.

### Application:  **Wavelets**[4] **I**

Wavelets are a technique for representing information that can lead to **compression** (a reduction in the amount of data needed to adequately represent the information) and to **approximation** (a model that closely resembles the original information). When we can achieve a high compression ratio (that is, when a large percentage of the original data can be omitted) and when we can build accurate models with the data that remains, then we can transmit vast amounts of information quickly and cost-effectively. (For instance, satellite images from space probes require high compression in transmission.)

For example, consider an $n$-vector that contains data. We seek a way to transform the information into a form that reveals intrinsic properties of the data that can be inspected for possible compression and ultimate approximation. It is truly amazing that a simple combination of **averaging** and **differencing** can prove to be an effective tool. We start with a 2-vector $\mathbf{x}$.

**Averaging:**   Let $\mathbf{x} = \begin{bmatrix} x_1 \\ x_2 \end{bmatrix}$; then the average is

$$a_1 = \frac{1}{2} \begin{bmatrix} 1 & 1 \end{bmatrix} \begin{bmatrix} x_1 \\ x_2 \end{bmatrix} = \frac{1}{2}(x_1 + x_2).$$

Thus to average we used a scalar multiple of a dot product.

**Differencing:**   Let $d_1 = x_1 - a_1$.

Then we claim that the vector $\mathbf{y} = \begin{bmatrix} a_1 \\ d_1 \end{bmatrix}$, which contains the average of the data and the difference of the first data entry and the average (that is, the distance between the first data and the average) contains the same information as the vector $\mathbf{x}$. (We say $\mathbf{y}$ is a transformation of the data in $\mathbf{x}$.) To verify this claim, we assume that we know the vector $\mathbf{y}$ and show how to obtain the vector $\mathbf{x}$. In fact, the entries of $\mathbf{x}$ are linear combinations of the entries of $\mathbf{y}$:

$$a_1 + d_1 = a_1 + (x_1 - a_1) = x_1$$

and

$$a_1 - d_1 = a_1 - (x_1 - a_1) = 2a_1 - x_1 = 2\left(\tfrac{1}{2}(x_1 + x_2)\right) - x_1 = x_2.$$

---

[4]This is a first look at the topic of wavelets. We will revisit it as we acquire more matrix tools to develop further steps of the process.

Since vector $\mathbf{y}$ can produce the original data $\mathbf{x}$, the information content of the two vectors is the same. Instead of performing the separate steps of averaging and differencing, we can perform them simultaneously by a matrix vector product. (See Exercise 71.) The transformation process can be extended to $n$-vectors of data using a partitioned matrix approach. (See Exercises 72 through 74.) In Section 2.4 we will show how compression is achieved and briefly discuss a technique for approximation.

### Application: Path Lengths

In Section 1.1 we introduced an incidence matrix as a mathematical representation of a graph. The graph in Figure 3 is represented by the incidence matrix

$$\mathbf{A} = \begin{bmatrix} 0 & 1 & 1 & 0 \\ 1 & 0 & 1 & 0 \\ 1 & 1 & 0 & 1 \\ 0 & 0 & 1 & 0 \end{bmatrix}.$$

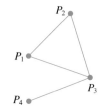

FIGURE 3

We say that there is a **path** between nodes $P_i$ and $P_j$ if we can move along edges of the graph to go from $P_i$ to $P_j$. For example, there is a path between $P_1$ and $P_4$ in Figure 3; traverse the edge from $P_1$ to $P_3$ and then from $P_3$ to $P_4$. However, there is no edge from $P_1$ to $P_4$. For the graph in Figure 3, there is a path connecting every pair of nodes. This is not always the case; see Figure 4. There is no path between $P_1$ and $P_4$. The graph in Figure 4 has two separate, or disjoint, pieces.

In some applications that can be modeled by using a graph there can be more than one edge between a pair of nodes. For the graph in Figure 5 there are two edges between $P_2$ and $P_3$. In this case the incidence matrix is

$$\mathbf{B} = \begin{bmatrix} 0 & 1 & 1 & 0 \\ 1 & 0 & 2 & 0 \\ 1 & 2 & 0 & 1 \\ 0 & 0 & 1 & 0 \end{bmatrix}.$$

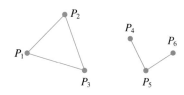

FIGURE 4

We use the term **path of length 1** to mean that there are no intervening nodes of the graph between the ends of the path. (The term *length* here does not mean a measure of the actual distance between nodes, rather it is a count of the number of edges traversed between the starting node and the ending node of a path.) The entries of the incidence matrix of a graph count the number of paths of length 1 between a pair of nodes.

In certain applications it is important to determine the number of paths of a certain length $k$ between each pair of nodes of a graph. Also, one frequently must calculate the longest path length that will ever be needed to travel between any pair of nodes. Both of these problems can be solved by manipulating the incidence matrix associated with a graph. Our manipulation tools will be matrix addition and multiplication.

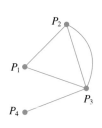

FIGURE 5

Since the incidence matrix $\mathbf{A}$ of a graph is square, its powers $\mathbf{A}^2$, $\mathbf{A}^3$, etc. are defined. We have

$$\mathrm{ent}_{ij}(\mathbf{A}^2) = \mathrm{row}_i(\mathbf{A}) \cdot \mathrm{col}_j(\mathbf{A}) = a_{i1}a_{1j} + a_{i2}a_{2j} + \cdots + a_{in}a_{nj}.$$

We know that

$$a_{i1} = \text{number of paths of length 1 from } P_i \text{ to } P_1$$
$$a_{1j} = \text{number of paths of length 1 from } P_1 \text{ to } P_j$$

so the only way there could be a path from $P_i$ to $P_j$ through node $P_1$ is if both $a_{i1}$ and $a_{1j}$ are not zero. Thus product $a_{i1}a_{1j}$ is the number of paths of length 2 between $P_i$ and $P_j$ through $P_1$. Similarly, $a_{is}a_{sj}$ is the number of paths of length 2 between $P_i$ and $P_j$ through $P_s$. Hence the dot product $\text{row}_i(\mathbf{A}) \cdot \text{col}_j(\mathbf{A})$ gives the total number of paths of length 2 between $P_i$ and $P_j$. It follows that matrix $\mathbf{A}^2$ provides a mathematical model whose entries count the number of paths of length 2 between nodes of the graph. By corresponding arguments, the number of paths of length $k$ from $P_i$ to $P_j$ will be given by the $(i, j)$-entry of matrix $\mathbf{A}^k$.

In order to calculate the longest path length that will ever be required to travel between any pair of nodes in a graph, we proceed as follows. The $(i, j)$-entry of matrix $\mathbf{A} + \mathbf{A}^2$ is the number of paths of length less than or equal to 2 from $P_i$ to $P_j$. Similarly, the entries of $\mathbf{A} + \mathbf{A}^2 + \mathbf{A}^3$ are the numbers of paths of length less than or equal to 3 between nodes of the graph. We continue this process until we find a value of $k$ such that all the nondiagonal entries of

$$\mathbf{A} + \mathbf{A}^2 + \mathbf{A}^3 + \cdots + \mathbf{A}^k$$

are nonzero. This implies that there is a path of length $k$ or less between distinct nodes of the graph. (Why can the diagonal entries be ignored?) The value of $k$ determined in this way is the length of the longest path we would ever need to traverse to go from one node of the graph to any other node of the graph.

## EXERCISES 1.3

*In Exercises 1–3, compute the dot product of each pair of vectors $\mathbf{v}$ and $\mathbf{w}$.*

**1.** $\mathbf{v} = \begin{bmatrix} 2 & -1 & 3 \end{bmatrix}$, $\mathbf{w} = \begin{bmatrix} 4 & 5 & 2 \end{bmatrix}$.

**2.** $\mathbf{v} = \begin{bmatrix} 5 \\ 2 \\ -3 \end{bmatrix}$, $\mathbf{w} = \begin{bmatrix} 0 \\ 3 \\ 2 \end{bmatrix}$.

**3.** $\mathbf{v} = \begin{bmatrix} 1 & -1 & 2 \end{bmatrix}$, $\mathbf{w} = \begin{bmatrix} 3 \\ 3 \\ 5 \end{bmatrix}$.

**4.** Determine the value of $x$ so that $\mathbf{v} \cdot \mathbf{w} = 0$, where
$$\mathbf{v} = \begin{bmatrix} 2 & x & -1 \end{bmatrix} \quad \text{and} \quad \mathbf{w} = \begin{bmatrix} 3 & 2 & 2 \end{bmatrix}.$$

*In Exercises 5–7, compute the dot product of each pair of vectors $\mathbf{v}$ and $\mathbf{w}$.*

**5.** $\mathbf{v} = \begin{bmatrix} 1 \\ 2 \\ -3 \\ 1 \end{bmatrix}$, $\mathbf{w} = \begin{bmatrix} 2 \\ -3 \\ 0 \\ 2 \end{bmatrix}$.

**6.** $\mathbf{v} = \begin{bmatrix} 1 \\ 2 \\ -3 \\ 1 \end{bmatrix}$, $\mathbf{w} = \begin{bmatrix} 5 & 0 & 0 & -6 \end{bmatrix}$.

**7.** $\mathbf{v} = \begin{bmatrix} 0 \\ 1 \\ 0 \\ 1 \end{bmatrix}$, $\mathbf{w} = \begin{bmatrix} 1 \\ -1 \\ 1 \\ 1 \end{bmatrix}$.

**8.** Determine the value of $x$ so that $\mathbf{v} \cdot \mathbf{w} = 0$, where
$$\mathbf{v} = \begin{bmatrix} 1 & -3 & 4 & x \end{bmatrix}$$

and
$$\mathbf{w} = \begin{bmatrix} x & 2 & -1 & 1 \end{bmatrix}.$$

**9.** Determine values for $x$ and $y$ so that $\mathbf{v} \cdot \mathbf{w} = 0$ and $\mathbf{v} \cdot \mathbf{u} = 0$ where $\mathbf{v} = \begin{bmatrix} x & 1 & y \end{bmatrix}$, $\mathbf{w} = \begin{bmatrix} 2 & -2 & 1 \end{bmatrix}$, and $\mathbf{u} = \begin{bmatrix} 1 & 8 & 2 \end{bmatrix}$.

**10.** Determine values for $x$ and $y$ so that $\mathbf{v} \cdot \mathbf{w} = 0$ and $\mathbf{v} \cdot \mathbf{u} = 0$ where $\mathbf{v} = \begin{bmatrix} x & 1 & y \end{bmatrix}$, $\mathbf{w} = \begin{bmatrix} x & -2 & 0 \end{bmatrix}$, and $\mathbf{u} = \begin{bmatrix} 0 & -9 & y \end{bmatrix}$.

**11.** Let $\mathbf{v} = \begin{bmatrix} a & b & c & d \end{bmatrix}$, where the entries are real numbers. If $\mathbf{v} \cdot \mathbf{v} = 0$, then what are the values of $a$, $b$, $c$, and $d$?

**12.** If $\mathbf{v}$ is an $n$-vector of real numbers, when is $\mathbf{v} \cdot \mathbf{v} = 0$? That is, what are the entries of vector $\mathbf{v}$?

**13.** Let $\mathbf{w} = \begin{bmatrix} \sin(\theta) & \cos(\theta) \end{bmatrix}$. Compute $\mathbf{w} \cdot \mathbf{w}$.

**14.** Let $\mathbf{v}$ and $\mathbf{w}$ be $n$-vectors of scalars; then explain why $\mathbf{v} \cdot \mathbf{w} = \mathbf{w} \cdot \mathbf{v}$.

**15.** Let $\mathbf{v}$, $\mathbf{w}$, and $\mathbf{u}$ be $n$-vectors of scalars.
   (a) Show that $\mathbf{v} \cdot (\mathbf{w} + \mathbf{u}) = \mathbf{v} \cdot \mathbf{w} + \mathbf{v} \cdot \mathbf{u}$.
   (b) Show that for any scalar $k$,
$$k(\mathbf{v} \cdot \mathbf{w}) = (k\mathbf{v}) \cdot \mathbf{w} = \mathbf{v} \cdot (k\mathbf{w}).$$

**16.** Use the properties of dot products shown in Exercise 15 to compute the following dot products involving linear combinations where we are given $\mathbf{v} \cdot \mathbf{w} = 2$, $\mathbf{v} \cdot \mathbf{u} = -3$, $\mathbf{v} \cdot \mathbf{x} = 4$, $\mathbf{x} \cdot \mathbf{w} = 1$, $\mathbf{x} \cdot \mathbf{u} = 5$.
   (a) $\mathbf{v} \cdot (\mathbf{w} + 5\mathbf{u})$.       (b) $\mathbf{v} \cdot (3\mathbf{w} - 5\mathbf{x})$.
   (c) $\mathbf{v} \cdot (2\mathbf{w} + 3\mathbf{u} - \mathbf{x})$.   (d) $(\mathbf{x} + \mathbf{v}) \cdot (\mathbf{w} + \mathbf{u})$.

**17.** Let $\mathbf{v} = \begin{bmatrix} a & b & c & d \end{bmatrix}$. Determine a vector $\mathbf{w}$ so that $\mathbf{v}\cdot\mathbf{w} = a + b + c + d$. If $\mathbf{v}$ were an $n$-vector, what is $\mathbf{w}$?

**18.** Use the result from Exercise 17 to develop a formula for the average of the entries in an $n$-vector $\mathbf{v} = \begin{bmatrix} v_1 & v_2 & \cdots & v_n \end{bmatrix}$ in terms of a ratio of dot products.

**19.** A furniture department stocks three sizes of bean bag chairs, small, large, and jumbo. A 3-vector contains the number of each size that was sold in a month, with the first, second, and third entries representing the sales figures for the small, large, and jumbo sizes, respectively. If

$$\mathbf{v1} = \begin{bmatrix} 25 & 20 & 19 \end{bmatrix}, \quad \mathbf{v2} = \begin{bmatrix} 22 & 31 & 24 \end{bmatrix},$$

and

$$\mathbf{v3} = \begin{bmatrix} 19 & 33 & 26 \end{bmatrix}$$

represent the sales vectors for a three-month period, then determine each of the following using linear combinations and dot products as appropriate.

(a) The total sales for the first two months.

(b) The sales for the three-month period.

(c) The average sales for given size of chair for the three-month period.

(d) If the stock at the beginning of the first month is given by vector $\mathbf{s1} = \begin{bmatrix} 120 & 95 & 87 \end{bmatrix}$ and no new shipments were received, determine the stock at the end of the three-month period.

(e) If a shipment of bean bags given by vector $\mathbf{s2} = \begin{bmatrix} 45 & 40 & 30 \end{bmatrix}$ was received during the second month, determine the stock at the end of the three-month period.

(f) If the price in dollars for the small, large, and jumbo bean bags is given by the 3-vector $\mathbf{p} = \begin{bmatrix} 23 & 28 & 35 \end{bmatrix}$, determine the total sales in dollars for each month.

(g) Using the price vector $\mathbf{p}$ from part (f), what is the sales total in dollars for jumbo bean bags over the three-month period?

**20.** In Section 1.2 we discussed how a polynomial of degree $k$ can be represented by its coefficient vector of $k+1$ entries $\mathbf{c} = \begin{bmatrix} c_k & c_{k-1} & \cdots & c_1 & c_0 \end{bmatrix}$, where the value of the subscript corresponds to the power of the unknown in the polynomial expression. If $\mathbf{v} = \begin{bmatrix} x^k & x^{k-1} & \cdots & x & 1 \end{bmatrix}$, then the polynomial is given by a dot product as $\mathbf{c}\cdot\mathbf{v}$. Use this procedure to evaluate each of the following polynomials at the specified values of $x$. In each case, also show the standard expression for the polynomial.

(a) $\mathbf{c} = \begin{bmatrix} 2 & 1 & -3 \end{bmatrix}, x = 4$.

(b) $\mathbf{c} = \begin{bmatrix} 5 & -6 & 2 & 1 \end{bmatrix}, x = -2$.

(c) $\mathbf{c} = \begin{bmatrix} -5 & 2 & 0 & 0 & -1 \end{bmatrix}, x = 3$ and $x = -4$.

**21.** In the case that the entries of $n$-vectors

$$\mathbf{a} = \begin{bmatrix} a_1 & a_2 & \cdots & a_n \end{bmatrix}$$

and

$$\mathbf{b} = \begin{bmatrix} b_1 & b_2 & \cdots & b_n \end{bmatrix}$$

are complex scalars, an alternative[5] to our definition of a dot product is often used. In words, take our dot product of the conjugate of the first vector with the second vector. We denote the conjugate of vector $\mathbf{a}$ as $\bar{\mathbf{a}}$ and it is computed as the conjugate of each entry. (See Appendix 1.) Thus this complex dot product of $\mathbf{a}$ and $\mathbf{b}$ is

$$\bar{\mathbf{a}}\cdot\mathbf{b} = \sum_{k=1}^{n} \bar{a_k}\, b_k.$$

(a) Let $\mathbf{v} = \begin{bmatrix} 2 - i & 4 + 2i \end{bmatrix}$ and

$$\mathbf{w} = \begin{bmatrix} 5 + 3i & 1 - i \end{bmatrix}.$$

Compute the complex dot product of $\mathbf{v}$ and $\mathbf{w}$ as $\bar{\mathbf{v}}\cdot\mathbf{w}$.

(b) Use the vectors in part (a) and compute the complex dot product of $\mathbf{w}$ and $\mathbf{v}$ as $\bar{\mathbf{w}}\cdot\mathbf{v}$.

(c) What is the relationship between the results from parts (a) and (b)?

(d) The result in part (c) generalizes to $n$-vectors of any size. Complete the following: the complex dot product of $\mathbf{v}$ and $\mathbf{w}$ is the _____ of the complex dot product of $\mathbf{w}$ and $\mathbf{v}$.

**22.** Referring to Exercise 21, if we consider $\mathbf{a}$ and $\mathbf{b}$ as column vectors, then the complex dot product is easily expressed in terms of a row-by-column product $\mathbf{a} * \mathbf{b}$, where $*$ is the conjugate transpose[6] as defined in the exercises of Section 1.1. Use this form to compute the dot products of the following pairs of vectors.

(a) $\mathbf{a} = \begin{bmatrix} 2 - 3i \\ 1 + 2i \\ 4 \end{bmatrix}, \mathbf{b} = \begin{bmatrix} 2i \\ 1 - i \\ 3 + 4i \end{bmatrix}$.

(b) $\mathbf{a} = \begin{bmatrix} 1 - 3i \\ 1 + 3i \end{bmatrix}, \mathbf{b} = \begin{bmatrix} 2i \\ 6 \end{bmatrix}$.

(c) $\mathbf{a} = \mathbf{b} = \begin{bmatrix} 2 + 2i \\ 3 \\ 1 - 2i \\ -4i \end{bmatrix}$.

*In Exercises 23–26, express the matrix-vector product $\mathbf{Ac}$ as a linear combination of the columns of $\mathbf{A}$.*

**23.** $\mathbf{A} = \begin{bmatrix} 5 & 4 \\ -1 & 3 \end{bmatrix}, \mathbf{c} = \begin{bmatrix} 2 & -7 \end{bmatrix}$.

[5] The dot product of complex $n$-vectors can take several different forms. The definition may be different in other books.
[6] This technique for the complex dot product is used in later topics.

**24.** $\mathbf{A} = \begin{bmatrix} 2 & 1 & 0 \\ 4 & -3 & 6 \end{bmatrix}$, $\mathbf{c} = \begin{bmatrix} 5 \\ -2 \\ 1 \end{bmatrix}$.

**25.** $\mathbf{A} = \begin{bmatrix} 4 & 2 \\ 2 & 0 \\ 1 & 1 \\ -1 & 0 \end{bmatrix}$, $\mathbf{c} = \begin{bmatrix} -3 \\ 5 \end{bmatrix}$.

**26.** $\mathbf{A} = \begin{bmatrix} 1+i & -2i \\ 3-i & 4+3i \\ 4 & -5+7i \end{bmatrix}$, $\mathbf{c} = \begin{bmatrix} 2+2i \\ 4-i \end{bmatrix}$.

*In Exercises 27–30, express each linear combination as a matrix-vector product* **Ac**.

**27.** $c_1 \begin{bmatrix} 4 \\ 1 \end{bmatrix} + c_2 \begin{bmatrix} 1 \\ -1 \end{bmatrix} + c_3 \begin{bmatrix} 2 \\ 2 \end{bmatrix}$.

**28.** $c_1 \begin{bmatrix} 1 \\ 0 \\ 1 \end{bmatrix} + c_2 \begin{bmatrix} 1 \\ 2 \\ 1 \end{bmatrix}$.

**29.** $\begin{bmatrix} 2 \\ 1 \\ 3 \end{bmatrix} - 4 \begin{bmatrix} -1 \\ 0 \\ 1 \end{bmatrix} + 3 \begin{bmatrix} 1 \\ 1 \\ 2 \end{bmatrix}$.

**30.** $4 \begin{bmatrix} 1+i \\ 2-i \end{bmatrix} - 2 \begin{bmatrix} 5 \\ 2+3i \end{bmatrix}$.

*In Exercises 31–33, express each set of expressions as a matrix-vector product.*

**31.** $5c_1 + 3c_2 - 2c_3$
$c_1 - 6c_2 + c_3$.

**32.** $c_1 - c_2 - 2c_3$
$c_2 + 4c_3$
$3c_1 - 5c_2$ .

**33.** $c_1 + c_2$
$2c_1 - c_2$
$3c_1 + 2c_2$
$-c_1 + 4c_2$.

*In Exercises 34–37, compute each matrix-vector product* **Ac** *using dot products of rows of* **A** *with the column* **c**.

**34.** $\mathbf{A} = \begin{bmatrix} 4 & 2 & -1 \\ 3 & 0 & 1 \end{bmatrix}$, $\mathbf{c} = \begin{bmatrix} 2 \\ 1 \\ -3 \end{bmatrix}$.

**35.** $\mathbf{A} = \begin{bmatrix} 1 & 2 \\ 1 & 0 \\ -2 & 3 \\ 4 & 1 \end{bmatrix}$, $\mathbf{c} = \begin{bmatrix} 2 \\ -1 \end{bmatrix}$.

**36.** $\mathbf{A} = \begin{bmatrix} 4 & 0 & 1 & 2 \\ 3 & -1 & 0 & 1 \end{bmatrix}$, $\mathbf{c} = \begin{bmatrix} 1 \\ 2 \\ -1 \\ 3 \end{bmatrix}$.

**37.** $\mathbf{A} = \begin{bmatrix} 3 & 2 & -1 \\ 0 & 0 & 0 \\ 1 & 0 & -1 \end{bmatrix}$, $\mathbf{c} = \begin{bmatrix} 1 \\ -1 \\ 1 \end{bmatrix}$.

*In Exercises 38–41, write each linear system of equations as a matrix equation* $\mathbf{Ax} = \mathbf{b}$, *where* **A** *is the coefficient matrix,* **x** *is a column of unknowns, and* **b** *is the right side.*

**38.** $4x_1 - 3x_2 = 1$
$x_1 + 2x_2 = 3.$

**39.** $x_1 + 2x_2 = 5$
$-2x_1 + 3x_2 = 4$
$4x_1 - x_2 = 2.$

**40.** $x_1 - x_2 + 2x_3 = 2$
$x_1 - x_3 = -1$
$x_1 + 3x_2 = 0.$

**41.** $x_1 + x_2 = 0$
$2x_1 - x_2 = 0$
$3x_1 + 2x_2 = 0$
$-x_1 + 4x_2 = 0.$

**42.** Verify each of the following properties.
(a) $\mathbf{A}(\mathbf{B} + \mathbf{C}) = \mathbf{AB} + \mathbf{AC}$.
(b) $\mathbf{A}(r\mathbf{x} + s\mathbf{y}) = r\mathbf{Ax} + s\mathbf{Ay}$.

*In Exercises 43–56, compute the indicated product, when possible, given that*

$$\mathbf{A} = \begin{bmatrix} 1 & 0 \\ 2 & 1 \\ -1 & 3 \end{bmatrix}, \quad \mathbf{B} = \begin{bmatrix} 3 & 1 \\ -1 & 2 \end{bmatrix},$$

$$\mathbf{C} = \begin{bmatrix} 0 & -2 \\ 2 & 4 \end{bmatrix}, \quad \mathbf{D} = \begin{bmatrix} 4 & 1 & 2 \\ 0 & -1 & 3 \end{bmatrix},$$

$$\mathbf{x} = \begin{bmatrix} 2 \\ 3 \end{bmatrix}, \quad \mathbf{y} = \begin{bmatrix} -1 \\ 4 \end{bmatrix}.$$

**43.** $\mathbf{AB}$.  **44.** $\mathbf{BA}$.  **45.** $\mathbf{AD}$.

**46.** $\mathbf{DA}$.  **47.** $\mathbf{BC}$.  **48.** $\mathbf{CB}$.

**49.** $(\mathbf{B} + \mathbf{C})\mathbf{D}$.  **50.** $\mathbf{D}(\mathbf{B} + \mathbf{C})$.

**51.** $\mathbf{A}(\mathbf{B} + \mathbf{C})$.  **52.** $(\mathbf{A} + \mathbf{D}^T)\mathbf{B}$.

**53.** $\mathbf{A}(-5\mathbf{x} + 3\mathbf{y})$.  **54.** $\mathbf{C}(2\mathbf{y} - \mathbf{x})$.

**55.** $\mathbf{D}(2\mathbf{x} + \mathbf{y})$.  **56.** $(\mathbf{C} + \mathbf{D})(3\mathbf{x} - 2\mathbf{y})$.

**57.** A hardware store stocks paint in $\frac{1}{2}$-pint, pint, quart, and gallon cans and carries both latex and enamel paints. A matrix of the following form is used to record the stock and weekly sales of each type of paint in each size of can.

| | $\frac{1}{2}$-pt. | pt. | qt. | gal. |
|---|---|---|---|---|
| latex | | | | |
| enamel | | | | |

The price in dollars for the various sizes of latex paint is given by the column vector

$$\mathbf{I} = \begin{bmatrix} 1.98 & 3.29 & 5.98 & 16.98 \end{bmatrix}^T,$$

and $\mathbf{e} = \begin{bmatrix} 2.29 & 3.98 & 6.38 & 18.98 \end{bmatrix}^T$ is the price vector for the enamel. Let the stock matrix at the beginning of a month be

$$S = \begin{bmatrix} 95 & 95 & 120 & 75 \\ 75 & 140 & 160 & 120 \end{bmatrix}.$$

Let sales matrices for a four-week period be, respectively,

$$\mathbf{w1} = \begin{bmatrix} 21 & 18 & 15 & 9 \\ 12 & 21 & 23 & 25 \end{bmatrix},$$

$$\mathbf{w2} = \begin{bmatrix} 18 & 14 & 10 & 16 \\ 15 & 23 & 28 & 40 \end{bmatrix},$$

$$\mathbf{w3} = \begin{bmatrix} 10 & 18 & 16 & 12 \\ 16 & 25 & 25 & 32 \end{bmatrix},$$

$$\mathbf{w4} = \begin{bmatrix} 15 & 21 & 13 & 15 \\ 10 & 18 & 31 & 22 \end{bmatrix}.$$

Determine each of the following using linear combinations and products.

(a) Sales matrix for the first two weeks.

(b) The stock remaining after the third week.

(c) The sales in dollars of latex paint during the third week.

(d) The sales in dollars of enamel paint during the first two weeks.

(e) The average sales in dollars per week of latex paint; of enamel paint.

(f) The value of the stock on hand at the end of the first week.

**58.** Let $\mathbf{A} = \begin{bmatrix} x & 0 \\ 0 & y \end{bmatrix}$. Compute $\mathbf{A}^2$, $\mathbf{A}^3$, and $\mathbf{A}^4$.

**59.** Let $\mathbf{A} = \begin{bmatrix} x & 0 & 0 \\ 0 & y & 0 \\ 0 & 0 & z \end{bmatrix}$. Compute $\mathbf{A}^2$, $\mathbf{A}^3$, and $\mathbf{A}^4$.

**60.** Using the results from Exercises 58 and 59, complete the following conjecture: If $\mathbf{A}$ is a diagonal matrix, then $A^k$ is a diagonal matrix with diagonal entries _____.

**61.** Let $\mathbf{A} = \begin{bmatrix} 0 & 1 & 2 \\ 0 & 0 & 3 \\ 0 & 0 & 0 \end{bmatrix}$. Compute $\mathbf{A}^2$, $\mathbf{A}^3$, and $\mathbf{A}^4$. What is $\mathbf{A}^k$ for $k \geq 5$? Explain.

**62.** If $\mathbf{A}$ has a row of all zeros, explain why $\mathbf{AB}$ will have the same row of all zeros.

**63.** If $\mathbf{B}$ has a column of all zeros, explain why $\mathbf{AB}$ will have the same column of all zeros.

**64.** If matrix $\mathbf{A}$ is square, then a linear combination like $5\mathbf{A}^3 + 3\mathbf{A}^2 - 2\mathbf{A} + 4\mathbf{I}$ is called a polynomial of degree 3 in the matrix $\mathbf{A}$. We view this as though polynomial $p(x) = 5x^3 + 3x^2 - 2x + 4$ is evaluated at the matrix $\mathbf{A}$; that is, $p(\mathbf{A}) = 5\mathbf{A}^3 + 3\mathbf{A}^2 - 2\mathbf{A} + 4\mathbf{I}$. We use the convention that the constant term 4 is $4x^0$, thus $4\mathbf{A}^0 = 4\mathbf{I}$, where $\mathbf{I}$ is the identity matrix of the same size as $\mathbf{A}$. For each of the following compute $p(\mathbf{A})$.

(a) $p(x) = 3x^2 + 2$, $\mathbf{A} = \begin{bmatrix} 1 & 2 \\ 0 & 1 \end{bmatrix}$.

(b) $p(x) = x^2 - x + 3$, $\mathbf{A} = \begin{bmatrix} 2 & 1 \\ -1 & 0 \end{bmatrix}$.

(c) $p(x) = -x^2 + 2x$, $\mathbf{A} = \begin{bmatrix} 1 & 2 & 3 \\ 0 & 0 & 1 \\ 0 & 0 & 0 \end{bmatrix}$.

**65.** Let $\mathbf{a}$ be an $m \times 1$ column vector and $\mathbf{b}$ a $1 \times n$ row vector. The product $\mathbf{ab}$ is an $m \times n$ matrix that is called an **outer product** of vectors $\mathbf{a}$ and $\mathbf{b}$. Compute each of the following outer products. (*Note*: Perform a matrix product not a dot product.)

(a) $\begin{bmatrix} 2 \\ -1 \end{bmatrix} \begin{bmatrix} 3 & 4 & 5 \end{bmatrix}$.

(b) $\begin{bmatrix} 1 \\ 2 \\ 1 \end{bmatrix} \begin{bmatrix} 1 & 2 & 1 \end{bmatrix}$.

(c) $\begin{bmatrix} 1 \\ 0 \\ -1 \end{bmatrix} \begin{bmatrix} 2 & 4 & 6 & 8 \end{bmatrix}$.

(d) $\begin{bmatrix} 1 \\ 1 \\ 1 \end{bmatrix} \begin{bmatrix} 1 & 1 & 1 \end{bmatrix}$.

**66.** Refer to Exercise 65.

(a) Compute the outer product $\mathbf{ab}$ where $\mathbf{a} = \begin{bmatrix} p \\ q \\ r \end{bmatrix}$ and $\mathbf{b} = \begin{bmatrix} x & y & z \end{bmatrix}$.

(b) Write the result in part (a) as a partitioned matrix whose rows are scalar multiples of the same row.

(c) Write the result in part (a) as a partitioned matrix whose columns are scalar multiples of the same column.

**67.** Refer to Exercise 65. Let $\mathbf{a} = \begin{bmatrix} 1 \\ 0 \\ 1 \end{bmatrix}$, $\mathbf{b} = \begin{bmatrix} 1 \\ 1 \\ 0 \end{bmatrix}$, and $\mathbf{c} = \begin{bmatrix} 1 \\ 1 \\ 1 \end{bmatrix}$.

(a) Compute the linear combination of outer products given by $2\mathbf{aa}^T + \mathbf{bb}^T + 3\mathbf{cc}^T$.

(b) Identify the type of square matrix that you computed in part (a).

**68.** Let $\mathbf{A}$ be an $m \times n$ matrix with real entries.

(a) Show that $\mathbf{AA}^T$ is $m \times m$ and symmetric.

(b) Show that $\mathbf{A}^T\mathbf{A}$ is $n \times n$ and symmetric.

(c) Let $\mathbf{v}_1, \mathbf{v}_2, \ldots, \mathbf{v}_k$ be column vectors with $n$ real entries and $c_1, c_2, \ldots, c_k$ be real scalars. Explain why the linear combination of outer products

$$c_1 \mathbf{v}_1 \mathbf{v}_1^T + c_2 \mathbf{v}_2 \mathbf{v}_2^T + \cdots + c_k \mathbf{v}_k \mathbf{v}_k^T = \sum_{j=1}^{k} c_j \mathbf{v}_j \mathbf{v}_j^T$$

is an $n \times n$ symmetric matrix.

**69.** In Example 7, change the transition matrix to

$$\mathbf{T} = \begin{bmatrix} \frac{2}{3} & \frac{1}{2} \\ \frac{1}{3} & \frac{1}{2} \end{bmatrix}.$$

Recompute the stock vectors for the ends of the first five weeks. What appears to be the steady state of this Markov process?

**70.** (*A Markov Chain*) Suppose that only two rival companies, $R$ and $S$, manufacture a certain computer component. Each year, company $R$ keeps $\frac{1}{4}$ of its customers while $\frac{3}{4}$ switch to $S$. Each year, $S$ keeps $\frac{2}{3}$ of its customers while $\frac{1}{3}$ switch to $R$. This information is displayed in the transition matrix $\mathbf{T}$ as

$$\begin{array}{cc} & R \quad\; S \end{array}$$
$$\mathbf{T} = \begin{bmatrix} \frac{1}{4} & \frac{1}{3} \\ \frac{3}{4} & \frac{2}{3} \end{bmatrix} \begin{array}{c} R \\ S \end{array}.$$

When the manufacture of the product first starts, $R$ has $\frac{3}{5}$ of the market (the market is the total number of customers) while $S$ has $\frac{2}{5}$ of the market. We denote the initial distribution of the market by

$$\mathbf{m}_0 = \begin{bmatrix} \frac{3}{5} \\ \frac{2}{5} \end{bmatrix}.$$

Compute the distribution of the market one year later; two years later; three years later. Rounding to two decimal places, what is the steady state market distribution?

**71.** In "Wavelets I" we saw that the entries of the original data vector $\mathbf{x}$ were linear combinations of the entries of the transformed vector $\mathbf{y}$ obtained by averaging and differencing. But in fact the reverse is also true since

$$a_1 = \frac{1}{2}(x_1 + x_2) = \frac{1}{2} \begin{bmatrix} 1 & 1 \end{bmatrix} \begin{bmatrix} x_1 \\ x_2 \end{bmatrix},$$

$$d_1 = x_1 - a_1 = x_1 - \frac{1}{2}(x_1 + x_2)$$

$$= \frac{1}{2}x_1 - \frac{1}{2}x_2 = \frac{1}{2} \begin{bmatrix} 1 & -1 \end{bmatrix} \begin{bmatrix} x_1 \\ x_2 \end{bmatrix}.$$

Use these linear combinations to determine a $2 \times 2$ matrix $\mathbf{M}$ so that $\mathbf{Mx} = \mathbf{y}$.

**72.** Use the matrix $\mathbf{M}$ from Exercise 71 to transform each of the following 2-vectors into the equivalent vector consisting of an average and a difference. In each case verify that the resulting vector $\mathbf{y}$ is correct.

(a) $\mathbf{x} = \begin{bmatrix} 2 \\ 6 \end{bmatrix}.$     (b) $\mathbf{x} = \begin{bmatrix} -3 \\ 6 \end{bmatrix}.$

(c) $\mathbf{x} = \begin{bmatrix} 8 \\ -3 \end{bmatrix}.$     (d) $\mathbf{x} = \begin{bmatrix} 5 \\ 1 \end{bmatrix}.$

**73.** If the information we wish to average and difference as in our discussion of wavelets is an $n$-vector, $n > 2$, where $n$ is even, then we reshape this vector into a matrix $\mathbf{D}_1$ of size $2 \times (n/2)$ and multiply it on the left by the matrix $\mathbf{M}$ determined in Exercise 71. Let

$$\mathbf{x} = \begin{bmatrix} 2 & 6 & -3 & 6 & 8 & -3 & 5 & 1 \end{bmatrix}$$

and define matrix $\mathbf{D}_1$ to be the $2 \times 4$ matrix

$$\mathbf{D}_1 = \begin{bmatrix} 2 & -3 & 8 & 5 \\ 6 & 6 & -3 & 1 \end{bmatrix}.$$

Show that the product $\mathbf{MD}_1$ gives the same results as the individual transformations in Exercise 72. Here the result is a $2 \times 4$ matrix whose columns are the average-difference pairs computed in Exercise 72. (The reshaping is not really a partitioning but is nonetheless a valuable way to rearrange the data in vector $\mathbf{x}$.)

**74.** Use the procedure of Exercise 73 to obtain the average-difference pairs for the vector

$$\mathbf{x} = \begin{bmatrix} 1 & 2 & 3 & 4 & 5 & 6 & 7 & 8 & 9 & 10 & 11 & 12 \end{bmatrix}.$$

**75.** In Exercise 73 we assumed that the $n$-vector had an even number of entries to perform the reshaping. If $n$ were odd, make a proposal about how to obtain the transformed information.

**76.** Do each of the following. (These results will be useful in Section 2.5.)

(a) Let $\mathbf{z}_i$ be an $n$-vector that has zero entries except possibly in the $i$th entry. Explain why $\mathbf{z}_i \cdot \mathbf{z}_j = 0$ provided $i \neq j$.

(b) Let $\mathbf{Z}$ be the $n \times n$ matrix of all zeros. Let $\mathbf{Z}_{ij}$ be the $n \times n$ matrix whose entries are all zero except possibly in the $(i, j)$-entry. Explain why $\mathbf{Z}_{ij} \mathbf{Z}_{rs} = \mathbf{Z}$ when $i \neq r$ or $j \neq s$.

(c) Let $\mathbf{E}_{ij} = \mathbf{I}_n + \mathbf{Z}_{ij}$, where $\mathbf{I}_n$ is the $n \times n$ identity matrix and $\mathbf{Z}_{ij}$ is as defined in part (b). Explain why $\mathbf{E}_{ij} \mathbf{E}_{rs} = \mathbf{I}_n + \mathbf{Z}_{ij} + \mathbf{Z}_{rs}$ when $i \neq r$ or $j \neq s$.

(d) In part (c), if both $\mathbf{E}_{ij}$ and $\mathbf{E}_{rs}$ are lower triangular, explain why $\mathbf{E}_{ij} \mathbf{E}_{rs}$ is lower triangular.

**77.** Use the graph in Figure 5.

(a) Determine the number of paths of length 2 from $P_2$ to $P_1$. Make a list of the paths of the form $P_2 P_j P_1$.

(b) Determine the number of paths of length 3 from $P_2$ to $P_2$. Make a list of the paths of the form $P_2 P_j P_i P_2$.

(c) Is there a path of length 2 from $P_3$ to $P_4$?

## In MATLAB

*In this section we define the dot product of any two n-vectors* **x** *and* **y**. *In* MATLAB *this computation is most easily performed when both vectors are columns, that is, $n \times 1$ matrices. Whether the vectors have real or complex entries the dot product is given by* **x**′ ∗ **y**. *In the real case this just performs a row (**x**′) by column (**y**) product. In the complex case* **x**′ *is a row of conjugates of the entries of column* **x** *so the result is our complex dot product.*

*The operation of matrix multiplication requires the explicit use of the multiplication operator* ∗, *as in* **A** ∗ **B**. *Similarly, powers of a square matrix require the explicit use of the exponentiation operator* ^.

**ML. 1.** Compute the dot product of each of the following pairs of vectors in MATLAB.

(a) $\begin{bmatrix} 1 \\ 2 \\ -1 \end{bmatrix}, \begin{bmatrix} 4 \\ -5 \\ 3 \end{bmatrix}.$  (b) $\begin{bmatrix} 1+i \\ 2i \end{bmatrix}, \begin{bmatrix} 2-i \\ 4+2i \end{bmatrix}.$

(c) $\begin{bmatrix} 4 \\ 3 \\ 0 \\ 1 \end{bmatrix}, \begin{bmatrix} i \\ 2 \\ 1+i \\ 1+2i \end{bmatrix}.$

**ML. 2.** Use MATLAB to check your work in Exercises 43–56.

**ML. 3.** Let $\mathbf{A} = \begin{bmatrix} 1 & 2 & 0 \\ 2 & -1 & 2 \\ 1 & 4 & 1 \end{bmatrix}$. Compute $p(\mathbf{A})$ where $p(x) = x^3 - 5x^2 + 2x$. (See Exercise 64.)

**ML. 4.** In Example 7 we introduced a Markov chain or Markov process. From some initial state we computed future states by successive matrix multiplication. Using MATLAB we can perform the necessary matrix algebra and provide a graphical display of the long-term behavior of the process. Consider the following situation, which can also be modeled by a Markov chain.

At "Perfect City" the weather is either sunny, cloudy, or rainy. The day-to-day change from one type of weather to another is controlled by a transition probability, which is an entry of a 3 by 3 matrix **A**, called a transition matrix. Once we know the vector **w** of probabilities for today's weather, we compute the probability vector for tomorrow's weather from the matrix product **A** ∗ **w**. We use this process repeatedly to determine the weather probabilities for future days.

Today's weather probability vector = **w**

Tomorrow's weather probability vector = **A** ∗ **w**

The next day's weather probability vector = **A** ∗ **Aw**

and so on

One such transition matrix **A** for this weather model is

Today
Sunny Cloudy Rainy
$\begin{bmatrix} \frac{1}{2} & \frac{1}{4} & \frac{1}{8} \\ \frac{1}{4} & \frac{3}{8} & \frac{5}{8} \\ \frac{1}{4} & \frac{3}{8} & \frac{1}{4} \end{bmatrix}$ Sunny
Cloudy  Tomorrow
Rainy

We interpret this as follows: the (2, 3)-entry implies that if it is rainy today then the probability that it will be cloudy tomorrow is $\frac{5}{8}$. Other entries are interpreted in a similar fashion. If the current day's weather is expressed as a vector of probabilities, like

$$\mathbf{w} = \begin{bmatrix} \frac{1}{4} \\ \frac{1}{2} \\ \frac{1}{4} \end{bmatrix} \begin{array}{l} \text{Sunny} \\ \text{Cloudy} \\ \text{Rainy} \end{array}$$

then the weather probability vector for the next day is computed as **A** ∗ **w**. In MATLAB enter the command **weather** and follow the directions using the matrix **A** and vector **w** given previously to see the long-term behavior for the probabilities of the three types of weather and a graphical display. For predicting eight days ahead, the graphical display of the behavior should appear as in Figure 6. From the graphs, estimate the numerical value of the long-term probability for each type of weather.

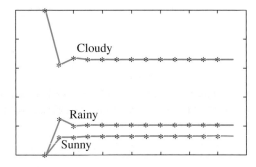

FIGURE 6

**ML. 5.** Use **weather** as in Exercise ML.4, but change the initial vector **w** to $\begin{bmatrix} \frac{1}{2} & \frac{1}{3} & \frac{1}{6} \end{bmatrix}'$. Write a short paragraph that compares the behavior after eight predictions in this case with that of Exercise ML.4. Experiment with other initial vectors. (The entries must be greater than or equal to zero and add up to 1.) Does changing the initial vector have any effect on the long-term behavior?

**ML. 6.** Repeat Exercise ML.4 using the transition matrix

$$\begin{bmatrix} \frac{2}{3} & \frac{1}{3} & \frac{1}{4} \\ \frac{1}{3} & \frac{1}{3} & \frac{1}{2} \\ 0 & \frac{1}{3} & \frac{1}{4} \end{bmatrix}.$$

You select the starting vector **w**. Describe the long-term behavior for the probabilities of the three types of weather.

**ML. 7.** The matrix **M** of Exercise 71 is $\frac{1}{2}\begin{bmatrix} 1 & 1 \\ 1 & -1 \end{bmatrix}$. In Exercise 73 we reshaped the vector

$$\mathbf{x} = \begin{bmatrix} 2 & 6 & -3 & 6 & 8 & -3 & 5 & 1 \end{bmatrix}$$

into the $2 \times 4$ matrix

$$\mathbf{D}_1 = \begin{bmatrix} 2 & -3 & 8 & 5 \\ 6 & 6 & -3 & 1 \end{bmatrix}$$

and then computed the product $\mathbf{MD}_1$ to obtain the averages and differences. MATLAB has a command that will reshape a matrix. The command is named **reshape**. Enter

$$\mathbf{x} = \begin{bmatrix} 2 & 6 & -3 & 6 & 8 & -3 & 5 & 1 \end{bmatrix}$$

into MATLAB and then use the command **reshape(x,2,4)** to obtain the matrix $\mathbf{D}_1$. Experiment

with the **reshape** command by investigating the output of each of the following.

**reshape(x,4,2)   reshape(x,8,1)**
**reshape(x,3,3)**

**ML. 8.** Construct the incidence matrix for the graph in Figure 7.

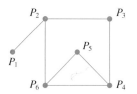

FIGURE 7

(a) Determine the number of paths of length 3 from $P_5$ to $P_2$.

(b) Determine the number of paths of length 2 from $P_4$ to $P_6$.

(c) Determine the length $k$ of the longest path needed to travel between any pair of distinct nodes.

## True/False Review Questions

*Determine whether each of the following statements is true or false.*

**1.** If **x** and **y** are both $3 \times 1$ matrices then $\mathbf{x}\cdot\mathbf{y}$ is the same as $\mathbf{x}^T\mathbf{y}$.

**2.** If **A** is $3 \times 2$ and **x** is $2 \times 1$ then $\mathbf{Ax}$ is a linear combination of the columns of **A**.

**3.** If **A** is $4 \times 2$ and **B** is $2 \times 4$, then $\mathbf{AB} = \mathbf{BA}$.

**4.** If **A** is $10 \times 7$ and **B** is $7 \times 8$, then $\text{ent}_{5,2}(\mathbf{AB}) = \text{row}_5(\mathbf{A})\text{col}_2(\mathbf{B})$.

**5.** Matrix multiplication is commutative.

**6.** If **A** is $3 \times 3$ and **b** is $3 \times 1$ so that $\mathbf{Ab} = \begin{bmatrix} 0 \\ 0 \\ 0 \end{bmatrix}$, then either **A** is a zero matrix or **b** is a zero column.

**7.** If **A** and **B** are the same size, then $\mathbf{AB} = \mathbf{BA}$.

**8.** Powers of a matrix are really successive matrix products.

**9.** The dot product of a pair of real $n$-vectors **x** and **y** is commutative; that is, $\mathbf{x}\cdot\mathbf{y} = \mathbf{y}\cdot\mathbf{x}$.

**10.** Let $\mathbf{x} = \begin{bmatrix} 1 & 1 & 1 \end{bmatrix}$, then $(\mathbf{x}^T\mathbf{x})^2 = 3\mathbf{x}^T\mathbf{x}$.

## Terminology

| | |
|---|---|
| Dot product; inner product | Matrix; Row-by-column product |
| Matrix-vector product | Matrix equation |
| Matrix product | Matrix powers |
| Markov chain; steady state | Angle between 2-vectors |

There is a connection between some of the terms listed in the preceding table. Understanding and being able to explicitly state such connections is important in later sections. For practice with the concepts of this section, formulate responses to the following questions and statements in your own words.

- How are dot products and row-by-column products related?
- Describe how matrix-vector products are computed using dot products.
- Explain how we compute a matrix-vector product as a linear combination of columns.
- Explain how we formulate every linear system of equations as a matrix equation. (Explicitly describe all matrices used in the equation.)
- Explain how to express the matrix product **AB** in terms of matrix-vector products.
- Explain how to express the matrix product **AB** in terms of dot products.
- For what type of matrix **A** can we compute powers, $\mathbf{A}^2, \mathbf{A}^3, \ldots$, etc. Explain why there is such a restriction.
- What do we mean when we say that matrix multiplication distributes over matrix addition?
- Complete the following: The transpose of a product of two matrices is _____.
- Explain what we mean by the steady state of a Markov chain. (Express this relationship in words and using a matrix equation.)

## 1.4 ■ GEOMETRY OF LINEAR COMBINATIONS

Sections 1.2 and 1.3 focused on the algebra of matrices and linear combinations. We showed how linear combinations could be used to build new vectors and matrices and how matrix products are constructed from linear combinations. In this section we explore a geometric representation of linear combinations of vectors, concentrating on 2-vectors and 3-vectors.

**Linear Combination Game**

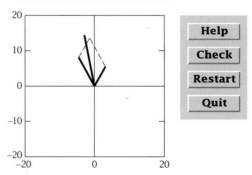

In MATLAB, **lincombo.m**

Let $R^n$ denote the set of all $n$-vectors with real entries. $R^n$ is called **$n$-space**[1]. Whether we regard a vector **a** in $R^n$ as a row or column will be determined by the

---

[1]We use $C^n$ to denote the set of all $n$-vectors with complex entries. $C^n$ is called complex $n$-space. We restrict our attention to $R^n$ in this section.

context in which we use the vector. Recall that we have shown that the following properties hold in $R^n$:

$R^n$ is closed under both addition of vectors and scalar multiplication.[2] (See Exercise 37 in Section 1.2.)

For **a**, **b** and **c** in $R^n$ and scalars $r$ and $k$ we have

$$\mathbf{a} + \mathbf{b} = \mathbf{b} + \mathbf{a} \qquad \textit{(commutativity of addition)}$$
(see Exercise 52(a) in Section 1.2.)

$$\mathbf{a} + (\mathbf{b} + \mathbf{c}) = (\mathbf{a} + \mathbf{b}) + \mathbf{c} \qquad \textit{(associativity of addition)}$$
(see Exercise 52(b) in Section 1.2.)

$$k(\mathbf{a} + \mathbf{b}) = k\mathbf{a} + k\mathbf{b} \qquad \text{(see Section 1.2.)}$$

$$(k + r)\mathbf{a} = k\mathbf{a} + r\mathbf{a} \qquad \text{(see Exercise 53(b) in Section 1.2.)}$$

$$k(r\mathbf{a}) = (kr)\mathbf{a} \qquad \text{(see Exercise 53(a) in Section 1.2.)}$$

These properties are the tools for manipulating linear combinations. There are three other properties of addition and scalar multiplication in $R^n$, which we list now:

- There is a unique vector **z**, so that $\mathbf{a} + \mathbf{z} = \mathbf{a}$, for any vector **a**. **z** is called the **zero vector** and $\mathbf{z} = \begin{bmatrix} 0 & 0 & \cdots & 0 \end{bmatrix}$ in $R^n$. We use **0** to denote the zero vector in $R^n$.
- For a vector **b**, there exists a unique vector **x**, so that $\mathbf{b} + \mathbf{x} = \mathbf{0}$, the zero vector. We call **x** the **negative** of **b** and denote it as $-\mathbf{b}$ or, equivalently, as $(-1)\mathbf{b}$.
- For any vector **b**, $1\mathbf{b} = \mathbf{b}$.

(See Exercises 1 and 2.)

Let **a** be in $R^2$. Then $\mathbf{a} = \begin{bmatrix} a_1 & a_2 \end{bmatrix}$ is a **matrix**, but we can also view **a** as an ordered pair representing the coordinates of a **point** $(a_1, a_2)$ in the $xy$-plane as in Figure 1. From a more geometric point of view, **a** is a directed line segment, a vector[3], from the origin $(0, 0)$ to the point $(a_1, a_2)$ as in Figure 2.

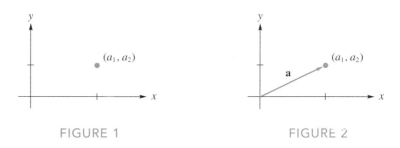

FIGURE 1  FIGURE 2

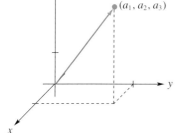

FIGURE 3

Hence **a** in $R^2$ can have three equivalent meanings: a matrix, a point, and a vector. The same is true for members of $R^3$ as shown in Figure 3. These representations are also valid in $R^n$, $n > 3$, but it is hard to provide a picture beyond the case $n = 3$.

Such multiple interpretations give us flexibility regarding the information within the members of $R^2$, $R^3$, and $R^n$. We can choose the form best suited for a particular situation.

[2]In this section all scalars are taken to be real numbers.

[3]Note that the term *vector* has another meaning. In Section 1.1 we called a row or column matrix a vector, and here we also call a directed line segment a vector. This merely shows that algebra and geometry are interchangeable and complement one another.

Next we develop geometric interpretations of linear combinations in $R^2$, which we generalize on an intuitive basis to $R^n$. Let **a** be in $R^2$. Then for a scalar $k$, $k\mathbf{a} = \begin{bmatrix} ka_1 & ka_2 \end{bmatrix}$. Figure 4 gives us a pictorial representation of the vector $k\mathbf{a}$ depending upon the value of $k$.

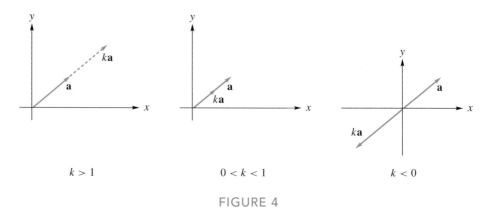

$$k > 1 \qquad\qquad 0 < k < 1 \qquad\qquad k < 0$$

FIGURE 4

From Figure 4 we see that scalar multiplication can change the length of a vector while maintaining the same direction when $k > 0$ or reversing the direction for $k < 0$. If $k \geq 1$ then $k\mathbf{a}$ is called a dilation (or a stretching) of **a**, while for $0 < k < 1$ we have that $k\mathbf{a}$ is called a contraction (or a shrinking) of **a**.

## Length of a Vector

To compute the **length** or **magnitude** of a vector **a** we use its representation as a directed line segment starting at the origin and terminating at the point determined by **a**. Thus in $R^2$ we compute the distance from $(0, 0)$ to $(a_1, a_2)$:

$$\text{distance from } (0, 0) \text{ to } (a_1, a_2) = \sqrt{(a_1 - 0)^2 + (a_2 - 0)^2} = \sqrt{a_1^2 + a_2^2}.$$

(Note the ease with which we switched from matrix to vector to point.) The length of vector **a** is denoted by $\|\mathbf{a}\|$. Thus we have

$$\|k\mathbf{a}\| = \sqrt{(ka_1 - 0)^2 + (ka_2 - 0)^2} = \sqrt{k^2 a_1^2 + k^2 a_2^2}$$

$$= \sqrt{k^2(a_1^2 + a_2^2)} = |k|\sqrt{a_1^2 + a_2^2} = |k|\,\|\mathbf{a}\|.$$

Hence scalar multiplication alters, or scales, the length of a vector by the absolute value of the scalar. The idea of length generalizes to $R^n$, where for **x** in $R^n$,

$$\|\mathbf{x}\| = \sqrt{x_1^2 + x_2^2 + \cdots + x_n^2} = \sqrt{\sum_{j=1}^{n} x_j^2}$$

and it follows that $\|k\mathbf{x}\| = |k|\,\|\mathbf{x}\|$.

EXAMPLE 1   Let
$$\mathbf{a} = \begin{bmatrix} 2 & 3 \end{bmatrix}, \quad \mathbf{b} = \begin{bmatrix} -1 & 0 & 4 \end{bmatrix}, \quad \text{and} \quad \mathbf{c} = \begin{bmatrix} 1 & 2 & -1 & -4 & 0 \end{bmatrix}.$$
Then
$$\|-3\mathbf{a}\| = |-3|\,\|\mathbf{a}\| = |-3|\sqrt{2^2 + 3^2} = 3\sqrt{13},$$
$$\|\mathbf{b}\| = \sqrt{(-1)^2 + 0^2 + 4^2} = \sqrt{17},$$

and

$$\|2\mathbf{c}\| = |2|\,\|\mathbf{c}\| = 2\sqrt{1^2 + 2^2 + (-1)^2 + (-4)^2 + 0^2} = 2\sqrt{22}. \qquad \blacksquare$$

The expression for the length or magnitude of a vector $\mathbf{x}$ in $R^n$ can be written in another way. From the definition of dot product in Section 1.2 we have that $\mathbf{x}\cdot\mathbf{x} = x_1^2 + x_2^2 + \cdots + x_n^2$, so

$$\|\mathbf{x}\| = \sqrt{\mathbf{x}\cdot\mathbf{x}}.$$

This is a convenient connection between the algebraic notion of the dot product and the geometric visualization of members of $R^n$ as vectors. This expression for the length of a vector will be used frequently.

We can give another interpretation of a scalar multiple of a member of $R^n$. We see from Figure 4 that if $\mathbf{b} = k\mathbf{a}$, then geometrically the vectors $\mathbf{a}$ and $\mathbf{b}$ are parallel; that is, they point either in the same direction or in opposite directions. Hence $\mathbf{a}$ and $\mathbf{b}$ are parallel vectors provided there is a scalar $k$ so that $\mathbf{b} = k\mathbf{a}$; that is, they are scalar multiples of one another.

## A Geometric Model for Sums and Differences of Vectors in $R^2$

If $\mathbf{a}$ and $\mathbf{b}$ are in $R^2$, then

$$\mathbf{a} + \mathbf{b} = \begin{bmatrix} a_1 & a_2 \end{bmatrix} + \begin{bmatrix} b_1 & b_2 \end{bmatrix} = \begin{bmatrix} a_1 + b_1 & a_2 + b_2 \end{bmatrix}.$$

A geometric interpretation is that the sum, $\mathbf{a} + \mathbf{b}$, is a new vector found by connecting $(0, 0)$ to the point $(a_1 + b_1, a_2 + b_2)$ as shown in Figure 5.

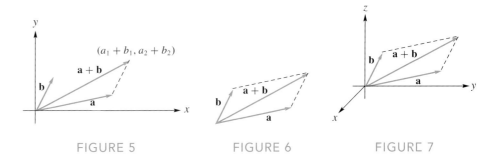

FIGURE 5     FIGURE 6     FIGURE 7

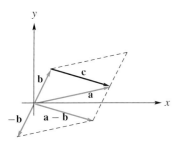

FIGURE 8

If we complete the drawing in Figure 5 into a parallelogram as in Figure 6, we see that vector $\mathbf{a} + \mathbf{b}$ is the diagonal of the parallelogram. Hence addition in $R^n$ is sometimes referred to as the **parallelogram rule**. In $R^3$ the parallelogram rule appears geometrically the same, but technically we are in the plane through the origin determined by the vectors $\mathbf{a}$ and $\mathbf{b}$. (See Figure 7.)

To visualize the difference $\mathbf{a} - \mathbf{b}$, recall that $\mathbf{a} - \mathbf{b} = \mathbf{a} + (-\mathbf{b}) = \mathbf{a} + (-1)\mathbf{b}$. Thus in $R^2$ we use the parallelogram rule on $\mathbf{a}$ and $-\mathbf{b}$ as shown in Figure 8. In terms of the parallelogram determined by $\mathbf{a}$ and $\mathbf{b}$, vector $\mathbf{a} - \mathbf{b}$ is in the same direction as directed line segment $\mathbf{c}$ and they have the same length. We call $\mathbf{c}$ the reverse diagonal of the parallelogram. In $R^3$ the result is identical and generalizes to $R^n$.

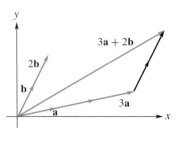

FIGURE 9

Geometrically a vector **a** is a directed line segment starting with the initial point at the origin and ending at the terminal point whose coordinates are the entries of **a**. In $R^2$, $(0, 0)$ is the initial point or **tail** and $(a_1, a_2)$ is the terminal point or **head** of **a**. To form Figure 5 we copied vector **b** onto the head of vector **a** and then the line segment representing **a** + **b** was drawn from $(0, 0)$ to the head of the copy of vector **b**. This **head-to-tail** construction is how we geometrically construct sums of vectors (Figures 5 and 7) and differences of vectors (Figure 8) in $R^2$ and $R^3$. Next we use this head-to-tail construction to form a linear combination. The linear combination combines the dilation/contraction notion from scalar multiplication with the parallelogram law. This is illustrated in Figure 9.

**EXAMPLE 2**    Use head-to-tail construction to illustrate $(\mathbf{a} + \mathbf{b}) + \mathbf{c}$ for the vectors given in Figure 10. The steps are shown in Figure 11.

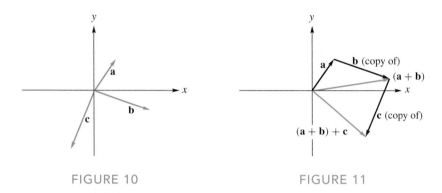

FIGURE 10                          FIGURE 11                          ■

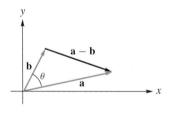

FIGURE 12

Behind the scenes in both scalar multiples and sums is the notion of the angle $\theta$ between a pair of vectors **a** and **b**. Again we use $R^2$ to model the general situation. From the parallelogram rule we have the triangle shown in Figure 12. Using the law of cosines[4], we have

$$\|\mathbf{a} - \mathbf{b}\| = \|\mathbf{a}\|^2 + \|\mathbf{b}\|^2 - 2\|\mathbf{a}\| \|\mathbf{b}\| \cos(\theta)$$

or

$$\cos(\theta) = \frac{\|\mathbf{a} - \mathbf{b}\|^2 - \|\mathbf{a}\|^2 - \|\mathbf{b}\|^2}{-2\|\mathbf{a}\| \|\mathbf{b}\|}.$$

Now we use the connection between lengths and dot products developed earlier so that

$$\|\mathbf{a} - \mathbf{b}\|^2 = (\mathbf{a} - \mathbf{b})\cdot(\mathbf{a} - \mathbf{b}) = \mathbf{a}\cdot\mathbf{a} - 2\mathbf{a}\cdot\mathbf{b} + \mathbf{b}\cdot\mathbf{b} = \|\mathbf{a}\|^2 - 2\mathbf{a}\cdot\mathbf{b} + \|\mathbf{b}\|^2$$

and thus

$$\cos(\theta) = \frac{\|\mathbf{a}\|^2 - 2\mathbf{a}\cdot\mathbf{b} + \|\mathbf{b}\|^2 - \|\mathbf{a}\|^2 - \|\mathbf{b}\|^2}{-2\|\mathbf{a}\| \|\mathbf{b}\|} = \frac{\mathbf{a}\cdot\mathbf{b}}{\|\mathbf{a}\| \|\mathbf{b}\|}. \qquad (1)$$

An almost identical argument can be carried out in $R^3$ to show that Equation (1) gives the cosine of the angle $\theta$ between 3-vectors **a** and **b**.

[4]In Section 1.3 we gave an alternative development in $R^2$. Here we emphasize the lengths of vectors to obtain the expression in (1).

## Angles in $R^n$

We also define the cosine of the angle $\theta$ between the vectors $\mathbf{a}$ and $\mathbf{b}$ in $R^n$ by the expression in (1). Since $-1 \le \cos(\theta) \le 1$, we must be sure that the expression

$$\frac{\mathbf{a}\cdot\mathbf{b}}{\|\mathbf{a}\|\,\|\mathbf{b}\|}$$

has a value in $[-1, 1]$ for $n > 3$. The key to showing this result is the Cauchy-Schwartz-Buniakovsky inequality, whose proof we omit at this time (see Section 6.2), which states that $|\mathbf{a}\cdot\mathbf{b}| \le \|\mathbf{a}\|\,\|\mathbf{b}\|$ for any vectors $\mathbf{a}$ and $\mathbf{b}$ in $R^n$. From this result it follows that if neither $\mathbf{a}$ nor $\mathbf{b}$ is the zero vector, then

$$-1 \le \frac{\mathbf{a}\cdot\mathbf{b}}{\|\mathbf{a}\|\,\|\mathbf{b}\|} \le 1.$$

EXAMPLE 3 Find the angle between the pairs of vectors in each of the following.

(a) For $\mathbf{a} = \begin{bmatrix} 2 & 1 \end{bmatrix}$ and $\mathbf{b} = \begin{bmatrix} -3 & 3 \end{bmatrix}$ we have

$$\cos(\theta) = \frac{-6+3}{\sqrt{5}\,\sqrt{18}} = \frac{-3}{\sqrt{5}\,\sqrt{18}}.$$

Thus

$$\theta = \arccos\left(\frac{-3}{\sqrt{5}\,\sqrt{18}}\right) \approx 1.89 \text{ radians} \approx 108.43°.$$

(b) For $\mathbf{u} = \begin{bmatrix} 1 & 3 & 2 & 1 \end{bmatrix}$ and $\mathbf{v} = \begin{bmatrix} 4 & -2 & 1 & 0 \end{bmatrix}$ we have

$$\cos(\theta) = \frac{4-6+2+0}{\sqrt{15}\,\sqrt{21}} = \frac{0}{\sqrt{15}\,\sqrt{21}} = 0.$$

Thus

$$\theta = \arccos(0) = \frac{\pi}{2} \text{ radians} = 90°. \qquad ■$$

If $\mathbf{a}$ and $\mathbf{b}$ are perpendicular, that is, if the angle between them is $\frac{\pi}{2}$ radians, then the law of cosines implies that

$$0 = \cos\left(\frac{\pi}{2}\right) = \frac{\mathbf{a}\cdot\mathbf{b}}{\|\mathbf{a}\|\,\|\mathbf{b}\|}$$

and thus it follows that $\mathbf{a}\cdot\mathbf{b} = 0$. (This assumes that neither $\mathbf{a}$ nor $\mathbf{b}$ is the zero vector, which has length 0.) In summary, we have that nonzero vectors $\mathbf{a}$ and $\mathbf{b}$ are **perpendicular** or **orthogonal** if and only if $\mathbf{a}\cdot\mathbf{b} = 0$. (Again notice the close connection between geometry and algebra.) We will use the term *orthogonal* from now on.

## Application:  Translations

In our head-to-tail construction for linear combinations, the step of copying vector $\mathbf{b}$ at the head of vector $\mathbf{a}$ as in Figure 5 can be viewed as a shift or translation of vector $\mathbf{b}$ by vector $\mathbf{a}$. By this we mean that the tail of vector $\mathbf{b}$ was translated to point $(a_1, a_2)$ and the head of vector $\mathbf{b}$ was translated to point $(a_1 + b_1, a_2 + b_2)$. This is shown in Figure 13, where we see the translation by following the dashed

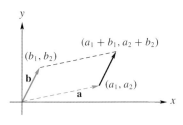

FIGURE 13

path "along" vector **a**. Algebraically we have

$$\text{New tail:} \quad (0, 0) + (a_1, a_2)$$
$$\text{New head:} \ (b_1, b_2) + (a_1, a_2)$$

where these expressions involve points, not vectors. The copy or result of the translation is a directed line segment that goes through point $(a_1, a_2)$ with the same direction and length as vector **b**, or equivalently the line segment from $(a_1, a_2)$ to $(a_1 + b_1, a_2 + b_2)$. (Note that we have been careful not to call the result of the translation a vector because that would imply that its tail is at the origin.) However, a translation or a shift of a set of vectors plays a fundamental role in understanding the structure of the solution set of a system of linear equations, so we want to relate it to vectors and vector addition.

We adopt the following approach. Consider $L$ to be the set of all vectors of the form $k\mathbf{b}$, where $k$ is any scalar and **b** is a nonzero vector in $R^2$. (Again we use $R^2$ as a model because the geometry and algebra here are easy, but the following development generalizes to $R^n$.) Every member of $L$ is parallel to every other member since each is a scalar multiple of vector **b**. The heads of the members of $L$ lie on a line through the origin. The line goes through points $(0, 0)$ and $(b_1, b_2)$ so its equation is

$$\mathbf{y} = \frac{b_2}{b_1}\mathbf{x} \quad \text{if } b_1 \neq 0$$

and $\mathbf{x} = \mathbf{0}$ if $b_1 = 0$. Note that an equivalent equation is

$$\mathbf{y} = \frac{kb_2}{kb_1}\mathbf{x}, \quad k \neq 0 \text{ and } b_1 \neq 0.$$

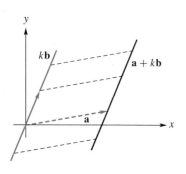

Hence we can say that the set of vectors $L$ determines a line through the origin. It is more convenient to use $k\mathbf{b}$ to represent such a line.

Returning to the notion of translation, we say that $\mathbf{a} + k\mathbf{b}$ is a translation of line $k\mathbf{b}$. (Note that here we are adding vectors **a** and $k\mathbf{b}$.) In $R^2$, $\mathbf{a} + k\mathbf{b}$ is a line parallel to line $k\mathbf{b}$ through point $(a_1, a_2)$ (see Figure 14). Hence we conclude that any line not through the origin is a translation of some line $k\mathbf{b}$.

FIGURE 14

## EXERCISES 1.4

1. Let **a** be any vector in $R^n$. Show that there exists a vector **z** in $R^n$ such that $\mathbf{a} + \mathbf{z} = \mathbf{a}$. Explicitly state the entries of vector **z**.

2. Show that vector **z** from Exercise 1 is unique. That is, show that there is one and only one vector **z** in $R^n$ such that $\mathbf{a} + \mathbf{z} = \mathbf{a}$ for every $n$-vector **a**. (*Hint*: Suppose that it is true that $\mathbf{a} + \mathbf{b} = \mathbf{a}$ for every vector **a** in $R^n$. Start with the fact that $\mathbf{b} = \mathbf{b} + \mathbf{z}$ and use Exercise 1 to finish the argument to show $\mathbf{b} = \mathbf{z}$.)

3. Let **b** be a vector in $n$-space. Show that there exists a unique $n$-vector **x** such that $\mathbf{b} + \mathbf{x} = \mathbf{z}$. (See Exercise 1.)

4. Let $\mathbf{a} = \begin{bmatrix} a_1 & a_2 \end{bmatrix}$ be a vector in 2-space, $R^2$. Assume that $a_1 > 0$ and $a_2 > 0$; then draw a right triangle that could be used to compute the formula for the length of **a** and carefully label its sides.

5. Repeat Exercise 4 for vector $\mathbf{a} = \begin{bmatrix} a_1 & a_2 & a_3 \end{bmatrix}$ in 3-space, when all of its entries are positive.

*In Exercises 6–9, determine the length of the given vector using the expression for length in terms of dot products.*

6. $\mathbf{a} = \begin{bmatrix} 5 \\ -1 \\ 3 \end{bmatrix}$.

7. $\mathbf{b} = \begin{bmatrix} 4 & -5 \end{bmatrix}$.

8. $\mathbf{c} = \begin{bmatrix} 0 & 2 & 1 & -3 \end{bmatrix}$.

9. $\mathbf{d} = \text{col}_3(\mathbf{I}_5)$.

10. Using dot products, explain why the length of a vector can never be negative.

*In Exercises 11–14, determine all values of the scalar $t$ so that the given condition is satisfied.*

11. $\|\mathbf{a}\| = 3$ for $\mathbf{a} = \begin{bmatrix} 2 & t & 0 \end{bmatrix}$.

12. $\|\mathbf{b}\| = 5$ for $\mathbf{b} = \begin{bmatrix} 1 & t & -1 & t \end{bmatrix}$.

**13.** $\|\mathbf{c}\| = 1$ for $\mathbf{c} = \dfrac{1}{t}\begin{bmatrix} 1 \\ 2 \\ -2 \end{bmatrix}$.

**14.** $\|\mathbf{d}\| = 1$ for $\mathbf{d} = \dfrac{1}{t}\begin{bmatrix} 0 & 1 & -1 & 2 \end{bmatrix}$.

**15.** An $n$-vector $\mathbf{a}$ is called a **unit vector** if $\|\mathbf{a}\| = 1$. For any $n$-vector except the zero vector $\mathbf{0}$, we can find a scalar $k \neq 0$ so that $\mathbf{u} = \dfrac{1}{k}\mathbf{a}$ is a unit vector. Determine a formula for $k$ and then use it in each of the following. (*Hint*: Compute $\|\mathbf{u}\| = \left\| \dfrac{1}{k}\mathbf{a} \right\|$ and see Exercises 13 and 14.)

(a) Determine the unit vector associated with $\mathbf{a} = \begin{bmatrix} 1 \\ 3 \\ 0 \end{bmatrix}$.

(b) Determine the unit vector associated with $\mathbf{b} = \begin{bmatrix} 1 & 4 & -3 & 1 \end{bmatrix}$.

(c) Determine the unit vector associated with $\mathbf{c} = \begin{bmatrix} \dfrac{1}{\sqrt{3}} & \dfrac{1}{\sqrt{3}} & \dfrac{-1}{\sqrt{3}} \end{bmatrix}$.

*In Exercises* 16–18, *use head-to-tail construction to determine geometrically* $\mathbf{a} + \mathbf{b}$ *and* $\mathbf{a} - \mathbf{b}$.

**16.**    **17.**

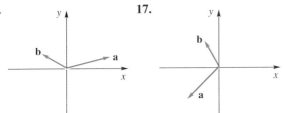

**18.**

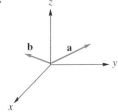

**19.** Use head-to-tail construction on the vectors in Figure 10 to show that $\mathbf{a} + (\mathbf{b} + \mathbf{c})$ is the same as $(\mathbf{a} + \mathbf{b}) + \mathbf{c}$, which is displayed in Figure 11. Figure 10 is repeated here for convenience; put your figure on the axes supplied.

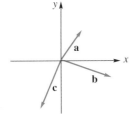

FIGURE 10 (again)          Put your figure here.

**20.** Use the vectors shown in Figure 15 with head-to-tail construction to determine the vector $2\mathbf{a} - 3\mathbf{b} + \mathbf{c}$. Put your figure on the axes supplied.

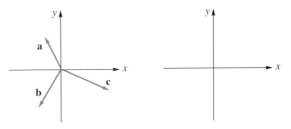

FIGURE 15          Put your figure here.

**21.** Determine which pairs of vectors from the following set of vectors in $R^4$ are orthogonal.

$$\mathbf{a} = \begin{bmatrix} 1 & 2 & 1 & 0 \end{bmatrix}, \quad \mathbf{b} = \begin{bmatrix} 0 & -1 & 2 & 4 \end{bmatrix}$$
$$\mathbf{c} = \begin{bmatrix} 5 & 1 & -7 & 1 \end{bmatrix}, \quad \mathbf{d} = \begin{bmatrix} -8 & 4 & 0 & 1 \end{bmatrix},$$
$$\mathbf{e} = \begin{bmatrix} -5 & 2 & 1 & 0 \end{bmatrix}.$$

**22.** Determine all values of the scalar $t$ such that $\mathbf{a} \cdot \mathbf{b} = 0$, where $\mathbf{a} = \begin{bmatrix} t & 2 & 1 \end{bmatrix}$ and $\mathbf{b} = \begin{bmatrix} 1 & t & 6 \end{bmatrix}$.

**23.** Determine all values of the scalar $t$ such that $\mathbf{a} \cdot \mathbf{b} = 0$, where $\mathbf{a} = \begin{bmatrix} t & 3 & 2 & t \end{bmatrix}$ and $\mathbf{b} = \begin{bmatrix} t & -2 & 3 & 1 \end{bmatrix}$.

**24.** Determine vector $\mathbf{c}$ so that both $\mathbf{a}$ and $\mathbf{b}$ are orthogonal to $\mathbf{c}$, where $\mathbf{a} = \begin{bmatrix} 1 & 2 & 3 \end{bmatrix}$, $\mathbf{b} = \begin{bmatrix} 0 & 1 & 4 \end{bmatrix}$, and $\mathbf{c} = \begin{bmatrix} t & s & 1 \end{bmatrix}$.

**25.** Determine vector $\mathbf{c}$ so that both $\mathbf{a}$ and $\mathbf{b}$ are orthogonal to $\mathbf{c}$, where

$$\mathbf{a} = \begin{bmatrix} 1 \\ -2 \\ 0 \\ 3 \end{bmatrix}, \quad \mathbf{b} = \begin{bmatrix} 0 \\ -1 \\ 4 \\ 0 \end{bmatrix}, \quad \text{and} \quad \mathbf{c} = \begin{bmatrix} 2 \\ t \\ 1 \\ s \end{bmatrix}.$$

**26.** Find a unit vector $\mathbf{c}$ of the form $\mathbf{c} = \begin{bmatrix} c_1 & 0 & c_3 \end{bmatrix}$ that is orthogonal to both $\mathbf{a} = \begin{bmatrix} -2 & 2 & 2 \end{bmatrix}$ and $\mathbf{b} = \begin{bmatrix} 1 & 8 & -1 \end{bmatrix}$.

**27.** Let $\mathbf{A} = \begin{bmatrix} 2 & 1 \\ -4 & -2 \end{bmatrix}$. Find a nonzero vector $\mathbf{c} = \begin{bmatrix} c_1 \\ c_2 \end{bmatrix}$ so that the matrix-vector product $\mathbf{Ac} = \begin{bmatrix} 0 \\ 0 \end{bmatrix}$. (*Hint*: Recall that

$$\mathbf{Ac} = \begin{bmatrix} \text{row}_1(\mathbf{A}) \cdot \mathbf{c} \\ \text{row}_2(\mathbf{A}) \cdot \mathbf{c} \end{bmatrix},$$

so $\mathbf{c}$ is to be orthogonal to each row of matrix $\mathbf{A}$.)

**28.** Let

$$\mathbf{A} = \begin{bmatrix} 2 & -2 & 1 \\ 1 & 0 & 1 \\ 0 & -2 & -1 \end{bmatrix}.$$

Find a vector $\mathbf{c}$ of the form $\mathbf{c} = \begin{bmatrix} c_1 \\ 1 \\ c_3 \end{bmatrix}$ so that the matrix-vector product $\mathbf{Ac} = \mathbf{0}$.

**29.** Show that it is not possible to find a vector $\mathbf{c} \neq \mathbf{0}$ so that $\mathbf{Ac} = \mathbf{0}$, where $\mathbf{A} = \begin{bmatrix} 2 & 1 \\ 1 & 1 \end{bmatrix}$.

**30.** Determine all vectors $\mathbf{c} = \begin{bmatrix} c_1 \\ c_2 \\ c_3 \\ c_4 \end{bmatrix}$ such that $\mathbf{I}_4 \mathbf{c} = \mathbf{0}$. Use this result to answer the following.

The only vector orthogonal to each of the rows of $\mathbf{I}_4$ is _____.

Explain why the word *row* can be replaced by the word *column* in the preceding statement.

*In Exercises* 31–34, *find the cosine of the angle between each pair of vectors* **a** *and* **b**.

**31.** $\mathbf{a} = \begin{bmatrix} 1 & -2 & 0 \end{bmatrix}$ and $\mathbf{b} = \begin{bmatrix} 3 & -2 & 1 \end{bmatrix}$.

**32.** $\mathbf{a} = \begin{bmatrix} 0 & 4 & 2 & 3 \end{bmatrix}$ and $\mathbf{b} = \begin{bmatrix} 0 & 2 & -1 & 0 \end{bmatrix}$.

**33.** $\mathbf{a} = \begin{bmatrix} 1 \\ 4 \\ 1 \\ 1 \end{bmatrix}$ and $\mathbf{b} = \begin{bmatrix} -1 \\ 1 \\ 2 \\ -2 \end{bmatrix}$.

**34.** $\mathbf{a} = \begin{bmatrix} 1 \\ -1 \\ 2 \end{bmatrix}$ and $\mathbf{b} = \begin{bmatrix} -2 \\ 0 \\ 1 \end{bmatrix}$.

---

In this section we restricted ourselves to $R^n$ so that we could use geometric illustrations from $R^2$ and $R^3$. However, the notions of length, unit vectors, and orthogonal vectors generalize to $C^n$ when we use the complex dot product as defined in Exercise 21 of Section 1.3. If $\mathbf{v}$ and $\mathbf{w}$ are in $C^n$, then

$$\|\mathbf{v}\| = \sqrt{\bar{\mathbf{v}} \cdot \mathbf{v}}, \qquad \frac{\mathbf{v}}{\|\mathbf{v}\|} \text{ is a unit vector,}$$

and $\mathbf{v}$ and $\mathbf{w}$ are orthogonal provided $\bar{\mathbf{v}} \cdot \mathbf{w} = 0$.

---

**35.** Determine the length of each of the following vectors in $C^n$.

(a) $\begin{bmatrix} 2 - 3i \\ 1 + 2i \\ 4 \end{bmatrix}$.  (b) $\begin{bmatrix} 2i \\ 6 \end{bmatrix}$.  (c) $\begin{bmatrix} 2 + 3i \\ 3 \\ 1 - 2i \\ -4i \end{bmatrix}$.

**36.** For each vector in Exercise 35, determine a unit vector parallel to it.

**37.** Determine whether each of the following pairs of vectors is orthogonal.

(a) $\begin{bmatrix} 2 + i \\ i \end{bmatrix}$, $\begin{bmatrix} 1 + 3i \\ 5 - 5i \end{bmatrix}$.

(b) $\begin{bmatrix} i \\ 2 - i \end{bmatrix}$, $\begin{bmatrix} 3 + 2i \\ i \end{bmatrix}$.

(c) $\begin{bmatrix} 1 + 2i \\ 4 \\ -i \end{bmatrix}$, $\begin{bmatrix} 2 \\ 1 + i \\ 6i \end{bmatrix}$.

**38.** We say that $\mathbf{v}$ is orthogonal to $\mathbf{w}$ provided $\bar{\mathbf{v}} \cdot \mathbf{w} = 0$. Show that if $\mathbf{v}$ is orthogonal to $\mathbf{w}$, then $\mathbf{w}$ is orthogonal to $\mathbf{v}$.

### In MATLAB

*The geometry of linear combinations can be illustrated using* MATLAB*'s graphics. In the following exercises we introduce three* MATLAB *routines that can be used for instruction and student experimentation. These routines display the geometric properties of the algebra developed in this section.*

**ML. 1.** In MATLAB type **help vecdemo**. This routine provides a graphical look at the basic operations of addition, subtraction, and scalar multiplication in 2-space and 3-space.

(a) Start the routine by typing **vecdemo**, then choose dimension 2 and select the built-in demo. Make sure you can identify each of the line segments that is displayed. Restart the routine choosing dimension 2, but this time select the option to enter you own vectors and the option to show the graphs on individual screens. Try $\mathbf{u} = \begin{bmatrix} -5 & -9 \end{bmatrix}$ and $\mathbf{v} = \begin{bmatrix} 1 & 15 \end{bmatrix}$.

(b) Use **vecdemo** with dimension 2 for a pair of vectors $\mathbf{u}$ and $\mathbf{v}$ where $\mathbf{v} = -2 * \mathbf{u}$. Before executing the routine, record a conjecture for the type of parallelograms that will be displayed. Run the routine and check your conjecture.

(c) Choose dimension 3 in **vecdemo** and experiment with several vectors. Why do the displays look like those for the two dimensional case, but tilted a bit? Explain.

**ML. 2.** Routine **lincombo** provides practice with linear combinations using the parallelogram rule in 2-space. The object is to use sliders to change the coefficients for a pair of vectors so that their linear combination is the third vector shown. Geometrically we change the size of a parallelogram so that its diagonal is this third vector. Once you get close you can have the routine check your linear combination. There is an option to type in the coefficients rather than change them using the sliders. To execute this routine just type **lincombo** in MATLAB. Experiment with this routine to get a feel for the interplay between the algebra and geometry of linear combinations.

**ML. 3.** A convex linear combination of vectors $\mathbf{u}$ and $\mathbf{v}$ has the form $t * \mathbf{u} + (1 - t) * \mathbf{v}$ where $t$ is in $[0, 1]$. The set of all possible convex linear combinations determines a familiar geometric figure together with the vectors $\mathbf{u}$ and $\mathbf{v}$. To illustrate aspects of convex linear combinations use the routine **convex** in MATLAB.

(a) Type **convex** in MATLAB, follow the screen directions, and choose one of the demos. To see the formation of the convex linear combinations choose values for the scalar $t = 0, 0.1, 0.2, \ldots, 1$. You will see a parallelogram formed for each of the linear combinations generated, and we leave an asterisk at the result of the linear combination. Record a conjecture for the geometric object determined by the set of asterisks. Execute **convex** again, this time choosing a spacing of 0.01 for the values of $t$. Check your conjecture with the display generated.

(b) Using the result from part (a), state a conjecture for the geometric figure determined by **u**, **v**, and the set of all asterisks that denote the end points of the convex linear combinations of **u** and **v**. Execute **convex** to check your conjecture.

(c) Experiment with routine **convex** by entering your own vectors and spacing for the scalar $t$.

(d) If you choose the option to select your own values for the scalar $t$, the routine **convex** lets you specify values other than those in [0, 1]. For example, choose one of the demos and the option to select your values for $t$. Then enter $[-2 : 0.1 : 5]$ which is the set of $t$'s from $-2$ in steps of 0.1 to 5. Inspect the geometric object determined by the set of asterisks. Execute the routine again, this time entering $[-10 : 0.2 : 15]$. From these two experiments and the result of part (a), form a conjecture for the object generated by the set of asterisks if $t$ could be selected to go from $-\infty$ to $+\infty$.

## True/False Review Questions

*Determine whether each of the following statements is true or false.*

1. The length of vector **x** in $R^n$ is **x·x**.

2. For $\mathbf{v} = \begin{bmatrix} v_1 \\ v_2 \end{bmatrix}$ in $R^2$, $\|\mathbf{v}\|$ is the same as the distance from the origin $(0, 0)$ to the point $(v_1, v_2)$.

3. In $R^2$, the sum of the vectors **a** and **b** is, geometrically, a diagonal of a parallelogram with sides **a** and **b**.

4. If vectors **x** and **y** in $R^n$ satisfy **x·y** $= 0$, then **x** is parallel to **y**.

5. A unit vector in the same direction as $\begin{bmatrix} 1 \\ 0 \\ 1 \end{bmatrix}$ is $\frac{1}{2} \begin{bmatrix} 1 \\ 0 \\ 1 \end{bmatrix}$.

6. Two vectors are orthogonal if the angle between them is $\pi$ radians.

7. If vectors **a** and **b** in $R^n$ are such that $(\mathbf{a} + \mathbf{b}) \cdot (\mathbf{a} - \mathbf{b}) = 0$, then $\|\mathbf{a}\| = \|\mathbf{b}\|$.

8. Non-zero vectors **a** and **b** in $R^n$ are parallel provided one is a scalar multiple of the other.

9. $\|k\mathbf{a}\| = k\|\mathbf{a}\|$.

10. $\|\mathbf{a} + \mathbf{b}\| = \|\mathbf{a}\| + \|\mathbf{b}\|$.

## Terminology

| | |
|---|---|
| $R^n$; $n$-space | Threefold interpretation of **a** in $R^2$ |
| Dilation; contraction | The length of an $n$-vector |
| Parallel vectors | Parallelogram rule |
| Angle between vectors in $R^n$ | Orthogonal vectors |
| Unit vector | |

Algebra and geometry are different sides of the same coin. The concepts in this section showed aspects of both sides of the coin. It will be convenient to switch from side-to-side in order to provide different perspectives on computations and concepts. Answer the following to help yourself see various interpretations of some of the concepts in this section.

- List the three interpretations of a vector **a** in $R^2$. Then give an example of a situation in which each interpretation has been used.

- Describe the relationship between scalar multiples and length in a dilation; in a contraction; for parallel vectors.
- Explain in your own words the parallelogram rule for adding two vectors (geometrically) in $R^2$.
- How is the difference of two vectors depicted in the parallelogram rule?
- How do we compute the angle between two $n$-vectors?
- What is the angle between parallel vectors?
- What is the angle between orthogonal vectors?
- Explain how to construct a unit vector from a nonzero vector $\mathbf{v}$ of $R^n$.

## 1.5 ■ MATRIX TRANSFORMATIONS

In Section 1.4 we showed that a linear combination of vectors in $R^n$ could be viewed geometrically as a head-to-tail construction. Now we use this geometric model to investigate a matrix-vector product. One result is that matrix-vector products define a function relationship. This point of view has far-ranging effects for upcoming topics. As previously[1], we use the geometry of $R^2$ and $R^3$ to provide visual models.

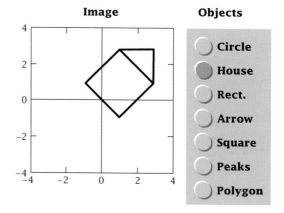

In MATLAB, **mapit.m**

In Section 1.3 we showed that a matrix-vector product $\mathbf{Ac}$ was a linear combination of the columns of $\mathbf{A}$ with coefficients that are the entries of $\mathbf{c}$:

$$\mathbf{Ac} = c_1 \text{col}_1(\mathbf{A}) + c_2 \text{col}_2(\mathbf{A}) + \cdots + c_n \text{col}_n(\mathbf{A}).$$

where $\mathbf{A}$ is $m \times n$ and $\mathbf{c}$ is an $n$-vector. This is the same as saying that $\mathbf{Ac}$ belongs to

$$\text{span}\{\text{col}_1(\mathbf{A}), \text{col}_2(\mathbf{A}), \dots, \text{col}_n(\mathbf{A})\}.$$

EXAMPLE 1    For $\mathbf{A} = \begin{bmatrix} 2 & 4 \\ 3 & 1 \end{bmatrix}$ and $\mathbf{c} = \begin{bmatrix} 2 \\ -1 \end{bmatrix}$ we have matrix-vector product

$$\mathbf{Ac} = \begin{bmatrix} 2 & 4 \\ 3 & 1 \end{bmatrix} \begin{bmatrix} 2 \\ -1 \end{bmatrix} = 2 \begin{bmatrix} 2 \\ 3 \end{bmatrix} - 1 \begin{bmatrix} 4 \\ 1 \end{bmatrix} = \begin{bmatrix} 0 \\ 5 \end{bmatrix},$$

[1]As in Section 1.4, $n$-vectors refer only to vectors from $R^n$ in this section.

which is shown in Figure 1. Also, for $\mathbf{c} = \begin{bmatrix} 1 \\ 2 \end{bmatrix}$,

$$\mathbf{Ac} = \begin{bmatrix} 2 & 4 \\ 3 & 1 \end{bmatrix} \begin{bmatrix} 1 \\ 2 \end{bmatrix} = 1 \begin{bmatrix} 2 \\ 3 \end{bmatrix} + 2 \begin{bmatrix} 4 \\ 1 \end{bmatrix} = \begin{bmatrix} 10 \\ 5 \end{bmatrix},$$

which is shown in Figure 2.

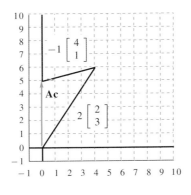

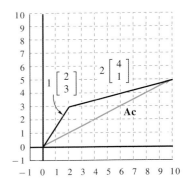

FIGURE 1                                    FIGURE 2

Both Figures 1 and 2 show the product $\mathbf{Ac}$ as a linear combination of the columns of $\mathbf{A}$.    ■

In Example 1 if we continued to select 2-vectors $\mathbf{c}$ and then compute $\mathbf{Ac}$, the resulting vectors would be in span $\left\{ \begin{bmatrix} 2 \\ 3 \end{bmatrix}, \begin{bmatrix} 4 \\ 1 \end{bmatrix} \right\}$. This association $\mathbf{c} \rightarrow \mathbf{Ac}$ defines a function, $f$, whose **domain** or inputs are 2-vectors $\mathbf{c}$ and whose **range** or outputs are vectors $\mathbf{Ac}$ in span $\left\{ \begin{bmatrix} 2 \\ 3 \end{bmatrix}, \begin{bmatrix} 4 \\ 1 \end{bmatrix} \right\}$. It follows that we can express $f$ as

$$f(\mathbf{c}) = \mathbf{Ac} = c_1 \begin{bmatrix} 2 \\ 3 \end{bmatrix} + c_2 \begin{bmatrix} 1 \\ 4 \end{bmatrix}.$$

In this case we say that $f$ takes $R^2$ into $R^2$ since the inputs are 2-vectors and the outputs are also 2-vectors. If matrix $\mathbf{A}$ is $2 \times 3$, then

$$f(\mathbf{c}) = \mathbf{Ac} = c_1 \mathrm{col}_1(\mathbf{A}) + c_2 \mathrm{col}_2(\mathbf{A}) + c_3 \mathrm{col}_3(\mathbf{A})$$

and we say that $f$ takes $R^3$ into $R^2$. Here $\mathbf{c}$ is a 3-vector and the outputs are in the span of the columns of $\mathbf{A}$, which are 2-vectors. Thus we are led to the following definition.

---

**Definition**    For an $m \times n$ matrix $\mathbf{A}$ the function $f$ defined by $f(\mathbf{c}) = \mathbf{Ac}$ is called a **matrix transformation** that takes $R^n$ into $R^m$. The domain is the set of all $n$-vectors $\mathbf{c}$ and the range is a set of $m$-vectors in span$\{\mathrm{col}_1(\mathbf{A}), \mathrm{col}_2(\mathbf{A}), \dots, \mathrm{col}_n(\mathbf{A})\}$. We call the range vectors **images**; $\mathbf{Ac}$ is the image of $\mathbf{c}$.

---

EXAMPLE 2

(a)  From Example 1, the matrix transformation is

$$f(\mathbf{c}) = \mathbf{Ac} = \begin{bmatrix} 2 & 4 \\ 3 & 1 \end{bmatrix} \begin{bmatrix} c_1 \\ c_2 \end{bmatrix}.$$

The image of $\begin{bmatrix} 2 \\ -1 \end{bmatrix}$ is $\begin{bmatrix} 0 \\ 5 \end{bmatrix}$, while the image of $\begin{bmatrix} 1 \\ 2 \end{bmatrix}$ is $\begin{bmatrix} 10 \\ 5 \end{bmatrix}$.

(b) For $\mathbf{A} = \begin{bmatrix} 1 & 2 & 0 \\ 1 & -1 & 1 \end{bmatrix}$, the matrix transformation is

$$f(\mathbf{c}) = \mathbf{Ac} = \begin{bmatrix} 1 & 2 & 0 \\ 1 & -1 & 1 \end{bmatrix} \begin{bmatrix} c_1 \\ c_2 \\ c_3 \end{bmatrix}.$$

The image of $\begin{bmatrix} 1 \\ 0 \\ 1 \end{bmatrix}$ is $\begin{bmatrix} 1 \\ 2 \end{bmatrix}$, the image of $\begin{bmatrix} 0 \\ 1 \\ 3 \end{bmatrix}$ is $\begin{bmatrix} 2 \\ 2 \end{bmatrix}$, and the image of $\begin{bmatrix} -2 \\ 1 \\ 3 \end{bmatrix}$ is $\begin{bmatrix} 0 \\ 0 \end{bmatrix}$. (Verify.)  ■

Observe that if $\mathbf{A}$ is an $m \times n$ matrix and $f(\mathbf{c}) = \mathbf{Ac}$ is a matrix transformation taking $R^n$ to $R^m$, then a vector $\mathbf{d}$ in $R^m$ is in the range of $f$ only if we can find a vector $\mathbf{c}$ in $R^n$ such that $f(\mathbf{c}) = \mathbf{d}$. That is, the range of $f$ is the span of the columns of matrix $\mathbf{A}$ and may not be all of $R^m$.

EXAMPLE 3   Let $\mathbf{A} = \begin{bmatrix} 1 & 2 \\ -2 & 3 \end{bmatrix}$ and consider the matrix transformation defined by $f(\mathbf{c}) = \mathbf{Ac}$. Determine if vector $\mathbf{d} = \begin{bmatrix} 4 \\ -1 \end{bmatrix}$ is in the range of $f$. That is, is there a vector $\mathbf{c} = \begin{bmatrix} c_1 \\ c_2 \end{bmatrix}$ such that $\mathbf{Ac} = \mathbf{d}$? We have

$$\mathbf{Ac} = \begin{bmatrix} c_1 + 2c_2 \\ -2c_1 + 3c_2 \end{bmatrix} = \mathbf{d} = \begin{bmatrix} 4 \\ -1 \end{bmatrix} \quad \text{only if} \quad \begin{matrix} c_1 + 2c_2 = 4 \\ -2c_1 + 3c_2 = -1. \end{matrix}$$

Solving this linear system of equations by the familiar method of elimination, we get $c_1 = 2$ and $c_2 = 1$. (Verify.) Thus $\mathbf{d}$ is in the range of $f$. In particular, if $\mathbf{c} = \begin{bmatrix} 2 \\ 1 \end{bmatrix}$ then $f(\mathbf{c}) = \mathbf{d}$.  ■

For certain matrix transformations we can display a simple geometric model. If $f(\mathbf{c}) = \mathbf{Ac}$ takes $R^2$ into $R^2$, $R^3$ into $R^3$, $R^3$ into $R^2$, or even $R^2$ into $R^3$, then we can draw pictures that illustrate the character of the effect of matrix $\mathbf{A}$. Examples 4 through 7 illustrate such geometric models. (Example 7 will be useful in several upcoming topics.)

EXAMPLE 4   The matrix transformation from $R^2$ to $R^2$ defined by

$$\mathbf{A} = \begin{bmatrix} 1 & 0 \\ 0 & -1 \end{bmatrix}$$

generates images

$$\mathbf{Ac} = \begin{bmatrix} c_1 \\ -c_2 \end{bmatrix}.$$

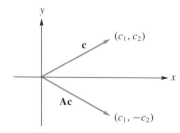

FIGURE 3  Reflection about the $x$-axis.

Using our interpretation of a 2-vector $\mathbf{c}$ as a point we have the model displayed in Figure 3. We call this matrix transformation a **reflection about the $x$-axis** in $R^2$.  ■

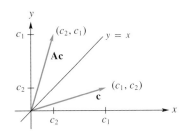

FIGURE 4  Reflection about the line $y = x$.

EXAMPLE 5    The matrix transformation from $R^2$ to $R^2$ defined by

$$\mathbf{A} = \begin{bmatrix} 0 & 1 \\ 1 & 0 \end{bmatrix}$$

generates images

$$\mathbf{Ac} = \begin{bmatrix} c_2 \\ c_1 \end{bmatrix}.$$

Using our interpretation of a 2-vector $\mathbf{c}$ as a point, we have the model displayed in Figure 4. We call this matrix transformation a **reflection about the line $y = x$ in** $R^2$. ■

EXAMPLE 6    The matrix transformation from $R^3$ to $R^2$ defined by

$$\mathbf{A} = \begin{bmatrix} 1 & 0 & 0 \\ 0 & 1 & 0 \end{bmatrix}$$

associates the 3-vector $\mathbf{c} = \begin{bmatrix} c_1 \\ c_2 \\ c_3 \end{bmatrix}$ with the 2-vector $\mathbf{Ac} = \begin{bmatrix} c_1 \\ c_2 \end{bmatrix}$. Figure 5 provides a model for this matrix transformation, which is called a **projection onto the xy-plane**. ($\mathbf{Ac}$ appears to be the shadow cast by $\mathbf{c}$ onto the $xy$-plane.) Here we use the three-dimensional $xyz$-coordinate system as a model for $R^3$ and the two-dimensional $xy$-coordinate system as a model for $R^2$. Note that if

$$\mathbf{b} = \begin{bmatrix} c_1 \\ c_2 \\ s \end{bmatrix}$$

where $s$ is any scalar, then

$$\mathbf{Ab} = \begin{bmatrix} c_1 \\ c_2 \end{bmatrix} = \mathbf{Ac}.$$

Hence there are infinitely many 3-vectors that have the same projection image. This is shown in Figure 6.

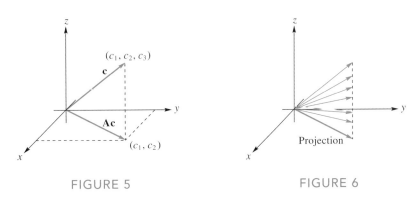

FIGURE 5                    FIGURE 6                    ■

Note that the matrix transformation from $R^3$ to $R^3$ defined by

$$\mathbf{B} = \begin{bmatrix} 1 & 0 & 0 \\ 0 & 1 & 0 \\ 0 & 0 & 0 \end{bmatrix}$$

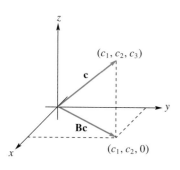

FIGURE 7

generates images of the form

$$\mathbf{Bc} = \begin{bmatrix} c_1 \\ c_2 \\ 0 \end{bmatrix}.$$

Figure 7 shows this transformation. The picture is almost the same as Figure 5 except that the head of $\mathbf{Bc}$ is the ordered triple $(c_1, c_2, 0)$.

EXAMPLE 7    The matrix transformation defined by

$$\mathbf{A} = \begin{bmatrix} \cos(\theta) & -\sin(\theta) \\ \sin(\theta) & \cos(\theta) \end{bmatrix}$$

where $\theta$ is an angle in degrees or radians generates images of the form

$$\mathbf{Ac} = c_1 \begin{bmatrix} \cos(\theta) \\ \sin(\theta) \end{bmatrix} + c_2 \begin{bmatrix} -\sin(\theta) \\ \cos(\theta) \end{bmatrix} = \begin{bmatrix} c_1 \cos(\theta) - c_2 \sin(\theta) \\ c_1 \sin(\theta) + c_2 \cos(\theta) \end{bmatrix}.$$

Recall that sines and cosines are called circular functions and that pairs of points $(\cos(\theta), \sin(\theta))$ are on the circumference of a circle of radius 1. Moreover, the entries of the image vector are linear combinations of sines and cosines. These observations lead us to conjecture that this matrix transformation may be associated with a rotation of the vector $\mathbf{c}$ through an angle $\theta$. (A conjecture is a guess based on limited evidence.) This conjecture is shown pictorially in Figure 8. Our goal will be to gather evidence to support this conjecture. We start by noting that if image $\mathbf{Ac}$ is the result of rotating vector $\mathbf{c}$, then the length of $\mathbf{Ac}$ should be the same as the length of $\mathbf{c}$. Computing $\|\mathbf{Ac}\|$, we have

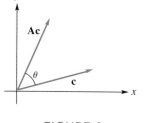

FIGURE 8

$$\|\mathbf{Ac}\| = \sqrt{(c_1 \cos(\theta) - c_2 \sin(\theta))^2 + (c_1 \sin(\theta) + c_2 \cos(\theta))^2}$$

$$= \sqrt{\begin{aligned} c_1^2 \cos^2(\theta) - 2c_1c_2 \cos(\theta)\sin(\theta) + c_2^2 \sin^2(\theta) \\ + c_1^2 \sin^2(\theta) + 2c_1c_2 \cos(\theta)\sin(\theta) + c_2^2 \cos^2(\theta) \end{aligned}}$$

$$= \sqrt{c_1^2 \cos^2(\theta) + c_1^2 \sin^2(\theta) + c_2^2 \cos^2(\theta) + c_2^2 \sin^2(\theta)}$$

$$= \sqrt{c_1^2(\cos^2(\theta) + \sin^2(\theta)) + c_2^2(\cos^2(\theta) + \sin^2(\theta))}$$

$$= \sqrt{c_1^2 + c_2^2} = \|\mathbf{c}\|.$$

Hence the length of the image is the same as the length of the original vector.

Next we compute the cosine of the angles that $\mathbf{c}$ and $\mathbf{Ac}$ make with a vector that is oriented in the direction of the positive $x$-axis, namely $\mathbf{i} = \begin{bmatrix} 1 \\ 0 \end{bmatrix}$; note that $\|\mathbf{i}\| = 1$. Let $\alpha$ be the angle between $\mathbf{c}$ and $\mathbf{i}$; then we have

$$\cos(\alpha) = \frac{\mathbf{c \cdot i}}{\|\mathbf{c}\| \, \|\mathbf{i}\|} = \frac{c_1}{\|\mathbf{c}\|}.$$

Let $\beta$ be the angle between $\mathbf{Ac}$ and $\mathbf{i}$; then we have

$$\cos(\beta) = \frac{\mathbf{Ac \cdot i}}{\|\mathbf{Ac}\| \, \|\mathbf{i}\|} = \frac{c_1 \cos(\theta) - c_2 \sin(\theta)}{\|\mathbf{Ac}\|} = \frac{c_1 \cos(\theta) - c_2 \sin(\theta)}{\|\mathbf{c}\|}.$$

Based on our conjecture, we want to show that $\cos(\beta) = \cos(\alpha + \theta)$. Here is where we use connections between trigonometry and geometry. The head of vector $\mathbf{c}$ is point $(c_1, c_2)$, which lies on a circle of radius $\|\mathbf{c}\|$ centered at the origin. Using the basic definitions of sine and cosine and Figure 9, we see that

$$\sin(\alpha) = \frac{c_2}{\|\mathbf{c}\|} \quad \text{and} \quad \cos(\alpha) = \frac{c_1}{\|\mathbf{c}\|}.$$

Substituting these into the previous expression for $\cos(\beta)$, we have

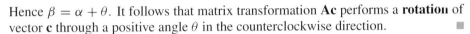

$$\cos(\beta) = \frac{c_1 \cos(\theta) - c_2 \sin(\theta)}{\|\mathbf{c}\|}$$

$$= \left(\frac{c_1}{\|\mathbf{c}\|}\right) \cos(\theta) - \left(\frac{c_2}{\|\mathbf{c}\|}\right) \sin(\theta)$$

$$= \cos(\alpha) \cos(\theta) - \sin(\alpha) \sin(\theta)$$

$$= \cos(\alpha + \theta).$$

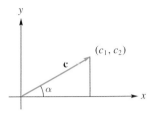

FIGURE 9

Hence $\beta = \alpha + \theta$. It follows that matrix transformation $\mathbf{Ac}$ performs a **rotation** of vector $\mathbf{c}$ through a positive angle $\theta$ in the counterclockwise direction. ■

Matrix transformations can be used to manipulate vectors graphically as shown in the preceding examples. However, they can do more—namely, manipulate figures. (Again we use $R^2$ to illustrate our ideas.) We can view a figure such as a triangle, square, or circle either as a set of points or vectors, each of which can be represented by a 2-vector. The manipulation is to compute the images of the set representing the figure and then display the resulting graph. Applying the matrix transformations from Example 4, 5, or 7, we can reflect about the $x$-axis, reflect about the line $y = x$, or rotate through an angle $\theta$, respectively. In Example 8 we introduce another matrix transformation that is quite useful in graphics displays.

EXAMPLE 8    Let $\mathbf{A}$ be the diagonal matrix given by

$$\mathbf{A} = \begin{bmatrix} h & 0 \\ 0 & k \end{bmatrix}$$

with $h$ and $k$ both nonzero. The matrix transformation from $R^2$ to $R^2$ defined by $\mathbf{A}$ generates images

$$\mathbf{Ac} = \begin{bmatrix} hc_1 \\ kc_2 \end{bmatrix}.$$

We illustrate the behavior of this transformation as follows.

(a) A unit square has each side of length 1. We describe the unit square by the points $(0, 0)$, $(1, 0)$, $(1, 1)$, $(0, 1)$, and $(0, 0)$, which when connected with straight line segments traces its outline. We record this information as the columns of the matrix

$$\mathbf{S} = \begin{bmatrix} 0 & 1 & 1 & 0 & 0 \\ 0 & 0 & 1 & 1 & 0 \end{bmatrix}.$$

Then the image of the unit square is described by the information in matrix

$$\mathbf{AS} = \begin{bmatrix} h & 0 \\ 0 & k \end{bmatrix} \begin{bmatrix} 0 & 1 & 1 & 0 & 0 \\ 0 & 0 & 1 & 1 & 0 \end{bmatrix} = \begin{bmatrix} 0 & h & h & 0 & 0 \\ 0 & 0 & k & k & 0 \end{bmatrix}.$$

We show this transformation in Figure 10, where both scalars $h$ and $k$ are taken to be positive.

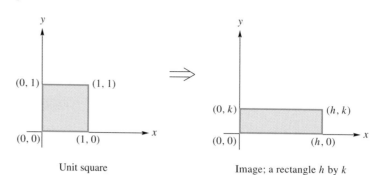

Unit square                                      Image; a rectangle $h$ by $k$

FIGURE 10

(b) A unit circle has radius 1, and we will take its center at the origin. Unfortunately, its circumference cannot be traced by specifying a few points as we did for the unit square. But each point on the unit circle is described by an ordered pair $(\cos(\theta), \sin(\theta))$, where the angle $\theta$ takes on values from 0 to $2\pi$ radians. So here we will represent an arbitrary point on the unit circle by the vector

$$\mathbf{c} = \begin{bmatrix} \cos(\theta) \\ \sin(\theta) \end{bmatrix}.$$

Hence the set of images of the unit circle that result from matrix transformation $\mathbf{Ac}$ are given by

$$\mathbf{Ac} = \begin{bmatrix} h & 0 \\ 0 & k \end{bmatrix}\begin{bmatrix} \cos(\theta) \\ \sin(\theta) \end{bmatrix} = \begin{bmatrix} h\cos(\theta) \\ k\sin(\theta) \end{bmatrix}.$$

We recall that a circle of radius 1 centered at the origin has equation $x^2 + y^2 = 1$, and this, of course, agrees with the points $(\cos(\theta), \sin(\theta))$ on the circumference by the Pythagoras identity, $\sin^2(\theta) + \cos^2(\theta) = 1$. We want to develop an equation describing the images of the unit circle. Let

$$\begin{bmatrix} x' \\ y' \end{bmatrix} = \mathbf{Ac} = \begin{bmatrix} h\cos(\theta) \\ k\sin(\theta) \end{bmatrix}.$$

Then we have $x' = h\cos(\theta)$ and $y' = k\sin(\theta)$, which is equivalent to

$$\frac{x'}{h} = \cos(\theta), \quad \frac{y'}{k} = \sin(\theta).$$

It follows that

$$\left(\frac{x'}{h}\right)^2 + \left(\frac{y'}{k}\right)^2 = 1$$

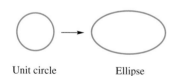

Unit circle        Ellipse

FIGURE 11

which is the equation of an ellipse. Thus the image of the unit circle by a matrix transformation with $\mathbf{A} = \begin{bmatrix} h & 0 \\ 0 & k \end{bmatrix}$ is an ellipse centered at the origin. See Figure 11. ■

The geometric nature of the preceding examples of matrix transformations hints at the use of such manipulations in animation and computer graphics. Simple geometric maneuvers such as those above are used one after another to achieve the type of sophisticated effects seen in arcade games and animated computer demonstrations. For example, to show a wheel spinning we can rotate the spokes through an angle $\theta_1$ followed by a second rotation through an angle $\theta_2$ and so on. Let the 2-vector **c** represent a spoke of the wheel, let $f$ be the matrix transformation defined by matrix

$$\mathbf{A} = \begin{bmatrix} \cos(\theta_1) & -\sin(\theta_1) \\ \sin(\theta_1) & \cos(\theta_1) \end{bmatrix},$$

and let $g$ be the matrix transformation defined by the matrix

$$\mathbf{B} = \begin{bmatrix} \cos(\theta_2) & -\sin(\theta_2) \\ \sin(\theta_2) & \cos(\theta_2) \end{bmatrix}.$$

We represent the succession of rotations of the spoke **c** by

$$g(f(\mathbf{c})) = g(\mathbf{Ac}) = \mathbf{B}(\mathbf{Ac}).$$

The matrix-vector product **Ac** is performed first and generates a rotation of **c** through angle $\theta_1$; then matrix-vector product **B(Ac)** generates the second rotation. We have

$$\mathbf{B}(\mathbf{Ac}) = \mathbf{B}(c_1 \text{col}_1(\mathbf{A}) + c_2 \text{col}_2(\mathbf{A})) = c_1 \mathbf{B}\,\text{col}_1(\mathbf{A}) + c_2 \mathbf{B}\,\text{col}_2(\mathbf{A})$$

and the final expression is a linear combination of column vectors $\mathbf{B}\text{col}_1(\mathbf{A})$ and $\mathbf{B}\,\text{col}_2(\mathbf{A})$ which we can write as the product

$$\begin{bmatrix} \mathbf{B}\,\text{col}_1(\mathbf{A}) & \mathbf{B}\,\text{col}_2(\mathbf{A}) \end{bmatrix} \begin{bmatrix} c_1 \\ c_2 \end{bmatrix}.$$

From the definition of matrix multiplication, $\begin{bmatrix} \mathbf{B}\,\text{col}_1(\mathbf{A}) & \mathbf{B}\,\text{col}_2(\mathbf{A}) \end{bmatrix} = \mathbf{BA}$, so we have

$$\mathbf{B}(\mathbf{Ac}) = (\mathbf{BA})\mathbf{c}$$

which says that instead of applying the transformations in succession, $f$ followed by $g$, we can achieve the same result by forming the matrix product **BA** and using it to define a matrix transformation on the spokes of the wheel. The expression $g(f(\mathbf{c}))$ is called a **composition** of matrix transformations, and its matrix is **BA**. Note that the rather bizarre rule of matrix multiplication, row-by-column products, makes things fit together so that the succession of transformations can be accomplished directly with matrix **BA**.

Suppose that we had a set of spokes represented by the columns of a matrix **C**. Since

$$\mathbf{AC} = \begin{bmatrix} \mathbf{A}\,\text{col}_1(\mathbf{C}) & \mathbf{A}\,\text{col}_2(\mathbf{C}) & \cdots & \mathbf{A}\,\text{col}_n(\mathbf{C}) \end{bmatrix}$$

we have

$$\begin{aligned} \mathbf{B}(\mathbf{AC}) &= \mathbf{B}\begin{bmatrix} \mathbf{A}\,\text{col}_1(\mathbf{C}) & \mathbf{A}\,\text{col}_2(\mathbf{C}) & \cdots & \mathbf{A}\,\text{col}_n(\mathbf{C}) \end{bmatrix} \\ &= \begin{bmatrix} (\mathbf{BA})\text{col}_1(\mathbf{C}) & (\mathbf{BA})\text{col}_2(\mathbf{C}) & \cdots & (\mathbf{BA})\text{col}_n(\mathbf{C}) \end{bmatrix} \\ &= (\mathbf{BA})\mathbf{C}. \end{aligned}$$

Expression $\mathbf{B}(\mathbf{AC}) = (\mathbf{BA})\mathbf{C}$ is called the **associative property** of matrix multiplication and is valid for matrices of any size for which the products are defined.

If $f$ is the matrix transformation defined by the matrix **A**, then $f(\mathbf{c}) = \mathbf{Ac}$. By properties developed in Section 1.3, we have

$$f(\mathbf{c} + \mathbf{d}) = \mathbf{A}(\mathbf{c} + \mathbf{d}) = \mathbf{Ac} + \mathbf{Ad} = f(\mathbf{c}) + f(\mathbf{d})$$

and

$$f(k\mathbf{c}) = \mathbf{A}(k\mathbf{c}) = k(\mathbf{Ac}) = kf(\mathbf{c}).$$

In Chapter 7, we investigate more general transformations and introduce the following terminology:

Any function $f$ from $R^n$ to $R^m$ that satisfies $f(\mathbf{c} + \mathbf{d}) = f(\mathbf{c}) + f(\mathbf{d})$ for every vector $\mathbf{c}$ and $\mathbf{d}$ in $R^n$ and $f(k\mathbf{c}) = kf(\mathbf{c})$ for any scalar $k$ is called a **linear transformation**.

Hence every matrix transformation is a linear transformation. We note that the two aforementioned properties are equivalent to the expression

$$f(k\mathbf{c} + r\mathbf{d}) = kf(\mathbf{c}) + rf(\mathbf{d})$$

which tells us the behavior of a linear transformation on a linear combination. Namely, a linear transformation "splits apart sums and bypasses scalar multiples."

EXAMPLE 9    Let $f$ be the matrix transformation defined by a $2 \times 2$ matrix $\mathbf{A}$ and let $\mathbf{a} + k\mathbf{b}$ be a translation of line $k\mathbf{b}$. (See the application in Section 1.4.) The image of the vector $\mathbf{a} + k\mathbf{b}$ under $f$ is

$$f(\mathbf{a} + k\mathbf{b}) = \mathbf{A}(\mathbf{a} + k\mathbf{b}) = \mathbf{Aa} + k\mathbf{Ab}.$$

Then $\mathbf{Aa}$ is a specific 2-vector and $k(\mathbf{Ab})$ is a line in $R^2$, so the image of $\mathbf{a} + k\mathbf{b}$ is a translation of line $k\mathbf{Ab}$. We say that **the image of a line is a line**. This provides another characterization of matrix transformations.  ■

Application:   **Orthogonal Projection**

In Example 6 we saw how to project a vector $\mathbf{c}$ in $R^3$ onto the $xy$-plane by dropping a perpendicular line segment from the head of $\mathbf{c}$ to the $xy$-plane. The resulting vector $\mathbf{p} = \mathbf{Ac}$ in the $xy$-plane can be thought of as the shadow that vector $\mathbf{c}$ casts upon the $xy$-plane (assuming there is a light immediately above the $xy$-plane). Alternatively, it can be shown that the projection $\mathbf{p}$ is the vector in the $xy$-plane closest to vector $\mathbf{c}$. The projection of a vector onto a plane is one type of projection that we investigate in more detail later, but for now we consider the projection of one vector onto another vector. As we have done previously, for convenience, we present our development in $R^2$, but the result is valid for any pair of nonzero vectors in $R^n$.

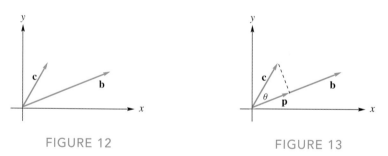

FIGURE 12                    FIGURE 13

Let $\mathbf{b}$ and $\mathbf{c}$ be nonzero vectors in $R^2$ as shown in Figure 12. From the head of vector $\mathbf{c}$ construct a line segment perpendicular to vector $\mathbf{b}$. Let $\mathbf{p}$ be the vector whose head is the point on $\mathbf{b}$ where the perpendicular intersects $\mathbf{b}$ (see Figure 13). Vector $\mathbf{p}$ is called the (orthogonal) projection of $\mathbf{c}$ onto $\mathbf{b}$ and we denote this as $\mathbf{p} = \text{proj}_\mathbf{b}\mathbf{c}$. Our goal is to determine an expression for $\mathbf{p}$ in terms of vectors $\mathbf{b}$ and $\mathbf{c}$. Our strategy will be to determine the length of $\mathbf{p}$ and then use the fact that $\mathbf{p}$ is parallel to $\mathbf{b}$.

• From our earlier development, the angle $\theta$ between vectors $\mathbf{b}$ and $\mathbf{c}$ satisfies

$$\cos(\theta) = \frac{\mathbf{b} \cdot \mathbf{c}}{\|\mathbf{b}\| \, \|\mathbf{c}\|}.$$

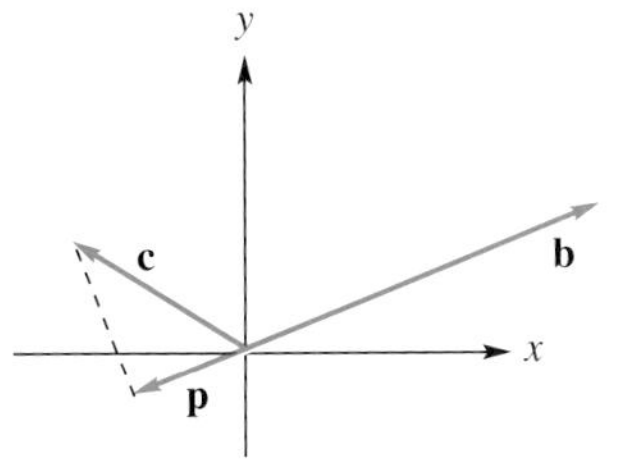

FIGURE 14

- Using trigonometry on the right triangle in Figure 13, we have that

$$\cos(\theta) = \frac{\|\mathbf{p}\|}{\|\mathbf{c}\|}.$$

(This assumes that $\theta$ is an acute angle. See Figure 14 for another look at a projection.)

- Setting the absolute value of these expressions equal and solving for $\|\mathbf{p}\|$, we get

$$\|\mathbf{p}\| = \frac{|\mathbf{b}\cdot\mathbf{c}|}{\|\mathbf{b}\|}. \tag{1}$$

- Since $\mathbf{p}$ is parallel to $\mathbf{b}$, these vectors are scalar multiples of one another. It follows that

$$\mathbf{p} = (\text{sign}) * \|\mathbf{p}\| * (\text{unit vector in the direction of } \mathbf{b}),$$

where sign is $+$ if $\mathbf{p}$ and $\mathbf{b}$ are in the same direction and $-$ if they are in opposite directions. The sign required is the same as the sign of $\mathbf{b}\cdot\mathbf{c}$ as long as $\mathbf{b}$ and $\mathbf{c}$ are not orthogonal. (When $\mathbf{b}$ and $\mathbf{c}$ are orthogonal, $\mathbf{p} = \mathbf{0}$.) Hence using (1), we have

$$\mathbf{p} = \frac{\mathbf{b}\cdot\mathbf{c}}{\|\mathbf{b}\|}\frac{\mathbf{b}}{\|\mathbf{b}\|} = \frac{\mathbf{b}\cdot\mathbf{c}}{\|\mathbf{b}\|^2}\mathbf{b} = \frac{\mathbf{b}\cdot\mathbf{c}}{\mathbf{b}\cdot\mathbf{b}}\mathbf{b}. \tag{2}$$

EXAMPLE 10   Determine $\mathbf{p} = \text{proj}_{\mathbf{b}}\mathbf{c}$ for each of the following pairs of vectors.

(a) $\mathbf{b} = \begin{bmatrix} 4 \\ 9 \end{bmatrix}$, $\mathbf{c} = \begin{bmatrix} 1 \\ 3 \end{bmatrix}$; $\text{proj}_{\mathbf{b}}\mathbf{c} = \dfrac{\mathbf{b}\cdot\mathbf{c}}{\mathbf{b}\cdot\mathbf{b}}\mathbf{b} = \dfrac{4+27}{16+81}\mathbf{b} = \dfrac{31}{97}\mathbf{b} = \begin{bmatrix} 124 \\ 876 \end{bmatrix}$.

(b) $\mathbf{b} = \begin{bmatrix} 1 \\ -2 \\ 0 \\ 2 \end{bmatrix}$, $\mathbf{c} = \begin{bmatrix} 0 \\ 4 \\ 1 \\ 1 \end{bmatrix}$;

$$\text{proj}_{\mathbf{b}}\mathbf{c} = \frac{\mathbf{b}\cdot\mathbf{c}}{\mathbf{b}\cdot\mathbf{b}}\mathbf{b} = \frac{0-8+0+2}{1+4+0+4}\mathbf{b} = -\frac{6}{9}\mathbf{b} = -\frac{2}{3}\mathbf{b} = \begin{bmatrix} -\frac{2}{3} \\ \frac{4}{3} \\ 0 \\ -\frac{4}{3} \end{bmatrix}. \quad ■$$

Let $\mathbf{n}$ denote the directed line segment from the head of $\mathbf{c}$ to the head of $\mathbf{p}$ ($\mathbf{n}$ is a directed line segment parallel to a vector; for convenience we will refer to it as a vector). We see that $\mathbf{n}$ and $\mathbf{b}$ are orthogonal (that is, $\mathbf{n}\cdot\mathbf{b} = 0$) and that $\mathbf{c} = \mathbf{p} - \mathbf{n}$. Hence any linear combination of vectors $\mathbf{b}$ and $\mathbf{c}$, say $r\mathbf{b} + s\mathbf{c}$, can be expressed as a linear combination of the orthogonal vectors $\mathbf{n}$ and $\mathbf{b}$ since

$$r\mathbf{b} + s\mathbf{c} = r\mathbf{b} + s(\mathbf{p} - \mathbf{n}) = r\mathbf{b} + s\left(\frac{\mathbf{b}\cdot\mathbf{c}}{\mathbf{b}\cdot\mathbf{b}}\mathbf{b} - \mathbf{n}\right) = \left(r + s\frac{\mathbf{b}\cdot\mathbf{c}}{\mathbf{b}\cdot\mathbf{b}}\right)\mathbf{b} - s\mathbf{n}$$

and the reverse is also true. Namely, any linear combination of $\mathbf{n}$ and $\mathbf{b}$ can be expressed in terms of $\mathbf{b}$ and $\mathbf{c}$. Thus we can say that $\text{span}\{\mathbf{b}, \mathbf{c}\} = \text{span}\{\mathbf{b}, \mathbf{n}\}$ and we call $\{\mathbf{b}, \mathbf{n}\}$ an **orthogonal spanning set**. As we will see later, there are advantages to having spanning sets of orthogonal vectors.

## EXERCISES 1.5

**1.** Determine a set of column vectors $\mathbf{u}$, $\mathbf{v}$, and $\mathbf{w}$ in $R^2$ such that every matrix-vector product $\mathbf{Ac}$ is in span$\{\mathbf{u}, \mathbf{v}, \mathbf{w}\}$ where $\mathbf{A} = \begin{bmatrix} 1 & -3 & -1 \\ 2 & 4 & 0 \end{bmatrix}$.

**2.** Determine a pair of column vectors $\mathbf{u}$ and $\mathbf{v}$ in $R^3$ such that every matrix-vector product $\mathbf{Ac}$ is in span$\{\mathbf{u}, \mathbf{v}\}$ where
$$\mathbf{A} = \begin{bmatrix} 4 & 2 \\ -2 & 1 \\ 1 & 1 \end{bmatrix}.$$

*In Exercises 3–8, compute $f(\mathbf{c}) = \mathbf{Ac}$ for each pair $\mathbf{A}$ and $\mathbf{c}$.*

**3.** $\mathbf{A} = \begin{bmatrix} 1 & 2 \\ -1 & 1 \end{bmatrix}, \mathbf{c} = \begin{bmatrix} 3 \\ -2 \end{bmatrix}.$

**4.** $\mathbf{A} = \begin{bmatrix} 1 & -1 \\ 2 & 0 \\ 1 & 3 \end{bmatrix}, \mathbf{c} = \begin{bmatrix} 4 \\ -1 \end{bmatrix}.$

**5.** $\mathbf{A} = \begin{bmatrix} 2 & -1 & 0 \\ 3 & 2 & -4 \end{bmatrix}, \mathbf{c} = \begin{bmatrix} 4 \\ -2 \\ 1 \end{bmatrix}.$

**6.** $\mathbf{A} = \begin{bmatrix} 2 & 3 & 1 \\ 1 & 2 & 1 \\ 0 & 1 & 1 \end{bmatrix}, \mathbf{c} = \begin{bmatrix} 1 \\ -1 \\ 1 \end{bmatrix}.$

**7.** $\mathbf{A} = \begin{bmatrix} 2 & 3 & 1 \\ 1 & 2 & 1 \\ 0 & 1 & 1 \end{bmatrix}, \mathbf{c} = \begin{bmatrix} 0 \\ 0 \\ 0 \end{bmatrix}.$

**8.** $\mathbf{A} = \begin{bmatrix} 2 & 3 & 1 \\ 1 & 2 & 1 \\ 0 & 1 & 1 \end{bmatrix}, \mathbf{c} = \begin{bmatrix} 2 \\ -2 \\ 2 \end{bmatrix}.$

**9.** Vector $\mathbf{b} = \begin{bmatrix} 0 \\ 1 \end{bmatrix}$ is in span$\{\mathbf{u}, \mathbf{v}\}$ where
$$\mathbf{A} = \begin{bmatrix} \mathbf{u} & \mathbf{v} \end{bmatrix} = \begin{bmatrix} 1 & 2 \\ 1 & 1 \end{bmatrix}.$$
Determine vector $\mathbf{c}$ in $R^2$ so that $\mathbf{Ac} = \mathbf{b}$.

**10.** Vector $\mathbf{b} = \begin{bmatrix} 0 \\ 8 \end{bmatrix}$ is in span$\{\mathbf{u}, \mathbf{v}, \mathbf{w}\}$ where
$$\mathbf{A} = \begin{bmatrix} \mathbf{u} & \mathbf{v} & \mathbf{w} \end{bmatrix} = \begin{bmatrix} 1 & 0 & 0 \\ 0 & 0 & 4 \end{bmatrix}.$$
Determine vector $\mathbf{c}$ in $R^3$ so that $\mathbf{Ac} = \mathbf{b}$.

**11.** Some matrix transformations $f$ have the property that $f(\mathbf{c}) = f(\mathbf{d})$, when $\mathbf{c} \neq \mathbf{d}$. That is, the images of different vectors can be the same. For each of the following matrix transformations, find two different vectors $\mathbf{c}$ and $\mathbf{d}$ such that $f(\mathbf{c}) = f(\mathbf{d}) = \mathbf{b}$.

(a) $\mathbf{A} = \begin{bmatrix} 1 & 2 & 0 \\ 0 & 1 & -1 \end{bmatrix}, \mathbf{b} = \begin{bmatrix} 0 \\ -1 \end{bmatrix}.$

(b) $\mathbf{A} = \begin{bmatrix} 2 & 1 & 0 \\ 0 & 2 & -1 \end{bmatrix}, \mathbf{b} = \begin{bmatrix} 4 \\ 4 \end{bmatrix}.$

*In Exercises 12–16, determine whether the given vector is in the range of matrix transformation $f(\mathbf{c}) = \mathbf{Ac}$. (See Example 3.) Let $\mathbf{A} = \begin{bmatrix} 1 & 2 \\ 0 & 1 \\ 1 & 1 \end{bmatrix}.$*

**12.** $\begin{bmatrix} 1 \\ -1 \\ 2 \end{bmatrix}.$     **13.** $\begin{bmatrix} 1 \\ 1 \\ 1 \end{bmatrix}.$     **14.** $\begin{bmatrix} 0 \\ 0 \\ 0 \end{bmatrix}.$

**15.** $\begin{bmatrix} 8 \\ 5 \\ 3 \end{bmatrix}.$     **16.** $\begin{bmatrix} 1 \\ 4 \\ 2 \end{bmatrix}.$

*In Exercises 17–19, give a geometric description of the images that result from the matrix transformation defined by the matrix $\mathbf{A}$ and then draw a diagram showing the domain vector and corresponding range vector. (Hint: Let $\mathbf{c} = \begin{bmatrix} c_1 \\ c_2 \end{bmatrix}$ and then compute $\mathbf{Ac}$.)*

**17.** $\mathbf{A} = \begin{bmatrix} 1 & 0 \\ 0 & 0 \end{bmatrix}.$     **18.** $\mathbf{A} = \begin{bmatrix} 0 & 0 \\ 0 & 1 \end{bmatrix}.$

**19.** $\mathbf{A} = \begin{bmatrix} 1 & 0 \\ 0 & 1 \end{bmatrix}.$

*In Exercises 20–22, give a geometric description of the images that result from the matrix transformation defined by the matrix $\mathbf{A}$ and then draw a diagram showing the domain vector and corresponding range vector. (Hint: Let $\mathbf{c} = \begin{bmatrix} c_1 \\ c_2 \\ c_3 \end{bmatrix}$ and then compute $\mathbf{Ac}$.)*

**20.** $\mathbf{A} = \begin{bmatrix} 1 & 0 & 0 \\ 0 & 0 & 0 \\ 0 & 0 & 1 \end{bmatrix}.$     **21.** $\mathbf{A} = \begin{bmatrix} 0 & 0 & 0 \\ 0 & 1 & 0 \\ 0 & 0 & 1 \end{bmatrix}.$

**22.** $\mathbf{A} = \begin{bmatrix} 1 & 0 & 0 \\ 0 & 1 & 0 \\ 0 & 0 & 1 \end{bmatrix}.$

**23.** The matrix transformation in $R^2$ defined by
$$\mathbf{A} = \begin{bmatrix} 1 & k \\ 0 & 1 \end{bmatrix},$$
where $k$ is any real scalar, is called a **shear in the $x$-direction by the factor $k$**. To investigate the geometric behavior of this transformation, we follow the procedure in Example 8(a) and construct the image of the unit square $S$.

(a) Compute the image of the unit square by the matrix transformation defined by **A**; that is, determine

$$\mathbf{AS} = \begin{bmatrix} 1 & k \\ 0 & 1 \end{bmatrix} \begin{bmatrix} 0 & 1 & 1 & 0 & 0 \\ 0 & 0 & 1 & 1 & 0 \end{bmatrix}.$$

(b) On separate figures with the same scale along the axes, draw the images of $S$ for the cases $k = 1, 2, 3, -1$, and $-2$. Under each figure give a brief geometric description of the image.

(c) Take images from part (b) and lay them on top of one another, aligning the axes and the origin. (You can sketch all the images on one set of axes to imitate this procedure.) If $k$ were any value other than those used in part (b), make a list of the properties that the image has in common with these superimposed sketches.

(d) If $k$ is large, say 25, describe the image of the shear.

(e) You are to make a physical model of the image resulting from a shear using two sticks of equal length and two rubber bands. Give directions on how to build the model.

**24.** Referring to Exercise 23, we know that the area of the unit square is 1 square unit. Determine the area of the image resulting from the shear. As a geometric aid, assume $k > 0$ and draw the image; then determine its area.

**25.** The matrix transformation in $R^2$ defined by $\mathbf{A} = \begin{bmatrix} 1 & 0 \\ k & 1 \end{bmatrix}$, where $k$ is any real scalar, is called a **shear in the $y$-direction by the factor $k$**. To investigate the geometric behavior of this transformation, we follow the procedure in Example 8(a) and construct the image of the unit square $S$.

(a) Compute the image of the unit square by the matrix transformation defined by **A**; that is, determine

$$\mathbf{AS} = \begin{bmatrix} 1 & 0 \\ k & 1 \end{bmatrix} \begin{bmatrix} 0 & 1 & 1 & 0 & 0 \\ 0 & 0 & 1 & 1 & 0 \end{bmatrix}.$$

(b) On separate figures with the same scale along the axes, draw the images of $S$ for the cases $k = 1, 2, 3, -1$, and $-2$. Under each figure give a brief geometric description of the image.

(c) Take images from part (b) and lay them on top of one another, aligning the axes and the origin. (You can sketch all the images on one set of axes to imitate this procedure.) If $k$ were any value other than those used in part (b), make a list of the properties that the image has in common with these superimposed sketches.

(d) If $k$ is large, say 25, describe the image of the shear.

(e) You are to make a physical model of the image resulting from a shear using two sticks of equal length and two rubber bands. Give directions on how to build the model.

**26.** Referring to the Exercise 25, we know that the area of the unit square is 1 square unit. Determine the area of the image resulting from the shear. As a geometric aid, assume $k > 0$ and draw the image; then determine its area.

**27.** Let $f$ be the matrix transformation in $R^2$ defined by $\mathbf{A} = \begin{bmatrix} 1 & 1 \\ k & k \end{bmatrix}$ where $k$ is any real scalar. From Example 8(a),

$$\mathbf{S} = \begin{bmatrix} 0 & 1 & 1 & 0 & 0 \\ 0 & 0 & 1 & 1 & 0 \end{bmatrix}$$

represents the vertices of the unit square.

(a) Describe the image $f(\mathbf{S})$ for each of the values $k = 2, 3, -4$.

(b) What is the area of the image in each case?

**28.** Let $f$ be the matrix transformation in $R^2$ that is a rotation by an angle $\theta$ as given in Example 7. Let

$$\mathbf{Q} = \begin{bmatrix} 1 & 0 & -1 & 0 & 1 \\ 0 & \sqrt{3} & 0 & -\sqrt{3} & 0 \end{bmatrix}$$

represent the vertices of a rhombus with side of length 2.

(a) Describe the image $f(\mathbf{Q})$ for each of the values $\theta = 30°, 45°$, and $90°$.

(b) What is the area of the image in each case?

**29.** Let $S$ be the unit square as described in Example 8(a) that we represent by the matrix

$$\mathbf{S} = \begin{bmatrix} 0 & 1 & 1 & 0 & 0 \\ 0 & 0 & 1 & 1 & 0 \end{bmatrix}$$

whose columns are the coordinates of the corners of the square. Let $f$ be the matrix transformation defined by $\mathbf{A} = \begin{bmatrix} a & b \\ c & d \end{bmatrix}$. Then the matrix representing the image $f(\mathbf{S})$ is

$$\mathbf{AS} = \begin{bmatrix} 0 & a & a+b & b & 0 \\ 0 & c & c+d & d & 0 \end{bmatrix}.$$

Geometrically the image is represented by plotting the coordinates of the points represented by each column and connecting them, in order, by straight line segments. We get the parallelogram shown. (For ease of drawing the figure, we have assumed that all the entries of **A** are positive; this has no bearing on what follows.)

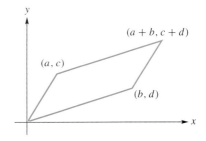

If $\mathbf{v} = \text{col}_1(\mathbf{A})$ and $\mathbf{w} = \text{col}_2(\mathbf{A})$, then we see that the image is a parallelogram determined by vectors $\mathbf{v}$ and $\mathbf{w}$. The area of a parallelogram is computed as area = base * height (see the following figure).

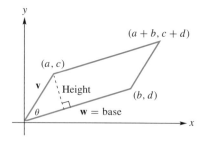

We have area $= \|\mathbf{w}\| * \text{height} = \|\mathbf{w}\| \sin(\theta)\|\mathbf{v}\|$. Our objective is develop a formula for the area of the parallelogram image in terms of the entries of the matrix $\mathbf{A}$.

(a) Express $\|\mathbf{v}\|$ and $\|\mathbf{w}\|$ in terms of the entries of $\mathbf{A}$.

(b) Express $\cos(\theta)$ in terms of vectors $\mathbf{v}$ and $\mathbf{w}$.

(c) Express $|\sin(\theta)|$ in terms of $\cos(\theta)$.

(d) Express $|\sin(\theta)|$ in terms of vectors $\mathbf{v}$ and $\mathbf{w}$.

(e) Express $|\sin(\theta)|$ in terms of the entries of $\mathbf{v}$ and $\mathbf{w}$.

(f) Express the area $\|\mathbf{v}\| \|\mathbf{w}\| \sin(\theta)$ in terms of the entries of $\mathbf{A}$ and simplify the expression to get $|ad - bc|$.

**30.** Use the result of Exercise 29(f) to compute the area of the image of the unit square in each of the following cases and draw the image.

(a) $\mathbf{A} = \begin{bmatrix} 2 & 0 \\ 0 & 3 \end{bmatrix}$.  (b) $\mathbf{A} = \begin{bmatrix} 2 & 5 \\ 0 & 3 \end{bmatrix}$.

(c) $\mathbf{A} = \begin{bmatrix} 2 & -2 \\ 0 & 3 \end{bmatrix}$.  (d) $\mathbf{A} = \begin{bmatrix} 2 & 3 \\ 2 & 3 \end{bmatrix}$.

**31.** Use the result of Exercise 29(f) to compute the area of the image of the unit square in each of the following cases and draw the image.

(a) $\mathbf{A} = \begin{bmatrix} 1 & 2 \\ 3 & 4 \end{bmatrix}$.  (b) $\mathbf{A} = \begin{bmatrix} 3 & 2 \\ 6 & 4 \end{bmatrix}$.

(c) $\mathbf{A} = \begin{bmatrix} 0 & 4 \\ -5 & 0 \end{bmatrix}$.  (d) $\mathbf{A} = \begin{bmatrix} 1 & 9 \\ 8 & 1 \end{bmatrix}$.

**32.** If $f$ is the matrix transformation that performs a rotation by an angle $\theta$ in $R^2$, then the associated matrix is

$$\mathbf{A} = \begin{bmatrix} \cos(\theta) & -\sin(\theta) \\ \sin(\theta) & \cos(\theta) \end{bmatrix}.$$

(See Example 7.) If $g$ is the matrix transformation that performs a rotation by an angle $\phi$ in $R^2$, then the associated matrix is

$$\mathbf{B} = \begin{bmatrix} \cos(\phi) & -\sin(\phi) \\ \sin(\phi) & \cos(\phi) \end{bmatrix}.$$

(*Hint*: Use trigonometric identities in the following.)

(a) Show that the composition $g(f(\mathbf{c}))$, for $\mathbf{c}$ in $R^2$, has matrix

$$\begin{bmatrix} \cos(\theta + \phi) & -\sin(\theta + \phi) \\ \sin(\theta + \phi) & \cos(\theta + \phi) \end{bmatrix}.$$

(b) Show that the composition $f(g(\mathbf{c}))$, for $\mathbf{c}$ in $R^2$, has matrix

$$\begin{bmatrix} \cos(\theta + \phi) & -\sin(\theta + \phi) \\ \sin(\theta + \phi) & \cos(\theta + \phi) \end{bmatrix}.$$

(c) Describe geometrically the meaning of the relation $g(f(\mathbf{c})) = f(g(\mathbf{c}))$.

(d) Describe in terms of the matrices $\mathbf{A}$ and $\mathbf{B}$ the meaning of the relation $g(f(\mathbf{c})) = f(g(\mathbf{c}))$.

**33.** If $f$ is the matrix transformation that reflects a vector about the $x$-axis in $R^2$, then the associated matrix is $\mathbf{A} = \begin{bmatrix} 1 & 0 \\ 0 & -1 \end{bmatrix}$ (see Example 4) and if $g$ is the matrix transformation that projects a vector onto the $x$-axis in $R^2$, then the associated matrix is $\mathbf{B} = \begin{bmatrix} 1 & 0 \\ 0 & 0 \end{bmatrix}$ (see Exercise 17).

(a) Determine the matrix $\mathbf{W}$ associated with the composition $g(f(\mathbf{c}))$ for $\mathbf{c}$ in $R^2$.

(b) Determine the matrix $\mathbf{Q}$ associated with the composition $f(g(\mathbf{c}))$ for $\mathbf{c}$ in $R^2$.

(c) Describe geometrically the meaning of the relation $g(f(\mathbf{c})) = f(g(\mathbf{c}))$.

(d) Describe in terms of the matrices $\mathbf{W}$ and $\mathbf{Q}$ the meaning of the relation $g(f(\mathbf{c})) = f(g(\mathbf{c}))$.

**34.** If $f$ is the matrix transformation that reflects a vector about the line $y = x$ in $R^2$, then the associated matrix $\mathbf{A} = \begin{bmatrix} 0 & 1 \\ 1 & 0 \end{bmatrix}$ (see Example 5) and if $g$ is the matrix transformation that projects a vector onto the $y$-axis in $R^2$, then the associated matrix $\mathbf{B} = \begin{bmatrix} 0 & 0 \\ 0 & 1 \end{bmatrix}$ (see Exercise 18).

(a) Determine the matrix $\mathbf{W}$ associated with the composition $g(f(\mathbf{c}))$ for $\mathbf{c}$ in $R^2$.

(b) Determine the matrix $\mathbf{Q}$ associated with the composition $f(g(\mathbf{c}))$ for $\mathbf{c}$ in $R^2$.

(c) Describe geometrically the meaning of the relation $g(f(\mathbf{c})) \neq f(g(\mathbf{c}))$.

(d) Describe in terms of the matrices $\mathbf{W}$ and $\mathbf{Q}$ the meaning of the relation $g(f(\mathbf{c})) \neq f(g(\mathbf{c}))$.

**35.** If $f$ is the matrix transformation that rotates a vector $30°$ in $R^2$, then the associated matrix is

$$\mathbf{A} = \begin{bmatrix} \cos(30°) & -\sin(30°) \\ \sin(30°) & \cos(30°) \end{bmatrix}.$$

If $g$ is the matrix transformation that projects a vector onto the $x$-axis in $R^2$, then the associated matrix is

$$\mathbf{B} = \begin{bmatrix} 1 & 0 \\ 0 & 0 \end{bmatrix}.$$

(a) Determine the matrix **W** associated with the composition $g(f(\mathbf{c}))$ for **c** in $R^2$.

(b) Determine the matrix **Q** associated with the composition $f(g(\mathbf{c}))$ for **c** in $R^2$.

(c) Describe geometrically the meaning of the relation $g(f(\mathbf{c})) \neq f(g(\mathbf{c}))$.

(d) Describe in terms of the matrices **W** and **Q** the meaning of the relation $g(f(\mathbf{c})) \neq f(g(\mathbf{c}))$.

**36.** If $f$ is the matrix transformation that performs the projection onto the $xy$-plane in $R^3$, then the associated matrix

$$\mathbf{A} = \begin{bmatrix} 1 & 0 & 0 \\ 0 & 1 & 0 \\ 0 & 0 & 0 \end{bmatrix}.$$

If $g$ is the matrix transformation that performs a reflection through the $xz$-plane in $R^3$, then the associated matrix

$$\mathbf{B} = \begin{bmatrix} -1 & 0 & 0 \\ 0 & -1 & 0 \\ 0 & 0 & 1 \end{bmatrix}.$$

(a) Determine the matrix **W** associated with the composition $g(f(\mathbf{c}))$ for **c** in $R^2$.

(b) Determine the matrix **Q** associated with the composition $f(g(\mathbf{c}))$ for **c** in $R^2$.

(c) Describe geometrically the meaning of the relation $g(f(\mathbf{c})) = f(g(\mathbf{c}))$.

(d) Describe in terms of the matrices **W** and **Q** the meaning of the relation $g(f(\mathbf{c})) = f(g(\mathbf{c}))$.

**37.** Let $\mathbf{i} = \begin{bmatrix} 1 & 0 \end{bmatrix}^T$ and $\mathbf{j} = \begin{bmatrix} 0 & 1 \end{bmatrix}^T$.

(a) Show that **i** and **j** are unit vectors and that they are orthogonal.

(b) Show that every 2-vector $\mathbf{c} = \begin{bmatrix} c_1 & c_2 \end{bmatrix}^T$ is a linear combination of **i** and **j**.

(c) Explain why span$\{\mathbf{i}, \mathbf{j}\}$ is $R^2$.

(d) Compute the projection of the 2-vector **c** onto the $x$-axis in terms of the vector **i**. Denote the resulting vector as $\operatorname{proj}_x \mathbf{c}$. (See Exercise 17.)

(e) Compute the projection of the 2-vector **c** onto the $y$-axis in terms of the vector **j**. Denote the resulting vector as $\operatorname{proj}_y \mathbf{c}$. (See Exercise 18.)

(f) Explain why we can express any vector **c** in $R^2$ as $\mathbf{c} = \operatorname{proj}_x \mathbf{c} + \operatorname{proj}_y \mathbf{c}$.

**38.** Let $\mathbf{i} = \begin{bmatrix} 1 & 0 & 0 \end{bmatrix}^T$, $\mathbf{j} = \begin{bmatrix} 0 & 1 & 0 \end{bmatrix}^T$, and $\mathbf{k} = \begin{bmatrix} 0 & 0 & 1 \end{bmatrix}^T$.

(a) Show that **i**, **j**, and **k** are unit vectors.

(b) Show that the following pairs of vectors are orthogonal: **i** and **j**; **j** and **k**; **i** and **k**.

(c) Show that every 3-vector $\mathbf{c} = \begin{bmatrix} c_1 & c_2 & c_3 \end{bmatrix}^T$ is a linear combination of **i**, **j**, and **k**.

(d) Explain why span$\{\mathbf{i}, \mathbf{j}, \mathbf{k}\}$ is $R^3$.

(e) Compute the projection of the 3-vector **c** onto the $x$-axis in terms of the vector **i**. Denote the resulting vector as $\operatorname{proj}_x \mathbf{c}$.

(f) Compute the projection of the 3-vector **c** onto the $y$-axis in terms of the vector **j**. Denote the resulting vector as $\operatorname{proj}_y \mathbf{c}$.

(g) Compute the projection of the 3-vector **c** onto the $z$-axis in terms of the vector **k**. Denote the resulting vector as $\operatorname{proj}_z \mathbf{c}$.

(h) Explain why we can express any vector **c** in $R^3$ as $\mathbf{c} = \operatorname{proj}_x \mathbf{c} + \operatorname{proj}_y \mathbf{c} + \operatorname{proj}_z \mathbf{c}$.

**39.** In Example 6 we showed that the projection of **c** in $R^3$ onto the $xy$-plane is computed as

$$f(\mathbf{c}) = \mathbf{Bc} = \begin{bmatrix} 1 & 0 & 0 \\ 0 & 1 & 0 \\ 0 & 0 & 0 \end{bmatrix} \begin{bmatrix} c_1 \\ c_2 \\ c_3 \end{bmatrix} = \begin{bmatrix} c_1 \\ c_2 \\ 0 \end{bmatrix}.$$

Express $f(\mathbf{c})$ in terms of the vectors **i** and **j** defined in Exercise 38.

*In the application* **Orthogonal projections** *we showed that the projection of vector* **c** *onto vector* **b** *is*

$$\mathbf{p} = \operatorname{proj}_{\mathbf{b}} \mathbf{c} = \frac{(\mathbf{b} \cdot \mathbf{c})}{(\mathbf{b} \cdot \mathbf{b})} \mathbf{b}.$$

*In Exercises* 40–42, *compute* **p** *for each pair of vectors in* $R^2$. *Sketch* **b**, **c**, *and* **p**.

**40.** $\mathbf{b} = \begin{bmatrix} 5 \\ 4 \end{bmatrix}$, $\mathbf{c} = \begin{bmatrix} 3 \\ 1 \end{bmatrix}$.

**41.** $\mathbf{b} = \begin{bmatrix} 3 \\ 7 \end{bmatrix}$, $\mathbf{c} = \begin{bmatrix} 1 \\ -1 \end{bmatrix}$.

**42.** $\mathbf{b} = \begin{bmatrix} -5 \\ 8 \end{bmatrix}$, $\mathbf{c} = \begin{bmatrix} -2 \\ -4 \end{bmatrix}$.

*In the application* **Orthogonal projections** *we showed how to compute the projection of one vector onto another vector. In Exercises* 43–46, *compute the indicated projections.*

**43.** $\operatorname{proj}_{\mathbf{w}} \mathbf{v}$, where $\mathbf{w} = \begin{bmatrix} 2 & 3 \end{bmatrix}$ and $\mathbf{v} = \begin{bmatrix} -2 & 5 \end{bmatrix}$.

**44.** $\operatorname{proj}_{\mathbf{w}} \mathbf{v}$, where $\mathbf{w} = \begin{bmatrix} 3 \\ 1 \\ -4 \end{bmatrix}$ and $\mathbf{v} = \begin{bmatrix} 0 \\ 3 \\ 1 \end{bmatrix}$.

**45.** $\operatorname{proj}_{\mathbf{r}} \mathbf{q}$, where $\mathbf{r} = \begin{bmatrix} -2 & 4 & 5 \end{bmatrix}$ and $\mathbf{q} = \begin{bmatrix} 1 & 4 & 1 \end{bmatrix}$.

**46.** $\operatorname{proj}_{\mathbf{r}} \mathbf{q}$, where $\mathbf{r} = \begin{bmatrix} 1 \\ 0 \\ 2 \\ 3 \end{bmatrix}$ and $\mathbf{q} = \begin{bmatrix} 0 \\ 3 \\ -3 \\ 2 \end{bmatrix}$.

**47.** True or false: $\operatorname{proj}_{\mathbf{b}} \mathbf{c} = \operatorname{proj}_{\mathbf{c}} \mathbf{b}$. Explain your answer.

**48.** Show that the projection of vector **c** onto **b** is a linear transformation. That is, show that $\text{proj}_\mathbf{b} k\mathbf{c} = k\,\text{proj}_\mathbf{b}\mathbf{c}$ and that $\text{proj}_\mathbf{b}(\mathbf{c} + \mathbf{d}) = \text{proj}_\mathbf{b}\mathbf{c} + \text{proj}_\mathbf{b}\mathbf{d}$.

**49.** Let **A** be an $n \times n$ matrix and let $\text{Tr}(\mathbf{A}) = a_{11} + a_{22} + \cdots + a_{nn}$. ($\text{Tr}(\mathbf{A})$ is called the **trace** of matrix **A**.) Show that this function is a linear transformation.

**50.** In order to provide a geometric display of the effect of a matrix transformation, we restricted much of our work in this section to $R^2$ and $R^3$. As our definition of a matrix transformation stated, no such restriction is needed. Moreover, there is no need to restrict the entries of the vector or matrix to be real numbers. If **A** is an $m \times n$ matrix with complex entries, then $f(\mathbf{c}) = \mathbf{Ac}$ determines a matrix transformation from $C^n$ to $C^n$. In addition, our development of orthogonal projections can be extended to vectors in $C^n$ where we use the complex dot product as given in Exercise 21 of Section 1.3; Equation (2) is then written as

$$\mathbf{p} = \frac{\overline{\mathbf{b} \cdot \mathbf{c}}}{\overline{\mathbf{b} \cdot \mathbf{b}}}\mathbf{b}.$$

(a) Let $\mathbf{A} = \begin{bmatrix} 2+i & 5 \\ 1-i & 3i \end{bmatrix}$. Determine the image of each of the following vectors for the matrix transformation determined by **A**.

$$\mathbf{c} = \begin{bmatrix} i \\ 1+i \end{bmatrix}, \quad \mathbf{d} = \begin{bmatrix} 1 \\ 2 \end{bmatrix}, \quad \mathbf{e} = \begin{bmatrix} 3+2i \\ -1+i \end{bmatrix}.$$

(b) For the matrix **A** in part (a), is $\begin{bmatrix} 2+i \\ 7 \end{bmatrix}$ in the range of the matrix transformation determined by **A**?

(c) Let $\mathbf{c} = \begin{bmatrix} 3+2i \\ 1-i \end{bmatrix}$ and $\mathbf{b} = \begin{bmatrix} 2i \\ 2 \end{bmatrix}$. Determine the orthogonal projection **p** of **c** onto **b**.

## In MATLAB

*This section introduced matrix transformations, which are functions whose input and output are vectors which are related by a matrix multiplication: $f(\mathbf{c}) = \mathbf{Ac}$. The input **c** can be a single vector or collection of vectors which represent a figure or object. (Note, we can view vectors as points and vice-versa.) Most of our geometric examples let **A** be a 2 × 2 matrix so that we could easily display the output, the image. In the following exercises we continue this practice and use MATLAB to construct and display images. The MATLAB routines in these exercises provide an opportunity for you to gain experience with the visualization of matrix transformations.*

**ML. 1.** A matrix transformation from $R^m$ to $R^n$ is often called a mapping or map for short. (This is just another name for a function.) We discussed a variety of mappings in this section. Using MATLAB routine **mapit**, we provide further illustrations and experimental capabilities. In MATLAB type **help mapit** and read the brief description. To start this routine type **mapit**. At the lower left is the Comment Window, which will provide directions for using this routine. (*Note*: In **mapit** all matrices are named **A** even if you perform a composition.)

(a) Select the 'Object' Square by clicking the mouse on the word. Look at the Comment Window. Next click on the View button. The unit square will appear on the left set of axes. Now click on the MATRIX button, then enter matrix

$$\mathbf{A} = \begin{bmatrix} 2 & 0 \\ 0 & 4 \end{bmatrix}.$$

by typing **[2 0;0 4]** followed by Enter. Click on the MAP IT button to see the image of the unit square determined by the matrix **A**. What is the area of the image? (See Example 8(a).)

(b) Next click on the Composite button and then the MATRIX button that appears. This time enter the matrix

$$\begin{bmatrix} \frac{1}{2} & 0 \\ 0 & \frac{1}{4} \end{bmatrix},$$

then click on MAP IT. What is the area of the composite image?

(c) If $\mathbf{A} = \begin{bmatrix} 2 & 0 \\ 0 & 4 \end{bmatrix}$ and $\mathbf{B} = \begin{bmatrix} \frac{1}{2} & 0 \\ 0 & \frac{1}{4} \end{bmatrix}$, then we say that

$$f(\text{unit square}) = \mathbf{A} * (\text{unit square})$$
$$= \text{first image}$$

and

$$g(\text{first image}) = \mathbf{B} * (\text{first image})$$
$$= \text{composite image}.$$

From our discussion of composition we have

$$g(f(\text{unit square})) = \mathbf{B} * (\mathbf{A} * (\text{unit square}))$$
$$= \text{composite image}.$$

Compute the matrix $\mathbf{B} * \mathbf{A}$ and explain how the result of this composition is related to the unit square. (Click on Quit to exit the routine **mapit**.)

**ML. 2.** In MATLAB type **mapit**.

(a) Select the 'Object' Circle by clicking the mouse on the word. Next click on the View button. The unit circle will appear on the left set of axes. Now click on the Matrix button, then enter matrix $\mathbf{A} = \begin{bmatrix} 3 & 0 \\ 0 & 5 \end{bmatrix}$ by typing **[3 0;0 5]** followed by Enter. Click on MAP IT to see the image. As predicted in Example 8(b), the image is an ellipse. What is the length of its major axis? Of its minor axis?

(b) Restart **mapit**, choose the Circle, and generate the image determined by the matrix $\mathbf{A} = \begin{bmatrix} 4 & 0 \\ 0 & 4 \end{bmatrix}$. Carefully describe the image and its relationship to the unit circle.

(c) Determine a matrix $\mathbf{A}$ so that the image of the unit circle is the ellipse shown in Figure 15. Verify your choice using **mapit**.

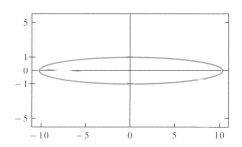

FIGURE 15

(d) Use **mapit** with the Circle and matrix $\mathbf{A} = \begin{bmatrix} 1 & 2 \\ 0 & 1 \end{bmatrix}$ and determine the corresponding image. How does the image in this case vary from those generated in parts (a) and (b)?

**ML. 3.** Let $\mathbf{A} = \begin{bmatrix} 3 & 1 \\ 0 & 2 \end{bmatrix}$ and $\mathbf{B} = \begin{bmatrix} -1 & 2 \\ 2 & 0 \end{bmatrix}$.

(a) Use **mapit** with the Rect. object and matrix $\mathbf{A}$. Choose the composite option, this time entering the matrix $\mathbf{B}$. The final image is $\mathbf{B}*(\mathbf{A}*\text{Rect.})$. Carefully make a sketch of this composite image or, if you have print capability, press Enter, type **print**, and return to the graph by typing **figure(gcf)**. Click the restart button before going to part (b).

(b) Use **mapit** with the Rect. object and matrix $\mathbf{B}$. Choose the composite option, this time entering the matrix $\mathbf{A}$. The final image is $\mathbf{A}*(\mathbf{B}*\text{Rect.})$. Carefully make a sketch of this composite image or, if you have print capability, press Enter, type **print**, and return to the graph by typing **figure(gcf)**. Click the Quit button to exit the routine.

(c) Are the composite images in parts (a) and (b) the same? Compute $\mathbf{B} * \mathbf{A}$ and then compute $\mathbf{A} * \mathbf{B}$ and use the results to provide an explanation for your response to this question.

**ML. 4.** Use **mapit** to perform each of the following.

(a) Select the Arrow object. Determine a matrix $\mathbf{A}$ so that the image is an arrow pointed in the opposite direction.

(b) Select the Arrow object. Determine a matrix $\mathbf{A}$ so that the image is an arrow pointed in the same direction but only half as long.

(c) Select the Arrow object and use the matrix

$$\begin{bmatrix} \cos(\text{pi}/4) & \sin(\text{pi}/4) \\ -\sin(\text{pi}/4) & \cos(\text{pi}/4) \end{bmatrix}.$$

Describe the resulting image. What angle does it make with the positive horizontal axis? To help answer this question, use the grid button and then inspect the grid generated on the mapped arrow.

(d) Using part (c), determine the coordinates of the top end of the arrow.

**ML. 5.** Use **mapit** to perform each of the following.

(a) Select the House object. Determine a matrix $\mathbf{A}$ so that the image is a house only half as wide.

(b) Select the House object. Determine a matrix $\mathbf{A}$ so that the image is a house only half as wide and half as tall.

(c) Select the House object. Use $\mathbf{A} = \begin{bmatrix} 1 & 1 \\ 1 & 1 \end{bmatrix}$, determine the image, and explain why the house is uninhabitable.

(d) Select the House object. What matrix will "shrink" the house to a single point? Verify your choice.

**ML. 6.** Matrix transformations from $R^2$ to $R^2$ are sometimes called plane linear transformations. The MATLAB routine **planelt** lets us experiment with such transformations by choosing the geometric operation we want to perform on a figure. The routine then uses the appropriate matrix to compute the image, displays the matrix, and keeps a graphical record of the original, the previous image, and the current image. Routine **planelt** is quite versatile since you can enter your own figure and/or matrix.

(a) To start this routine type **planelt**. Read the descriptions that appear and follow the on-screen directions until you get to FIGURE CHOICES. There select the triangle, choose to 'See the Triangle', and then the option 'Use this Figure. Go to select transformations.' Select the rotation and use a 45° angle. After the figures are displayed, press Enter. You will see the Plane Linear Transformation menu of options again. If you choose a transformation at this point it will be used "compositely" with the transformation just performed. Try this by choosing to reflect the current figure about the $y$-axis. The figures displayed show the original triangle, this triangle rotated through 45°, and then this image reflected about the $y$-axis. Record a sketch of the composite figure.

(b) Reverse the order of the transformations in part (a). Record a sketch of the figure obtained from this composition. Compare this sketch with that from part (a). If $\mathbf{A}$ is the matrix of the 45° rotation and $\mathbf{B}$ is the matrix of the reflection about the $y$-axis then explain how we know that $\mathbf{BA} \neq \mathbf{AB}$.

**ML. 7.** Use **planelt** with the parallelogram. Choose a composition of transformations so that the final figure is that shown in Figure 16.

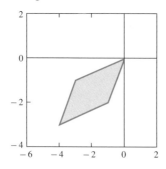

FIGURE 16

**ML. 8.** Orthogonal projections will play a fundamental role in a variety of situations for us later. While this section developed the algebra for computing projec-

tions, MATLAB can help present the geometric aspects. In MATLAB type **help project** and read the description. To start this routine type **project** and then choose the demo for a first look at this routine. Use **project** to determine $\text{proj}_w\mathbf{u}$ for each of the following pairs of vectors. As the figure is generated, note whether the projection is longer than vector $\mathbf{w}$ and whether it is in the same or opposite direction.

(a) $\mathbf{u} = \begin{bmatrix} 5 \\ 4 \end{bmatrix}$, $\mathbf{w} = \begin{bmatrix} 3 \\ 1 \end{bmatrix}$.

(b) $\mathbf{u} = \begin{bmatrix} 1 \\ -4 \end{bmatrix}$, $\mathbf{w} = \begin{bmatrix} 3 \\ 7 \end{bmatrix}$.

(c) $\mathbf{u} = \begin{bmatrix} 5 \\ 3 \\ 1 \end{bmatrix}$, $\mathbf{w} = \begin{bmatrix} 3 \\ 1 \\ -4 \end{bmatrix}$.

(d) $\mathbf{u} = \begin{bmatrix} 4 \\ 6 \\ 0 \end{bmatrix}$, $\mathbf{w} = \begin{bmatrix} 2 \\ 3 \\ 8 \end{bmatrix}$.

## True/False Review Questions

*Determine whether each of the following statements is true or false.*

1. If $f$ is a matrix transformation with associated matrix $\mathbf{A}$, then $f(\mathbf{x} + \mathbf{y}) = f(\mathbf{x}) + f(\mathbf{y})$.

2. If $f$ is a matrix transformation with associated matrix $\mathbf{A}$, then $f(k\mathbf{x}) = kf(\mathbf{x})$.

3. The image from a matrix transformation with associated matrix $\mathbf{A}$ is a linear combination of the columns of $\mathbf{A}$.

4. If $f$ and $g$ are matrix transformations from $R^2$ to $R^2$, then $f(g(\mathbf{x})) = g(f(\mathbf{x}))$ for all vectors $\mathbf{x}$ in $R^2$.

5. If $f$ is the matrix transformation from $R^2$ to $R^2$ with associated matrix $\mathbf{A} = \begin{bmatrix} 5 & 0 \\ 0 & 3 \end{bmatrix}$, then the image of the unit square has area 15 square units.

6. If $f$ is the matrix transformation from $R^2$ to $R^2$ with associated matrix $\mathbf{A} = \begin{bmatrix} -2 & 0 \\ 0 & 4 \end{bmatrix}$, then the image of the unit circle is an ellipse.

7. If $f$ and $g$ are rotation matrix transformations from $R^2$ to $R^2$, then $f(g(\mathbf{x})) = g(f(\mathbf{x}))$.

8. Let $\mathbf{v} = \begin{bmatrix} 0 \\ 5 \end{bmatrix}$ and $\mathbf{w} = \begin{bmatrix} -1 \\ 2 \end{bmatrix}$, then $\text{proj}_w\mathbf{v} = 2\mathbf{w}$.

9. $\text{proj}_w\mathbf{v}$ is the same as $\text{proj}_v\mathbf{w}$.

10. In $R^2$ the composition of a reflection about the $x$-axis followed by a reflection about the $y$-axis is the same as a reflection about the line $y = x$.

## Terminology

| | |
|---|---|
| Matrix transformation; domain and range | Images |
| Reflections | Projections onto axes or coordinate planes |
| Rotations | Composition of matrix transformations |
| Linear transformation | Projection of vector **c** onto **b** |
| Orthogonal spanning set | |

The concepts in this section connect the algebraic operations of matrix-vector products and dot products to geometric notions. Hence we establish further bonds between the algebra of matrices and geometry in $R^2$ and $R^3$. Answer the following in your own words to help you form visualizations of the terms listed in the preceding table.

- Let $f(\mathbf{c}) = \mathbf{Ac} = \mathbf{b}$. We say _____ is the image of _____. Vector $\mathbf{b}$ is in the _____ of the matrix transformation and vector $\mathbf{c}$ is in the _____.
- Explain both algebraically and geometrically what it means to reflect vector $\begin{bmatrix} c_1 \\ c_2 \end{bmatrix}$ about the $x$-axis; about the $y$-axis. Draw a figure in each case.
- Explain both algebraically and geometrically what it means to project vector $\begin{bmatrix} c_1 \\ c_2 \end{bmatrix}$ onto the $x$-axis; onto the $y$-axis. Draw a figure in each case.
- Explain both algebraically and geometrically what it means to project vector $\begin{bmatrix} c_1 \\ c_2 \\ c_3 \end{bmatrix}$ onto the $xy$-plane; onto the $y$-axis. Draw a figure.
- If matrix transformation $f$ is a rotation through $\alpha$ degrees that uses matrix $\mathbf{A}$ and matrix transformation $g$ is another rotation through $\beta$ degrees that uses matrix $\mathbf{B}$, then

    What is the matrix corresponding to $f(g(\mathbf{c}))$?

    What is the matrix corresponding to $g(f(\mathbf{c}))$?

    How are the images $f(g(\mathbf{c}))$ and $g(f(\mathbf{c}))$ related? Explain your answer.

    How are the matrices corresponding to $f(g(\mathbf{c}))$ and $g(f(\mathbf{c}))$ related? Explain your answer.

    How may the results change if $f$ and $g$ are not both rotations?
- The projection of vector $\mathbf{c}$ onto vector $\mathbf{b}$ is parallel to vector _____.
- Explain why a matrix transformation is a particular example of a linear transformation.
- State an orthogonal spanning set for the plane $R^2$ and for $R^3$.

## CHAPTER TEST

*For questions 1–6, use the following matrices.* (*None of the answers involve the size of the matrices.*)

$$\mathbf{A} = \begin{bmatrix} 2 & 4 \\ 0 & 1 \end{bmatrix}, \quad \mathbf{b} = \begin{bmatrix} 3 \\ -1 \end{bmatrix}, \quad \mathbf{c} = \begin{bmatrix} 2 \\ 4 \end{bmatrix}, \quad \mathbf{D} = \begin{bmatrix} 3 & 0 \\ 0 & 5 \end{bmatrix}, \quad \mathbf{E} = \begin{bmatrix} 3 & 2 \\ 0 & 5 \end{bmatrix}, \quad \mathbf{u} = \begin{bmatrix} 1 \\ 1 \end{bmatrix}.$$

1. Compute $\mathbf{A} + 2\mathbf{E}^T =$ _____. The resulting matrix is _____.

2. Solve the matrix equation $\mathbf{Ax} = \mathbf{b}$ where $\mathbf{x} = \begin{bmatrix} x_1 \\ x_2 \end{bmatrix}$.

3. Find the projection of $\mathbf{c}$ onto $\mathbf{b}$.

4. Compute $2\mathbf{Dc} - (\mathbf{cb}^T)\mathbf{u}$.

5. Find the cosine of the angle between $\mathbf{Eu}$ and $\mathbf{c}$.

6. Let $f$ be the matrix transformation $f(\mathbf{v}) = \mathbf{Dv}$ and $g$ be the matrix transformation $g(\mathbf{v}) = \mathbf{Av}$, where $\mathbf{v}$ is a 2-vector (a column).

   (a) Compute $f(g(\mathbf{w}))$ where $\mathbf{w} = \begin{bmatrix} 2 \\ -2 \end{bmatrix}$.

   (b) Using matrices $\mathbf{D}$ and $\mathbf{A}$ show that $f(g(\mathbf{v})) \neq g(f(\mathbf{v}))$.

7. Let $\mathbf{A}$ and $\mathbf{B}$ be matrices of appropriate sizes so that their product $\mathbf{AB}$ is defined.

   (a) If $\mathbf{A}$ is $m \times 3$, then what size is $\mathbf{B}$?

   (b) Is it true that we are guaranteed that $\mathbf{BA}$ will be defined? Explain your answer.

   (c) If $\mathbf{A} = \begin{bmatrix} 2 & -1 \\ -6 & 3 \end{bmatrix}$. Find matrix $\mathbf{B} = \begin{bmatrix} b_1 \\ b_2 \end{bmatrix}$, whose entries are *not* both zero, so that $\mathbf{AB} = \begin{bmatrix} 0 \\ 0 \end{bmatrix}$.

8. (a) If $\mathbf{w} = k\mathbf{v}$, where $k$ is a scalar, then we say vectors $\mathbf{v}$ and $\mathbf{w}$ are _____.

   (b) If $\mathbf{v} \cdot \mathbf{w} = 0$, then we say vectors $\mathbf{v}$ and $\mathbf{w}$ are _____.

   (c) If $\mathbf{v} = 5\mathbf{w} - 8\mathbf{u}$, then we say that vector $\mathbf{v}$ is a _____ of $\mathbf{w}$ and $\mathbf{u}$.

   (d) Matrix $\mathbf{A}$ is 5 by 5 and $\mathbf{x}$ is 5 by 1. We know $\mathbf{Ax} = 3\mathbf{x}$. Compute $\mathbf{y} = (-4\mathbf{A}^2 + \mathbf{A} + 2\mathbf{I}_3)\mathbf{x}$ as a scalar multiple of $\mathbf{x}$.

9. Let $S$ be the set of all 3-vectors that are orthogonal to $\mathbf{v} = \begin{bmatrix} v_1 \\ v_2 \\ v_3 \end{bmatrix}$. This means that if $\mathbf{w}$ is in $S$, then $\mathbf{w} \cdot \mathbf{v} = 0$.

   (a) Show that if vectors $\mathbf{u}$ and $\mathbf{w}$ are in $S$, then so is $\mathbf{u} + \mathbf{w}$.

   (b) Show that if vector $\mathbf{u}$ is in $S$, then so is vector $k\mathbf{u}$, for any scalar $k$.

   (c) Explain why $S$ is a closed set.

10. Matrix $\mathbf{A}$ is $4 \times 3$ and column $\mathbf{c}$ is $3 \times 1$. Express the matrix-vector product $\mathbf{Ac}$ in two distinct ways.

    (a) In terms of a linear combination.

    (b) Using dot products.

11. Let $\mathbf{A}$ be a square matrix. Show that $\mathbf{A} + \mathbf{A}^T$ is symmetric.

12. Matrix $\mathbf{A}$ is $9 \times 9$ and $\mathbf{B}$ is $9 \times 5$. State an expression that computes the $(7, 3)$-entry of $\mathbf{AB}$.

13. Let $T$ be the set of vectors $\{\mathbf{v}_1, \mathbf{v}_2, \mathbf{v}_3, \mathbf{v}_4\}$. Define the span$\{T\}$.

14. Let $S$ be the set of all 4-vectors of the form $\begin{bmatrix} t \\ r \\ t \\ r \end{bmatrix}$ where $t$ and $r$ are scalars. State a set of vectors that will *generate* the set $S$.

    (That is, find vectors $\mathbf{v}_1$ and $\mathbf{v}_2$ such that span$\{\mathbf{v}_1, \mathbf{v}_2\} = S$.)

# CHAPTER

# 2

# *LINEAR SYSTEMS AND THEIR SOLUTION*

A system of linear equations expresses a relationship between a set of quantities. It is the purpose of this chapter to establish procedures for determining whether the relationship is valid, and if it is, to determine values for the quantities involved. We will see that when the relationship is valid there can, in certain instances, be more than one set of values that make the relationship true.

We show that the data within a system of linear equations can be represented in matrix form and can be manipulated by simple arithmetic procedures to reveal information about the quantities involved. Our goal is to develop a systematic way to represent and analyze the behavior of linear systems. As we establish the computational process, we develop accompanying properties that can be generalized to more abstract settings in later chapters. The theoretical concepts developed in this chapter are fundamental to our study of linear algebra and provide a rigorous validation of the computational models we develop. The mastery of these concepts will make it easy to assimilate later topics. In addition, we introduce a variety of applications that use linear systems to construct mathematical models that represent information or solve mathematical models of particular physical situations.

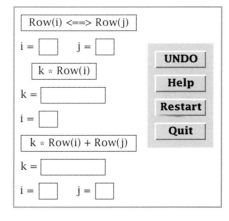

In MATLAB, **rowop.m**

## 2.1 ■ LINEAR SYSTEMS OF EQUATIONS

A linear system of $m$ equations in $n$ unknowns $x_1, x_2, \ldots, x_n$ is expressed as

$$
\begin{aligned}
a_{11}x_1 + a_{12}x_2 + \cdots + a_{1n}x_n &= b_1 \\
a_{21}x_1 + a_{22}x_2 + \cdots + a_{2n}x_n &= b_2 \\
&\ \ \vdots \\
a_{i1}x_1 + a_{i2}x_2 + \cdots + a_{in}x_n &= b_i \\
&\ \ \vdots \\
a_{j1}x_1 + a_{j2}x_2 + \cdots + a_{jn}x_n &= b_j \\
&\ \ \vdots \\
a_{m1}x_1 + a_{m2}x_2 + \cdots + a_{mn}x_n &= b_m.
\end{aligned}
\tag{1}
$$

The coefficient matrix is

$$
\mathbf{A} = \begin{bmatrix}
a_{11} & a_{12} & \cdots & \cdots & a_{1n} \\
a_{21} & a_{22} & \cdots & \cdots & a_{2n} \\
\vdots & \vdots & \cdots & \cdots & \vdots \\
\vdots & \vdots & \cdots & \cdots & \vdots \\
a_{m1} & a_{m2} & \cdots & \cdots & a_{mn}
\end{bmatrix}
$$

and the right side is

$$
\mathbf{b} = \begin{bmatrix}
b_1 \\
b_2 \\
\vdots \\
\vdots \\
b_m
\end{bmatrix}.
$$

As we showed in Example 4 of Section 1.3, this linear system can be written as the matrix equation $\mathbf{Ax} = \mathbf{b}$ where

$$
\mathbf{x} = \begin{bmatrix}
x_1 \\
x_2 \\
\vdots \\
\vdots \\
x_n
\end{bmatrix}.
$$

In Examples 4 and 5 in Section 1.1 we noted that the information contained in the system can be represented by the partitioned matrix $\begin{bmatrix} \mathbf{A} \mid \mathbf{b} \end{bmatrix}$, which we call the **augmented matrix** of the linear system. The right side $\mathbf{b}$ is called the **augmented column** of the partitioned matrix. From Section 1.3 we know that the matrix product $\mathbf{Ax}$ is a linear combination of the columns of $\mathbf{A}$; that is,

$$
\mathbf{Ax} = \sum_{i=1}^{n} x_i \mathrm{col}_i(\mathbf{A}).
$$

Hence when we are asked to solve the matrix equation $\mathbf{Ax} = \mathbf{b}$ we seek a vector $\mathbf{x}$ so that vector $\mathbf{b} = \sum_{i=1}^{n} x_i \mathrm{col}_i(\mathbf{A})$. This is the same as saying that linear system $\mathbf{Ax} = \mathbf{b}$ has a solution if and only if $\mathbf{b}$ is in span$\{\mathrm{col}_1(\mathbf{A}), \mathrm{col}_2(\mathbf{A}), \ldots, \mathrm{col}_n(\mathbf{A})\}$.

**Definition**    The linear system $\mathbf{Ax} = \mathbf{b}$ is called **consistent** provided that $\mathbf{b}$ is in span$\{\text{col}_1(\mathbf{A}), \text{col}_2(\mathbf{A}), \ldots, \text{col}_n(\mathbf{A})\}$; otherwise the linear system is called **inconsistent**.

Thus a consistent linear system has a solution while an inconsistent linear system has no solution, and conversely.

A major goal is to be able to identify whether a linear system $\mathbf{Ax} = \mathbf{b}$ is consistent or inconsistent and, for those that are consistent, to determine a solution vector $\mathbf{x}$ so that the product $\mathbf{Ax}$ yields $\mathbf{b}$. Thus merely being presented with the matrix equation $\mathbf{Ax} = \mathbf{b}$ does not imply that the linear system is consistent.

**Definition**    A linear system $\mathbf{Ax} = \mathbf{b}$ is called **homogeneous** provided the right side $\mathbf{b} = \mathbf{0}$. Every homogeneous linear system can be represented as a matrix equation of the form $\mathbf{Ax} = \mathbf{0}$.

A linear system $\mathbf{Ax} = \mathbf{b}$ is called **nonhomogeneous** provided that $\mathbf{b} \neq \mathbf{0}$.

A homogeneous linear system is always consistent since

$$0\,\text{col}_1(\mathbf{A}) + 0\,\text{col}_2(\mathbf{A}) + \cdots + 0\,\text{col}_n(\mathbf{A}) = \mathbf{A0} = \mathbf{0}$$

That is, $\mathbf{x} = \mathbf{0}$ is always a solution of a homogeneous linear system; we call $\mathbf{0}$ the **trivial solution**. (Recall the trivial linear combination introduced in Section 1.2.) As we will see, some homogeneous linear systems have solutions other than the trivial solution; such solutions are called **nontrivial**.

A linear system is either consistent or inconsistent. However, a consistent linear system can be one of two types:

1.  We say that $\mathbf{Ax} = \mathbf{b}$ has a **unique solution** if there is one and only one vector $\mathbf{x}$ so that the product $\mathbf{Ax}$ gives $\mathbf{b}$. This is equivalent to saying that there is one and only one way to express $\mathbf{b}$ as a linear combination of the columns of $\mathbf{A}$.
2.  We say that $\mathbf{Ax} = \mathbf{b}$ has multiple solutions if there is more than one vector $\mathbf{x}$ so that the product $\mathbf{Ax}$ gives $\mathbf{b}$. This is equivalent to saying that there is more than one set of coefficients that can be used to express $\mathbf{b}$ as a linear combination of the columns of $\mathbf{A}$.

A geometric illustration of the possibilities for the **solution set**[1] of a linear system of two equations in two unknowns is given by considering the possible intersection of a pair of straight lines $L_1: ax + by = c$ and $L_2: dx + ey = f$ in 2-space.

Let

$$\mathbf{A} = \begin{bmatrix} a & b \\ d & e \end{bmatrix}, \quad \mathbf{x} = \begin{bmatrix} x \\ y \end{bmatrix}, \quad \text{and} \quad \mathbf{b} = \begin{bmatrix} c \\ f \end{bmatrix}.$$

1.  If $L_1$ and $L_2$ are parallel (see Figure 1), then $\mathbf{Ax} = \mathbf{b}$ is inconsistent. There is no point $(x, y)$ that lies on both $L_1$ and $L_2$.
2.  If the lines intersect in a single point, then $\mathbf{Ax} = \mathbf{b}$ is consistent and the linear system has a unique solution. (See Figure 2.)
3.  If the lines are coincident (that is, they are the same line), then $\mathbf{Ax} = \mathbf{b}$ is consistent and there are multiple solutions. In fact, there are infinitely many solutions since $L_1$ and $L_2$ share all their points. (See Figure 3.)

---

[1]The solution set of a linear system is sometimes called the **general solution** of the linear system.

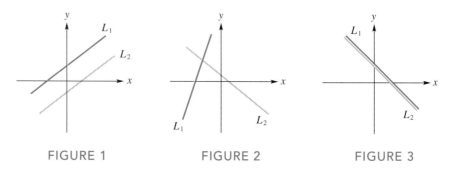

FIGURE 1                    FIGURE 2                    FIGURE 3

The case of multiple solutions is important to investigate further. We do this in Example 1, where we use the matrix equation representing a linear system together with properties of matrix multiplication.

EXAMPLE 1    Suppose that the linear system $\mathbf{Ax} = \mathbf{b}$ has two different solutions, vectors $\mathbf{y}$ and $\mathbf{s}$. Thus $\mathbf{Ay} = \mathbf{b}$ and $\mathbf{As} = \mathbf{b}$ where $\mathbf{y} \neq \mathbf{s}$. Adding the two expressions $\mathbf{Ay} = \mathbf{b}$ and $\mathbf{As} = \mathbf{b}$ gives the following:

$$\mathbf{Ay} + \mathbf{As} = \mathbf{b} + \mathbf{b} \quad \text{or, equivalently,} \quad \tfrac{1}{2}(\mathbf{Ay} + \mathbf{As}) = \mathbf{A}(\tfrac{1}{2}\mathbf{y} + \tfrac{1}{2}\mathbf{s}) = \mathbf{b}.$$

Thus we see that $\tfrac{1}{2}\mathbf{y} + \tfrac{1}{2}\mathbf{s}$ is another solution of $\mathbf{Ax} = \mathbf{b}$, which is different from either $\mathbf{y}$ or $\mathbf{s}$ since $\mathbf{y} \neq \mathbf{s}$. Let's also note that $\tfrac{1}{3}\mathbf{y} + \tfrac{2}{3}\mathbf{s}$ is another distinct solution. (Verify.)  In fact, for any scalar $t$, the linear combination $(1 - t)\mathbf{y} + t\mathbf{s}$ is also a solution. (Verify.) If follows that if a linear system has multiple solutions, then it has infinitely many solutions.    ■

The following summarizes the types of solution sets for a linear system $\mathbf{Ax} = \mathbf{b}$:

No solution.    $\Leftarrow$    Inconsistent linear system.

A unique solution or
infinitely many solutions.    $\Leftarrow$    Consistent linear system.

A basic strategy in determining the solution set of a linear system $\mathbf{Ax} = \mathbf{b}$ is to manipulate the information in the coefficient matrix $\mathbf{A}$ and right side $\mathbf{b}$ to obtain a **simpler system** without changing the solution set. By *simpler system* we mean a linear system from which it is easier to determine the solution set. Given the intimate connection of linear systems to linear combinations, we suspect that the manipulations that preserve the solution set will involve linear combinations. We next investigate three fundamental manipulations that will form the basis for our solution techniques.

Given a linear system $\mathbf{Ax} = \mathbf{b}$ for a particular $m \times n$ coefficient matrix $\mathbf{A}$ and a particular $m \times 1$ right side $\mathbf{b}$, an obviously inefficient search procedure to determine solutions is to take every possible $n \times 1$ vector $\mathbf{x}$ and see if the product $\mathbf{Ax}$ gives $\mathbf{b}$. The system will be inconsistent if $\mathbf{Ax} = \mathbf{b}$ is false for *all* choices of $\mathbf{x}$ and it is consistent if there are *some* choices of $\mathbf{x}$ for which $\mathbf{Ax} = \mathbf{b}$ is true. For a consistent system, the solution set of $\mathbf{Ax} = \mathbf{b}$ is the set of $\mathbf{x}$'s that make $\mathbf{Ax} = \mathbf{b}$ true. Or we say the solution set is the set of all vectors $\mathbf{x}$ that "satisfy" the matrix equation $\mathbf{Ax} = \mathbf{b}$. So for a vector $\mathbf{x}$, the truth value of $\mathbf{Ax} = \mathbf{b}$ is false if $\mathbf{x}$ is not a solution and true if $\mathbf{x}$ is a solution.

Our solution technique, which we describe later, replaces system $\mathbf{Ax} = \mathbf{b}$ with a related, but different, system $\mathbf{Cx} = \mathbf{d}$ so that the truth value of $\mathbf{Cx} = \mathbf{d}$ is the same as the truth value of $\mathbf{Ax} = \mathbf{b}$ for every vector $\mathbf{x}$. Thus the solution set of the new system $\mathbf{Cx} = \mathbf{d}$ is the same as the solution set of the original system $\mathbf{Ax} = \mathbf{b}$. The remaining questions are what operations on the linear system $\mathbf{Ax} = \mathbf{b}$ preserve the solution set and how do we organize them to determine the solution set. We first investigate operations that preserve solution sets.

Recall that the $m \times n$ matrix $\mathbf{A}$ can be partitioned into its rows as

$$\mathbf{A} = \begin{bmatrix} \text{row}_1(\mathbf{A}) \\ \text{row}_2(\mathbf{A}) \\ \vdots \\ \vdots \\ \text{row}_m(\mathbf{A}) \end{bmatrix}$$

and so the matrix equation $\mathbf{Ax} = \mathbf{b}$ gives the set of relations

$$\text{row}_t(\mathbf{A}) \cdot \mathbf{x} = b_t \quad \text{for } t = 1, 2, \ldots, m. \tag{2}$$

The expressions in (2) are equivalent to the individual equations in (1), but the properties of the dot product give us easy access to manipulations with linear combinations. We make the following observations.

**I.** If we multiply the $i$th equation in (1) by a scalar $k \neq 0$, then the relations in (2) remain unchanged except for $t = i$. The new relation is $k\,\text{row}_i(\mathbf{A}) \cdot \mathbf{x} = kb_i$. For every vector $\mathbf{x}$ we must determine if the truth value of this new relation is the same as the original relation, $\text{row}_i(\mathbf{A}) \cdot \mathbf{x} = b_i$. Of course, the answer is yes since we can multiply both sides of the new relation by $1/k$. Hence multiplying an equation of (1) by a nonzero scalar preserves the solution set. This is the same as saying that we can multiply a row of the augmented matrix $\begin{bmatrix} \mathbf{A} \mid \mathbf{b} \end{bmatrix}$ by $k \neq 0$ and the resulting linear system has the same solution set as $\mathbf{Ax} = \mathbf{b}$.

**II.** If we replace the $i$th equation of (1) by the linear combination ($j$th equation) + ($i$th equation), then we have a linear system with all equations the same as in (1) except that the $i$th equation is now

$$(a_{j1}x_1 + a_{j2}x_2 + \cdots + a_{jn}x_n) + (a_{i1}x_1 + a_{i2}x_2 + \cdots + a_{in}x_n) = b_j + b_i.$$

The relations in (2) remain unchanged except for $t = i$. The new relation is

$$(\text{row}_j(\mathbf{A}) + \text{row}_i(\mathbf{A})) \cdot \mathbf{x} = b_j + b_i. \tag{3}$$

For every vector $\mathbf{x}$ we must determine if the truth value of this new relation (3) is the same as the truth value of the original relation, $\text{row}_i(\mathbf{A}) \cdot \mathbf{x} = b_i$. Using properties of the dot product, we have that (3) is equivalent to

$$\text{row}_j(\mathbf{A}) \cdot \mathbf{x} + \text{row}_i(\mathbf{A}) \cdot \mathbf{x} = b_j + b_i. \tag{4}$$

For a given vector $\mathbf{x}$, if all the relations in (2) are true, then (4) must be true. Similarly, if all the relations in (2) are true with the $i$th relation replaced by (4), it follows that all the original relations in (2) are true. For a given vector $\mathbf{x}$ if one of the relations in (2) is false, say when $t = s$, then we need only check the truth value of (4) when $i = s$. If $i = s$, then $\text{row}_i(\mathbf{A}) \cdot \mathbf{x} \neq b_i$. If $\text{row}_j(\mathbf{A}) \cdot \mathbf{x} = b_j$, then (4) is false and if $\text{row}_j(\mathbf{A}) \cdot \mathbf{x} \neq b_j$, then the truth value of all the relations in (2) with the $i$th relation replaced by (4) is false. A corresponding argument shows that if (4) is false, then the truth value of the relations in (2) is false. Hence the replacement of one equation by the sum of itself and another equation preserves the solution set. This is the same as saying that we can replace one row of augmented matrix $\begin{bmatrix} \mathbf{A} \mid \mathbf{b} \end{bmatrix}$ by the sum of itself and another row and the resulting linear system has the same solution set.

**III.** If we interchange the positions of two equations in (1), then the truth value of the set of expressions in (2) is unchanged. This is the same as saying that interchanging two rows of the augmented matrix $\left[\,\mathbf{A}\mid\mathbf{b}\,\right]$ gives a new linear system with the same solution set.

Operations I and II above can be combined to say that replacing the $i$th equation by the linear combination $k * (j$th equation$) + (i$th equation$)$ yields a linear system with the same solution set. It is simpler to apply these operations to the augmented matrix, which is a more efficient data structure for the information in a linear system. Thus we summarize these operations as follows. The solution set of a linear system $\mathbf{Ax} = \mathbf{b}$ will be unchanged if

two rows of $\left[\,\mathbf{A}\mid\mathbf{b}\,\right]$ are interchanged

or

a row of $\left[\,\mathbf{A}\mid\mathbf{b}\,\right]$ is multiplied by a nonzero scalar

or

a given row is replaced by a scalar times any row added to the given row.

We call these **row operations** and abbreviate them as

$R_i \leftrightarrow R_j$, meaning interchange rows $i$ and $j$

$kR_i$, meaning multiply row $i$ by a nonzero scalar $k$

$kR_i + R_j$, meaning linear combination $kR_i + R_j$ replaces row $j$.

---

**Definition**    A pair of matrices is said to be **row equivalent** provided that one matrix can be obtained from the other by a finite sequence of row operations.

---

Thus if $\left[\,\mathbf{A}\mid\mathbf{b}\,\right]$ is row equivalent to $\left[\,\mathbf{C}\mid\mathbf{d}\,\right]$, then the linear systems $\mathbf{Ax} = \mathbf{b}$ and $\mathbf{Cx} = \mathbf{d}$ have the same solution set. Thus we say that $\mathbf{Ax} = \mathbf{b}$ and $\mathbf{Cx} = \mathbf{d}$ are **equivalent linear systems**.

Since the row operations applied to an augmented matrix preserve the solution set, using a sequence of such operations generates a sequence of row-equivalent augmented matrices. We want to develop a strategy for choosing row operations that leads to simpler equivalent linear systems. However, it is important that we determine the types of linear systems that are easy to solve, or the type of linear system that reveals a solution by inspection. Keep in mind that such a strategy should first reveal if the linear system is consistent or not, and if it is consistent then we may be required to use more row operations to find the set of all solutions.

**EXAMPLE 2**    If $\mathbf{A} = \begin{bmatrix} 2 & 0 & 0 \\ 0 & -4 & 0 \\ 0 & 0 & 5 \end{bmatrix}$ and $\mathbf{b} = \begin{bmatrix} 3 \\ 20 \\ 35 \end{bmatrix}$, then the linear system $\mathbf{Ax} = \mathbf{b}$ is

$$\begin{aligned} 2x_1 \qquad\qquad &= 3 \\ -4x_2 \qquad &= 20 \\ 5x_3 &= 35. \end{aligned}$$

We see that each equation contains exactly one unknown, so we can solve for it directly. We get the unique solution

$$x_1 = \frac{3}{2}, \quad x_2 = \frac{20}{-4} = -5, \quad x_3 = \frac{35}{5} = 7.$$

■

If the linear system had $n$ equations in $n$ unknowns with a diagonal coefficient matrix, then we would expect to obtain a solution in a correspondingly easy fashion. However, if

$$\mathbf{C} = \begin{bmatrix} 2 & 0 & 0 \\ 0 & 0 & 0 \\ 0 & 0 & 5 \end{bmatrix},$$

then the linear system $\mathbf{Cx} = \mathbf{b}$ is

$$\begin{aligned} 2x_1 \quad\quad\quad &= 3 \\ 0x_2 \quad &= 20 \\ 5x_3 &= 35. \end{aligned}$$

We see that the second equation implies that $0 = 20$, which is absurd. Such an expression is inconsistent with known facts, and we are led to the conclusion that the linear system $\mathbf{Cx} = \mathbf{b}$ has no solution, that is, it is inconsistent. The augmented matrices of the linear systems $\mathbf{Ax} = \mathbf{b}$ and $\mathbf{Cx} = \mathbf{b}$ have a structure worth noting, in addition to the fact that the coefficient matrix is diagonal. We have

$$\begin{bmatrix} \mathbf{A} \mid \mathbf{b} \end{bmatrix} = \left[ \begin{array}{ccc|c} 2 & 0 & 0 & 3 \\ 0 & -4 & 0 & 20 \\ 0 & 0 & 5 & 35 \end{array} \right] \quad \text{and} \quad \begin{bmatrix} \mathbf{C} \mid \mathbf{b} \end{bmatrix} = \left[ \begin{array}{ccc|c} 2 & 0 & 0 & 3 \\ 0 & 0 & 0 & 20 \\ 0 & 0 & 5 & 35 \end{array} \right].$$

The inconsistent system $\mathbf{Cx} = \mathbf{b}$ has a row of the form $\begin{bmatrix} \text{all zeros} \mid \neq 0 \end{bmatrix}$, which leads to the contradiction of a known fact. However, regardless of the row operations that may be applied to $\begin{bmatrix} \mathbf{A} \mid \mathbf{b} \end{bmatrix}$, no such row could appear. For $\mathbf{Ax} = \mathbf{b}$ we can obtain an even simpler system: (We indicate the row operation that is applied to a matrix by writing it on the lower right).

$$\begin{bmatrix} \mathbf{A} \mid \mathbf{b} \end{bmatrix} = \left[ \begin{array}{ccc|c} 2 & 0 & 0 & 3 \\ 0 & -4 & 0 & 20 \\ 0 & 0 & 5 & 35 \end{array} \right]_{\left(\frac{1}{2}\right)R_1} \Rightarrow \left[ \begin{array}{ccc|c} 1 & 0 & 0 & \frac{3}{2} \\ 0 & -4 & 0 & 20 \\ 0 & 0 & 5 & 35 \end{array} \right]_{\left(-\frac{1}{4}\right)R_2}$$

$$\Rightarrow \left[ \begin{array}{ccc|c} 1 & 0 & 0 & \frac{3}{2} \\ 0 & 1 & 0 & -5 \\ 0 & 0 & 5 & 35 \end{array} \right]_{\left(\frac{1}{5}\right)R_3} \Rightarrow \left[ \begin{array}{ccc|c} 1 & 0 & 0 & \frac{3}{2} \\ 0 & 1 & 0 & -5 \\ 0 & 0 & 1 & 7 \end{array} \right].$$

The final equivalent matrix reveals by inspection the unique solution to $\mathbf{Ax} = \mathbf{b}$, which is $x_1 = \frac{3}{2}$, $x_2 = -5$, $x_3 = 7$. Note that the coefficient matrix of this final augmented matrix is the identity matrix $\mathbf{I}_3$.

EXAMPLE 3    If $\mathbf{A} = \begin{bmatrix} 2 & 1 & -2 \\ 0 & -4 & 2 \\ 0 & 0 & 5 \end{bmatrix}$ and $\mathbf{b} = \begin{bmatrix} 0 \\ 10 \\ 5 \end{bmatrix}$, then the linear system $\mathbf{Ax} = \mathbf{b}$ is

$$\begin{aligned} 2x_1 + x_2 - 2x_3 &= 0 \\ -4x_2 + 2x_3 &= 10 \\ 5x_3 &= 5. \end{aligned}$$

We see that the third equation contains only the one unknown $x_3$, so we can solve for it directly to get $x_3 = \frac{5}{5} = 1$. Next we substitute this value of $x_3$ into the second equation to obtain the following equation with the single unknown $x_2$:

$$-4x_2 + 2(1) = 10.$$

We solve directly for $x_2$, obtaining

$$x_2 = \frac{10 - 2(1)}{-4} = -2.$$

Since we have values for both $x_3$ and $x_2$, we use them in the first equation to get the next equation in the single variable $x_1$:

$$2x_1 + (-2) - 2(1) = 0.$$

We solve directly for $x_1$, obtaining

$$x_1 = \frac{0 - (-2) + 2(1)}{2},$$

which implies that $x_1 = 2$. Hence the unique solution vector is

$$\mathbf{x} = \begin{bmatrix} x_1 \\ x_2 \\ x_3 \end{bmatrix} = \begin{bmatrix} 2 \\ -2 \\ 1 \end{bmatrix}.$$   ■

The coefficient matrix $\mathbf{A}$ in Example 3 is upper triangular (see Example 10 in Section 1.1) so we were able to start with the last equation and work our way up toward the first equation, obtaining a value for an entry of the solution vector at each step. This procedure is called **back substitution** and can be used on a linear system when its coefficient matrix is upper triangular. (A general expression for back substitution on a square linear system is given in Exercise 23.) However, note that the diagonal entries of the coefficient matrix $\mathbf{A}$ were used in the denominators of the expressions for $x_3$, $x_2$, and $x_1$. Hence if $\mathbf{A}$ is upper triangular with some diagonal entry zero, then we are warned that there may be difficulties encountered in the back substitution process. We illustrate two important cases in Examples 4 and 5. Both of these examples point to more general situations that arise when solving linear systems. Pay close attention to the observations cited in the examples and the terminology introduced, particularly in Example 5.

EXAMPLE 4   Let $\mathbf{C} = \begin{bmatrix} 2 & 1 & -2 \\ 0 & 0 & 2 \\ 0 & 0 & 5 \end{bmatrix}$ and $\mathbf{b} = \begin{bmatrix} 0 \\ 10 \\ 5 \end{bmatrix}$. Then the linear system $\mathbf{Cx} = \mathbf{b}$ is

$$\begin{aligned} 2x_1 + x_2 - 2x_3 &= 0 \\ 2x_3 &= 10 \\ 5x_3 &= 5. \end{aligned}$$

Using back substitution, the third equation gives us $x_3 = \frac{5}{5} = 1$ and from the second equation $x_3 = \frac{10}{2} = 5$. But $x_3$ cannot have two distinct values, so $\mathbf{Cx} = \mathbf{b}$ must be inconsistent.   ■

EXAMPLE 5   The linear system $\mathbf{Dx} = \mathbf{b}$, where

$$\mathbf{D} = \begin{bmatrix} 2 & 1 & -2 \\ 0 & 0 & 10 \\ 0 & 0 & 5 \end{bmatrix} \quad \text{and} \quad \mathbf{b} = \begin{bmatrix} 0 \\ 10 \\ 5 \end{bmatrix},$$

corresponds to

$$2x_1 + x_2 - 2x_3 = 0$$
$$10x_3 = 10$$
$$5x_3 = 5.$$

Using back substitution, we have from the third equation $x_3 = 1$, from the second equation $x_3 = 1$, and from the first equation

$$x_1 = \frac{0 + 2(1) - x_2}{2} = \frac{2 - x_2}{2}.$$

It follows that $x_3$ must be 1, $x_1$ depends on the value of $x_2$, and there are no restrictions on $x_2$ since it does not appear on the left side of one of the equations obtained from the back substitution process. Because there are no restrictions on unknown $x_2$, it could be any scalar value t. We say $t$ is an **arbitrary constant** or, equivalently, that $x_2$ is a **free variable**. Thus we have

$$\mathbf{x} = \begin{bmatrix} x_1 \\ x_2 \\ x_3 \end{bmatrix} = \begin{bmatrix} \dfrac{2 - x_2}{2} \\ t \\ 1 \end{bmatrix} = \begin{bmatrix} \dfrac{2 - t}{2} \\ t \\ 1 \end{bmatrix}.$$

Since $t$ is arbitrary, we have a solution for each choice of $t$, so the linear system $\mathbf{Dx} = \mathbf{b}$ is consistent with infinitely many solutions. Using properties of linear combinations, we have that the infinite set of solutions to $\mathbf{Dx} = \mathbf{b}$ can be expressed in the form

$$\mathbf{x} = \begin{bmatrix} 1 - \frac{1}{2}t \\ t \\ 1 \end{bmatrix} = \begin{bmatrix} 1 \\ 0 \\ 1 \end{bmatrix} + \begin{bmatrix} -\frac{1}{2}t \\ t \\ 0 \end{bmatrix} = \begin{bmatrix} 1 \\ 0 \\ 1 \end{bmatrix} + t \begin{bmatrix} -\frac{1}{2} \\ 1 \\ 0 \end{bmatrix}.$$

Recall from Section 1.4 the pattern of a translation: vector $+ k$ (another vector). This pattern generalizes to $R^n$, so the solution set of $\mathbf{Dx} = \mathbf{b}$ is a translation of line

$$t \begin{bmatrix} -\frac{1}{2} \\ 1 \\ 0 \end{bmatrix}.$$

■

So far our observations about determining solution sets of linear systems with square coefficient matrices include the following:

1. If the coefficient matrix is diagonal and all the diagonal entries are different from zero, then there is a unique solution and it is easy to compute.
2. If the coefficient matrix is square upper triangular and all the diagonal entries are different from zero, then there is a unique solution and we can compute it with little work by back substitution.
3. If the coefficient matrix is diagonal or square upper triangular with some diagonal entry zero, then the linear system may be inconsistent or may be consistent with infinitely many solutions. To determine which case applies, we must carefully perform the algebraic steps.

Examples 2 through 5 involved square matrices. They do not represent all the linear systems that are easy to solve, but they do suggest the following approach.

> **Strategy:**    For $\mathbf{Ax} = \mathbf{b}$, the augmented matrix is $\begin{bmatrix} \mathbf{A} \mid \mathbf{b} \end{bmatrix}$. Apply row operations to obtain row-equivalent matrices until you have an upper triangular form[2] for the coefficient matrix; then apply back substitution.

We call this strategy (row) **reduction to (upper) triangular** form followed by back substitution or (row) **reduction** for short. In addition to this strategy we make the following observations:

- If at any step of the reduction to triangular form there appears a row of the form $\begin{bmatrix} 0 & 0 & \cdots & 0 \mid \neq 0 \end{bmatrix}$, then the linear system is inconsistent and we need proceed no further.
- We may continue the reduction process until we get a diagonal or nearly a diagonal coefficient matrix, which often simplifies the back substitution process.

We illustrate this strategy and use these observations, when appropriate, in Examples 6 through 9. In these examples we chose to illustrate certain features of the reduction process, so our approach may not be the most efficient since there are many ways to choose row operations to achieve upper triangular form.

**EXAMPLE 6**    Determine the solution set of the following linear system by reducing the augmented matrix to upper triangular form.

$$
\begin{aligned}
3x_1 - x_2 + 2x_3 &= 12 \\
x_1 + 2x_2 + 3x_3 &= 11 \\
2x_1 - 2x_2 + x_3 &= 2.
\end{aligned}
$$

We form the augmented matrix and start the reduction to upper triangular form. The row operations are selected to generate zeros below diagonal entries. (Algebraically this eliminates unknowns from the following equations, so the process is sometimes called (row) **elimination**.)

$$
\begin{bmatrix} \mathbf{A} \mid \mathbf{b} \end{bmatrix} = \begin{bmatrix} 3 & -1 & 2 & \mid & 12 \\ 1 & 2 & 3 & \mid & 11 \\ 2 & -2 & 1 & \mid & 2 \end{bmatrix}.
$$

The arithmetic involved with the row operations is often simplified if we arrange to have a 1 in the diagonal position below which we want to generate zeros. (We will use the term **zero-out** or **eliminate** to describe this set of actions.) We start with the $(1, 1)$-entry and plan how to zero-out below it. One row operation that will produce a 1 in the $(1, 1)$-entry is $\left(\frac{1}{3}\right) R_1$. Another is to interchange rows 1 and 2; $R_1 \leftrightarrow R_2$. We can use either operation to start our reduction process; however, since we are doing the calculations by hand, we can avoid the use of fractions by doing the row interchange. We choose this as a first step.

$$
\begin{bmatrix} \mathbf{A} \mid \mathbf{b} \end{bmatrix} = \begin{bmatrix} 3 & -1 & 2 & \mid & 12 \\ 1 & 2 & 3 & \mid & 11 \\ 2 & -2 & 1 & \mid & 2 \end{bmatrix}_{R_1 \leftrightarrow R_2} \Rightarrow \begin{bmatrix} 1 & 2 & 3 & \mid & 11 \\ 3 & -1 & 2 & \mid & 12 \\ 2 & -2 & 1 & \mid & 2 \end{bmatrix}.
$$

[2]By *upper triangular form* we mean that all entries below the diagonal entries are zero. See Section 1.1.

The next row operation is chosen to produce a zero in the $(2, 1)$-entry. We see that $-3R_1 + R_2$ will give a new second row of $\begin{bmatrix} 0 & -7 & -7 & | & -21 \end{bmatrix}$. We get

$$\Rightarrow \begin{bmatrix} 1 & 2 & 3 & | & 11 \\ 3 & -1 & 2 & | & 12 \\ 2 & -2 & 1 & | & 2 \end{bmatrix}_{-3R_1+R_2} \Rightarrow \begin{bmatrix} 1 & 2 & 3 & | & 11 \\ 0 & -7 & -7 & | & -21 \\ 2 & -2 & 1 & | & 2 \end{bmatrix}.$$

Now we choose a row operation that will produce a zero in the $(3, 1)$-entry. We see that $-2R_1 + R_3$ will give a new third row of $\begin{bmatrix} 0 & -6 & -5 & | & -20 \end{bmatrix}$. We get

$$\Rightarrow \begin{bmatrix} 1 & 2 & 3 & | & 11 \\ 0 & -7 & -7 & | & -21 \\ 2 & -2 & 1 & | & 2 \end{bmatrix}_{-2R_1+R_3} \Rightarrow \begin{bmatrix} 1 & 2 & 3 & | & 11 \\ 0 & -7 & -7 & | & -21 \\ 0 & -6 & -5 & | & -20 \end{bmatrix}.$$

We say that this is the end of the first stage of the reduction process since we have zeroed-out below the $(1, 1)$-entry. The second stage chooses row operations to zero-out below the $(2, 2)$-entry. Following our guideline that a 1 in the diagonal position below which we want to generate zeros may simplify the algebra, we use the row operation $\left(-\frac{1}{7}\right) R_2$. We get

$$\Rightarrow \begin{bmatrix} 1 & 2 & 3 & | & 11 \\ 0 & -7 & -7 & | & -21 \\ 0 & -6 & -5 & | & -20 \end{bmatrix}_{\left(-\frac{1}{7}\right)R_2} \Rightarrow \begin{bmatrix} 1 & 2 & 3 & | & 11 \\ 0 & 1 & 1 & | & 3 \\ 0 & -6 & -5 & | & -20 \end{bmatrix}.$$

To eliminate below the $(2, 2)$-entry, we use $6R_2 + R_3$. We get

$$\Rightarrow \begin{bmatrix} 1 & 2 & 3 & | & 11 \\ 0 & 1 & 1 & | & 3 \\ 0 & -6 & -5 & | & -20 \end{bmatrix}_{6R_2+R_3} \Rightarrow \begin{bmatrix} 1 & 2 & 3 & | & 11 \\ 0 & 1 & 1 & | & 3 \\ 0 & 0 & 1 & | & -2 \end{bmatrix}.$$

We note that the coefficient matrix in this final augmented matrix is upper triangular and all the diagonal entries are different from zero. Hence it follows that the linear system is consistent and has a unique solution. We use back substitution to find the solution. The following linear system has the same solution set as the original system but is easier to solve:

$$\begin{aligned} x_1 + 2x_2 + 3x_3 &= 11 \\ x_2 + x_3 &= 3 \\ x_3 &= -2. \end{aligned}$$

Using back substitution, we get

$$\begin{aligned} x_3 &= -2, \\ x_2 &= 3 - x_3 = 3 - (-2) = 5, \\ x_1 &= 11 - 2x_2 - 3x_3 = 11 - 2(5) - 3(-2) = 7. \end{aligned}$$

Thus the unique solution vector is

$$\mathbf{x} = \begin{bmatrix} 7 \\ 5 \\ -2 \end{bmatrix}.$$

We deliberately maneuvered to keep the arithmetic simple in the preceding reduction. An alternate sequence of row operations is given in the following equations, where we do make the diagonal entries 1 before zeroing-out below them. Note the early introduction of fractions.

$$\left[\mathbf{A} \mid \mathbf{b}\right] = \begin{bmatrix} 3 & -1 & 2 & \big| & 12 \\ 1 & 2 & 3 & \big| & 11 \\ 2 & -2 & 1 & \big| & 2 \end{bmatrix}_{\left(-\frac{1}{3}\right)R_1+R_2} \Rightarrow \begin{bmatrix} 3 & -1 & 2 & \big| & 12 \\ 0 & \frac{7}{3} & \frac{7}{3} & \big| & 7 \\ 2 & -2 & 1 & \big| & 2 \end{bmatrix}_{\left(-\frac{2}{3}\right)R_1+R_3}$$

$$\Rightarrow \begin{bmatrix} 3 & -1 & 2 & \big| & 12 \\ 0 & \frac{7}{3} & \frac{7}{3} & \big| & 7 \\ 0 & -\frac{4}{3} & -\frac{1}{3} & \big| & -6 \end{bmatrix}_{\left(\frac{4}{7}\right)R_2+R_3} \Rightarrow \begin{bmatrix} 3 & -1 & 2 & \big| & 12 \\ 0 & \frac{7}{3} & \frac{7}{3} & \big| & 7 \\ 0 & 0 & 1 & \big| & -2 \end{bmatrix}.$$

The coefficient matrix of this final augmented matrix is again upper triangular with all its diagonal entries different from zero, so the linear system is consistent and back substitution will give exactly the same unique solution. (Verify.)    ■

Note that in the reduction process in the second method used in Example 6 we did not use row interchanges, $R_i \leftrightarrow R_j$, or scalar multiples of a row, $kR_i$. If you carefully inspect the pattern of the linear combinations used, you will see the form

$$\frac{(-1)*(\text{entry to be zeroed out})}{(\text{diagonal entry})} * R_{\text{row containing diagonal entry}} + R_{\text{row containing entry to be zeroed-out}}$$

which of course assumes that the diagonal entry is not zero. The diagonal entry is called the **pivot**[3] and the row containing it is called the **pivot row**. Thus we can express the preceding pattern as

$$\frac{(-1)*(\text{entry to be zeroed out})}{(\text{pivot})} * R_{\text{pivot row}} + R_{\text{row containing entry to be zeroed-out}}$$

Thus we may, in other instances, need to use row interchanges to ensure that a pivot is not zero.

EXAMPLE 7    Determine the solution set of the following linear system by performing the reduction process.

$$\begin{aligned} 2x_1 - 4x_2 + 6x_3 &= 4 \\ -4x_1 + x_2 - 16x_3 &= -5 \\ 2x_1 + 3x_2 + 10x_3 &= 3. \end{aligned}$$

We form the augmented matrix and then choose row operations to generate zeros

[3] We will see that pivots are not restricted to being diagonal entries. In general, a pivot is an entry of a coefficient matrix below which we want to zero out; that is, we choose row operations to generate equivalent linear systems whose entries below the pivot are zero. The elimination process, zeroing out of entries (above or below pivots), can be applied to square or nonsquare linear systems.

below the pivots.

$$[\mathbf{A} \mid \mathbf{b}] = \begin{bmatrix} 2 & -4 & 6 & | & 4 \\ -4 & 1 & -16 & | & -5 \\ 2 & 3 & 10 & | & 3 \end{bmatrix}_{\left(\frac{1}{2}\right)R_1} \Rightarrow \begin{bmatrix} 1 & -2 & 3 & | & 2 \\ -4 & 1 & -16 & | & -5 \\ 2 & 3 & 10 & | & 3 \end{bmatrix}_{4R_1+R_2}$$

$$\Rightarrow \begin{bmatrix} 1 & -2 & 3 & | & 2 \\ 0 & -7 & -4 & | & 3 \\ 2 & 3 & 10 & | & 3 \end{bmatrix}_{-2R_1+R_3} \Rightarrow \begin{bmatrix} 1 & -2 & 3 & | & 2 \\ 0 & -7 & -4 & | & 3 \\ 0 & 7 & 4 & | & -1 \end{bmatrix}_{1R_2+R_3}$$

$$\Rightarrow \begin{bmatrix} 1 & -2 & 3 & | & 2 \\ 0 & -7 & -4 & | & 3 \\ 0 & 0 & 0 & | & 2 \end{bmatrix}.$$

The sequence of row operations employed here starts by generating a 1 in the first pivot position, the $(1, 1)$ position, and then we finish the first stage of the reduction by using linear combinations of the rows to zero-out below the pivot. The second stage of the reduction used the $(2, 2)$-entry of $-7$ as the second pivot and zeroed-out below it to obtain the last augmented matrix displayed. Inspecting this final matrix, we see that the last row has all zeros in the coefficient matrix part and a nonzero entry in the last row of the augmented column. Hence by our previous observations, this linear system is inconsistent.

If the right side of the original linear system had been

$$\begin{bmatrix} 4 \\ -5 \\ 1 \end{bmatrix},$$

then the linear system would be consistent (verify). If we have a pair of linear systems with the same coefficient matrix $\mathbf{A}$, but different right sides $\mathbf{b}$ and $\mathbf{c}$, respectively, we can form the partitioned matrix $\begin{bmatrix} \mathbf{A} \mid \mathbf{b} \mid \mathbf{c} \end{bmatrix}$ and apply row operations to determine the solution set of each of the systems $\mathbf{Ax} = \mathbf{b}$ and $\mathbf{Ax} = \mathbf{c}$. ∎

EXAMPLE 8   Determine the solution set of the following linear system by performing the reduction process without making the pivots 1 before zeroing out below them.

$$\begin{aligned} 4x_1 + x_2 \qquad\quad + 2x_4 &= 5 \\ -2x_1 + 2x_2 + x_3 + 3x_4 &= -1 \\ 8x_1 - 3x_2 - 2x_3 - 4x_4 &= 7 \\ x_3 - x_4 &= 2. \end{aligned}$$

Forming the augmented matrix and performing the row operations gives the following sequence of equivalent augmented matrices:

$$[\mathbf{A} \mid \mathbf{b}] = \begin{bmatrix} 4 & 1 & 0 & 2 & | & 5 \\ -2 & 2 & 1 & 3 & | & -1 \\ 8 & -3 & -2 & -4 & | & 7 \\ 0 & 0 & 1 & -1 & | & 2 \end{bmatrix}_{\substack{\left(\frac{1}{2}\right)R_1 + R_2 \\ -2R_1 + R_3}}$$

$$\Rightarrow \begin{bmatrix} 4 & 1 & 0 & 2 & | & 5 \\ 0 & 2.5 & 1 & 4 & | & 1.5 \\ 0 & -5 & -2 & -8 & | & -3 \\ 0 & 0 & 1 & -1 & | & 2 \end{bmatrix}_{2R_2+R_3}$$

$$\Rightarrow \begin{bmatrix} 4 & 1 & 0 & 2 & | & 5 \\ 0 & 2.5 & 1 & 4 & | & 1.5 \\ 0 & 0 & 0 & 0 & | & 0 \\ 0 & 0 & 1 & -1 & | & 2 \end{bmatrix}_{R_3 \leftrightarrow R_4}$$

$$\Rightarrow \begin{bmatrix} 4 & 1 & 0 & 2 & | & 5 \\ 0 & 2.5 & 1 & 4 & | & 1.5 \\ 0 & 0 & 1 & -1 & | & 2 \\ 0 & 0 & 0 & 0 & | & 0 \end{bmatrix}.$$

The pivots have values 4, 2.5, and 1, respectively; there was no need for a fourth pivot since the row interchange made the row of all zeros come last. We see that there is no row of the form $\begin{bmatrix} 0 & 0 & 0 & 0 & | & \neq 0 \end{bmatrix}$ so the linear system is consistent. This final augmented matrix corresponds to the linear system

$$\begin{aligned} 4x_1 + \quad x_2 \quad\quad + 2x_4 &= 5 \\ 2.5x_2 + x_3 + 4x_4 &= 1.5 \\ x_3 - \quad x_4 &= 2. \end{aligned}$$

This linear system has three equations in four unknowns, so we suspect that one of the unknowns can be chosen arbitrarily. Applying back substitution, we start with the nonzero row nearest the bottom of the final augmented matrix, which corresponds to the last equation shown. The usual procedure is to solve the equation for the unknown with the lowest subscript, here 3, since that is the position occupied by the pivot. We get $x_3 = 2 + x_4$. In the second equation, solving for $x_2$ and substituting for $x_3$ gives

$$x_2 = \frac{1.5 - x_3 - 4x_4}{2.5} = \frac{1.5 - (2 + x_4) - 4x_4}{2.5} = \frac{-0.5 - 5x_4}{2.5} = -0.2 - 2x_4.$$

Similarly, from the first equation we get

$$x_1 = \frac{5 - x_2 - 2x_4}{4} = \frac{5 - (-0.2 - 2x_4) - 2x_4}{4} = \frac{5.2}{4} = 1.3.$$

From the back substitution we see that there is no restriction on $x_4$, so we set it equal to an arbitrary constant $t$. Hence we conclude that there are infinitely many solutions and the set of all solutions is given by the expression

$$\mathbf{x} = \begin{bmatrix} x_1 \\ x_2 \\ x_3 \\ x_4 \end{bmatrix} = \begin{bmatrix} 1.3 \\ -0.2 - 2x_4 \\ 2 + x_4 \\ x_4 \end{bmatrix} = \begin{bmatrix} 1.3 \\ -0.2 - 2t \\ 2 + t \\ t \end{bmatrix} = \begin{bmatrix} 1.3 \\ -0.2 \\ 2 \\ 0 \end{bmatrix} + t \begin{bmatrix} 0 \\ -2 \\ 1 \\ 1 \end{bmatrix}.$$

We see that the solution set is the translation of the line $t \begin{bmatrix} 0 & -2 & 1 & 1 \end{bmatrix}^T$ in $R^4$. We say that the solution set has one free variable since there is one arbitrary constant.

■

The number of unknowns that can be chosen arbitrarily (or the number of free variables) in the formation of the solution set of a linear system is called the number of **degrees of freedom** in the linear system. If a system has its degree of freedom $\geq 1$, then it has infinitely many solutions. The following important example extends the procedure introduced in Example 5.

EXAMPLE 9   Determine the solution set of the homogeneous linear system having the following augmented matrix.

$$[\,\mathbf{A}\mid\mathbf{b}\,] = \begin{bmatrix} 2 & -4 & 6 & -2 & \bigm| & 0 \\ 0 & 3 & -9 & 0 & \bigm| & 0 \\ 0 & 0 & 0 & 0 & \bigm| & 0 \end{bmatrix}.$$

We see that the coefficient matrix is in upper triangular form and hence no reduction is required. Furthermore, since we have a homogeneous linear system, the system is consistent. We see that there are just two nonzero rows and four columns in the coefficient matrix part; hence we have two equations in four unknowns. Thus we suspect that there will be infinitely many solutions. Instead of solving by back substitution at this point, we will perform several row operations that simplify the back substitution process:

$$[\,\mathbf{A}\mid\mathbf{b}\,] = \begin{bmatrix} 2 & -4 & 6 & -2 & \bigm| & 0 \\ 0 & 3 & -9 & 0 & \bigm| & 0 \\ 0 & 0 & 0 & 0 & \bigm| & 0 \end{bmatrix} \begin{array}{l} \left(\tfrac{1}{2}\right)R_1 \\ \left(\tfrac{1}{3}\right)R_2 \end{array}$$

$$\Rightarrow \begin{bmatrix} 1 & -2 & 3 & -1 & \bigm| & 0 \\ 0 & 1 & -3 & 0 & \bigm| & 0 \\ 0 & 0 & 0 & 0 & \bigm| & 0 \end{bmatrix}_{2R_2+R_1}$$

$$\Rightarrow \begin{bmatrix} 1 & 0 & -3 & -1 & \bigm| & 0 \\ 0 & 1 & -3 & 0 & \bigm| & 0 \\ 0 & 0 & 0 & 0 & \bigm| & 0 \end{bmatrix}.$$

The first two row operations made the pivots have value 1, and the third row operation generated a zero "above" the second pivot. Hence the first two columns look like columns from an identity matrix. From back substitution we have $x_2 = 3x_3$, and $x_1 = 3x_3 + x_4$ from which we conclude that both unknowns $x_3$ and $x_4$ can be chosen arbitrarily. We assign a different arbitrary constant to each of these, say $x_3 = t$ and $x_4 = s$. Then the set of solutions is given by

$$\mathbf{x} = \begin{bmatrix} x_1 \\ x_2 \\ x_3 \\ x_4 \end{bmatrix} = \begin{bmatrix} 3x_3 + x_4 \\ 3x_3 \\ x_3 \\ x_4 \end{bmatrix} = \begin{bmatrix} 3t + s \\ 3t \\ t \\ s \end{bmatrix} = t \begin{bmatrix} 3 \\ 3 \\ 1 \\ 0 \end{bmatrix} + s \begin{bmatrix} 1 \\ 0 \\ 0 \\ 1 \end{bmatrix}. \qquad ■$$

   In the solution set of Example 9 we say that there are two degrees of freedom since there are two arbitrary constants. Note that no translation is involved, but the solution set consists of all possible linear combinations of the vectors

$$\begin{bmatrix} 3 & 3 & 1 & 0 \end{bmatrix}^T \quad \text{and} \quad \begin{bmatrix} 1 & 0 & 0 & 1 \end{bmatrix}^T.$$

In fact, the set of all solutions is

$$\text{span} \left\{ \begin{bmatrix} 3 \\ 3 \\ 1 \\ 0 \end{bmatrix}, \begin{bmatrix} 1 \\ 0 \\ 0 \\ 1 \end{bmatrix} \right\}.$$

Recall that the span of a set of vectors is closed under both scalar multiplication and addition, so the solution set in this case is quite special. We investigate this further in the next section.

Application:   **Quadratic Interpolation**

Various approximation techniques in science and engineering use a parabola constructed through three data points $\{(x_1, y_1), (x_2, y_2), (x_3, y_3)\}$ where $x_i \neq x_j$ for $i \neq j$. We call these **distinct points**, since the $x$-coordinates are all different. The graph of a quadratic polynomial $p(x) = ax^2 + bx + c$ is a parabola and we use the data points to determine the coefficients $a$, $b$, and $c$ as follows. Requiring that $p(x_i) = y_i$, $i = 1, 2, 3$ gives us three equations:

$$p(x_1) = y_1 \quad \Leftrightarrow \quad ax_1^2 + bx_1 + c = y_1$$
$$p(x_2) = y_2 \quad \Leftrightarrow \quad ax_2^2 + bx_2 + c = y_2$$
$$p(x_3) = y_3 \quad \Leftrightarrow \quad ax_3^2 + bx_3 + c = y_3.$$

Let

$$\mathbf{A} = \begin{bmatrix} x_1^2 & x_1 & 1 \\ x_2^2 & x_2 & 1 \\ x_3^2 & x_3 & 1 \end{bmatrix}, \quad \mathbf{v} = \begin{bmatrix} a \\ b \\ c \end{bmatrix}, \quad \text{and} \quad \mathbf{y} = \begin{bmatrix} y_1 \\ y_2 \\ y_3 \end{bmatrix}.$$

Then we have the matrix equation $\mathbf{Av} = \mathbf{y}$, or equivalently, the augmented matrix

$$\begin{bmatrix} \mathbf{A} \mid \mathbf{y} \end{bmatrix} = \begin{bmatrix} x_1^2 & x_1 & 1 & y_1 \\ x_2^2 & x_2 & 1 & y_2 \\ x_3^2 & x_3 & 1 & y_3 \end{bmatrix}.$$

We solve for the vector $\mathbf{v}$ using our reduction process. It can be shown that there is a unique solution if and only if the points are distinct. The construction of the parabola that matches the points of the data set is called **quadratic interpolation** and the parabola is called the **quadratic interpolant**. This process can be generalized to distinct data sets of $n + 1$ points and polynomials of degree $n$. We illustrate the construction of the quadratic interpolant in Example 10.

**EXAMPLE 10**   For the three distinct points $\{(1, -5), (-1, 1), (2, 7)\}$ we obtain a linear system for the quadratic interpolant whose augmented matrix is

$$\begin{bmatrix} \mathbf{A} \mid \mathbf{y} \end{bmatrix} = \begin{bmatrix} 1 & 1 & 1 & -5 \\ 1 & -1 & 1 & 1 \\ 4 & 2 & 1 & 7 \end{bmatrix}.$$

For the reduction process we perform the following row operations:

$$\begin{bmatrix} \mathbf{A} \mid \mathbf{y} \end{bmatrix} = \begin{bmatrix} 1 & 1 & 1 & -5 \\ 1 & -1 & 1 & 1 \\ 4 & 2 & 1 & 7 \end{bmatrix} \underset{\substack{-1R_1 + R_2 \\ -4R_1 + R_3}}{\Rightarrow} \begin{bmatrix} 1 & 1 & 1 & -5 \\ 0 & -2 & 0 & 6 \\ 0 & -2 & -3 & 27 \end{bmatrix} \underset{-1R_2 + R_3}{}$$

$$\Rightarrow \begin{bmatrix} 1 & 1 & 1 & -5 \\ 0 & -2 & 0 & 6 \\ 0 & 0 & -3 & 21 \end{bmatrix}.$$

Back substitution gives

$$\mathbf{v} = \begin{bmatrix} a \\ b \\ c \end{bmatrix} = \begin{bmatrix} 5 \\ -3 \\ -7 \end{bmatrix}.$$

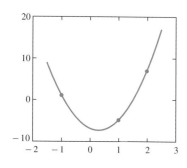

FIGURE 4

(Verify.) Thus the quadratic interpolant is $p(x) = 5x^2 - 3x - 7$ and its graph is given in Figure 4. The solid dots represent the three data points.   ■

## Application: Temperature Distribution

A simple model for estimating the temperature distribution on a square plate uses a linear system of equations. To construct the appropriate linear system, we use the following information. The square plate is perfectly insulated on its top and bottom so that the only heat flow is through the plate itself. The four edges are held at various temperatures. To estimate the temperature at an interior point on the plate, we use the rule that it is the average of the temperatures at its four compass point neighbors, to the west, north, east, and south. Let's construct such a linear system to estimate the temperatures $T_i$, $i = 1, 2, 3, 4$ at the four equispaced interior points on the plate shown in Figure 5. The points at which we need the temperature of the plate for this model are indicated by dots. Using our averaging rule, we obtain the equations

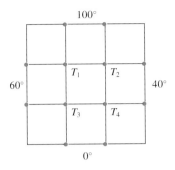

FIGURE 5

$$T_1 = \frac{60 + 100 + T_2 + T_3}{4} \quad \Rightarrow \quad 4T_1 - T_2 - T_3 \qquad = 160$$

$$T_2 = \frac{T_1 + 100 + 40 + T_4}{4} \quad \Rightarrow \quad -T_1 + 4T_2 \qquad - T_4 = 140$$

$$T_3 = \frac{60 + T_1 + T_4 + 0}{4} \quad \Rightarrow \quad -T_1 \qquad + 4T_3 - T_4 = 60$$

$$T_4 = \frac{T_3 + T_2 + 40 + 0}{4} \quad \Rightarrow \qquad - T_2 - T_3 + 4T_4 = 40.$$

Constructing the augmented matrix for the linear system and choosing row operations to obtain an upper triangular form, we get the following sequence of row-equivalent matrices. (Many such sequences are possible. The row operations were chosen to avoid the use of fractions.)

$$[\mathbf{A} \mid \mathbf{b}] = \begin{bmatrix} 4 & -1 & -1 & 0 & | & 160 \\ -1 & 4 & 0 & -1 & | & 140 \\ -1 & 0 & 4 & -1 & | & 60 \\ 0 & -1 & -1 & 4 & | & 40 \end{bmatrix}_{R_1 \leftrightarrow R_2}$$

$$\Rightarrow \begin{bmatrix} -1 & 4 & 0 & -1 & | & 140 \\ 4 & -1 & -1 & 0 & | & 160 \\ -1 & 0 & 4 & -1 & | & 60 \\ 0 & -1 & -1 & 4 & | & 40 \end{bmatrix}_{\substack{4R_1 + R_2 \\ -1R_1 + R_3}}$$

$$\Rightarrow \begin{bmatrix} -1 & 4 & 0 & -1 & | & 140 \\ 0 & 15 & -1 & -4 & | & 720 \\ 0 & -4 & 4 & 0 & | & -80 \\ 0 & -1 & -1 & 4 & | & 40 \end{bmatrix}_{R_2 \leftrightarrow R_4}$$

$$\Rightarrow \begin{bmatrix} -1 & 4 & 0 & -1 & | & 140 \\ 0 & -1 & -1 & 4 & | & 40 \\ 0 & -4 & 4 & 0 & | & -80 \\ 0 & 15 & -1 & -4 & | & 720 \end{bmatrix}_{\substack{-4R_2 + R_3 \\ 15R_2 + R_4}}$$

$$\Rightarrow \begin{bmatrix} -1 & 4 & 0 & -1 & | & 140 \\ 0 & -1 & -1 & 4 & | & 40 \\ 0 & 0 & 8 & -16 & | & -240 \\ 0 & 0 & -16 & 56 & | & 1320 \end{bmatrix}_{2R_3 + R_4}$$

$$\Rightarrow \begin{bmatrix} -1 & 4 & 0 & -1 & | & 140 \\ 0 & -1 & -1 & 4 & | & 40 \\ 0 & 0 & 8 & -16 & | & -240 \\ 0 & 0 & 0 & 24 & | & 840 \end{bmatrix}.$$

Performing back substitution, we get the unique solution $T_1 = 65°$, $T_2 = 60°$, $T_3 = 40°$, and $T_4 = 35°$. (Verify.) Note that the original coefficient matrix is strictly diagonally dominant. (See the Exercises for Section 1.1.) It can be shown that a linear system with a strictly diagonally dominant coefficient matrix will always have a unique solution. This is one instance when properties or the pattern of the entries of a coefficient matrix provide immediate insight into the solution set of the linear system. We will see other such cases later.

**Application:   Parametric Cubic Curves in $R^3$ (Calculus Required)**

The parametric form of a cubic curve $g$ in $R^3$ is given by

$$\begin{aligned} x &= a_1 t^3 + b_1 t^2 + c_1 t + d_1 \\ y &= a_2 t^3 + b_2 t^2 + c_2 t + d_2 \\ z &= a_3 t^3 + b_3 t^2 + c_3 t + d_3. \end{aligned} \tag{5}$$

(See Exercise 65 in Section 1.2.) Let

$$\mathbf{a} = \begin{bmatrix} a_1 \\ a_2 \\ a_3 \end{bmatrix}, \quad \mathbf{b} = \begin{bmatrix} b_1 \\ b_2 \\ b_3 \end{bmatrix}, \quad \mathbf{c} = \begin{bmatrix} c_1 \\ c_2 \\ c_3 \end{bmatrix}, \quad \mathbf{d} = \begin{bmatrix} d_1 \\ d_2 \\ d_3 \end{bmatrix}$$

then define

$$\mathbf{p}(t) = t^3 \mathbf{a} + t^2 \mathbf{b} + t \mathbf{c} + \mathbf{d}. \tag{6}$$

For a value of parameter $t$, $\mathbf{p}(t)$ produces a point on curve $g$. An alternative form of $\mathbf{p}(t)$, useful in geometric settings, is to express $\mathbf{p}(t)$ as a linear combination of information at two points that lie on the curve. The location of the two points will be involved as well as the value of the parametric derivative,

$$\frac{d}{dt}\mathbf{p}(t) = \mathbf{p}'(t),$$

at the points. It follows that

$$\mathbf{p}'(t) = 3t^2 \mathbf{a} + 2t \mathbf{b} + \mathbf{c}. \tag{7}$$

We now use a system of equations to convert the expression for the cubic curve in (6) into an alternative form that is a linear combination of $\mathbf{p}(0)$, $\mathbf{p}'(0)$, $\mathbf{p}(1)$, and $\mathbf{p}'(1)$ rather than $\mathbf{a}$, $\mathbf{b}$, $\mathbf{c}$, and $\mathbf{d}$. We see that $\mathbf{p}(0)$ and $\mathbf{p}(1)$ are points on $g$, while $\mathbf{p}'(0)$ and $\mathbf{p}'(1)$ are values of the parametric derivative at the respective points. [The terms of the equations involved are vectors; hence we do not use a matrix form for the system of equations developed in (8).] From (6) and (7) we have (verify)

$$\begin{aligned} \mathbf{p}(0) &= \mathbf{d} \\ \mathbf{p}'(0) &= \mathbf{c} \\ \mathbf{p}(1) &= \mathbf{a} + \mathbf{b} + \mathbf{c} + \mathbf{d} = \mathbf{a} + \mathbf{b} + \mathbf{p}'(0) + \mathbf{p}(0) \\ \mathbf{p}'(1) &= 3\mathbf{a} + 2\mathbf{b} + \mathbf{c} = 3\mathbf{a} + 2\mathbf{b} + \mathbf{p}'(0). \end{aligned} \tag{8}$$

Solving the two equations

$$\begin{aligned} \mathbf{p}(1) &= \mathbf{a} + \mathbf{b} + \mathbf{p}'(0) + \mathbf{p}(0) \\ \mathbf{p}'(1) &= 3\mathbf{a} + 2\mathbf{b} + \mathbf{p}'(0) \end{aligned}$$

simultaneously for **a** and **b**, we find that (verify)

$$\mathbf{a} = 2\mathbf{p}(0) - 2\mathbf{p}(1) + \mathbf{p}'(0) + \mathbf{p}'(1)$$
$$\mathbf{b} = -3\mathbf{p}(0) + 3\mathbf{p}(1) - 2\mathbf{p}'(0) - \mathbf{p}'(1).$$

Substituting the expression for **a**, **b**, **c**, and **d** into (6) and collecting terms, we get

$$\begin{aligned}
\mathbf{p}(t) &= t^3\mathbf{a} + t^2\mathbf{b} + t\mathbf{c} + \mathbf{d} \\
&= t^3(2\mathbf{p}(0) - 2\mathbf{p}(1) + \mathbf{p}'(0) + \mathbf{p}'(1)) \\
&\quad + t^2(-3\mathbf{p}(0) + 3\mathbf{p}(1) - 2\mathbf{p}'(0) - \mathbf{p}'(1)) + t\mathbf{p}'(0) + \mathbf{p}(0) \\
&= (2t^3 - 3t^2 + 1)\mathbf{p}(0) + (-2t^3 + 3t^2)\mathbf{p}(1) \\
&\quad + (t^3 - 2t^2 + t)\mathbf{p}'(0) + (t^3 - t^2)\mathbf{p}'(1).
\end{aligned}$$

Hence points along the cubic curve $g$ are expressed in terms of information at the points that correspond to $\mathbf{p}(0)$ and $\mathbf{p}(1)$. We call this form of the cubic curve in $R^3$ the **geometric equation**. The preceding expression may seem limiting in some sense. However, we are free to choose the points $\mathbf{p}(0)$ and $\mathbf{p}(1)$ and the parametric derivatives, which are all 3-vectors, in order to generate curves of various types. Thus we have the flexibility to develop cubic models with very simple choices of information. We illustrate this in Example 11.

EXAMPLE 11    Let

$$\mathbf{p}(0) = \begin{bmatrix} 0 \\ 0 \\ 0 \end{bmatrix} \quad \text{and} \quad \mathbf{p}(1) = \begin{bmatrix} 1 \\ 1 \\ 1 \end{bmatrix}.$$

By choosing different 3-vectors for the parametric derivatives at these points, we can generate a variety of cubic curves in 3-space. Figure 6(a) was generated with the parametric derivatives given by

$$\mathbf{p}'(0) = \begin{bmatrix} 20 \\ 1 \\ 0 \end{bmatrix} \quad \text{and} \quad \mathbf{p}'(1) = \begin{bmatrix} -1 \\ 1 \\ 0 \end{bmatrix}.$$

while Figure 6(b) was generated with

$$\mathbf{p}'(0) = \begin{bmatrix} 20 \\ 1 \\ 0 \end{bmatrix} \quad \text{and} \quad \mathbf{p}'(1) = \begin{bmatrix} 1 \\ 1 \\ -5 \end{bmatrix}.$$

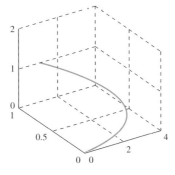

FIGURE 6(a)

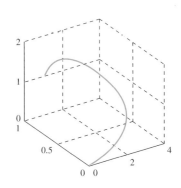

FIGURE 6(b)

These two curves have quite different shapes. To experiment with the geometric form of a parametric cubic curve in MATLAB, see Exercise ML.7.    ■

We used a system of equations to transform the parametric form of the cubic equation in (5) into the geometric equation. Linear algebra plays an important role in developing alternative representations of equations and information.

## EXERCISES 2.1

*In Exercises 1–3, determine whether the augmented matrix represents a consistent linear system. If it does not, explain why not.*

**1.** $\begin{bmatrix} 2 & 1 & -1 & | & 3 \\ 0 & -1 & 4 & | & 1 \\ 0 & 0 & 0 & | & 0 \end{bmatrix}$.    **2.** $\begin{bmatrix} 4 & 2 & | & 1 \\ 0 & 0 & | & -3 \end{bmatrix}$.

**3.** $\begin{bmatrix} 0 & 6 & -3 & | & 2 \\ 1 & -4 & 5 & | & 3 \\ 0 & 0 & 8 & | & 4 \end{bmatrix}$.

*In Exercises 4–8, each given augmented matrix represents a consistent linear system. Determine whether there is a unique solution or infinitely many solutions. Explain why your answer is correct.*

**4.** $\begin{bmatrix} 3 & 4 & | & -1 \\ 0 & 2 & | & 6 \end{bmatrix}$.    **5.** $\begin{bmatrix} 0 & 0 & | & 0 \\ 1 & 2 & | & 6 \end{bmatrix}$.

**6.** $\begin{bmatrix} 1 & 3 & 4 \\ 0 & 2 & 0 \\ 0 & 0 & 0 \end{bmatrix}$.    **7.** $\begin{bmatrix} 2 & 4 & 5 & | & -1 \\ 0 & 0 & 1 & | & 3 \\ 0 & 0 & 0 & | & 0 \end{bmatrix}$.

**8.** $\begin{bmatrix} 2 & -3 & 5 & | & 2 \\ 0 & -1 & 4 & | & 1 \\ 0 & 0 & 6 & | & 3 \end{bmatrix}$.

*In Exercises 9–12, each given augmented matrix represents a linear system with infinitely many solutions. Determine the number of free variables or arbitrary constants in the solution.*

**9.** $\begin{bmatrix} 2 & -3 & | & 5 \\ 0 & 0 & | & 0 \end{bmatrix}$.    **10.** $\begin{bmatrix} 3 & -4 & 5 & | & 0 \\ 0 & 0 & 2 & | & 0 \\ 0 & 0 & 0 & | & 0 \end{bmatrix}$.

**11.** $\begin{bmatrix} 1 & -2 & 0 & 4 & | & 1 \\ 0 & 0 & 1 & -5 & | & 6 \\ 0 & 0 & 0 & 0 & | & 0 \\ 0 & 0 & 0 & 0 & | & 0 \end{bmatrix}$.

**12.** $\begin{bmatrix} 2 & 8 & -4 & 1 & | & 0 \\ 0 & 0 & 0 & 0 & | & 0 \\ 0 & 0 & 0 & 0 & | & 0 \end{bmatrix}$.

**13.** Let $P_1$, $P_2$, and $P_3$ be planes in 3-space where $P_i$ is represented by the equation

$$a_i x + b_i y + c_i z = d_i \quad \text{for } i = 1, 2, 3$$

(a) Construct the augmented matrix of a linear system that can be used to determine the intersection points, if any, of these planes.

(b) Label each of the following figures with a phrase that describes the number of points in the solution set of the linear system from part (a).

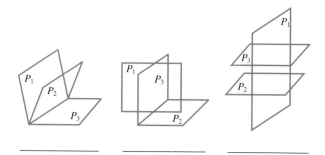

_____    _____    _____

*In Exercises 14–21, perform row operations on the augmented matrix to obtain upper triangular form; then determine the solution set of the linear system using back substitution. Use unknowns $x_1, x_2, \dots$ as needed.*

**14.** $\begin{bmatrix} 0 & 0 & 2 & | & 6 \\ 0 & 1 & 0 & | & 1 \\ 4 & 3 & 2 & | & 1 \end{bmatrix}$.    **15.** $\begin{bmatrix} 2 & 3 & 5 & | & 0 \\ 0 & 1 & 2 & | & 1 \\ 4 & 2 & 2 & | & -4 \end{bmatrix}$.

**16.** $\begin{bmatrix} 1 & 0 & 2 & | & -3 \\ -1 & 1 & -2 & | & 0 \\ -1 & 3 & -2 & | & 4 \end{bmatrix}$.

**17.** $\begin{bmatrix} 2 & 1 & | & 2 \\ 1 & 0 & | & 3 \\ 3 & 1 & | & 5 \end{bmatrix}$.    **18.** $\begin{bmatrix} 1 & 2 & 0 & 0 & | & 0 \\ 2 & 1 & 2 & 0 & | & 0 \\ 0 & 2 & 1 & 2 & | & 0 \\ 0 & 0 & 2 & 1 & | & 2 \end{bmatrix}$.

**19.** $\begin{bmatrix} 1 & 0 & 0 & 0 & | & 2 \\ 3 & 1 & 0 & 0 & | & 1 \\ 0 & 3 & 1 & 0 & | & -1 \\ 0 & 0 & 3 & 1 & | & 2 \end{bmatrix}$.

**20.** $\begin{bmatrix} 1 & 0 & 0 & 2 & | & 0 \\ 2 & 1 & 0 & 4 & | & 0 \\ -3 & 2 & 1 & 0 & | & 0 \end{bmatrix}$.    **21.** $\begin{bmatrix} 1 & 1 & 0 & 3 & | & 4 \\ 2 & 2 & 0 & 5 & | & 2 \\ 3 & 3 & 0 & 7 & | & 1 \end{bmatrix}$.

**22.** Each of the augmented matrices in Exercises 9–12 is in upper triangular form. Use back substitution to construct the solution set. Use unknowns $x_1, x_2, \dots$ as needed.

**23.** Let $\mathbf{A}$ be an $n \times n$ matrix that is upper triangular with nonzero diagonal entries; $a_{ii} \neq 0$, $i = 1, 2, \ldots, n$. Then $\mathbf{Ax} = \mathbf{b}$ has a unique solution that we can determine by back substitution. The augmented matrix of the linear system is

$$\left[ \begin{array}{ccccccc|c}
a_{11} & a_{12} & \cdots & \cdots & \cdots & a_{1\,n-1} & a_{1n} & b_1 \\
0 & a_{22} & \cdots & \cdots & \cdots & a_{2\,n-1} & a_{2n} & b_2 \\
\vdots & \vdots & \vdots & \vdots & \vdots & \vdots & \vdots & \vdots \\
0 & 0 & \cdots & a_{ii} & \cdots & a_{i\,n-1} & a_{in} & b_i \\
\vdots & \vdots & \vdots & \vdots & \vdots & \vdots & \vdots & \vdots \\
0 & 0 & \cdots & \cdots & \cdots & a_{n-1\,n-1} & a_{n-1\,n} & b_{n-1} \\
0 & 0 & \cdots & \cdots & \cdots & 0 & a_{nn} & b_n
\end{array} \right].$$

(a) Construct the expression for $x_n$.
(b) Construct the expression for $x_{n-1}$.
(c) Construct the expression for $x_{n-2}$.
(d) Construct the expression for $x_i$.
(e) Verify that your expression for $x_i$ is equivalent to

$$x_i = \frac{b_i - \displaystyle\sum_{j=i+1}^{n} a_{ij} x_j}{a_{ii}}.$$

*A linear system $\mathbf{Ax} = \mathbf{b}$ is said to be in lower triangular form if its coefficient matrix $\mathbf{A}$ is lower triangular. (See Section 1.1.) In such a case we can solve the linear system by **forward substitution**, which begins with the first equation to determine $x_1$, then uses the value of $x_1$ to determine $x_2$ from the second equation, and so on. Use forward substitution to determine the solution set of each of the linear systems in Exercises 24–26.*

**24.** $\begin{aligned} 2x_1 &= 6 \\ x_1 + x_2 &= 4 \\ 2x_1 + x_2 - x_3 &= 5. \end{aligned}$

**25.** $\begin{aligned} 5x_1 &= 3 \\ 2x_1 + 4x_2 &= -2. \end{aligned}$

**26.** $\begin{aligned} -3x_1 &= 9 \\ x_1 + 2x_2 &= 7 \\ 2x_1 + 4x_2 &= 14. \end{aligned}$

**27.** Let $\mathbf{A}$ be an $n \times n$ matrix that is lower triangular with nonzero diagonal entries; $a_{ii} \neq 0$, $i = 1, 2, \ldots, n$. Then $\mathbf{Ax} = \mathbf{b}$ has a unique solution that we can determine by forward substitution. The augmented matrix of the linear system is

$$\left[ \begin{array}{ccccccc|c}
a_{11} & 0 & \cdots & \cdots & \cdots & 0 & 0 & b_1 \\
a_{21} & a_{22} & 0 & \cdots & \cdots & 0 & 0 & b_2 \\
\vdots & \vdots & \vdots & \vdots & \vdots & \vdots & \vdots & \vdots \\
a_{i1} & a_{i2} & \cdots & a_{ii} & 0 & 0 & 0 & b_i \\
\vdots & \vdots & \vdots & \vdots & \vdots & \vdots & \vdots & \vdots \\
a_{n-1\,1} & a_{n-1\,2} & \cdots & \cdots & \cdots & a_{n-1\,n-1} & 0 & b_{n-1} \\
a_{n1} & a_{n2} & \cdots & \cdots & \cdots & a_{n\,n-1} & a_{nn} & b_n
\end{array} \right].$$

(a) Construct the expression for $x_1$.
(b) Construct the expression for $x_2$.
(c) Construct the expression for $x_3$.
(d) Construct the expression for $x_i$.
(e) Verify that your expression for $x_i$ is equivalent to

$$x_i = \frac{b_i - \displaystyle\sum_{j=1}^{i-1} a_{ij} x_j}{a_{ii}}.$$

**28.** If the square linear system $\mathbf{Ax} = \mathbf{b}$ is in lower triangular form, then explain how to recognize that there will be a unique solution.

*In Exercises 29–31, vector $\mathbf{x}$ represents the solution set of a linear system with infinitely many solutions. Express $\mathbf{x}$ as a translation of a vector (see Section 1.4). Scalar $t$ represents an arbitrary constant.*

**29.** $\mathbf{x} = \begin{bmatrix} 2+t \\ 3 \\ t \end{bmatrix}.$       **30.** $\mathbf{x} = \begin{bmatrix} 1+2t \\ t \\ 0 \end{bmatrix}.$

**31.** $\mathbf{x} = \begin{bmatrix} -1-t \\ 2t \\ 3 \\ t \end{bmatrix}.$

*In Exercises 32–34, the vector $\mathbf{x}$ represents the solution set of a homogeneous linear system with infinitely many solutions. Express vector $\mathbf{x}$ as the span of a set of vectors that contain no arbitrary constants. Scalars $s$ and $t$ represent arbitrary constants.*

**32.** $\mathbf{x} = \begin{bmatrix} 2s-t \\ s \\ -s+3t \\ t \end{bmatrix}.$       **33.** $\mathbf{x} = \begin{bmatrix} 2t \\ -3t \\ t \end{bmatrix}.$

**34.** $\mathbf{x} = \begin{bmatrix} 0 \\ s+t \\ 3s-4t \\ s \\ t \end{bmatrix}.$

**35.** Let $\mathbf{A} = \begin{bmatrix} 1 & -1 \\ -2 & 2 \end{bmatrix}$. Determine which of the following vectors $\mathbf{b}$ are in the range of the matrix transformation determined by $\mathbf{A}$. (That is, is there a vector $\mathbf{x} = \begin{bmatrix} x_1 \\ x_2 \end{bmatrix}$ such that $\mathbf{Ax} = \mathbf{b}$?)

(a) $\mathbf{b} = \begin{bmatrix} 0 \\ 0 \end{bmatrix}.$       (b) $\mathbf{b} = \begin{bmatrix} 1 \\ 2 \end{bmatrix}.$

(c) $\mathbf{b} = \begin{bmatrix} 4 \\ 4 \end{bmatrix}.$

**36.** Let $A = \begin{bmatrix} 1 & 1 & 1 \\ 1 & -1 & 0 \\ 2 & 0 & 1 \end{bmatrix}$. Determine which of the following vectors **b** are in the range of the matrix transformation determined by **A**. (That is, is there a vector $x = \begin{bmatrix} x_1 \\ x_2 \\ x_3 \end{bmatrix}$ such that $Ax = b$?)

(a) $b = \begin{bmatrix} 4 \\ 0 \\ 4 \end{bmatrix}$.　(b) $b = \begin{bmatrix} 0 \\ -2 \\ -2 \end{bmatrix}$.

(c) $b = \begin{bmatrix} 1 \\ 1 \\ 1 \end{bmatrix}$.

*In Exercises 37–40, construct a lower triangular matrix* **L** *and an upper triangular matrix* **U** *so that* $LU = A$ *for each given matrix* **A**. *We view this as factoring* **A** *into a product of a lower triangular matrix and an upper triangular matrix and call this an* **LU-factorization** *of* **A**.

**37.** Let $A = \begin{bmatrix} 3 & 2 \\ 6 & 3 \end{bmatrix}$, $L = \begin{bmatrix} 1 & 0 \\ t & 1 \end{bmatrix}$, and $U = \begin{bmatrix} 3 & 2 \\ 0 & -1 \end{bmatrix}$.
Find a scalar $t$ so that $LU = A$.

**38.** Let $A = \begin{bmatrix} 2 & 6 \\ 1 & 1 \end{bmatrix}$, $L = \begin{bmatrix} 2 & 0 \\ t & -2 \end{bmatrix}$, and $U = \begin{bmatrix} 1 & 3 \\ 0 & s \end{bmatrix}$.
Find scalars $s$ and $t$ so that $LU = A$.

**39.** Let

$$A = \begin{bmatrix} 6 & 2 & 8 \\ 9 & 5 & 11 \\ 3 & 1 & 6 \end{bmatrix}, \quad L = \begin{bmatrix} 2 & 0 & 0 \\ t & s & 0 \\ 1 & 0 & -1 \end{bmatrix},$$

and

$$U = \begin{bmatrix} r & 1 & 4 \\ 0 & 2 & -1 \\ 0 & 0 & p \end{bmatrix}.$$

Find scalars $r$, $s$, $t$, and $p$ so that $LU = A$.

**40.** Let $A = \begin{bmatrix} 2 & 6 \\ 1 & 1 \end{bmatrix}$, $L = \begin{bmatrix} 1 & 0 \\ r & 1 \end{bmatrix}$, and $U = \begin{bmatrix} s & t \\ 0 & p \end{bmatrix}$.
Find scalars $r$, $s$, $t$, and $p$ so that $LU = A$.

*Let* **A** *be a square matrix with an LU-factorization,* $A = LU$. *(See the discussion prior to Exercise 37). To solve the linear system* $Ax = b$ *no row operations are required. We use a forward substitution followed by a back substitution in the following way:*

*Since* $Ax = b$ *is the same as* $LUx = b$, *let* $Ux = z$. *Then we have the lower triangular linear system* $Lz = b$, *which we can solve for* **z** *by forward substitution. Once we have computed* **z**, *we then solve the upper triangular linear system* $Ux = z$ *for* **x** *by back substitution.*

*Use this procedure for the linear systems in Exercises 41 and 42.*

**41.** Solve $Ax = b$, where

$$A = \begin{bmatrix} 2 & 8 & 0 \\ 2 & 2 & -3 \\ 1 & 2 & 7 \end{bmatrix}, \quad b = \begin{bmatrix} 18 \\ 3 \\ 12 \end{bmatrix},$$

$$L = \begin{bmatrix} 2 & 0 & 0 \\ 2 & -3 & 0 \\ 1 & -1 & 4 \end{bmatrix}, \quad \text{and} \quad U = \begin{bmatrix} 1 & 4 & 0 \\ 0 & 2 & 1 \\ 0 & 0 & 2 \end{bmatrix}.$$

**42.** Solve $Ax = b$, where

$$A = \begin{bmatrix} 8 & 12 & -4 \\ 6 & 5 & 7 \\ 2 & 1 & 6 \end{bmatrix}, \quad b = \begin{bmatrix} -36 \\ 11 \\ 16 \end{bmatrix},$$

$$L = \begin{bmatrix} 4 & 0 & 0 \\ 3 & 2 & 0 \\ 1 & 1 & 1 \end{bmatrix}, \quad \text{and} \quad U = \begin{bmatrix} 2 & 3 & -1 \\ 0 & -2 & 5 \\ 0 & 0 & 2 \end{bmatrix}.$$

**43.** Let **A** have an LU-factorization, $A = LU$. By inspecting the lower triangular matrix **L** and the upper triangular matrix **U**, explain how to claim that the linear system $Ax = LUx = b$ does not have a unique solution.

*In Exercises 44 and 45, determine the quadratic interpolant to each given data set. Follow the procedure in Example 10.*

**44.** $\{(0, 2), (1, 5), (2, 14)\}$.

**45.** $\{(-1, 2), (3, 14), (0, -1)\}$.

**46.** (Calculus Required) Construct a linear system of equations to determine a quadratic polynomial $p(x) = ax^2 + bx + c$ that satisfies the conditions $p(0) = f(0)$, $p'(0) = f'(0)$, and $p''(0) = f''(0)$, where $f(x) = e^{2x}$.

**47.** (Calculus Required) Construct a linear system of equations to determine a quadratic polynomial $p(x) = ax^2 + bx + c$ that satisfies the conditions $p(1) = f(1)$, $p'(1) = f'(1)$, and $p''(1) = f''(1)$, where $f(x) = xe^{x-1}$.

**48.** Determine the temperatures at the interior points $T_i, i = 1, 2, 3, 4$ for the plate shown in Figure 7. (See the application on Temperature Distribution.)

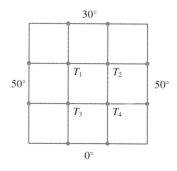

FIGURE 7

*Our strategy for solving linear systems of equations with complex entries is the same as that for real entries that was emphasized in this section. Use row operations with complex arithmetic to reduce each given augmented matrix in Exercises 49–52 to upper triangular form and then determine the solution set.*

**49.** $\begin{bmatrix} 1-i & 2+i & | & 2+2i \\ 2 & 1-2i & | & 1+3i \end{bmatrix}$.

**50.** $\begin{bmatrix} i & 2 & 1-i & | & 1-2i \\ 0 & 2i & 2+i & | & -2+i \\ 0 & -i & 1 & | & -1-i \end{bmatrix}$.

**51.** $\begin{bmatrix} 1-i & 2+2i & | & 1 \\ 1+i & -2+2i & | & i \end{bmatrix}$.

**52.** $\begin{bmatrix} 1-i & 2+2i & | & i \\ 1+i & -2+2i & | & -2 \end{bmatrix}$.

## In MATLAB

*This section shows how to obtain row equivalent linear systems that are in upper triangular form and can then be solved by back substitution. Exercises* ML.1 *through* ML.3 *use* MATLAB *to provide a sense of the reduction process as an algorithm. These three exercises provide opportunities for students experiencing difficulties with the stages of the process to see the actions of reduction to upper triangular form and back substitution and may not be needed by all students. In addition, in Exercises* ML.4 *and* ML.5 *we use computational tools that focus on choosing the row operations for the reduction while letting* MATLAB *handle the arithmetic involved. These exercises are recommended for all students since they focus on the strategy for the reduction process, while* MATLAB *handles the arithmetic of the row operations.*

**ML. 1.** For the linear system

$$\begin{aligned} 3x_1 + 2x_2 + 4x_3 - x_4 &= 20 \\ 2x_1 \quad\quad + x_3 - 3x_4 &= 19 \\ x_1 + x_2 \quad\quad + 2x_4 &= 0 \\ 5x_1 + 2x_2 + x_3 + x_4 &= 24 \end{aligned}$$

enter the coefficient matrix into MATLAB and name it **A**. Enter the right side and name it **b**. Form the augmented matrix, naming it **C**, using the command **C = [A b]**. (There is a space between **A** and **b** in the command.)

(a) To get a sense of the reduction of this linear system to upper triangular form, use command **utriquik(C)**. Choose the rational display and original pivot value options. Watch carefully as row operations are applied automatically to obtain an equivalent upper triangular augmented matrix. The process appears to "sweep out" the lower triangular entries. Rerun this example and note the order in which quantities are swept out.

(b) To see the "sweep out" process with an element-by-element dialog, enter command **utriview(C)** and choose the same options as in part (a).

(c) To see the specific row operations used to "sweep out" the lower triangular entries, use **utriview(C)** and choose the same options as in part (a).

(d) Write a description of the path through the lower triangular entries used by the reduction process for the preceding parts.

**ML. 2.** The result of each of parts (a), (b), and (c) in Exercise ML.1 is the upper triangular linear system

$$\begin{aligned} 3x_1 + 2x_2 + 4x_3 - x_4 &= 20 \\ -\tfrac{4}{3}x_2 - \tfrac{5}{3}x_3 - \tfrac{7}{3}x_4 &= \tfrac{17}{3} \\ -\tfrac{7}{4}x_3 + \tfrac{7}{4}x_4 &= \tfrac{-21}{4} \\ x_4 &= -3. \end{aligned}$$

(a) Solve this linear system by hand using back substitution.

(b) Repeat Exercise ML.1(b), this time choose the option to force the pivots to be 1. Solve the resulting upper triangular linear system by hand using back substitution.

(c) The results in parts (a) and (b) must be identical. (Explain why.) It is often said that with hand calculations of the solution of an upper triangular linear system, the form obtained in part (b) is easier than that in part (a). Explain why this is the case.

**ML. 3.** The back substitution process is used quite frequently. To gain experience and confidence in using it, we can practice it using MATLAB as a smart tutor. In MATLAB type **help bksubpr** and read the description displayed. This routine will provide a way to check your step-by-step use of back substitution. For example, let **A** be the upper triangular coefficient matrix and **b** be the right side of the linear system

$$\begin{aligned} 5x_1 + x_2 + 2x_3 &= 12 \\ -2x_2 + 3x_3 &= -8 \\ 4x_3 &= -8 \end{aligned}$$

$$\Leftrightarrow \begin{bmatrix} 5 & 1 & 2 \\ 0 & -2 & 3 \\ 0 & 0 & 4 \end{bmatrix} \begin{bmatrix} x_1 \\ x_2 \\ x_3 \end{bmatrix} = \begin{bmatrix} 12 \\ -8 \\ -8 \end{bmatrix}$$

$$\Leftrightarrow \mathbf{Ax = b}.$$

Back substitution starts with the last equation and moves toward the first equation, solving for one unknown at a time. Enter the matrices **A** and **b** into MATLAB and then type command

**bksubpr(A,b);**

[*Note*: In MATLAB entries of a vector **c** are denoted **c(1)**, **c(2)**, ... rather than with subscripts.] For the first component of the solution you are asked to enter a value or expression for **x(3)**. From the last equation we can see that **x(3)** $= -8/4 = -2$ or, equivalently, **x(3)** = **b(3)/A(3,3)**. (Respond with either the value $-2$ or the expression **b(3)/A(3,3)**.) If the response is not correct, then options to 'Try again' or 'Help' can be chosen. Once the response for **x(3)** is correct, you are prompted to enter a value or expression for **x(2)**. From equation $-2x_2 + 3x_3 = -8$, we see that

$$x_2 = \frac{-8 - 3x_3}{-2}.$$

We can compute this value and enter it directly or have MATLAB compute it by entering

### (b(2)–A(2,3)*x(3))/A(2,2)

Finish the back substitution on this system in the MATLAB routine **bksubpr**. For further practice use routine **bksubpr** with the following matrices **A** and **b**.

(a) $\mathbf{A} = \begin{bmatrix} 2 & 3 & -1 \\ 0 & -2 & 4 \\ 0 & 0 & 5 \end{bmatrix}, \mathbf{b} = \begin{bmatrix} -3 \\ 14 \\ 10 \end{bmatrix}.$

(b) $\mathbf{A} = \begin{bmatrix} 5 & 1 & -3 \\ 0 & 0 & 2 \\ 0 & 0 & 4 \end{bmatrix}, \mathbf{b} = \begin{bmatrix} 6 \\ 8 \\ 1 \end{bmatrix}.$

(c) $\mathbf{A} = \begin{bmatrix} 3 & 2 & 1 & 1 \\ 0 & -1 & 4 & -2 \\ 0 & 0 & 6 & 1 \\ 0 & 0 & 0 & 5 \end{bmatrix}, \mathbf{b} = \begin{bmatrix} 7.5 \\ 7.5 \\ -13 \\ -5 \end{bmatrix}.$

(d) $\mathbf{A} = \begin{bmatrix} 2 & -1 & 2 & 0 & 0 \\ 0 & 3 & -1 & 2 & 0 \\ 0 & 0 & 4 & -1 & 2 \\ 0 & 0 & 0 & 5 & -1 \\ 0 & 0 & 0 & 0 & 2 \end{bmatrix}, \mathbf{b} = \begin{bmatrix} 0 \\ 19 \\ -12 \\ 9 \\ 2 \end{bmatrix}.$

**ML. 4.** In order to perform the reduction of the augmented matrix of linear system $\mathbf{Ax} = \mathbf{b}$ to upper triangular form, we use row operations. The routine **reduce** in MATLAB provides a tool for performing row operations. In **reduce** you must specify the row operation and MATLAB will perform the associated arithmetic.

Enter the command **reduce** and select the option to use a built-in demo. Select the second linear system. When the menu appears, turn on the rational display, but do not perform any row interchanges or multiply a row by a nonzero scalar. Use just three row operations of the form

$$k * Row(i) + Row(j) ==> Row(j)$$

to obtain the upper triangular form shown next.

---

**"REDUCE" a Matrix by Row Reduction**

The current matrix is:

$\mathbf{A} =$

| 3 | 4 | 5 | −8 |
|---|---|---|----|
| −2 | 1 | 0 | 3 |
| −1 | 3 | 1 | 5 |

<< OPTIONS >>

<1>  Row(i) <==> Row(j)
<2>  k * Row(i)       (k not zero)
<3>  k * Row(i) + Row(j)  ==>  Row(j)

<4>  Turn on rational display.
<5>  Turn off rational display.
<−1>  "Undo" previous row operation.
<0>  Quit reduce !

ENTER your choice  ===>

| 3 | 4 | 5 | −8 |
|---|---|---|----|
| 0 | 11/3 | 10/3 | −7/3 |
| 0 | 0 | −14/11 | 56/11 |

To obtain the solution, use back substitution. Use **reduce** to practice your strategy for reduction to upper triangular form. Check your solutions to the linear systems in Exercises 14–21, 49–52, or use any of the following linear systems.

(a) $\left[\begin{array}{ccc|c} 1 & 0 & 1 & 2 \\ -1 & 1 & 0 & -3 \\ 0 & 1 & -1 & -3 \end{array}\right].$

(b) $\left[\begin{array}{ccc|c} 1 & 2 & -3 & -5 \\ -3 & 1 & 2 & 1 \\ 2 & -3 & 1 & 4 \end{array}\right].$

(c) $\left[\begin{array}{ccc|c} 2 & -2 & 1 & 2 \\ 1 & 2 & -1 & 1 \\ 3 & 0 & 0 & 1 \end{array}\right].$

(d) $\left[\begin{array}{cccc|c} 1 & 2 & -2 & 1 & -15 \\ 1 & 1 & 2 & -2 & 16 \\ -2 & 1 & 1 & 2 & -13 \\ 2 & -2 & 1 & 1 & 12 \end{array}\right].$

**ML. 5.** An alternative to routine **reduce** (see Exercise ML.4) is routine **rowop**. Routine **rowop** is similar to **reduce**, but it uses the mouse to click on fields where information for row operations is to be typed. To execute this routine, type **rowop** and follow the directions that appear. Practice the reduction steps with the linear systems in Exercise ML.4.

**ML. 6.** Determine the temperature at the interior points $T_i$, $i = 1, 2, \cdots, 6$ for the plate shown in Figure 8. Use either **reduce** to perform row reduction to upper triangular form on the linear system you construct. Then solve the system by back substitution.

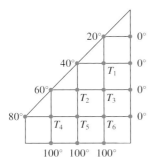

FIGURE 8

**ML. 7.** (*Calculus Required*) The geometric form of a parametric cubic curve was developed in Example 11. The MATLAB routine **paracub** lets you experiment with drawing cubic segments between points $\mathbf{p}(0) = \mathbf{Q}$ and $\mathbf{p}(1) = \mathbf{R}$ while you specify the parametric derivatives at these points. Keeping points $\mathbf{Q}$ and $\mathbf{R}$ fixed and varying the parametric derivatives lets you change the shape of the cubic segment. The input to **paracub** is a set of four 3-vectors, $\mathbf{p}(0)$, $\mathbf{p}(1)$, $\mathbf{p}'(0)$, and $\mathbf{p}'(1)$, in the order specified. (The entries of these 3-vectors can be any real numbers.) To test this, enter the following MATLAB commands; the result should appear like Figure 6(b) in this section.

**Q = [0 0 0]';R = [1 1 1]';Qd = [20 1 0]';**
**Rd = [1 1 −5]';paracub(Q,R,Qd,Rd)**

(a) Enter the following commands and describe the curve displayed.

**Q=[0 0 0]';R=[1 1 1]';Qd=[0 0 0]';**
**Rd=[0 0 0]';paracub(Q,R,Qd,Rd)**

(b) Enter the following commands and describe the curve displayed.

**k=1;Q=[0 0 0]';R=[1 1 1]';Qd=[1 k 1]';**
**Rd=[1 k 1]';paracub(Q,R,Qd,Rd)**

Now repeat the preceding commands (use your up arrow), but change $k$ to 2; then 4; then 8. Describe how the curve in the successive graphs is changing.

(c) Enter the following commands and describe the curve displayed.

**k=1;Q=[0 0 0]';R=[1 1 1]';**
**Qd=k*[1 0 1]';Rd=k*[1 1 1]';**
**paracub(Q,R,Qd,Rd)**

Now repeat the preceding command (use your up arrow), but change $k$ to 3; then 5; then 7; then 9. Describe how the curve in the successive graphs is changing.

(d) Experiment with the following commands by changing the value of $k$ until you get a cubic curve similar to that shown in Figure 9.

**k=1;Q=[0 0 0]';R=[1 1 1]';Qd=[1 k 1]';**
**Rd=[1 1 1]';paracub(Q,R,Qd,Rd),grid**

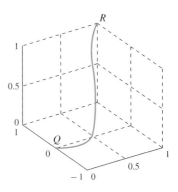

FIGURE 9

## True/False Review Questions

*Determine whether each of the following statements is true or false.*

1. A linear system which has no solution is called consistent.

2. There are only three possible types of solution sets to $\mathbf{Ax} = \mathbf{b}$; no solution, a unique solution, or infinitely many solutions.

3. A homogeneous linear system always has at least one solution.

4. Two matrices $\mathbf{A}$ and $\mathbf{B}$ are row equivalent provided there exist row operations which when applied to $\mathbf{A}$ will produce $\mathbf{B}$.

5. Row operations applied to an augmented matrix $\begin{bmatrix} \mathbf{A} \mid \mathbf{b} \end{bmatrix}$ yield a new linear system with exactly the same solution set.

6. The appearance of a free variable in the general solution of a linear system implies it is inconsistent.

7. Applying back substitution to the linear system with augmented matrix $\begin{bmatrix} 2 & 3 & 0 & 7 & | & 1 \\ 0 & 0 & 1 & 1 & | & 3 \\ 0 & 0 & 0 & 0 & | & 0 \end{bmatrix}$ shows that its solution set has 2 degrees of freedom.

8. The linear system with augmented matrix $\begin{bmatrix} 2 & 1 & | & 4 \\ -6 & -3 & | & 5 \end{bmatrix}$ is inconsistent.

9. When we are reducing an augmented matrix $\begin{bmatrix} A & | & b \end{bmatrix}$ to upper triangular form, the appearance of a row of the form $\begin{bmatrix} 0 & 0 & \cdots & 0 & | & \neq 0 \end{bmatrix}$ implies the system has infinitely many solutions.

10. The linear system with augmented matrix $\begin{bmatrix} a & 0 & 0 & | & 5 \\ 0 & b & 0 & | & 2 \\ 0 & 0 & c & | & 1 \end{bmatrix}$ always has a unique solution for any choice of real numbers $a$, $b$, and $c$.

## Terminology

| | |
|---|---|
| Augmented matrix | Consistent system; inconsistent system |
| Homogeneous system | Trivial solution |
| Solution set of a linear system; general solution | Row operations |
| Row equivalent matrices | Back substitution |
| Arbitrary constant; free variable | Solution strategy for linear systems |
| Row reduction | Pivot; pivot row |
| Forward substitution | Parametric cubic curve |

This section provides us with the basic algebraic tools to solve linear systems. While the overall strategy is quite simple, there are fundamental theoretical principles that justify the process. By answering the following questions and responding to the statements included, you can review the basic reasons why the mechanical process of solving linear systems is valid. These concepts will be extended in succeeding sections.

- Describe the augmented matrix associated with a linear system.
- If we have two linear systems with the same coefficient matrix but different right sides, how can we extend the idea of an augmented matrix to solve both linear systems at the same time?
- What is the solution set of an inconsistent system?
- Describe two different types of consistent linear systems.
- Why can we say that every homogeneous linear system is consistent?
- What does it mean to say that two matrices are row equivalent?
- If the two linear systems described by augmented matrices $\begin{bmatrix} A & | & b \end{bmatrix}$ and $\begin{bmatrix} C & | & d \end{bmatrix}$ are row equivalent, how are their solution sets related?
- Carefully outline our basic strategy for solving a linear system $Ax = b$.
- Describe the solution set of a linear system with three free variables.
- When performing row reduction, we produce equivalent linear systems by using row operations to eliminate entries in the coefficient matrix. Describe the role of a pivot and a pivot row in this process.
- Back substitution is applied to systems in _____ triangular form, while forward substitution is applied to systems in _____ triangular form.

## 2.2 ■ ECHELON FORMS

Section 2.1 showed how to determine the solution set of a linear system $\mathbf{Ax} = \mathbf{b}$. We formed the augmented matrix $\begin{bmatrix} \mathbf{A} \mid \mathbf{b} \end{bmatrix}$ and used row operations to generate equivalent linear systems that were easier to solve. The procedure reduced $\begin{bmatrix} \mathbf{A} \mid \mathbf{b} \end{bmatrix}$ to $\begin{bmatrix} \mathbf{U} \mid \mathbf{c} \end{bmatrix}$ where $\mathbf{U}$ is upper triangular and then the solution of the equivalent linear system $\mathbf{Ux} = \mathbf{c}$ was obtained using back substitution. This reduction process can be quite efficient when we carefully choose row operations to eliminate entries systematically below the pivots, but we must complete the process with the algebraic manipulations of back substitution. For even moderately sized linear systems, using hand calculations with this procedure can become cumbersome. In this section we explore alternatives that employ more row operations so that the back substitution step becomes less complicated or becomes completely unnecessary. Such procedures can be visually appealing since, by appropriate choices of row operations, we force the equivalent linear systems to have many zero entries in the coefficient matrix. However, such approaches are rarely used in computer algorithms, many of which use a variation of reduction to upper triangular form and LU-factorization. (See Exercises 37–42 in Section 2.1.) This section also develops a characteriza-

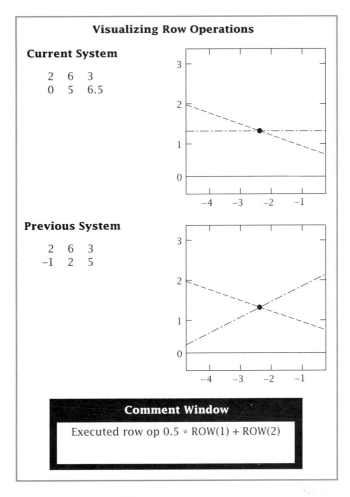

In MATLAB, **vizrowop.m**

tion or general pattern for the set of solutions of any consistent linear system. It is the recognition that there is a pattern that reveals the power of the matrix structure. The pattern we develop makes visible the information contained in the matrix and provides a means to express it in terms of fundamental building blocks.

The techniques that we discuss in this section have a common first step:

$$\text{reduce } [\,\mathbf{A} \mid \mathbf{b}\,] \text{ to upper triangular form } [\,\mathbf{U} \mid \mathbf{c}\,].$$

There are alternatives for succeeding steps, but the basic idea is to continue to use row operations to introduce zeros above the pivots. Zeroing out above the pivots eliminates more unknowns from the equations so that back substitution is simpler. For efficiency it is recommended that we use the rows of $[\,\mathbf{U} \mid \mathbf{c}\,]$ from the bottom upward to perform eliminations. We will employ this strategy, and we illustrate this in Example 1.

**EXAMPLE 1**    Let

$$[\,\mathbf{A} \mid \mathbf{b}\,] = \begin{bmatrix} 6 & -6 & 7 & -10 & -18 \\ 2 & -3 & 2 & -4 & -6 \\ 4 & -15 & 1 & -13 & -10 \end{bmatrix}.$$

This augmented matrix represents a linear system with three equations in four unknowns. We use row operations

$$\left(-\tfrac{2}{6}\right) R_1 + R_2, \quad \left(-\tfrac{4}{6}\right) R_1 + R_3, \quad \text{and} \quad \left(-\tfrac{-11}{-1}\right) R_2 + R_3$$

to obtain the equivalent upper triangular linear system

$$[\,\mathbf{U} \mid \mathbf{c}\,] = \begin{bmatrix} 6 & -6 & 7 & -10 & -18 \\ 0 & -1 & -\tfrac{1}{3} & -\tfrac{2}{3} & 0 \\ 0 & 0 & 0 & 1 & 2 \end{bmatrix} \qquad \text{(verify).} \qquad (1)$$

The pivots are in the $(1, 1)$, $(2, 2)$, and $(3, 4)$ entries and none of these entries are zero, so the linear system is consistent since back substitution would not fail. (If there is a row of the form $[\,0 \;\; 0 \;\; 0 \;\; 0 \mid \neq 0\,]$, then there is a zero in a pivot position and back substitution would fail, indicating that the linear system was inconsistent. If that were the case we would stop at this point, since further row operations could not change the situation.) Since the linear system has solutions, we can continue to use row operations to obtain simpler equivalent systems. Because we are using hand calculations, we clear the fractions in row 2 using operation $3R_2$ to obtain

$$\begin{bmatrix} 6 & -6 & 7 & -10 & -18 \\ 0 & -3 & -1 & -2 & 0 \\ 0 & 0 & 0 & 1 & 2 \end{bmatrix}. \qquad (2)$$

Starting with the third pivot, the $(3, 4)$-entry, row operations $2R_3 + R_2$ and $10R_3 + R_1$ zero-out the entries above it. (Verify.) We obtain

$$\begin{bmatrix} 6 & -6 & 7 & 0 & 2 \\ 0 & -3 & -1 & 0 & 4 \\ 0 & 0 & 0 & 1 & 2 \end{bmatrix}.$$

Next we zero-out above the second pivot, the $(2, 2)$-entry, using operation

$$\left(-\tfrac{-6}{-3}\right) R_2 + R_1$$

to obtain

$$\left[\begin{array}{cccc|c} 6 & 0 & 9 & 0 & -6 \\ 0 & -3 & -1 & 0 & 4 \\ 0 & 0 & 0 & 1 & 2 \end{array}\right] \qquad \text{(verify).} \qquad (3)$$

Since there is no pivot in column 3, we cannot eliminate the $(1, 3)$-entry. From (3) we see that this consistent linear system has three (nontrivial) equations in four unknowns, call them $x_1, x_2, x_3, x_4$, so one variable will be free. Thus there will be infinitely many solutions. Applying back substitution to (3) is simpler than using back substitution on (2). From (3) we obtain

$$x_4 = 2, \quad -3x_2 - x_3 = 4, \quad 6x_1 + 9x_3 = -6 \qquad \text{(verify)}$$

which implies that $x_3$ is a free variable. An equivalent form is

$$x_4 = 2, \quad x_2 = \frac{4 + x_3}{-3}, \quad x_1 = \frac{-6 - 9x_3}{6}.$$

Letting $x_3 = t$, an arbitrary constant, we have

$$\mathbf{x} = \left[\begin{array}{c} x_1 \\ x_2 \\ x_3 \\ x_4 \end{array}\right] = \left[\begin{array}{c} \dfrac{-6 - 9x_3}{6} \\ \dfrac{4 + x_3}{-3} \\ x_3 \\ 2 \end{array}\right] = \left[\begin{array}{c} \dfrac{-6 - 9t}{6} \\ \dfrac{4 + t}{-3} \\ t \\ 2 \end{array}\right] = \left[\begin{array}{c} -1 \\ -\frac{4}{3} \\ 0 \\ 2 \end{array}\right] + t \left[\begin{array}{c} -\frac{3}{2} \\ -\frac{1}{3} \\ 1 \\ 0 \end{array}\right] \qquad (4)$$

so the solution set is a translate of a line in $R^4$. While back substitution in (3) is easy, an alternate procedure is to make the pivot entries have numerical value 1 before eliminating the entries above the pivot. We use this variation on (1) next.

$$\left[\mathbf{U} \mid \mathbf{c}\right] = \left[\begin{array}{cccc|c} 6 & -6 & 7 & -10 & -18 \\ 0 & -1 & -\frac{1}{3} & -\frac{2}{3} & 0 \\ 0 & 0 & 0 & 1 & 2 \end{array}\right] \underset{-1R_2}{\overset{(\frac{1}{6})R_1}{\Rightarrow}} \left[\begin{array}{cccc|c} 1 & -1 & \frac{7}{6} & -\frac{5}{3} & -3 \\ 0 & 1 & \frac{1}{3} & \frac{2}{3} & 0 \\ 0 & 0 & 0 & 1 & 2 \end{array}\right].$$

Since the pivots are all 1, it is easy to construct the row operations to zero out the entries above the pivot.

$$\left[\begin{array}{cccc|c} 1 & -1 & \frac{7}{6} & -\frac{5}{3} & -3 \\ 0 & 1 & \frac{1}{3} & \frac{2}{3} & 0 \\ 0 & 0 & 0 & 1 & 2 \end{array}\right] \underset{\substack{(\frac{5}{3})R_3 + R_1}}{\overset{(-\frac{2}{3})R_3 + R_2}{\Rightarrow}} \left[\begin{array}{cccc|c} 1 & -1 & \frac{7}{6} & 0 & \frac{1}{3} \\ 0 & 1 & \frac{1}{3} & 0 & -\frac{4}{3} \\ 0 & 0 & 0 & 1 & 2 \end{array}\right]_{1R_2 + R_1}$$

$$\Rightarrow \left[\begin{array}{cccc|c} 1 & 0 & \frac{3}{2} & 0 & -1 \\ 0 & 1 & \frac{1}{3} & 0 & -\frac{4}{3} \\ 0 & 0 & 0 & 1 & 2 \end{array}\right]. \qquad (5)$$

Performing back substitution on (5) gives the solution set displayed in (4). In (5) the columns containing the pivots are columns of $\mathbf{I}_3$, the $3 \times 3$ identity matrix. ■

There are other variations than those developed in Example 1. However, there are two, so called **standard forms**, that are used frequently. We define these next.

---

**Definition**    A matrix is in **reduced row echelon form (RREF)** if

1. Rows of all zeros, if there are any, appear at the bottom of the matrix.
2. The first nonzero entry of a nonzero row is a 1. This is called a **leading 1**.
3. For each nonzero row, the leading 1 appears to the right and below any leading 1's in preceding rows.
4. Any column in which a leading 1 appears has zeros in every other entry.

---

A matrix in reduced row echelon form appears as a staircase ("echelon") pattern of leading 1's descending from the upper left corner of the matrix. The columns containing the leading 1's are columns of an identity matrix. This is reminiscent of the upper triangular form for linear systems. See Figure 1 for examples of RREF of matrices, square and rectangular.

$$\begin{bmatrix} 1 & 0 & 0 \\ 0 & 1 & 0 \\ 0 & 0 & 1 \end{bmatrix} \quad \begin{bmatrix} 1 & x & 0 \\ 0 & 0 & 1 \\ 0 & 0 & 0 \end{bmatrix} \quad \begin{bmatrix} 1 & 0 & x \\ 0 & 1 & y \\ 0 & 0 & 0 \end{bmatrix} \quad \begin{bmatrix} 1 & x & y & 0 \\ 0 & 0 & 0 & 1 \\ 0 & 0 & 0 & 0 \end{bmatrix}$$

$$\begin{bmatrix} 1 & 0 & x & 0 \\ 0 & 1 & y & 0 \\ 0 & 0 & 0 & 1 \end{bmatrix} \quad \begin{bmatrix} 1 & 0 & x & y \\ 0 & 1 & z & w \\ 0 & 0 & 0 & 0 \\ 0 & 0 & 0 & 0 \end{bmatrix} \quad \begin{bmatrix} 1 & x & 0 & 0 \\ 0 & 0 & 1 & 0 \\ 0 & 0 & 0 & 1 \\ 0 & 0 & 0 & 0 \end{bmatrix} \quad \begin{bmatrix} 1 & 0 & x \\ 0 & 1 & y \\ 0 & 0 & 0 \\ 0 & 0 & 0 \end{bmatrix}$$

$x$, $y$, $z$, or $w$ can be nonzero

FIGURE 1

---

**Definition**    A matrix is in **row echelon form (REF)** if properties 1, 2, and 3 in the definition of reduced row echelon form are satisfied.

---

A matrix is in row echelon form, REF, provided that zero rows come last, the nonzero rows have leading 1's, and the leading 1's form a staircase descending to the right. We do not require that the columns containing the leading 1's have zeros above the leading 1's. See Figure 2 for examples of REF matrices, square and rectangular.

$$\begin{bmatrix} 1 & x & y \\ 0 & 1 & z \\ 0 & 0 & 1 \end{bmatrix} \quad \begin{bmatrix} 1 & x & y \\ 0 & 0 & 1 \\ 0 & 0 & 0 \end{bmatrix} \quad \begin{bmatrix} 1 & z & x \\ 0 & 1 & y \\ 0 & 0 & 0 \end{bmatrix} \quad \begin{bmatrix} 1 & x & y \\ 0 & 1 & z \end{bmatrix}$$

$$\begin{bmatrix} 1 & v & x & y \\ 0 & 1 & z & w \\ 0 & 0 & 0 & 0 \\ 0 & 0 & 0 & 0 \end{bmatrix} \quad \begin{bmatrix} 1 & x & y & z \\ 0 & 0 & 1 & w \\ 0 & 0 & 0 & 1 \\ 0 & 0 & 0 & 0 \end{bmatrix} \quad \begin{bmatrix} 1 & x & y \\ 0 & 0 & 1 \\ 0 & 0 & 0 \\ 0 & 0 & 0 \end{bmatrix}$$

$x$, $y$, $z$, $v$, or $w$ can be nonzero

FIGURE 2

REF is upper triangular form with leading 1's. Such a form is convenient when we use hand calculations to solve linear systems. It follows that any matrix in RREF is also in REF. The RREF is unique, but the REF of a matrix and the upper triangular form are not.

In Section 2.1 we reduced the augmented matrix of a linear system of equations to upper triangular form before applying back substitution. Upper triangular form can be changed to REF by using row operations to convert the pivots to leading 1's. The matrix in (2) is in upper triangular form and the following row operations convert it to REF:

$$
\begin{bmatrix}
6 & -6 & 7 & -10 & | & -18 \\
0 & -3 & -1 & -2 & | & 0 \\
0 & 0 & 0 & 1 & | & 2
\end{bmatrix}
\begin{array}{l}(\frac{1}{6}) R_1 \\ (-\frac{1}{3}) R_2\end{array}
\Rightarrow
\begin{bmatrix}
1 & -1 & \frac{7}{6} & -\frac{5}{3} & | & -3 \\
0 & 1 & \frac{1}{3} & \frac{2}{3} & | & 0 \\
0 & 0 & 0 & 1 & | & 2
\end{bmatrix}.
\qquad (6)
$$

The matrix in (5) is in RREF and, of course, eliminating above the pivots in (6) will yield (5).

**EXAMPLE 2**    Determine which of the following matrices are in upper triangular form, REF, RREF or none of these forms.

$$
\mathbf{A} = \begin{bmatrix} 1 & 2 & 5 \\ 0 & 1 & 3 \\ 0 & 0 & 0 \end{bmatrix}, \quad
\mathbf{B} = \begin{bmatrix} 1 & 0 & 0 & -3 & 4 \\ 0 & 0 & 1 & 2 & -5 \\ 0 & 0 & 0 & 0 & 0 \\ 0 & 0 & 0 & 0 & 0 \end{bmatrix}
$$

$$
\mathbf{C} = \begin{bmatrix} 1 & 3 & 0 & 0 \\ 0 & 0 & 0 & 1 \\ 0 & 1 & 1 & 0 \end{bmatrix}, \quad
\mathbf{D} = \begin{bmatrix} 1 & 0 & 1 & -4 \\ 0 & 2 & 3 & 0 \\ 0 & 0 & 0 & 0 \end{bmatrix}.
$$

We will summarize the results in the following matrix.

|   | Upper Triangular | REF | RREF |
|---|---|---|---|
| **A** | Yes | Yes | No, since there is a nonzero entry above a pivot. |
| **B** | Yes | Yes | Yes |
| **C** | No, we need to interchange rows 2 and 3. | No, because it is not in upper triangular form. | No, because it is not in upper triangular form or REF. |
| **D** | Yes | No, since the second pivot is not 1. | No, since it is not in REF. |

Any of the three forms, upper triangular, REF, or RREF, provide the same information about the solution set of a linear system. In particular, we can determine whether the linear system is consistent or inconsistent and for a consistent system whether there is a unique solution or infinitely many solutions. We illustrate this in Examples 3 and 4.

EXAMPLE 3    Determine which of the following linear systems is consistent, and for the consistent systems determine whether or not there is a unique solution.

(a) $\begin{bmatrix} 1 & 2 & | & 5 \\ 0 & 2 & | & 4 \\ 0 & 0 & | & 0 \end{bmatrix}$ is in upper triangular form.

We see that there are two nonzero equations in two unknowns with a nonzero pivot in each of the nonzero rows. Hence back substitution will be successful, so the system is consistent. Using back substitution, we could solve for both unknowns with no free variables so there is a unique solution. Verify that the unique solution is $x_1 = 1$, $x_2 = 2$.

(b) $\begin{bmatrix} 1 & 3 & 0 & | & 0 \\ 0 & 0 & 1 & | & 0 \\ 0 & 0 & 0 & | & 1 \end{bmatrix}$ is in RREF.

We see that the third row is of the form $\begin{bmatrix} 0 & 0 & 0 & | & \neq 0 \end{bmatrix}$, which indicates that the linear system is inconsistent.

(c) $\begin{bmatrix} 1 & -2 & 0 & -3 & | & 4 \\ 0 & 0 & 1 & 2 & | & -5 \\ 0 & 0 & 0 & 0 & | & 0 \end{bmatrix}$ is in RREF.

We see that there is no row of the form $\begin{bmatrix} 0 & 0 & 0 & 0 & | & \neq 0 \end{bmatrix}$, which indicates that the linear system is consistent. Furthermore, there are two nonzero equations in four unknowns, so back substitution will lead to expressions for the unknowns corresponding to the columns containing the leading 1's in terms of the other unknowns. Hence there will be infinitely many solutions with two degrees of freedom. Verify that $x_3 = -5 - 2x_4$ and $x_1 = 4 + 2x_2 + 3x_4$; then determine the solution set.

(d) $\begin{bmatrix} 1 & 2 & 1 & | & -4 \\ 0 & 1 & 3 & | & 1 \\ 0 & 0 & 0 & | & 0 \end{bmatrix}$ is in REF.

We see that there is no row of the form $\begin{bmatrix} 0 & 0 & 0 & | & \neq 0 \end{bmatrix}$, which indicates that the linear system is consistent. Furthermore, there are two nonzero equations in three unknowns, so back substitution will lead to expressions for the two unknowns corresponding to the columns containing the leading 1's in terms of the other unknown. Hence there will be infinitely many solutions with one degree of freedom. Verify that $x_2 = 1 - 3x_3$ and $x_1 = -4 - x_3 - 2x_2$; then determine the solution set.    ■

In a wide variety of applications we encounter a linear system with a square coefficient matrix. If an equivalent upper triangular form of the augmented matrix has a nonzero pivot in each row or if the REF or RREF has a leading 1 in each row, then back substitution cannot fail. But even more is true; there are no free variables, hence each unknown will have a particular value so the linear system will have a unique solution. (See Examples 3 and 6 in Section 2.1, the two applications at the end of Section 2.1, and the application on cubic splines at the end of this section.) We provide a characterization of this important class of linear systems in observation 3, which follows Example 4. However, rectangular linear systems can also have unique solutions, as illustrated in Example 4.

EXAMPLE 4    The linear system of four equations in three unknowns represented by the augmented matrix

$$\left[\begin{array}{ccc|c} 2 & 2 & 3 & 2 \\ 1 & 2 & 1 & 4 \\ 1 & 0 & 2 & -2 \\ 1 & 1 & 1 & 1 \end{array}\right]$$

has RREF

$$\left[\begin{array}{ccc|c} 1 & 0 & 0 & -2 \\ 0 & 1 & 0 & 3 \\ 0 & 0 & 1 & 0 \\ 0 & 0 & 0 & 0 \end{array}\right]$$

(verify). Thus the unique solution is

$$\mathbf{x} = \left[\begin{array}{c} -2 \\ 3 \\ 0 \end{array}\right].$$

As we show in Section 2.3, this situation corresponds to one of the original equations containing redundant information, hence not necessary for determining the solution set.    ■

The examples in Section 2.1 together with Examples 3 and 4 above lead us to the following set of observations, which are phrased in terms of the RREF but can be applied to REF and upper triangular form with minor modifications.

**Observations**

1. If the RREF of $\left[\begin{array}{c|c} \mathbf{A} & \mathbf{b} \end{array}\right]$ has a row of the form $\left[\begin{array}{cccc|c} 0 & 0 & \cdots & 0 & \neq 0 \end{array}\right]$, then $\mathbf{Ax} = \mathbf{b}$ is *inconsistent* and no solution exists. (Actually, if at any time during manipulations with row operations there is a row of this type, then the linear system is inconsistent so the reduction process can be stopped.)

2. If the RREF of $\left[\begin{array}{c|c} \mathbf{A} & \mathbf{b} \end{array}\right]$ represents a consistent linear system, then there are *infinitely many solutions* if the number of columns in the coefficient matrix (the number of unknowns) is greater than the number of leading 1's (the number of pivots). Otherwise there is a *unique solution* and it appears in the augmented column. (If we are examining a system in REF or upper triangular form, then we will need to apply back substitution to get the unique solution or convert the augmented matrix to RREF.) We also note that

$$\left(\begin{array}{c} \text{the number of} \\ \text{free variables} \end{array}\right) = \left(\begin{array}{c} \text{the number} \\ \text{of unknowns} \end{array}\right) - \left(\begin{array}{c} \text{the number} \\ \text{of pivots} \end{array}\right).$$

3. The following is an important special case. Let $\mathbf{A}$ be an $n \times n$ matrix. If the RREF of $\left[\begin{array}{c|c} \mathbf{A} & \mathbf{b} \end{array}\right]$ is $\left[\begin{array}{c|c} \mathbf{I}_n & \mathbf{c} \end{array}\right]$, where $\mathbf{I}_n$ is the $n \times n$ identity matrix, then the linear system has a unique solution $\mathbf{x} = \mathbf{c}$. In fact, in this case we can say even more: The linear system $\mathbf{Ax} = \mathbf{b}$ has a unique solution for any $n \times 1$ vector $\mathbf{b}$ if and only if $\mathbf{rref}(\mathbf{A}) = \mathbf{I}_n$.

EXAMPLE 5    Each of the following augmented matrices is in RREF and consistent. Determine whether there is a unique solution or infinitely many solutions. For those with infinitely many solutions, determine the number of degrees of freedom.

(a)
$$\begin{bmatrix} 1 & 0 & 2 & 0 & | & 4 \\ 0 & 1 & -2 & 0 & | & -3 \\ 0 & 0 & 0 & 1 & | & 2 \\ 0 & 0 & 0 & 0 & | & 0 \end{bmatrix}.$$

The linear system has four unknowns and three leading 1's; hence there are infinitely many solutions with $4 - 3 = 1$ degree of freedom.

(b)
$$\begin{bmatrix} 1 & -4 & 0 & -1 & 0 & | & -6 \\ 0 & 0 & 1 & 5 & 0 & | & 2 \\ 0 & 0 & 0 & 0 & 1 & | & -3 \\ 0 & 0 & 0 & 0 & 0 & | & 0 \end{bmatrix}.$$

The linear system has five unknowns and three leading 1's; hence there are infinitely many solutions with $5 - 3 = 2$ degrees of freedom.

(c)
$$\begin{bmatrix} 1 & 0 & 0 & | & 7 \\ 0 & 1 & 0 & | & -2 \\ 0 & 0 & 1 & | & 4 \end{bmatrix}.$$

The linear system has three unknowns and three leading 1's; hence there are no degrees of freedom. This implies that there is a unique solution. The solution appears in the augmented column.                                           ■

We have seen that the solution set of a nonhomogeneous linear system with one degree of freedom can be thought of as the translation of a line [see (4) or the material in Section 2.1]. If the system has more than one degree of freedom, then the solution set is the translation of a span of a set of vectors. Using back substitution on Example 5b gives $x_5 = -3$, $x_3 = 2 - 5x_4$, and $x_1 = -6 + x_4 + 4x_2$. Thus the set of solutions to the linear system is given by the following where we set $x_2 = s$ and $x_4 = t$:

$$\mathbf{x} = \begin{bmatrix} x_1 \\ x_2 \\ x_3 \\ x_4 \\ x_5 \end{bmatrix} = \begin{bmatrix} -6 + x_4 + 4x_2 \\ x_2 \\ 2 - 5x_4 \\ x_4 \\ -3 \end{bmatrix} = \begin{bmatrix} -6 + t + 4s \\ s \\ 2 - 5t \\ t \\ -3 \end{bmatrix}$$

$$= \begin{bmatrix} -6 \\ 0 \\ 2 \\ 0 \\ -3 \end{bmatrix} + t \begin{bmatrix} 1 \\ 0 \\ -5 \\ 1 \\ 0 \end{bmatrix} + s \begin{bmatrix} 4 \\ 1 \\ 0 \\ 0 \\ 0 \end{bmatrix} \qquad (7)$$

and we say this is a translation of

$$\text{span} \left\{ \begin{bmatrix} 1 \\ 0 \\ -5 \\ 1 \\ 0 \end{bmatrix}, \begin{bmatrix} 4 \\ 1 \\ 0 \\ 0 \\ 0 \end{bmatrix} \right\} \text{ by vector } \begin{bmatrix} -6 \\ 0 \\ 2 \\ 0 \\ -3 \end{bmatrix}.$$

The set of vectors which are multiplied by the free variables as determined from the RREF has a special property, which we investigate in Section 2.3.

Application:   **Street Networks**

The central business area of many large cities is a network of one-way streets. Any repairs to these thoroughfares, closings for emergency or civic functions, or even traffic accidents disrupt the normal flow of traffic. In addition, many cities do not permit parking on such busy streets during working hours. The daily flow of traffic through these downtown street networks can be studied by local governments to assist with road repairs, city planning, and the planning of emergency evacuation routes. For one-way street networks there is a simple rule: Vehicles entering an intersection from a street must also exit the intersection by another street. So for each intersection we have an **equilibrium equation** or, put simply, an **input-equals-output equation**. Thus after some data collection involving entry and exit volumes at intersections, a city traffic commission can construct network models for traffic flow patterns.

EXAMPLE 6    The city traffic commission has collected data for vehicles entering and exiting the network shown in Figure 3. The direction of the traffic flow between intersections is indicated by the arrows. The intersections are labeled A through D, and the average number of vehicles per hour on portions of the streets is indicated by $x_1$ through $x_5$. The average number of vehicles that enter or exit on a street appears near the street. For instance, at intersection A, 300 vehicles exit and at intersection C a total of $x_3 + x_4 + 200$ vehicles enter. In Table 2.1 we develop an input-equals-output equation for each intersection. The resulting linear system models the traffic flow for the street network shown in Figure 3.

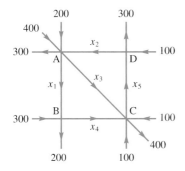

FIGURE 3

TABLE 1

| Intersection | Input | Output | Equation |
|---|---|---|---|
| A | $x_2 + 600$ | $x_1 + x_3 + 300$ | $-x_1 + x_2 - x_3 = -300$ |
| B | $x_1 + 300$ | $x_4 + 200$ | $x_1 - x_4 = -100$ |
| C | $x_3 + x_4 + 200$ | $x_5 + 400$ | $x_3 + x_4 - x_5 = 200$ |
| D | $x_5 + 100$ | $x_2 + 300$ | $-x_2 + x_5 = 200$ |

From this set of equations we construct the augmented matrix and compute the reduced row echelon form to obtain

$$\mathbf{rref}\left(\begin{bmatrix} -1 & 1 & -1 & 0 & 0 & -300 \\ 1 & 0 & 0 & -1 & 0 & -100 \\ 0 & 0 & 1 & 1 & -1 & 200 \\ 0 & -1 & 0 & 0 & 1 & 200 \end{bmatrix}\right)$$

$$= \begin{bmatrix} 1 & 0 & 0 & -1 & 0 & -100 \\ 0 & 1 & 0 & 0 & -1 & -200 \\ 0 & 0 & 1 & 1 & -1 & 200 \\ 0 & 0 & 0 & 0 & 0 & 0 \end{bmatrix}.$$

Solving for the unknowns corresponding to the leading 1's, we find that $x_4$ and $x_5$

are free variables:

$$x_1 = x_4 - 100$$
$$x_2 = x_5 - 200$$
$$x_3 = 200 - x_4 + x_5.$$

Since each of the traffic flows $x_i \geq 0$, we have that $x_1 = x_4 - 100 \geq 0$ so $x_4 \geq 100$ and $x_2 = x_5 - 200 \geq 0$ so $x_5 \geq 200$. If the streets that carry traffic flows $x_4$ and $x_5$ cannot accommodate this level of traffic per hour or had to be closed, then the road network might experience gridlock. This information can be useful for a traffic commission's planning efforts.                                                                    ■

## Application:  Delivery Planning

EXAMPLE 7    The delivery of products to retail stores from multiple warehouses requires a plan to ensure that each store receives the required number of units (cases, barrels, tons, etc.).  Suppose that a certain product is stored in two warehouses containing 30 and 50 units, respectively.  The product must be delivered to three stores whose requirements are 20, 25, and 35 units, respectively.  Let the matrix in Figure 4 represent this situation, where $x_1$ is the number of units shipped from warehouse 1 to store 1, $x_2$ is the number of units shipped from warehouse 2 to store 1, and so on.

| Warehouse ⇒ Store ⇓ | #1 | #2 | Store Requirement |
|---|---|---|---|
| #1 | $x_1$ | $x_2$ | 20 |
| #2 | $x_3$ | $x_4$ | 25 |
| #3 | $x_5$ | $x_6$ | 35 |
| Warehouse Stock ⇒ | 30 | 50 | |

FIGURE 4

We form two types of equations:

the sum of the $x$'s in a row = a store requirement

and

the sum of the $x$'s in a column = warehouse stock.

This gives the set of equations

$$x_1 + x_2 = 20$$
$$x_3 + x_4 = 25$$
$$x_5 + x_6 = 35 \qquad (8)$$
$$x_1 + x_3 + x_5 = 30$$
$$x_2 + x_4 + x_6 = 50.$$

We form the augmented matrix for the linear system in (8) and compute its reduced row echelon form to obtain

$$\textbf{rref}\left(\left[\begin{array}{cccccc|c} 1 & 1 & 0 & 0 & 0 & 0 & 20 \\ 0 & 0 & 1 & 1 & 0 & 0 & 25 \\ 0 & 0 & 0 & 0 & 1 & 1 & 35 \\ 1 & 0 & 1 & 0 & 1 & 0 & 30 \\ 0 & 1 & 0 & 1 & 0 & 1 & 50 \end{array}\right]\right)$$

$$=\left[\begin{array}{cccccc|c} 1 & 0 & 0 & -1 & 0 & -1 & -30 \\ 0 & 1 & 0 & 1 & 0 & 1 & 50 \\ 0 & 0 & 1 & 1 & 0 & 0 & 25 \\ 0 & 0 & 0 & 0 & 1 & 1 & 35 \\ 0 & 0 & 0 & 0 & 0 & 0 & 0 \end{array}\right].$$

Solving for the unknowns corresponding to the leading 1's, we find that $x_4$ and $x_6$ are free variables:

$$x_1 = x_4 + x_6 - 30$$
$$x_2 = -x_4 - x_6 + 50$$
$$x_3 = -x_4 + 25$$
$$x_5 = -x_6 + 35.$$

Since each of the $x_i \geq 0$, we have

$$\begin{aligned} x_1 = &\ x_4 + x_6 - 30 \geq 0 &\Rightarrow&\quad x_4 + x_6 \geq 30 \\ x_2 = &-x_4 - x_6 + 50 \geq 0 &\Rightarrow&\quad x_4 + x_6 \leq 50 \\ x_3 = &\quad -x_4 + 25 \geq 0 &\Rightarrow&\quad x_4 \leq 25 \\ x_5 = &\quad -x_6 + 35 \geq 0 &\Rightarrow&\quad x_6 \leq 35. \end{aligned} \qquad (9)$$

| Warehouse ⇒<br>Store ⇓ | #1 | #2 |
|---|---|---|
| #1 | 10 | 10 |
| #2 | 5 | 20 |
| #3 | 15 | 20 |

FIGURE 5

One delivery plan is obtained by setting $x_4 = 20$ and $x_6 = 20$ in (9). (Verify that all the restrictions are satisfied.) It is convenient to display the plan as a (schedule) matrix as in Figure 5. We see that warehouse 1 will ship 10 units to store 1, 5 units to store 2, and 15 units to store 3, while warehouse number 2 will ship 10 units to store 1, 20 units to store 2, and 20 units to store 3.    ■

Associated with each delivery is a cost that includes salaries, truck expenses, and stocking and storage costs. Since many alternative delivery plans are possible, an obvious business question is which plan minimizes the total delivery costs, provided we know the individual costs of shipping between a warehouse and store. The linear system in (8), together with the constraint of minimizing the shipping cost, requires the use of techniques in linear programming, a topic which is not discussed in this book.

Application:    **Cubic Splines (Calculus Required)**

In Example 4 in Section 1.1 and the application "Quadratic Interpolation" in Section 2.1, we illustrated how to compute a polynomial that was guaranteed to go through all points of a specified data set. If the number of data points is large, sometimes the resulting polynomial has many relative maxima and minima. This is referred to as the "polynomial wiggle" phenomenon.[1]  To avoid this type of behavior another approach is used to construct a function that goes through the data. Instead of

---

[1] A more technical term for the oscillations that occur is the Runge phenomena. A detailed investigation of such behavior is part of a first course in numerical analysis.

constructing a single polynomial expression, a piecewise cubic polynomial, called a **cubic spline**, is developed. This procedure computes a separate cubic polynomial between successive data points so that the cubic pieces join together smoothly. By *smooth* we mean that the cubic pieces to the left and right of a data point both go through the point and have continuous first and second derivatives at the point. Using these requirements, the coefficients of the cubic pieces are determined from a linear system of equations. We illustrate this construction in Example 8.

EXAMPLE 8    For the data set

$$D = \{(1, 2), (2, 5), (3, 4)\}$$

a cubic spline $S(x)$ that goes through each point is a pair of cubic polynomials

$$S(x) = \begin{cases} S_1(x) = a_1 + b_1(x - 1) + c_1(x - 1)^2 + d_1(x - 1)^3 & \text{on } [1, 2] \\ S_2(x) = a_2 + b_2(x - 2) + c_2(x - 2)^2 + d_2(x - 2)^3 & \text{on } [2, 3]. \end{cases}$$

The cubic spline $S(x)$ must go through each point of data set $D$, so we have the requirements

$$S_1(1) = 2, \quad S_1(2) = 5, \quad S_2(2) = 5, \quad S_2(3) = 4.$$

It follows that $S_1(x)$ and $S_2(x)$ both go through point $(2, 5)$ since $S_1(2) = S_2(2)$. To join the cubic pieces smoothly at the points they share, in this case just $(2, 5)$, we must further require that the slopes (the values of the first derivative) and the values of the second derivative agree; that is

$$S_1'(2) = S_2'(2) \quad \text{and} \quad S_1''(2) = S_2''(2).$$

Each of the six requirements corresponds to an equation involving the coefficients of the cubic pieces. The following table summarizes this information.

| Requirement | Corresponding Equation |
|---|---|
| $S_1(1) = 2$ | $a_1 = 2$ |
| $S_1(2) = 5$ | $a_1 + b_1 + c_1 + d_1 = 5 \Rightarrow b_1 + c_1 + d_1 = 3$ |
| $S_2(2) = 5$ | $a_2 = 5$ |
| $S_2(3) = 4$ | $a_2 + b_2 + c_2 + d_2 = 4 \Rightarrow b_2 + c_2 + d_2 = -1$ |
| $S_1'(2) = S_2'(2)$ | $b_1 + 2c_1 + 3d_1 = b_2$ |
| $S_1''(2) = S_2''(2)$ | $2c_1 + 6d_1 = 2c_2$ |

Since $a_1$ and $a_2$ are determined directly, we have four equations in the six unknowns $b_1$, $b_2$, $c_1$, $c_2$, $d_1$, and $d_2$. We expect that there will be two free variables. Forming the augmented matrix and computing the reduced row echelon form, we have

$$\mathbf{rref}\left(\begin{bmatrix} 1 & 0 & 1 & 0 & 1 & 0 & 3 \\ 0 & 1 & 0 & 1 & 0 & 1 & -1 \\ 1 & -1 & 2 & 0 & 3 & 0 & 0 \\ 0 & 0 & 2 & -2 & 6 & 0 & 0 \end{bmatrix}\right)$$

$$= \begin{bmatrix} 1 & 0 & 0 & 0 & -1.5 & -0.5 & 5 \\ 0 & 1 & 0 & 0 & 0.5 & 0.5 & 1 \\ 0 & 0 & 1 & 0 & 2.5 & 0.5 & -2 \\ 0 & 0 & 0 & 1 & -0.5 & 0.5 & -2 \end{bmatrix}$$

which confirms our suspicion about the free variables. Hence there are many cubic splines that go through the data in set D. You can construct your own personal cubic spline for this data set by choosing two of the unknowns $b_1$, $b_2$, $c_1$, $c_2$, $d_1$, and $d_2$.

■

Cubic splines are really a mathematical model for a drafter's spline, which is used to draw "smooth curves" through points in a variety of applications. The drafter's spline is a thin, flexible rod that assumes its natural straight configuration to the left of the first point and to the right of the last point of the data involved. Mathematically we incorporate this straight configuration by requiring that spline $S(x)$ have its second derivative equal to zero at the first and last data points. We formalize this in the following definition.

---

**Definition**    A **natural cubic spline** is a cubic spline that has its second derivative equal to zero at the first and last data points.

---

For our data in D in Example 8 the second derivative requirement at the first and last data point for a natural spline gives us the following additional two equations:

$$S_1''(1) = 0 \quad \Rightarrow \quad 2c_1 = 0$$

$$S_2''(3) = 0 \quad \Rightarrow \quad 2c_2 + 6d_2 = 0.$$

These two equations with the four equations from the preceding table give us six equations in six unknowns. It can be shown that there is a unique solution to the system of equations derived for a natural cubic spline regardless of the number of data points with distinct $x$-values. For our data we get the following information.

$$\mathbf{rref} \left( \left[ \begin{array}{cccccc|c} 1 & 0 & 1 & 0 & 1 & 0 & 3 \\ 0 & 1 & 0 & 1 & 0 & 1 & -1 \\ 1 & -1 & 2 & 0 & 3 & 0 & 0 \\ 0 & 0 & 2 & -2 & 6 & 0 & 0 \\ 0 & 0 & 1 & 0 & 0 & 0 & 0 \\ 0 & 0 & 0 & 2 & 0 & 6 & 0 \end{array} \right] \right)$$

$$= \left[ \begin{array}{cccccc|c} 1 & 0 & 0 & 0 & 0 & 0 & 4 \\ 0 & 1 & 0 & 0 & 0 & 0 & 1 \\ 0 & 0 & 1 & 0 & 0 & 0 & 0 \\ 0 & 0 & 0 & 1 & 0 & 0 & -3 \\ 0 & 0 & 0 & 0 & 1 & 0 & -1 \\ 0 & 0 & 0 & 0 & 0 & 1 & 1 \end{array} \right].$$

Thus the natural cubic spline for the data in set $D$ of Example 8 is

$$S(x) = \begin{cases} S_1(x) = 2 + 4(x-1) - 1(x-1)^3 & \text{on } [1, 2] \\ S_2(x) = 5 + (x-2) - 3(x-2)^2 + (x-2)^3 & \text{on } [2, 3]. \end{cases}$$

For sets with more than three points the linear system required to solve for the coefficients of a natural cubic spline is correspondingly larger than that of Example 8. However, there are more efficient ways to compute the spline coefficients that result in much smaller linear systems. Such alternative approaches are usually discussed in a course in numerical analysis and will not be pursued further here. If there is more information about the data available (for instance, if we knew the slopes required at the first and last data points) then alternatives to the natural spline can be computed by adjoining a different pair of equations to the original system that has two degrees of freedom. Splines are extremely versatile for a wide variety of applications and are used frequently to provide approximations to data.

## EXERCISES 2.2

*In Exercises 1–5, determine whether the given matrix is in REF and/or RREF or neither of these forms. Write your response on the line below the matrix.*

1. $\begin{bmatrix} 1 & -2 & 3 & 5 \\ 0 & 0 & 1 & 4 \\ 0 & 0 & 0 & 1 \end{bmatrix}$.    2. $\begin{bmatrix} 0 & 1 & 0 & 5 \\ 1 & 3 & 2 & -1 \\ 0 & 1 & -2 & 4 \\ 0 & 0 & 0 & 0 \end{bmatrix}$.

_____        _____

3. $\begin{bmatrix} 1 & 6 & 0 \\ 0 & 0 & 1 \\ 0 & 0 & 0 \end{bmatrix}$.    4. $\begin{bmatrix} 2 & 0 & 3 & 2 \\ 0 & 1 & 1 & -5 \\ 0 & 0 & 0 & 0 \end{bmatrix}$.

_____        _____

5. $\begin{bmatrix} 1 & 0 & 3 \\ 0 & 1 & 4 \\ 0 & 0 & 0 \\ 0 & 0 & 0 \end{bmatrix}$.

_____

*In Exercises 6–8, determine a REF for each given matrix. (Note: There are many REFs for a matrix.)*

6. $\begin{bmatrix} 1 & -2 & 4 \\ 2 & -3 & 7 \\ -3 & 6 & -11 \end{bmatrix}$.    7. $\begin{bmatrix} 1 & 3 & 2 & 1 \\ -4 & 12 & -7 & -2 \\ 2 & 6 & 4 & 3 \end{bmatrix}$.

8. $\begin{bmatrix} 2 & 8 & 1 \\ 1 & 4 & 0 \\ -3 & -12 & -2 \end{bmatrix}$.

9. Find the RREF of each given matrix in Exercises 6–8.

10. Let **A** be a $3 \times 3$ matrix without any column of all zeros. The linear system $\mathbf{Ax} = \mathbf{b}$ has augmented matrix $[\, \mathbf{A} \mid \mathbf{b} \,]$ and we will denote $\mathbf{rref}([\, \mathbf{A} \mid \mathbf{b} \,])$ by $[\, \mathbf{W} \mid \mathbf{c} \,]$. There are only four possible forms for $[\, \mathbf{W} \mid \mathbf{c} \,]$, and these are displayed here. ($c_j$, $j = 1, 2, 3$ are the entries of vector **c** and scalars p and q represent possibly nonzero values that result from row reduction steps.)

(P) $\begin{bmatrix} 1 & 0 & 0 & c_1 \\ 0 & 1 & 0 & c_2 \\ 0 & 0 & 1 & c_3 \end{bmatrix}$,    (Q) $\begin{bmatrix} 1 & 0 & p & c_1 \\ 0 & 1 & q & c_2 \\ 0 & 0 & 0 & c_3 \end{bmatrix}$,

(R) $\begin{bmatrix} 1 & p & 0 & c_1 \\ 0 & 0 & 1 & c_2 \\ 0 & 0 & 0 & c_3 \end{bmatrix}$,    (S) $\begin{bmatrix} 1 & p & q & c_1 \\ 0 & 0 & 0 & c_2 \\ 0 & 0 & 0 & c_3 \end{bmatrix}$.

List the letter, P, Q, R, or S, next to a statement if the form has the property described. (More than one or none of the forms may have the property described in a statement.)

(a) _____ Has a unique solution.

(b) _____ Has zero degrees of freedom.

(c) _____ Has 1 degree of freedom, if consistent.

(d) _____ Has 2 degrees of freedom, if consistent.

(e) _____ Has 3 degrees of freedom, if consistent.

(f) _____ Can be inconsistent.

(g) _____ The solution set is a translation of a span of one vector, if consistent.

(h) _____ The solution set is a translation of a span of two vectors, if consistent.

11. Let **A** be a $2 \times 2$ matrix with no column of all zeros. Write out all possible forms for $\mathbf{rref}([\, \mathbf{A} \mid \mathbf{b} \,])$ using the style shown in Exercise 10. (See also Figure 1.)

12. Let **A** be a $3 \times 2$ matrix with no column of all zeros. Write out all possible forms for $\mathbf{rref}([\, \mathbf{A} \mid \mathbf{b} \,])$ using the style shown in Exercise 10.

13. Let **A** be a $3 \times 4$ matrix with no column of all zeros. Write out all possible forms for $\mathbf{rref}([\, \mathbf{A} \mid \mathbf{b} \,])$ using the style shown in Exercise 10. (*Hint*: There are seven possible forms in this case.)

14. Each of the following is a REF or the RREF of a linear system $\mathbf{Ax} = \mathbf{b}$. Identify each linear system as consistent or inconsistent, and for those that are consistent state the number of vectors in its solution set.

Part A.

(a) $\begin{bmatrix} 1 & 0 & 3 & 4 \\ 0 & 1 & -2 & -5 \\ 0 & 0 & 0 & 0 \end{bmatrix}$.

(b) $\begin{bmatrix} 1 & 3 & 0 & 1 \\ 0 & 1 & 0 & -1 \\ 0 & 0 & 1 & 2 \\ 0 & 0 & 0 & 4 \end{bmatrix}$.

(c) $\begin{bmatrix} 1 & 7 & 2 & 4 \\ 0 & 1 & -3 & 5 \\ 0 & 0 & 1 & 0 \end{bmatrix}$.

(d) $\begin{bmatrix} 1 & 0 & 4 & 3 & 0 \\ 0 & 1 & 6 & 5 & 0 \\ 0 & 0 & 0 & 0 & 0 \end{bmatrix}$.

(e) $\begin{bmatrix} 1 & 2 & 8 & 4 & 1 \\ 0 & 0 & 1 & -1 & -1 \\ 0 & 0 & 0 & 0 & 2 \\ 0 & 0 & 0 & 0 & 0 \end{bmatrix}$.

**Part B.** For each of the consistent linear systems in Part A, record the information requested in the following table.

| Number of leading 1's | Number of free variables | Number of columns in the coefficient matrix |
|---|---|---|
| (a) | | |
| (b) | | |
| (c) | | |
| (d) | | |
| (e) | | |

From this limited sample of information, formulate a conjecture concerning the relationship among the quantities appearing in the columns of the table.

*In Exercises 15–17, determine the solution set of each given linear system and write it in the form of a translation of the span of a set of vectors.*

**15.** $x_1 + x_2 + 2x_3 = -1$
$x_1 - 2x_2 + x_3 = -5$
$x_1 + 4x_2 + 3x_3 = 3.$

**16.** $x_1 + x_2 + 3x_3 + 2x_4 = 7$
$2x_1 - x_2 + 4x_4 = 8$
$3x_2 + 6x_4 = 6.$

**17.** $x_1 + x_2 + x_3 + x_4 = 6$
$2x_1 + x_2 - x_3 = 3$
$3x_1 + x_2 + 2x_4 = 6$
$6x_1 + 3x_2 + 3x_4 = 15.$

*In Exercises 18 and 19, find all the values of the scalar a for which each given linear system has (1) no solution, (2) a unique solution, and (3) infinitely many solutions.*

**18.** $x_1 + x_2 = 3$
$x_1 + (a^2 - 8)x_2 = a.$

**19.** $x_1 + x_2 - x_3 = 2$
$x_1 + 2x_2 + x_3 = 3$
$x_1 + x_2 + (a^2 - 5)x_3 = a.$

**20.** Determine all values of scalars $p$ and $q$ so that the linear system is consistent;

$$4x_1 + 6x_2 = p$$
$$-2x_1 - 3x_2 = q.$$

**21.** Determine all values of scalars $p$, $q$, and $r$ so that the linear system is consistent;

$$x_1 + 2x_2 - 3x_3 = p$$
$$2x_1 + 3x_2 + 3x_3 = q$$
$$5x_1 + 9x_2 - 6x_3 = r.$$

**22.** In Table 2.1 the right side of the equations is the sum of the known outputs − the sum of the known inputs. For instance, at intersection A we have $300 - 600 = -300$ and at intersection D we have $300 - 100 = 200$. If we did not have specific values for the external inputs and outputs at each intersection, then the resulting linear system for the network in Example 6 has the form

$$-x_1 + x_2 - x_3 = a$$
$$x_1 - x_4 = b$$
$$x_3 + x_4 - x_5 = c$$
$$- x_2 + x_5 = d.$$

The RREF of the augmented matrix for this linear system is

$$\begin{bmatrix} 1 & 0 & 0 & -1 & 0 & | & b \\ 0 & 1 & 0 & 0 & -1 & | & a+b+c \\ 0 & 0 & 1 & 1 & -1 & | & c \\ 0 & 0 & 0 & 0 & 0 & | & a+b+c+d \end{bmatrix}.$$

(a) What is the meaning of the last row of this RREF?

(b) Determine a set of values for $a$, $b$, $c$, and $d$ so that the linear system modeling the traffic flow in Example 6 has a solution. Then fill in the boxes in Figure 6 with values that will generate your choice of $a$, $b$, $c$, and $d$.

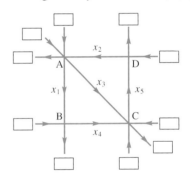

FIGURE 6

**23.** Follow the procedure in Example 6 to determine the traffic flows for the network shown in Figure 7.

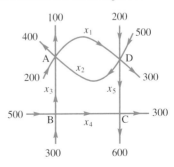

FIGURE 7

**24.** Determine a set of constant or known input and output values for intersections in Figure 7 that render the associated linear system inconsistent. Describe what such a situation could mean to the traffic flow in the network.

**25.** For the delivery planning application of Example 7, use the requirements developed in (9) to find a delivery schedule different from that shown in Figure 5.

**26.** Use the technique developed in Example 7 to find a delivery schedule for the warehouse/store information given in Figure 8.

| Warehouse ⇒<br>Store ⇓ | #1 | #2 | Store<br>Requirement |
|---|---|---|---|
| #1 | $x_1$ | $x_2$ | 50 |
| #2 | $x_3$ | $x_4$ | 30 |
| #3 | $x_5$ | $x_6$ | 20 |
| #4 | $x_7$ | $x_8$ | 40 |
| **Warehouse**<br>**Stock ⇒** | 60 | 80 | |

FIGURE 8

**27.** (a) From Example 7, use the delivery plan displayed in Figure 5. Assume that the cost of delivering one unit of the product from warehouse $i$ to store $j$ is given by the cost matrix displayed in Figure 9. Determine the total cost of meeting the delivery plan of Figure 5.

| Warehouse ⇒<br>Store ⇓ | #1 | #2 |
|---|---|---|
| #1 | $3.00 | $1.50 |
| #2 | $2.50 | $4.50 |
| #3 | $3.50 | $3.00 |

FIGURE 9

(b) Determine the total cost of meeting the delivery plan you constructed for Exercise 25 using the cost matrix in Figure 9.

**28.** Use the technique developed in Example 8 to determine the natural cubic spline for the data in $D = \{(2, 12), (4, 0), (5, 6)\}$.

**29.** Use the technique developed in Example 8 to determine the natural cubic spline for the data in $D = \{(2, 4), (4, 6), (5, 4)\}$.

*REF and RREF apply to linear systems involving complex entries as well as to systems with real entries. In Exercises 30–33, use row operations with complex arithmetic to find the RREF of each given augmented matrix and then determine the solution set.*

**30.** $\begin{bmatrix} 1 - i & 2 + i & | & 2 + 2i \\ 2 & 1 + 2i & | & 1 + 3i \end{bmatrix}.$

**31.** $\begin{bmatrix} i & 2 & 1 - i & | & 1 - 2i \\ 0 & 2i & 2 + i & | & -2 + i \\ 0 & -i & 1 & | & -1 - i \end{bmatrix}.$

**32.** $\begin{bmatrix} 1 - i & 2 + 2i & | & 1 \\ 1 + i & -2 + 2i & | & i \end{bmatrix}.$

**33.** $\begin{bmatrix} 1 - i & 2 + 2i & | & i \\ 1 + i & -2 + 2i & | & -2 \end{bmatrix}.$

## In MATLAB

*In this section we show how to extend the reduction process of a linear system to get equivalent linear systems that are simpler to solve than if we stopped at upper triangular form. The following exercises use MATLAB as a computational tool for the arithmetic involved in the row operations to get to REF or RREF. Exercise ML.2 provides the opportunity to use routines that can give you a sense of the sweeping out process above and below pivots. (This exercise can be skipped if you are proficient determining the RREF.) The reduction to RREF is more than just algebraic. There is also a geometric aspect which we can explore and provides a foundation for material on bases that arises in Chapter 4.*

**ML. 1.** Use **reduce** or **rowop** (see Exercises ML.4 and ML.5 in Section 2.1) to find the RREF of each of the following linear systems $\mathbf{Ax} = \mathbf{b}$; then determine the solution set of the system.

(a) $\mathbf{A} = \begin{bmatrix} 2 & 3 & 4 \\ -1 & 2 & 0 \\ 5 & 4 & 8 \end{bmatrix}, \mathbf{b} = \begin{bmatrix} 1 \\ -1 \\ 3 \end{bmatrix}.$

(b) $\mathbf{A} = \begin{bmatrix} 2 & 4 & 5 & 6 \\ -1 & 0 & 2 & 1 \\ 0 & 2 & 2 & 1 \\ 3 & 1 & 3 & 3 \end{bmatrix}, \mathbf{b} = \begin{bmatrix} 8 \\ 0 \\ 1 \\ 6 \end{bmatrix}.$

(c) $\mathbf{A} = \begin{bmatrix} 4 & 1 & -2 \\ 5 & 2 & 0 \\ 1 & 1 & 3 \\ 2 & 1 & -4 \end{bmatrix}, \mathbf{b} = \begin{bmatrix} 12 \\ 9 \\ -6 \\ 16 \end{bmatrix}.$

(d) $\mathbf{A} = \begin{bmatrix} -2 & 1 & 4 & 0 \\ 6 & -3 & -12 & 0 \\ 8 & 3 & 0 & -4 \\ 3 & 2 & 2 & -2 \end{bmatrix}, \mathbf{b} = \begin{bmatrix} -7 \\ 21 \\ 47 \\ 20 \end{bmatrix}.$

**ML. 2.** For the linear system

$$3x_1 + 2x_2 + 4x_3 - x_4 = 20$$
$$2x_1 \qquad + x_3 - 3x_4 = 19$$
$$x_1 + x_2 \qquad + 2x_4 = 0$$
$$5x_1 + 2x_2 + x_3 + x_4 = 24.$$

enter the coefficient matrix into MATLAB and name it **A**. Enter the right side and name it **b**. Form the augmented matrix and name it **C** using the command **C = [A b]**. (There is a space between the **A** and **b** in the command.)

(a) To get a sense of the reduction of this linear system to RREF, use the command **rrefquik(C)**. The process appears to "sweep out" the lower triangular entries and then the upper triangular entries. Rerun this example (as needed) and note the order in which quantities are swept out.

(b) To see the "sweep-out" process with an element-by-element dialog, enter the command **rrefview(C)**.

(c) To see the specific row operations used to obtain RREF, use the command **rrefstep(C)**.

Write a description of the path swept out in the reduction process in each of the preceding parts.
*Note*: If you need to practice the reduction of a matrix to RREF, use routine **rrefstep** as follows. Before each row operation is to be performed, write out the row operation that you think should be used. Check your work by selecting the option to 'Explain the Step'.

**ML. 3.** In some situations we want to compute the RREF of a matrix quickly, without seeing the progressive effect of the reduction as displayed in **rrefquik**, **rrefview**, or **rrefstep**. The MATLAB command **rref** does just that. Use **help rref** for directions before using **rref** to check your results for Exercises 6–8 and 30–33.

**ML. 4.** The MATLAB routines for row operations, **reduce** and **rowop**, and those for RREF in this section, focus on the algebraic manipulations to obtain equivalent linear systems. Next we provide a geometric setting for elementary row operations using routine **vizrowop**. This routine displays two linear equations in two unknowns as a pair of lines. The algebraic action of the row operations preserves the solution set but can change geometrically the pair of lines involved. We illustrate the routine as follows. To see the "geometric" elimination of the linear system

$$4x_1 + 7x_2 = 36$$
$$3x_1 + \ x_2 = 10$$

enter the command **vizrowop** and follow the directions that appear on the screen. Once you have entered the augmented matrix and pressed Enter, you will see a palette of buttons, and the augmented matrix is displayed as well as the graph of the pair of lines corresponding to the equations of the linear system. Note the color coordination between the rows of the augmented matrix and the lines in the graphics display.

(a) Make the appropriate selections to perform row operation $\left(\frac{-3}{4}\right)\text{Row}(1) + \text{Row}(2)$. You will see that the second row of the augmented matrix will appear as

$$0 \ -4.25 \ -17$$

Use this algebraic change to explain the geometric change that occurs.

(b) Next click on the 'Rational' option. Now the second row of the augmented matrix is shown as

$$0 \ -17/4 \ -17$$

Make the appropriate selections to perform the row operation $\left(-\frac{4}{17}\right)\text{Row}(2)$. Explain the relationship between the algebraic and geometric changes that occur.

(c) Next perform the row operation $-7\text{Row}(2) + \text{Row}(1)$. Explain the relationship between the algebraic and geometric changes that occur.

(d) Finally, perform row operation $\left(\frac{1}{4}\right)\text{Row}(1)$. Explain the relationship between the algebraic and geometric changes that occur.

(e) The linear system here has a unique solution. For such systems of two equations in two unknowns, form a conjecture for the graphical display of its RREF.

**ML. 5.** Use **vizrowop** to obtain the RREF of each of the following linear systems. Explain how the geometry of the lines displayed predicts the form of the RREF.

(a)   $4x_1 - \ x_2 = \ 3$
    $-8x_1 + 2x_2 = -6.$

(b)  $-2x_1 + 3x_2 = 5$
    $4x_1 - 6x_2 = 1.$

**ML. 6.** The routine **vizrowop** dealt with lines in 2-space. However, the ideas generalize to sets of equations in 3-space and $n$-space. Visualizing solution sets in $n$-space, $n > 2$, is not easy. For practice with these ideas in 3-space, execute the routine **vizplane**. There are five demos available. You will see the graphs of three planes one at a time and then the picture with the three planes on the same set of axes. For each demo record the equations of the three planes and find the solution set of this linear system by hand.

After the picture of the three planes on the same set of axes there appears the message 'Press Enter'. Pressing Enter will show the three planes in a larger window. You can change the view of the picture by clicking on the 'Next View' button, which behaves as though you were moving around the $z$-axis. As you change the view, look for the solution set you determined algebraically. Record the number of the view that, in your opinion, best shows the solution set.

## True/False Review Questions

*Determine whether each of the following statements is true or false.*

1. $\begin{bmatrix} 1 & 3 & 0 & 0 & | & 5 \\ 0 & 0 & 1 & 2 & | & 4 \\ 0 & 0 & 0 & 0 & | & 0 \end{bmatrix}$ is in reduced row echelon form.

2. There can be many different reduced row echelon forms for a linear system $\mathbf{Ax} = \mathbf{b}$.

3. It usually requires more row operations to find the RREF than to find the REF.

4. If we use row operations to reduce $\begin{bmatrix} \mathbf{A} & | & \mathbf{b} \end{bmatrix}$ to upper triangular form, the resulting augmented matrix is the REF of the system.

5. If the REF of $\begin{bmatrix} \mathbf{A} & | & \mathbf{b} \end{bmatrix}$ has a zero row, then the system has infinitely many solutions.

6. If $\mathbf{A}$ is $3 \times 3$ and the RREF of $\begin{bmatrix} \mathbf{A} & | & \mathbf{b} \end{bmatrix}$ is $\begin{bmatrix} \mathbf{I}_3 & | & \mathbf{c} \end{bmatrix}$, then the only solution of $\mathbf{Ax} = \mathbf{b}$ is $\mathbf{x} = \mathbf{c}$.

7. You can always convert a REF for a system to a RREF.

8. If $\mathbf{A}$ is $3 \times 5$ and $\mathbf{b}$ is $3 \times 1$ with two leading 1's in the RREF of $\begin{bmatrix} \mathbf{A} & | & \mathbf{b} \end{bmatrix}$, then such a consistent system has 3 degrees of freedom in its solution set.

9. A linear system whose RREF of its augmented matrix is $\begin{bmatrix} 1 & 5 & 0 & | & 3 \\ 0 & 0 & 1 & | & -4 \\ 0 & 0 & 0 & | & 0 \end{bmatrix}$ has a solution set which is a translation of the span of one vector.

10. If $\mathbf{A}$ is $3 \times 3$ and $\mathbf{b} = \begin{bmatrix} r \\ s \\ t \end{bmatrix}$ with RREF of $\begin{bmatrix} \mathbf{A} & | & \mathbf{b} \end{bmatrix}$ given by $\begin{bmatrix} 1 & -2 & 0 & | & r+s \\ 0 & 0 & 1 & | & s+2t \\ 0 & 0 & 0 & | & r-s+t \end{bmatrix}$, then the system is consistent if and only if $r - s + t = 0$.

## Terminology

| Standard forms; REF & RREF | Leading 1 |
|---|---|

There are very few new pieces of terminology introduced in this section. All of the items in the preceding table have been used in one form or another in preceding sections. Here we organized previous reduction processes into a collection of steps that produce an equivalent system with certain common features. The so-called standard forms reveal detailed information about the solution set of the linear system whose augmented matrix was input to the reduction process. We will continue to use these standard forms in succeeding sections to elicit structural information about the solution set of a linear system and the coefficient matrix of the system. Hence a firm understanding of the terms listed in the preceding table is crucial. To review these concepts in detail, respond to the following.

- Carefully describe the REF of $\begin{bmatrix} \mathbf{A} & | & \mathbf{b} \end{bmatrix}$.
- Carefully describe the RREF of $\begin{bmatrix} \mathbf{A} & | & \mathbf{b} \end{bmatrix}$ and explain how it differs from the REF.
- When computing either REF or RREF, how do we recognize that a system is inconsistent?
- Suppose we have computed the RREF of $\begin{bmatrix} \mathbf{A} & | & \mathbf{b} \end{bmatrix}$. How do we recognize that the system has a unique solution?

- Suppose we have computed the RREF of $\begin{bmatrix} \mathbf{A} \mid \mathbf{b} \end{bmatrix}$. How do we recognize that the system has infinitely many solutions?
- How do we compute the number of pivots from the RREF?
- How do we compute the number of free variables from the RREF?
- Suppose that we have a square system with the RREF of $\begin{bmatrix} \mathbf{A} \mid \mathbf{b} \end{bmatrix}$ equal to $\begin{bmatrix} \mathbf{I}_n \mid \mathbf{c} \end{bmatrix}$. Without any further work, how do we determine the solution set of the system? Does the system have a unique solution?

Construct the following, if possible.

- An example of a $3 \times 3$ linear system that has an "empty" solution set.
- An example of a $3 \times 3$ linear system that has exactly one vector in its solution set.
- An example of a $3 \times 3$ linear system that has exactly two vectors in its solution set.
- An example of a $3 \times 3$ linear system that has infinitely many vectors in its solution set.

## 2.3 ■ PROPERTIES OF SOLUTION SETS

In Section 2.2 we developed the reduced row echelon form (RREF) and the row echelon form (REF) of a matrix. These so-called standard forms of an augmented matrix $\begin{bmatrix} \mathbf{A} \mid \mathbf{b} \end{bmatrix}$ are used to determine the nature of the set of solutions to the linear system $\mathbf{Ax} = \mathbf{b}$. If the system is consistent, then the pattern of the entries of either the RREF or the REF enables us to determine whether there is a unique solution or infinitely many solutions. To write out the solution set often requires further algebra in the form of back substitution.

In this section we use the RREF to develop additional properties of the solution set of a linear system. The case of no solutions, an inconsistent system, requires nothing further. The case of a unique solution will provide information about the rows and columns of the original coefficient matrix. However, *our primary focus will be on the nature of the solution set of linear systems that have at least one free variable.* For such systems we develop a geometric view of the solution set. In addition, we obtain information about relationships among the rows of the coefficient matrix as well as relationships among the columns. For these results keep in mind that row operations produce equivalent linear systems. That is, row operations do not change the solution set. Hence the information obtained from the RREF (or REF) applies to the original linear system and, of course, to its coefficient matrix.

### Solution Set Structure

In Example 5(b) of Section 2.2 we showed that the linear system $\mathbf{Ax} = \mathbf{b}$ which has augmented matrix

$$\begin{bmatrix} 1 & -4 & 0 & -1 & 0 & \mid & -6 \\ 0 & 0 & 1 & 5 & 0 & \mid & 2 \\ 0 & 0 & 0 & 0 & 1 & \mid & -3 \\ 0 & 0 & 0 & 0 & 0 & \mid & 0 \end{bmatrix}$$

has a solution set consisting of all vectors in $R^5$ of the form

$$\mathbf{x} = \begin{bmatrix} x_1 \\ x_2 \\ x_3 \\ x_4 \\ x_5 \end{bmatrix} = \begin{bmatrix} -6 \\ 0 \\ 2 \\ 0 \\ -3 \end{bmatrix} + t \begin{bmatrix} 1 \\ 0 \\ -5 \\ 1 \\ 0 \end{bmatrix} + s \begin{bmatrix} 4 \\ 1 \\ 0 \\ 0 \\ 0 \end{bmatrix}. \tag{1}$$

Geometrically the solution set is a translation of

$$\text{span} \left\{ \begin{bmatrix} 1 \\ 0 \\ -5 \\ 1 \\ 0 \end{bmatrix}, \begin{bmatrix} 4 \\ 1 \\ 0 \\ 0 \\ 0 \end{bmatrix} \right\}$$

by vector $\begin{bmatrix} -6 \\ 0 \\ 2 \\ 0 \\ -3 \end{bmatrix}$. We will develop a corresponding algebraic way to view such a solution set.

The vectors associated with the free variables in (1), namely

$$\begin{bmatrix} 1 \\ 0 \\ -5 \\ 1 \\ 0 \end{bmatrix} \quad \text{and} \quad \begin{bmatrix} 4 \\ 1 \\ 0 \\ 0 \\ 0 \end{bmatrix},$$

have a particular property that is important with regard to the set of all solutions of the associated homogeneous linear system $\mathbf{Ax} = \mathbf{0}$. We first investigate this property and then introduce concepts that provide a characterization, or a way of considering, the set of all solutions to $\mathbf{Ax} = \mathbf{0}$. We will show that the set of all solutions to $\mathbf{Ax} = \mathbf{0}$ has a structure (a pattern of behavior) that is easy to understand and yields insight into the nature of the solution set of homogeneous linear systems. Ultimately this will provide a foundation for describing algebraically the solution set of the nonhomogeneous linear system $\mathbf{Ax} = \mathbf{b}$.

The ideas we develop in this section will be encountered again in Chapter 4 and are applied to a more general setting in Chapter 5. **It is important that you master this material since it forms a core of ideas for further developments in linear algebra.**

---

**Definition**    A set of vectors is called **linearly independent** provided the *only* way that a linear combination of the vectors produces the zero vector is when all the coefficients are zero. If there is some set of coefficients, not all zero, for which a linear combination of the vectors is zero, then the set of vectors is called **linearly dependent**. That is, $\{\mathbf{v}_1, \mathbf{v}_2, \ldots, \mathbf{v}_k\}$ is linearly dependent if we can find constants $c_1, c_2, \ldots, c_k$ *not all zero* such that

$$c_1\mathbf{v}_1 + c_2\mathbf{v}_2 + \cdots + c_k\mathbf{v}_k = \mathbf{0}.$$

---

EXAMPLE 1    Show that the set

$$S = \{\mathbf{v}_1, \mathbf{v}_2\} = \left\{ \begin{bmatrix} 1 \\ 0 \\ -5 \\ 1 \\ 0 \end{bmatrix}, \begin{bmatrix} 4 \\ 1 \\ 0 \\ 0 \\ 0 \end{bmatrix} \right\}$$

is linearly independent.

We form a linear combination of $\mathbf{v}_1$ and $\mathbf{v}_2$ with unknown coefficients $c_1$ and $c_2$, set it equal to the zero vector, and then proceed to show that the only time this is true is when both coefficients are zero.

$$c_1\mathbf{v}_1 + c_2\mathbf{v}_2 = \mathbf{0} \Leftrightarrow c_1 \begin{bmatrix} 1 \\ 0 \\ -5 \\ 1 \\ 0 \end{bmatrix} + c_2 \begin{bmatrix} 4 \\ 1 \\ 0 \\ 0 \\ 0 \end{bmatrix} = \begin{bmatrix} 0 \\ 0 \\ 0 \\ 0 \\ 0 \end{bmatrix} \Leftrightarrow \begin{bmatrix} c_1 + 4c_2 \\ c_2 \\ -5c_1 \\ c_1 \\ 0 \end{bmatrix} = \begin{bmatrix} 0 \\ 0 \\ 0 \\ 0 \\ 0 \end{bmatrix}.$$

Equating corresponding entries shows that $c_1 = c_2 = 0$, so $\mathbf{v}_1$ and $\mathbf{v}_2$ are linearly independent.    ■

In Example 1 the expression

$$c_1 \begin{bmatrix} 1 \\ 0 \\ -5 \\ 1 \\ 0 \end{bmatrix} + c_2 \begin{bmatrix} 4 \\ 1 \\ 0 \\ 0 \\ 0 \end{bmatrix}$$

is the same as

$$t \begin{bmatrix} 1 \\ 0 \\ -5 \\ 1 \\ 0 \end{bmatrix} + s \begin{bmatrix} 4 \\ 1 \\ 0 \\ 0 \\ 0 \end{bmatrix}$$

from (7) in Section 2.2 except that the coefficients have different names. Since $t$ and $s$ are arbitrary constants for different free variables, it follows that in the linear combination there will be one entry that contains $t$ (or $c_1$) alone and another entry that contains $s$ (or $c_2$) alone. Equating the linear combination to the zero vector then must give $t = s = 0$ (or $c_1 = c_2 = 0$). It follows *that any consistent linear system with infinitely many solutions can be expressed as a translation of the span of a set of vectors S.* **When we obtain the vectors in $S$ from the RREF (or the REF) they will be a linearly independent set** [1] This result is awkward to prove in a general case, so we do not pursue a formal proof here. However, we will use this result for $m \times n$ systems of linear equations when the need arises. Hence we summarize this as follows:

> The solution set of a linear system with infinitely many solutions is the translation of the span of a set $S$ of linearly independent vectors. The number of linearly independent vectors in $S$ is the same as the number of free variables in the solution set of the linear system.

---

[1] Note that we have adopted the procedure that solves for the unknowns that correspond to the pivots. Hence the free variables correspond to the unknowns associated with columns not containing a leading 1. If another procedure is used, then the set $S$ may change.

If the linear system is homogeneous and has infinitely many solutions, then the solution set is just the span of a set $S$ of linearly independent vectors, since the augmented column consists entirely of zeros and remains unchanged by the row operations that are used. Example 2 illustrates such a case.

EXAMPLE 2    Determine the solution set of the homogeneous linear system $\mathbf{Ax} = \mathbf{0}$ when the RREF of $\begin{bmatrix} \mathbf{A} & | & \mathbf{0} \end{bmatrix}$ is

$$\left[\begin{array}{ccccc|c} 1 & 0 & 2 & 4 & 0 & 0 \\ 0 & 1 & -3 & 1 & 0 & 0 \\ 0 & 0 & 0 & 0 & 1 & 0 \\ 0 & 0 & 0 & 0 & 0 & 0 \end{array}\right].$$

Using the unknowns $x_1, x_2, \ldots, x_5$ we get from back substitution

$$x_5 = 0, \quad x_2 = 3x_3 - x_4, \quad x_1 = -2x_3 - 4x_4. \quad \text{(Verify.)}$$

Thus we have

$$\mathbf{x} = \begin{bmatrix} -2x_3 - 4x_4 \\ 3x_3 - x_4 \\ x_3 \\ x_4 \\ 0 \end{bmatrix} = \begin{bmatrix} -2s - 4t \\ 3s - t \\ s \\ t \\ 0 \end{bmatrix} = s \begin{bmatrix} -2 \\ 3 \\ 1 \\ 0 \\ 0 \end{bmatrix} + t \begin{bmatrix} -4 \\ -1 \\ 0 \\ 1 \\ 0 \end{bmatrix}$$

and the solution set is

$$\text{span}\{S\} = \text{span} \left\{ \begin{bmatrix} -2 \\ 3 \\ 1 \\ 0 \\ 0 \end{bmatrix}, \begin{bmatrix} -4 \\ -1 \\ 0 \\ 1 \\ 0 \end{bmatrix} \right\}.$$

If we had the same coefficient matrix $\mathbf{A}$, but a nonhomogeneous linear system for which the RREF is

$$\left[\begin{array}{ccccc|c} 1 & 0 & 2 & 4 & 0 & 5 \\ 0 & 1 & -3 & 1 & 0 & 2 \\ 0 & 0 & 0 & 0 & 1 & -3 \\ 0 & 0 & 0 & 0 & 0 & 0 \end{array}\right]$$

then the solution set is (verify)

$$\mathbf{x} = \begin{bmatrix} 5 - 2x_3 - 4x_4 \\ 2 + 3x_3 - x_4 \\ x_3 \\ x_4 \\ -3 \end{bmatrix} = \begin{bmatrix} 5 - 2s - 4t \\ 2 + 3s - t \\ s \\ t \\ -3 \end{bmatrix} = \begin{bmatrix} 5 \\ 2 \\ 0 \\ 0 \\ -3 \end{bmatrix} + s \begin{bmatrix} -2 \\ 3 \\ 1 \\ 0 \\ 0 \end{bmatrix} + t \begin{bmatrix} -4 \\ -1 \\ 0 \\ 1 \\ 0 \end{bmatrix}.$$

Since $s$ and $t$ could be chosen to be 0, it follows that vector

$$\begin{bmatrix} 5 & 2 & 0 & 0 & -3 \end{bmatrix}^T$$

is a solution of the nonhomogeneous linear system $\mathbf{Ax} = \mathbf{b}$. We have seen that all the vectors in span$\{S\}$ are solutions of $\mathbf{Ax} = \mathbf{0}$. Such a split of the solution set is valid in general.    ■

Example 2 leads us to the following general principle which reveals the pattern for solution sets of consistent linear systems.

---

If the linear system $\mathbf{Ax} = \mathbf{b}$ has infinitely many solutions, then the solution set is the translation of the span of a set $S$ of linearly independent vectors. We express this as $\mathbf{x} = \mathbf{x}_p + \text{span}\{S\}$, where $\mathbf{x}_p$ is a **particular solution** of the nonhomogeneous system $\mathbf{Ax} = \mathbf{b}$ and the vectors in $\text{span}\{S\}$ form the solution set of the associated homogeneous linear system $\mathbf{Ax} = \mathbf{0}$. If we let $\mathbf{x}_h$ represent an arbitrary vector in $\text{span}\{S\}$, then the set of solutions is given by

$$\mathbf{x} = \mathbf{x}_p + \mathbf{x}_h.$$

---

We use the term *particular solution* for $\mathbf{x}_p$ since its entries are completely determined from the algorithm presented here; that is, the entries do not depend on any of the arbitrary constants. It is important to keep in mind that $\mathbf{x}_h$ represents one vector in the span of a set of vectors. Note that $\mathbf{A}(\mathbf{x}_p + \mathbf{x}_h) = \mathbf{Ax}_p + \mathbf{Ax}_h = \mathbf{b} + \mathbf{0} = \mathbf{b}$.

The solution set of a homogeneous linear system $\mathbf{Ax} = \mathbf{0}$ arises so frequently that we refer to it as the **solution space** or as the **null space**[2] **of matrix A**. In addition, we use the notation $\mathbf{ns}(\mathbf{A})$ to denote this special set of vectors. Since every homogeneous linear system is consistent, $\mathbf{ns}(\mathbf{A})$ consists either of a single vector or infinitely many vectors. If $\mathbf{Ax} = \mathbf{0}$ has a unique solution, then $\mathbf{ns}(\mathbf{A}) = \{\mathbf{0}\}$; that is, the only solution is the zero vector. As we have shown, if $\mathbf{Ax} = \mathbf{0}$ has infinitely many solutions, then $\mathbf{ns}(\mathbf{A}) = \text{span}\{S\}$, where $S$ is the set of vectors obtained from the information contained in an equivalent linear system in upper triangular form, REF, or RREF. In addition, it was argued that these vectors are linearly independent by virtue of choosing the arbitrary constants for the free variables.

We can illustrate the general solution of a linear system $\mathbf{Ax} = \mathbf{b}$ along with $\mathbf{ns}(\mathbf{A})$ in 3-space. In the case that $\mathbf{ns}(\mathbf{A})$ is the span of a pair of linearly independent vectors, this set can be represented by a plane through the origin. Then the general solution is the set of vectors $\mathbf{x} = \mathbf{x}_p + \mathbf{x}_h = \mathbf{x}_p + \mathbf{ns}(\mathbf{A})$, which is a translation of the plane representing $\mathbf{ns}(\mathbf{A})$. Example 3 depicts this situation. While we cannot draw the corresponding situation for higher dimensions, we can use the 3-space illustration as a model for more general situations.

EXAMPLE 3   Determine the general solution to $\mathbf{Ax} = \mathbf{b}$ where

$$\mathbf{A} = \begin{bmatrix} 3 & -6 & 12 \\ 1 & -2 & 4 \\ -2 & 4 & -8 \end{bmatrix} \quad \text{and} \quad \mathbf{b} = \begin{bmatrix} 9 \\ 3 \\ -6 \end{bmatrix}$$

and represent it as a translation of $\mathbf{ns}(\mathbf{A})$. Computing $\mathbf{rref}([\,\mathbf{A}\,|\,\mathbf{b}\,])$ we get

$$\begin{bmatrix} 1 & -2 & 4 & | & 3 \\ 0 & 0 & 0 & | & 0 \\ 0 & 0 & 0 & | & 0 \end{bmatrix}.$$

Hence it follows that

$$\mathbf{x} = \begin{bmatrix} 3 + 2x_2 - 4x_3 \\ x_2 \\ x_3 \end{bmatrix}.$$

---

[2]The term *null space* refers to the set of vectors $\mathbf{x}$ such that multiplication by $\mathbf{A}$ produces the null or zero vector.

If we let $x_2 = s$ and $x_3 = t$ we have the general solution

$$\mathbf{x} = \mathbf{x}_p + \mathbf{x}_h = \begin{bmatrix} 3 \\ 0 \\ 0 \end{bmatrix} + s \begin{bmatrix} 1 \\ 1 \\ 0 \end{bmatrix} + t \begin{bmatrix} -4 \\ 0 \\ 1 \end{bmatrix}.$$

Thus

$$\mathbf{ns}(\mathbf{A}) = \text{span} \left\{ \begin{bmatrix} 1 \\ 1 \\ 0 \end{bmatrix}, \begin{bmatrix} -4 \\ 0 \\ 1 \end{bmatrix} \right\}.$$

Figure 1 shows the general solution as the set of vectors obtained by translating the null space of $\mathbf{A}$ by the vector $\begin{bmatrix} 3 \\ 0 \\ 0 \end{bmatrix}$.    ■

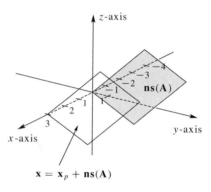

$$\mathbf{x} = \mathbf{x}_p + \mathbf{ns}(\mathbf{A})$$

FIGURE 1

## Additional Structural Properties

Next we develop ideas that will assist us in determining the nature of the set $\mathbf{ns}(\mathbf{A})$ and other sets of vectors. The span of a set $T$, span$\{T\}$, is the set that consists of all possible linear combinations of the members of $T$. (See Section 1.2.) We say span$\{T\}$ is **generated** by the members of $T$. In Section 1.2, we said that a set is **closed** provided every linear combination of its members also belongs to the set. It follows that span$\{T\}$ is a closed set since a linear combination of the members of span$\{T\}$ is just another linear combination of the members of $T$. (See Exercise 53.) Thus $\mathbf{ns}(\mathbf{A}) = $ span$\{S\}$ is closed.

---

**Definition**    A subset of $R^n$ ($C^n$) that is closed is called a **subspace** of $R^n$ ($C^n$).

---

Observe that if $\mathbf{A}$ is an $m \times n$ real matrix, then any solution of $\mathbf{Ax} = \mathbf{0}$ is an $n$-vector in $R^n$. It then follows that $\mathbf{ns}(\mathbf{A})$ is a subspace of $R^n$. For the linear system $\mathbf{Ax} = \mathbf{0}$, this means that any linear combination of solutions is another solution. But the nature of $\mathbf{ns}(\mathbf{A})$ is even simpler because of the way we have obtained the set $S$, which generates $\mathbf{ns}(\mathbf{A})$; recall that $\mathbf{ns}(\mathbf{A}) = $ span$\{S\}$. To develop the necessary ideas, we next investigate the nature of linearly independent and linearly dependent sets of spanning vectors. We begin with a set $T$ of two distinct vectors in $R^n$: $T = \{\mathbf{v}_1, \mathbf{v}_2\}$. We have

1. If $T$ is linearly dependent, then there is a pair of scalars $c$ and $d$, not both zero, so that $c\mathbf{v}_1 + d\mathbf{v}_2 = \mathbf{0}$. If $c \neq 0$, we then have

$$\mathbf{v}_1 = \left(-\frac{d}{c}\right)\mathbf{v}_2.$$

   This says that **if** $T = \{\mathbf{v}_1, \mathbf{v}_2\}$ **is linearly dependent, then one vector is a scalar multiple of the other**.

2. If $T$ is such that one vector is a scalar multiple of the other, say $\mathbf{v}_2 = k\mathbf{v}_1$, then $k\mathbf{v}_1 + (-1)\mathbf{v}_2 = \mathbf{0}$. This is a linear combination with a nonzero coefficient, $(-1)$, which produces the zero vector. Hence $T$ is a linearly dependent set. Thus we have that **if** $T = \{\mathbf{v}_1, \mathbf{v}_2\}$ **is such that one vector is a scalar multiple of the other then,** $T$ **is a linearly dependent set**.

Combining 1 and 2, we can say that $T = \{\mathbf{v}_1, \mathbf{v}_2\}$ **is a linearly dependent set if and only if one vector is a scalar multiple of the other**. This is equivalent to the statement that $T = \{\mathbf{v}_1, \mathbf{v}_2\}$ **is a linearly independent set if and only if neither vector is a scalar multiple of the other**.

3. If $T$ is linearly independent, then neither $\mathbf{v}_1$ nor $\mathbf{v}_2$ can be the zero vector because if $\mathbf{v}_1 = \mathbf{0}$, then $k\mathbf{v}_1 + 0\mathbf{v}_2 = \mathbf{0}$ for any scalar $k \neq 0$, which contradicts the assumption that $T$ is linearly independent. It follows that **any set containing the zero vector is linearly dependent**.

4. Span$\{T\}$ is the set of all linear combinations $c\mathbf{v}_1 + d\mathbf{v}_2$. If $T$ is linearly dependent, then we also know that one of $\mathbf{v}_1$ or $\mathbf{v}_2$ is a scalar multiple of the other, say $\mathbf{v}_2 = k\mathbf{v}_1$. Thus

$$c\mathbf{v}_1 + d\mathbf{v}_2 = c\mathbf{v}_1 + d(k\mathbf{v}_1) = (c + dk)\mathbf{v}_1.$$

   Hence span$\{T\}$ is just the set of all scalar multiples of the single vector $\mathbf{v}_1$. It follows that **if** $T$ **is a linearly dependent set of vectors, then span$\{T\}$ can be generated by fewer vectors from** $T$. It also follows that **if** $T$ **is a linearly independent set of vectors, then span$\{T\}$ cannot be generated by fewer vectors from** $T$.

For a set $T$ with more than two vectors, $T = \{\mathbf{v}_1, \mathbf{v}_2, \mathbf{v}_3, \ldots, \mathbf{v}_r\}$, essentially the same properties hold. We need only revise the combined statement of properties 1 and 2 as follows (see Exercise 52):

   $T$ is linearly dependent if and only if (at least) one of the vectors in $T$ is a linear combination of the other vectors.

These properties of linearly independent and dependent sets have far-reaching implications. We summarize the observations above in terms of the information required to construct all the members of the subspace span$\{T\}$.

   If $T$ is a linearly dependent set, then the subspace span$\{T\}$ can be generated by using a subset of $T$.

That is, the information given by the vectors of $T$ is redundant since we can use fewer vectors and generate the same subspace. (However, to determine which vectors can be omitted we have to do more work. We discuss a procedure for omitting redundant information later.) It also follows that

If $T$ is a linearly independent set, then any subset of fewer vectors cannot generate the whole subspace span$\{T\}$.

That is, *a linearly independent spanning set contains the minimal amount of information to generate a subspace.* To indicate this we introduce the following terminology.

---

**Definition**    A linearly independent spanning set of a subspace is called a **basis**.

---

**EXAMPLE 4**    Let $T = \{\mathbf{v}_1, \mathbf{v}_2\}$ be a linearly independent set in $R^2$. Then we know from properties 1 and 2 that $\mathbf{v}_2$ is not a scalar multiple of $\mathbf{v}_1$ and from property 3 that neither $\mathbf{v}_1$ nor $\mathbf{v}_2$ is the zero vector. We claim that span$\{T\}$ is $R^2$. That is, if $\mathbf{w}$ is any vector in $R^2$, then there exist scalars $c_1$ and $c_2$ such that

$$c_1\mathbf{v}_1 + c_2\mathbf{v}_2 = \mathbf{w}. \tag{2}$$

Let $\mathbf{v}_1 = \begin{bmatrix} a \\ b \end{bmatrix}$, $\mathbf{v}_2 = \begin{bmatrix} c \\ d \end{bmatrix}$, and $\mathbf{w} = \begin{bmatrix} e \\ f \end{bmatrix}$. Then (2) is equivalent to the linear system

$$\begin{bmatrix} a & c \\ b & d \end{bmatrix} \begin{bmatrix} c_1 \\ c_2 \end{bmatrix} = \begin{bmatrix} e \\ f \end{bmatrix} \tag{3}$$

whose augmented matrix is $\left[\begin{array}{cc|c} a & c & e \\ b & d & f \end{array}\right]$. To substantiate our claim we need only show that (3) is always consistent for any choice of entries $e$ and $f$. Working with the augmented matrix and row operations, we proceed as follows. Since $\mathbf{v}_1 \neq \mathbf{0}$ either $a$ or $b$ is not zero. Suppose $a \neq 0$.[3] Then

$$\left[\begin{array}{cc|c} a & c & e \\ b & d & f \end{array}\right]_{-\frac{b}{a}R_1 + R_2} \Rightarrow \left[\begin{array}{cc|c} a & c & e \\ 0 & d - \dfrac{bc}{a} & f - \dfrac{be}{a} \end{array}\right].$$

This system will always be consistent provided $d - \dfrac{bc}{a} \neq 0$. (Explain why this is true.) Next we note that

$$d - \frac{bc}{a} \neq 0 \quad \Leftrightarrow \quad d \neq \frac{bc}{a},$$

that is, if and only if

$$\begin{bmatrix} c \\ d \end{bmatrix} \neq \begin{bmatrix} c \\ \dfrac{c}{a}b \end{bmatrix} = \frac{c}{a}\begin{bmatrix} a \\ b \end{bmatrix}.$$

But it must be true that

$$\begin{bmatrix} c \\ d \end{bmatrix} \neq \frac{c}{a}\begin{bmatrix} a \\ b \end{bmatrix}$$

since $\mathbf{v}_2$ is not a scalar multiple of $\mathbf{v}_1$. Hence it follows that $d - \dfrac{bc}{a} \neq 0$. So (3) is always consistent and thus $T$ is a basis for $R^2$. That is, **any pair of linearly independent vectors in $R^2$ is a basis**. See also Exercise 49. ∎

---

[3]See Exercise 48 for an alternative approach if $a = 0$.

For a homogeneous linear system $\mathbf{Ax} = \mathbf{0}$ with infinitely many solutions, our construction of the set $S$ produces a basis for the subspace $\mathbf{ns}(\mathbf{A})$. We continue the investigation of linearly independent sets, and in particular bases, by defining two other subspaces that are associated with a matrix.

---

**Definition**    For a matrix $\mathbf{A}$ the span of the rows of $\mathbf{A}$ is called the **row space of $\mathbf{A}$**, denoted $\mathbf{row}(\mathbf{A})$, and the span of the columns of $\mathbf{A}$ is called the column space of $\mathbf{A}$, denoted $\mathbf{col}(\mathbf{A})$.

---

If $\mathbf{A}$ is $m \times n$, then the members of $\mathbf{row}(\mathbf{A})$ are vectors in $R^n$ and the members of $\mathbf{col}(\mathbf{A})$ are vectors in $R^m$. Since the span of any set of matrices is closed,[4] $\mathbf{row}(\mathbf{A})$ is a subspace of $R^n$ and $\mathbf{col}(\mathbf{A})$ is a subspace of $R^m$. Our goal now is to determine a basis for each of these subspaces associated with matrix $\mathbf{A}$.

A member of the row space of $\mathbf{A}$ is a linear combination of the rows of $\mathbf{A}$. If $\mathbf{y}$ is a $1 \times m$ vector, then the **vector-matrix product**

$$\mathbf{yA} = \begin{bmatrix} y_1 & y_2 & \cdots & y_m \end{bmatrix} \begin{bmatrix} \mathbf{row}_1(\mathbf{A}) \\ \mathbf{row}_2(\mathbf{A}) \\ \vdots \\ \mathbf{row}_m(\mathbf{A}) \end{bmatrix}$$

$$= y_1 \, \mathbf{row}_1(\mathbf{A}) + y_2 \, \mathbf{row}_2(\mathbf{A}) + \cdots + y_m \, \mathbf{row}_m(\mathbf{A})$$

produces a member of $\mathbf{row}(\mathbf{A})$. Row operations applied to $\mathbf{A}$ manipulate its rows and (except for interchanges) produce linear combinations of its rows. If matrix $\mathbf{B}$ is row equivalent to $\mathbf{A}$, then $\mathbf{row}(\mathbf{B})$ is the same as $\mathbf{row}(\mathbf{A})$ since a row of $\mathbf{B}$ is a linear combination of rows of $\mathbf{A}$. (See Exercise 47.) It follows that

$$\mathbf{row}(\mathbf{A}) = \mathbf{row}(\mathbf{rref}(\mathbf{A})).$$

Suppose that $\mathbf{rref}(\mathbf{A})$ has $k$ nonzero rows. These $k$ rows have a leading 1 in different columns. When we form a linear combination of these rows with coefficients $c_1, c_2, \ldots, c_k$ each of the c's will appear by itself in an entry of the resulting vector. Equating the resulting vector to the zero (row) vector must then give $c_1 = c_2 = \cdots = c_k = 0$. Hence the nonzero rows of $\mathbf{rref}(\mathbf{A})$ are linearly independent. Since $\mathbf{row}(\mathbf{rref}(\mathbf{A})) = \mathbf{row}(\mathbf{A})$, it follows that the nonzero rows of $\mathbf{rref}(\mathbf{A})$ form a basis for $\mathbf{row}(\mathbf{A})$. [This argument is similar to that given for the basis of $\mathbf{ns}(\mathbf{A})$.]

A member of the column space of $\mathbf{A}$ is a linear combination of the columns of $\mathbf{A}$. If $\mathbf{x}$ is an $n \times 1$ vector, then the **matrix-vector product**

$$\mathbf{Ax} = x_1 \, \mathbf{col}_1(\mathbf{A}) + x_2 \, \mathbf{col}_2(\mathbf{A}) + \cdots + x_n \, \mathbf{col}_n(\mathbf{A})$$

produces a member of $\mathbf{col}(\mathbf{A})$. Thus we see that **a linear system $\mathbf{Ax} = \mathbf{b}$ is consistent if and only if $\mathbf{b}$ is in the column space of $\mathbf{A}$**. It follows that a vector $\mathbf{x}$ is in the solution set of a linear system if and only if its entries are coefficients for the columns of $\mathbf{A}$ that produce vector $\mathbf{b}$. To determine a basis for $\mathbf{col}(\mathbf{A})$, we first recall that the transpose converts columns to rows. Hence we have that $\mathbf{col}(\mathbf{A}) = \mathbf{row}(\mathbf{A}^T)$ and since $\mathbf{row}(\mathbf{A}^T) = \mathbf{row}(\mathbf{rref}(\mathbf{A}^T))$, it follows that the nonzero columns of $(\mathbf{rref}(\mathbf{A}^T))^T$ form a basis for $\mathbf{col}(\mathbf{A})$.

---

[4] See Example 12 in Section 1.2 and the comments immediately following it. See also Exercise 37 in Section 1.2.

EXAMPLE 5   Determine a basis for **row(A)**, **col(A)**, and **ns(A)** where

$$A = \begin{bmatrix} 1 & -2 & 0 & 3 \\ 3 & 2 & 8 & 1 \\ 2 & 3 & 7 & 2 \\ -1 & 2 & 0 & 4 \end{bmatrix}.$$

Applying row operations, we obtain (verify)

$$\mathbf{rref(A)} = \begin{bmatrix} 1 & 0 & 2 & 0 \\ 0 & 1 & 1 & 0 \\ 0 & 0 & 0 & 1 \\ 0 & 0 & 0 & 0 \end{bmatrix}$$

so a basis for **row(A)** is

$$\left\{ \begin{bmatrix} 1 & 0 & 2 & 0 \end{bmatrix}, \begin{bmatrix} 0 & 1 & 1 & 0 \end{bmatrix}, \begin{bmatrix} 0 & 0 & 0 & 1 \end{bmatrix} \right\}.$$

Also, we have that the solution set of $\mathbf{Ax} = \mathbf{0}$ is given by $x_4 = 0$, $x_2 = -x_3$, $x_1 = -2x_3$. (Verify.) Thus

$$\mathbf{x} = \begin{bmatrix} x_1 \\ x_2 \\ x_3 \\ x_4 \end{bmatrix} = \begin{bmatrix} -2x_3 \\ -x_3 \\ x_3 \\ 0 \end{bmatrix} = \begin{bmatrix} -2t \\ -t \\ t \\ 0 \end{bmatrix} = t \begin{bmatrix} -2 \\ -1 \\ 1 \\ 0 \end{bmatrix}$$

and hence a basis for **ns(A)** is $\left\{ \begin{bmatrix} -2 & -1 & 1 & 0 \end{bmatrix}^T \right\}$. Computing $\mathbf{rref(A}^T)$ and taking the transpose of the result gives (verify)

$$(\mathbf{rref(A)}^T)^T = \left( \begin{bmatrix} 1 & 0 & 0 & \frac{11}{24} \\ 0 & 1 & 0 & -\frac{49}{24} \\ 0 & 0 & 1 & \frac{7}{3} \\ 0 & 0 & 0 & 0 \end{bmatrix} \right)^T = \begin{bmatrix} 1 & 0 & 0 & 0 \\ 0 & 1 & 0 & 0 \\ 0 & 0 & 1 & 0 \\ \frac{11}{24} & -\frac{49}{24} & \frac{7}{3} & 0 \end{bmatrix}.$$

Thus a basis for **col**(A) is

$$\left\{ \begin{bmatrix} 1 \\ 0 \\ 0 \\ \frac{11}{24} \end{bmatrix}, \begin{bmatrix} 0 \\ 1 \\ 0 \\ -\frac{49}{24} \end{bmatrix}, \begin{bmatrix} 0 \\ 0 \\ 1 \\ \frac{7}{3} \end{bmatrix} \right\}. \qquad ■$$

The preceding development and Example 5 show that the reduced row echelon form of a matrix provides a pattern that reveals the information needed to construct bases for the three subspaces, row, column, and null, associated with a matrix. We will see in Section 2.4 that the reduced row echelon form can be used to determine more information about a matrix and the solution set of a linear system $\mathbf{Ax} = \mathbf{b}$. Hence it follows that row operations are a powerful computational tool not only to construct solution sets, but also to yield information related to these fundamental subspaces associated with a matrix.

We can use the techniques for computing bases for **row(A)** and **col(A)** just developed to determine if a set $T = \{\mathbf{v}_1, \mathbf{v}_2, \dots, \mathbf{v}_k\}$ in $R^n$ is linearly independent or linearly dependent. Applying the definition, we form the linear combination

$\sum\limits_{j=1}^{k} c_j \mathbf{v}_j$, set it equal to the zero vector, $\mathbf{0}$, and then see if we can find nonzero values for the coefficients $c_j$, $j = 1, 2, \ldots, k$ so that

$$\sum_{j=1}^{k} c_j \mathbf{v}_j = \mathbf{0}.$$

This leads to a homogeneous linear system for which we determine whether there is a nontrivial solution or only the trivial solution. (See Examples 1 and 2.) We compute the reduced row echelon form of the linear system to make this determination.

A shortcut that provides the same structure is to form the matrix $\mathbf{A}$ whose columns are the vectors in $T$, taking transposes if vectors $\mathbf{v}_j$ are rows, and compute $\mathbf{rref}(\mathbf{A})$. There will be a nontrivial solution, implying that

> $T$ is a linearly dependent set if (the number of columns of $\mathbf{A}$) - (the number of leading 1's in $\mathbf{rref}(\mathbf{A})$ is positive.

This is equivalent to saying that $\mathbf{Ac} = \mathbf{0}$ has a nontrivial solution.

> If (the number of columns of $\mathbf{A}$) − (the number of leading 1's) is zero, then $T$ is a linearly independent set.

EXAMPLE 6    Determine whether the set $T$ is linearly independent or linearly dependent.

(a) $T = \{\mathbf{v}_1, \mathbf{v}_2, \mathbf{v}_3\} = \left\{ \begin{bmatrix} 1 \\ 0 \\ 1 \\ 2 \end{bmatrix}, \begin{bmatrix} 0 \\ 1 \\ 1 \\ 2 \end{bmatrix}, \begin{bmatrix} 1 \\ 1 \\ 1 \\ 3 \end{bmatrix} \right\}.$

Form the matrix

$$\mathbf{A} = \begin{bmatrix} \mathbf{v}_1 & \mathbf{v}_2 & \mathbf{v}_3 \end{bmatrix} = \begin{bmatrix} 1 & 0 & 1 \\ 0 & 1 & 1 \\ 1 & 1 & 1 \\ 2 & 2 & 3 \end{bmatrix}.$$

Then

$$\mathbf{rref}(\mathbf{A}) = \begin{bmatrix} 1 & 0 & 0 \\ 0 & 1 & 0 \\ 0 & 0 & 1 \\ 0 & 0 & 0 \end{bmatrix}. \qquad \text{(Verify.)}$$

Since (the number of columns of $\mathbf{A}$) − (the number of leading 1's) is zero, we conclude that $T$ is a linearly independent set. It follows that $T$ is a basis for span$\{T\}$, which is a subspace of $R^4$.

(b) $T = \{\mathbf{v}_1, \mathbf{v}_2, \mathbf{v}_3, \mathbf{v}_4\}$
$= \{\begin{bmatrix} 1 & 2 & -1 \end{bmatrix}, \begin{bmatrix} 1 & -2 & 1 \end{bmatrix}, \begin{bmatrix} -3 & 2 & -1 \end{bmatrix}, \begin{bmatrix} 2 & 0 & 0 \end{bmatrix}\}.$

Form the matrix

$$\mathbf{A} = \begin{bmatrix} \mathbf{v}_1^T & \mathbf{v}_2^T & \mathbf{v}_3^T & \mathbf{v}_4^T \end{bmatrix} = \begin{bmatrix} 1 & 1 & -3 & 2 \\ 2 & -2 & 2 & 0 \\ -1 & 1 & -1 & 0 \end{bmatrix}.$$

Then

$$\mathbf{rref(A)} = \begin{bmatrix} 1 & 0 & -1 & 1 \\ 0 & 1 & -2 & 1 \\ 0 & 0 & 0 & 0 \end{bmatrix}. \qquad \text{(Verify.)}$$

Since (the number of columns of $\mathbf{A}$) $-$ (the number of leading 1's) is 2, we conclude that $T$ is a linearly dependent set. It follows that $T$ is not a basis for span$\{T\}$, which is a subspace of $R^3$. To determine a basis for span$\{T\}$, we form the matrix $\mathbf{B}$ whose rows are the $\mathbf{v}_j$; that is,

$$\mathbf{B} = \begin{bmatrix} 1 & 2 & -1 \\ 1 & -2 & 1 \\ -3 & 2 & -1 \\ 2 & 0 & 0 \end{bmatrix}.$$

Then it follows that span$\{T\} = \mathbf{row(B)}$, so the nonzero rows of $\mathbf{rref(B)}$ form a basis for span$\{T\}$. We have

$$\mathbf{rref(B)} = \begin{bmatrix} 1 & 0 & 0 \\ 0 & 1 & -0.5 \\ 0 & 0 & 0 \\ 0 & 0 & 0 \end{bmatrix};$$

thus $\{\begin{bmatrix} 1 & 0 & 0 \end{bmatrix}, \begin{bmatrix} 0 & 1 & -0.5 \end{bmatrix}\}$ is a basis for span$\{T\}$. ■

A consistent linear system $\mathbf{Ax} = \mathbf{b}$ in which some of the unknowns can be chosen arbitrarily is called **underdetermined**. Example 4 in Section 2.2 contains an underdetermined nonhomogeneous linear system and Example 2 contains an underdetermined homogeneous linear system.

A simple geometrical example of a consistent underdetermined linear system is the intersection of two planes in 3-space. The solution set represents a line. Consistent underdetermined linear systems appear in such areas as resource allocation, network flows, and the analysis of closed economies. In such applications it is often of interest to determine which variables are free and how the choice of these free variables affects the other variables in the model. Such an analysis can help to develop alternative plans for allocation of resources, changes in street or highway systems, and equilibrium pricing. (See Section 2.2.)

We next introduce two concepts that provide a count of the number of linearly independent rows of a matrix and a count of the largest number of linearly independent vectors that can be present in a subset of a subspace.

---

**Definition**    The **rank** of a matrix $\mathbf{A}$, denoted rank$(\mathbf{A})$, is the number of leading 1's in $\mathbf{rref(A)}$.

---

EXAMPLE 7    From Example 5, we have that

$$\mathbf{rref(A)} = \mathbf{rref}\left(\begin{bmatrix} 1 & -2 & 0 & 3 \\ 3 & 2 & 8 & 1 \\ 2 & 3 & 7 & 2 \\ -1 & 2 & 0 & 4 \end{bmatrix}\right) = \begin{bmatrix} 1 & 0 & 2 & 0 \\ 0 & 1 & 1 & 0 \\ 0 & 0 & 0 & 1 \\ 0 & 0 & 0 & 0 \end{bmatrix}.$$

Hence rank$(\mathbf{A}) = 3$. It follows that

$$\{\begin{bmatrix} 1 & 0 & 2 & 0 \end{bmatrix}, \begin{bmatrix} 0 & 1 & 1 & 0 \end{bmatrix}, \begin{bmatrix} 0 & 0 & 0 & 1 \end{bmatrix}\}$$

is a basis for $\mathbf{row(A)}$; hence $\mathbf{A}$ has three linearly independent rows. ■

In Example 4 we showed that any pair of linearly independent vectors is a basis for $R^2$. (See also Exercise 49.) It will be shown in Chapter 5 that all bases for a particular subspace contain the same number of vectors. Thus we make the following definition.

**Definition**    The **dimension** of a subspace $V$ of $R^n$, denoted dim($V$), is the number of vectors in a basis for $V$.

EXAMPLE 8    From Example 2 we have that **ns(A)** consists of all vectors of the form

$$s\mathbf{v}_1 + t\mathbf{v}_2, \quad \text{where } \mathbf{v}_1 = \begin{bmatrix} -2 \\ 3 \\ 1 \\ 0 \\ 0 \end{bmatrix} \text{ and } \mathbf{v}_2 = \begin{bmatrix} -4 \\ -1 \\ 0 \\ 1 \\ 0 \end{bmatrix}.$$

Since $\mathbf{v}_1$ is not a scalar multiple of $\mathbf{v}_2$, these vectors are linearly independent. It follows that $\{\mathbf{v}_1, \mathbf{v}_2\}$ is a basis for **ns(A)**, and so dim(**ns(A)**) $= 2$. We note that if $\mathbf{v}_3$ is any other vector in **ns(A)** then $\{\mathbf{v}_1, \mathbf{v}_2, \mathbf{v}_3\}$ is linearly dependent since $\mathbf{v}_3$ is a linear combination of $\mathbf{v}_1$ and $\mathbf{v}_2$.    ■

## Extending a Basis for a Subspace to a Basis for $R^k$

Let matrix **A** be $m \times n$; then we have three subspaces associated with **A**:

1. **ns(A)** is the solution space of $\mathbf{Ax} = \mathbf{0}$ and is a subspace of $R^n$.
2. **row(A)** is the row space of **A** and is a subspace of $R^n$.
3. **col(A)** is the column space of **A** and is a subspace of $R^m$.

Since $R^n$ and $R^m$ are closed sets, they are subspaces. From Exercise 58, it follows that dim($R^n$) $= n$ and dim($R^m$) $= m$. We also have

$$\dim(\mathbf{ns(A)}) = \text{the number of free variables in the}$$
$$\text{general solution of } \mathbf{Ax} = \mathbf{0} \le n$$
$$\dim(\mathbf{row(A)}) = \text{the number of leading 1's in } \mathbf{rref(A)} \le n$$

and

$$\dim(\mathbf{col(A)}) = \text{the number of leading 1's in } \mathbf{rref(A}^T) \le m.$$

An important idea that is used in Chapter 4 is that a basis for a subspace $V$ of $R^k$ can be extended to a basis for $R^k$. Here we show how to determine vectors that, when adjoined to a basis for subspace $V$ of $R^k$, yield a basis for $R^k$. Justification for the following technique is discussed in Section 5.3.

Let $V$ be a subspace of $R^k$ with dim($V$) $= r < k$. Denote the basis for $V$ as the set $\{\mathbf{v}_1, \mathbf{v}_2, \ldots, \mathbf{v}_r\}$. It is convenient to consider the vectors $\mathbf{v}_j$, $j = 1, 2, \ldots, r$ as columns of size $k \times 1$. Next let $\mathbf{e}_j = \text{col}_j(\mathbf{I}_k)$, $j = 1, 2, \ldots, k$. Form the matrix

$$\mathbf{B} = \begin{bmatrix} \mathbf{v}_1 & \mathbf{v}_2 & \cdots & \mathbf{v}_r & \mathbf{e}_1 & \mathbf{e}_2 & \cdots & \mathbf{e}_k \end{bmatrix}.$$

Compute **rref(B)**. There will be $k$ leading 1's in **rref(B)** since **col(B)** $= R^k$. (Explain why.) Let the column numbers of **rref(B)** that contain leading 1's be denoted $i_1, i_2, \ldots, i_k$. It can be shown that $i_1 = 1$, $i_2 = 2$, $\ldots$, $i_r = r$ and that this set of columns of **B** form a basis $S$ for $R^k$. It follows that basis $S$ contains vectors

$\mathbf{v}_1, \mathbf{v}_2, \ldots, \mathbf{v}_r$ and some subset of $k - r$ columns of $\mathbf{I}_k$. Thus we say that basis $\{\mathbf{v}_1, \mathbf{v}_2, \ldots, \mathbf{v}_r\}$ of V has been extended to a basis for $R^k$.

EXAMPLE 9    Suppose that linear system $\mathbf{Ax} = \mathbf{0}$ has basis for $\mathbf{ns(A)}$ given by

$$T = \{\mathbf{v}_1, \mathbf{v}_2\} = \left\{ \begin{bmatrix} -3 \\ 1 \\ 0 \\ 0 \end{bmatrix}, \begin{bmatrix} 2 \\ 0 \\ 1 \\ 0 \end{bmatrix} \right\}.$$

To extend this basis for subspace $\mathbf{ns(A)}$ to a basis for $R^4$, we form matrix

$$\mathbf{B} = \begin{bmatrix} -3 & 2 & 1 & 0 & 0 & 0 \\ 1 & 0 & 0 & 1 & 0 & 0 \\ 0 & 1 & 0 & 0 & 1 & 0 \\ 0 & 0 & 0 & 0 & 0 & 1 \end{bmatrix}$$

and compute its RREF. We obtain (verify)

$$\mathbf{rref(B)} = \begin{bmatrix} 1 & 0 & 0 & 1 & 0 & 0 \\ 0 & 1 & 0 & 0 & 1 & 0 \\ 0 & 0 & 1 & 3 & -2 & 0 \\ 0 & 0 & 0 & 0 & 0 & 1 \end{bmatrix}.$$

There are four leading 1's, in columns 1, 2, 3, and 6. It follows from the preceding discussion that the corresponding columns of $\mathbf{B}$ are a basis $S$ for $R^4$ where

$$S = \left\{ \mathbf{v}_1 = \begin{bmatrix} -3 \\ 1 \\ 0 \\ 0 \end{bmatrix}, \mathbf{v}_2 = \begin{bmatrix} 2 \\ 0 \\ 1 \\ 0 \end{bmatrix}, \begin{bmatrix} 1 \\ 0 \\ 0 \\ 0 \end{bmatrix}, \begin{bmatrix} 0 \\ 0 \\ 0 \\ 1 \end{bmatrix} \right\}.$$

Thus the $T$ basis for $\mathbf{ns(A)}$ has been extended to a basis for $R^4$.    ■

## EXERCISES 2.3

*In Exercises 1–6, determine whether the given set is linearly independent.*

**1.** $T = \left\{ \begin{bmatrix} 1 \\ 2 \end{bmatrix}, \begin{bmatrix} -1 \\ 3 \end{bmatrix}, \begin{bmatrix} 3 \\ 1 \end{bmatrix} \right\}.$

**2.** $T = \left\{ \begin{bmatrix} 4 \\ -2 \end{bmatrix}, \begin{bmatrix} -2 \\ 1 \end{bmatrix} \right\}.$

**3.** $T = \left\{ \begin{bmatrix} 2 \\ -2 \end{bmatrix}, \begin{bmatrix} -2 \\ 1 \end{bmatrix} \right\}.$

**4.** $T = \left\{ \begin{bmatrix} 1 \\ 2 \\ 1 \end{bmatrix}, \begin{bmatrix} -1 \\ 3 \\ 0 \end{bmatrix}, \begin{bmatrix} 3 \\ 1 \\ 0 \end{bmatrix} \right\}.$

**5.** $T = \left\{ \begin{bmatrix} 0 \\ 1 \\ 2 \end{bmatrix}, \begin{bmatrix} 2 \\ 1 \\ 0 \end{bmatrix}, \begin{bmatrix} 0 \\ 0 \\ 0 \end{bmatrix}, \begin{bmatrix} 0 \\ 0 \\ 1 \end{bmatrix} \right\}.$

**6.** $T = \left\{ \begin{bmatrix} 2 \\ 1 \\ 1 \end{bmatrix}, \begin{bmatrix} 1 \\ 2 \\ 2 \end{bmatrix}, \begin{bmatrix} 2 \\ 1 \\ 2 \end{bmatrix} \right\}.$

**7.** Let $T = \{\mathbf{v}_1, \mathbf{v}_2, \mathbf{v}_3\}$ be any three vectors in $R^3$. It is true that $0\mathbf{v}_1 + 0\mathbf{v}_2 + 0\mathbf{v}_3 = \mathbf{0}$. Does this guarantee that $T$ is always linearly independent? Explain your answer.

**8.** Let $T = \{\mathbf{v}_1, \mathbf{v}_2, \mathbf{v}_3, \mathbf{v}_4\}$ be a subset of $R^n$ such that $\mathbf{v}_3 = 2\mathbf{v}_1 - \mathbf{v}_2 + 5\mathbf{v}_4$. Is $T$ linearly independent or linearly dependent? Explain your answer.

*In Exercises 9–16, determine $\mathbf{ns(A)}$ for each given matrix.*

**9.** $A = \begin{bmatrix} 2 & 4 \\ -1 & -2 \end{bmatrix}.$

**10.** $A = \begin{bmatrix} 2 & 4 \\ 1 & -2 \end{bmatrix}.$

**11.** $A = \begin{bmatrix} 1 & 0 & 2 \\ 2 & 1 & 3 \\ 3 & 2 & 4 \end{bmatrix}.$

**12.** $A = \begin{bmatrix} 6 & 0 & 3 \\ 2 & 0 & 1 \\ 3 & 2 & 4 \end{bmatrix}.$

**13.** $\mathbf{A} = \begin{bmatrix} 0 & 1 & 1 \\ 1 & -2 & 1 \\ 1 & 1 & 4 \\ -1 & 6 & 3 \end{bmatrix}$.

**14.** $\mathbf{A} = \begin{bmatrix} 2 & 1 & 3 \\ 1 & 2 & 3 \\ 3 & 1 & 1 \\ 1 & 1 & 1 \end{bmatrix}$.

**15.** $\mathbf{A} = \begin{bmatrix} 1 & 2 & 2 & -1 & 1 \\ 0 & 2 & 2 & -2 & -1 \\ 2 & 6 & 2 & -4 & 1 \\ 1 & 4 & 0 & -3 & 0 \end{bmatrix}$.

**16.** $\mathbf{A} = \begin{bmatrix} 1 & 2 & -1 & 3 & -1 \\ 2 & 2 & -1 & 2 & 0 \\ 1 & 0 & 3 & 3 & 1 \end{bmatrix}$.

*In Exercises 17–19, determine the null space of the homogeneous linear system $(t\mathbf{I}_n - \mathbf{A})\mathbf{x} = \mathbf{0}$ for the given scalar t and the given matrix* $\mathbf{A}$.

**17.** $t = 1; \mathbf{A} = \begin{bmatrix} 3 & 2 \\ 1 & 2 \end{bmatrix}$.

**18.** $t = 2; \mathbf{A} = \begin{bmatrix} 1 & 1 & -2 \\ -1 & 2 & 1 \\ 0 & 1 & -1 \end{bmatrix}$.

**19.** $t = 3; \mathbf{A} = \begin{bmatrix} 1 & 1 & -2 \\ -1 & 2 & 1 \\ 0 & 1 & -1 \end{bmatrix}$.

*In Exercises 20–22, find a nonzero vector* $\mathbf{x}$ *so that* $\mathbf{A}\mathbf{x} = t\mathbf{x}$ *for the given scalar t and the given matrix* $\mathbf{A}$. *(Hint: See the statement preceding Exercise 17.)*

**20.** $t = 3; \mathbf{A} = \begin{bmatrix} 2 & 1 \\ 1 & 2 \end{bmatrix}$.

**21.** $t = 0; \mathbf{A} = \begin{bmatrix} 1 & 1 \\ 1 & 1 \end{bmatrix}$.

**22.** $t = 1, \mathbf{A} = \begin{bmatrix} 1 & 2 & -1 \\ 1 & 0 & 1 \\ 4 & -4 & 5 \end{bmatrix}$.

**23.** $S = \left\{ \begin{bmatrix} 1 \\ 2 \end{bmatrix}, \begin{bmatrix} 2 \\ -2 \end{bmatrix} \right\}$ is a basis for $R^2$. Express each of the following vectors as a linear combination of the vectors in $S$.

(a) $\mathbf{v} = \begin{bmatrix} 0 \\ 6 \end{bmatrix}$.         (b) $\mathbf{v} = \begin{bmatrix} 10 \\ 2 \end{bmatrix}$.

**24.** $S = \left\{ \begin{bmatrix} 1 \\ -1 \\ 0 \end{bmatrix}, \begin{bmatrix} 2 \\ 1 \\ 1 \end{bmatrix}, \begin{bmatrix} 3 \\ 0 \\ 2 \end{bmatrix} \right\}$ is a basis for $R^3$. Express each of the following vectors as a linear combination of the vectors in $S$.

(a) $\mathbf{v} = \begin{bmatrix} 0 \\ 0 \\ -1 \end{bmatrix}$.         (b) $\mathbf{v} = \begin{bmatrix} 0 \\ -3 \\ -1 \end{bmatrix}$.

*In Exercises 25–32, find a basis for* $\mathbf{row}(\mathbf{A})$ *for each of the given matrices.*

**25.** $\mathbf{A} = \begin{bmatrix} 2 & 4 \\ -1 & -2 \end{bmatrix}$.

**26.** $\mathbf{A} = \begin{bmatrix} 2 & 4 \\ 1 & -2 \end{bmatrix}$.

**27.** $\mathbf{A} = \begin{bmatrix} 1 & 0 & 2 \\ 2 & 1 & 3 \\ 3 & 2 & 4 \end{bmatrix}$.

**28.** $\mathbf{A} = \begin{bmatrix} 6 & 0 & 3 \\ 2 & 0 & 1 \\ 3 & 2 & 4 \end{bmatrix}$.

**29.** $\mathbf{A} = \begin{bmatrix} 1 & -2 & 4 \\ -3 & 6 & -11 \\ -1 & 2 & -3 \end{bmatrix}$.

**30.** $\mathbf{A} = \begin{bmatrix} 0 & 1 & 1 \\ 1 & -2 & 1 \\ 1 & 1 & 4 \\ -1 & 6 & 3 \end{bmatrix}$.

**31.** $\mathbf{A} = \begin{bmatrix} 1 & 0 & 5 \\ 1 & 1 & 1 \\ 0 & 1 & -4 \end{bmatrix}$.

**32.** $\mathbf{A} = \begin{bmatrix} 1 & 3 & 0 & 5 & -4 \\ 3 & 8 & 1 & 17 & -10 \\ 4 & 10 & 1 & 20 & -12 \end{bmatrix}$.

**33.** Find a basis for $\mathbf{col}(\mathbf{A})$ for each of the matrices in Exercises 25–32.

*In Exercises 34–39, find a basis for* $\mathbf{row}(\mathbf{A})$, $\mathbf{col}(\mathbf{A})$, *and* $\mathbf{ns}(\mathbf{A})$.

**34.** $\mathbf{A} = \begin{bmatrix} 1 & 1 \\ 1 & 1 \end{bmatrix}$.

**35.** $\mathbf{A} = \begin{bmatrix} 2 & 1 & 3 \\ 1 & 2 & 3 \\ 3 & 1 & 1 \\ 1 & 1 & 1 \end{bmatrix}$.

**36.** $\mathbf{A} = \begin{bmatrix} 4 & 3 & 0 \\ 1 & 1 & 2 \\ 3 & 2 & -2 \end{bmatrix}$.

**37.** $\mathbf{A} = \begin{bmatrix} -1 & 1 & 0 \\ 1 & 1 & 2 \\ 2 & -1 & 1 \\ 4 & -2 & 2 \end{bmatrix}$.

**38.** $A = \begin{bmatrix} 1 & 2 & 0 & 3 & -1 \\ -2 & -4 & 1 & -7 & 2 \\ -1 & -2 & 0 & -5 & 0 \\ 1 & 2 & 1 & 4 & 0 \end{bmatrix}$.

**39.** $A = \begin{bmatrix} -1 & 2 & 2 & -1 \\ -3 & 1 & 1 & -5 \\ 6 & -2 & -2 & 11 \\ 2 & 1 & 1 & 5 \\ 3 & -1 & -1 & 6 \end{bmatrix}$.

**40.** If **A** is symmetric, how are the subspaces **row(A)** and **col(A)** related?

**41.** How are the subspaces **row(A)** and **row(kA)**, $k \neq 0$, related?

*By observation only, label each of the sets in Exercises 42–46 as linearly independent (LI) or linearly dependent (LD) and give a reason for your choice.*

**42.** $\left\{ \begin{bmatrix} 1 \\ 2 \\ 0 \end{bmatrix}, \begin{bmatrix} 0 \\ 4 \\ -1 \end{bmatrix} \right\}$.

**43.** $\left\{ \begin{bmatrix} 1 \\ 0 \\ 1 \\ 0 \end{bmatrix}, \begin{bmatrix} 3 \\ 0 \\ 3 \\ 0 \end{bmatrix} \right\}$.

**44.** $\{ \begin{bmatrix} 1 & -2 \end{bmatrix}, \begin{bmatrix} 0 & 0 \end{bmatrix} \}$.

**45.** $\{ \begin{bmatrix} 2 & 0 & -3 \end{bmatrix}, \begin{bmatrix} \frac{1}{2} & 0 & -\frac{3}{4} \end{bmatrix} \}$.

**46.** $\left\{ \begin{bmatrix} 1 \\ 0 \\ 0 \end{bmatrix}, \begin{bmatrix} 1 \\ 1 \\ 0 \end{bmatrix}, \begin{bmatrix} 0 \\ 0 \\ 0 \end{bmatrix} \right\}$.

**47.** Explain why if **A** is row equivalent to **B**, then **row(A) = row(B)**.

**48.** In Example 4 we showed that any set of two linearly independent vectors in $R^2$ is a basis for $R^2$. As part of the development we used $a \neq 0$. If $a = 0$, then we could interchange rows as follows:

$$\begin{bmatrix} a & c & | & e \\ b & d & | & f \end{bmatrix}_{R_1 \leftrightarrow R_2} \Rightarrow \begin{bmatrix} b & d & | & f \\ 0 & c & | & e \end{bmatrix}.$$

(a) Explain why $c \neq 0$.
(b) Explain why the linear system in this case must always be consistent.

**49.** Using Example 4 as a model, construct a linear system of equations that must be consistent so that any set of three linearly independent vectors in $R^3$ is a basis for $R^3$. (You are not asked to show that the linear system is always consistent. Developments in later sections can be used to show easily that the linear system is always consistent.) This idea generalizes, so that any set of $n$ linearly independent vectors in $R^n$ is a basis for $R^n$.

**50.** Let **A** be $n \times n$ with a row of all zeros. Explain why we can say that linear system $Ax = b$, for any $n \times 1$ vector **b**, does not have a unique solution. (*Hint*: Consider cases of inconsistent and consistent systems.)

**51.** Let **A** be $m \times n$ with $n > m$. Explain why we can say that $Ax = 0$ has infinitely many solutions. (*Note*: This property says that a homogeneous linear system with more unknowns than equations is underdetermined.)

**52.** (a) Let $T = \{v_1, v_2, \ldots, v_k\}$ be a linearly dependent set of vectors in $R^n$. Then we know that there exist scalars $c_1, c_2, \ldots, c_k$, not all zero, such that

$$c_1 v_1 + c_2 v_2 + \cdots + c_k v_k = 0.$$

Suppose that we have arranged the vectors in $T$ so that vector $v_1$ has a nonzero coefficient $c_1$. Explain why we can say $v_1$ must be a linear combination of $v_2, v_3, \ldots, v_k$. (The general statement we derive from this is often stated as follows: If $T$ is a linearly dependent set, then one of its members is a linear combination of the other members in $T$.)

(b) Let $T = \{v_1, v_2, \ldots, v_k\}$ be a set of vectors in $R^n$ in which one of the vectors is a linear combination of the other vectors. For instance, suppose $v_1$ is a linear combination of $v_2, v_3, \ldots, v_k$. Show that $T$ is a linearly dependent set.

**53.** We want to show that a linear combination of vectors from span$\{T\}$, where $T = \{v_1, v_2, \ldots, v_k\}$ is a subset of $R^m$, is a vector that is also in span$\{T\}$. Let $w_1, w_2, \ldots, w_n$ be in span$\{T\}$; then each $w_j$ is a linear combination of members of $T$. Hence let

$$w_1 = c_{11}v_1 + c_{12}v_2 + \cdots + c_{1k}v_k$$
$$w_2 = c_{21}v_1 + c_{22}v_2 + \cdots + c_{2k}v_k$$
$$\vdots$$
$$w_n = c_{n1}v_1 + c_{n2}v_2 + \cdots + c_{nk}v_k$$

and let $u = p_1 w_1 + p_2 w_2 + \cdots + p_n w_n$. Next use substitution to show that $u$ can be expressed as a linear combination of the members of $T$.

**54.** If we assume that the members of $T$ in Exercise 53 are $m \times 1$, then an alternate way to verify the statement in Exercise 53 is to use the fact that a matrix-vector product is a linear combination of the columns of the matrix. Proceed as follows: Let $A = \begin{bmatrix} v_1 & v_2 & \cdots & v_k \end{bmatrix}$, $C = \begin{bmatrix} c_{ij} \end{bmatrix}$, $i = 1, 2, \ldots, n, j = 1, 2, \ldots, k$,

$$W = \begin{bmatrix} w_1 & w_2 & \cdots & w_n \end{bmatrix},$$

and $p = \begin{bmatrix} p_1 & p_2 & \cdots & p_n \end{bmatrix}^T$.
(a) List the sizes of **A, C, W**, and **p**.
(b) Explain why $W = AC$.
(c) Explain why $u = Wp$.
(d) Explain why $u = ACp$ and why we can say **u** belongs to span$\{T\}$.

**55.** (a) If $\mathbf{A}$ is $m \times n$, what is the largest value of rank($\mathbf{A}$)?

(b) Explain why rank($\mathbf{A}$) is always the same as rank$\left(\left[\ \mathbf{A}\ |\ \mathbf{0}\ \right]\right)$.

(c) For a linear system $\mathbf{Ax} = \mathbf{b}$, explain why rank($\mathbf{A}$) $\neq$ rank$\left(\left[\ \mathbf{A}\ |\ \mathbf{b}\ \right]\right)$ implies that the system is inconsistent.

**56.** Give a basis for $R^2$. What is dim($R^2$)?

**57.** Give a basis for $R^3$. What is dim($R^3$)?

**58.** Give a basis for $R^k$. What is dim($R^k$)?

**59.** What is dim($V$) for $V = \mathrm{span}\left\{\left[\begin{matrix} 1 & 2 & 1 \end{matrix}\right]\right\}$?

**60.** What is dim($V$) for $V = \mathrm{span}\left\{\begin{bmatrix} 1 \\ -1 \end{bmatrix}, \begin{bmatrix} 0 \\ 0 \end{bmatrix}\right\}$?

**61.** What is dim($V$) for $V = \mathrm{span}\left\{\begin{bmatrix} 1 \\ 0 \\ 1 \end{bmatrix}, \begin{bmatrix} 0 \\ 1 \\ 0 \end{bmatrix}\right\}$?

**62.** Explain why dim(**row**($\mathbf{A}$)) is the number of leading 1's in **rref**($\mathbf{A}$).

**63.** Explain why dim(**col**($\mathbf{A}$)) is the number of leading 1's in **rref**($\mathbf{A}^T$).

**64.** Use the results of Exercises 25–33 to form a conjecture about the relationship between dim(**row**($\mathbf{A}$)) and dim(**col**($\mathbf{A}$)).

**65.** Let $\mathbf{A}$ be a matrix so that rank($\mathbf{A}$) $= r$.

(a) What is dim(**row**($\mathbf{A}$))? Explain.

(b) Form a conjecture for dim(**col**($\mathbf{A}$)). (*Hint*: use Exercise 64.)

**66.** Explain why it is reasonable to claim rank($\mathbf{A}$) $=$ rank($\mathbf{A}^T$). [*Hint*: Use Exercise 65 and that **col**($\mathbf{A}$) $=$ **row**($\mathbf{A}^T$).]

**67.** Let $\mathbf{A}$ be an $n \times n$ matrix such that **rref**($\mathbf{A}$) $= \mathbf{I}_n$.

(a) Describe **ns**($\mathbf{A}$).

(b) What is dim(**col**($\mathbf{A}$))?

(c) What is dim(**row**($\mathbf{A}$))?

(d) What is rank($\mathbf{A}$)?

(e) How many linearly independent rows does $\mathbf{A}$ have?

(f) How many linearly independent columns does $\mathbf{A}$ have?

(g) In solving $\mathbf{Ax} = \mathbf{0}$, how many free variables are there?

(h) Use the answers to parts (a) and (g) to formulate a conjecture for dim(**ns**($\mathbf{A}$)). Explain your conjecture.

**68.** Let $\mathbf{A}$ be the $4 \times 4$ matrix in Example 5.

(a) From Example 5, determine dim(**row**($\mathbf{A}$)).

(b) Use Exercise 58 to deduce dim($R^4$).

**(c)** Explain why the basis for **row**($\mathbf{A}$) determined in Example 4 cannot be a basis for $R^4$. [*Hint*: Let the basis for **row**($\mathbf{A}$) be $\{\mathbf{v}_1, \mathbf{v}_2, \mathbf{v}_3\}$ and $\mathbf{v} = \left[\begin{matrix} a & b & c & d \end{matrix}\right]$ be an arbitrary vector in $R^4$. Show that you cannot always find scalars $k_1$, $k_2$, and $k_3$ so that $k_1\mathbf{v}_1 + k_2\mathbf{v}_2 + k_3\mathbf{v}_3 = \mathbf{v}$.]

**69.** A homogeneous linear system $\mathbf{Bx} = \mathbf{0}$ is such that

$$\mathbf{rref}\left(\left[\ \mathbf{B}\ |\ \mathbf{0}\ \right]\right) = \left[\begin{array}{cccc|c} 1 & -2 & 0 & 1 & 0 \\ 0 & 0 & 1 & 2 & 0 \\ 0 & 0 & 0 & 0 & 0 \\ 0 & 0 & 0 & 0 & 0 \end{array}\right].$$

(a) Find a basis for **ns**($\mathbf{B}$).

(b) What is dim(**ns**($\mathbf{B}$))?

(c) Use Exercise 58 to deduce dim($R^4$).

**70.** Let $4 \times 1$ vectors from part (a) of Exercise 69 be denoted $\mathbf{v}$ and $\mathbf{w}$, and the columns of $\mathbf{I}_4$ be denoted $\mathbf{e}_1, \mathbf{e}_2, \mathbf{e}_3$, and $\mathbf{e}_4$ respectively.

(a) Show that $\{\mathbf{v}, \mathbf{w}, \mathbf{e}_1, \mathbf{e}_3\}$ is linearly independent.

(b) Show that $\{\mathbf{v}, \mathbf{w}, \mathbf{e}_1, \mathbf{e}_3\}$ spans $R^4$.

(c) Why is $\{\mathbf{v}, \mathbf{w}, \mathbf{e}_1, \mathbf{e}_3\}$ a basis for $R^4$?

**71.** Let $4 \times 1$ vectors from part (a) of Exercise 69 be denoted $\mathbf{v}$ and $\mathbf{w}$, and the columns of $\mathbf{I}_4$ be denoted $\mathbf{e}_1, \mathbf{e}_2, \mathbf{e}_3$, and $\mathbf{e}_4$ respectively.

(a) Show that $\{\mathbf{v}, \mathbf{w}, \mathbf{e}_1, \mathbf{e}_2\}$ is linearly dependent.

(b) Can $\{\mathbf{v}, \mathbf{w}, \mathbf{e}_1, \mathbf{e}_2\}$ be a basis for $R^4$? Explain.

*Find a basis for* **ns**($\mathbf{A}$) *and extend it to a basis for* $R^3$ *for the matrices in Exercises* 72 *and* 73.

**72.** $\mathbf{A} = \begin{bmatrix} 4 & 3 & 0 \\ 1 & 1 & 2 \\ 3 & 2 & -2 \end{bmatrix}$.

**73.** $\mathbf{A} = \begin{bmatrix} 1 & 2 & 1 \\ -2 & -4 & -2 \\ 0 & 0 & 0 \end{bmatrix}$.

*In Exercises* 74 *and* 75, *find a basis for* **row**($\mathbf{A}$) *and extend it to a basis for* $R^4$ *for each given matrix.* (*Write the basis vectors for* **row**($\mathbf{A}$) *as columns, then append matrix* $\mathbf{I}_4$.)

**74.** $\mathbf{A} = \begin{bmatrix} 1 & -2 & 0 & 1 \\ 1 & 0 & 0 & 1 \\ 2 & -2 & 0 & 2 \\ 3 & -2 & 0 & 3 \end{bmatrix}$.

**75.** $\mathbf{A} = \begin{bmatrix} 1 & 0 & 2 & 2 \\ 1 & 1 & 0 & 1 \\ 0 & 0 & 1 & 0 \end{bmatrix}$.

*In Exercises* 76 *and* 77, *find a basis for* **row**($\mathbf{A}$), **col**($\mathbf{A}$), *and* **ns**($\mathbf{A}$) *for each given matrix.*

**76.** $\mathbf{A} = \begin{bmatrix} 1+i & 2-i \\ 1 & 1-i \end{bmatrix}$.

**77.** $\mathbf{A} = \begin{bmatrix} 2-i & 2 & 1+2i \\ 0 & i & 1+i \\ 2-i & 2-2i & -1 \end{bmatrix}$.

## In MATLAB

*The techniques of this section rely heavily on the RREF of a matrix. In MATLAB, command* **rref** *was introduced in Section 2.2 and we will use it here. The steps to determine* **ns**(A), **row**(A), *and* **col**(A) *are identical to those used with hand calculations, except MATLAB does the arithmetic of the reduction. You must still formulate the appropriate conclusion.*

**ML. 1.** Determine which of the following sets is linearly dependent.

(a) $\left\{ \begin{bmatrix} 1 \\ 2 \\ 1 \\ 0 \\ 1 \end{bmatrix}, \begin{bmatrix} 0 \\ 1 \\ 2 \\ 1 \\ 1 \end{bmatrix}, \begin{bmatrix} 1 \\ 1 \\ -1 \\ -1 \\ 0 \end{bmatrix} \right\}$.

(b) $\left\{ \begin{bmatrix} 2 & 1 & 4 & 3 \end{bmatrix}, \begin{bmatrix} 1 & 2 & 3 & 4 \end{bmatrix}, \begin{bmatrix} 4 & 3 & 2 & 1 \end{bmatrix} \right\}$.

(c) $\left\{ \begin{bmatrix} 1 \\ 2 \\ 1 \\ 1 \end{bmatrix}, \begin{bmatrix} 1 \\ 4 \\ 1 \\ 1 \end{bmatrix}, \begin{bmatrix} 1 \\ 8 \\ 1 \\ 1 \end{bmatrix}, \begin{bmatrix} 1 \\ 16 \\ 1 \\ 1 \end{bmatrix} \right\}$.

(d) $\left\{ \begin{bmatrix} 1 & 2 \\ 3 & 4 \end{bmatrix}, \begin{bmatrix} 2 & 3 \\ 1 & 0 \end{bmatrix}, \begin{bmatrix} 1 & 1 \\ -2 & -4 \end{bmatrix}, \begin{bmatrix} 1 & 0 \\ 0 & 1 \end{bmatrix} \right\}$.

**ML. 2.** Use MATLAB to check your work in Exercises 1–6.

**ML. 3.** Determine **ns**(A) for each of the following.

(a) $\mathbf{A} = \begin{bmatrix} 2 & 1 & 3 \\ 3 & 2 & 1 \\ 1 & 3 & 2 \\ 1 & 1 & 1 \\ 1 & 3 & 1 \end{bmatrix}$.

(b) $\mathbf{A} = \begin{bmatrix} 4 & 2 & 1 \\ 1 & 0 & 1 \\ 3 & 2 & 0 \\ 8 & 4 & 2 \end{bmatrix}$.

(c) $\mathbf{A} = \begin{bmatrix} 1 & -4 & 3 & 0 & 2 \\ 2 & 1 & -4 & 3 & 0 \\ 0 & 2 & 1 & -4 & 3 \end{bmatrix}$.

**ML. 4.** Determine a nonzero vector **x** so that $\mathbf{Ax} = t\mathbf{x}$ for each of the following pairs $t$, **A**.

(a) $t = 2$, $\mathbf{A} = \begin{bmatrix} -6 & -3 & -5 \\ 5 & 2 & 5 \\ 3 & 3 & 2 \end{bmatrix}$.

(b) $t = 6$, $\mathbf{A} = \begin{bmatrix} 4 & 2 & 0 & 0 \\ 3 & 3 & 0 & 0 \\ 0 & 0 & 5 & -1 \\ 0 & 0 & 2 & 2 \end{bmatrix}$.

(c) $t = 3$, $\mathbf{A} = \begin{bmatrix} 4 & 1.4 & 1.6 & 1 \\ -1 & 0.4 & -0.4 & -1 \\ -2 & -1.6 & -1.4 & -2 \\ 1 & 1.4 & 1.6 & 4 \end{bmatrix}$.

**ML. 5.** Find a basis for **row**(A), **col**(A), and **ns**(A).

(a) $\mathbf{A} = \begin{bmatrix} 4 & 1 & 3 & 8 \\ 2 & 0 & 2 & 4 \\ 1 & 1 & 0 & 2 \end{bmatrix}$.

(b) $\mathbf{A} = \begin{bmatrix} 2 & 4 & 1 & 1 & 2 \\ 1 & 1 & 4 & 2 & 4 \\ 4 & 2 & 2 & 4 & 1 \end{bmatrix}$.

(c) $\mathbf{A} = \begin{bmatrix} 2 & -1 & 3 & 4 & 6 \\ 0 & 2 & -1 & 1 & -3 \\ 0 & 0 & 2 & 2 & 2 \\ -1 & 0 & -2 & -3 & -3 \\ -1 & -1 & -2 & -4 & -2 \end{bmatrix}$.

**ML. 6.** In MATLAB, command **rank(A)** computes the rank of matrix **A**. Determine the rank of each of the matrices in Exercises 25–32.

**ML. 7.** An $n \times n$ matrix is called a **circulant** provided that each row, after the first, is obtained from the preceding one by shifting all the entries one position to the right, with the $n$th entry becoming the first entry. For example,

$$\mathbf{A} = \begin{bmatrix} 1 & 2 & 3 \\ 3 & 1 & 2 \\ 2 & 3 & 1 \end{bmatrix}$$

and

$$\mathbf{B} = \begin{bmatrix} 5 & 8 & 9 & -6 \\ -6 & 5 & 8 & 9 \\ 9 & -6 & 5 & 8 \\ 8 & 9 & -6 & 5 \end{bmatrix}$$

are circulants. In MATLAB we can use command **circ** to construct circulants. Matrix **A** is obtained from **circ([1 2 3])** and **B** from **circ([5 8 9 −6])**. Use command **circ** in each of the following. (There can be more than one solution to some of the following.)

(a) Determine a $5 \times 5$ circulant that has rank 1.

(b) Determine a $5 \times 5$ circulant **C** so that dim(**ns**(C)) = 1.

(c) For your matrix **C** in part (b), determine a nonzero vector such that $\mathbf{Cx} = \mathbf{0}$.

(d) Determine a $5 \times 5$ circulant **D** that has dim(**ns**(D)) = 0.

(e) Besides the pattern of the entries used to define a circulant, what other pattern(s) of the entries do you observe in a circulant? Carefully describe your observations.

## True/False Review Questions

*Determine whether each of the following statements is true or false.*

1. If $\mathbf{v}$ and $\mathbf{w}$ are 3-vectors so that $\mathbf{v} = -4\mathbf{w}$, then $\{\mathbf{v}, \mathbf{w}\}$ is linearly independent.

2. If linear system $\mathbf{Ax} = \mathbf{0}$ has two degrees of freedom, then its solution set is the span of a pair of linearly independent vectors.

3. If linear system $\mathbf{Ax} = \mathbf{b}$ is consistent, then its general solution is the sum of a particular solution and the solution set of $\mathbf{Ax} = \mathbf{0}$.

4. The span of any set is a closed set.

5. $\mathbf{ns(A)}$ is the set of all solutions to $\mathbf{Ax} = \mathbf{0}$.

6. If $\mathbf{A}$ is a real $m \times n$ matrix, then $\mathbf{ns(A)}$ is a subspace of $R^n$.

7. The set $\left\{ \begin{bmatrix} 1 \\ 2 \\ 1 \end{bmatrix}, \begin{bmatrix} 1 \\ 0 \\ 1 \end{bmatrix}, \begin{bmatrix} 0 \\ 0 \\ 0 \end{bmatrix} \right\}$ is linearly dependent.

8. If $\{\mathbf{v}_1, \mathbf{v}_2, \mathbf{v}_3, \mathbf{v}_4\}$ is a linearly dependent set, then at least one of the $\mathbf{v}$'s can be expressed as a linear combination of the other vectors in the set.

9. A basis of a subspace is the smallest set of vectors that will generate every vector in the subspace.

10. If $T$ is a linearly dependent set, then span$\{T\}$ can be generated with fewer vectors than those in $T$.

11. If $\mathbf{Ax} = \mathbf{0}$ has 3 degrees of freedom, then a basis for $\mathbf{ns(A)}$ will have 3 vectors.

12. If $5\mathbf{v}_1 - 2\mathbf{v}_2 + 6\mathbf{v}_3 + 2\mathbf{v}_4 = \mathbf{0}$, then $\{\mathbf{v}_1, \mathbf{v}_2, \mathbf{v}_3, \mathbf{v}_4\}$ is a linearly dependent set.

13. $\mathbf{Row(A)}$ is the span of the rows of matrix $\mathbf{A}$.

14. If $\mathbf{rref(A)} = \begin{bmatrix} 1 & 3 & 0 & 2 & 0 \\ 0 & 0 & 1 & -1 & 0 \\ 0 & 0 & 0 & 0 & 1 \\ 0 & 0 & 0 & 0 & 0 \end{bmatrix}$, then $\dim(\mathbf{row(A)}) = 3$.

15. If $\mathbf{rref(A)} = \begin{bmatrix} 1 & 3 & 0 & 2 & 0 \\ 0 & 0 & 1 & -1 & 0 \\ 0 & 0 & 0 & 0 & 1 \\ 0 & 0 & 0 & 0 & 0 \end{bmatrix}$, then the rows of $\mathbf{A}$ are a linearly dependent set.

16. A linear system $\mathbf{Ax} = \mathbf{b}$ is consistent if and only if $\mathbf{b}$ is in $\mathbf{col(A)}$.

17. Rank $(\mathbf{A}) = \dim(\mathbf{row(A)})$.

18. If $S = \{\mathbf{v}_1, \mathbf{v}_2, \mathbf{v}_3\}$ is a basis for a subspace $V$ of $R^5$, then every vector $\mathbf{w}$ in $V$ is uniquely expressible as a linear combination of the vectors in $S$.

19. The nonzero rows of $\mathbf{rref(A}^T)$, when written as columns, are a basis for $\mathbf{col(A)}$.

20. If linear system $\mathbf{Ax} = \mathbf{b}$ has $\mathbf{rref}\left( \begin{bmatrix} \mathbf{A} & | & \mathbf{b} \end{bmatrix} \right) = \left[ \begin{array}{cccc|c} 1 & 0 & -4 & 0 & 0 \\ 0 & 1 & 2 & 0 & 0 \\ 0 & 0 & 0 & 1 & 0 \\ 0 & 0 & 0 & 0 & 1 \end{array} \right]$, then $\mathbf{b}$ is not in $\mathbf{col(A)}$.

## Terminology

| | |
|---|---|
| Linear independence and dependence | General solution: $\mathbf{x}_p + \mathbf{x}_h$ |
| Null space | Subspace |
| Basis | Row space and column space |
| Underdetermined system | Rank of a matrix |
| Dimension of a subspace | Extending a basis |

The concepts introduced in this section were used to provide information about the structure of solution sets and also the structure of the coefficient matrix. Thus it is important to have a solid understanding of them. Formulate responses to the following in your own words.

- Explain what it means to say that the set of vectors $\{\mathbf{v}_1, \mathbf{v}_2, \cdots, \mathbf{v}_k\}$ is linearly independent.
- In terms of linear combinations, what does it mean to say that the set of vectors $\{\mathbf{w}_1, \mathbf{w}_2, \cdots, \mathbf{w}_k\}$ is linearly dependent?
- For a set containing exactly two vectors, explain how to determine if it is linearly independent or not.
- Any set containing the _____ vector is linearly dependent.
- If the set of vectors $\{\mathbf{w}_1, \mathbf{w}_2, \cdots, \mathbf{w}_k\}$ is linearly dependent, why can we drop at least one of the vectors from this set without a loss of information?
- Define a subspace of $R^n$.
- What is $\mathbf{ns}(\mathbf{A})$? Explain why it is a subspace.
- True or false: The span of the rows of $\mathbf{A}$ is the same as the span of the rows of $\mathbf{rref}(\mathbf{A})$. Give a reason for your choice.
- A linearly independent spanning set of a subspace is called a _____.
- Explain how to find a basis for $\mathbf{ns}(\mathbf{A})$.
- If we compute $\mathbf{rref}(\mathbf{A})$, then the nonzero rows form a _____ for _____.
- Explain how to find a basis for $\mathbf{col}(\mathbf{A})$.
- What does the rank of a matrix count?
- What does the dimension of a subspace count?
- Describe the solution set of $\mathbf{Ax} = \mathbf{b}$ if $\dim(\mathbf{ns}(\mathbf{A})) = 0$.

The focus of this section was on the nature of the solution set of a consistent linear system $\mathbf{Ax} = \mathbf{b}$ which had at least one free variable. In this regard we showed that the general solution, the set of all solutions, was described by $\mathbf{x}_p + \mathbf{x}_h$.

- What does $\mathbf{x}_p$ represent? What does $\mathbf{x}_h$ represent?
- The term $\mathbf{x}_h$ really came from the span of a set $S$ of vectors associated with the solution set of the linear system _____. Since it is our practice to use RREF, the vectors in the set $S$ are guaranteed to be _____. This meant we had the _____ amount of information needed in order to select $\mathbf{x}_h$ correctly.
- In using RREF to reduce $\begin{bmatrix} \mathbf{A} \mid \mathbf{b} \end{bmatrix}$ we found that there were $k \geq 1$ free variables. This told us $\dim(\mathbf{ns}(\mathbf{A})) = $ _____ and, if $\mathbf{A}$ were $m \times n$, $\dim(\mathbf{row}(\mathbf{A})) = $ _____.

## 2.4 ■ NONSINGULAR LINEAR SYSTEMS

In Sections 2.1, 2.2, and 2.3 we considered linear systems of $m$ equations in $n$ unknowns. We found that for consistent linear systems the solution set contained either infinitely many vectors or a single vector. We showed that the contents of the solution set could be determined by using row operations to obtain equivalent linear systems in upper triangular form or in either REF or RREF. In this section we focus on square linear systems $\mathbf{Ax} = \mathbf{b}$ whose solution set is a single vector; that is, $\mathbf{Ax} = \mathbf{b}$ has a unique solution. Our goal is to develop properties of the coefficient matrix $\mathbf{A}$ of such systems. We will show that such systems can be manipulated like algebraic equations in order to determine a formula for their solution. Such a capability provides great flexibility in dealing with a wide range of questions that involve matrix equations.

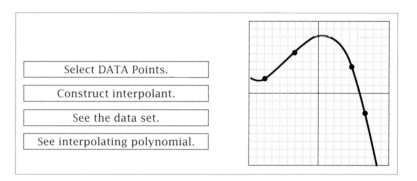

In MATLAB, **pinterp.m**

Let $\mathbf{A}$ be $n \times n$. From Observation 3 immediately before Example 5 in Section 2.2 (see page 111), we have the following important result.

$\mathbf{Ax} = \mathbf{b}$ has a unique solution for *any* $n \times 1$ right side $\mathbf{b}$ if and only if $\mathbf{rref}(\mathbf{A}) = \mathbf{I}_n$, the $n \times n$ identity matrix.

It follows that if $\mathbf{rref}(\mathbf{A}) = \mathbf{I}_n$, then $\mathbf{rref}\left(\left[\,\mathbf{A} \mid \mathbf{b}\,\right]\right) = \left[\,\mathbf{I}_n \mid \mathbf{c}\,\right]$ and $\mathbf{x} = \mathbf{c}$ is the unique solution to $\mathbf{Ax} = \mathbf{b}$. The case where $\mathbf{b} = \mathbf{0}$ is a homogeneous linear system, $\mathbf{x} = \mathbf{c} = \mathbf{0}$ which implies that the linear system has only the trivial solution. (That is, there are no arbitrary constants, and hence zero degrees of freedom in the solution set of such a system.) Coefficient matrices of square linear systems with unique solutions are important in their own right, and we make the following definition.

---

**Definition**   An $n \times n$ matrix $\mathbf{A}$ is called **nonsingular**[1] or **invertible** provided that $\mathbf{rref}(\mathbf{A}) = \mathbf{I}_n$.

---

Thus we say that a **nonsingular linear system** has a unique solution, meaning that the coefficient matrix of the linear system is square and its RREF is an identity matrix.

---

[1]Historically the word *singular* came to mean "out of the ordinary, troublesome." Thus *nonsingular* is the opposite, meaning that some ordinary or algebraically natural behavior is to be expected. See *The Words of Mathematics* by S. Schwartzman (Washington, D.C.: The Mathematical Association of America, 1994).

Next we develop an important property of nonsingular matrices. We will use linear combinations, linear systems, RREF, and properties of partitioned matrices from Section 1.3. Recall that a matrix product $\mathbf{PQ}$ can be viewed as $\mathbf{P}$ times $\mathbf{Q}$ partitioned into its columns, that is, $\mathbf{Q} = \begin{bmatrix} \mathbf{q}_1 & \mathbf{q}_2 & \cdots & \mathbf{q}_n \end{bmatrix}$, so

$$\mathbf{PQ} = \mathbf{P}\begin{bmatrix} \mathbf{q}_1 & \mathbf{q}_2 & \cdots & \mathbf{q}_n \end{bmatrix} = \begin{bmatrix} \mathbf{Pq}_1 & \mathbf{Pq}_2 & \cdots & \mathbf{Pq}_n \end{bmatrix}.$$

Thus each column of the product is really a linear combination of the columns of matrix $\mathbf{P}$.

If $\mathbf{A}$ is an $n \times n$ nonsingular matrix, then for any $n \times 1$ matrix $\mathbf{b}$, the linear system $\mathbf{Ax} = \mathbf{b}$ has a unique solution. Partition $\mathbf{I}_n$ into its columns which we denote as $\mathbf{e}_j$, $j = 1, 2, \ldots, n$; $\mathbf{I}_n = \begin{bmatrix} \mathbf{e}_1 & \mathbf{e}_2 & \cdots & \mathbf{e}_n \end{bmatrix}$. Then $\mathbf{Ax} = \mathbf{e}_j$ has a unique solution for each $j$. Denote the solution as $\mathbf{c}_j$. Hence we have

$$\mathbf{Ac}_1 = \mathbf{e}_1, \quad \mathbf{Ac}_2 = \mathbf{e}_2, \quad \ldots, \quad \mathbf{Ac}_n = \mathbf{e}_n. \tag{1}$$

Writing (1) in matrix form, we obtain

$$\begin{bmatrix} \mathbf{Ac}_1 & \mathbf{Ac}_2 & \cdots & \mathbf{Ac}_n \end{bmatrix} = \begin{bmatrix} \mathbf{e}_1 & \mathbf{e}_2 & \cdots & \mathbf{e}_n \end{bmatrix} = \mathbf{I}_n.$$

From our observation on partitioned matrices previously, this equation can be written in the form

$$\mathbf{A}\begin{bmatrix} \mathbf{c}_1 & \mathbf{c}_2 & \cdots & \mathbf{c}_n \end{bmatrix} = \mathbf{I}_n.$$

Letting $\mathbf{C} = \begin{bmatrix} \mathbf{c}_1 & \mathbf{c}_2 & \cdots & \mathbf{c}_n \end{bmatrix}$, we have just shown the following result.

**Statement 1**   If $\mathbf{A}$ is nonsingular then there exists a unique matrix $\mathbf{C}$ such that $\mathbf{AC} = \mathbf{I}_n$.

Note that vector $\mathbf{c}_j = \mathbf{col}_j(\mathbf{C})$ is the unique solution of $\mathbf{Ax} = \mathbf{e}_j$; hence it follows that matrix $\mathbf{C}$ is unique. It is also true that

**Statement 2**   If $\mathbf{A}$ is nonsingular, then there exists a unique matrix $\mathbf{D}$ such that $\mathbf{DA} = \mathbf{I}_n$.

This is verified at the end of this section. There is no loss of continuity by omitting a discussion of the verification of Statement 2.

Next we show that $\mathbf{C} = \mathbf{D}$ as follows:

$$\mathbf{C} = \mathbf{I}_n\mathbf{C} = (\mathbf{DA})\mathbf{C} = \mathbf{D}(\mathbf{AC}) = \mathbf{DI}_n = \mathbf{D}. \tag{2}$$

Thus combining Statements 1 and 2, we have

**Statement 3**   If $\mathbf{A}$ is nonsingular, then there exists a unique matrix $\mathbf{C}$ such that $\mathbf{AC} = \mathbf{CA} = \mathbf{I}_n$.

Statement 3 motivates the following definition.

---

**Definition**   For an $n \times n$ nonsingular matrix $\mathbf{A}$, the unique matrix $\mathbf{C}$ such that $\mathbf{AC} = \mathbf{CA} = \mathbf{I}_n$ is called the **inverse** of $\mathbf{A}$. We denote the inverse of $\mathbf{A}$ by the symbol $\mathbf{A}^{-1}$.

---

Now you see why nonsingular matrices are also called invertible matrices; they have an inverse.

**WARNING:** $A^{-1}$ is not to be interpreted as a reciprocal[2]; it is just the notation to denote the inverse of matrix $A$.

To complete this characterization of nonsingular matrices we also need to verify that if $A$ has an inverse then $\mathbf{rref}(A) = I_n$. We do this at the end of this section.

To summarize, we present the following set of equivalent statements:

1. $A$ is nonsingular.
2. $\mathbf{rref}(A) = I_n$.
3. $A$ has an inverse.

We have tied together the unique solution of a square linear system, RREF, and the matrix multiplication property $AA^{-1} = A^{-1}A = I_n$. This leads us to the following observation.

A nonsingular linear system $Ax = b$ has the unique solution $x = A^{-1}b$.

This follows from the equivalent statements given next:

$$Ax = b \quad \text{or} \quad A^{-1}Ax = A^{-1}b \quad \text{or} \quad I_n x = A^{-1}b \quad \text{or} \quad x = A^{-1}b. \qquad (3)$$

The relations in (3) say that we have a formula for the unique solution of a nonsingular linear system and it can be computed in two steps:

1. Find $A^{-1}$.
2. Compute the matrix-vector product $A^{-1}x$.

The final result in (3) is advantageous for algebraic manipulation of expressions and theoretical purposes, but the solution of $Ax = b$ is more efficiently computed using the fact that $\mathbf{rref}\left(\begin{bmatrix} A & | & b \end{bmatrix}\right) = \begin{bmatrix} I_n & | & c \end{bmatrix}$ implies that the unique solution is $x = c$ or by using LU-factorization, which is discussed in Section 2.5. These comments imply that from a computational point of view we should (try to) avoid the direct computation of the inverse of a matrix. To do this we use the connection that the inverse can be avoided by solving a linear system, which is the meaning of (3) read from right to left. We illustrate this in Example 1.

EXAMPLE 1    Determine the vector $p$ that is given by $p = \begin{bmatrix} A^{-1} + W \end{bmatrix} b$ where

$$A = \begin{bmatrix} 1 & 4 & 5 \\ 2 & 9 & 10 \\ 3 & 13 & 16 \end{bmatrix}, \quad W = \begin{bmatrix} 1 & 2 & 1 \\ -3 & 4 & 1 \\ -1 & 8 & 3 \end{bmatrix}, \quad \text{and} \quad b = \begin{bmatrix} 2 \\ -3 \\ 0 \end{bmatrix}.$$

One approach is to compute $A^{-1}$, then form the sum $A^{-1} + W$, and finally multiply $b$ by this sum. However, so far we have not given a procedure for computing the inverse of a matrix (but we will, shortly) so we have no alternative at this time but to take another approach. We see that the computation of $p$ is equivalent to the sum $A^{-1}b + Wb$. From (3) it follows that if we let $x = A^{-1}b$, then this expression is the same as saying solve the linear system $Ax = b$. Thus we compute $x$ from $\mathbf{rref}\left(\begin{bmatrix} A & | & b \end{bmatrix}\right) = \begin{bmatrix} I_n & | & c \end{bmatrix}$; that is, $x = c$. Hence we compute $p$ from the expression $p = c + Wb$. The computational details are as follows:

[2]In the real or complex numbers, the term *reciprocal* means "inverse."

(1) $\mathbf{rref}\left(\begin{bmatrix} \mathbf{A} \mid \mathbf{b} \end{bmatrix}\right) = \mathbf{rref}\left(\begin{bmatrix} 1 & 4 & 5 & 2 \\ 2 & 9 & 10 & -3 \\ 3 & 13 & 16 & 0 \end{bmatrix}\right) = \begin{bmatrix} 1 & 0 & 0 & 25 \\ 0 & 1 & 0 & -7 \\ 0 & 0 & 1 & 1 \end{bmatrix}.$

(2) $\mathbf{c} = \begin{bmatrix} 25 \\ -7 \\ 1 \end{bmatrix}$, so

$$\mathbf{p} = \mathbf{c} + \mathbf{Wb} = \begin{bmatrix} 25 \\ -7 \\ 1 \end{bmatrix} + \begin{bmatrix} 1 & 2 & 1 \\ -3 & 4 & 1 \\ -1 & 8 & 3 \end{bmatrix} \begin{bmatrix} 2 \\ -3 \\ 0 \end{bmatrix}$$

$$= \begin{bmatrix} 25 \\ -7 \\ 1 \end{bmatrix} + \begin{bmatrix} -4 \\ -18 \\ -26 \end{bmatrix} = \begin{bmatrix} 21 \\ -25 \\ -25 \end{bmatrix}. \quad ■$$

The important message in Example 1 is that we need not know how to compute the inverse of a matrix directly in order to work with it in matrix expressions. We can often perform matrix algebra so we can use the computation of the solution of a linear system instead of computing the inverse.

Next we show how to compute the inverse of a matrix. The procedure we develop simultaneously determines if a square matrix has an inverse, and if it does, computes it. From (1) we have that the columns of the inverse of $\mathbf{A}$ are the solutions of the linear systems

$$\mathbf{Ax} = \mathrm{col}_j(\mathbf{A}) = \mathbf{e}_j, \quad j = 1, 2, \ldots, n.$$

The corresponding set of augmented matrices can be combined into the following partitioned matrix:

$$\begin{bmatrix} \mathbf{A} \mid \mathbf{e}_1 \mid \mathbf{e}_2 \mid \cdots \mid \mathbf{e}_n \end{bmatrix} = \begin{bmatrix} \mathbf{A} \mid \mathbf{I}_n \end{bmatrix}.$$

From our previous arguments we have the following statement:

**Statement 4**  For an $n \times n$ matrix, $\mathbf{rref}\left(\begin{bmatrix} \mathbf{A} \mid \mathbf{I}_n \end{bmatrix}\right) = \begin{bmatrix} \mathbf{I}_n \mid \mathbf{C} \end{bmatrix}$ if and only if $\mathbf{A}$ is nonsingular and $\mathbf{C} = \mathbf{A}^{-1}$.

It follows that if $\mathbf{rref}\left(\begin{bmatrix} \mathbf{A} \mid \mathbf{I}_n \end{bmatrix}\right) \neq \begin{bmatrix} \mathbf{I}_n \mid \mathbf{C} \end{bmatrix}$, then $\mathbf{A}$ has no inverse. We call an $n \times n$ matrix with no inverse a **singular** matrix. Recall that our focus here is on square matrices. The terms *singular* and *nonsingular* do not apply to matrices that are not square.

EXAMPLE 2    For each of the following matrices determine if it is nonsingular, and if it is, find its inverse.

(a) $\mathbf{A} = \begin{bmatrix} 1 & 4 & 5 \\ 2 & 9 & 10 \\ 3 & 13 & 16 \end{bmatrix}$. We compute $\mathbf{rref}\left(\begin{bmatrix} \mathbf{A} \mid \mathbf{I}_3 \end{bmatrix}\right)$ and get

$$\begin{bmatrix} 1 & 0 & 0 & 14 & 1 & -5 \\ 0 & 1 & 0 & -2 & 1 & 0 \\ 0 & 0 & 1 & -1 & -1 & 1 \end{bmatrix}.$$

(Verify.) Hence $\mathbf{A}$ is nonsingular and

$$\mathbf{A}^{-1} = \begin{bmatrix} 14 & 1 & -5 \\ -2 & 1 & 0 \\ -1 & -1 & 1 \end{bmatrix}.$$

(b) $\mathbf{W} = \begin{bmatrix} 1 & 2 & 1 \\ -3 & 4 & 1 \\ -1 & 8 & 3 \end{bmatrix}$. We compute $\mathbf{rref}\left(\left[\ \mathbf{W} \mid \mathbf{I}_3\ \right]\right)$ and get

$$\left[\begin{array}{ccc|ccc} 1 & 0 & 0 & 0.2 & -0.4 & 0.2 \\ 0 & 1 & 0.4 & 0 & -0.5 & 0.15 \\ 0 & 0 & 0 & 1 & 0.5 & -0.5 \end{array}\right].$$

(Verify.) Hence $\mathbf{W}$ is singular.  ■

**EXAMPLE 3**   Let $\mathbf{A} = \begin{bmatrix} a & b \\ c & d \end{bmatrix}$ be nonsingular. By carefully using row operations on $\left[\ \mathbf{A} \mid \mathbf{I}_2\ \right]$ we can develop an expression for $\mathbf{A}^{-1}$ as follows. To start, we know that $\mathbf{rref}(\mathbf{A}) = \mathbf{I}_2$; hence one of the entries in column 1 of $\mathbf{A}$ is not zero. Assume that $a \neq 0$; if $a = 0$, then we can interchange rows 1 and 2 and proceed. Use the entry $a$ as the first pivot and perform the following sequence of row operations:

$$\left[\begin{array}{cc|cc} a & b & 1 & 0 \\ c & d & 0 & 1 \end{array}\right]_{\left(\frac{1}{a}\right)R_1} \Rightarrow \left[\begin{array}{cc|cc} 1 & \dfrac{b}{a} & \dfrac{1}{a} & 0 \\ c & d & 0 & 1 \end{array}\right]_{-cR_1+R_2}$$

$$\Rightarrow \left[\begin{array}{cc|cc} 1 & \dfrac{b}{a} & \dfrac{1}{a} & 0 \\ 0 & d - \dfrac{bc}{a} & -\dfrac{c}{a} & 1 \end{array}\right].$$

Since $\mathbf{rref}(\mathbf{A}) = \mathbf{I}_2$ entry $d - \dfrac{bc}{a} = \dfrac{ad - bc}{a} \neq 0$ (Why must this be true?) so we can use it as our second pivot to continue the reduction to RREF.

$$\Rightarrow \left[\begin{array}{cc|cc} 1 & \dfrac{b}{a} & \dfrac{1}{a} & 0 \\ 0 & d - \dfrac{bc}{a} & -\dfrac{c}{a} & 1 \end{array}\right]_{\left(\frac{a}{ad-bc}\right)R_2} \Rightarrow \left[\begin{array}{cc|cc} 1 & \dfrac{b}{a} & \dfrac{1}{a} & 0 \\ 0 & 1 & \dfrac{-c}{ad - bc} & \dfrac{a}{ad - bc} \end{array}\right].$$

Next we eliminate above the leading 1 in the second row:

$$\Rightarrow \left[\begin{array}{cc|cc} 1 & \dfrac{b}{a} & \dfrac{1}{a} & 0 \\ 0 & 1 & \dfrac{-c}{ad - bc} & \dfrac{a}{ad - bc} \end{array}\right]_{\left(-\frac{b}{a}\right)R_2+R_1}$$

$$\Rightarrow \left[\begin{array}{cc|cc} 1 & 0 & \dfrac{d}{ad - bc} & \dfrac{-b}{ad - bc} \\ 0 & 1 & \dfrac{-c}{ad - bc} & \dfrac{a}{ad - bc} \end{array}\right].$$

Thus

$$\mathbf{A}^{-1} = \begin{bmatrix} \dfrac{d}{ad-bc} & \dfrac{-b}{ad-bc} \\[2ex] \dfrac{-c}{ad-bc} & \dfrac{a}{ad-bc} \end{bmatrix} = \dfrac{1}{ad-bc} \begin{bmatrix} d & -b \\ -c & a \end{bmatrix}.$$

To use this formula we note that we compute the reciprocal of $(ad-bc)$, interchange the diagonal entries, and change the sign of the off-diagonal entries. Following this prescription for

$$\mathbf{A} = \begin{bmatrix} 3 & -1 \\ 2 & 0 \end{bmatrix},$$

we have

$$\begin{bmatrix} 3 & -1 \\ 2 & 0 \end{bmatrix}^{-1} = \dfrac{1}{(3)(0)-(2)(-1)} \begin{bmatrix} 0 & 1 \\ -2 & 3 \end{bmatrix} = \dfrac{1}{2} \begin{bmatrix} 0 & 1 \\ -2 & 3 \end{bmatrix}.$$

We also note that if $ad-bc=0$, then the $2 \times 2$ matrix has no inverse; that is, it is singular. This computation, $ad-bc$, is often used as a quick method to test if a $2 \times 2$ matrix is nonsingular or not. We will see this in more detail in Chapter 3. ■

For $3 \times 3$ matrices a development similar to that in Example 3 can be made. The result is a good deal more complicated. Let $w = iae - afh - idb + dch + gbf - gce$. Then

$$\begin{bmatrix} a & b & c \\ d & e & f \\ g & h & i \end{bmatrix}^{-1} = \dfrac{1}{w} \begin{bmatrix} ie-fh & -(ib-ch) & bf-ce \\ -id+fg & ia-cg & -(af-cd) \\ -(-dh+eg) & -(ah-bg) & ae-bd \end{bmatrix}.$$

If the value of $w=0$, then the $3 \times 3$ matrix is singular. Computationally it is more realistic to compute $\mathbf{rref}\left(\begin{bmatrix} \mathbf{A} & | & \mathbf{I}_n \end{bmatrix}\right)$ and extract $\mathbf{A}^{-1}$ as we did in Example 2(a).

The computation of the inverse of a nonsingular matrix can be viewed as an operation on the matrix and need not be linked to the solution of a linear system. The inverse operation has algebraic properties in the same way that the transpose of a matrix had properties. We adopt the convention that if we write the symbol $\mathbf{A}^{-1}$, then it is understood that $\mathbf{A}$ is nonsingular. Using this convention, we next list properties of the inverse.

### Algebraic Properties of the Inverse

1. $(\mathbf{AB})^{-1} = \mathbf{B}^{-1}\mathbf{A}^{-1}$     The inverse of a product is the product of the inverses in reverse order.

2. $(\mathbf{A}^T)^{-1} = (\mathbf{A}^{-1})^T$     When $\mathbf{A}$ is nonsingular, so is $\mathbf{A}^T$ and the inverse of $\mathbf{A}^T$ is the transpose of $\mathbf{A}^{-1}$.

3. $(\mathbf{A}^{-1})^{-1} = \mathbf{A}$     When $\mathbf{A}$ is nonsingular, so is $\mathbf{A}^{-1}$ and its inverse is $\mathbf{A}$.

4. $(k\mathbf{A})^{-1} = \dfrac{1}{k}\mathbf{A}^{-1}$   $k \neq 0$     When $\mathbf{A}$ is nonsingular, so is any nonzero scalar multiple of it.

5. $(\mathbf{A}_1\mathbf{A}_2 \cdots \mathbf{A}_r)^{-1} = \mathbf{A}_r^{-1} \cdots \mathbf{A}_2^{-1}\mathbf{A}_1^{-1}$     The inverse of a product extends to any number of factors.

The verification of the preceding properties uses the fact that a matrix $\mathbf{C}$ is the inverse of an $n \times n$ matrix $\mathbf{D}$ provided that $\mathbf{CD} = \mathbf{I}_n$, or $\mathbf{DC} = \mathbf{I}_n$. This is the alternative to the definition of nonsingular matrices given in Statement 3 earlier. We illustrate the use of this equivalent characterization of nonsingular matrices in Example 4, where we verify property 1 from the preceding list and leave the verification of the remaining properties as exercises.

**EXAMPLE 4**   To show that $(\mathbf{AB})^{-1} = \mathbf{B}^{-1}\mathbf{A}^{-1}$, we need only verify that when we multiply $\mathbf{AB}$ by $\mathbf{B}^{-1}\mathbf{A}^{-1}$, we get the identity matrix. We have

$$(\mathbf{AB})(\mathbf{B}^{-1}\mathbf{A}^{-1}) = \mathbf{A}(\mathbf{BB}^{-1})\mathbf{A}^{-1} = \mathbf{A}(\mathbf{I}_n)\mathbf{A}^{-1} = \mathbf{AA}^{-1} = \mathbf{I}_n$$

Thus the inverse of $\mathbf{AB}$ is $\mathbf{B}^{-1}\mathbf{A}^{-1}$.   ∎

A common thread that has linked our topics in this chapter is linear combinations. It is unfortunate that the operation of computing the inverse does not follow one of the principles of linear combinations. Namely, $(\mathbf{A} + \mathbf{B})^{-1}$ is *not* guaranteed to be $\mathbf{A}^{-1} + \mathbf{B}^{-1}$. We illustrate this failure and more in Example 5.

**EXAMPLE 5**

(a) Let

$$\mathbf{A} = \begin{bmatrix} 1 & 2 \\ 0 & 1 \end{bmatrix} \quad \text{and} \quad \mathbf{B} = \begin{bmatrix} -1 & 6 \\ 0 & 1 \end{bmatrix}.$$

Both $\mathbf{A}$ and $\mathbf{B}$ are nonsingular (verify), but

$$\mathbf{A} + \mathbf{B} = \begin{bmatrix} 0 & 8 \\ 0 & 2 \end{bmatrix}$$

is not since its RREF is $\begin{bmatrix} 0 & 1 \\ 0 & 0 \end{bmatrix} \neq \mathbf{I}_2$. So $\mathbf{A} + \mathbf{B}$ is singular.

(b) Let

$$\mathbf{A} = \begin{bmatrix} 1 & 0 \\ 0 & 0 \end{bmatrix} \quad \text{and} \quad \mathbf{B} = \begin{bmatrix} 0 & 0 \\ 0 & 1 \end{bmatrix}.$$

Both $\mathbf{A}$ and $\mathbf{B}$ are singular. (Verify.) However, $\mathbf{A} + \mathbf{B} = \mathbf{I}_2$ is nonsingular. (Verify.)

Hence nonsingular matrices are not closed under addition, nor are singular matrices.   ∎

As we have seen, the construction of the inverse of a matrix can require quite a few arithmetic steps, which are carried out using row operations. However, there are several types of matrices for which it is easy to determine by observation if they are nonsingular. We present several such results next.

**Statement 5**   A diagonal matrix, a square lower triangular matrix, and a square upper triangular matrix are nonsingular if and only if all the diagonal entries are different from zero.

A verification of Statement 5 can be constructed using an argument that relies only on the definition of a nonsingular matrix; namely, its RREF is an identity matrix. (See Exercises 33 through 35.) For a nonsingular diagonal matrix $\mathbf{D}$ we can explicitly construct its inverse, $\mathbf{D}^{-1}$, as we show in Example 6.

EXAMPLE 6   Let $\mathbf{D}$ be an $n \times n$ nonsingular diagonal matrix. By Statement 5 we have

$$\mathbf{D} = \begin{bmatrix} d_{11} & 0 & \cdots & 0 & 0 \\ 0 & d_{22} & \cdots & 0 & 0 \\ 0 & 0 & \ddots & \vdots & \vdots \\ \vdots & \vdots & \cdots & \ddots & 0 \\ 0 & 0 & \cdots & 0 & d_{nn} \end{bmatrix}$$

where $d_{jj} \neq 0$, $j = 1, 2, \ldots, n$. To construct $\mathbf{D}^{-1}$ we compute $\mathbf{rref}\left(\left[\, \mathbf{D} \mid \mathbf{I}_n \,\right]\right)$. Since the only nonzero entries are the diagonal entries, we use row operations to produce leading 1's. It follows that the $n$ row operations

$$\left(\tfrac{1}{d_{11}}\right) R_1, \quad \left(\tfrac{1}{d_{22}}\right) R_2, \quad \ldots, \quad \left(\tfrac{1}{d_{nn}}\right) R_n$$

applied to $\left[\, \mathbf{D} \mid \mathbf{I}_n \,\right]$ give

$$\mathbf{rref}\left(\left[\, \mathbf{D} \mid \mathbf{I}_n \,\right]\right) = \left[\, \mathbf{I}_n \mid \mathbf{D}^{-1} \,\right]$$

$$= \left[\begin{array}{ccccc|ccccc} 1 & 0 & \cdots & 0 & 0 & \dfrac{1}{d_{11}} & 0 & \cdots & 0 & 0 \\ 0 & 1 & \cdots & 0 & 0 & 0 & \dfrac{1}{d_{22}} & \cdots & 0 & 0 \\ 0 & 0 & \ddots & \vdots & \vdots & 0 & 0 & \ddots & \vdots & \vdots \\ \vdots & \vdots & \cdots & \ddots & 0 & \vdots & \vdots & \cdots & \ddots & 0 \\ 0 & 0 & \cdots & 0 & 1 & 0 & 0 & \cdots & 0 & \dfrac{1}{d_{nn}} \end{array}\right].$$

Hence $\mathbf{D}^{-1}$ is a diagonal matrix whose diagonal entries are the reciprocals of the diagonal entries of $\mathbf{D}$. ■

We next investigate the subspaces associated with a nonsingular matrix. The *key* to verifying each of the following statements is that the RREF of a nonsingular matrix is an identity matrix. Let $\mathbf{A}$ be an $n \times n$ nonsingular matrix; then we have the following properties.

### Properties of a Nonsingular Matrix

1. A basis for $\mathbf{row(A)}$ ($\mathbf{col(A)}$) are the rows (columns) of $\mathbf{I}_n$.
2. $\mathbf{ns(A)} = \mathbf{0}$.
3. The rows (columns) of $\mathbf{A}$ are linearly independent.
4. $\dim(\mathbf{row(A)}) = n$ and $\dim(\mathbf{col(A)}) = n$.
5. $\mathrm{rank}(\mathbf{A}) = n$.

It is often difficult to determine by inspection whether a square matrix $\mathbf{A}$ is singular or nonsingular. For us, at this time, the determination depends upon looking at the RREF. However, since the RREF of a nonsingular matrix must be an identity matrix, sometimes an inspection of $\mathbf{A}$ can tell us that this is not possible, and hence we can claim that $\mathbf{A}$ is singular. The following is a list of occurrences that immediately imply that $\mathbf{A}$ is singular.

## Ways to Recognize Singular Matrices

1. **A** has a zero row or zero column.
2. **A** has a pair of rows (columns) that are multiples of one another.
3. As we row reduce **A**, a zero row appears.
4. The rows (columns) of **A** are linearly dependent.
5. **ns(A)** contains a nonzero vector.

The algebraic properties of the inverse and properties of nonsingular matrices together with the recognition information given above regarding singular matrices give us a wide variety of tools for use in many different situations. The key is to be able to choose the correct tool for the situation. Acquiring such a skill requires practice, and the exercises are designed as a first step in this regard.

Nonsingular linear systems and nonsingular matrices appear in many different settings. The accompanying applications are just a sample of topics that use such systems and matrices. We have also included an important computational instance where singular matrices arise in Section 2.6.

### Verification of Statement 2 (Optional)

If **A** is nonsingular then there exists a matrix **D** such that $\mathbf{DA} = \mathbf{I}_n$.

If **A** is nonsingular and $n \times n$, then we know that $\mathbf{rref(A)} = \mathbf{I}_n$. This says **A** is row equivalent to $\mathbf{I}_n$. Thus there is a sequence of row operations that when applied to **A** reduces it to $\mathbf{I}_n$. However, each row operation applied to **A** can be performed by a matrix multiplication using the following device.

---

**Definition**   An $n \times n$ **elementary matrix** is a matrix obtained by performing a single row operation on $\mathbf{I}_n$.

---

Exercise 54 shows that if **E** is the elementary matrix corresponding to a row operation, then matrix product **EA** gives the same result as performing the row operation directly on **A**. Thus we have the following:

If $\mathbf{E}_j$ is the elementary matrix that corresponds to the $j$th row operation that is performed on **A** in the process of computing $\mathbf{rref(A)}$, and there are $N$ such operations, then

$$\mathbf{E}_N \cdots \mathbf{E}_2 \mathbf{E}_1 \mathbf{A} = \mathbf{rref(A)}. \tag{4}$$

Let $\mathbf{D} = \mathbf{E}_N \cdots \mathbf{E}_2 \mathbf{E}_1$. Then we have an $n \times n$ matrix **D** such that $\mathbf{DA} = \mathbf{rref(A)}$.

### Verification of the claim "If **A** has an inverse, then $\mathbf{rref(A)} = \mathbf{I}_n$." (Optional)

If **A** has an inverse, then there exists a unique matrix **C** such that $\mathbf{AC} = \mathbf{I}_n$. Partitioning **C** and $\mathbf{I}_n$ into their columns, we have

$$\mathbf{AC} = \mathbf{A} \begin{bmatrix} \mathbf{c}_1 & \mathbf{c}_2 & \cdots & \mathbf{c}_n \end{bmatrix} = \begin{bmatrix} \mathbf{Ac}_1 & \mathbf{Ac}_2 & \cdots & \mathbf{Ac}_n \end{bmatrix}$$
$$= \mathbf{I}_n = \begin{bmatrix} \mathbf{e}_1 & \mathbf{e}_2 & \cdots & \mathbf{e}_n \end{bmatrix}$$

and equating corresponding columns, it follows that $\mathbf{Ac}_j = \mathbf{e}_j$, $j = 1, 2, \ldots, n$. Showing that $\mathbf{rref(A)} = \mathbf{I}_n$ is equivalent to showing that the linear system $\mathbf{Ax} = \mathbf{b}$ has a unique solution for any right side **b**. For any $n \times 1$ vector **b** we have

$$\mathbf{b} = \sum_{j=1}^{n} b_j \mathbf{e}_j.$$

Now we observe that

$$\mathbf{A}\left(\sum_{j=1}^{n} b_j \mathbf{c}_j\right) = \sum_{j=1}^{n} b_j (\mathbf{A}\mathbf{c}_j) = \sum_{j=1}^{n} b_j \mathbf{e}_j = \mathbf{b}.$$

Thus $\mathbf{x} = \sum_{j=1}^{n} b_j \mathbf{c}_j$ is a solution to $\mathbf{A}\mathbf{x} = \mathbf{b}$, and since the $\mathbf{c}_j$ are unique, so is $\mathbf{x}$. By the equivalence cited previously, we have that $\mathbf{rref}(\mathbf{A}) = \mathbf{I}_n$.

### Application:    Polynomial Interpolation

For a set of $n + 1$ distinct data points, $D = \{(x_0, y_0), (x_1, y_1), \ldots, (x_n, y_n)\}$ with $x_i \neq x_j$ when $i \neq j$, there exists a unique polynomial $P_n(x)$ of degree $n$ or less such that

$$P_n(x_0) = y_0, \quad P_n(x_1) = y_1, \quad \ldots, \quad P_n(x_n) = y_n. \tag{5}$$

We say that $P_n(x)$ interpolates the data in $D$ and call $P_n(x)$ the **polynomial interpolant** of the data in $D$. We have previously used special cases of polynomial interpolation in Example 4 and Exercise 4 in Section 1.1 and in the application "Quadratic Interpolation" and Exercises 44 and 45 in Section 2.1. Here we investigate it in general.

The process of polynomial interpolation constructs a polynomial to match (interpolate) the data in $D$. This is expressed algebraically by (5), which we use to determine the coefficients of the polynomial $P_n(x)$. For the polynomial

$$P_n(x) = a_n x^n + a_{n-1} x^{n-1} + \cdots + a_1 x + a_0$$

we show that we can determine the $n + 1$ coefficients $a_n, a_{n-1}, \ldots, a_1, a_0$ as the solution of a nonsingular linear system. Each of the requirements in (5) gives us an equation:

$$P_n(x_0) = y_0 \Rightarrow a_n x_0^n + a_{n-1} x_0^{n-1} + \cdots + a_2 x_0^2 + a_1 x_0 + a_0 = y_0$$
$$P_n(x_1) = y_1 \Rightarrow a_n x_1^n + a_{n-1} x_1^{n-1} + \cdots + a_2 x_1^2 + a_1 x_1 + a_0 = y_1$$
$$\vdots \qquad\qquad\qquad\qquad \vdots \qquad\qquad\qquad\qquad \vdots$$
$$P_n(x_n) = y_n \Rightarrow a_n x_n^n + a_{n-1} x_n^{n-1} + \cdots + a_2 x_n^2 + a_1 x_n + a_0 = y_n.$$

This is a linear system of $n+1$ equations in the $n+1$ unknowns $a_n, a_{n-1}, \ldots, a_1, a_0$. Let

$$\mathbf{A} = \begin{bmatrix} x_0^n & x_0^{n-1} & \cdots & x_0 & 1 \\ x_1^n & x_1^{n-1} & \cdots & x_1 & 1 \\ \vdots & \vdots & \vdots & \vdots & \vdots \\ \vdots & \vdots & \vdots & \vdots & \vdots \\ x_n^n & x_n^{n-1} & \cdots & x_n & 1 \end{bmatrix}, \quad \mathbf{x} = \begin{bmatrix} a_n \\ a_{n-1} \\ \vdots \\ \vdots \\ a_0 \end{bmatrix}, \quad \text{and} \quad \mathbf{b} = \begin{bmatrix} y_0 \\ y_1 \\ \vdots \\ \vdots \\ y_n \end{bmatrix}.$$

Then we can write the linear system in matrix form as $\mathbf{A}\mathbf{x} = \mathbf{b}$.

Note the pattern of the entries in matrix $\mathbf{A}$; such a matrix is called a **Vandermonde matrix** in $x_0, x_1, \ldots, x_n$. If any two $x$-values are the same, then there will be two identical rows, which implies that the matrix is singular. (Explain why.) We must show that $\mathbf{A}$ is nonsingular when the $x$'s are distinct. We outline how to do this for $n = 2$, that is, when $\mathbf{A}$ is $3 \times 3$, in Exercise 62. The general case follows similarly but requires further row operations.

Polynomial interpolation is important for constructing a variety of approximation procedures. We discuss one of these in Example 7.

EXAMPLE 7  (Calculus Required) Let $f(x) = \sin(\pi x^2)$. Unfortunately, we cannot compute the value of

$$\int_0^1 \sin(\pi x^2)\, dx$$

directly since $f(x)$ does not have an antiderivative in terms of elementary functions. However, we can estimate this integral by first approximating $f(x)$ by an interpolant constructed to match information about $f(x)$ in $[0, 1]$. Let

$$x_0 = 0, \quad x_1 = \frac{1}{2}, \quad x_2 = \frac{\sqrt{2}}{2}, \quad x_3 = 1.$$

Then, computing the corresponding values of $f(x)$, we obtain the data set

$$D - \left\{ (0, 0), \left(\frac{1}{2}, \frac{\sqrt{2}}{2}\right), \left(\frac{\sqrt{2}}{2}, 1\right), (1, 0) \right\}.$$

To construct the interpolating polynomial to these data, we form the corresponding Vandermonde matrix $\mathbf{A}$ and solve the nonsingular linear system $\mathbf{Ax} = \mathbf{b}$, where $\mathbf{b}$ is the vector of $y$-coordinates from $D$. We obtain the linear system whose augmented matrix is

$$[\, \mathbf{A} \mid \mathbf{b} \,] = \begin{bmatrix} 0 & 0 & 0 & 1 & 0 \\ \frac{1}{8} & \frac{1}{4} & \frac{1}{2} & 1 & \frac{\sqrt{2}}{2} \\ \frac{2\sqrt{2}}{8} & \frac{1}{2} & \frac{\sqrt{2}}{2} & 1 & 1 \\ 1 & 1 & 1 & 1 & 0 \end{bmatrix}.$$

Using row operations, we can obtain the equivalent upper triangular form

$$\begin{bmatrix} 1 & 1 & 1 & 1 & 0 \\ 0 & 1 & 3 & 7 & 4\sqrt{2} \\ 0 & 0 & 1 & 3 + \sqrt{2} & -2 \\ 0 & 0 & 0 & 1 & 0 \end{bmatrix} \tag{6}$$

and it follows by back substitution that

$$a_0 = 0, \quad a_1 = -2, \quad a_2 = 6 + 4\sqrt{2}, \quad a_3 = -4 - 4\sqrt{2}.$$

(See Exercise 63.) Thus the interpolating polynomial is

$$P_3(x) = (-4 - 4\sqrt{2})x^3 + (6 + 4\sqrt{2})x^2 - 2x.$$

Hence we have

$$\int_0^1 \sin(\pi x^2)\, dx \approx \int_0^1 P_3(x)\, dx.$$

Figure 1 shows a graphical comparison of $\sin(\pi x^2)$ and $P_3(x)$, and Exercise 64 asks you to complete the numerical approximation. Inspection of Figure 1 reveals that $P_3(x)$ is quite good on $[0.5, 1]$ but differs substantially from $f(x)$ on $[0, 0.5]$. One possible way to improve the approximation is to choose another value of $x$ in $[0, 0.5]$ and hence adjoin another point to our data set. This requires that we recompute the interpolant, this time obtaining a polynomial of degree 4. Such topics are explored in more detail in a course on numerical analysis.  ■

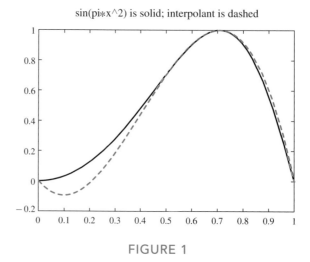

sin(pi∗x^2) is solid; interpolant is dashed

FIGURE 1

## Application:   **Cryptography**

Cryptography is the technique of coding and decoding messages; it goes back to the time of the ancient Greeks. Today a primary use of cryptography is to ensure the security of electronic transmissions of information. A simple cryptographic scheme assigns a different number to each character of the alphabet used, partitions the now digitized message into columns, and encodes it by multiplying the columns by a nonsingular matrix **C**, called the **encoding matrix**. To ensure security the encoding matrix is kept secret but must be known to the receiver so that the message can be decoded. We illustrate this process in Example 8.

EXAMPLE 8    Associate the letters of our standard alphabet with the numbers 1 through 26 as follows:

$$
\begin{array}{ccccccc}
A & B & C & \cdots & X & Y & Z \\
\updownarrow & \updownarrow & \updownarrow & \cdots & \updownarrow & \updownarrow & \updownarrow \\
1 & 2 & 3 & \cdots & 24 & 25 & 26.
\end{array}
\tag{7}
$$

The message to be sent is

LASTDAYFORTHISCODE

Its digitized form is

12  1  19  20  4  1  25  6  15  18  20  8  9  19  3  15  4  5

The current encoding matrix is

$$
\mathbf{C} = \begin{bmatrix} 1 & -1 & -2 \\ 1 & 1 & -1 \\ -1 & 2 & 2 \end{bmatrix}.
$$

Hence we partition the digitized message into columns of three entries each, padding the number of entries with any numerical value if the length is not divisible evenly by 3. In this case no padding is needed and the partitioned message has the form

$$
\mathbf{M} = \begin{bmatrix} 12 & 20 & 25 & 18 & 9 & 15 \\ 1 & 4 & 6 & 20 & 19 & 4 \\ 19 & 1 & 15 & 8 & 3 & 5 \end{bmatrix}.
$$

To encode the message we compute

$$\mathbf{CM} = \begin{bmatrix} 1 & -1 & -2 \\ 1 & 1 & -1 \\ -1 & 2 & 2 \end{bmatrix} \begin{bmatrix} 12 & 20 & 25 & 18 & 9 & 15 \\ 1 & 4 & 6 & 20 & 19 & 4 \\ 19 & 1 & 15 & 8 & 3 & 5 \end{bmatrix}$$

$$= \begin{bmatrix} -27 & 14 & -11 & -18 & -16 & 1 \\ -6 & 23 & 16 & 30 & 25 & 14 \\ 28 & -10 & 17 & 38 & 35 & 3 \end{bmatrix}.$$

The encoded message is now transmitted by sending the string of numbers

$$-27 \;\; -6 \;\; 28 \;\; 14 \;\; 23 \;\; -10 \;\; -11 \;\; 16 \;\; 17 \;\; -18 \;\; 30 \;\; 38 \;\; -16 \;\; 25 \;\; 35 \;\; 1 \;\; 14 \;\; 3$$

The receiver decodes the message by partitioning this string to form columns with three entries, multiplying by $\mathbf{C}^{-1}$, and then reversing the digitizing process. If the encoding matrix $\mathbf{C}$ were of a different size, then the message would be partitioned so the matrix product $\mathbf{CM}$ could be computed and the contents of $\mathbf{M}$ padded if needed.

■

Simple coding schemes like those in Example 8 can be cracked by a number of techniques, including the analysis of frequency of letters. Many more elaborate coding techniques have been developed, but they all basically follow the pattern encode-transmit-decode with variations on how to perform the encoding. Designers must ensure that the process is reversible, so something akin to a nonsingular matrix is needed.

Application:     **Wavelets II**

In Section 1.3 in Wavelets I, we indicated how a vector of information can be transformed into an equivalent form involving averages and differences. Exercise 71 in Section 1.3 showed that matrix

$$\mathbf{M} = \frac{1}{2} \begin{bmatrix} 1 & 1 \\ 1 & -1 \end{bmatrix}$$

transformed the information in a 2-vector $\begin{bmatrix} a \\ b \end{bmatrix}$ into the equivalent average-difference vector

$$\begin{bmatrix} c \\ d \end{bmatrix} = \begin{bmatrix} \dfrac{a+b}{2} \\ a - c \end{bmatrix}.$$

Exercise 73 in Section 1.3 showed how to extend this transformation to an $n$-vector $\mathbf{x}$, where $n$ is even. The procedure was to reshape the vector $\mathbf{x}$ into a $2 \times (n/2)$ matrix $\mathbf{D}_1$, and then compute the matrix product $\mathbf{Q}_1 = \mathbf{MD}_1$, which is the same size as $\mathbf{D}_1$. Since $\mathbf{M}$ is nonsingular (verify), the transformation process is completely reversible and hence no information is lost. That is, if we know the matrix $\mathbf{Q}_1$, then the original information in vector $\mathbf{x}$ is recoverable from the matrix product $\mathbf{M}^{-1}\mathbf{Q}_1$, which can be reshaped into an $n$-vector by stringing together its columns.

EXAMPLE 9     An 8-vector $\mathbf{x}$ has been transformed by the average-difference process in Exercise 73 of Section 1.3. The result is

$$\mathbf{Q}_1 = \begin{bmatrix} 4 & 1.5 & 2.5 & 3 \\ -2 & -4.5 & 5.5 & 3 \end{bmatrix}.$$

To find the original vector of information $\mathbf{x}$, we compute $\mathbf{M}^{-1}\mathbf{Q}_1$, which gives (verify)

$$\begin{bmatrix} 2 & -3 & 8 & 5 \\ 6 & 6 & -3 & 1 \end{bmatrix}.$$

Reshaping this matrix column-wise into an 8-vector gives (verify)

$$\begin{bmatrix} 2 & 6 & -3 & 6 & 8 & -3 & 5 & 1 \end{bmatrix}$$

which is the original vector in Exercise 73 of Section 1.3.    ■

A lesson to be learned here is that if we wish to develop a matrix transformation that is reversible with no information loss, then we should use a nonsingular matrix.

We can obtain a second level of transformation by averaging and differencing the averages obtained in the matrix product $\mathbf{Q}_1 = \mathbf{MD}_1$. The averages of the columns of matrix $\mathbf{D}_1$ appear in the first row of $\mathbf{Q}_1$. We save the second row of $\mathbf{Q}_1$, the differences, for later use. To perform the second level of transformation we reshape $\text{row}_1(\mathbf{Q}_1)$ into a matrix $\mathbf{D}_2$ of size $2 \times (n/2^2)$. Then the product $\mathbf{Q}_2 = \mathbf{MD}_2$ yields the second level of the transformation. This transformation process is used recursively[3] until we have a single average term and a set of $n - 1$ differences that have been saved from the transformation levels. We illustrate this in Example 10.

EXAMPLE 10    Let $\mathbf{x} = \begin{bmatrix} 2 & 6 & -3 & 6 & 8 & -3 & 5 & 1 \end{bmatrix}$ as in Exercise 73 of Section 1.3. There we showed that $\mathbf{Q}_1 = \mathbf{MD}_1$ gave the first level of the transformation:

$$\mathbf{Q}_1 = \begin{bmatrix} 4 & 1.5 & 2.5 & 3 \\ -2 & -4.5 & 5.5 & 2 \end{bmatrix}.$$

Applying the average-difference procedure recursively, we let $\mathbf{x}_1 = \text{row}_1(\mathbf{Q}_1)$, the pairwise averages of the data in $\mathbf{x}$. Label the first-level differences

$$\mathbf{d}_1 = \begin{bmatrix} -2 & -4.5 & 5.5 & 2 \end{bmatrix}.$$

We refer to the differences as detail coefficients. Now we reshape $\mathbf{x}_1$ into a $2 \times 2$ matrix $\mathbf{D}_2$ and multiply by $\mathbf{M}$ to compute the second level of transformation:

$$\mathbf{Q}_2 = \mathbf{MD}_2 = \frac{1}{2}\begin{bmatrix} 1 & 1 \\ 1 & -1 \end{bmatrix}\begin{bmatrix} 4 & 2.5 \\ 1.5 & 3 \end{bmatrix} = \begin{bmatrix} 2.75 & 2.75 \\ 1.25 & -0.25 \end{bmatrix}.$$

Let $\mathbf{x}_2 = \text{row}_1(\mathbf{Q}_2)$ be the second-level averages and $\mathbf{d}_2 = \text{row}_2(\mathbf{Q}_2)$ the second-level detail coefficients. Continuing, we apply the process again. Reshape $\mathbf{x}_2$ into a $2 \times 1$ matrix $\mathbf{D}_3$, and compute the third level of transformation:

$$\mathbf{Q}_3 = \mathbf{MD}_3 = \frac{1}{2}\begin{bmatrix} 1 & 1 \\ 1 & -1 \end{bmatrix}\begin{bmatrix} 2.75 \\ 2.75 \end{bmatrix} = \begin{bmatrix} 2.75 \\ 0 \end{bmatrix}.$$

Let $\mathbf{x}_3 = \text{row}_1(\mathbf{Q}_3)$ be the third-level averages and $\mathbf{d}_3 = \text{row}_2(\mathbf{Q}_3)$ the third-level detail coefficients. Since $\mathbf{x}_3$ contains a single entry, this is the final level of the transformation. The information in the original vector $\mathbf{x}$ is now represented by the "final average" $\mathbf{x}_3$ and the three sets of detail coefficients $\mathbf{d}_3$, $\mathbf{d}_2$, and $\mathbf{d}_1$. Table 2.1 summarizes the recursive process step-by-step notationally, while Table 2.2 provides a corresponding summary showing the average-difference data at each level.

---

[3]"Recursion is a technique in which the result of a mathematical process is 'run back through' the process again to be further refined." See *The Words of Mathematics* by S. Schwartzman (Washington, D.C.: The Mathematical Association of America, 1994).

TABLE 1

| x | | |
|---|---|---|
| $x_1$ | | $d_1$ |
| $x_2$ | $d_2$ | $d_1$ |
| $x_3$ | $d_3$ $d_2$ | $d_1$ |

TABLE 2

| 2 | 6 | −3 | 6 | 8 | −3 | 5 | 1 |
|---|---|---|---|---|---|---|---|
| 4 | 1.5 | 2.5 | 3 | −2 | −4.5 | 5.5 | 2 |
| 2.75 | 2.75 | 1.25 | −0.25 | −2 | −4.5 | 5.5 | 2 |
| 2.75 | 0 | 1.25 | −0.25 | −2 | −4.5 | 5.5 | 2 |

■

Once we reach the final level of the transformation, the original data in $n$-vector **x** is now represented by the "final average" and $n - 1$ detail coefficients; see the last row of Table 2.2 for the case $n = 8$. At this time the next step of a wavelet is to initiate an approximation. Namely, if the absolute value of a detail coefficient is smaller than a specified value, called a **threshold value**[4], then it is set to zero. This generates an approximation to the information represented by the "final average" and detail coefficients. Let $\tilde{d}_i$ represent the detail coefficients at level $i$ as a result of the approximation step. Thus we have the approximate information

| $x_3$ | $\tilde{d}_3$ | $\tilde{d}_2$ | $\tilde{d}_1$ |
|---|---|---|---|

In wavelet applications with huge data sets, the next step is to compress the information for transmission by not sending the zeros that have been inserted by the use of the threshold value. Upon receipt of the compressed/approximate data, the transformation process is reversed by using $\mathbf{M}^{-1}$ and the result is a data vector $\tilde{\mathbf{x}}$ that approximates the original data in **x**. The wavelet process becomes economical when the data that are actually sent are a small fraction of the original data and the approximate data in $\tilde{\mathbf{x}}$ that is reconstructed from the transmitted data provide a good representation of the information contained in the original data vector **x**. An area that has profited from the wavelet process is the transmission of photographic information. We cannot explore this area at this time, but Example 11 illustrates the use of a threshold value. Several exercises also explore the wavelet process.

EXAMPLE 11  To illustrate the wavelet process and the use of a threshold value, we sample $f(x) = x^3$ on $[0, 1]$ at the points $x_i = i/7$, $i = 0, 1, 2, \ldots, 7$. To the corresponding function values $f(x_i)$ we apply the averaging and differencing procedure to generate the information in Table 2.3. (We have recorded the data to 4 decimal places.) The first row is the data $f(x_i)$ and the last seven entries in row 4 are the detail coefficients. (See Table 2.1 for a description of the regions of the table.)

TABLE 3

| 0 | 0.0029 | 0.0233 | 0.0787 | 0.1866 | 0.3644 | 0.6297 | 1.0000 |
|---|---|---|---|---|---|---|---|
| 0.0015 | 0.0510 | 0.2755 | 0.8149 | **−0.0015** | **−0.0277** | **−0.0889** | **−0.1851** |
| 0.0262 | 0.5452 | **−0.0248** | **−0.2697** | **−0.0015** | **−0.0277** | **−0.0889** | **−0.1851** |
| 0.2857 | **−0.2595** | **−0.0248** | **−0.2697** | **−0.0015** | **−0.0277** | **−0.0889** | **−0.1851** |

[4]The threshold value is dependent upon the data and the application from which it came. There are still many unanswered questions about the choice of a threshold value.

If we choose a threshold value of 0.1, then the approximate detail coefficients are

| −0.2595 | 0 | −0.2697 | 0 | 0 | 0 | −0.1851 |
|---|---|---|---|---|---|---|

Reversing the averaging and differencing gives us approximate function values

| 0.0262 | 0.0262 | 0.0262 | 0.0262 | 0.2755 | 0.2755 | 0.6297 | 1.0000 |
|---|---|---|---|---|---|---|---|

We compare the original data and the approximation generated by the wavelet process in Figure 2. The solid curve is $y = f(x) = x^3$ and the dashed curve is obtained from the approximate function values generated by the wavelet process.

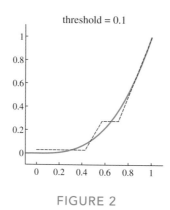

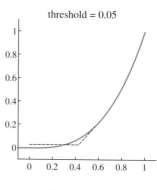

FIGURE 2

FIGURE 3

Choosing a smaller threshold, like 0.05, gives us approximate detail coefficients containing fewer zeros. Figure 3 shows the comparison in this case.    ■

# EXERCISES 2.4

*In Exercises 1–5, show that each given linear system has a unique solution and find it.*

**1.** $x_1 - 2x_2 + 4x_3 = 4$
$3x_2 + 5x_3 = 8$
$- 2x_3 = -2.$

**2.** $2x_1 - x_2 = 9$
$3x_1 + 2x_2 = 17.$

**3.** $x_1 + x_2 + x_3 = 3$
$2x_1 + 3x_2 + x_3 = 5$
$x_1 + 2x_2 + x_3 = 4.$

**4.** $2x_1 - x_2 = 1$
$-x_1 + 2x_2 - x_3 = 0$
$- x_2 + 2x_3 - x_4 = 0$
$- x_3 + 2x_4 = 0.$

**5.** $2ix_1 + (1+i)x_2 = 3+i$
$x_1 + (1+2i)x_2 = 1+i.$

*In Exercises 6–10, show that each given matrix is nonsingular.*

**6.** $\begin{bmatrix} 2 & 0 & 0 \\ 3 & -1 & 0 \\ 4 & 0 & 2 \end{bmatrix}.$     **7.** $\begin{bmatrix} 1 & 1 & 1 \\ 2 & 3 & 1 \\ 1 & 2 & 1 \end{bmatrix}.$

**8.** $\begin{bmatrix} 1 & 2 \\ 3 & 4 \end{bmatrix}.$     **9.** $\begin{bmatrix} 3 & 4 & -1 & 0 \\ 1 & -2 & 2 & 0 \\ -1 & 0 & 1 & 0 \\ 0 & 0 & 0 & 8 \end{bmatrix}.$

**10.** $\begin{bmatrix} 1 & 1 & 0 \\ i & 1 & 1 \\ 0 & 1 & i \end{bmatrix}.$

**11.** Find the inverse of each of the matrices in Exercises 6–10.

*In Exercises 12–15, determine the unique solution of $\mathbf{Ax} = \mathbf{b}$ using just the given information.*

**12.** $\mathbf{A}^{-1} = \begin{bmatrix} -1 & 3 \\ 1 & -2 \end{bmatrix}, \mathbf{b} = \begin{bmatrix} 1 \\ 4 \end{bmatrix}.$

**13.** $\mathbf{A}^{-1} = \begin{bmatrix} 1 & 1 & -2 \\ -1 & 0 & 1 \\ 1 & -1 & 1 \end{bmatrix}$, $\mathbf{b} = \begin{bmatrix} 3 \\ 5 \\ 4 \end{bmatrix}$.

**14.** $\mathbf{A}^{-1} = \dfrac{1}{25} \begin{bmatrix} 11 & 4 & 1 & -1 & -4 \\ 4 & 11 & 4 & 1 & -1 \\ 1 & 4 & 11 & 4 & 1 \\ -1 & 1 & 4 & 11 & 4 \\ -4 & -1 & 1 & 4 & 11 \end{bmatrix}$, $\mathbf{b} = \begin{bmatrix} 1 \\ 1 \\ 1 \\ 1 \\ 1 \end{bmatrix}$.

**15.** $\mathbf{A}^{-1} = \begin{bmatrix} 14 & 1 & -5 \\ -2 & 1 & 0 \\ -1 & -1 & 1 \end{bmatrix}$, $\mathbf{b} = \begin{bmatrix} 0 \\ 2 \\ 1 \end{bmatrix}$.

**16.** For each of the Exercises 12–15, explain how to find the original coefficient matrix $\mathbf{A}$.

*In Exercises 17–20, show that each given pair of matrices are inverses of one another.*

**17.** $\begin{bmatrix} 3 & 4 \\ 1 & 1 \end{bmatrix}$, $\begin{bmatrix} -1 & 4 \\ 1 & -3 \end{bmatrix}$.

**18.** $\begin{bmatrix} 1 & 0 & 1 \\ 1 & 1 & 0 \\ 0 & 1 & 1 \end{bmatrix}$, $\dfrac{1}{2}\begin{bmatrix} 1 & 1 & -1 \\ -1 & 1 & 1 \\ 1 & -1 & 1 \end{bmatrix}$.

**19.** $\begin{bmatrix} 1 & a & 0 & 0 \\ 0 & 1 & a & 0 \\ 0 & 0 & 1 & a \\ 0 & 0 & 0 & 1 \end{bmatrix}$, $\begin{bmatrix} 1 & -a & a^2 & -a^3 \\ 0 & 1 & -a & a^2 \\ 0 & 0 & 1 & -a \\ 0 & 0 & 0 & 1 \end{bmatrix}$.

**20.** $\begin{bmatrix} 0 & 1 & 2 \\ 2 & 0 & 1 \\ 1 & 0 & 2 \end{bmatrix}$, $\dfrac{1}{3}\begin{bmatrix} 0 & 2 & -1 \\ 3 & 2 & -4 \\ 0 & -1 & 2 \end{bmatrix}$.

**21.** As part of a project, two students must determine the inverse of a given $7 \times 7$ matrix $\mathbf{A}$. Each performs the required calculation and returns their results $\mathbf{A}_1$ and $\mathbf{A}_2$, respectively, to the instructor.

(a) What must be true about the two results? Why?

(b) How does the instructor check their work without repeating the calculations the students used?

*In Exercises 22–24, compute the vector $\mathbf{w}$ for each given expression without computing the inverse of any matrix given that*

$$\mathbf{A} = \begin{bmatrix} 1 & 0 & -2 \\ 1 & 1 & 0 \\ 0 & 1 & 1 \end{bmatrix}, \quad \mathbf{C} = \begin{bmatrix} 1 & 1 & 1 \\ 2 & 3 & 1 \\ 1 & 2 & 1 \end{bmatrix},$$

$$\mathbf{F} = \begin{bmatrix} 2 & 1 & 0 \\ -3 & 0 & 2 \\ -1 & 1 & 2 \end{bmatrix}, \quad \mathbf{v} = \begin{bmatrix} 6 \\ 7 \\ -3 \end{bmatrix}.$$

**22.** $\mathbf{w} = \mathbf{A}^{-1}(\mathbf{C} + \mathbf{F})\mathbf{v}$.

**23.** $\mathbf{w} = (\mathbf{F} + 2\mathbf{A})\mathbf{C}^{-1}\mathbf{v}$.

**24.** $\mathbf{w} = \mathbf{C}^{-1}(\mathbf{A} - 2\mathbf{F})\mathbf{v}$.

**25.** Determine all values of $s$ so that $\mathbf{A} = \begin{bmatrix} 0 & 1 & 2 \\ 2 & 1 & 1 \\ 1 & s & 2 \end{bmatrix}$ is nonsingular.

**26.** Determine all values of $s$ so that $\mathbf{A} = \begin{bmatrix} s & 1 & 0 \\ 1 & s & 1 \\ 0 & 1 & s \end{bmatrix}$ is nonsingular.

**27.** (a) For complex numbers $a$ and $b$, it is not generally true that
$$\frac{1}{a+b} = \frac{1}{a} + \frac{1}{b}.$$
Find a pair of values $a$ and $b$ for which this relation is true.

(b) Use the result from part a to find a pair of $2 \times 2$ matrices $\mathbf{A}$ and $\mathbf{B}$ such that $(\mathbf{A} + \mathbf{B})^{-1} = \mathbf{A}^{-1} + \mathbf{B}^{-1}$.

*In Exercises 28–31, verify each indicated algebraic property of the inverse.*

**28.** $(\mathbf{A}^T)^{-1} = (\mathbf{A}^{-1})^T$.

**29.** $(\mathbf{A}^{-1})^{-1} = \mathbf{A}$.

**30.** $(k\mathbf{A})^{-1} = \dfrac{1}{k}\mathbf{A}^{-1}$ $k \neq 0$.

**31.** $(\mathbf{ABC})^{-1} = \mathbf{C}^{-1}\mathbf{B}^{-1}\mathbf{A}^{-1}$.

**32.** Let $\mathbf{A}$ be a symmetric nonsingular matrix. Show that $\mathbf{A}^{-1}$ is symmetric. (*Hint*: Use Exercise 28.)

**33.** Show that the $n \times n$ diagonal matrix

$$D = \begin{bmatrix} d_{11} & 0 & \cdots & 0 & 0 \\ 0 & d_{22} & \cdots & 0 & 0 \\ 0 & 0 & \ddots & \vdots & \vdots \\ \vdots & \vdots & \cdots & \ddots & 0 \\ 0 & 0 & \cdots & 0 & d_{nn} \end{bmatrix}$$

is nonsingular if and only if all its diagonal entries are nonzero. [*Hint*: Show that $\mathbf{rref}(D) = \mathbf{I}_n$ if and only if $d_{jj} \neq 0$. Use ideas from Example 6.]

**34.** Show that the $n \times n$ lower triangular matrix

$$\mathbf{L} = \begin{bmatrix} l_{11} & 0 & \cdots & 0 & 0 \\ l_{21} & l_{22} & \cdots & 0 & 0 \\ 0 & 0 & \ddots & \vdots & \vdots \\ \vdots & \vdots & \cdots & \ddots & 0 \\ l_{n1} & l_{n2} & \cdots & l_{n\,n-1} & l_{nn} \end{bmatrix}$$

is nonsingular if and only if all its diagonal entries are nonzero.

**35.** Show that the $n \times n$ upper triangular matrix

$$\mathbf{U} = \begin{bmatrix} u_{11} & u_{12} & \cdots & u_{1\,n-1} & u_{1n} \\ 0 & u_{22} & \cdots & u_{2\,n-1} & u_{2n} \\ \vdots & \vdots & \cdots & \ddots & 0 \\ \vdots & \vdots & \cdots & \ddots & u_{n-1\,n} \\ 0 & 0 & \cdots & 0 & u_{nn} \end{bmatrix}$$

is nonsingular if and only if all its diagonal entries are nonzero.

*In Exercises 36–42, determine whether each given matrix is nonsingular just by inspecting the structure of the matrix. (See Exercises 33 through 35.) If it is nonsingular, find the inverse.*

**36.** $\begin{bmatrix} 4 & 0 & 0 \\ 3 & 0 & 0 \\ -5 & 1 & 2 \end{bmatrix}$.

**37.** $\begin{bmatrix} 5 & 0 & 0 \\ 0 & -6 & 0 \\ 0 & 0 & 3 \end{bmatrix}$.

**38.** $\begin{bmatrix} 5 & 5 \\ 0 & 5 \end{bmatrix}$.

**39.** $\begin{bmatrix} 3 & 0 & 0 \\ 2 & 1 & 0 \\ 0 & 2 & 1 \end{bmatrix}$.

**40.** $\begin{bmatrix} 2 & 0 & 0 \\ 0 & 0 & 0 \\ 0 & 0 & 8 \end{bmatrix}$.

**41.** $\begin{bmatrix} 1 & 0 & 0 & 0 \\ 0 & \frac{1}{2} & 0 & 0 \\ 0 & 0 & \frac{1}{3} & 0 \\ 0 & 0 & 0 & \frac{1}{4} \end{bmatrix}$.

**42.** $\begin{bmatrix} i & 0 & 0 \\ 0 & 1+i & 0 \\ 0 & 0 & 2 \end{bmatrix}$.

*For each of the matrices in Exercises 43–47, determine the following information.*

(i) A basis for its row space.

(ii) A basis for its column space.

(iii) Its null space.

(iv) Its rank.

**43.** $\begin{bmatrix} 2 & 0 & 0 \\ 3 & -1 & 0 \\ 4 & 0 & 2 \end{bmatrix}$.

**44.** $\begin{bmatrix} 1 & 1 & 1 \\ 2 & 3 & 1 \\ 1 & 2 & 1 \end{bmatrix}$.

**45.** $\begin{bmatrix} 1 & 2 \\ 3 & 4 \end{bmatrix}$.

**46.** $\begin{bmatrix} 3 & 4 & -1 & 0 \\ 1 & -2 & 2 & 0 \\ -1 & 0 & 1 & 0 \\ 0 & 0 & 0 & 8 \end{bmatrix}$.

**47.** $\begin{bmatrix} 1 & 1 & 0 \\ i & 1 & 1 \\ 0 & 1 & i \end{bmatrix}$.

**48.** Let **B** be any $5 \times 5$ nonsingular matrix.

(a) Describe a common feature of **rref(B)**.

(b) What is the largest number of vectors possible in a basis for **row(B)**?

(c) Describe a common feature of **ns(B)**.

(d) What is the largest possible value for rank(**B**)?

*In Exercises 49–53, explain why you can say that each given matrix is singular without computing its RREF.*

**49.** $\begin{bmatrix} 3 & -2 \\ -6 & 4 \end{bmatrix}$.

**50.** $\begin{bmatrix} 1 & 2 & 0 \\ 3 & -1 & 0 \\ 8 & 7 & 0 \end{bmatrix}$.

**51.** $\begin{bmatrix} 1 & 0 & 4 \\ 0 & 2 & -3 \\ 1 & 2 & 1 \end{bmatrix}$.

**52.** $\begin{bmatrix} -1 & 2 & 0 \\ 3 & 7 & 2 \\ 4 & -8 & 0 \end{bmatrix}$.

**53.** $\begin{bmatrix} 1+2i & -i \\ -2+i & 1 \end{bmatrix}$.

**54.** Let $\mathbf{A} = \begin{bmatrix} a & b & c \\ d & e & f \\ g & h & i \end{bmatrix}$.

(a) Let $\mathbf{E}_1 = \mathbf{I}_{R_2 \leftrightarrow R_3}$. Compute $\mathbf{E}_1\mathbf{A}$. Compare this result with $\begin{bmatrix} a & b & c \\ d & e & f \\ g & h & i \end{bmatrix}_{R_2 \leftrightarrow R_3}$.

(b) Let $\mathbf{E}_2 = \mathbf{I}_{t R_2}$. Compute $\mathbf{E}_2\mathbf{A}$. Compare this result with $\begin{bmatrix} a & b & c \\ d & e & f \\ g & h & i \end{bmatrix}_{t R_2}$.

(c) Let $\mathbf{E}_3 = \mathbf{I}_{t R_1 + R_3}$. Compute $\mathbf{E}_3\mathbf{A}$. Compare this result with $\begin{bmatrix} a & b & c \\ d & e & f \\ g & h & i \end{bmatrix}_{t R_1 + R_3}$. (See note on next page.)

(*Note*: We chose matrix **A** to be $3 \times 3$ and we chose particular row operations for convenience. Hence this is not a proof that multiplication by elementary matrices mimics the action of row operations. However, these results are valid in general.)

**55.** Let $\mathbf{A} = \begin{bmatrix} 2 & 4 & 0 \\ 2 & 3 & 2 \\ -3 & 8 & 1 \end{bmatrix}$.

  (a) Perform the row operations $\left(\frac{1}{2}\right) R_1$, $-2R_1 + R_2$, and $3R_1 + R_3$ in succession on **A** and record the resulting matrix.

  (b) Construct the elementary matrix $\mathbf{E}_1$ that corresponds to $\left(\frac{1}{2}\right) R_1$.

  (c) Construct the elementary matrix $\mathbf{E}_2$ that corresponds to $-2R_1 + R_2$.

  (d) Construct the elementary matrix $\mathbf{E}_3$ that corresponds to $3R_1 + R_3$.

  (e) Compute the product $\mathbf{E}_3(\mathbf{E}_2(\mathbf{E}_1\mathbf{A}))$ and compare this result with that from part (a).

**56.** Let $\mathbf{A} = \begin{bmatrix} 1 & 0 \\ 4 & 2 \end{bmatrix}$. Compute $\mathbf{rref}(\mathbf{A})$ using elementary matrices.

**57.** Let $\mathbf{A} = \begin{bmatrix} 2 & -4 & 0 \\ 0 & 1 & 3 \\ 0 & 0 & 6 \end{bmatrix}$. Compute $\mathbf{rref}(\mathbf{A})$ using elementary matrices.

**58.** Let **A** be an $n \times n$ matrix.

  (a) If $\mathbf{B} = \mathbf{A}_{kR_j}$, then determine a row operation that when applied to **B** gives **A**.

  (b) If $\mathbf{C} = \mathbf{A}_{R_j \leftrightarrow R_i}$, then determine a row operation that when applied to **C** gives **A**.

  (c) If $\mathbf{D} = \mathbf{A}_{kR_j + R_i}$, then determine a row operation that when applied to **D** gives **A**.

**59.** Exercise 58 implies that the effect of any row operation can be reversed by another (suitable) row operation.

  (a) Let $\mathbf{E}_1 = \mathbf{I}_{kR_j}$, $k \neq 0$. Explain why $\mathbf{E}_1$ is nonsingular.

  (b) Let $\mathbf{E}_2 = \mathbf{I}_{R_j \leftrightarrow R_i}$. Explain why $\mathbf{E}_2$ is nonsingular.

  (c) Let $\mathbf{E}_3 = \mathbf{I}_{kR_j + R_i}$, $i \neq j$. Explain why $\mathbf{E}_3$ is nonsingular.

**60.** Explain why we can say, "a product of elementary matrices is a nonsingular matrix." (*Hint*: Use Exercise 59 and property 5 listed under "Algebraic Properties of the Inverse," page 148.)

**61.** Explain why we can say, "for any matrix **A** there exists a nonsingular matrix **F** such that $\mathbf{FA} = \mathbf{rref}(\mathbf{A})$." [*Hint*: See (4).]

**62.** For distinct values $x_0$, $x_1$, and $x_2$, show that the $3 \times 3$ Vandermonde matrix

$$\mathbf{A} = \begin{bmatrix} x_0^2 & x_0 & 1 \\ x_1^2 & x_1 & 1 \\ x_2^2 & x_2 & 1 \end{bmatrix}$$

is nonsingular. From the definition of *nonsingular* we need to show $\mathbf{rref}(\mathbf{A}) = \mathbf{I}_3$. However, the row operations will be simpler if we compute $\mathbf{rref}(\mathbf{A}^T)$ instead. Recall that **A** is nonsingular if and only $\mathbf{A}^T$ is nonsingular; see the "Algebraic Properties of the Inverse" and Exercise 28. Proceed as follows.

  (a) Compute $\mathbf{B} = \mathbf{A}^T = \begin{bmatrix} x_0^2 & x_1^2 & x_2^2 \\ x_0 & x_1 & x_2 \\ 1 & 1 & 1 \end{bmatrix}_{R_1 \leftrightarrow R_3}$.

  (b) Use the leading 1 in row 1 as the pivot and eliminate the entries below it in matrix **B** to obtain a new matrix, which we call **C**.

  (c) Why is the $(2, 2)$-entry of **C** guaranteed to be nonzero? Use the $(2, 2)$-entry as a pivot, eliminate below it, and call the resulting matrix **D**.

  (d) Why is the $(3, 3)$-entry guaranteed to be nonzero? (*Hint*: Factor it into a product.)

  (e) The matrix **D** is an upper triangular matrix. Why can it be reduced to $\mathbf{I}_3$ and hence why does it follow that **A** is nonsingular?

**63.** Verify that the solution to the linear system corresponding to the augmented matrix in (6) is $a_0 = 0$, $a_1 = -2$, $a_2 = 6 + 4\sqrt{2}$, $a_3 = -4 - 4\sqrt{2}$.

**64.** Compute the approximation to $\int_0^1 \sin(\pi x^2)\, dx$ developed in Example 7 by integrating the interpolant $P_3(x)$.

**65.** Use the digitizing scheme in (7) and the encoding matrix **C** in Example 8 to decode the message

$$-40\ 4\ 55\ -13\ 24\ 31\ -25\ 29\ 46\ -37\ 12\ 57\ -37\ 10\ 51$$

**66.** Use the digitizing scheme in (7) and the encoding matrix $\mathbf{C} = \begin{bmatrix} 5 & 3 \\ 2 & 1 \end{bmatrix}$.

  (a) Encode the message WHYMATH for transmission.

  (b) Decode the reply

$$68\ 27\ 124\ 48\ 68\ 27\ 70\ 27\ 134\ 51\ 117\ 44\ 100\ 35$$

**67.** In Example 7 and Exercise 66 the inverse of the encoding matrix **C** has only integer entries. Explain why this is convenient for the simple coding scheme we have used.

**68.** The encoding matrix

$$\mathbf{C} = \begin{bmatrix} 1 & 2 & 1 \\ 1 & 0 & 2 \\ 2 & 2 & 3 \end{bmatrix}$$

was used to encode a message. The coded message was sent and received. What problems will be encountered in decoding this message?

**69.** Determine the inverse of matrix $\mathbf{M} = \dfrac{1}{2}\begin{bmatrix} 1 & 1 \\ 1 & -1 \end{bmatrix}$.

**70.** Let $\mathbf{x} = \begin{bmatrix} 87 & 81 & 62 & 64 \end{bmatrix}$. Determine $\mathbf{x}_2$, the final average, and the two levels of detail coefficients $\mathbf{d}_1$ and $\mathbf{d}_2$ by following the procedure in Example 10.

**71.** Let $\mathbf{x} = \begin{bmatrix} 87 & 81 & 62 & 64 & 76 & 78 & 68 & 54 \end{bmatrix}$. Determine $\mathbf{x}_3$, the final average, and the three levels of detail coefficients $\mathbf{d}_1$, $\mathbf{d}_2$, and $\mathbf{d}_3$ by following the procedure in Example 10.

**72.** The following table is a sample of a function $\mathbf{x}(t)$ taken at equispaced values of $t$.

| $t$ | 1 | 2 | 3 | 4 | 5 | 6 | 7 | 8 |
|-----|----|----|---|----|----|----|----|---|
| $\mathbf{x}$ | 37 | 33 | 6 | 16 | 26 | 28 | 18 | 4 |

If we graph the ordered pairs and connected successive points with straight line segments, we obtain what is called a piecewise linear approximation to the full graph of $\mathbf{x}(t)$. The larger the data set we use the better the piecewise linear approximation obtained. (We are deliberately using only a small sample, 8 points.) To illustrate the use of wavelets to approximate graphical data, perform the following steps.

(a) Determine $\mathbf{x}_3$, the final average, and the three levels of detail coefficients $\mathbf{d}_1$, $\mathbf{d}_2$ and $\mathbf{d}_3$ by following the procedure in Example 10.

(b) To perform the approximation step, set a detail coefficient to zero if its absolute value is less than or equal to 3. That is, we are using a threshold value of 3. You should get the following:

| $\mathbf{x}_3$ | $\tilde{\mathbf{d}}_3$ | $\tilde{\mathbf{d}}_2$ | | $\tilde{\mathbf{d}}_1$ | | |
|-----|---|----|---|---|----|---|
| 21 | 0 | 12 | 8 | 0 | −5 | 0 | 7 |

(c) Using the information in part (b), we reverse the wavelet transformation to obtain an approximation to the original data sample as follows.

Compute $\begin{bmatrix} a \\ b \end{bmatrix} = \mathbf{M}^{-1}\begin{bmatrix} 21 \\ 0 \end{bmatrix}$.

Compute $\begin{bmatrix} c & d \\ e & f \end{bmatrix} = \mathbf{M}^{-1}\begin{bmatrix} a & b \\ 12 & 8 \end{bmatrix}$.

Compute $\begin{bmatrix} g & h & i & j \\ k & l & m & n \end{bmatrix}$

$= \mathbf{M}^{-1}\begin{bmatrix} c & e & d & f \\ 0 & -5 & 0 & 7 \end{bmatrix}$.

(Note that we used $\begin{bmatrix} c & d \\ e & f \end{bmatrix}$ columnwise.)

Reshaping $\begin{bmatrix} g & h & i & j \\ k & l & m & n \end{bmatrix}$ columnwise, you should get the wavelet approximation to the values

of $\mathbf{x}(t)$ as

$\begin{bmatrix} g & k & h & l & i & m & j & n \end{bmatrix}$
$= \begin{bmatrix} 33 & 33 & 4 & 14 & 29 & 29 & 20 & 6 \end{bmatrix}$.

The result of the wavelet process gives the approximate data as

| $t$ | 1 | 2 | 3 | 4 | 5 | 6 | 7 | 8 |
|-----|----|----|---|----|----|----|----|---|
| $\mathbf{x}$ | 33 | 33 | 4 | 14 | 29 | 29 | 20 | 6 |

To visually compare the original data with the model developed as a result of the wavelet process, we plot the original data set and the approximate data set.

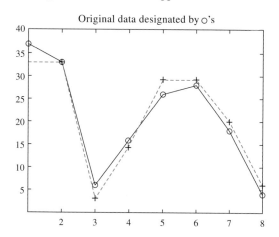

Original data designated by ○'s

The data designated by the + signs approximate the original data designated by the ○'s. This wavelet approximation is surprisingly good, considering how few data points we used.

**73.** Repeat Exercise 72 using

$$\mathbf{x}(t) = \begin{bmatrix} 87 & 81 & 62 & 64 & 76 & 78 & 68 & 54 \end{bmatrix}.$$

### In MATLAB

*In this section we considered square linear systems $\mathbf{Ax} = \mathbf{b}$ whose coefficient matrix $\mathbf{A}$ was nonsingular or invertible. Such systems have a unique solution, and the matrix $\mathbf{A}$ has an inverse. We solved these linear systems by finding the RREF of the augmented matrix $\begin{bmatrix} \mathbf{A} \mid \mathbf{b} \end{bmatrix}$ and could find $\mathbf{A}^{-1}$ by computing the RREF of the matrix $\begin{bmatrix} \mathbf{A} \mid \mathbf{I}_n \end{bmatrix}$. All of these computations can be performed in MATLAB using the **rref** command (see Exercise ML.3. in Section 2.2). However, linear systems arise so frequently that not only do we need to obtain a solution, but we must do it as quickly and accurately as possible. In most cases this is done in a computing environment that uses an arithmetic that is just an approximation of our exact computations. Unfortunately, finding the RREF of a system is not the best way to solve linear systems, so MATLAB has other commands that are more efficient and preferred by scientists* *and engineers. We introduce these commands and present several brief experiments using them, but we relegate the details regarding these commands to courses in numerical analysis.*

**ML. 1.** Use MATLAB's **rref** command to solve each of the following linear systems. From the result displayed by MATLAB, explain how you know that the system you solved is nonsingular.

(a) $2x_1 + 3x_2 - 2x_3 = 10$
$x_1 + 4x_2 + 5x_3 = -13$
$2x_2 - 6x_3 = 18.$

(b) $-2x_1 + x_2 - x_3 + 2x_4 = 5$
$2x_1 - 2x_2 + x_3 - x_4 = -3$
$-x_1 + 2x_2 - 2x_3 + x_4 = 1$
$- x_2 + 2x_3 - 2x_4 = -3.$

(c) $2x_1 + 6x_2 + (5+4i)x_3 = 37 + 4i$
$(5i)x_1 + (9+3i)x_2 + (8+6i)x_3 = 49 + 13i$
$(6+6i)x_1 + 3x_2 + (5i)x_3 = 16 + 27i.$

**ML. 2.** In MATLAB we can generate an identity matrix of size $3 \times 3$ using command **eye(3)**; **eye(4)** produces a $4 \times 4$ identity matrix; **eye(k)** produces a $k \times k$ identity matrix. Use the **rref** command and the **eye** command to obtain the inverse of each matrix **A** by computing the RREF of $\begin{bmatrix} \mathbf{A} & | & \mathbf{I}_n \end{bmatrix}$.

(a) $\mathbf{A} = \begin{bmatrix} 1 & 0 & -1 \\ -1 & 1 & 0 \\ 1 & -1 & 1 \end{bmatrix}.$

(b) $\mathbf{A} = \begin{bmatrix} 1 & \frac{1}{2} & \frac{1}{3} \\ \frac{1}{2} & \frac{1}{3} & \frac{1}{4} \\ \frac{1}{3} & \frac{1}{4} & \frac{1}{5} \end{bmatrix}.$

(c) $\mathbf{A} = \begin{bmatrix} \frac{1}{3} & \frac{1}{4} & \frac{1}{5} & \frac{1}{6} \\ \frac{1}{4} & \frac{1}{5} & \frac{1}{6} & \frac{1}{7} \\ \frac{1}{5} & \frac{1}{6} & \frac{1}{7} & \frac{1}{8} \\ \frac{1}{6} & \frac{1}{7} & \frac{1}{8} & \frac{1}{9} \end{bmatrix}.$

The inverses of each of these matrices have a common property. What is it? Do not expect all inverses to have this property.

**ML. 3.** The command **inv** in MATLAB will compute the inverse of a nonsingular matrix directly. However, this command does not use RREF. The procedure used is related to matrix factorization, which is discussed in Section 2.5.

(a) Use **inv** to compute the inverse of each matrix in Exercise ML.2. Use **help inv** for directions.

(b) A matrix that arises in a variety of instances in mathematics is known as the Hilbert matrix. The following are the $3 \times 3$ and $4 \times 4$ Hilbert matrices:

$$\begin{bmatrix} 1 & \frac{1}{2} & \frac{1}{3} \\ \frac{1}{2} & \frac{1}{3} & \frac{1}{4} \\ \frac{1}{3} & \frac{1}{4} & \frac{1}{5} \end{bmatrix}, \begin{bmatrix} 1 & \frac{1}{2} & \frac{1}{3} & \frac{1}{4} \\ \frac{1}{2} & \frac{1}{3} & \frac{1}{4} & \frac{1}{5} \\ \frac{1}{3} & \frac{1}{4} & \frac{1}{5} & \frac{1}{6} \\ \frac{1}{4} & \frac{1}{5} & \frac{1}{6} & \frac{1}{7} \end{bmatrix}.$$

To compare the efficiency of computing inverses using commands **inv** and **rref**, we will count the arithmetic operations performed by MATLAB when we invert Hilbert matrices of sizes 3 through 10. To count the arithmetic operations, we use a command called **flops**. (The term **flops** stands for floating point operations. Type **help flops** for more information.)

(c) To see this comparison, enter the MATLAB command **compinv**. We show the comparison in tabular and graphic forms.

(d) Based on the information displayed by routine **compinv**, which command, **rref** or **inv**, is more efficient and why?

(e) Using part (d), estimate how much more efficient. Explain your estimation procedure.

**ML. 4.** MATLAB has a command that will solve nonsingular linear systems directly provided we enter the coefficient matrix **A** and the right side **b**. The command is denoted by the symbol \ and to execute it we type $\mathbf{x} = \mathbf{A} \backslash \mathbf{b}$. This command does not use RREF but uses a procedure involving matrix factorization, which we discuss in Section 2.5. If the matrix **A** is detected to be singular, a warning is displayed. Use the backslash, \, command on each of the following linear systems. Determine which are nonsingular systems.

(a) $x_1 - 2x_2 + 4x_3 = 0$
$3x_2 + 5x_3 = 8$
$- 2x_3 = 1.$

(b) $2x_1 - x_2 = 0$
$3x_1 + 2x_2 = 17.$

(c) $x_1 + x_2 + x_3 = 3$
$2x_1 + 3x_2 + 2x_3 = 5$
$x_1 + 2x_2 + x_3 = 4.$

(d) $2x_1 - x_2 = 1$
$-x_1 + 2x_2 - x_3 = -1$
$-x_2 + 2x_3 - x_4 = 2$
$- x_3 + 2x_4 = 0.$

(e) $2ix_1 + (1+i)x_2 = 3 + i$
$x_1 + (1+2i)x_2 = 1 + i.$

**ML. 5.** We introduced elementary matrices in this section as a tool to develop a matrix equation that relates a matrix and its RREF. Elementary matrices are used in succeeding topics. The MATLAB routine **elemmat** contains a summary of the properties of elementary matrices and can be used for review or further reading on this topic. In MATLAB type **help elemmat** for further directions.

**ML. 6.** In our discussion of polynomial interpolation we constructed a nonsingular linear system of equations whose solution provided a set of coefficients for a polynomial whose graph went through each of the distinct points of the specified data set. The construction of an interpolation polynomial is the foundation of a number of topics that often appear in a numerical analysis course. The MATLAB routine **pinterp** provides computational support for obtaining and graphing interpolation polynomials. Use the following data sets with **pinterp**. Type **help pinterp** for more information and directions.

(a) Find the interpolation polynomial $\mathbf{p}_1(x)$ to the set $S_1 = \{(1, 1), (2, 0), (3, 0)\}$. Record the equation for $\mathbf{p}_1(x)$ and inspect its graph.

(b) Find the interpolation polynomial $\mathbf{p}_2(x)$ to the set $S_2 = \{(1, 0), (2, 1), (3, 0)\}$. Record the equation for $\mathbf{p}_2(x)$ and inspect its graph.

(c) Find the interpolation polynomial $\mathbf{p}_3(x)$ to the set $S_3 = \{(1, 0), (2, 0), (3, 1)\}$. Record the equation for $\mathbf{p}_3(x)$ and inspect its graph.

(d) Find the interpolation polynomial $\mathbf{p}_4(x)$ to the set $S_4 = \{(1, 4), (2, -3), (3, 5)\}$. Record the equation for $\mathbf{p}_4(x)$ and inspect its graph.

(e) Determine the values of $a$, $b$, $c$ such that $a * \mathbf{p}_1(\mathbf{x}) + b * \mathbf{p}_2(\mathbf{x}) + c * \mathbf{p}_3(\mathbf{x}) = \mathbf{p}_4(\mathbf{x})$.

(f) What is the relationship between the values of $a$, $b$, and $c$ from part (e) and the data set $S_4$? Explain.

(g) Look at the data sets $S_1$, $S_2$, $S_3$, and $S_4$. Note that the $x$-coordinates are the same. The $y$-coordinates in $S_1$ are $\mathbf{row}_1(\mathbf{I}_3)$. The $y$-coordinates in $S_2$ are $\mathbf{row}_2(\mathbf{I}_3)$. The $y$-coordinates in $S_3$ are $\mathbf{row}_3(\mathbf{I}_3)$. It follows that the $y$-coordinates in $S_4$ are $4 * \mathbf{row}_1(\mathbf{I}_3) - 3 * \mathbf{row}_2(\mathbf{I}_3) + 5 * \mathbf{row}_3(\mathbf{I}_3)$. How is this linear combination related to that in part (e)?

(h) To see a graphical display in MATLAB of the interpolation polynomials and the linear combination in part (e), type **pintdemo**. Write a brief description of the action that takes place in **pintdemo**.

**ML. 7.** An option in the routine **pinterp** is to choose your data set using the mouse. Use this option for each of the following tasks.

(a) Choose a data set of five points so that the interpolant has approximately the shape in the figure below.

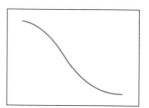

(b) Choose a data set of seven points so that the interpolant has approximately the shape in the figure below.

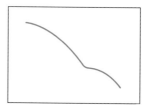

(c) Experiment with the option of choosing data using the mouse and generate the interpolant. Draw a curve on paper and try to choose data points so the interpolant looks like the curve you drew.

## True/False Review Questions

*Determine whether each of the following statements is true or false.*

1. If $\mathbf{rref}(\mathbf{A}) = \begin{bmatrix} 1 & 0 & 2 \\ 0 & 1 & -1 \\ 0 & 0 & 0 \end{bmatrix}$, then $\mathbf{A}$ is nonsingular.

2. If $\mathbf{A}$ is square and the linear system $\mathbf{A}\mathbf{x} = \mathbf{0}$ has two degrees of freedom, then $\mathbf{A}$ is nonsingular.

3. If $\mathbf{A}$ is nonsingular, then the linear system $\mathbf{A}\mathbf{x} = \mathbf{b}$ has the unique solution $\mathbf{x} = \mathbf{A}^{-1}\mathbf{b}$.

4. The $2 \times 2$ matrix $\mathbf{A} = \begin{bmatrix} a & b \\ c & d \end{bmatrix}$ is singular if and only if $ad - bc \neq 0$.

5. If $\mathbf{A}$ and $\mathbf{B}$ are the same size and nonsingular, then $(\mathbf{A}\mathbf{B})^{-1} = \mathbf{A}^{-1}\mathbf{B}^{-1}$.

6. The inverse of $\mathbf{A}^{-1}$ is $\mathbf{A}$.

7. Every diagonal matrix is nonsingular.

8. If $\mathbf{A}$ is square and $\mathbf{ns}(\mathbf{A}) = \mathbf{0}$, then $\mathbf{A}$ is nonsingular.

9. A nonsingular matrix has linearly independent columns.

10. A matrix with a zero entry must be singular.

11. A nonsingular matrix can have many inverses.

12. If $\mathbf{rref}\left(\left[\, \mathbf{A} \mid \mathbf{I} \,\right]\right) = \left[\, \mathbf{I} \mid \mathbf{B} \,\right]$ then $\mathbf{B} = \mathbf{A}^{-1}$.

13. If $\mathbf{A}$ is nonsingular, then $\mathrm{row}_j(\mathbf{A})\,\mathrm{col}_j(\mathbf{A}^{-1}) = 1$.

14. The product of nonsingular matrices (of the same size) is another nonsingular matrix.

15. If $\mathbf{A}$ is square and $\mathbf{rref}(\mathbf{A})$ has a zero row, then $\mathbf{A}$ has no inverse.

## Terminology

| | |
|---|---|
| Nonsingular or invertible matrix | The inverse of a matrix |
| Singular matrix | Computing an inverse |
| Properties of inverses | Properties of nonsingular matrices related to linear systems |
| Recognizing singular matrices | Elementary matrices |

Section 2.3 focused on linear systems with at least one degree of freedom. This section focused on linear systems with zero degrees of freedom, in particular square linear systems with a unique solution. We characterized such systems in terms of properties of the coefficient matrix. The properties of such coefficient matrices involve many of the structural components of linear systems that we have studied in this chapter. Your responses to the following questions and statements provide a review of the ideas in this section and show the connections to previously discussed topics.

- Describe a nonsingular matrix in terms of RREF.
- If $\mathbf{A}$ is nonsingular, describe the solution set of $\mathbf{Ax} = \mathbf{0}$.
- If $\mathbf{A}$ is nonsingular, state a formula for the solution to $\mathbf{Ax} = \mathbf{b}$.
- In the list of terms we used the phrase *The inverse of a matrix*. Explain why we can say *The inverse*.
- If $\mathbf{A}$ is nonsingular, then it has an _____.
- If we must compute an inverse (and in most instances we don't), explain the technique we developed using RREF.
- Determine whether the following statements are true or false.
   1. The inverse of the inverse of $\mathbf{A}$ is $\mathbf{A}$.
   2. The inverse of $\mathbf{A}$ transpose is the transpose of $\mathbf{A}$ inverse.
   3. The inverse of a product $\mathbf{AB}$ is the product of the respective inverses; that is, $\mathbf{A}^{-1}\mathbf{B}^{-1}$.
   4. If $\mathbf{A}$ is nonsingular, then the rows of $\mathbf{A}$ are linearly independent.
   5. If $\mathbf{A}$ has a zero row, then $\mathbf{A}$ is nonsingular.
   6. The null space of a singular matrix contains a nonzero vector.
   7. If $\mathbf{A}$ is $n \times n$ and nonsingular, then for any $n$-vector $\mathbf{b}$, the system $\mathbf{Ax} = \mathbf{b}$ has exactly one solution.
   8. If $\mathbf{A}$ is singular, then its RREF is not the identity matrix.
   9. If $\mathbf{A}$ is nonsingular, then as we compute $\mathbf{rref}(\mathbf{A})$ we will not encounter a row of all zeros.
   10. If $f$ is a matrix transformation using a nonsingular matrix, then there exists a matrix transformation $g$ that undoes $f$.

## 2.5 ■ LU-FACTORIZATION

In Sections 2.2 through 2.4 we showed how to use row operations to obtain a solution to a nonsingular linear system $\mathbf{Ax} = \mathbf{b}$ by reducing $\begin{bmatrix} \mathbf{A} \mid \mathbf{b} \end{bmatrix}$ to upper triangular form $\begin{bmatrix} \mathbf{U} \mid \mathbf{c} \end{bmatrix}$ and then applying back substitution. To be as economical as possible we do not make the diagonal entries of $\mathbf{U}$ have value 1 as in REF, nor do we eliminate the entries above the pivots as in RREF. The "bare-bones" reduction procedure to upper triangular form $\begin{bmatrix} \mathbf{U} \mid \mathbf{c} \end{bmatrix}$ followed by back substitution is commonly called **Gaussian elimination**. Both REF and RREF are variants of Gaussian elimination that are convenient for small linear systems and are used with hand calculations. REF and RREF simplify the back substitution process; however, both require more effort than Gaussian elimination when implemented on a computer. For this reason Gaussian elimination is widely used for solving linear systems on a computer.[1]

---

<< Find an LU-FACTORIZATION by Row Reduction >>

**L** =
```
   1   0   0
   0   1   0
   0   0   1
```
**U** =
```
   4   5      -2
   0   1.75   0.5
  -2   3      1
```

You just performed operation $-0.25 * \text{Row}(1) + \text{Row}(2)$

Insert a value in **L** in the position you just eliminated in U. Let the multiplier you just used be called num. It has the value $-0.25$.

Enter row number of **L** to change.  2

Enter column number of **L** to change.  1

Value of L(2,1) = ____

---

In MATLAB, **lupr.m**

In this section we develop a modification of Gaussian elimination that factors a nonsingular matrix $\mathbf{A}$ into a product of a lower triangular matrix $\mathbf{L}$ and an upper triangular matrix $\mathbf{U}$; that is, $\mathbf{A} = \mathbf{LU}$. We call this an **LU-factorization** of matrix $\mathbf{A}$. We briefly explored the use of LU-factorizations in Exercises 37 through 43 of Section 2.1. For completeness we include part of that information here and identify specific steps that utilize this factorization to solve the linear system $\mathbf{Ax} = \mathbf{L}(\mathbf{Ux}) = \mathbf{b}$.

For a nonsingular linear system $\mathbf{Ax} = \mathbf{b}$ for which we know (or can find) an LU-factorization, $\mathbf{A} = \mathbf{LU}$, we proceed as follows:

1. Replace $\mathbf{Ax} = \mathbf{b}$ with $\mathbf{L}(\mathbf{Ux}) = \mathbf{b}$.
2. Let $\mathbf{Ux} = \mathbf{z}$. Then $\mathbf{L}(\mathbf{Ux}) = \mathbf{Lz} = \mathbf{b}$ is a linear system with a lower triangular coefficient matrix that can be solved for $\mathbf{z}$ using forward substitution. (See Exercise 27 in Section 2.1.)

---

[1]MATLAB uses a variation of Gaussian elimination that incorporates LU-factorization, the topic of this section, in its command that automatically solves nonsingular linear systems.

**3.** Once we have $\mathbf{z}$, we solve the upper triangular linear system $\mathbf{Ux} = \mathbf{z}$ by back substitution for the solution $\mathbf{x}$. (See Exercise 23 in Section 2.1.)

In summary, once we have an LU-factorization of $\mathbf{A}$, solve $\mathbf{Ax} = \mathbf{L}(\mathbf{Ux}) = \mathbf{b}$ using forward substitution followed by a back substitution. (See Exercises 37 through 43 in Section 2.1.)

Our focus in this section is to show how to obtain an LU-factorization using the row operations from Gaussian elimination. The procedure we develop involves the same amount of effort as Gaussian elimination but organizes the information from the reduction steps in a different way. Instead of reducing the augmented matrix $\begin{bmatrix} \mathbf{A} \mid \mathbf{b} \end{bmatrix}$ we reduce just the coefficient matrix $\mathbf{A}$ to upper triangular form $\mathbf{U}$ and save information from the row operations to construct the lower triangular matrix $\mathbf{L}$. In the preceding list, step 2 corresponds to using the row operations on the augmented column and step 3 is the same as the back substitution in Gaussian elimination.

We develop a simplified form of LU-factorization. In this form we permit only row operations of the form $k R_i + R_j$ which are specifically constructed to eliminate an entry in $\mathbf{A}$. Furthermore, we will adopt a systematic reduction scheme that eliminates the entries below a pivot. Unfortunately, this restrictive procedure can fail if a row interchange is needed to obtain a nonzero pivot. It is possible to incorporate row interchanges, but doing so requires that we introduce a bookkeeping scheme to keep track of row interchanges. This more general method of LU-factorization is often covered in a numerical analysis course and will not be pursued here.

As we use row operations of the form $k R_i + R_j$ to reduce $\mathbf{A}$ to upper triangular form $\mathbf{U}$, we simultaneously build the lower triangular matrix $\mathbf{L}$ as follows:

**1.** $\mathbf{L}$ is lower triangular with its diagonal entries equal to 1. Initially the entries below the diagonal entries are blank. (We designate these by $*$.) As we use row operations the blanks are filled in. The initial form of $\mathbf{L}$ is

$$\mathbf{L} = \begin{bmatrix} 1 & 0 & \cdots & 0 & 0 \\ * & 1 & \cdots & 0 & 0 \\ * & \cdots & \ddots & \vdots & \vdots \\ * & \cdots & * & 1 & 0 \\ * & \cdots & * & * & 1 \end{bmatrix}.$$

**2.** When row operation $k R_i + R_j$ eliminates or zeros out the $(j, i)$-entry in $\mathbf{A}$, then we insert the value $-k$ into the $(j, i)$-entry of $\mathbf{L}$. (We call this **storage of multipliers**.)

At the end of this section we discuss the reasons why this construction of $\mathbf{L}$ is valid. We illustrate the mechanics of the construction of $\mathbf{L}$ and $\mathbf{U}$ in Example 1.

EXAMPLE 1   Determine an LU-factorization of

$$\mathbf{A} = \begin{bmatrix} 6 & -2 & -4 & 4 \\ 3 & -3 & -6 & 1 \\ -12 & 8 & 21 & -8 \\ -6 & 0 & -10 & 7 \end{bmatrix}$$

using the storage of multipliers technique. We proceed to zero out the entries below the diagonal entries using only row operations of the form $k R_i + R_j$ . We let $\mathbf{U}_j$ and $\mathbf{L}_j$ be matrices on their way to becoming $\mathbf{U}$ and $\mathbf{L}$, respectively. Note that we set the entries of $\mathbf{L}$ equal to the *negatives of the multipliers* used to eliminate the corresponding entry in the $\mathbf{U}$-matrices.

$$\mathbf{U}_1 = \mathbf{A}_{\substack{(-\frac{1}{2})R_1 + R_2 \\ 2R_1 + R_3 \\ 1R_1 + R_4}} = \begin{bmatrix} 6 & -2 & -4 & 4 \\ 0 & -2 & -4 & -1 \\ 0 & 4 & 13 & 0 \\ 0 & -2 & -14 & 11 \end{bmatrix} \qquad \mathbf{L}_1 = \begin{bmatrix} 1 & 0 & 0 & 0 \\ \frac{1}{2} & 1 & 0 & 0 \\ -2 & * & 1 & 0 \\ -1 & * & * & 1 \end{bmatrix}.$$

$$\mathbf{U}_2 = \mathbf{U}_{1\substack{2R_2 + R_3 \\ -1R_2 + R_4}} = \begin{bmatrix} 6 & -2 & -4 & 4 \\ 0 & -2 & -4 & -1 \\ 0 & 0 & 5 & -2 \\ 0 & 0 & -10 & 12 \end{bmatrix} \qquad \mathbf{L}_2 = \begin{bmatrix} 1 & 0 & 0 & 0 \\ \frac{1}{2} & 1 & 0 & 0 \\ -2 & -2 & 1 & 0 \\ -1 & 1 & * & 1 \end{bmatrix}.$$

$$\mathbf{U}_3 = \mathbf{U}_{2\,2R_3 + R_4} = \begin{bmatrix} 6 & -2 & -4 & 4 \\ 0 & -2 & -4 & -1 \\ 0 & 0 & 5 & -2 \\ 0 & 0 & 0 & 8 \end{bmatrix} \qquad \mathbf{L}_3 = \begin{bmatrix} 1 & 0 & 0 & 0 \\ \frac{1}{2} & 1 & 0 & 0 \\ -2 & -2 & 1 & 0 \\ -1 & 1 & -2 & 1 \end{bmatrix}.$$

Now we define $\mathbf{L} = \mathbf{L}_3$ and $\mathbf{U} = \mathbf{U}_3$ so that $\mathbf{A} = \mathbf{LU}$. (Verify that we have an LU-factorization by showing that the product $\mathbf{L} * \mathbf{U}$ gives $\mathbf{A}$.) ■

**EXAMPLE 2**   Let $\mathbf{A}$ be specified as in Example 1. Solve the linear system $\mathbf{Ax} = \mathbf{b}$ by using the LU-factorization developed in Example 1 when

$$\mathbf{b} = \begin{bmatrix} 2 \\ -4 \\ 8 \\ -43 \end{bmatrix}.$$

We have $\mathbf{Ax} = \mathbf{L}(\mathbf{Ux}) = \mathbf{b}$. Let $\mathbf{z} = \mathbf{Ux}$ and then solve $\mathbf{Lz} = \mathbf{b}$ by forward substitution:

$$\begin{bmatrix} 1 & 0 & 0 & 0 \\ \frac{1}{2} & 1 & 0 & 0 \\ -2 & -2 & 1 & 0 \\ -1 & 1 & -2 & 1 \end{bmatrix} \begin{bmatrix} z_1 \\ z_2 \\ z_3 \\ z_4 \end{bmatrix} = \begin{bmatrix} 2 \\ -4 \\ 8 \\ -43 \end{bmatrix} \quad \Leftrightarrow$$

$$\begin{aligned} z_1 &= 2 \\ z_2 &= -4 - \tfrac{1}{2}z_1 = -5 \\ z_3 &= 8 + 2z_1 + 2z_2 = 2 \\ z_4 &= -43 + z_1 - z_2 + 2z_3 = -32 \end{aligned} \qquad \Leftrightarrow \quad \mathbf{z} = \begin{bmatrix} 2 \\ -5 \\ 2 \\ -32 \end{bmatrix}.$$

Next we solve $\mathbf{Ux} = \mathbf{z}$ by back substitution:

$$\begin{bmatrix} 6 & -2 & -4 & 4 \\ 0 & -2 & -4 & -1 \\ 0 & 0 & 5 & -2 \\ 0 & 0 & 0 & 8 \end{bmatrix} \begin{bmatrix} x_1 \\ x_2 \\ x_3 \\ x_4 \end{bmatrix} = \begin{bmatrix} 2 \\ -5 \\ 2 \\ -32 \end{bmatrix} \Leftrightarrow$$

$$x_4 = \frac{-32}{8} = -4$$

$$x_3 = \frac{2 + 2x_4}{5} = -1.2$$

$$x_2 = \frac{-5 + 4x_3 + x_4}{-2} = 6.9 \qquad \Leftrightarrow \quad \mathbf{x} = \begin{bmatrix} 4.5 \\ 6.9 \\ -1.2 \\ -4 \end{bmatrix}.$$

$$x_1 = \frac{2 + 2x_2 + 4x_3 - 4x_4}{6} = 4.5$$

It is important to note that if the right side $\mathbf{b}$ were changed in Example 2, then we need only perform the forward and back substitution to solve the linear system. Since $\mathbf{A}$ did not change, we use the same LU-factorization. Thus if we have a set of linear systems with the same coefficient matrix $\mathbf{A}$ but different right sides, we compute the LU-factorization once and save it. Then we use the matrices $\mathbf{L}$ and $\mathbf{U}$ with the various right sides to solve each new linear system. This is an advantage of LU-factorization over Gaussian elimination.

In the step-by-step LU-factorization process shown in Example 1, if the $(j + 1, j + 1)$-entry of $\mathbf{U}_j$ is zero, the method fails to produce a factorization. A possible remedy, assuming that the coefficient matrix is indeed nonsingular, is to rearrange the order of the original system of equations and begin again. If it happens that $\mathbf{A}$ is singular, then no reordering of the equations will produce a successful conclusion to the factorization process. In computer algorithms that incorporate row interchanges in the factorization process, the failure to obtain a nonzero pivot at any stage is often used to indicate that the matrix is singular. Hence in such computer programs the LU-factorization process can be used as a test to see if a matrix is nonsingular. (Because of the use of computer arithmetic, which is not the same as exact arithmetic, an extremely small pivot is a warning that the matrix may be singular.)

**Verification that the storage of multipliers technique is valid (Optional)**
Assume that $\mathbf{A}$ is nonsingular and that we can reduce $\mathbf{A}$ to upper triangular form $\mathbf{U}$ by using only row operations of the form $kR_i + R_j$ to eliminate the entries below the pivots. If $\mathbf{A}$ is $n \times n$ then $n - 1$ such row operations are needed to eliminate below the first pivot, $n - 2$ to eliminate below the second pivot, and so on until we need only 1 to eliminate below the $(n-1)$st pivot. [It also follows that the $n$th pivot, the $(n, n)$-entry of $\mathbf{U}$, is nonzero.] Hence we require

$$N = (n - 1) + (n - 2) + \cdots + 2 + 1 = \frac{n(n - 1)}{2}$$

such row operations. Corresponding to each of these $N$ row operations is an elementary matrix (see Section 2.4) that is lower triangular with (at most) one nonzero entry below the diagonal.[2] For row operation $kR_i + R_j$ of this type the corresponding elementary matrix is

---

[2] We assume that the row operations eliminate successive entries below the pivots.

$$i$$
$$\downarrow$$

$$\mathbf{E}_p = \begin{bmatrix} 1 & & & & & \\ & \ddots & & & & \\ & & \ddots & & & \\ & & & \ddots & & \\ j \rightarrow & & k & & \ddots & \\ & & & & & 1 \end{bmatrix} \qquad \text{where } j > i. \tag{1}$$

That is, the $(j, i)$-entry of elementary matrix $\mathbf{E}_p$ is $k$ and all the other nondiagonal entries are zero.

From the discussion about elementary matrices in Section 2.4, it follows that we have the matrix equation

$$\mathbf{E}_N \cdots \mathbf{E}_2 \mathbf{E}_1 \mathbf{A} = \mathbf{U} \tag{2}$$

where the $\mathbf{E}$'s are the elementary matrices of the form shown in (1). By Exercises 59 and 60 in Section 2.4, each of the $\mathbf{E}$'s is nonsingular and moreover

$$i$$
$$\downarrow$$

$$\mathbf{E}_p^{-1} = \begin{bmatrix} 1 & & & & & \\ & \ddots & & & & \\ & & \ddots & & & \\ & & & \ddots & & \\ j \rightarrow & & -k & & \ddots & \\ & & & & & 1 \end{bmatrix} \qquad \text{where } j > i. \tag{3}$$

Using these observations, we have that (2) is equivalent to

$$\mathbf{A} = \mathbf{E}_1^{-1} \mathbf{E}_2^{-1} \cdots \mathbf{E}_N^{-1} \mathbf{U}. \tag{4}$$

Let $\mathbf{Q} = \mathbf{E}_1^{-1} \mathbf{E}_2^{-1} \cdots \mathbf{E}_N^{-1}$. Then $\mathbf{Q}$ is nonsingular since it is the product of nonsingular matrices. (See property 5 given in the list of "Algebraic Properties of the Inverse" in Section 2.4.) In addition, each of the matrices $\mathbf{E}_p^{-1}$ is lower triangular with 1's on the diagonal and so is their product $\mathbf{Q}$. (See Exercise 76 in Section 1.3.) It only remains to verify that the entries below the diagonal in $\mathbf{Q}$ are the negatives of the multipliers of the row operations from the reduction to upper triangular form. A general argument for this is quite cumbersome, so instead we indicate how to observe that we have the desired form. Consider computing the product of a pair of distinct matrices of the form given in (3). The nonzero entries below the diagonal in each of these matrices are in different positions. Because the other rows and columns are from an identity matrix, the row-by-column product will produce a matrix with 1's on the diagonal, the pair of nonzero entries will appear in the same positions as in the individual matrices, and there will be zeros everywhere else.[3]

---

[3]It is assumed that the elimination process proceeds to zero out below the $(1, 1)$-entry, then below the $(2, 2)$-entry, and so on. With this assumption we get the behavior of products of the inverses of the matrices $\mathbf{E}_j$ stated here.

This behavior is valid for a product of more than two matrices of the form in (3). For example,

$$\begin{bmatrix} 1 & 0 & 0 & 0 \\ -k_1 & 1 & 0 & 0 \\ 0 & 0 & 1 & 0 \\ 0 & 0 & 0 & 1 \end{bmatrix} \begin{bmatrix} 1 & 0 & 0 & 0 \\ 0 & 1 & 0 & 0 \\ 0 & 0 & 1 & 0 \\ 0 & -k_2 & 0 & 1 \end{bmatrix} = \begin{bmatrix} 1 & 0 & 0 & 0 \\ -k_1 & 1 & 0 & 0 \\ 0 & 0 & 1 & 0 \\ 0 & -k_2 & 0 & 1 \end{bmatrix}$$

and

$$\begin{bmatrix} 1 & 0 & 0 & 0 \\ -k_1 & 1 & 0 & 0 \\ 0 & 0 & 1 & 0 \\ 0 & 0 & 0 & 1 \end{bmatrix} \begin{bmatrix} 1 & 0 & 0 & 0 \\ 0 & 1 & 0 & 0 \\ 0 & 0 & 1 & 0 \\ 0 & -k_2 & 0 & 1 \end{bmatrix} \begin{bmatrix} 1 & 0 & 0 & 0 \\ 0 & 1 & 0 & 0 \\ -k_3 & 0 & 1 & 0 \\ 0 & 0 & 0 & 1 \end{bmatrix}$$

$$= \begin{bmatrix} 1 & 0 & 0 & 0 \\ -k_1 & 1 & 0 & 0 \\ 0 & 0 & 1 & 0 \\ 0 & -k_2 & 0 & 1 \end{bmatrix} \begin{bmatrix} 1 & 0 & 0 & 0 \\ 0 & 1 & 0 & 0 \\ -k_3 & 0 & 1 & 0 \\ 0 & 0 & 0 & 1 \end{bmatrix} = \begin{bmatrix} 1 & 0 & 0 & 0 \\ -k_1 & 1 & 0 & 0 \\ -k_3 & 0 & 1 & 0 \\ 0 & -k_2 & 0 & 1 \end{bmatrix}.$$

Thus $\mathbf{Q}$ is lower triangular with 1's on its diagonal and has the negatives of the multipliers appearing in the appropriate positions in the entries below the diagonal. Hence we define $\mathbf{L} = \mathbf{Q}$ and (4) becomes $\mathbf{A} = \mathbf{LU}$ as asserted.

# EXERCISES 2.5

*In Exercises 1 through 4, solve the linear system $\mathbf{Ax} = \mathbf{b}$ with the given LU-factorization using a forward substitution followed by a back substitution.*

**1.** $\mathbf{A} = \begin{bmatrix} 2 & 8 & 0 \\ 2 & 2 & -3 \\ 1 & 2 & 7 \end{bmatrix}$, $\mathbf{b} = \begin{bmatrix} 18 \\ 3 \\ 12 \end{bmatrix}$,

$\mathbf{L} = \begin{bmatrix} 2 & 0 & 0 \\ 2 & -3 & 0 \\ 1 & -1 & 4 \end{bmatrix}$, $\mathbf{U} = \begin{bmatrix} 1 & 4 & 0 \\ 0 & 2 & 1 \\ 0 & 0 & 2 \end{bmatrix}$.

**2.** $\mathbf{A} = \begin{bmatrix} 8 & 12 & -4 \\ 6 & 5 & 7 \\ 2 & 1 & 6 \end{bmatrix}$, $\mathbf{b} = \begin{bmatrix} -36 \\ 11 \\ 16 \end{bmatrix}$,

$\mathbf{L} = \begin{bmatrix} 4 & 0 & 0 \\ 3 & 2 & 0 \\ 1 & 1 & 1 \end{bmatrix}$, $\mathbf{U} = \begin{bmatrix} 2 & 3 & -1 \\ 0 & -2 & 5 \\ 0 & 0 & 2 \end{bmatrix}$.

**3.** $\mathbf{A} = \begin{bmatrix} 2 & 3 & 0 & 1 \\ 4 & 5 & 3 & 3 \\ -2 & -6 & 7 & 7 \\ 8 & 9 & 5 & 21 \end{bmatrix}$, $\mathbf{b} = \begin{bmatrix} -2 \\ -2 \\ -16 \\ -66 \end{bmatrix}$,

$\mathbf{L} = \begin{bmatrix} 1 & 0 & 0 & 0 \\ 2 & 1 & 0 & 0 \\ -1 & 3 & 1 & 0 \\ 4 & 3 & 2 & 1 \end{bmatrix}$, $\mathbf{U} = \begin{bmatrix} 2 & 3 & 0 & 1 \\ 0 & -1 & 3 & 1 \\ 0 & 0 & -2 & 5 \\ 0 & 0 & 0 & 4 \end{bmatrix}$.

**4.** $\mathbf{A} = \begin{bmatrix} 2 & 1 & 0 \\ 2i & 2i & 1 \\ 2+2i & 2+i & 2-i \end{bmatrix}$, $\mathbf{b} = \begin{bmatrix} 2+i \\ -1+2i \\ 3+3i \end{bmatrix}$,

$\mathbf{L} = \begin{bmatrix} 1 & 0 & 0 \\ i & 1 & 0 \\ 1+i & -i & 1 \end{bmatrix}$, $\mathbf{U} = \begin{bmatrix} 2 & 1 & 0 \\ 0 & i & 1 \\ 0 & 0 & 2 \end{bmatrix}$.

*In Exercises 5 through 9, find an LU-factorization of the coefficient matrix $\mathbf{A}$ of the linear system $\mathbf{Ax} = \mathbf{b}$. Solve the linear system using a forward substitution followed by a back substitution.*

**5.** $\mathbf{A} = \begin{bmatrix} 2 & 3 & 4 \\ 4 & 5 & 10 \\ 4 & 8 & 2 \end{bmatrix}$, $\mathbf{b} = \begin{bmatrix} 6 \\ 16 \\ 2 \end{bmatrix}$.

**6.** $\mathbf{A} = \begin{bmatrix} -3 & 1 & -2 \\ -12 & 10 & -6 \\ 15 & 13 & 12 \end{bmatrix}$, $\mathbf{b} = \begin{bmatrix} 15 \\ 82 \\ -5 \end{bmatrix}$.

**7.** $\mathbf{A} = \begin{bmatrix} -5 & 4 & 0 & 1 \\ -30 & 27 & 2 & 7 \\ 5 & 2 & 0 & 2 \\ 10 & 1 & -2 & 1 \end{bmatrix}$, $\mathbf{b} = \begin{bmatrix} -17 \\ -102 \\ 7 \\ -6 \end{bmatrix}$.

**8.** $\mathbf{A} = \begin{bmatrix} 1 & -2 & 0 & 0 & 0 \\ -2 & 1 & -2 & 0 & 0 \\ 0 & -2 & 1 & -2 & 0 \\ 0 & 0 & -2 & 1 & -2 \\ 0 & 0 & 0 & -2 & 1 \end{bmatrix}$, $\mathbf{b} = \begin{bmatrix} 1 \\ -2 \\ 0 \\ -2 \\ 1 \end{bmatrix}$.

**9.** $\mathbf{A} = \begin{bmatrix} 2 & 1 & i \\ 4 & 3 & 2i \\ 2i & 0 & 1 \end{bmatrix}$, $\mathbf{b} = \begin{bmatrix} 1+3i \\ 3+6i \\ -1 \end{bmatrix}$.

In MATLAB

*Use the* MATLAB *routine **lupr** to practice generating an LU-factorization. The routine requires explicit responses to the* necessary steps for obtaining the factorization and checks that you have made correct choices. Use ***help lupr*** for directions. Practice on the preceding exercises.

## True/False Review Questions

*Determine whether each of the following statements is true or false.*

**1.** $\mathbf{L} = \begin{bmatrix} 1 & 0 \\ 2 & 1 \end{bmatrix}$ and $\mathbf{U} = \begin{bmatrix} 3 & -1 \\ 0 & 2 \end{bmatrix}$ is an LU-factorization of $\mathbf{A} = \begin{bmatrix} 3 & -1 \\ 0 & 6 \end{bmatrix}$.

**2.** Given an LU-factorization of nonsingular matrix $\mathbf{A}$, we solve $\mathbf{Ax} = \mathbf{b}$ using a forward substitution followed by a back substitution.

**3.** The technique we described for finding an LU-factorization will always work.

### Terminology

| LU-factorization | Storage of multipliers |
| --- | --- |

LU-factorization is an alternative procedure for solving nonsingular linear systems. It is actually more efficient than using RREF in a number of ways and hence is used in many computer algorithms. As a review for the ideas in this section, develop responses to the following questions and statements in your own words.

- Given that we have an LU-factorization of the coefficient matrix of linear system $\mathbf{Ax} = \mathbf{b}$, tell how we use it to solve the system.
- For the technique described in this chapter, explain the structure of the lower triangular matrix $\mathbf{L}$ and how it is obtained as we eliminate below the pivots.
- If we have linear systems with the same coefficient matrix but different right sides,
  1. Explain how to solve them at the same time using RREF.
  2. Explain how to solve them using an LU-factorization.
- The method described in this section for LU-factorization can fail. Why can this happen, and what is the remedy?

## 2.6 ■ THE (LEAST SQUARES) LINE OF BEST FIT

Deviations

In MATLAB, **lsqgame.m**

In applications discussed so far, consistent linear systems have provided an important mathematical model. In Section 2.4 we discussed nonsingular linear systems that have a unique solution. In Sections 2.2 and 2.3 we discussed underdetermined linear systems that have infinitely many solutions. In each of these cases the linear system $\mathbf{Ax} = \mathbf{b}$ is consistent or, equivalently, $\mathbf{b}$ is in the column space of $\mathbf{A}$. That is, $\mathbf{b}$ is a linear combination of the columns of $\mathbf{A}$.

It would be convenient if we could ignore all inconsistent systems; however, we cannot. Inconsistent systems do occur, and we must deal with them. In order to accomplish this, we need to alter our notion that a solution to a matrix problem satisfies the equality requirement that $\mathbf{A}$ times $\mathbf{x}$ gives $\mathbf{b}$.

Consider the case of an **overdetermined** linear system; that is, a linear system of equations in which there are more equations than unknowns. Such linear systems arise naturally from experiments or collections of data that consist of a large number of observations that are used to estimate a few unknowns in a mathematical model. Examples include computing the orbit of a satellite or path of a projectile, determining rate constants of various types, and, in general, calculating coefficients in a proposed model of a physical phenomenon or process. Since errors will invariably be included in such observational data, it is expected that such overdetermined linear systems will be inconsistent. Whether the resulting linear system is inconsistent or not, values of unknown coefficients of the model are required. If the linear system $\mathbf{Ax} = \mathbf{b}$ is inconsistent, then the use of row operations will fail to yield a result. In such cases an alternative is to seek a vector $\mathbf{z}$ so that the product $\mathbf{Az}$ is as close to the right side $\mathbf{b}$ as possible. Since $\mathbf{Az}$ is a linear combination of the columns of $\mathbf{A}$ and hence in $\mathbf{col}(\mathbf{A})$, we can rephrase the situation as follows:

Determine the vector in $\mathbf{col}(\mathbf{A})$ that is closest to $\mathbf{b}$.

This implies that we must solve a minimization problem; namely, determine the minimum distance from $\mathbf{b}$ to the subspace $\mathbf{col}(\mathbf{A})$. Note that if the linear system is consistent, then the standard solution $\mathbf{x}$ to $\mathbf{Ax} = \mathbf{b}$ is in $\mathbf{col}(\mathbf{A})$ and hence the distance between $\mathbf{b}$ and $\mathbf{col}(\mathbf{A})$ is zero. Thus the minimization problem includes the solution of consistent linear systems as a special case.

A matrix approach to solving this minimization problem requires concepts we have not yet developed, but an alternative is to formulate the minimization problem in terms of calculus. Here we consider the special case in which there are just two unknown coefficients in the linear system. We use a geometric model to construct the appropriate minimization statement and then show how to reformulate the result in terms of matrices.

## Geometric Model (Special Case)

Let $D = \{(x_1, y_1), (x_2, y_2), \ldots, (x_n, y_n)\}$ be a data set in which the points are distinct in the sense that all the $x$-coordinates are different from one another. We say there is a "linear" relationship between the $x$- and $y$-coordinates provided that there is some line $y = mx + b$ that either goes through each point $(x_i, y_i)$ or comes close to all the ordered pairs in $D$. If each point of $D$ lies on the same line, then the values of $m$ and $b$ can be determined from the solution of a nonsingular linear system. If there is no line that goes through all the points in $D$, then an inconsistent linear system arises in trying to determine $m$ and $b$. In either case we have the linear system in the unknowns $m$ and $b$ given by

$$
\begin{aligned}
mx_1 + b &= y_1 \\
mx_2 + b &= y_2 \\
&\vdots \\
&\vdots \\
mx_n + b &= y_n
\end{aligned}
\quad \text{or} \quad
\begin{bmatrix} x_1 & 1 \\ x_2 & 1 \\ \vdots & \vdots \\ \vdots & \vdots \\ x_n & 1 \end{bmatrix}
\begin{bmatrix} m \\ b \end{bmatrix}
=
\begin{bmatrix} y_1 \\ y_2 \\ \vdots \\ \vdots \\ y_n \end{bmatrix}.
\tag{1}
$$

Let

$$
\mathbf{A} = \begin{bmatrix} x_1 & 1 \\ x_2 & 1 \\ \vdots & \vdots \\ \vdots & \vdots \\ x_n & 1 \end{bmatrix}, \quad
\mathbf{x} = \begin{bmatrix} m \\ b \end{bmatrix}, \quad \text{and} \quad
\begin{bmatrix} y_1 \\ y_2 \\ \vdots \\ \vdots \\ y_n \end{bmatrix}.
$$

Then the linear system in (1) is $\mathbf{Ax} = \mathbf{y}$.

Here we investigate the case in which $\mathbf{Ax} = \mathbf{y}$ is inconsistent. Thus geometrically, the data in $D$ is a set of noncollinear points as shown in Figure 1. The line

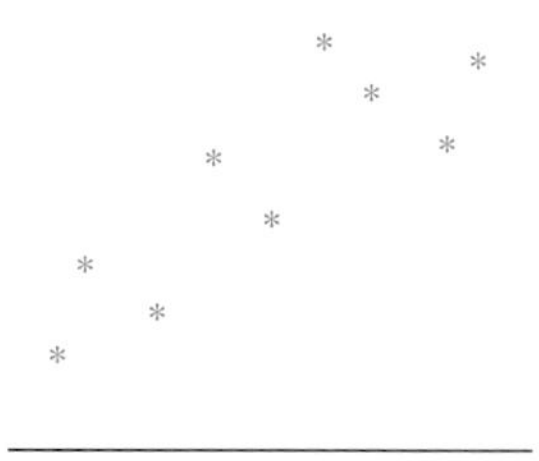

FIGURE 1

that comes closest to all the data in $D$ is called the **line of best fit**. The technique to determine the line of best fit is known as the **method of least squares** since we adopt the criterion that we want to minimize the sum of the squares of the vertical distances from the data points to the line $y = mx + b$. The vertical distance, often called the **deviation**, from a point $(x_i, y_i)$ to the line $y = mx + b$ is given by the expression $(mx_i + b - y_i)$, which is just the difference of the $y$-coordinates of the data point and the point on the line when $x = x_i$. Computing this difference for each point in $D$, squaring the quantities, and adding them gives the expression $E(m, b)$ shown next:

$$E(m, b) = (mx_1 + b - y_1)^2 + (mx_2 + b - y_2)^2 + \cdots + (mx_n + b - y_n)^2.$$

This expression is called the **sum of the squares of the deviations**. The line of best fit is obtained by determining the values of $m$ and $b$ that minimize the sum of the squares of the deviations given in the expression $E(m, b)$.

Using summation notation, we have

$$E(m, b) = \sum_{i=1}^{n} (mx_i + b - y_i)^2.$$

To use calculus[1] to obtain the values that minimize $E(m, b)$, we proceed as follows. Compute the partial derivative of $E(m, b)$ with respect to $m$ and the partial with respect to $b$, set them equal to zero, and solve for $m$ and $b$. We obtain

$$\frac{\partial E(m, b)}{\partial m} = 2 \sum_{i=1}^{n} (mx_i + b - y_i) \cdot x_i = 0,$$

$$\frac{\partial E(m, b)}{\partial b} = 2 \sum_{i=1}^{n} (mx_i + b - y_i) = 0.$$

This gives us two equations in the unknowns $m$ and $b$, which can be simplified and rearranged into the following form:

$$m \sum_{i=1}^{n} x_i^2 + b \sum_{i=1}^{n} x_i = \sum_{i=1}^{n} x_i y_i,$$

$$m \sum_{i=1}^{n} x_i + b \sum_{i=1}^{n} 1 = \sum_{i=1}^{n} y_i.$$

Note that $\sum_{i=1}^{n} 1 = n$; thus this linear system in matrix form is given by

$$\begin{bmatrix} \sum_{i=1}^{n} x_i^2 & \sum_{i=1}^{n} x_i \\ \sum_{i=1}^{n} x_i & n \end{bmatrix} \begin{bmatrix} m \\ b \end{bmatrix} = \begin{bmatrix} \sum_{i=1}^{n} x_i y_i \\ \sum_{i=1}^{n} y_i \end{bmatrix}. \tag{2}$$

[1] If you are unfamiliar with partial derivatives, then the material in the next display can be skipped. You can proceed to the linear system given in (2). This is the linear system that must be solved in order to determine the slope and $y$-intercept of the line of best fit.

Let

$$
\mathbf{C} = \begin{bmatrix} \displaystyle\sum_{i=1}^{n} x_i^2 & \displaystyle\sum_{i=1}^{n} x_i \\ \displaystyle\sum_{i=1}^{n} x_i & n \end{bmatrix}, \quad \mathbf{d} = \begin{bmatrix} \displaystyle\sum_{i=1}^{n} x_i y_i \\ \displaystyle\sum_{i=1}^{n} y_i \end{bmatrix}.
$$

Then (2) is given by matrix equation $\mathbf{Cx} = \mathbf{d}$. The matrix $\mathbf{C}$ is nonsingular (see Exercise 9) so (2) has a unique solution. The resulting values of $m$ and $b$ are, respectively, the slope and $y$-intercept of the line of best fit. The linear system $\mathbf{Cx} = \mathbf{d}$ is called the **normal system of equations** for the line of best fit. When we want to compute the line of best fit for a particular data set, we can immediately construct the normal system of equations and then find its solution. There is no need to repeat the minimization steps employed to obtain (2). We illustrate the technique in Example 1.

EXAMPLE 1    Various airlines publish a table showing how the temperature (in °F) outside an airplane changes as the altitude (in 1000 feet) changes. This data is given in Table 1. Determine the line of best fit to this data set.

TABLE 1

| x (Altitude in 1000 ft) | 1 | 5 | 10 | 15 | 20 | 30 | 36 |
|---|---|---|---|---|---|---|---|
| y (Temperature in °F) | 56 | 41 | 23 | 5 | −15 | −47 | −69 |

Using (2), we construct the normal system of equations:

$$n = 7$$

$$\sum_{i=1}^{7} x_i = 1 + 5 + 10 + 15 + 20 + 30 + 36 = 117$$

$$\sum_{i=1}^{7} x_i^2 = 1^2 + 5^2 + 10^2 + 15^2 + 20^2 + 30^2 + 36^2 = 2947$$

$$\sum_{i=1}^{7} y_i = 56 + 41 + 23 + 5 - 15 - 47 - 69 = -6$$

$$\sum_{i=1}^{7} x_i y_i = (1)(56) + (5)(41) + (10)(23) + (15)(5) + (20)(-15)$$
$$+ (30)(-47) + (36)(-69) = -3628$$

$$
\begin{bmatrix} \displaystyle\sum_{i=1}^{7} x_i^2 & \displaystyle\sum_{i=1}^{7} x_i \\ \displaystyle\sum_{i=1}^{7} x_i & 7 \end{bmatrix} \begin{bmatrix} m \\ b \end{bmatrix} = \begin{bmatrix} \displaystyle\sum_{i=1}^{7} x_i y_i \\ \displaystyle\sum_{i=1}^{7} y_i \end{bmatrix}
$$

or

$$
\begin{bmatrix} 2947 & 117 \\ 117 & 7 \end{bmatrix} \begin{bmatrix} m \\ b \end{bmatrix} = \begin{bmatrix} -3628 \\ -6 \end{bmatrix}.
$$

Solving for $m$ and $b$, we find that (to four decimal places) $m = -3.5582$ and $b = 58.6159$. Hence the line of best fit is

$$y = -3.5582x + 58.6159. \qquad (3)$$

Figure 2 shows both the data from Table 1 and the line of best fit. Note that this line comes very close to all the data, which is a strong indication that there is a linear relationship between altitude and temperature.

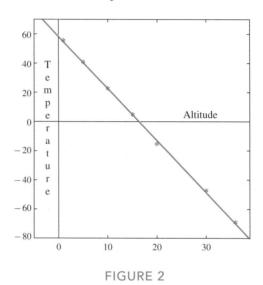

FIGURE 2

The line of best fit can also be used as a mathematical model to estimate either the temperature at a given altitude or the altitude at which a specific temperature occurs. For instance, to estimate the temperature at an altitude of 40,000 ft, we set $x = 40$ in (3) to obtain

$$y = -3.5582(40) + 58.6159 = -83.7121,$$

or approximately $-84°$F. In a similar fashion, if we want to estimate the altitude at which the temperature is $-30°$F, we set $y = -30$ and solve for $x$. We find that

$$x = \frac{-30 - 58.6159}{-3.5582} = \frac{-88.6159}{-3.5582} \approx 24.9047$$

so a temperature of $-30°$F occurs at approximately 25,000 ft.                    ■

Another mathematical model can be obtained by interchanging the roles of the $x$ and $y$ data. Thus geometrically this would amount to plotting the temperature horizontally and the altitude vertically. The model obtained in this manner is different from that obtained in Example 1.

It is quite possible that a given data set is not well approximated by a line. In such instances the method of least squares[2] can be extended to obtain functions of other shapes that come close to all the points in the set. We do not investigate this area in detail here, but the special case of a parabola of best fit is discussed in Exercise 8.

---

[2]Recall that we have considered only a special case: a line of best fit to a data set.

## Matrix Computations to Obtain the Normal Equations in (2) (Optional)

The normal system of equations in (2) was derived using calculus. Here we show how to develop the normal equations directly from the original system of equations

$$
\begin{bmatrix} x_1 & 1 \\ x_2 & 1 \\ \vdots & \vdots \\ \vdots & \vdots \\ x_n & 1 \end{bmatrix} \begin{bmatrix} m \\ b \end{bmatrix} = \begin{bmatrix} y_1 \\ y_2 \\ \vdots \\ \vdots \\ y_n \end{bmatrix} \qquad \Leftrightarrow \qquad \mathbf{Ax} = \mathbf{y} \tag{4}
$$

given in (1). The development we present is not an alternate verification that we have obtained the line closest to all the data in the least squares sense; rather it provides an easy matrix formulation of the normal system of equations. We proceed with a set of observations regarding the entries of the matrix $\mathbf{C}$ and right side $\mathbf{d}$ of the normal system of equations. Each of the entries of $\mathbf{C}$ and $\mathbf{d}$ can be expressed as a dot product:

$$
\sum_{i=1}^{n} x_i^2 = \begin{bmatrix} x_1 & x_2 & \cdots & x_n \end{bmatrix} \cdot \begin{bmatrix} x_1 \\ x_2 \\ \vdots \\ x_n \end{bmatrix}, \quad \sum_{i=1}^{n} x_i y_i = \begin{bmatrix} x_1 & x_2 & \cdots & x_n \end{bmatrix} \cdot \begin{bmatrix} y_1 \\ y_2 \\ \vdots \\ y_n \end{bmatrix},
$$

$$
\sum_{i=1}^{n} x_i = \begin{bmatrix} 1 & 1 & \cdots & 1 \end{bmatrix} \cdot \begin{bmatrix} x_1 \\ x_2 \\ \vdots \\ x_n \end{bmatrix}, \qquad \sum_{i=1}^{n} y_i = \begin{bmatrix} 1 & 1 & \cdots & 1 \end{bmatrix} \cdot \begin{bmatrix} y_1 \\ y_2 \\ \vdots \\ y_n \end{bmatrix},
$$

$$
n = \begin{bmatrix} 1 & 1 & \cdots & 1 \end{bmatrix} \cdot \begin{bmatrix} 1 \\ 1 \\ \vdots \\ 1 \end{bmatrix}.
$$

Comparing these expressions to the entries of $\mathbf{C}$ and $\mathbf{d}$ and recalling that the entries of the product of a pair of matrices can be expressed as dot products, we see that

$$
\mathbf{A}^T \mathbf{A} = \begin{bmatrix} \sum_{i=1}^{n} x_i^2 & \sum_{i=1}^{n} x_i \\ \sum_{i=1}^{n} x_i & n \end{bmatrix} = \mathbf{C}, \quad \mathbf{A}^T \mathbf{y} = \begin{bmatrix} \sum_{i=1}^{n} x_i y_i \\ \sum_{i=1}^{n} y_i \end{bmatrix} = \mathbf{d}.
$$

Hence the normal system of equations is obtained from the original system $\mathbf{Ax} = \mathbf{y}$ by multiplying both sides by $\mathbf{A}^T$ on the left to give $\mathbf{A}^T \mathbf{Ax} = \mathbf{A}^T \mathbf{y}$. This matrix formulation of the normal system of equations, whose solution leads to the slope and y-intercept of the line of best fit, is useful in computer calculations.

We show in Section 5.5 that solution $\mathbf{x}$ to the normal system $\mathbf{A}^T \mathbf{Ax} = \mathbf{A}^T \mathbf{y}$ gives the least squares solution of inconsistent systems in general. From that development it follows that the vector in $\mathbf{col}(\mathbf{A})$ that is closest to $\mathbf{y}$ is $\mathbf{Ax}$. (See Exercises 5 through 7.)

## EXERCISES 2.6

**1.** Determine the line of best fit to the data $D = \{(3, 2),$ $(4, 3), (5, 2), (6, 4), (7, 3)\}$.

**2.** In an experiment designed to determine the extent of a person's natural spatial orientation, a subject is put in a special room and kept there for a certain length of time. He is then asked to find a way out of a maze and a record is made of the time it takes the subject to accomplish this task. The following data are obtained:

| Time in Room (in Hours) | Time to Find Way Out of Maze (in Minutes) |
|:---:|:---:|
| 1 | 0.8 |
| 2 | 2.1 |
| 3 | 2.6 |
| 4 | 2.0 |
| 5 | 3.1 |
| 6 | 3.3 |

Let **x** be the vector denoting the data for the number of hours in the room and let **y** be the vector denoting the data for the number of minutes that it takes the subject to find his way out.

(a) Determine the line of best fit to the data in vectors **x** and **y**.

(b) Use the equation of the line of best fit to estimate the time it will take the subject to find his way out of the maze after 8 hours in the room.

**3.** The U.S. Government gave the following data on personal health care expenditures for the years listed.

| Year | Personal Health Care Expenditures (in billions of $) |
|:---:|:---:|
| 1985 | 376.4 |
| 1990 | 614.7 |
| 1991 | 676.2 |
| 1992 | 739.8 |
| 1993 | 786.5 |
| 1994 | 831.7 |

Let $\mathbf{x} = \begin{bmatrix} 85 & 90 & 91 & 92 & 93 & 94 \end{bmatrix}$ be the vector of data for the given years and

$$\mathbf{y} = \begin{bmatrix} 376.4 & 614.7 & 676.2 & 739.8 & 786.5 & 831.7 \end{bmatrix}$$

be the vector of data for the expenditures.

(a) Determine the line of best fit to the data in vectors **x** and **y**.

(b) Use the equation of the line of best fit to estimate the expenditures in 1999. [*Note*: Set $x = 99$ in the equation from part (a).]

**4.** A real estate agent was looking over her sales information for the past year and made a table of selling price of a house and the number of bedrooms in the house. A sample of this data is as follows.

| Number of Bedrooms | Selling Price (in thousands of $) |
|:---:|:---:|
| 3 | 110 |
| 3 | 114 |
| 2 | 87 |
| 4 | 131 |
| 4 | 125 |
| 3 | 120 |

Let $\mathbf{x} = \begin{bmatrix} 3 & 3 & 2 & 4 & 4 & 3 \end{bmatrix}$ be the vector of data for the number of bedrooms and

$$\mathbf{y} = \begin{bmatrix} 110 & 114 & 87 & 131 & 125 & 120 \end{bmatrix}$$

be the vector of data for the selling price of the house.

(a) Determine the line of best fit to the data in vectors **x** and **y**.

(b) Use the equation of the line of best fit to estimate the selling price for a house with five bedrooms.

(c) To get a crude average of the selling price for a house with $t$ bedrooms the agent computes $y = mt + b$, where $m$ and $b$ are derived from the line of best fit. What are these crude averages for $t = 3$ and $t = 4$?

*The line of best to a set of data is just one application of what is known as the least squares solution of an inconsistent linear system* $\mathbf{Ax} = \mathbf{b}$. *Note that the term* solution *refers to the solution of the associated normal system of equations, which can be obtained from the matrix formulation* $\mathbf{A}^T\mathbf{Ax} = \mathbf{A}^T\mathbf{b}$. *In Exercises 5–7, obtain the least squares solution of each given linear system.*

**5.** $\mathbf{A} = \begin{bmatrix} 2 & 1 \\ 1 & 0 \\ -1 & 1 \end{bmatrix}, \mathbf{b} = \begin{bmatrix} 3 \\ 1 \\ 4 \end{bmatrix}.$

**6.** $\mathbf{A} = \begin{bmatrix} 3 & -2 \\ 2 & -3 \\ 1 & -1 \\ 2 & 3 \end{bmatrix}, \mathbf{b} = \begin{bmatrix} 2 \\ -1 \\ 0 \\ 0 \end{bmatrix}.$

**7.** $\mathbf{A} = \begin{bmatrix} 1 & 3 & 2 \\ 2 & 5 & 3 \\ 2 & 0 & 0 \\ 3 & 1 & 1 \end{bmatrix}, \mathbf{b} = \begin{bmatrix} 2 \\ 0 \\ 1 \\ -2 \end{bmatrix}.$

**8.** To determine a best fit quadratic $y = ax^2 + bx + c$ to a set of data $D = \{(x_1, y_1), (x_2, y_2), \ldots, (x_n, y_n)\}$ we first construct the linear system corresponding to the equations $y_i = ax_i^2 + bx_i + c$.. In matrix form this linear system is

$$\begin{bmatrix} x_1^2 & x_1 & 1 \\ x_2^2 & x_2 & 1 \\ \vdots & \vdots & \vdots \\ x_n^2 & x_n & 1 \end{bmatrix} \begin{bmatrix} a \\ b \\ c \end{bmatrix} = \begin{bmatrix} y_1 \\ y_2 \\ \vdots \\ \vdots \\ y_n \end{bmatrix}.$$

Let $\mathbf{A}$ denote the coefficient matrix, $\mathbf{x}$ denote the column of coefficients $\begin{bmatrix} a & b & c \end{bmatrix}^T$, and $\mathbf{y}$ denote the right side. Then the least squares quadratic is determined from the solution of the normal system of equations given by $\mathbf{A}^T\mathbf{A}\mathbf{x} = \mathbf{A}^T\mathbf{y}$. (This is verified in Section 6.4.) Find the least squares quadratic for the data in the following table.

| $x_i$ | 1 | 2 | 3 | 4 | 5 |
|-------|-----|-----|-----|-----|-----|
| $y_i$ | 4.5 | 5.1 | 4.3 | 2.5 | 1 |

**9.** The normal system of equation in (2) has a unique solution provided that the matrix

$$\mathbf{C} = \begin{bmatrix} \sum_{i=1}^{n} x_i^2 & \sum_{i=1}^{n} x_i \\ \sum_{i=1}^{n} x_i & n \end{bmatrix}$$

is nonsingular. To verify that this is indeed true when all the $x_i$ are distinct, we use the matrix formulation, $\mathbf{C} = \mathbf{A}^T\mathbf{A}$, where

$$\mathbf{A} = \begin{bmatrix} x_1 & 1 \\ x_2 & 1 \\ \vdots & \vdots \\ \vdots & \vdots \\ x_n & 1 \end{bmatrix}.$$

Let $\mathbf{k} = \begin{bmatrix} k_1 \\ k_2 \end{bmatrix}$. Then we know the following: The homogeneous linear system $\mathbf{C}\mathbf{k} = \mathbf{0}$ has only the trivial solution if and only if matrix $\mathbf{C}$ is nonsingular. Verify the following to show that $\mathbf{C}$ is nonsingular.

(a) Explain why the columns of $\mathbf{A}$ are linearly independent.

(b) Explain why $\mathbf{A}\mathbf{k}$ can only be equal to the zero vector $\mathbf{0}$ if $\mathbf{k} = \mathbf{0}$.

(c) Explain why $\mathbf{k}^T\mathbf{A}^T = (\mathbf{A}\mathbf{k})^T$.

(d) Assume $\mathbf{C}\mathbf{k} = \mathbf{0}$. Explain why $(\mathbf{A}\mathbf{k})^T(\mathbf{A}\mathbf{k}) = 0$. (*Note*: The right side is the zero scalar.)

(e) Explain why $(\mathbf{A}\mathbf{k})^T(\mathbf{A}\mathbf{k}) = 0$ is the same as the dot product $(\mathbf{A}\mathbf{k}) \cdot (\mathbf{A}\mathbf{k}) = 0$.

(f) When can the dot product of a vector with itself be zero?

(g) Use parts (b), (e), and (f) to explain why $\mathbf{k} = \mathbf{0}$ and hence $\mathbf{C}$ is nonsingular.

**10.** For a given data set $D$, let line $L$ be the least squares line and $E_L$ be the minimum value of the sum of the squares of the deviations. For the same data set $D$ let quadratic $Q$ be the least squares quadratic (see Exercise 8) and let $E_Q$ be the minimum value of the sum of the squares of its deviations. Why must $E_Q \leq E_L$? (*Hint*: We can think that the least squares process for determining $L$ looks over all polynomials of degree 1 or less to determine the one that minimizes the sum of the squares of the deviations. Similarly, the least squares process for determining $Q$ looks over all polynomials of degree 2 or less.)

In MATLAB

*This section gives us an algebraic technique for determining the least squares line. We used terminology like "the line that comes closest to all the points" and "the line that minimizes the sum of the squares of the deviations." Such phrases have a geometric flavor, and we can use MATLAB to emphasize this geometric context. To get a feel for "the line that comes closest to all the points," we display the data set graphically and ask you to choose two points that determine a line that you think will be close to the least squares line. We even provide an option for a bit of competition in determining such lines.*

**ML. 1.** In MATLAB type **help lsqgame**. Once you have read this overview you are ready to test your estimation capabilities for the graphical determination of lines of best fit.

(a) The following table gives the winning distance of the Women's Shot Put (8 lb, 13 oz) in the summer Olympics for the years 1976 to 1996.

| Year | Distance | Distance (in inches) |
|------|----------|----------------------|
| 1976 | 69′ 5.25″ | 833.25″ |
| 1980 | 73′ 6.25″ | 882.25″ |
| 1984 | 67′ 2.25″ | 806.25″ |
| 1988 | 72′ 11.5″ | 875.5″ |
| 1992 | 69′ 1.5″ | 829.5″ |
| 1996 | 67′ 5.5″ | 809.5″ |

In MATLAB enter the following vectors:

**x = [76  80  84  88  92  96]**
**y = [833.25  882.25  806.25  875.5**
    **829.5  809.5]**

Next enter command **lsqgame**. Select the option for one player and option 2 from the data entry menu. In response to 'Data Matrix =' type **[x;y]'** (be sure to type the ' since it means transpose in MATLAB) and press enter. You should then see a screen with the shot put data plotted and buttons for selecting (Least Square) lines. Click the mouse on the button for Line #1. Move the cursor to the plot of the data and decide where to click the mouse to select two points for your estimate of the line of best fit to this data. Once the first line is generated, if you think you can do better, click on the button for Line #2 and select two points for your second estimate of the line of best fit. A record of the sum of the squares of the deviations and the equations of the lines you select will be displayed. Check your estimates by clicking on the button that says Show LSQ Solution. (As a check, the line of best fit should have a sum of the squares of the deviations of about 4651. There is nothing that says the line of best fit must have a small sum of squares of the deviations. In fact, from the plot of the data we see that this data set is not well approximated by a straight line.) When you are finished inspecting the graphs and information on the screen, click on the Quit button.

(b) The following table gives the winning distance of the Men's Discus Throw in the summer Olympics for the years 1976 to 1996.

| Year | Distance | Distance (in inches) |
|------|----------|----------------------|
| 1976 | 221′ 5.4″ | 2657.4″ |
| 1980 | 218′ 8″ | 2624″ |
| 1984 | 218′ 6″ | 2622″ |
| 1988 | 225′ 9.25″ | 2709.25″ |
| 1992 | 213′ 7.75″ | 2563.75″ |
| 1996 | 227′ 8″ | 2732″ |

Follow the directions given in part (a) to estimate the line of best for this data.

(c) If you inspect either the table or the plot of the data in part (b), you see that the discus throw in 1992 is quite a bit shorter than in other years shown. Omit this data pair and use **lsqgame** to estimate the line of best fit to the discus throw data from years '76, '80, '84, '88, and '96. (A data point that seems quite different from the other data in a set is called an outlier. The inclusion of such points can often dramatically effect the line of best fit.) Were you better able to estimate the line of best fit for this data than for the data in part (b)? Explain your response.

**ML. 2.** To practice your estimation of the line of best fit for data sets, you can use **lsqgame** as a competitive game. With a partner (or opponent) flip a coin to see who will choose the data set. Start **lsqgame**, select two players, and select the data input option for using the mouse. The player who is to select the data chooses about 10 points in the graphics box (press q to quit selecting data). The player who did not choose the data then decides whether to estimate the first or second line of best fit. After both players have made their selection for the least squares line, click the Show LSQ Solution button. The player whose sum of squares of the deviations is closest to that given for the least squares line wins the game. Repeat this process, reversing the role of who chooses the data set.

## True/False Review Questions

*Determine whether each of the following statements is true or false.*

1. A least squares line computation minimizes the sum of the squares of the (vertical) deviations from data points to the line $y = mx + b$.

2. A least squares line is also called the line of best fit.

3. To determine the slope and $y$-intercept for a least squares line we solve a $2 \times 2$ system, called the normal system of equations.

## Terminology

| | |
|---|---|
| Overdetermined linear system | Line of best fit to a data set |
| Method of least squares | Deviations |
| Normal system of equations | |

The material in this section is an introduction to a valuable and versatile approximation procedure that appears in many forms in business, science, engineering, and applied mathematics. It shows that linear algebra can be used to obtain approximations when a so-called exact solution does not exist. The ideas here can be generalized in a number of directions and are the foundation for further studies. Make sure you understand the change of perspective that was used here to handle the special kind of problem discussed in this section. To aid in this regard, formulate responses to the following.

- What is an overdetermined system of equations?
- Give an example of an overdetermined system in two unknowns that has a unique solution.
- Give an example of an overdetermined system in two unknowns that is inconsistent.
- For an inconsistent system $\mathbf{A}\mathbf{x} = \mathbf{b}$, there is no vector $\mathbf{x}$ such that $\mathbf{A}$ times $\mathbf{x}$ produces $\mathbf{b}$. Describe how we change our point of view to handle this situation.
- In geometric terms, what do we mean by the *line of best fit*?
- In technical terms, how is the line of best fit computed? Why is the phrase *method of least squares* used in this regard?
- An original inconsistent system $\mathbf{A}\mathbf{x} = \mathbf{y}$ is replaced by a nonsingular linear system. Describe this nonsingular linear system and how it arises.
- When you solve the normal system $\mathbf{C}\mathbf{x} = \mathbf{d}$ for the special case of determining a line of best fit to a data set, what is the physical meaning of the unknowns in vector $\mathbf{x}$?

## CHAPTER TEST

1. Explain how to construct any vector in span$\{\mathbf{w}_1, \mathbf{w}_2, \mathbf{w}_3\}$.

2. Let $V$ be the set of all $2 \times 2$ matrices of the form $\begin{bmatrix} a & b \\ c & 5 \end{bmatrix}$, where $a$, $b$, and $c$ are any real numbers. Is $V$ a closed set? Explain your answer.

3. Let $V = \{\mathbf{v}_1, \mathbf{v}_2, \mathbf{v}_3\}$ be in $R^5$. Explain what it means to say $V$ is a linearly dependent set.

4. Explain the meaning of $\mathbf{ns}(\mathbf{A})$.

5. Suppose that the linear system $\mathbf{A}\mathbf{x} = \mathbf{b}$ is consistent and its general solution has 2 degrees of freedom. Carefully describe the general solution and explain any notation you use.

6. Find the RREF of $\mathbf{A} = \begin{bmatrix} 2 & -1 & 0 & 1 \\ 1 & 0 & 2 & 2 \\ -1 & 0 & -1 & 2 \end{bmatrix}$.

7. Determine whether the set $S = \left\{ \begin{bmatrix} 1 \\ 0 \\ 1 \end{bmatrix}, \begin{bmatrix} 0 \\ 1 \\ 1 \end{bmatrix}, \begin{bmatrix} 3 \\ -2 \\ 1 \end{bmatrix} \right\}$ is linearly independent or linearly dependent.

8. Determine the general solution of $\mathbf{A}\mathbf{x} = \mathbf{b}$ where $\mathbf{A} = \begin{bmatrix} 1 & 1 & 3 & -3 \\ 0 & 2 & 1 & -3 \\ 1 & 0 & 2 & -1 \end{bmatrix}$ and $\mathbf{b} = \begin{bmatrix} 0 \\ 3 \\ -1 \end{bmatrix}$.

9. Let $T = \left\{ \begin{bmatrix} 1 \\ 2 \\ 1 \end{bmatrix}, \begin{bmatrix} 2 \\ 0 \\ 1 \end{bmatrix}, \begin{bmatrix} 4 \\ 4 \\ 3 \end{bmatrix} \right\}$ and $\mathbf{v} = \begin{bmatrix} 1 \\ 1 \\ 1 \end{bmatrix}$. Is $\mathbf{v}$ in span$(T)$? Explain your answer.

**10.** Let $A = \begin{bmatrix} 2 & 1 & 3 \\ 1 & -2 & 3 \\ 0 & 5 & -3 \\ 3 & 4 & 3 \end{bmatrix}$.

    (a) Find a basis for $\mathbf{row}(A)$.

    (b) Find a basis for $\mathbf{col}(A)$.

    (c) Find a basis for $\mathbf{ns}(A)$.

**11.** Let $A = \begin{bmatrix} 2 & 9 & 10 \\ 3 & 13 & 16 \\ 1 & 4 & 5 \end{bmatrix}$. Find $A^{-1}$.

**12.** Let $A$ be $4 \times 4$ and nonsingular.

    (a) What is $\text{row}_2(A)\,\text{col}_3(A^{-1})$?

    (b) What is $\text{row}_3(A)\,\text{col}_3(A^{-1})$?

    (c) State a formula for the solution of $Ax = b$.

    (d) What is $\dim(\mathbf{row}(A))$?

    (e) What is $\dim(\mathbf{ns}(A))$?

    (f) Are the rows of $A$ linearly dependent? Explain your answer.

**13.** Let $A$ be a $2 \times 2$ real matrix and $f(\mathbf{x}) = A\mathbf{x}$ be the matrix transformation with associated matrix $A$.

    (a) If $\mathbf{x}$ belongs to $\mathbf{ns}(A)$, what is the image $f(\mathbf{x})$?

    (b) Let $\mathbf{y}$ be the 2-vector $\begin{bmatrix} 1 \\ 2 \end{bmatrix}$ and $A$ be nonsingular. If $f(\mathbf{x}) = \mathbf{y}$, then what is $\mathbf{x}$?

    (c) If $A = \begin{bmatrix} 2 & 3 \\ -2 & -3 \end{bmatrix}$ and $S = \begin{bmatrix} 0 & 1 & 1 & 0 & 0 \\ 0 & 0 & 1 & 1 & 0 \end{bmatrix}$ are the vertices of the unit square, what is the area of $f(S)$?

# 3

# *THE DETERMINANT, A GEOMETRIC APPROACH*

The focus of Chapter 2 was to determine information about the solution set of the linear system of equations given as the matrix equation $\mathbf{Ax} = \mathbf{b}$. We saw that in general, both the coefficient matrix $\mathbf{A}$ and right side $\mathbf{b}$ contributed to the specific nature of the solution set. This followed since the linear system $\mathbf{Ax} = \mathbf{b}$ is consistent if and only if $\mathbf{b}$ is a linear combination of the columns of $\mathbf{A}$. A goal was to determine the coefficients of such linear combinations, and to do so we used row operations. However, in the special case of a nonsingular matrix $\mathbf{A}$ it was an intrinsic quality of the matrix $\mathbf{A}$ that the linear system $\mathbf{Ax} = \mathbf{b}$ was consistent regardless of the contents of $\mathbf{b}$.

In this chapter we develop an intrinsic number associated only with a square matrix. We use geometric arguments to provide a foundation that such a number has merit. We then make a formal definition that we use for $2 \times 2$ and $3 \times 3$ matrices with an indication of how to generalize the concept to $n \times n$ matrices. (We postpone a detailed investigation of the general case until a later section.) We develop properties of the concept and show how it is related to linear systems that have a unique solution. Next, we show how to use row operations to compute this number efficiently. We also present an overview of a recursive algebraic manipulation that involves the entries of a square matrix that yields the same result. Finally, we give several computational procedures that use the ideas developed herein.

In this chapter we rely on preformal justifications of the principal ideas. That is, we use arguments based on geometrically intuitive basic concepts. We feel this is a reasonable starting point for learning this material. We will not give general step-by-step verifications; rather we develop a sense that the topics in this chapter are natural extensions of the ideas we have presented previously.

## 3.1 ■ THE MAGNIFICATION FACTOR

In Section 1.5 we discussed the matrix transformation determined by a matrix $\mathbf{A}$. For an $m \times n$ matrix $\mathbf{A}$ the function $f(\mathbf{c}) = \mathbf{Ac}$ provides a correspondence between vectors in $R^n$ and $R^m$. In this chapter we consider the matrix transformation for a square matrix from a geometric point of view. Initially we take $\mathbf{A}$ to be $2 \times 2$ and determine how the matrix transformation $f(\mathbf{c}) = \mathbf{Ac}$ changes the area of a closed polygon.

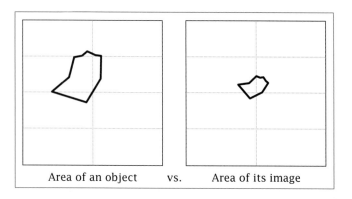

Area of an object     vs.     Area of its image

In MATLAB, **mapit.m**

In Example 8 of Section 1.5 we showed that the image of the unit square when transformed by the diagonal matrix $\begin{bmatrix} h & 0 \\ 0 & k \end{bmatrix}$ was a rectangle with sides of length $|h|$ and $|k|$. In Exercise 29 of Section 1.5 you were asked to show that the image of the unit square transformed by the matrix $\begin{bmatrix} a & b \\ c & d \end{bmatrix}$ is a parallelogram and to show how to compute the area of the parallelogram. In Example 1 we determine the image of an arbitrary rectangle and compute its area.

EXAMPLE 1    Let $\mathbf{A} = \begin{bmatrix} a & b \\ c & d \end{bmatrix}$. To find the image of the rectangle $R$ shown in Figure 1 by the matrix transformation determined by $\mathbf{A}$, we follow the procedure used in Example 8 of Section 1.5.

Let $\mathbf{S}$ be the matrix whose columns are the vertices of the rectangle

$$\mathbf{S} = \begin{bmatrix} 0 & 0 & x & x & 0 \\ 0 & y & y & 0 & 0 \end{bmatrix}.$$

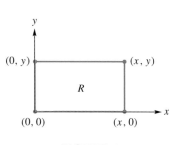

FIGURE 1

Then the image of the matrix transformation is

$$f(\mathbf{S}) = \mathbf{AS} = \begin{bmatrix} a & b \\ c & d \end{bmatrix} \begin{bmatrix} 0 & 0 & x & x & 0 \\ 0 & y & y & 0 & 0 \end{bmatrix} = \begin{bmatrix} 0 & by & ax+by & ax & 0 \\ 0 & dy & cx+dy & cx & 0 \end{bmatrix}.$$

The vertices in $f(\mathbf{S})$ define the parallelogram that is shown in Figure 2.

Next we determine the relationship between the areas of the rectangle $R$ in Figure 1 and the parallelogram $P$ in Figure 2. From Figure 3 we see that

1.  The area of the two rectangles labeled I is $2|(cx)(by)|$.
2.  The area of the two triangles labeled II is $|(by)(dy)|$.
3.  The area of the two triangles labeled III is $|(ax)(cx)|$.

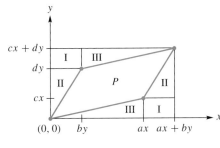

FIGURE 2                          FIGURE 3

**4.** The area of the rectangle enclosing P and the regions labeled I, II, and III is $|(ax + by)(cx + dy)|$.

Thus

$$\text{area}(P) = |(ax + by)(cx + dy)| - 2|(cx)(by)| - |(by)(dy)| - |(ax)(cx)|$$
$$= |ad - bc|\,|xy|$$

so the area of the image $P$ is $|ad - bc| * \text{area}(R)$. It follows that

$$\frac{\text{area}(P)}{\text{area}(R)} = \frac{|ad - bc|xy}{xy} = |ad - bc|. \tag{1}$$

(Exercise 29 in Section 1.5 obtains a similar result using an algebraic approach.)  ■

From (1) we see that the factor by which the area of the image changes depends only on the entries of the matrix **A**. We call this the **magnification factor** of the matrix transformation determined by **A**.

A natural question is as follows:

Does the magnification factor for a matrix transformation $f(\mathbf{c}) = \mathbf{Ac}$ change if we change the geometric figure?

To examine this question we investigate several cases for familiar figures.

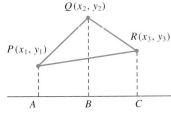

FIGURE 4

EXAMPLE 2    Let $T$ be the triangle with vertices $P$, $Q$, and $R$ in Figure 4. The area of triangle $T$ is a linear combination of the areas of trapezoids $APQB$, $BQRC$, and $APRC$ as follows:

$$\text{area}(T) = \text{area}(APQB) + \text{area}(BQRC) \quad \text{area}(APRC) \tag{2}$$

The area of a trapezoid is $\frac{1}{2}$ times "its height" times "the sum of the lengths of its parallel sides". We have (verify)

$$\text{area}(APQB) = \frac{(x_2 - x_1)(y_1 + y_2)}{2}$$

$$\text{area}(BQRC) = \frac{(x_3 - x_2)(y_2 + y_3)}{2}$$

$$\text{area}(APRC) = \frac{(x_3 - x_1)(y_1 + y_3)}{2}.$$

Substituting these expressions into (2) and collecting terms, we obtain (verify)

$$\text{area}(T) = \tfrac{1}{2}\left[x_1(y_3 - y_2) + x_2(y_1 - y_3) + x_3(y_2 - y_1)\right]$$

$$= \tfrac{1}{2}\begin{bmatrix} x_1 \\ x_2 \\ x_3 \end{bmatrix} \cdot \begin{bmatrix} y_3 - y_2 \\ y_1 - y_3 \\ y_2 - y_1 \end{bmatrix}.$$

To make this expression independent of the order in which the vertices of the triangle are labeled, we take the absolute value of the right side of the preceding expression:

$$\text{area}(T) = \frac{1}{2}\left| \begin{bmatrix} x_1 \\ x_2 \\ x_3 \end{bmatrix} \cdot \begin{bmatrix} y_3 - y_2 \\ y_1 - y_3 \\ y_2 - y_1 \end{bmatrix} \right|. \tag{3}$$

Before we compute the magnification factor for the matrix transformation determined by $\mathbf{A} = \begin{bmatrix} a & b \\ c & d \end{bmatrix}$, it is convenient to express (3) in an equivalent vector form. Let

$$\mathbf{C} = \begin{bmatrix} 0 & -1 & 1 \\ 1 & 0 & -1 \\ -1 & 1 & 0 \end{bmatrix}.$$

Then

$$\begin{bmatrix} y_3 - y_2 \\ y_1 - y_3 \\ y_2 - y_1 \end{bmatrix} = \mathbf{C}\begin{bmatrix} y_1 \\ y_2 \\ y_3 \end{bmatrix}.$$

(Verify.) With

$$\mathbf{x} = \begin{bmatrix} x_1 \\ x_2 \\ x_3 \end{bmatrix} \quad \text{and} \quad \mathbf{y} = \begin{bmatrix} y_1 \\ y_2 \\ y_3 \end{bmatrix},$$

(3) can be expressed in vector-matrix form as

$$\text{area}(T) = \tfrac{1}{2}\left|\mathbf{x}\cdot\mathbf{Cy}\right| = \tfrac{1}{2}\left|\mathbf{x}^T\mathbf{Cy}\right|. \tag{4}$$

The image of $T$ by the matrix transformation determined by $\mathbf{A} = \begin{bmatrix} a & b \\ c & d \end{bmatrix}$ is

$$f(\mathbf{S}) = \mathbf{A}\mathbf{S} = \begin{bmatrix} a & b \\ c & d \end{bmatrix}\begin{bmatrix} x_1 & x_2 & x_3 \\ y_1 & y_2 & y_3 \end{bmatrix}$$

$$= \begin{bmatrix} ax_1 + by_1 & ax_2 + by_2 & ax_3 + by_3 \\ cx_1 + dy_1 & cx_2 + dy_2 & cx_3 + dy_3 \end{bmatrix}.$$

Geometrically this image is a triangle $T_A$ with vertices

$$\{(ax_1 + by_1, cx_1 + dy_1), (ax_2 + by_2, cx_2 + dy_2), (ax_3 + by_3, cx_3 + dy_3)\},$$

which correspond to the columns of $f(\mathbf{S})$. The area of the image $T_A$ is computed using (4) with $\mathbf{x}$ replaced by

$$\begin{bmatrix} ax_1 + by_1 \\ ax_2 + by_2 \\ ax_3 + by_3 \end{bmatrix} = a\mathbf{x} + b\mathbf{y}$$

and $\mathbf{y}$ replaced by

$$\begin{bmatrix} cx_1 + dy_1 \\ cx_2 + dy_2 \\ cx_3 + dy_3 \end{bmatrix} = c\mathbf{x} + d\mathbf{y}.$$

Thus

$$\text{area}(T_A) = \tfrac{1}{2}\left|(a\mathbf{x} + b\mathbf{y})^T \mathbf{C}(c\mathbf{x} + d\mathbf{y})\right|$$
$$= \tfrac{1}{2}\left|ac(\mathbf{x}^T\mathbf{Cx}) + ad(\mathbf{x}^T\mathbf{Cy}) + bc(\mathbf{y}^T\mathbf{Cx}) + bd(\mathbf{y}^T\mathbf{Cy})\right|.$$

From Exercises 10 and 11 we have $\mathbf{x}^T\mathbf{Cx} = 0$, $\mathbf{y}^T\mathbf{Cx} = -\mathbf{x}^T\mathbf{Cy}$, and $\mathbf{y}^T\mathbf{Cy} = 0$. Hence it follows that

$$\text{area}(T_A) = \tfrac{1}{2}\left|ad(\mathbf{x}^T\mathbf{Cy}) - bc(\mathbf{x}^T\mathbf{Cy})\right|$$
$$= \tfrac{1}{2}\left|(ad - bc)(\mathbf{x}^T\mathbf{Cy})\right| = |(ad - bc)|\,\text{area}(T).$$

Therefore, the magnification factor in this case is also $|ad - bc|$.    ■

EXAMPLE 3    Let $P$ be the parallelogram shown in Figure 5. Since a diagonal of a parallelogram divides it into two congruent triangles we have from (4) that

$$\text{area}(P) = 2 * \tfrac{1}{2}\left|\mathbf{x}^T\mathbf{Cy}\right| = \left|\mathbf{x}^T\mathbf{Cy}\right|.$$

The image of $P$ by the matrix transformation determined by

$$\mathbf{A} = \begin{bmatrix} a & b \\ c & d \end{bmatrix}$$

is another parallelogram, hence a pair of congruent triangles. From Example 2 it follows that the area of the image of $P$ is $|ad - bc|$ times the area of $P$. Hence the magnification factor is again $|ad - bc|$.    ■

Next we investigate the magnification factor for the image of an arbitrary closed polygon by the matrix transformation determined by $\mathbf{A} = \begin{bmatrix} a & b \\ c & d \end{bmatrix}$. Since any closed polygon can be subdivided into a set of nonoverlapping triangles, as shown in Figures 6(a) and 6(b), we can use the results established so far. The process depicted in Figure 6(b) is called **triangulation**[1]. (Many different triangulations are possible for a given polygon. ) Once we have performed a triangulation we apply the results of Example 2. It follows that the magnification factor is again $|ad - bc|$.

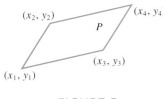

$(x_2, y_2)$  $(x_4, y_4)$  $P$  $(x_3, y_3)$  $(x_1, y_1)$

FIGURE 5

FIGURE 6(a)

FIGURE 6(b)

EXAMPLE 4    To determine the area of the polygon shown in Figure 7, we perform a triangulation as shown in Figure 8. Next we use the results of Example 2.

$$\text{area}(T_1) = \tfrac{1}{2}\left|\begin{bmatrix} -2 \\ 0 \\ 1 \end{bmatrix}^T \mathbf{C} \begin{bmatrix} 0 \\ 3 \\ 1 \end{bmatrix}\right| = 3.5,$$

[1]"Triangulation is the partitioning of a polygon into nonoverlapping triangles all of whose vertices are vertices of the polygon" from *The Words of Mathematics* by S. Schwartzman, Washington, D.C.: The Mathematical Association of America, 1994. Note that an edge is shared by the adjoining triangles.

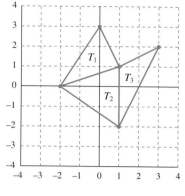

FIGURE 7                                         FIGURE 8

$$\text{area}(T_2) = \frac{1}{2} \left| \begin{bmatrix} -2 \\ 1 \\ 1 \end{bmatrix}^T \mathbf{C} \begin{bmatrix} 0 \\ 1 \\ -2 \end{bmatrix} \right| = 4.5,$$

$$\text{area}(T_3) = \frac{1}{2} \left| \begin{bmatrix} 1 \\ 1 \\ 3 \end{bmatrix}^T \mathbf{C} \begin{bmatrix} 1 \\ -2 \\ 2 \end{bmatrix} \right| = 3.$$

Thus we have

$$\text{area(polygon in Figure 7)} = \text{area}(T_1) + \text{area}(T_2) + \text{area}(T_3)$$
$$= 11 \text{ square units.} \quad ■$$

A remaining question is as follows:

What is the magnification factor if the geometric figure is a nonpolygonal region (that is, a closed region with curved boundaries)?

Many nonpolygonal regions can be approximated as closely as desired by a polygon, possibly with many sides (see Example 5). Thus it follows that the magnification factor for the matrix transformation determined by $\mathbf{A} = \begin{bmatrix} a & b \\ c & d \end{bmatrix}$ of a nonpolygonal region will also be $|ad - bc|$.

EXAMPLE 5    A circle can be approximated as closely as desired by inscribed polygons, as indicated in Figures 9(a), (b), and (c).

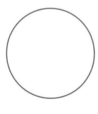    

FIGURE 9(a)           FIGURE 9(b)           FIGURE 9(c)           ■

In Example 3 in Section 2.4 we showed that a $2 \times 2$ matrix $\mathbf{A} = \begin{bmatrix} a & b \\ c & d \end{bmatrix}$ was nonsingular provided $ad - bc \neq 0$ and then

$$\mathbf{A}^{-1} = \frac{1}{ad - bc} \begin{bmatrix} d & -b \\ -a & c \end{bmatrix}.$$

From our development in this section we have that the magnification factor for the matrix transformation determined by $\mathbf{A} = \begin{bmatrix} a & b \\ c & d \end{bmatrix}$ is $|ad - bc|$. Geometrically, we can see that the matrix transformation is "reversible" if and only if $\mathbf{A}$ is nonsingular. If $\mathbf{A}$ were singular, then its rows are linearly dependent; that is, multiples of one another. In that case $\mathbf{A} = \begin{bmatrix} a & b \\ ka & kb \end{bmatrix}$ and the magnification factor is $|a(kb) - b(ka)| = 0$. Hence the area of the image that results from such a matrix transformation is zero. The image could be a line segment or even a single point.

The notion of magnification factor associated with a matrix transformation generalizes to $n \times n$ matrices. In the case of $3 \times 3$ matrices, the magnification factor is the ratio of the volume of the image divided by the volume of the original three-dimensional figure. We can display the images as surfaces in this case as in Example 6. However, if $n > 3$, we have no geometric model to view, but we will still use the notion of a magnification factor for a matrix transformation. Fortunately, the properties developed for the $2 \times 2$ case hold in general. However, we do not yet have a computational procedure for computing the magnification factor for $n > 2$. We investigate such issues in Section 3.2.

EXAMPLE 6    Let $R$ be the rectangular solid depicted in Figure 10. Its image from the matrix transformation determined by

$$\mathbf{A} = \begin{bmatrix} 2 & 1 & 0 \\ 3 & -1 & 2 \\ 1 & -2 & 1 \end{bmatrix}$$

is shown in Figure 11, which is a parallelepiped.

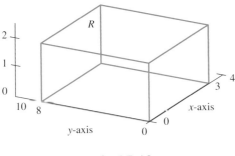

FIGURE 10

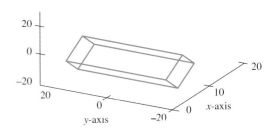

FIGURE 11

Note the change in scale between the figures. We can show that the magnification factor is 5; see Example 2 in Section 3.2. What is the volume of $R$? What is the volume of the image in Figure 11? ■

The magnification factor of a matrix transformation determined by a square matrix $\mathbf{A}$ is an intrinsic number associated with $\mathbf{A}$. We have shown that this number is important geometrically and also appears naturally when we wish to determine if a $2 \times 2$ matrix is nonsingular. (See Example 3 in Section 2.4.) In the next section we connect these ideas to larger matrices and introduce a number related to the magnification factor.

## EXERCISES 3.1

In Exercises 1 and 2, use Equation (4) to compute the area of each given triangle.

**1.**

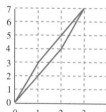

**2.**

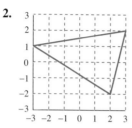

In Exercises 3 and 4, use Equation (4) to compute the area of each given parallelogram.

**3.**

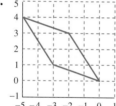

**4.**

In Exercises 5 and 6, use the process of triangulation to compute the area of each given polygonal region.

**5.**

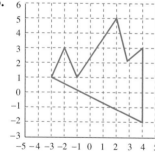

**6.**

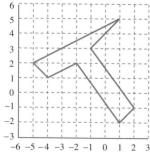

In Exercises 7 and 8, find the coordinates of the vertices of the image of the given figure determined by the matrix transfor-mation whose associated matrix is **A**. Sketch the image on the grid supplied.

**7.** $\mathbf{A} = \begin{bmatrix} 2 & -1 \\ 1 & 1 \end{bmatrix}$

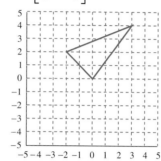

**8.** $\mathbf{A} = \begin{bmatrix} 2 & -1 \\ -2 & 1 \end{bmatrix}$

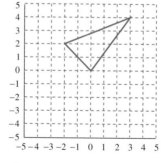

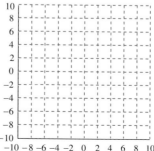

For Exercises 9–11, let $\mathbf{C} = \begin{bmatrix} 0 & -1 & 1 \\ 1 & 0 & -1 \\ -1 & 1 & 0 \end{bmatrix}$.

9. Show that $\mathbf{C}$ is skew symmetric; that is, $\mathbf{C}^T = -\mathbf{C}$. (See Exercise 31, Section 1.2.)

10. Let $\mathbf{x} = \begin{bmatrix} x_1 \\ x_2 \\ x_3 \end{bmatrix}$. Show that $\mathbf{x}^T \mathbf{C} \mathbf{x} = 0$. (*Hint*: $\mathbf{x}^T \mathbf{C} \mathbf{x}$ is a scalar so it is equal to its transpose.)

11. Let $\mathbf{x} = \begin{bmatrix} x_1 \\ x_2 \\ x_3 \end{bmatrix}$ and $\mathbf{y} = \begin{bmatrix} y_1 \\ y_2 \\ y_3 \end{bmatrix}$. Show that $\mathbf{y}^T \mathbf{C} \mathbf{x} = -\mathbf{x}^T \mathbf{C} \mathbf{y}$.

## In MATLAB

*This section showed that the magnification factor of a matrix transformation determined by a square matrix $\mathbf{A}$ is an intrinsic property of the matrix $\mathbf{A}$. Although we only considered cases for $2 \times 2$ and $3 \times 3$ matrices, the result is true for $n \times n$ matrices. The following MATLAB exercises focus on the geometric and algebraic relationships of the area of transformed planar regions. Matrix transformations of closed polygonal regions were discussed in Section 1.5, and we will use the routine **mapit** introduced in Section 1.5.*

ML. 1. Start MATLAB routine **mapit**. Select the house object, view it, and then click on the 'Grid On' button. Use the techniques of this section to compute the area of the house figure. Use each of the following matrices to "map" the house figure in routine **mapit** and determine the area of the image. (Note that there may be a change of scale when the image is displayed. Also, $\pi/4$ in MATLAB is entered as pi/4.)

(a) $\mathbf{A} = \begin{bmatrix} 2 & 1 \\ 1 & 2 \end{bmatrix}$.

(b) $\mathbf{A} = \begin{bmatrix} -1 & 3 \\ 1 & 2 \end{bmatrix}$.

(c) $\mathbf{A} = \begin{bmatrix} \cos(\pi/4) & \sin(\pi/4) \\ -\sin(\pi/4) & \cos(\pi/4) \end{bmatrix}$.

ML. 2. Start MATLAB routine **mapit**. Select the polygon object, view it, and then click on the 'Grid On' button. Use the techniques of this section to compute the area of the figure. Use each of the following matrices to "map" the figure in routine **mapit** and determine the area of the image. (Note that there may be a change of scale when the image is displayed. Also, $\pi/6$ in MATLAB is entered as pi/6.)

(a) $\mathbf{A} = \begin{bmatrix} 1 & -2 \\ 1 & 1 \end{bmatrix}$.

(b) $\mathbf{A} = \begin{bmatrix} 2 & 0 \\ 1 & 3 \end{bmatrix}$.

(c) $\mathbf{A} = \begin{bmatrix} \cos(\pi/6) & \sin(\pi/6) \\ -\sin(\pi/6) & \cos(\pi/6) \end{bmatrix}$.

ML. 3. Start MATLAB routine **mapit**. Select the peaks object, view it, and then click on the 'Grid On' button. Use the techniques of this section to compute the area of the peaks figure. Use each of the following pairs of matrices to form a composite "map" of the peaks figure. Determine the area of the final image. (Note that there may be a change of scale when the image is displayed.)

(a) $\mathbf{A} = \begin{bmatrix} 1 & 1 \\ -1 & 1 \end{bmatrix}$, $\mathbf{B} = \begin{bmatrix} 2 & 1 \\ 2 & 0 \end{bmatrix}$.

(b) $\mathbf{A} = \begin{bmatrix} 0 & 2 \\ 3 & 0 \end{bmatrix}$, $\mathbf{B} = \begin{bmatrix} 1 & 2 \\ 1 & 1 \end{bmatrix}$.

(c) $\mathbf{A} = \begin{bmatrix} 0 & -3 \\ 2 & 1 \end{bmatrix}$, $\mathbf{B} = \begin{bmatrix} 1 & 2 \\ -1 & -2 \end{bmatrix}$.

## True/False Review Questions

*Determine whether each of the following statements is true or false.*
*In each of the following, let $T$ be the matrix transformation from $R^2$ to $R^2$ with associated matrix $\mathbf{A}$; that is, $T(\mathbf{x}) = \mathbf{A}\mathbf{x}$ for $\mathbf{x}$ in $R^2$.*

1. Let $P$ be a closed polygon in $R^2$. Then the magnification factor of $T$ is $\dfrac{|\text{area}(T(P))|}{|\text{area}(P)|}$.

2. If $\mathbf{A} = \begin{bmatrix} h & 0 \\ 0 & k \end{bmatrix}$ and $\mathbf{S}$ is the unit square, then $\text{area}(T(\mathbf{S})) = |hk|$.

3. If $\mathbf{S}$ is the unit square and $T(\mathbf{S}) = P$, a parallelogram, then $\text{area}(T(\mathbf{S})) = |\text{area}(P)|$.

4. The matrix $\mathbf{C}$ determined in Example 2 is skew symmetric.

5. If $\mathbf{A} = \begin{bmatrix} 2 & 5 \\ -2 & -5 \end{bmatrix}$ and $\mathbf{S}$ is the unit square, then area $T(\mathbf{S})$ is zero.

Terminology

| | |
|---|---|
| Magnification factor | Triangulation |

In this section we investigated how a matrix transformation $f(\mathbf{c}) = \mathbf{Ac}$ applied to a closed figure in $R^2$, like a polygon, changed the area of the figure. We argued, primarily from geometric and basic algebra relations, that we could characterize the change independently of the figure to which the transformation was applied. This gave us an intrinsic number that we can associate with the matrix $\mathbf{A}$ of the transformation. We indicated that the ideas established here could be extended to an $n \times n$ real matrix, but we could not provide a visualization of the action for $n > 3$.

- What is the magnification factor of the matrix transformation $f(\mathbf{c}) = \mathbf{Ac}$?
- Describe the procedure for computing the area of a triangle whose vertices are $(x_1, y_1)$, $(x_2, y_2)$, $(x_3, y_3)$.
- Describe how to compute the area of a parallelogram whose vertices are $(x_1, y_1)$, $(x_2, y_2)$, $(x_3, y_3)$, $(x_4, y_4)$.
- Explain how to find the area of the region in Figure 12.
- If the area of the region in Figure 12 is denoted by $K$, explain how to determine the area of the image of this figure under the matrix transformation whose associated matrix is

$$\mathbf{A} = \begin{bmatrix} p & s \\ t & z \end{bmatrix}.$$

FIGURE 12                          FIGURE 13

- Explain how to use the techniques of this section to approximate the area of the region in Figure 13.

## 3.2 ■ THE DETERMINANT OF A MATRIX

In Section 3.1 we saw that the magnification factor of a matrix transformation determined by a square matrix $\mathbf{A}$ was an intrinsic number associated with the matrix. For a $2 \times 2$ matrix $\mathbf{A}$ we showed how to compute this number. However, for larger matrices we need a method for computing the magnification factor that does not rely exclusively on the geometric arguments. In this section we define a number associated with a square matrix that is related to the magnification factor. Here we focus on $2 \times 2$ and $3 \times 3$ matrices and indicate how to generalize the computation to any square matrix. We combine ideas from Section 3.1 with the concepts of this section in order to develop some properties. We return to computational strategies in Sections 3.3 and 3.4.

**Definition**    The **determinant of**

$$\mathbf{A} = \begin{bmatrix} a & b \\ c & d \end{bmatrix},$$

denoted $\det(\mathbf{A})$[1], is $ad - bc$.

We saw in Section 3.1 that the magnification factor associated with the matrix transformation determined by $\mathbf{A} = \begin{bmatrix} a & b \\ c & d \end{bmatrix}$ was $|ad - bc|$. Hence the magnification factor is $|\det(\mathbf{A})|$. We note that

$$\det\left(\begin{bmatrix} a & b \\ c & d \end{bmatrix}\right) = \left(\begin{array}{c} \text{product of} \\ \text{diagonal entries} \end{array}\right) - \left(\begin{array}{c} \text{product of reverse} \\ \text{diagonal entries} \end{array}\right).$$

This computation is depicted as

$$\det\left(\begin{bmatrix} a & b \\ c & d \end{bmatrix}\right) = ad - bc \tag{1}$$

and we refer to it as the **2×2 device**.

EXAMPLE 1    Let $\mathbf{A} = \begin{bmatrix} 5 & 2 \\ 4 & 1 \end{bmatrix}$. To compute $\det(\mathbf{A})$ we use the $2 \times 2$ device as shown in Equation (1):

$$\det\left(\begin{bmatrix} 5 & 2 \\ 4 & 1 \end{bmatrix}\right) = (5)(1) - (4)(2) = -3. \qquad \blacksquare$$

In Example 3 in Section 2.4 we showed that a $2 \times 2$ matrix $\mathbf{A} = \begin{bmatrix} a & b \\ c & d \end{bmatrix}$ was nonsingular provided $ad - bc \neq 0$ and then

$$\mathbf{A}^{-1} = \frac{1}{ad - bc} \begin{bmatrix} d & -b \\ -a & c \end{bmatrix}.$$

From our development here we have that **a 2 × 2 matrix A is nonsingular if and only if $\det(\mathbf{A}) \neq \mathbf{0}$.**

There is another way to look at the formula for the determinant of a $2 \times 2$ matrix that expresses it in a form that permits generalization to larger matrices. For this purpose we adopt the following:

The determinant of a $1 \times 1$ matrix is the value of its single entry.

We use this fact in:

$$\det(\mathbf{A}) = \det\left(\begin{bmatrix} a & b \\ c & d \end{bmatrix}\right) = ad - bc = a * \det\left(\begin{bmatrix} d \end{bmatrix}\right) - b * \det\left(\begin{bmatrix} c \end{bmatrix}\right)$$

where we see that $\begin{bmatrix} d \end{bmatrix}$ is the submatrix of $\mathbf{A}$ obtained from $\mathbf{A}$ by omitting its first row and first column and $\begin{bmatrix} c \end{bmatrix}$ is the submatrix of $\mathbf{A}$ obtained from $\mathbf{A}$ by omitting

---

[1]An alternate notation used for the determinant is $|\mathbf{A}|$. We will use $\det(\mathbf{A})$.

its first row and second column. Next we incorporate the row number and column number of the omitted row and column as follows (note the exponents of $(-1)$):

$$\det(\mathbf{A}) = \det\left(\begin{bmatrix} a & b \\ c & d \end{bmatrix}\right) = ad - bc = a * \det\left(\begin{bmatrix} d \end{bmatrix}\right) - b * \det\left(\begin{bmatrix} c \end{bmatrix}\right)$$

$$= (-1)^{1+1}a * \det\left(\begin{bmatrix} d \end{bmatrix}\right) + (-1)^{1+2}b * \det\left(\begin{bmatrix} c \end{bmatrix}\right).$$

We see that the computation of the determinant here used the entries from the first row as coefficients together with the determinants of submatrices obtained by omitting the row and column containing the coefficient. We call this **expansion about the first row**. What is rather surprising is that we get the same value when we use expansion about any row or column in a similar fashion. (See Exercises 1 through 4.)

We next consider $3 \times 3$ matrices.

---

**Definition**    The **determinant of**

$$\mathbf{A} = \begin{bmatrix} a & b & c \\ d & e & f \\ g & h & i \end{bmatrix},$$

denoted $\det(\mathbf{A})$, is given by the expression $aei + bfg + dch - gec - hfa - idb$. (Do not memorize this!)

---

Rather than compute $\det(\mathbf{A})$ using the expression given above, we present the following scheme for generating the products required. Adjoin to $\mathbf{A}$ its first two columns, then compute the products along the lines shown in Figure 1.

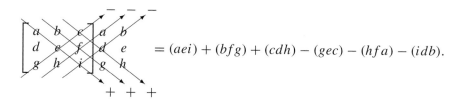

$$= (aei) + (bfg) + (cdh) - (gec) - (hfa) - (idb).$$

FIGURE 1

The products are then combined incorporating the sign that appears at the end of the line. Those products obtained along the lines parallel to the diagonal have a plus sign associated with them and those not parallel to the diagonal have a minus sign.

The computational procedure shown in Figure 1 is called the **3×3 device** for the determinant of a $3 \times 3$ matrix.

**WARNING:**    There are no such shortcuts or "devices" for larger matrices. (See Sections 3.3 and 3.4.)

**EXAMPLE 2**    Let $\mathbf{A} = \begin{bmatrix} 2 & 1 & 0 \\ 3 & -1 & 2 \\ 1 & -2 & 1 \end{bmatrix}$. To compute $\det(\mathbf{A})$ we use the $3 \times 3$ device as depicted in Figure 2.    ■

The expression for the determinant of a $3 \times 3$ matrix can be written as an expansion involving determinants of $2 \times 2$ submatrices and the entries from the first

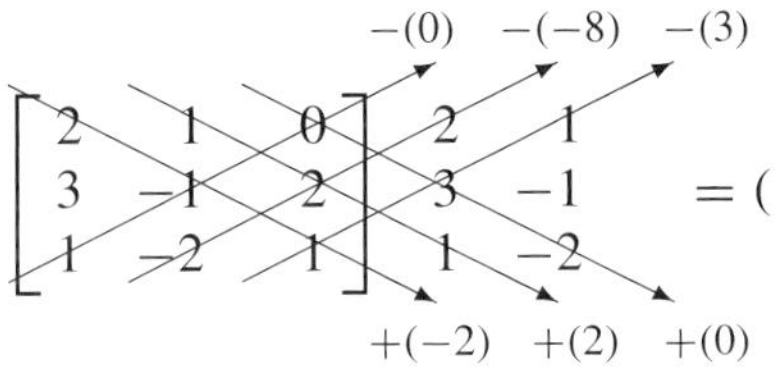

$$-(0) \quad -(-8) \quad -(3)$$

$$\begin{bmatrix} 2 & 1 & 0 \\ 3 & -1 & 2 \\ 1 & -2 & 1 \end{bmatrix} \begin{matrix} 2 & 1 \\ 3 & -1 \\ 1 & -2 \end{matrix} = (-2) + (2) + (0) - (0) - (-8) - (-3) = 5.$$

$$+(-2) \quad +(2) \quad +(0)$$

FIGURE 2

row. We have (verify by using the $2 \times 2$ device and simplifying)

$$\det \left( \begin{bmatrix} a & b & c \\ d & e & f \\ g & h & i \end{bmatrix} \right) = a * \det \left( \begin{bmatrix} e & f \\ h & i \end{bmatrix} \right)$$

$$- b * \det \left( \begin{bmatrix} d & f \\ g & i \end{bmatrix} \right) + c * \det \left( \begin{bmatrix} d & e \\ g & h \end{bmatrix} \right).$$

Note that

$\begin{bmatrix} e & f \\ h & i \end{bmatrix}$ is the submatrix obtained by omitting the first row and first column,

$\begin{bmatrix} d & f \\ g & i \end{bmatrix}$ is the submatrix obtained by omitting the first row and second column,

and

$\begin{bmatrix} d & e \\ g & h \end{bmatrix}$ is the submatrix obtained by omitting the first row and third column.

In addition, we can incorporate the row and column numbers of the rows and columns omitted by writing the expression in the form. (Note the exponents of $(-1)$.)

$$\det \left( \begin{bmatrix} a & b & c \\ d & e & f \\ g & h & i \end{bmatrix} \right) = (-1)^{1+1} a * \det \left( \begin{bmatrix} e & f \\ h & i \end{bmatrix} \right)$$

$$+ (-1)^{1+2} b * \det \left( \begin{bmatrix} d & f \\ g & i \end{bmatrix} \right)$$

$$+ (-1)^{1+3} c * \det \left( \begin{bmatrix} d & e \\ g & h \end{bmatrix} \right).$$

As in the $2 \times 2$ case, we call this expression **the expansion of the determinant about the first row**. In fact, we can use any row or column for the expansion with appropriate powers of $(-1)$ multiplying the entries and submatrices selected by omitting a row and column. (See Exercise 1 through 4.)

Following Example 3 in Section 2.4 we displayed the general formula for the inverse of a $3 \times 3$ matrix

$$\mathbf{A} = \begin{bmatrix} a & b & c \\ d & e & f \\ g & h & i \end{bmatrix}$$

as

$$\begin{bmatrix} a & b & c \\ d & e & f \\ g & h & i \end{bmatrix}^{-1} = \frac{1}{w} \begin{bmatrix} ie - fh & -(ib - ch) & bf - ce \\ -id + fg & ia - cg & -(af - cd) \\ -(-dh + eg) & -(ah - bg) & ae - bd \end{bmatrix}$$

where $w = aei + bfg + dch - gec - hfa - idb \neq 0$. It follows that

$$w = \det \left( \begin{bmatrix} a & b & c \\ d & e & f \\ g & h & i \end{bmatrix} \right)$$

and that $\mathbf{A}$ is nonsingular if and only if $\det(\mathbf{A}) \neq 0$.

The definition we give for the determinant of an $n \times n$ matrix is an expansion about the first row following the pattern of the $3 \times 3$ case as given previously. However, the expression involves determinants of $(n-1) \times (n-1)$ matrices, and hence is recursive. We explore the details of evaluating such expressions in Section 3.4.

---

**Definition**    The determinant of an $n \times n$ matrix $\mathbf{A} = \begin{bmatrix} a_{ij} \end{bmatrix}$, denoted $\det(\mathbf{A})$, is given by the expression

$$\det(\mathbf{A}) = (-1)^{1+1} a_{11} \det(\mathbf{A}_{11}) + (-1)^{1+2} a_{12} \det(\mathbf{A}_{12})$$
$$+ (-1)^{1+3} a_{13} \det(\mathbf{A}_{13}) + \cdots + (-1)^{1+n} a_{1n} \det(\mathbf{A}_{1n})$$

where $\mathbf{A}_{1k}$ is the submatrix obtained from $\mathbf{A}$ by omitting the first row and $k$th column. (We call this **expansion about the first row**.)

---

We state without verification that expansion about any row or column will yield the same value. See Section 3.4 for further details.

The relationship between the magnification factor and the determinant is

$$|\det(\mathbf{A})| = \text{the magnification factor}$$

We have shown this for $2 \times 2$ matrices, and we claim that it also holds for $n \times n$ matrices. Hence we have an algebraic procedure that computes the magnification factor of a matrix transformation determined by a square matrix. We next develop properties of the determinant.

In Section 1.5 we discussed the composition of matrix transformations and showed that the matrix corresponding to the composition was the product of the matrices of the individual matrix transformations. If $f(\mathbf{c}) = \mathbf{Ac}$ and $g(\mathbf{c}) = \mathbf{Bc}$, then $g(f(\mathbf{c})) = g(\mathbf{Ac}) = (\mathbf{BA})\mathbf{c}$. From geometrical considerations it follows that the magnification factor of the composition of the matrix transformations is the product of the magnification factors of the individual matrix transformations. This corresponds to the following important property of determinants, which we state without verification:

$$\det(\mathbf{BA}) = \det(\mathbf{B}) * \det(\mathbf{A}). \tag{2}$$

Since the determinant is a scalar, (2) implies the following result:

$$\det(\mathbf{BA}) = \det(\mathbf{B}) * \det(\mathbf{A}) = \det(\mathbf{A}) * \det(\mathbf{B}) = \det(\mathbf{AB}). \tag{3}$$

This does not imply that matrix $\mathbf{BA}$ is the same as matrix $\mathbf{AB}$, nor does it imply that the images of a given figure by the matrix transformations determined by $\mathbf{BA}$ and $\mathbf{AB}$ are identical. Rather, it does say that the magnification factors are the same. These ideas are illustrated in Example 3.

EXAMPLE 3    Let $\mathbf{A} = \begin{bmatrix} 2 & 1 \\ -2 & 0 \end{bmatrix}$ and $\mathbf{B} = \begin{bmatrix} -1 & 0 \\ 0 & 1 \end{bmatrix}$. In Figure 3 we show the results of the compositions of the matrix transformations determined by matrices $\mathbf{A}$ and $\mathbf{B}$ on the unit square.

**Unit Square**

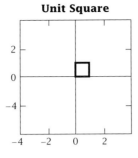

**A * (Unit Square)**

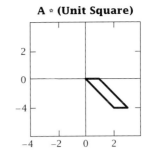

**B * A * (Unit Square)**

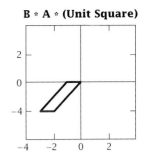

**B * (Unit Square)**

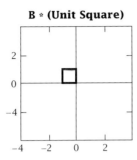

**A * B * (Unit Square)**

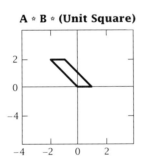

FIGURE 3

The images $\mathbf{B}*\mathbf{A}*$(unit square) and $\mathbf{A}*\mathbf{B}*$(unit square) are quite different yet their areas are the same. In order for the images to be the same, the matrices $\mathbf{A}$ and $\mathbf{B}$ must commute; that is, $\mathbf{AB} = \mathbf{BA}$. This can happen, but since matrix multiplication is not commutative, we cannot expect this situation to occur in general. ■

The computation of the determinant of $n \times n$ matrices, $n > 3$, from another point of view is discussed in Section 3.3, where we show how to use row operations to perform the calculation.

We conclude this section with an important result that connects row operations, nonsingular matrices, and determinants. In Exercises 15, 19, and 20 we show the effect that a row operation has on the value of a determinant. In summary,

$$\det(\mathbf{A} \text{ after a row operation}) = (\text{nonzero scalar}) * \det(\mathbf{A}).$$

Hence either $\det(\mathbf{A}$ after a row operation) and $\det(\mathbf{A})$ are both nonzero or they are both zero. From Section 2.4 it follows that every row operation on $\mathbf{A}$ can be performed by multiplying $\mathbf{A}$ on the left by an appropriate elementary matrix and that

$$\mathbf{rref}(\mathbf{A}) = (\text{product of elementary matrices}) * \mathbf{A}.$$

As shown in Exercises 15, 19, and 20, the determinant of an elementary matrix is not zero so, using Equation (2), we have

$$\det(\mathbf{rref}(\mathbf{A})) = \det(\text{product of elementary matrices}) * \det(\mathbf{A}) \qquad (4)$$

Thus either $\det(\mathbf{rref}(\mathbf{A}))$ and $\det(\mathbf{A})$ are both nonzero or they are both zero. However, from Section 2.4, $\mathbf{A}$ is nonsingular if and only if $\mathbf{rref}(\mathbf{A}) = \mathbf{I}_n$ and from Exercise 12 $\det(\mathbf{I}_n) = 1$. Putting this information together with (4) gives us the property

$$\mathbf{A} \text{ is nonsingular if and only if } \det(\mathbf{A}) \neq 0.$$

An equivalent form of the preceding statement follows:

$$\mathbf{A} \text{ is singular if and only if } \det(\mathbf{A}) = 0.$$

Combining information in Section 2.4 and the developments in this section, we obtain the following list of equivalent statements.

---

**Equivalent Statements for an $n \times n$ Nonsingular Matrix A**

1.  **A** is nonsingular.
2.  $\mathbf{rref}(\mathbf{A}) = \mathbf{I}_n$.
3.  **A** has an inverse.
4.  $\mathbf{Ax} = \mathbf{b}$ has a unique solution for every right side **b**.
5.  $\det(\mathbf{A}) \neq 0$.
6.  $\mathbf{ns}(\mathbf{A}) = \mathbf{0}$.
7.  The rows (columns) of **A** are linearly independent.
8.  $\dim(\mathbf{row}(\mathbf{A})) = \dim(\mathbf{col}(\mathbf{A})) = n$.
9.  $\text{rank}(\mathbf{A}) = n$.

---

There are a number of properties of the determinant that have been presented in this section and more that are obtained from the exercises that follow. Many of these properties will be used in succeeding sections. For ease of reference we list these properties in Table 1, which appears after the exercises.

# EXERCISES 3.2

1. Let $\mathbf{A} = \begin{bmatrix} a & b \\ c & d \end{bmatrix}$.

   (a) The expression

   $$(-1)^{1+1}a * \det\left(\begin{bmatrix} d \end{bmatrix}\right) + (-1)^{1+2}c * \det\left(\begin{bmatrix} b \end{bmatrix}\right)$$

   is called the expansion about the first column. Show that this expression is the same as $\det(\mathbf{A})$.

   (b) The expression

   $$(-1)^{1+2}c * \det\left(\begin{bmatrix} b \end{bmatrix}\right) + (-1)^{2+2}d * \det\left(\begin{bmatrix} a \end{bmatrix}\right)$$

   is called the expansion about the second row. Show that this expression is the same as $\det(\mathbf{A})$.

2. Construct the expansion about the second column for the matrix in Exercise 1. Show that the expression is the same as $\det(\mathbf{A})$.

3. Let $\mathbf{A} = \begin{bmatrix} a & b & c \\ d & e & f \\ g & h & i \end{bmatrix}$.

   (a) The expression

   $$(-1)^{3+1}g * \det\left(\begin{bmatrix} b & c \\ e & f \end{bmatrix}\right)$$

   $$+ (-1)^{3+2}h * \det\left(\begin{bmatrix} a & c \\ d & f \end{bmatrix}\right)$$

   $$+ (-1)^{3+3}i * \det\left(\begin{bmatrix} a & b \\ d & e \end{bmatrix}\right)$$

   is called the expansion about the third row. Show that this expression is the same as $\det(\mathbf{A})$.

(b) The expression

$$(-1)^{1+2}b * \det\left(\begin{bmatrix} d & f \\ g & i \end{bmatrix}\right)$$

$$+ (-1)^{2+2}e * \det\left(\begin{bmatrix} a & c \\ g & i \end{bmatrix}\right)$$

$$+ (-1)^{3+2}h * \det\left(\begin{bmatrix} a & c \\ d & f \end{bmatrix}\right)$$

is called the expansion about the second column. Show that this expression is the same as $\det(\mathbf{A})$.

4. Construct the expansion about the second row for the matrix in Exercise 3. Show that the expression is the same as $\det(\mathbf{A})$.

*In Exercises 5–11, compute the determinant of each given matrix.*

5. $\begin{bmatrix} 1 & 2 \\ 3 & 4 \end{bmatrix}$.

6. $\begin{bmatrix} -2 & 3 \\ 1 & 1 \end{bmatrix}$.

7. $\begin{bmatrix} 1 & 2 & 3 \\ 4 & 5 & 6 \\ 7 & 8 & 9 \end{bmatrix}$.

8. $\begin{bmatrix} 2 & 1 & 0 \\ 1 & 2 & 1 \\ 0 & 1 & 2 \end{bmatrix}$.

9. $\begin{bmatrix} \cos(\theta) & \sin(\theta) \\ -\sin(\theta) & \cos(\theta) \end{bmatrix}$.

10. $\begin{bmatrix} 1 & 2 & 3 \\ 4 & 5 & 6 \\ 7 & 8 & 0 \end{bmatrix}$.

11. $\begin{bmatrix} 0 & 1 & 0 \\ 1 & 0 & 0 \\ 0 & 0 & 1 \end{bmatrix}$.

12. Let $\mathbf{A} = \mathbf{I}_n$, the $n \times n$ identity matrix. Explain why the magnification factor for the matrix transformation determined by $\mathbf{A}$ is 1. What is $\det(\mathbf{I}_2)$? What is $\det(\mathbf{I}_3)$? Make a conjecture for $\det(\mathbf{I}_n)$.

13. If $\mathbf{A}$ is nonsingular, show that

$$\det(\mathbf{A}^{-1}) = \frac{1}{\det(\mathbf{A})}.$$

[*Hint*: Use Equation (3).]

14. (a) Compute the determinants of each of the following diagonal matrices·

$$\begin{bmatrix} 5 & 0 \\ 0 & 2 \end{bmatrix}, \quad \begin{bmatrix} 4 & 0 \\ 0 & -3 \end{bmatrix},$$

$$\begin{bmatrix} 2 & 0 & 0 \\ 0 & -5 & 0 \\ 0 & 0 & 3 \end{bmatrix}, \quad \begin{bmatrix} 1 & 0 & 0 \\ 0 & 0 & 0 \\ 0 & 0 & 7 \end{bmatrix}.$$

(b) Complete the following conjecture: The determinant of a diagonal matrix is the product of _____.

(c) If $\mathbf{A}$ is an $n \times n$ diagonal matrix, explain geometrically why the magnification factor of the matrix transformation determined by $\mathbf{A}$ agrees with the conjecture formulated in part (b).

(d) Does your conjecture from part (b) agree with the conjecture made in Exercise 12? Explain.

15. Let $\mathbf{A}$ be an $n \times n$ matrix. To obtain $\mathbf{A}_{kR_i}$ we can form the product $\mathbf{EA}$ where $\mathbf{E} = (\mathbf{I}_n)_{kR_i}$. (See the discussion about elementary matrices in Section 2.4.)

(a) $\mathbf{E}$ is a _____ matrix.

(b) $\det(\mathbf{E}) =$ _____.

(c) Compute $\det(\mathbf{EA})$ in terms of $\det(\mathbf{A})$; $\det(\mathbf{EA}) =$ _____.

16. Let $\mathbf{A}$ be an $n \times n$ matrix and $\mathbf{B} = k\mathbf{A}$. Then $b_{ij} = ka_{ij}$.

(a) Determine an $n \times n$ diagonal matrix $\mathbf{Q}$ so that $\mathbf{QA} = \mathbf{B}$. (*Hint*: See Exercise 15.)

(b) Compute $\det(k\mathbf{A})$ in terms of $\det(\mathbf{A})$; $\det(k\mathbf{A}) =$ _____.

17. (a) Compute the determinants of each of the following upper triangular matrices.

$$\begin{bmatrix} 5 & 3 \\ 0 & 2 \end{bmatrix}, \quad \begin{bmatrix} 4 & 9 \\ 0 & -3 \end{bmatrix},$$

$$\begin{bmatrix} 2 & 1 & 5 \\ 0 & -5 & 8 \\ 0 & 0 & 3 \end{bmatrix}, \quad \begin{bmatrix} -1 & 6 & 2 \\ 0 & 0 & 1 \\ 0 & 0 & 7 \end{bmatrix}.$$

(b) Complete the following conjecture: The determinant of an upper triangular matrix is the product of _____.

18. (a) Compute the determinants of each of the following lower triangular matrices.

$$\begin{bmatrix} 5 & 0 \\ 3 & 2 \end{bmatrix}, \quad \begin{bmatrix} 4 & 0 \\ 9 & -3 \end{bmatrix},$$

$$\begin{bmatrix} 2 & 0 & 0 \\ 1 & -5 & 0 \\ 5 & 8 & 3 \end{bmatrix}, \quad \begin{bmatrix} -1 & 0 & 0 \\ 6 & 0 & 0 \\ 2 & 1 & 7 \end{bmatrix}.$$

(b) Complete the following conjecture: The determinant of a lower triangular matrix is the product of _____.

19. Let $\mathbf{A}$ be an $n \times n$ matrix. To obtain $\mathbf{A}_{kR_i+R_j}$ we can form the product $\mathbf{EA}$ where $\mathbf{E} = (\mathbf{I}_n)_{kR_i+R_j}$. (See the discussion about elementary matrices in Section 2.4.)

(a) If $i > j$, then $\mathbf{E}$ is a _____ triangular matrix.

(b) If $i < j$, then $\mathbf{E}$ is a _____ triangular matrix.

(c) $\det(\mathbf{E}) =$ _____.

(d) Compute $\det(\mathbf{A}_{kR_i+R_j})$ in terms of $\det(\mathbf{A})$; $\det(\mathbf{A}_{kR_i+R_j}) =$ _____.

20. (a) Let $\mathbf{A} = \begin{bmatrix} a & b \\ c & d \end{bmatrix}$ and $\mathbf{B} = \mathbf{A}_{R_1 \leftrightarrow R_2}$. Compute $\det(\mathbf{B})$ in terms of $\det(\mathbf{A})$.

(b) Let $A = \begin{bmatrix} a & b & c \\ d & e & f \\ g & h & i \end{bmatrix}$, $B = A_{R_1 \leftrightarrow R_3}$, $C = A_{R_2 \leftrightarrow R_3}$,

and $D = A_{R_1 \leftrightarrow R_2}$. Compute $\det(B)$, $\det(C)$, and $\det(D)$ in terms of $\det(A)$.

(c) Complete the following conjecture for $i \neq j$:

$$\det(A_{R_i \leftrightarrow R_j}) = \underline{\qquad} \det(A).$$

**21.** (a) Let $A = \begin{bmatrix} a & b \\ c & d \end{bmatrix}$. Show by direct computation that $\det(A^T) = \det(A)$.

(b) Let $A = \begin{bmatrix} a & b & c \\ d & e & f \\ g & h & i \end{bmatrix}$. Show by direct computation that $\det(A^T) = \det(A)$.

(c) Complete the following conjecture:

For any square matrix $A$, $\det(A^T) = \underline{\qquad}$.

**22.** Equations (3) and (4) in Section 3.1 provide us with an expression for the area of a triangle in terms of a dot product. Show that

$$\frac{1}{2} \left| \det \left( \begin{bmatrix} x_1 & y_1 & 1 \\ x_2 & y_2 & 1 \\ x_3 & y_3 & 1 \end{bmatrix} \right) \right|$$

yields the same result.

**23.** Develop an expression for the area of the parallelogram in Figure 5 of Section 3.1 in terms of the determinant of an appropriate $3 \times 3$ matrix.

**24.** Let $P(x_1, y_1)$ and $Q(x_2, y_2)$ be two points in $R^2$. Show that the equation of the line through $P$ and $Q$ is given by the expression

$$\det \left( \begin{bmatrix} x & y & 1 \\ x_1 & y_1 & 1 \\ x_2 & y_2 & 1 \end{bmatrix} \right) = 0.$$

**25.** Using the expression in Exercise 24, explain how to decide if points $P(x_1, y_1)$, $Q(x_2, y_2)$, and $R(x_3, y_3)$ are collinear.

---

TABLE 1 **Properties of the Determinant**

---

$\det(AB) = \det(A)\det(B) = \det(B)\det(A) = \det(BA)$

$\det(A) \neq 0$ if and only if $A$ is nonsingular.

$\det(A) = 0$ if and only if $A$ is singular.

$\det(A^{-1}) = \dfrac{1}{\det(A)}$

$\det(I_n) = 1$

$\det(\text{diagonal matrix}) = \text{product of its diagonal entries}$

$\det(\text{upper triangular matrix}) = \text{product of its diagonal entries}$

$\det(\text{lower triangular matrix}) = \text{product of its diagonal entries}$

$\det(A^T) = \det(A)$

$\det(kA) = k^n \det(A)$, if $A$ is $n \times n$

$\det(A_{kR_i}) = k \det(A)$

$\det(A_{R_i \leftrightarrow R_j}) = -\det(A)$, $i \neq j$

$\det(A_{kR_i + R_j}) = \det(A)$, $i \neq j$

---

## True/False Review Questions

*Determine whether each of the following statements is true or false.*

**1.** $\det \left( \begin{bmatrix} 5 & 2 \\ -1 & 1 \end{bmatrix} \right) = 7$.

**2.** If $T$ is the matrix transformation from $R^2$ to $R^2$ with associated matrix $A = \begin{bmatrix} 4 & 3 \\ 2 & 1 \end{bmatrix}$, then for a closed polygon $P$, area $(T(P)) = 2\,\text{area}(P)$.

**3.** If $A$ is singular, then $\det(A) = 0$.

4. If **A** is singular, matrix transformation $T$ from $R^2$ to $R^2$ has associated matrix **A**, and $P$ is a closed polygon, then area$(T(P)) = 0$.

5. $\det\left(\begin{bmatrix} 0 & -1 & 1 \\ 1 & 0 & -1 \\ -1 & 1 & 0 \end{bmatrix}\right) = 0.$

6. $\det(\mathbf{AB}) = \det(\mathbf{A})\,\det(\mathbf{B})$.

7. $\det(\mathbf{A}^{-1}) = \dfrac{1}{\det(\mathbf{A})}$.

8. $\det(\mathbf{A}^T) = \det(\mathbf{A})$.

9. If **A** is nonsingular and **B** is singular, then **AB** is singular.

10. $\det(\mathbf{A} + \mathbf{B}) = \det(\mathbf{A}) + \det(\mathbf{B})$.

## Terminology

| Determinant | Determinant by expansion |
| --- | --- |

The determinant is an intrinsic number associated with a square matrix. In this section we showed how to compute the determinant for $2 \times 2$ and $3 \times 3$ matrices in a purely algebraic fashion. We then gave a general definition of the determinant that involved the determinant of submatrices of the original matrix. We also developed properties of the determinants as they related to concepts studied earlier.

- Explain the connection between det(**A**) and the magnification factor associated with the matrix transformation $f(\mathbf{c}) = \mathbf{Ac}$.
- In words, describe how to compute the determinant of a $2 \times 2$ matrix; of a $3 \times 3$ matrix.
- In words, explain what we mean by the expansion about the first row for a $3 \times 3$ matrix; for an $n \times n$ matrix.
- Describe how the determinant of a matrix changes when we perform a single row operation on it. Treat all three cases.
- Describe how the determinant of a matrix changes when we multiply the matrix by a scalar $k$.

Complete the following in words.

- The determinant of a product is _____.
- The determinant of a nonsingular matrix is _____.
- The determinant of a singular matrix is _____.
- The determinant of the inverse of a matrix is _____.
- The determinant of a diagonal matrix is _____.
- The determinant of the transpose of a matrix is _____.

## 3.3 ■ COMPUTING DET(A) USING ROW OPERATIONS

In Sections 3.1 and 3.2 we showed that det(**A**) is closely related to the magnification factor of the matrix transformation determined by **A** and is an intrinsic number associated with a square matrix. In effect, "det" defines a function whose domain is the set of square matrices and whose range is the set of scalars, possibly real or complex. In Section 3.2 and its exercises we developed properties of the "det" function which are listed in Table 1 at the end of Section 3.2.

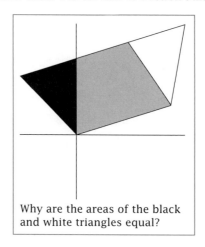

Why are the areas of the black and white triangles equal?

MATLAB routine, **areapara.m**

In addition, we developed computational procedures, or "devices," for computing the determinants of $2 \times 2$ and $3 \times 3$ matrices. These procedures are displayed in Figures 1 and 2, respectively. The determinant is a linear combination of the products of the entries along the lines shown in the figures using the sign that appears at the end of the line.

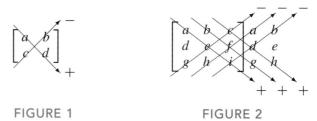

FIGURE 1                              FIGURE 2

Unfortunately, there are no such computational shortcuts for the determinants of $n \times n$ matrices for $n > 3$. In this section we show how to use row operations and the properties listed in Table 1 at the end of Section 3.2 to compute the determinant of any square matrix. An alternative recursive procedure is developed in Section 3.4.

Our strategy for computing det(**A**) using row operations is to apply row operations to matrix **A** to produce an upper triangular matrix while keeping track of the effect the row operations have on the determinant. We use the properties

$$\det(\mathbf{A}_{kR_i}) = k \det(\mathbf{A}),$$
$$\det(\mathbf{A}_{R_i \leftrightarrow R_j}) = -\det(\mathbf{A}), \quad i \neq j$$
$$\det(\mathbf{A}_{kR_i + R_j}) = \det(\mathbf{A}), \quad i \neq j.$$

It is convenient to rewrite these properties in terms of $\det(\mathbf{A})$:

1. $\det(\mathbf{A}) = \dfrac{1}{k} \det(\mathbf{A}_{kR_i})$, $k \neq 0$.
2. $\det(\mathbf{A}) = - \det(\mathbf{A}_{R_i \leftrightarrow R_j})$, $i \neq j$.
3. $\det(\mathbf{A}) = \det(\mathbf{A}_{kR_i + R_j})$, $i \neq j$.

These properties are interpreted as follows:

1. If we apply row operation $kR_i$, $k \neq 0$, to $\mathbf{A}$, then $\det(\mathbf{A})$ is $(1/k)$ times the determinant of the matrix that results from the row operation.
2. If we apply row operation $R_i \leftrightarrow R_j$, $i \neq j$, to $\mathbf{A}$, then $\det(\mathbf{A})$ is $(-1)$ times the determinant of the matrix that results from the row operation.
3. If we apply row operation $kR_i + R_j$, $i \neq j$, to $\mathbf{A}$, then $\det(\mathbf{A})$ is the same as the determinant of the matrix that results from the row operation.

We illustrate the procedure in Example 1 on a $3 \times 3$ matrix and compare the result with that obtained using the $3 \times 3$ device shown in Figure 2.

EXAMPLE 1   Let

$$\mathbf{A} = \begin{bmatrix} 4 & 3 & 2 \\ 3 & -2 & 5 \\ 2 & 4 & 6 \end{bmatrix}.$$

There are many sets of row operations that will lead us to an upper triangular form. The operations we choose are not the most efficient, but we do avoid fractions during the first few steps.

$$\det(\mathbf{A}) = 2 \det \left( \mathbf{A}_{\frac{1}{2}R_3} \right) = 2 \det \left( \begin{bmatrix} 4 & 3 & 2 \\ 3 & -2 & 5 \\ 1 & 2 & 3 \end{bmatrix} \right) \qquad \text{[multiply row 3 by } \tfrac{1}{2}\text{]}$$

$$= (-1)2 \det \left( \begin{bmatrix} 4 & 3 & 2 \\ 3 & -2 & 5 \\ 1 & 2 & 3 \end{bmatrix}_{R_1 \leftrightarrow R_3} \right) = -2 \det \left( \begin{bmatrix} 1 & 2 & 3 \\ 3 & -2 & 5 \\ 4 & 3 & 2 \end{bmatrix} \right) \qquad \text{[interchange rows 1 and 3]}$$

$$= -2 \det \left( \begin{bmatrix} 1 & 2 & 3 \\ 3 & -2 & 5 \\ 4 & 3 & 2 \end{bmatrix}_{\substack{-3R_1 + R_2 \\ -4R_1 + R_3}} \right) = -2 \det \left( \begin{bmatrix} 1 & 2 & 3 \\ 0 & -8 & -4 \\ 0 & -5 & -10 \end{bmatrix} \right) \qquad \text{[zero out below the (1, 1) entry]}$$

$$= -2 \det \left( \begin{bmatrix} 1 & 2 & 3 \\ 0 & -8 & -4 \\ 0 & -5 & -10 \end{bmatrix}_{\frac{-5}{8}R_2 + R_3} \right) = -2 \det \left( \begin{bmatrix} 1 & 2 & 3 \\ 0 & -8 & -4 \\ 0 & 0 & \frac{-30}{4} \end{bmatrix} \right) \qquad \text{[zero out below the (2, 2) entry]}$$

[Next we compute the determinant of the upper triangular matrix.] $\qquad = -2(1)(-8) \left( \dfrac{-30}{4} \right) = -120 \qquad$ [See Exercise 17 in Section 3.2]

The $3 \times 3$ device yields

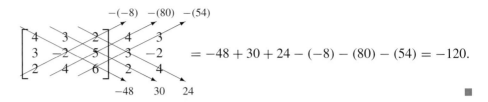

$$= -48 + 30 + 24 - (-8) - (80) - (54) = -120.$$

For $n \times n$ matrices, $n > 3$, we have no "devices," so the row operation procedure of Example 1 provides a computational technique for finding the determinant.

(The following discussion and example assumes familiarity with Section 2.5.) In Section 2.5 we developed a form of LU-factorization of a matrix $\mathbf{A}$ that uses only row operations of the form $kR_i + R_j$. When such a factorization is possible, the diagonal entries of the lower triangular matrix $\mathbf{L}$ are all 1's. In such a case we have

$$\det(\mathbf{A}) = \det(\mathbf{LU}) = \det(\mathbf{L}) \det(\mathbf{U}) = \det(\mathbf{U}).$$

**EXAMPLE 2**   In Example 1 in Section 2.5 we showed that

$$\mathbf{A} = \begin{bmatrix} 6 & -2 & -4 & 4 \\ 3 & -3 & -6 & 1 \\ -12 & 8 & 21 & -8 \\ -6 & 0 & -10 & 7 \end{bmatrix} = \mathbf{LU}$$

$$= \begin{bmatrix} 1 & 0 & 0 & 0 \\ \frac{1}{2} & 1 & 0 & 0 \\ -2 & -2 & 1 & 0 \\ -1 & 1 & -2 & 1 \end{bmatrix} \begin{bmatrix} 6 & -2 & -4 & 4 \\ 0 & -2 & -4 & -1 \\ 0 & 0 & 5 & -2 \\ 0 & 0 & 0 & 8 \end{bmatrix}.$$

Thus $\det(\mathbf{A}) = \det(\mathbf{U}) = (6)(-2)(5)(8) = -480.$ ■

Having an LU-factorization makes the determinant computation easy. Without an LU-factorization we could compute $\det(\mathbf{A})$ as in Example 1. If we try using only row operations of the form $kR_i + R_j$, then we run the risk of encountering a zero pivot and not being able to complete the reduction to upper triangular form. This is the same risk encountered in the special LU-factorization developed in Section 2.5 .

# EXERCISES 3.3

*In Exercises 1–4, compute the determinant of each given matrix using row operations as illustrated in Example 1. Compare your result with the value obtained by using the $3 \times 3$ device shown in Figure 2.*

1. $\begin{bmatrix} 4 & 1 & 3 \\ 2 & 3 & 0 \\ 1 & 3 & 2 \end{bmatrix}.$

2. $\begin{bmatrix} 1 & 3 & -4 \\ -2 & 1 & 2 \\ -9 & 15 & 0 \end{bmatrix}.$

3. $\begin{bmatrix} 4 & -3 & 5 \\ 5 & 2 & 0 \\ 2 & 0 & 4 \end{bmatrix}.$

4. $\begin{bmatrix} 1+i & 0 & i \\ 2 & 2+i & 1 \\ 1 & 0 & 2 \end{bmatrix}.$

*In Exercises 5–8, compute the determinant of each given matrix using row operations as illustrated in Example 1.*

5. $\begin{bmatrix} 2 & 0 & 1 & 4 \\ 1 & 2 & -6 & -10 \\ 2 & 3 & -1 & 0 \\ 11 & 8 & -4 & 6 \end{bmatrix}.$

6. $\begin{bmatrix} 4 & 2 & 3 & -4 \\ -1 & -4 & -2 & 9 \\ -2 & 0 & 1 & -3 \\ 8 & -2 & 6 & 4 \end{bmatrix}.$

**7.** $\begin{bmatrix} 2 & 1 & 2 & 1 \\ 4 & 0 & 1 & 2 \\ 0 & 2 & 0 & 1 \\ 0 & 0 & 5 & 5 \end{bmatrix}.$

**8.** $\begin{bmatrix} 1 & 1 & 0 & i \\ i & 0 & 1 & 0 \\ 0 & 1 & i & 0 \\ -i & 0 & 0 & 1 \end{bmatrix}.$

**9.** Use the $2 \times 2$ device to evaluate each of the following.

(a) $\det\left( \begin{bmatrix} \lambda - 1 & 2 \\ 3 & \lambda - 2 \end{bmatrix} \right).$

(b) $\det(\lambda \mathbf{I}_2 - \mathbf{A})$, where $\mathbf{A} = \begin{bmatrix} 4 & 2 \\ -1 & 1 \end{bmatrix}.$

**10.** For each of the computations in Exercise 9, find all the values of $\lambda$ for which the determinant is zero.

**11.** Use the $3 \times 3$ device to evaluate each of the following.

(a) $\det\left( \begin{bmatrix} \lambda - 1 & -2 & -1 \\ 0 & \lambda - 1 & 0 \\ 1 & -3 & \lambda - 2 \end{bmatrix} \right).$

(b) $\det(\lambda \mathbf{I}_3 - \mathbf{A})$, where $\mathbf{A} = \begin{bmatrix} -1 & 0 & 1 \\ -2 & 0 & -1 \\ 0 & 0 & 1 \end{bmatrix}.$

**12.** For each of the computations in Exercise 11, find all the values of $\lambda$ for which the determinant is zero.

**13.** Show that $\det\left( \begin{bmatrix} a^2 & a & 1 \\ b^2 & b & 1 \\ c^2 & c & 1 \end{bmatrix} \right) = (b-a)(c-a)(b-c).$

*For Exercises* 14 *through* 19, *use the properties of determinants listed in Table* 1 *of Section* 3.2.

**14.** If $\det(\mathbf{A}) = -3$, find

(a) $\det(\mathbf{A}^2)$.

(b) $\det(\mathbf{A}^5)$.

(c) $\det(\mathbf{A}^{-1})$.

**15.** If $\mathbf{A}$ and $\mathbf{B}$ are $n \times n$ matrices with $\det(\mathbf{A}) = 4$ and $\det(\mathbf{B}) = -3$, calculate $\det(\mathbf{A}^{-1}\mathbf{B}^T)$.

**16.** Show that for $n \times n$ matrices $\mathbf{A}$ and $\mathbf{B}$ if $\det(\mathbf{AB}) = 0$, then either $\mathbf{A}$ or $\mathbf{B}$ is singular.

**17.** Show that if $\det(\mathbf{A}^5) = 0$, then $\mathbf{A}$ is singular.

**18.** If $\mathbf{A}$ is upper (lower) triangular and nonsingular, describe the diagonal entries of $\mathbf{A}$.

**19.** Show that if $\mathbf{A}^T = \mathbf{A}^{-1}$, then $\det(\mathbf{A}) = \pm 1$.

*In Exercises* 20 *through* 23, *use properties* 1, 2, *and* 3 (*see page* 203), *which show the effect of row operations on the determinant, to formulate your explanation.*

**20.** Let $\mathbf{A}$ be an $n \times n$ matrix with two identical rows. Explain why $\det(\mathbf{A}) = 0$.

**21.** Let $\mathbf{A}$ be an $n \times n$ matrix with two identical columns. Explain why $\det(\mathbf{A}) = 0$.

**22.** Let $\mathbf{A}$ be an $n \times n$ matrix with a row of all zeros. Explain why $\det(\mathbf{A}) = 0$.

**23.** Let $\mathbf{A}$ be an $n \times n$ matrix with a column of all zeros. Explain why $\det(\mathbf{A}) = 0$.

**24.** (a) Find a pair of $3 \times 3$ nonzero matrices $\mathbf{A}$ and $\mathbf{B}$ such that $\det(\mathbf{A} + \mathbf{B}) = \det(\mathbf{A}) + \det(\mathbf{B})$.

(b) Find a pair of $3 \times 3$ nonzero matrices $\mathbf{A}$ and $\mathbf{B}$ such that $\det(\mathbf{A} + \mathbf{B}) \neq \det(\mathbf{A}) + \det(\mathbf{B})$.

**25.** Explain why "det" is not a linear transformation from the set of square matrices to the set of scalars. (See Section 1.5.)

**26.** Explain why $\det(\mathbf{A}^k) = (\det(\mathbf{A}))^k$, for $k$ a positive integer. (See Exercise 14.)

In MATLAB

*In Sections* 3.1 *and* 3.2 *we showed that the number we call the determinant is related to the magnification factor for a matrix transformation that corresponds to a square matrix. In this section we showed how to use properties of the determinant associated with row operations to develop a computational procedure for obtaining the value of the determinant. We use* MATLAB *as a smart tutor for practicing this method of finding* $\det(\mathbf{A})$. *In addition we provide a preformal proof that row reductions using only row operations of the form* $kR_i + R_j$ *preserve the magnification factor of a* $2 \times 2$ *matrix transformation. (See Exercise* ML.2 *through* ML.4.)

**ML. 1.** In MATLAB type **help detpr** and read the description. This routine acts like a smart tutor when using row operations to compute the determinant of a matrix. Enter each of the following matrices into MATLAB and then begin determinant practice by executing command **detpr(A)**. Choose row operations to reduce $\mathbf{A}$ to upper triangular form, keeping track of the effect of the row operations, as in Example 1. The routine will prompt you to develop the appropriate relationships.

(a) $\mathbf{A} = \begin{bmatrix} 5 & -2 & 1 \\ 4 & 3 & 0 \\ -1 & 7 & 1 \end{bmatrix}.$

(b) $\mathbf{A} = \begin{bmatrix} 2 & 4 & -3 \\ 6 & 5 & 2 \\ -2 & 3 & -8 \end{bmatrix}.$

(c) $\mathbf{A} = \begin{bmatrix} 10 & 9 & 9 & 10 \\ 3 & 8 & 5 & 8 \\ 7 & 5 & 7 & 2 \\ 5 & 1 & 8 & 5 \end{bmatrix}.$

(d) $\mathbf{A} = \begin{bmatrix} 3 & 4 & 5 & 3 & 2 \\ 8 & 1 & 2 & 2 & 4 \\ 3 & -5 & 7 & 0 & 0 \\ 3 & 7 & 4 & 0 & -5 \\ 7 & 0 & 6 & 5 & 2 \end{bmatrix}.$

(e) $\mathbf{A} = \begin{bmatrix} 1+4i & -23 & 2i \\ -5+4i & -11+i & -1 \\ i & 3 & 0 \end{bmatrix}$.

**ML. 2.** In MATLAB type **help areapara** and read the brief description. Next execute the command **areapara** and choose option 1 so that vectors $\mathbf{u} = \begin{bmatrix} 8 & 2 \end{bmatrix}$ and $\mathbf{v} = \begin{bmatrix} -4 & 5 \end{bmatrix}$ are used.

(a) Vectors $\mathbf{u}$ and $\mathbf{v}$, the rows of a matrix, will be displayed. Press ENTER to see the corresponding parallelogram. Press ENTER to begin row reduction using only row operations of the form $k R_i + R_j$.

In (b) through (f), formulate your explanations in terms of geometric relationships.

(b) After the First Elimination there appears a black triangle and a white triangle. Explain why these triangles have the same area.

(c) The black triangle is cleaved off and the white triangle is kept to form a new parallelogram. Explain why the area of the original parallelogram is equal to the area of the current parallelogram. (Press ENTER after you develop your explanation.)

(d) After the Second Elimination there is another pair of triangles, one black and one white. Explain why these triangles have the same area. (Press ENTER after you develop your explanation.)

(e) The black triangle is cleaved off and the white triangle is kept to form a new parallelogram. Explain why the area of the original parallelogram is equal to the area of the current parallelogram. (Press ENTER after you develop your explanation.)

(f) Explain why the area of the original parallelogram is equal to the area of the rectangle that is displayed. (To quit this routine press ENTER.)

**ML. 3.** In MATLAB, execute command **areapara** and choose option 2 so that vectors $\mathbf{u} = \begin{bmatrix} -7 & 6 \end{bmatrix}$ and $\mathbf{v} = \begin{bmatrix} 5 & -9 \end{bmatrix}$ are used. Follow the steps in Exercise ML.2 but be on the lookout for a different set of displays in parts (d) through (f). Give a reason that each display is correct before going to the next step.

**ML. 4.** Use routine **areapara** with a pair of (linearly independent) vectors that you supply. Give a reason for the geometric steps as in Exercises ML.2 and ML.3.

**ML. 5.** MATLAB command **det** computes the determinant of a square matrix. This routine first determines an LU-factorization where the diagonal entries of $\mathbf{L}$ are all ones, so that $\det(\mathbf{A}) = \det(\mathbf{U})$.

(a) What role do the diagonal entries of $\mathbf{U}$ play in the row reduction process that obtains the LU-factorization? See Section 2.5.

(b) Use command **det** to check your work in Exercises 1 through 8.

## True/False Review Questions

*Determine whether each of the following statements is true or false. (Assume $\mathbf{A}$, $\mathbf{B}$, and $\mathbf{C}$ are square matrices of the same size.)*

**1.** $\det(\mathbf{A}_{k R_i + R_j}) = \det(\mathbf{A})$.

**2.** For $i \neq j$, $\det(\mathbf{A}_{R_i \leftrightarrow R_j}) = \det(\mathbf{A})$.

**3.** $\det(k\mathbf{A}) = k \det(\mathbf{A})$.

**4.** $\det(\mathbf{A}^3) = 3 \det(\mathbf{A})$.

**5.** If $\mathbf{A}$ has two rows that are the same, then $\det(\mathbf{A}) = 0$.

**6.** Let $\mathbf{T} = \begin{bmatrix} 2 & 1 & 2 \\ 0 & 3 & 1 \\ 0 & 0 & -1 \end{bmatrix}$ be the result of applying row operations $R_1 \leftrightarrow R_3$, $2R_1 + R_2$, $-3R_1 + R_3$, $\frac{1}{2}R_2$ to $\mathbf{A}$.

Then $\det(\mathbf{A}) = 6$.

**7.** If $\det(\mathbf{A}) = 3$ and $\det(\mathbf{B}^{-1}) = 4$, then $\det(\mathbf{AB}) = 12$.

**8.** If $\det(\mathbf{A}) = 4$, $\det(\mathbf{B}) = 3$, and $\det(\mathbf{C}) = 8$, then $\det(\mathbf{A}^{-1}\mathbf{BC}) = 6$.

**9.** If $\det(\mathbf{AB}) \neq 0$, then both $\mathbf{A}$ and $\mathbf{B}$ are nonsingular.

**10.** If $\mathbf{A}$ is $3 \times 3$ and skew symmetric, then $\det(\mathbf{A}) = 0$.

## Terminology

Determinants using row operations

The use of row operations within a determinant calculation gives us a way to compute the determinant of any square matrix using familiar manipulations. In fact, this procedure forms the basis of determinant computations in most scientific computer software programs.

- Give a general description of how to use row operations to assist in the computation of a determinant.
- Describe how to compute the determinant of $\mathbf{A}^{100}$ without performing the exponentiation.
- In Section 1.5 we stated that a function $f$ was called a **linear transformation**, provided $f(\mathbf{A} + \mathbf{B}) = f(\mathbf{A}) + f(\mathbf{B})$ and $f(k\mathbf{A}) = kf(\mathbf{A})$, for any scalar $k$. Determine whether the function det is a linear transformation. Explain your reasoning.

## 3.4 ■ RECURSIVE COMPUTATION OF THE DETERMINANT (OPTIONAL)

In Section 3.3 we showed how to compute the determinant of an $n \times n$ matrix $\mathbf{A}$ using row operations. The process kept track of the effects of the row operations on the value of the determinant until we obtained an upper triangular matrix $\mathbf{U}$ that had the same size as $\mathbf{A}$. The final step computed $\det(\mathbf{U})$ as the product of its diagonal entries.

**Determinants, RECURSIVELY**

$$\mathbf{A} = \begin{bmatrix} 9 & 6 & 7 & 2 \\ -8 & 0 & 3 & 1 \\ 5 & -4 & 0 & 2 \\ 6 & -2 & -7 & -1 \end{bmatrix}$$

$$\mathbf{M}(1,1) = \begin{bmatrix} 0 & 3 & 1 \\ -4 & 0 & 2 \\ -2 & -7 & -1 \end{bmatrix} \quad \mathbf{M}(2,1) = \begin{bmatrix} 6 & 7 & 2 \\ -4 & 0 & 2 \\ -2 & -7 & -1 \end{bmatrix}$$

$$\mathbf{M}(3,1) = \begin{bmatrix} 6 & 7 & 2 \\ 0 & 3 & 1 \\ -2 & -7 & -1 \end{bmatrix} \quad \mathbf{M}(4,1) = \begin{bmatrix} 6 & 7 & 2 \\ 0 & 3 & 1 \\ -4 & 0 & 2 \end{bmatrix}$$

$$\det(\mathbf{A}) = (9)*\det(\mathbf{M}(1,1)) - (-8)*\det(\mathbf{M}(2,1)) + (5)*\det(\mathbf{M}(3,1)) - (6)*\det(M(4,1))$$

From MATLAB routine, **detrecur.m, det(A)** in terms of determinants of submatrices with coefficients from the first column of **A**.

FIGURE 1

In this section we develop a recursive procedure for computing $\det(\mathbf{A})$. Here we show how to compute $\det(\mathbf{A})$ as a linear combination of the determinants of submatrices of size $(n - 1) \times (n - 1)$. (See Figure 1.) We can then repeat the process for these $(n - 1) \times (n - 1)$ matrices until we obtain $\det(\mathbf{A})$ as a linear combination of the determinants of $2 \times 2$ submatrices.[1] The final step computes the determinants of the $2 \times 2$ matrices directly using our "$2 \times 2$ device". This process

---

[1] Sometimes it is convenient to write $\det(\mathbf{A})$ as a linear combination of the determinants of $3 \times 3$ submatrices and then use the "$3 \times 3$ device".

is not as computationally efficient as that developed in Section 3.3, but it does lead to an important formula for the inverse of a matrix.

In Section 3.2 we indicated a recursive procedure for computing the determinants of $2 \times 2$ and $3 \times 3$ matrices. Here we give a formal development for square matrices of arbitrary size. We begin with several definitions and then state, without verification, the recursive procedure.

---

**Definition**    Let **A** be an $n \times n$ matrix. The $(i, j)$-**minor** of **A**, denoted $\mathbf{A}_{ij}$, is the $(n-1) \times (n-1)$ submatrix of **A** obtained by deleting its $i$th row and $j$th column.

---

EXAMPLE 1    Let
$$\mathbf{A} = \begin{bmatrix} 4 & 3 & 2 \\ 3 & -2 & 5 \\ 2 & 4 & 6 \end{bmatrix}.$$
Then there is a minor associated with each row-column position of **A**. The minors corresponding to the positions across the first row are
$$\mathbf{A}_{11} = \begin{bmatrix} -2 & 5 \\ 4 & 6 \end{bmatrix}, \quad \mathbf{A}_{12} = \begin{bmatrix} 3 & 5 \\ 2 & 6 \end{bmatrix}, \quad \text{and} \quad \mathbf{A}_{13} = \begin{bmatrix} 3 & -2 \\ 2 & 4 \end{bmatrix}.$$
The minors corresponding to the positions in the third column are
$$\mathbf{A}_{13} = \begin{bmatrix} 3 & -2 \\ 2 & 4 \end{bmatrix}, \quad \mathbf{A}_{23} = \begin{bmatrix} 4 & 3 \\ 2 & 4 \end{bmatrix}, \quad \text{and} \quad \mathbf{A}_{33} = \begin{bmatrix} 4 & 3 \\ 3 & -2 \end{bmatrix}. \quad ■$$

---

**Definition**    Let **A** be an $n \times n$ matrix. The $(i, j)$-**cofactor** of **A**, denoted $C_{ij}$, is the determinant of the $(i, j)$-minor of **A** times $(-1)^{i+j}$; $C_{ij} = (-1)^{i+j} \det(\mathbf{A}_{ij})$.

---

The numerical value of the $(i, j)$-cofactor of a matrix **A** is independent of the values of the entries in row $i$ and column $j$ of **A**. Hence different matrices can have the same value for their $(i, j)$-cofactor. However, there is a pattern to the signs associated with cofactors. See Exercise 21.

EXAMPLE 2    Use the information in Example 1 to compute the cofactors associated with the positions in the first row of **A**.
$$C_{11} = (-1)^{1+1} \det(\mathbf{A}_{11}) = \det\left(\begin{bmatrix} -2 & 5 \\ 4 & 6 \end{bmatrix}\right) = -12 - 20 = -32,$$

$$C_{12} = (-1)^{1+2} \det(\mathbf{A}_{12}) = (-1) \det\left(\begin{bmatrix} 3 & 5 \\ 2 & 6 \end{bmatrix}\right) = (-1)(18 - 10) = -8,$$

$$C_{13} = (-1)^{1+3} \det(\mathbf{A}_{13}) = \det\left(\begin{bmatrix} 3 & -2 \\ 2 & 4 \end{bmatrix}\right) = 12 - (-4) = 16. \quad ■$$

The **cofactor expansion for det(A) about the $i$th row** is
$$\det(\mathbf{A}) = \sum_{j=1}^{n} a_{ij} C_{ij}$$
$$= a_{i1}(-1)^{i+1} \det(\mathbf{A}_{i1}) + a_{i2}(-1)^{i+2} \det(\mathbf{A}_{i2})$$
$$+ \cdots + a_{in}(-1)^{i+n} \det(\mathbf{A}_{in}). \tag{1}$$

Thus to compute $\det(\mathbf{A})$ using (1) we choose a row of $\mathbf{A}$, compute the cofactors for that row, multiply each cofactor by the corresponding entry of $\mathbf{A}$, and add the resulting values.

EXAMPLE 3   Compute

$$\det\left(\begin{bmatrix} 4 & 3 & 2 \\ 3 & -2 & 5 \\ 2 & 4 & 6 \end{bmatrix}\right)$$

using cofactor expansion about the first row. Using (1) and Examples 1 and 2, we have

$$\det\left(\begin{bmatrix} 4 & 3 & 2 \\ 3 & 2 & 5 \\ 2 & 4 & 6 \end{bmatrix}\right) = 4C_{11} + 3C_{12} + 2C_{13}$$

$$= 4(-1)^{1+1}\det(\mathbf{A}_{11}) + 3(-1)^{1+2}\det(\mathbf{A}_{12})$$

$$+ 2(-1)^{1+3}\det(\mathbf{A}_{13})$$

$$= 4(-32) + 3(-8) + 2(16) = -120.$$

This result is the same as that obtained in Example 1 in Section 3.2, where we used row operations.                                                                                       ■

The cofactor expansion procedure works using any row or any column. The **cofactor expansion about the $k$th column of $\mathbf{A}$** is

$$\det(\mathbf{A}) = \sum_{j=1}^{n} a_{jk}C_{jk}$$

$$= a_{1k}(-1)^{1+k}\det(\mathbf{A}_{1k}) + a_{2k}(-1)^{2+k}\det(\mathbf{A}_{2k})$$

$$+ \cdots + a_{nk}(-1)^{n+k}\det(\mathbf{A}_{nk}).$$

$$(2)$$

The coefficients of the linear combinations in both (1) and (2) are entries of $\mathbf{A}$. It follows that the optimal course of action is to expand about a row or column that has the largest number of zeros, since we need not compute the $(i, j)$-cofactors if $a_{ij} = 0$.

EXAMPLE 4   Let

$$\mathbf{A} = \begin{bmatrix} 1 & 2 & -3 & 4 \\ -1 & 2 & 1 & 3 \\ 3 & 0 & 0 & -3 \\ 2 & 0 & 2 & 0 \end{bmatrix}.$$

To compute $\det(\mathbf{A})$ we choose to use cofactor expansion about column 2. We get

$$\det(\mathbf{A}) = 2C_{12} + 2C_{22} + 0C_{32} + 0C_{42}$$

$$= 2(-1)^3 \det\left(\begin{bmatrix} -1 & 1 & 3 \\ 3 & 0 & -3 \\ 2 & 2 & 0 \end{bmatrix}\right) + 2(-1)^4 \det\left(\begin{bmatrix} 1 & -3 & 4 \\ 3 & 0 & -3 \\ 2 & 2 & 0 \end{bmatrix}\right).$$

Using the cofactor expansion about row 2 for each of the $3 \times 3$ determinants, we have

$$\det(\mathbf{A}) = -2\left(3(-1)^3 \det\left(\begin{bmatrix} 1 & 3 \\ 2 & 0 \end{bmatrix}\right) + 0(-1)^4 \det\left(\begin{bmatrix} -1 & 3 \\ 2 & 0 \end{bmatrix}\right)\right.$$

$$\left. -3(-1)^5 \det\left(\begin{bmatrix} -1 & 1 \\ 2 & 2 \end{bmatrix}\right)\right)$$

$$+2\left(3(-1)^3 \det\left(\begin{bmatrix} -3 & 4 \\ 2 & 0 \end{bmatrix}\right) + 0(-1)^4 \det\left(\begin{bmatrix} 1 & 4 \\ 2 & 0 \end{bmatrix}\right)\right.$$

$$\left. -3(-1)^5 \det\left(\begin{bmatrix} 1 & -3 \\ 2 & 2 \end{bmatrix}\right)\right)$$

$$= -2(3(-1)(-6) - 3(-1)(-4)) + 2(3(-1)(-8) - 3(-1)(8))$$

$$= -2(6) + 2(48) = 84. \qquad ■$$

In Section 2.4 we showed how to compute the inverse of a matrix $\mathbf{A}$ using row operations. Next we show how to use cofactors to obtain a formula for $\mathbf{A}^{-1}$ in terms of the entries of $\mathbf{A}$.

---

**Definition**    Let $\mathbf{A}$ be an $n \times n$ matrix. The $n \times n$ matrix adj($\mathbf{A}$), called the **adjoint of A**, is the matrix whose $(i, j)$-entry is the $(j, i)$-cofactor of $\mathbf{A}$; ent$_{ij}$(adj($\mathbf{A}$)) $= C_{ji}$.

---

An alternative way to describe adj($\mathbf{A}$) is to say that it is the transpose of the matrix whose entries are the cofactors of $\mathbf{A}$.

EXAMPLE 5    Let

$$\mathbf{A} = \begin{bmatrix} 3 & -2 & 1 \\ 5 & 6 & 2 \\ 1 & 0 & -3 \end{bmatrix}.$$

Then adj($\mathbf{A}$) is obtained by first determining the cofactors of $\mathbf{A}$:

$$C_{11} = (-1)^{1+1} \det\left(\begin{bmatrix} 6 & 2 \\ 0 & -3 \end{bmatrix}\right) = -18,$$

$$C_{12} = (-1)^{1+2} \det\left(\begin{bmatrix} 5 & 2 \\ 1 & -3 \end{bmatrix}\right) = 17,$$

$$C_{13} = (-1)^{1+3} \det\left(\begin{bmatrix} 5 & 6 \\ 1 & 0 \end{bmatrix}\right) = -6;$$

$$C_{21} = (-1)^{2+1} \det\left(\begin{bmatrix} -2 & 1 \\ 0 & -3 \end{bmatrix}\right) = -6,$$

$$C_{22} = (-1)^{2+2} \det\left(\begin{bmatrix} 3 & 1 \\ 1 & -3 \end{bmatrix}\right) = -10,$$

$$C_{23} = (-1)^{2+3} \det\left(\begin{bmatrix} 3 & -2 \\ 1 & 0 \end{bmatrix}\right) = -2;$$

$$C_{31} = (-1)^{3+1} \det \left( \begin{bmatrix} -2 & 1 \\ 6 & 2 \end{bmatrix} \right) = -10,$$

$$C_{32} = (-1)^{3+2} \det \left( \begin{bmatrix} 3 & 1 \\ 5 & 2 \end{bmatrix} \right) = -1,$$

$$C_{33} = (-1)^{3+3} \det \left( \begin{bmatrix} 3 & -2 \\ 5 & 6 \end{bmatrix} \right) = 28.$$

Hence

$$\mathrm{adj}(\mathbf{A}) = \begin{bmatrix} C_{11} & C_{12} & C_{13} \\ C_{21} & C_{22} & C_{23} \\ C_{31} & C_{32} & C_{33} \end{bmatrix}^T = \begin{bmatrix} -18 & -6 & -10 \\ 17 & -10 & -1 \\ -6 & -2 & 28 \end{bmatrix}.$$ ∎

In Exercise 20 of Section 3.3 you were asked to show that a matrix with a pair of identical rows has a determinant of zero. The cofactor expansion in (1) can be expressed as expansion about the $i$th row as

$$\det(\mathbf{A}) = \sum_{j=1}^{n} a_{ij} C_{ij} = \mathrm{row}_i(\mathbf{A}) * \mathrm{col}_i(\mathrm{adj}(\mathbf{A})) \tag{3}$$

and about the $i$th column as

$$\det(\mathbf{A}) = \sum_{j=1}^{n} a_{ji} C_{ji} = \mathrm{row}_i(\mathrm{adj}(\mathbf{A})) * \mathrm{col}_i(\mathbf{A}). \tag{4}$$

If we take cofactors from row $k$ and entries from row $i$, $i \neq k$, then we have

$$\mathrm{row}_i(\mathbf{A}) * \mathrm{col}_k(\mathrm{adj}(\mathbf{A})) = \sum_{j=1}^{n} a_{ij} C_{kj}, \quad k \neq i. \tag{5}$$

The corresponding expression with cofactors from column $k$ and entries from column $i$, $i \neq k$, is

$$\mathrm{row}_k(\mathrm{adj}(\mathbf{A})) * \mathrm{col}_i(\mathbf{A}) = \sum_{j=1}^{n} a_{ji} C_{jk}, \quad k \neq i. \tag{6}$$

The expression in (5) corresponds to the determinant of a matrix with two identical rows (the $i$th and $k$th); hence the result in (5) is zero. Similarly, for the expression in (6), only now the matrix has two identical columns (the $i$th and $k$th). (See Exercises 16 through 20.) The information in (3) through (6) leads us to the following relationship:

$$\mathbf{A}\,\mathrm{adj}(\mathbf{A}) = \mathrm{adj}(\mathbf{A})\mathbf{A} = \det(\mathbf{A})\mathbf{I}_n. \tag{7}$$

If $\det(\mathbf{A}) \neq 0$, then we have

$$\frac{1}{\det(\mathbf{A})}\,\mathrm{adj}(\mathbf{A})\mathbf{A} = \mathbf{I}_n$$

or, equivalently,

$$\mathbf{A}^{-1} = \frac{1}{\det(\mathbf{A})} \operatorname{adj}(\mathbf{A}). \tag{8}$$

The expression in (8) provides a formula for the inverse of **A** in terms of the entries of **A**, in contrast to the computational procedure employing row operations given in Section 3.3. This expression is not computationally efficient, but it is useful in the development of certain expressions for applications.

EXAMPLE 6    Let

$$\mathbf{A} = \begin{bmatrix} 3 & -2 & 1 \\ 5 & 6 & 2 \\ 1 & 0 & -3 \end{bmatrix}.$$

Using the adjoint computed in Example 5 we obtain

$$\mathbf{A} \operatorname{adj}(\mathbf{A}) = \begin{bmatrix} -94 & 0 & 0 \\ 0 & -94 & 0 \\ 0 & 0 & -94 \end{bmatrix}$$

so from (7) we have $\det(\mathbf{A}) = -94$ and from (8)

$$\mathbf{A}^{-1} = \frac{1}{\det(\mathbf{A})} \operatorname{adj}(\mathbf{A}) = \frac{1}{-94} \begin{bmatrix} -18 & -6 & -10 \\ 17 & -10 & -1 \\ -6 & -2 & 28 \end{bmatrix}.$$ ■

EXERCISES 3.4

1. Let $\mathbf{A} = \begin{bmatrix} 2 & 0 & 1 \\ 1 & 3 & 4 \\ 5 & 1 & -2 \end{bmatrix}$.

   (a) Find the following minors: $\mathbf{A}_{31}, \mathbf{A}_{22},$ and $\mathbf{A}_{23}$.

   (b) Find the following cofactors: $C_{31}, C_{22},$ and $C_{23}$.

2. Let $\mathbf{A} = \begin{bmatrix} 0 & 2 & 1 & 3 \\ -2 & 4 & 0 & -1 \\ 5 & -7 & 3 & 6 \\ 8 & -4 & 0 & -2 \end{bmatrix}$.

   (a) Find the following minors: $\mathbf{A}_{11}, \mathbf{A}_{33}, \mathbf{A}_{24},$ and $\mathbf{A}_{14}$.

   (b) Find the following cofactors: $C_{11}, C_{33}, C_{24},$ and $C_{14}$.

In Exercises 3–5, use cofactor expansion to find the determinant of each given matrix.

3. $\mathbf{A} = \begin{bmatrix} -1 & -2 & 0 \\ 2 & 5 & 1 \\ 3 & 4 & -3 \end{bmatrix}$.

4. $\mathbf{A} = \begin{bmatrix} 4 & 2 & 1 \\ -2 & 1 & 0 \\ 2 & 3 & 1 \end{bmatrix}$.

5. $\mathbf{A} = \begin{bmatrix} i & 0 & 1 \\ 1 & 1+i & 2 \\ 2 & 0 & i \end{bmatrix}$.

In Exercises 6–8, use cofactor expansion to find the determinant of each given matrix.

6. $\mathbf{A} = \begin{bmatrix} 2 & 0 & 1 & 3 \\ -1 & 1 & 2 & 4 \\ 0 & 5 & 1 & 0 \\ 2 & 6 & 7 & -3 \end{bmatrix}$.

7. $\mathbf{A} = \begin{bmatrix} 2 & -1 & 0 & 0 \\ -1 & 2 & -1 & 0 \\ 0 & -1 & 2 & -1 \\ 0 & 0 & -1 & 2 \end{bmatrix}$.

8. $\mathbf{A} = \begin{bmatrix} 1 & i & 0 & -i \\ 1 & 0 & 1 & 0 \\ 0 & 1 & i & 0 \\ i & 0 & 0 & 1 \end{bmatrix}$.

In Exercises 9–11, find the adjoint of each given matrix.

9. $\mathbf{A} = \begin{bmatrix} 3 & 2 & 0 \\ 5 & -1 & 1 \\ 4 & 3 & -3 \end{bmatrix}$.

10. $\mathbf{A} = \begin{bmatrix} 6 & 2 & 8 \\ -3 & 4 & 1 \\ 4 & -4 & 5 \end{bmatrix}$.

11. $\mathbf{A} = \begin{bmatrix} i & 0 & 1+i \\ -i & 1 & 0 \\ 1 & i & 2 \end{bmatrix}$.

*In Exercises* 12–14, *use the results from Exercises* 9–11 *to find* $\det(\mathbf{A})$ *by computing* $\mathbf{A}\,\mathrm{adj}(\mathbf{A})$, *then find* $\mathbf{A}^{-1}$, *if it exists.*

12. $\mathbf{A}$ is the matrix in Exercise 9.

13. $\mathbf{A}$ is the matrix in Exercise 10.

14. $\mathbf{A}$ is the matrix in Exercise 11.

15. Let $\mathbf{A}$ be a $3 \times 3$ matrix with all integer entries.
    (a) Explain why $\det(\mathbf{A})$ is an integer.
    (b) Explain why $\mathrm{adj}(\mathbf{A})$ has all integer entries.
    (c) If $\mathbf{A}$ were nonsingular, under what conditions would $\mathbf{A}^{-1}$ have all integer entries?

16. Let $\mathbf{A}$ be a $3 \times 3$ matrix with minors
    $$\mathbf{A}_{11} = \begin{bmatrix} e & f \\ h & i \end{bmatrix}, \quad \mathbf{A}_{12} = \begin{bmatrix} d & f \\ g & i \end{bmatrix},$$
    and
    $$\mathbf{A}_{13} = \begin{bmatrix} d & e \\ g & h \end{bmatrix}.$$
    Determine as many entries of matrix $\mathbf{A}$ as you can.

17. Let $\mathbf{A}$ be a $3 \times 3$ matrix with $\mathrm{row}_2(\mathbf{A}) = \begin{bmatrix} 4 & 8 & -7 \end{bmatrix}$. Given the following minors, determine matrix $\mathbf{A}$.
    $$\mathbf{A}_{21} = \begin{bmatrix} 6 & 1 \\ 0 & 3 \end{bmatrix}, \quad \mathbf{A}_{22} = \begin{bmatrix} 2 & 1 \\ 9 & 3 \end{bmatrix}.$$

18. Given
    $$\det(\mathbf{A}) = 4C_{11} + 1C_{12} + 8C_{13}$$
    $$= 4(-1)^2 \det\left(\begin{bmatrix} 3 & -5 \\ 0 & 7 \end{bmatrix}\right)$$
    $$+ 1(-1)^3 \det\left(\begin{bmatrix} 2 & -5 \\ 6 & 7 \end{bmatrix}\right)$$
    $$+ 8(-1)^4 \det\left(\begin{bmatrix} 2 & 3 \\ 6 & 0 \end{bmatrix}\right).$$
    (a) Compute $\det(\mathbf{A})$.
    (b) Find $\mathbf{A}$.

19. Given
    $$\det(\mathbf{B}) = 2C_{11} + 3C_{12} - 5C_{13}$$
    $$= 2(-1)^2 \det\left(\begin{bmatrix} 3 & -5 \\ 0 & 7 \end{bmatrix}\right)$$
    $$+ 3(-1)^3 \det\left(\begin{bmatrix} 2 & -5 \\ 6 & 7 \end{bmatrix}\right)$$
    $$- 5(-1)^4 \det\left(\begin{bmatrix} 2 & 3 \\ 6 & 0 \end{bmatrix}\right).$$

(a) Compute $\det(\mathbf{B})$.
(b) Find $\mathbf{B}$.

20. In Exercises 18 and 19 the cofactors (and minors) used are the same. The expressions in Exercises 18 and 19 are different linear combinations of the same cofactors. The expression in Exercise 18 used coefficients from the first row of $\mathbf{A}$ and that in Exercise 19 used coefficients from the second row of $\mathbf{A}$.
    (a) Describe the major difference between matrices $\mathbf{A}$ and $\mathbf{B}$.
    (b) Complete the following conjecture:
        A linear combination of cofactors from $\mathrm{row}_i(\mathbf{A})(\mathrm{col}_i(\mathbf{A}))$ with coefficients from $\mathrm{row}_k(\mathbf{A})(\mathrm{col}_k(\mathbf{A}))$, $i \neq k$, has numerical value _____ since it represents the determinant of a matrix with _____ rows (columns).

21. From the definition of a cofactor of a matrix $\mathbf{A}$, there is a power of $(-1)$ attached to the determinant of the associated minor.
    (a) For a $3 \times 3$ matrix, construct a display of the sign pattern associated with the cofactors of the entries.
    (b) For a $4 \times 4$ matrix, construct a display of the sign pattern associated with the cofactors of the entries.

## In MATLAB

*To illustrate the recursive computation of a determinant in* MATLAB *type* **detrecur**. *You will be presented with a set of options for selecting a* $4 \times 4$ *matrix to expand in terms of its cofactors. Choose an input matrix and then follow the directions on the screen or press the help button for an overview of the aspects of this demonstration. This routine can be used for exploration of the ideas of this section on an individual or group basis and by instructors for demonstrating the concepts involved.*

ML. 1. Use **detrecur** to compute the determinant of
$$\mathbf{A} = \begin{bmatrix} 4 & 0 & 0 & 4 \\ 0 & 1 & -1 & 0 \\ 2 & 1 & 2 & 1 \\ 1 & 2 & -1 & 2 \end{bmatrix}$$
so that you need only compute the determinant of two of the minors.

ML. 2. Use **detrecur** to compute the determinant of
$$\mathbf{A} = \begin{bmatrix} 1 & 2 & 4 & 1 & 0 \\ 0 & 1 & 3 & 0 & 2 \\ -1 & -1 & 2 & 0 & 3 \\ 0 & 0 & 2 & 1 & 5 \\ -7 & 4 & 3 & 0 & -6 \end{bmatrix}$$
as efficiently as possible.

**ML. 3.** In MATLAB type **help cofactor** and carefully read the information displayed.

(a) Use **cofactor** to compute the cofactors of the elements in the second row of

$$A = \begin{bmatrix} 1 & 5 & 0 \\ 2 & -1 & 3 \\ 3 & 2 & 1 \end{bmatrix}.$$

(b) Find the determinant of

$$A = \begin{bmatrix} 4 & 0 & -1 \\ -2 & 2 & -1 \\ 0 & 4 & 3 \end{bmatrix}$$

using the recursive technique with the aid of routine **cofactor**.

(c) Find the determinant of

$$A = \begin{bmatrix} -1 & 2 & 0 & 0 \\ 2 & -1 & 2 & 0 \\ 0 & 2 & -1 & 2 \\ 0 & 0 & 2 & -1 \end{bmatrix}$$

using the recursive technique with the aid of routine **cofactor**.

**ML. 4.** In MATLAB type **help adjoint** and carefully read the information displayed. Use this routine to compute the adjoint of each of the following matrices.

(a) $A = \begin{bmatrix} 2 & 3 \\ -1 & 2 \end{bmatrix}.$

(b) $A = \begin{bmatrix} 1 & 2 & -3 \\ -4 & -5 & 2 \\ -1 & 1 & -7 \end{bmatrix}.$

(c) $A = \begin{bmatrix} -1 & 2 & 0 & 0 \\ 2 & -1 & 2 & 0 \\ 0 & 2 & -1 & 2 \\ 0 & 0 & 2 & -1 \end{bmatrix}.$

## True/False Review Questions

*Determine whether each of the following statements is true or false.*

**1.** $\text{adj}(I_3) = I_3$.

**2.** If $\det(A) = 2 \det(A_{11}) - 3 \det(A_{12}) = 2(7) - 3(5)$, then $A = \begin{bmatrix} 2 & 3 \\ 5 & 7 \end{bmatrix}$.

**3.** If $A_{11} = \begin{bmatrix} 5 & 4 \\ 2 & 1 \end{bmatrix}$ and $A_{12} = \begin{bmatrix} 6 & 4 \\ 3 & 1 \end{bmatrix}$, then $\text{row}_2(A) = \begin{bmatrix} 6 & 5 & 4 \end{bmatrix}$.

**4.** The $(4, 4)$ cofactor of $A$ is $\det(A_{44})$.

**5.** If $A$ is nonsingular, $\text{adj}(A) = \det(A)A^{-1}$.

**6.** If we replace each entry of $A$ by its cofactor, then the resulting matrix is $\text{adj}(A)$.

**7.** If $A = \begin{bmatrix} 5 & 4 & 0 & 1 \\ 3 & -2 & 0 & 3 \\ 6 & 1 & 1 & 8 \\ 7 & 2 & 0 & 9 \end{bmatrix}$, then $\det(A) = -124$.

**8.** If $A$ is symmetric, then the $(3, 1)$ minor is the same as the $(1, 3)$ minor.

**9.** If $A$ is $4 \times 4$, then using cofactor expansions we write $\det(A)$ as a linear combination of twelve $2 \times 2$ submatrices of $A$.

**10.** If $\text{adj}(A)$ has a zero row, then $A$ is nonsingular.

## Terminology

| Minor | Cofactor |
|---|---|
| Cofactor expansion | Adjoint |

This section describes a recursive procedure for computing the determinant of any square matrix.

- Describe how to find a minor of a matrix.
- If the matrix is $n \times n$, how many minors does it have?
- How is a cofactor related to a minor?
- For a $4 \times 4$ matrix, give a detailed description of its cofactor expansion about the third row.
- If $\mathbf{A}$ is an $8 \times 8$ matrix, describe a strategy for starting the computation of its determinant using cofactor expansion that will help shorten the computational steps.
- Describe the adjoint of a matrix.
- How is the adjoint used to determine a formula for the inverse of a matrix? Explain why the computation of the inverse using row operations is preferred over this formula.

## 3.5 ■ CRAMER'S RULE (OPTIONAL)

In Section 3.4 we developed the following formula for the inverse of a matrix $\mathbf{A}$ using the determinant and the adjoint:

$$\mathbf{A}^{-1} = \frac{1}{\det(\mathbf{A})} \operatorname{adj}(\mathbf{A}). \tag{1}$$

In this section we show how to use the result in (1) to develop another method for solving a nonsingular linear system. The method, known as **Cramer's rule**, provides a formula for each of the unknowns of the linear system in terms of determinants. Cramer's rule is not applicable to linear systems with singular coefficient matrices. Why?

Let $\mathbf{Ax} = \mathbf{b}$ be a nonsingular linear system of $n$ equations in $n$ unknowns. Next let $\mathbf{B}_i$ denote the $n \times n$ matrix obtained from $\mathbf{A}$ by replacing its $i$th column by the right side $\mathbf{b}$:

$$\mathbf{B}_i = \begin{bmatrix} \operatorname{col}_1(\mathbf{A}) & \operatorname{col}_2(\mathbf{A}) & \cdots & \operatorname{col}_{i-1}(\mathbf{A}) & \mathbf{b} & \operatorname{col}_{i+1}(\mathbf{A}) & \cdots & \operatorname{col}_n(\mathbf{A}) \end{bmatrix}. \tag{2}$$

Using the cofactor expansion of $\mathbf{B}_i$ about its $i$th column, we have

$$\det(\mathbf{B}_i) = b_1 C_{1i} + b_2 C_{2i} + \cdots + b_n C_{ni} = \sum_{j=1}^{n} b_j C_{ji}. \tag{3}$$

Since $\mathbf{A}$ is nonsingular, using (1) we have that the solution of the linear system is given by

$$\mathbf{x} = \mathbf{A}^{-1}\mathbf{b} = \frac{1}{\det(\mathbf{A})} \operatorname{adj}(\mathbf{A})\mathbf{b} = \frac{1}{\det(\mathbf{A})} \begin{bmatrix} C_{11} & C_{21} & \cdots & C_{n1} \\ C_{12} & C_{22} & \cdots & C_{n2} \\ \vdots & \vdots & \vdots & \vdots \\ C_{1n} & C_{2n} & \cdots & C_{nn} \end{bmatrix} \begin{bmatrix} b_1 \\ b_2 \\ \vdots \\ b_n \end{bmatrix}.$$

It follows that

$$x_i = \frac{1}{\det(\mathbf{A})} \operatorname{row}_i(\operatorname{adj}(\mathbf{A})) \cdot \mathbf{b} = \frac{1}{\det(\mathbf{A})} \begin{bmatrix} C_{1i} & C_{2i} & \cdots & C_{ni} \end{bmatrix} \begin{bmatrix} b_1 \\ b_2 \\ \vdots \\ b_n \end{bmatrix}$$

$$= \frac{1}{\det(\mathbf{A})} \sum_{j=1}^{n} b_j C_{ji} = \frac{\det(\mathbf{B}_i)}{\det(\mathbf{A})}$$

where in the last step we used (3). Hence the solution of the linear system $\mathbf{Ax} = \mathbf{b}$ is given by

$$x_1 = \frac{\det(\mathbf{B}_1)}{\det(\mathbf{A})}, \quad x_2 = \frac{\det(\mathbf{B}_2)}{\det(\mathbf{A})}, \quad \ldots, \quad x_n = \frac{\det(\mathbf{B}_n)}{\det(\mathbf{A})}. \quad (4)$$

We refer to (4) as Cramer's rule.

EXAMPLE 1    To solve the linear system $\mathbf{Ax} = \mathbf{b}$ where

$$\mathbf{A} = \begin{bmatrix} -2 & -1 & 1 \\ -2 & 3 & -1 \\ 1 & 2 & -1 \end{bmatrix} \quad \text{and} \quad \mathbf{b} = \begin{bmatrix} 3 \\ 1 \\ 4 \end{bmatrix}$$

using Cramer's rule, we first form the matrices $\mathbf{B}_1$, $\mathbf{B}_2$, and $\mathbf{B}_3$ as shown in (2):

$$\mathbf{B}_1 = \begin{bmatrix} 3 & -1 & 1 \\ 1 & 3 & -1 \\ 4 & 2 & 1 \end{bmatrix}, \quad \mathbf{B}_2 = \begin{bmatrix} -2 & 3 & 1 \\ -2 & 1 & -1 \\ 1 & 4 & -1 \end{bmatrix}, \quad \mathbf{B}_3 = \begin{bmatrix} -2 & -1 & 3 \\ -2 & 3 & 1 \\ 1 & 2 & 4 \end{bmatrix}.$$

Next we compute $\det(\mathbf{A})$, $\det(\mathbf{B}_1)$, $\det(\mathbf{B}_2)$, and $\det(\mathbf{B}_3)$, obtaining (verify)

$$\det(\mathbf{A}) = -2, \quad \det(\mathbf{B}_1) = -10, \quad \det(\mathbf{B}_2) = -24, \quad \text{and} \quad \det(\mathbf{B}_3) = -50.$$

Hence from Cramer's rule in (4) we have

$$x_1 = \frac{\det(\mathbf{B}_1)}{\det(\mathbf{A})} = \frac{-10}{-2} = 5, \quad x_2 = \frac{\det(\mathbf{B}_2)}{\det(\mathbf{A})} = \frac{-24}{-2} = 12,$$

$$x_3 = \frac{\det(\mathbf{B}_3)}{\det(\mathbf{A})} = \frac{-50}{-2} = 25.$$ ■

Cramer's rule is not computationally efficient for systems larger than $3 \times 3$. It also has the major drawback that it is only applicable to nonsingular linear systems. In contrast, the row reduction techniques developed in Chapter 2 are applicable to any linear system.

## EXERCISES 3.5

1. If possible, solve the following linear system using Cramer's rule.

$$2x_1 + 4x_2 + 6x_3 = 2$$
$$x_1 \qquad + 2x_3 = 0$$
$$2x_1 + 3x_2 - x_3 = -5.$$

2. If possible, solve the following linear system using Cramer's rule.

$$2x_1 + x_2 - x_3 = 4$$
$$x_1 + 2x_2 + 3x_3 = -8$$
$$-x_1 + 2x_2 + x_3 = 16.$$

3. If possible, solve the following linear system using Cramer's rule.

$$2x_1 - x_2 + 3x_3 = 0$$
$$x_1 + 2x_2 - 3x_3 = 0$$
$$4x_1 + 2x_2 + x_3 = 0.$$

4. Solve the following linear system for $x_3$ using Cramer's rule.

$$2x_1 + x_2 + x_3 = 6$$
$$3x_1 + 2x_2 - 2x_3 = -2$$
$$x_1 + x_2 + 2x_3 = -4.$$

5. In Example 5 in Section 1.1 we had considered the following situation. The director of a trust fund has $100,000 to invest. The rules of the trust state that both a certificate of deposit (CD) and a long-term bond must be used. The director's goal is to have the trust yield $7800 on its investments for the year. The CD chosen returns 5% per annum and the bond 9%. The director determines the amount $x$ to invest in the CD and the amount $y$ to invest in the bond using a linear system:

$$x + y = 100{,}000$$
$$0.05x + 0.09y = 7{,}800.$$

Use Cramer's rule to determine $y$, the amount invested in the bond.

**6.** In Example 4 in Section 1.1 we were to use the linear system

$$a + b + c = -5$$
$$a - b + c = 1$$
$$4a + 2b + c = 7$$

to determine the coefficients $a$, $b$, and $c$ of a quadratic polynomial $p(x) = ax^2 + bx + c$ that went through the three points $(1, -5), (-1, 1), (2, 7)$.

(a) Use Cramer's rule to determine coefficient $b$ of polynomial $p(x)$.

(b) Use Cramer's rule to determine coefficient $c$ of polynomial $p(x)$.

**7.** An inheritance of \$24,000 is divided into three parts, with the second part twice as much as the first. The three parts are invested and earn interest at the annual rates of 9%, 10%, and 6%, respectively, and return \$2210 at the end of the first year. Use Cramer's rule to determine the part of the inheritance that is invested at a rate of 10%.

**8.** Solve the linear system $\mathbf{Ax} = \mathbf{b}$ for $x_2$ where

$$\mathbf{A} = \begin{bmatrix} 1 & 2 & -5 & 4 \\ 0 & 1 & 6 & 1 \\ 2 & 0 & 1 & 0 \\ 0 & 1 & 1 & 1 \end{bmatrix} \quad \text{and} \quad \mathbf{b} = \begin{bmatrix} 0 \\ 2 \\ 0 \\ 0 \end{bmatrix}$$

**9.** Four items were purchased at a hardware store and the receipt was lost. In order to get reimbursed by his company, an employee needs to determine the total cost. He recalls that cost of items 1 and 2 were the same, item 3 cost \$2 more than item 1, and item 4 cost \$5 less that the sum of the costs of items 2 and 3.

(a) Let $T$ be the total cost of the four items. Construct a linear system of equations that represents the situation given.

(b) After completing part (a), the employee recalls that the cost of the fourth item was \$10. Use Cramer's rule with the system in part (a) to determine $T$.

**10.** Use Cramer's rule to solve the following system of equations.

$$\frac{3}{x} - \frac{5}{y} + \frac{2}{z} = -5$$
$$\frac{2}{x} + \frac{3}{y} + \frac{1}{z} = 4$$
$$\frac{4}{x} - \frac{6}{y} + \frac{3}{z} = -7.$$

**11.** Determine all possible values of $a$, $b$, and $c$ so that the following linear system has a unique solution.

$$3x_1 \qquad + ax_3 = 2$$
$$-2x_1 + bx_2 \qquad = 1$$
$$cx_1 \qquad + x_3 = 2.$$

**12.** If possible, solve the following linear system using Cramer's rule. (Here $i = \sqrt{-1}$.)

$$x_1 + x_2 + ix_3 = 1$$
$$ix_1 \qquad + x_3 = 0$$
$$x_2 + ix_3 = i.$$

## True/False Review Questions

*Determine whether each of the following statements is true or false.*

**1.** Cramer's rule will find all the solutions of any square system of equations.

**2.** If $\mathbf{A} = \begin{bmatrix} 2 & -2 & a \\ 2 & -2 & b \\ 1 & 0 & c \end{bmatrix}$, where $a$, $b$, and $c$ are real numbers so that $\det(\mathbf{A}) \neq 0$, then linear system $\mathbf{Ax} = \begin{bmatrix} 0 \\ 0 \\ 1 \end{bmatrix}$ has $x_3 = 0$.

**3.** The linear system $\begin{bmatrix} 1 & 0 & 1 \\ 1 & 1 & -1 \\ 0 & 1 & 1 \end{bmatrix} \mathbf{x} = \begin{bmatrix} 1 \\ 1 \\ 1 \end{bmatrix}$ has a unique solution in which $x_2 = 2$.

**4.** If nonsingular system $\mathbf{Ax} = \mathbf{b}$ is such that $\text{col}_2(\mathbf{A}) = \mathbf{b}$, then $x_2 = 1$.

**5.** If nonsingular system $\mathbf{Ax} = \mathbf{b}$ is such that $\text{col}_2(\mathbf{A}) = \mathbf{b}$, then $x_1 = 0$.

## Terminology

### Cramer's Rule

This section provided a method to solve a nonsingular linear system using determinants, rather than row operations directly.

- Describe the process used in Cramer's rule.
- Can we use Cramer's rule to solve a rectangular system of equations? Explain.
- Explain why Cramer's rule is not computationally efficient for large linear systems.
- For certain linear systems we need the values of a few of the system's variables. Explain why Cramer's rule appears to be attractive for such situations.

## 3.6 ■ THE CROSS PRODUCT (OPTIONAL)

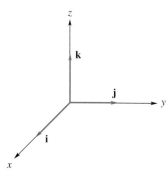

FIGURE 1

In this section we discuss an operation on vectors that is meaningful only in $R^3$. We show how to determine a vector that is orthogonal to a given pair of vectors using the determinant as a computational aid. This calculation arises in a number of situations involving areas, volumes, planes, and distances between objects in $R^3$.

In Section 1.5 we introduced the vectors

$$\mathbf{i} = \begin{bmatrix} 1 & 0 & 0 \end{bmatrix}, \quad \mathbf{j} = \begin{bmatrix} 0 & 1 & 0 \end{bmatrix}, \quad \text{and} \quad \mathbf{k} = \begin{bmatrix} 0 & 0 & 1 \end{bmatrix}$$

which are unit vectors that are mutually orthogonal in $R^3$. That is,

$$\|\mathbf{i}\| = \|\mathbf{j}\| = \|\mathbf{k}\| = 1, \quad \mathbf{i}\cdot\mathbf{j} = 0, \quad \mathbf{j}\cdot\mathbf{k} = 0, \quad \text{and} \quad \mathbf{k}\cdot\mathbf{i} = 0.$$

These vectors are depicted in Figure 1. In Section 2.3 we showed that $\{\mathbf{i}, \mathbf{j}, \mathbf{k}\}$ formed a basis for $R^3$; thus any vector in $R^3$ can be expressed uniquely as a linear combination of $\mathbf{i}$, $\mathbf{j}$, and $\mathbf{k}$. In addition we recognize these three vectors as the rows of $\mathbf{I}_3$.

---

**Definition**    For vectors $\mathbf{u}$ and $\mathbf{v}$ in $R^3$ given by

$$\mathbf{u} = \begin{bmatrix} u_1 & u_2 & u_3 \end{bmatrix} = u_1\mathbf{i} + u_2\mathbf{j} + u_3\mathbf{k} \quad \text{and}$$
$$\mathbf{v} = \begin{bmatrix} v_1 & v_2 & v_3 \end{bmatrix} = v_1\mathbf{i} + v_2\mathbf{j} + v_3\mathbf{k}$$

their **cross product**, denoted $\mathbf{u} \times \mathbf{v}$, is the vector

$$\mathbf{u} \times \mathbf{v} = (u_2 v_3 - u_3 v_2)\mathbf{i} + (u_3 v_1 - u_1 v_3)\mathbf{j} + (u_1 v_2 - u_2 v_1)\mathbf{k}. \qquad (1)$$

---

Carefully inspecting (1) and using the $2 \times 2$ device for determinants, we have

$$\mathbf{u} \times \mathbf{v} = \det\left( \begin{bmatrix} u_2 & u_3 \\ v_2 & v_3 \end{bmatrix} \right)\mathbf{i} - \det\left( \begin{bmatrix} u_1 & u_3 \\ v_1 & v_3 \end{bmatrix} \right)\mathbf{j} + \det\left( \begin{bmatrix} u_1 & u_2 \\ v_1 & v_2 \end{bmatrix} \right)\mathbf{k} \qquad (2)$$

which is the pattern of a cofactor expansion of the matrix

$$\begin{bmatrix} \mathbf{i} & \mathbf{j} & \mathbf{k} \\ u_1 & u_2 & u_3 \\ v_1 & v_2 & v_3 \end{bmatrix} \qquad (3)$$

about the first row. However, the matrix in (3) is a bit unusual since its first row consists of vectors, while rows 2 and 3 contain scalars. For our purposes we consider it a data structure that provides a way to obtain the cross product of a pair of vectors in terms of a determinant. Thus for computations we use the expression

$$\mathbf{u} \times \mathbf{v} = \det\left( \begin{bmatrix} \mathbf{i} & \mathbf{j} & \mathbf{k} \\ u_1 & u_2 & u_3 \\ v_1 & v_2 & v_3 \end{bmatrix} \right) \qquad (4)$$

as a shortcut to obtain the cross product of vectors $\mathbf{u}$ and $\mathbf{v}$.

EXAMPLE 1    Let $\mathbf{u} = 3\mathbf{i} + \mathbf{j} + 4\mathbf{k}$ and $\mathbf{v} = 2\mathbf{i} - \mathbf{j} - \mathbf{k}$. Then

$$\mathbf{u} \times \mathbf{v} = \det\left(\begin{bmatrix} \mathbf{i} & \mathbf{j} & \mathbf{k} \\ 3 & 1 & 4 \\ 2 & -1 & -1 \end{bmatrix}\right).$$

Using our $3 \times 3$ device for determinants we have

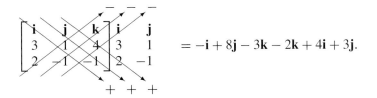

$$= -\mathbf{i} + 8\mathbf{j} - 3\mathbf{k} - 2\mathbf{k} + 4\mathbf{i} + 3\mathbf{j}.$$

Collecting like terms gives

$$\mathbf{u} \times \mathbf{v} = 3\mathbf{i} + 11\mathbf{j} - 5\mathbf{k}.$$ ■

Using (4) and properties of the determinant from Table 1 of Section 3.2 along with the Exercises in Section 3.3, we can show (see Exercise 4) that the cross product has the properties listed in Table 1.

TABLE 1

1. If $\mathbf{u}$ or $\mathbf{v}$ is the zero vector, then $\mathbf{u} \times \mathbf{v} = \mathbf{0}$.
2. If $\mathbf{u}$ and $\mathbf{v}$ are parallel, then $\mathbf{u} \times \mathbf{v} = \mathbf{0}$.
3. $\mathbf{u} \times \mathbf{u} = \mathbf{0}$.
4. $\mathbf{u} \times \mathbf{v} = -(\mathbf{v} \times \mathbf{u})$.
5. $c(\mathbf{u} \times \mathbf{v}) = (c\mathbf{u}) \times \mathbf{v} = \mathbf{u} \times (c\mathbf{v})$.

EXAMPLE 2    The vectors $\mathbf{i}$, $\mathbf{j}$, and $\mathbf{k}$ have the following properties. From Property 3 in Table 1, $\mathbf{i} \times \mathbf{i} = \mathbf{j} \times \mathbf{j} = \mathbf{k} \times \mathbf{k} = \mathbf{0}$. Using (3), we have

$$\mathbf{i} \times \mathbf{j} = \mathbf{k}, \quad \mathbf{j} \times \mathbf{k} = \mathbf{i}, \quad \mathbf{k} \times \mathbf{i} = \mathbf{j}. \tag{5}$$

From Property 4 in Table 1 we also have

$$\mathbf{j} \times \mathbf{i} = -\mathbf{k}, \quad \mathbf{k} \times \mathbf{j} = -\mathbf{i}, \quad \mathbf{i} \times \mathbf{k} = -\mathbf{j}. \tag{6}$$ ■

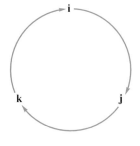

FIGURE 2

The results in (5) and (6) can be remembered by using Figure 2. Moving around the circle in a clockwise direction, we see that the cross product of the two vectors taken in the order indicated by the arrows is the third vector; moving in the counter clockwise direction, we see that the cross product taken in the opposite order from the arrows is the negative of the third vector.

Next we investigate geometric properties of the cross product. From (2) we have

$$(\mathbf{u} \times \mathbf{v}) \cdot \mathbf{w} = \left( \det \left( \begin{bmatrix} u_2 & u_3 \\ v_2 & v_3 \end{bmatrix} \right) \mathbf{i} - \det \left( \begin{bmatrix} u_1 & u_3 \\ v_1 & v_3 \end{bmatrix} \right) \mathbf{j} \right.$$

$$\left. + \det \left( \begin{bmatrix} u_1 & u_2 \\ v_1 & v_2 \end{bmatrix} \right) \mathbf{k} \right) \cdot (w_1 \mathbf{i} + w_2 \mathbf{j} + w_3 \mathbf{k})$$

$$= w_1 \det \left( \begin{bmatrix} u_2 & u_3 \\ v_2 & v_3 \end{bmatrix} \right) - w_2 \det \left( \begin{bmatrix} u_1 & u_3 \\ v_1 & v_3 \end{bmatrix} \right)$$

$$+ w_3 \det \left( \begin{bmatrix} u_1 & u_2 \\ v_1 & v_2 \end{bmatrix} \right).$$

It follows by direct calculation or from the pattern for cofactor expansion that the preceding expression can be written as

$$(\mathbf{u} \times \mathbf{v}) \cdot \mathbf{w} = \det \left( \begin{bmatrix} u_1 & u_2 & u_3 \\ v_1 & v_2 & v_3 \\ w_1 & w_2 & w_3 \end{bmatrix} \right). \tag{7}$$

From (7) and properties of determinants it follows that vector $\mathbf{u} \times \mathbf{v}$ is orthogonal to both $\mathbf{u}$ and $\mathbf{v}$; that is,

$$(\mathbf{u} \times \mathbf{v}) \cdot \mathbf{u} = 0 \quad \text{and} \quad (\mathbf{u} \times \mathbf{v}) \cdot \mathbf{v} = 0. \tag{8}$$

(See Exercise 8.)

EXAMPLE 3    Let $\mathbf{u}$ and $\mathbf{v}$ be as in Example 1 and let $\mathbf{n} = \mathbf{u} \times \mathbf{v}$ be as computed in Example 1. Then

$$\mathbf{u} \cdot \mathbf{n} = \begin{bmatrix} 3 & 1 & 4 \end{bmatrix} \cdot \begin{bmatrix} 3 & 11 & -5 \end{bmatrix} = 0$$

and

$$\mathbf{v} \cdot \mathbf{n} = \begin{bmatrix} 2 & -1 & -1 \end{bmatrix} \cdot \begin{bmatrix} 3 & 11 & -5 \end{bmatrix} = 0. \qquad ■$$

It also follows that if $\mathbf{w}$ is in span$\{\mathbf{u}, \mathbf{v}\}$ then $\mathbf{n} = \mathbf{u} \times \mathbf{v}$ is orthogonal to $\mathbf{w}$. (See Exercise 9.) Geometrically span$\{\mathbf{u}, \mathbf{v}\}$ is a plane in $R^3$ so $\mathbf{n}$ is orthogonal to this plane; see Figure 3. A vector orthogonal to a plane is called a **normal** to the plane, and we call such a vector a **normal vector**. Thus the cross product of $\mathbf{u}$ and $\mathbf{v}$ yields a normal vector to the plane spanned by $\mathbf{u}$ and $\mathbf{v}$.

The length of $\mathbf{u} \times \mathbf{v}$ is related to the parallelogram determined by vectors $\mathbf{u}$ and $\mathbf{v}$. From (1) we have the following after expanding and collecting terms:

$$\|\mathbf{u} \times \mathbf{v}\|^2 = (u_2 v_3 - u_3 v_2)^2 + (u_3 v_1 - u_1 v_3)^2 + (u_1 v_2 - u_2 v_1)^2$$

$$= (u_1^2 + u_2^2 + u_3^2)(v_1^2 + v_2^2 + v_3^2) - (u_1 v_1 + u_2 v_2 + u_3 v_3)^2 \tag{9}$$

$$= \|\mathbf{u}\| \|\mathbf{v}\|^2 - (\mathbf{u} \cdot \mathbf{v})^2.$$

Since $\mathbf{u} \cdot \mathbf{v} = \|\mathbf{u}\| \|\mathbf{v}\| \cos(\theta)$, where $\theta$ is the angle between $\mathbf{u}$ and $\mathbf{v}$ with $0 \le \theta \le \pi$, we get

$$\|\mathbf{u} \times \mathbf{v}\|^2 = \|\mathbf{u}\|^2 \|\mathbf{v}\|^2 - \|\mathbf{u}\|^2 \|\mathbf{v}\|^2 \cos^2(\theta)$$

$$= \|\mathbf{u}\|^2 \|\mathbf{v}\|^2 (1 - \cos^2(\theta))$$

$$= \|\mathbf{u}\|^2 \|\mathbf{v}\|^2 \sin^2(\theta)$$

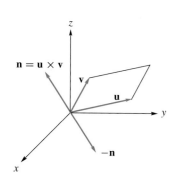

FIGURE 3

and thus

$$\|\mathbf{u} \times \mathbf{v}\| = \|\mathbf{u}\| \, \|\mathbf{v}\| \sin(\theta).$$

From Exercise 29 in Section 1.5 we know that $\|\mathbf{u}\| \, \|\mathbf{v}\| \sin(\theta)$ is the area of the parallelogram determined by $\mathbf{u}$ and $\mathbf{v}$. Hence the length of the cross product is the area of the parallelogram determined by the vectors $\mathbf{u}$ and $\mathbf{v}$ and $\frac{1}{2}\|\mathbf{u} \times \mathbf{v}\|$ is the area of the triangle determined by $\mathbf{u}$ and $\mathbf{v}$.

The volume $V$ of a parallelepiped whose edges are determined by vectors $\mathbf{u}$, $\mathbf{v}$, and $\mathbf{w}$ is the area of its base times its height; see Figure 4. In Figure 4 the base is determined by $\mathbf{u}$ and $\mathbf{v}$ and the height $h$ is the (perpendicular) distance from the tip of $\mathbf{w}$ to the plane determined by $\mathbf{u}$ and $\mathbf{v}$. We have

$$\text{area of base} = \|\mathbf{u} \times \mathbf{v}\|$$
$$\text{height} = h = |\cos(\theta)| \, \|\mathbf{w}\|$$

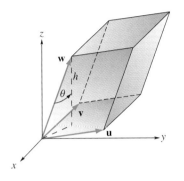

FIGURE 4

and thus

$$\text{volume} = V = \|\mathbf{u} \times \mathbf{v}\| \, \|\mathbf{w}\| \, |\cos(\theta)|.$$

Since vector $\mathbf{u} \times \mathbf{v}$ is orthogonal to the plane determined by $\mathbf{u}$ and $\mathbf{v}$, $\theta$ is also the angle between $\mathbf{u} \times \mathbf{v}$ and $\mathbf{w}$. Because a dot product is equal to the product of the lengths of the vectors times the cosine of the angle between them, we have from the preceding expression that

$$V = |(\mathbf{u} \times \mathbf{v}) \cdot \mathbf{w}| \qquad (10)$$

Then from (7) we have

$$V = \left| \det\left( \begin{bmatrix} u_1 & u_2 & u_3 \\ v_1 & v_2 & v_3 \\ w_1 & w_2 & w_3 \end{bmatrix} \right) \right| \qquad (11)$$

which gives the volume of the parallelepiped as a determinant with entries from the vectors determining its edges.

### Application:   Planes and Projections from a Plane

In $R^3$ three noncollinear points $P(x_1, y_1, z_1)$, $Q(x_2, y_2, z_2)$, and $R(x_3, y_3, z_3)$ determine a plane. In order to use easily the correspondence between vectors and points, we make the following definition.

---

**Definition**   The vector $\mathbf{u}$ that corresponds to the directed line segment with initial point at $P(x_1, y_1, z_1)$ and terminal point at $Q(x_2, y_2, z_2)$ with the same length as the line segment from $P$ to $Q$ is

$$\mathbf{u} = \begin{bmatrix} u_1 & u_2 & u_3 \end{bmatrix} = (x_2 - x_1)\mathbf{i} + (y_2 - y_1)\mathbf{j} + (z_2 - z_1)\mathbf{k}.$$

---

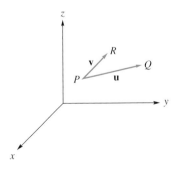

FIGURE 5

Let $\mathbf{u}$ the vector that corresponds to the segment from $P$ to $Q$ and

$$\mathbf{v} = \begin{bmatrix} v_1 & v_2 & v_3 \end{bmatrix} = (x_3 - x_1)\mathbf{i} + (y_3 - y_1)\mathbf{j} + (z_3 - z_1)\mathbf{k}$$

be the vector that corresponds to the segment from $P$ to $R$; see Figure 5. The plane determined by points $P$, $Q$, and $R$ is the same as that determined by vectors $\mathbf{u}$ and $\mathbf{v}$. A point $S(x, y, z)$ will lie in the plane determined by $P$, $Q$ and $R$ provided the vector $\mathbf{w}$ corresponding to the line segment from $P$ to $S$ is a linear combination of vectors $\mathbf{u}$ and $\mathbf{v}$. That is, $\mathbf{w} = (x - x_1)\mathbf{i} + (y - y_1)\mathbf{j} + (z - z_1)\mathbf{k}$ is in span$\{\mathbf{u}, \mathbf{v}\}$.

We have shown that $\mathbf{n} = \mathbf{u} \times \mathbf{v}$ is orthogonal to the plane determined by $\mathbf{u}$ and $\mathbf{v}$ and hence orthogonal to any vector in span$\{\mathbf{u}, \mathbf{v}\}$. It follows that if $\mathbf{w} \cdot \mathbf{n} = 0$, then $\mathbf{w}$ is in span$\{\mathbf{u}, \mathbf{v}\}$, and hence point $S$ is in the plane determined by points $P$, $Q$, and $R$. Let $\mathbf{n} = \mathbf{u} \times \mathbf{v}$, the normal to the plane determined by $\mathbf{u}$ and $\mathbf{v}$ be $\mathbf{n} = a\mathbf{i} + b\mathbf{j} + c\mathbf{k}$. Then

$$\mathbf{w} \cdot \mathbf{n} = 0 \Leftrightarrow a(x - x_1) + b(y - y_1) + c(z - z_1) = 0 \Leftrightarrow ax + by + cz = d$$

where $d = ax_1 + by_1 + cz_1$. The expression

$$ax + by + cz = d \tag{12}$$

for scalars $a$, $b$, $c$, and $d$ provides an equation for a plane in $R^3$ whose normal vector is $\mathbf{n} = \begin{bmatrix} a & b & c \end{bmatrix}$.

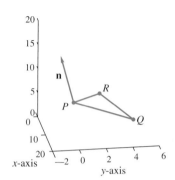

FIGURE 6

**EXAMPLE 4**    Determine an equation of the plane through points $P(2, 1, 2)$, $Q(0, 3, 4)$, and $R(5, 5, 0)$. The vector that corresponds to segment $PQ$ is $\mathbf{u} = \begin{bmatrix} -2 & 2 & 2 \end{bmatrix}$ and the vector that corresponds to segment $PR$ is $\mathbf{v} = \begin{bmatrix} 3 & 4 & -2 \end{bmatrix}$. (See Figure 6.) Then the normal to the plane determined by $P$, $Q$, and $R$ is

$$\mathbf{n} = \mathbf{u} \times \mathbf{v} = \begin{bmatrix} -12 & 2 & -14 \end{bmatrix}.$$

Thus an equation of the plane is determined as

$$\mathbf{n} \cdot ((x - 2)\mathbf{i} + (y - 1)\mathbf{j} + (z - 2)\mathbf{k}) = 0$$
$$\Leftrightarrow \quad -12(x - 2) + 2(y - 1) - 14(z - 2) = 0$$
$$\Leftrightarrow \quad -12x + 2y - 14z = -50. \qquad ■$$

Whenever we have an equation of a plane in the form in (12), we immediately can determine its normal vector from the coefficients of $x$, $y$, and $z$. This information provides us with the orientation of the plane, and by choosing values of $x$, $y$, and $z$ that satisfy (12) we can determine points that lie in the plane. Hence we can generate a sketch of the plane.

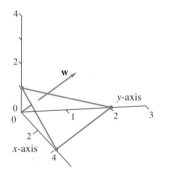

FIGURE 7

**EXAMPLE 5**    Sketch the plane $2x + 3y + 6z = 6$. We have that the normal vector is $\mathbf{n} = \begin{bmatrix} 2 & 3 & 6 \end{bmatrix}$. It follows that points $(3, 0, 0)$, $(0, 2, 0)$, and $(0, 0, 1)$ lie in the plane. With this information we can make the sketch that appears in Figure 7. As an aid for seeing how the plane is positioned, the vector w goes through the origin and is parallel to $\mathbf{n}$. ■

A pair of nonparallel planes $P_1$ and $P_2$ intersect in a line as in Figure 8. The angle of intersection of the planes is denoted by $\theta$ and is taken to be the acute angle between their normal vectors $\mathbf{n}_1$ and $\mathbf{n}_2$.

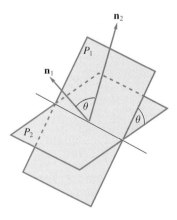

FIGURE 8

**EXAMPLE 6**    Determine the angle $\theta$ between planes $P_1$, given by $3x + 3y - z = 5$, and $P_2$, given by $-2x + 4y + 3z = 12$. From (12) we have that their respective normals are $\mathbf{n}_1 = \begin{bmatrix} 3 & 3 & -1 \end{bmatrix}$, and $\mathbf{n}_2 = \begin{bmatrix} -2 & 4 & 3 \end{bmatrix}$. Then

$$\cos(\theta) = \frac{\mathbf{n}_1 \cdot \mathbf{n}_2}{\|\mathbf{n}_1\| \, \|\mathbf{n}_2\|} = \frac{3}{\sqrt{19}\sqrt{29}} \approx 0.1278$$

and

$$\theta \approx \arccos(0.1278) \approx 1.4426 \text{ rad} \approx 82.6°. \qquad ■$$

In Section 1.5 we found that the projection $\mathbf{p}$ of a vector $\mathbf{v}$ onto a vector $\mathbf{w}$ is given by

$$\mathbf{p} = \text{proj}_{\mathbf{w}}\mathbf{v} = \frac{\mathbf{w}\cdot\mathbf{v}}{\|\mathbf{w}\|^2}\mathbf{w}$$

and it follows that

$$\|\mathbf{p}\| = \|\text{proj}_{\mathbf{w}}\mathbf{v}\| = \|\mathbf{v}\|\,|\cos(\theta)| \tag{13}$$

where $\theta$ is the angle between vectors $\mathbf{v}$ and $\mathbf{w}$. From (13) we see that the length of the projection varies with the angle $\theta$; that is, if $\theta$ is small, then $\|\mathbf{p}\|$ is close to $\|\mathbf{v}\|$, but if $\theta$ is near $\pi/2$, then $\|\mathbf{p}\|$ can be quite small. A corresponding result is true for planar regions. We consider the case of a closed region $R$ in a plane $P$ projected onto the $xy$-plane; we denote this projection by $\text{proj}_{xy}(R)$. (See Figure 9.) In this case we have

$$\left|\text{area}(\text{proj}_{xy}R)\right| = \left|\text{area}(R)\cos(\gamma)\right| \tag{14}$$

where $\gamma$ is the acute angle between plane $P$ and the $xy$-plane. If $P$ has equation $ax + by + cz = d$, then its normal is $\mathbf{n} = \begin{bmatrix} a & b & c \end{bmatrix}$ and the normal to the $xy$-plane is $\mathbf{k} = \begin{bmatrix} 0 & 0 & 1 \end{bmatrix}$. Thus

$$\cos(\gamma) = \frac{\mathbf{n}\cdot\mathbf{k}}{\|\mathbf{n}\|\,\|\mathbf{k}\|} = \frac{c}{\|\mathbf{n}\|} = \frac{c}{\sqrt{a^2 + b^2 + c^2}}. \qquad ■$$

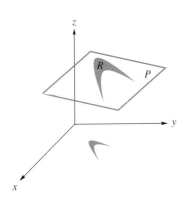

FIGURE 9

EXAMPLE 7   Let $R$ be the triangle determined by the points $(0, 0, 0)$, $(2, 4, 5)$, and $(3, -2, 6)$. To find $\text{area}(\text{proj}_{xy}R)$ we first note that $R$ is determined by the vectors $\mathbf{u} = \begin{bmatrix} 2 & 4 & 5 \end{bmatrix}$ and $\mathbf{v} = \begin{bmatrix} 3 & -2 & 6 \end{bmatrix}$. Then the area of this triangle is obtained from

$$\text{area}(R) = \tfrac{1}{2}\|\mathbf{n}\| = \tfrac{1}{2}\|\mathbf{u}\times\mathbf{v}\| = \tfrac{1}{2}\left\|\begin{bmatrix} 34 & 3 & -16 \end{bmatrix}\right\| \approx 18.848 \text{ square units.}$$

We also have

$$\cos(\gamma) = \frac{\begin{bmatrix} 34 & 3 & -16 \end{bmatrix}\cdot\begin{bmatrix} 0 & 0 & 1 \end{bmatrix}}{\left\|\begin{bmatrix} 34 & 3 & -16 \end{bmatrix}\right\|} = \frac{-16}{\sqrt{1421}}.$$

Thus

$$\left|\text{area}(\text{proj}_{xy}R)\right| = \left|\frac{1}{2}\|\mathbf{n}\|\cos(\gamma)\right| = \left|\frac{1}{2}\|\mathbf{n}\|\frac{-16}{\|\mathbf{n}\|}\right| = 8 \text{ square units.} \qquad ■$$

## EXERCISES 3.6

*In Exercises* 1–3, *use Equation* (4) *to compute* $\mathbf{u}\times\mathbf{v}$.

1. $\mathbf{u} = 3\mathbf{i} + 2\mathbf{j} - \mathbf{k}$, $\mathbf{v} = -\mathbf{i} + 3\mathbf{j} + 2\mathbf{k}$.

2. $\mathbf{u} = \begin{bmatrix} 1 & 0 & 2 \end{bmatrix}$, $\mathbf{v} = \begin{bmatrix} 3 & 1 & -4 \end{bmatrix}$.

3. $\mathbf{u} = \begin{bmatrix} -1 & -2 & 5 \end{bmatrix}$, $\mathbf{v} = 3\mathbf{u}$.

4. The following properties are listed in Table 1.
   (a) Show Property 1.
   (b) Show Property 2.
   (c) Explain why Property 3 follows from Property 2.
   (d) Use properties of the determinant to explain why Property 4 is valid.

   (e) Use properties of the determinant to explain why Property 5 is valid.

*In Exercises* 5–7, *compute* $(\mathbf{u}\times\mathbf{v})\cdot\mathbf{w}$.

5. $\mathbf{u} = 2\mathbf{i} - \mathbf{j} + 3\mathbf{k}$, $\mathbf{v} = 2\mathbf{j} - \mathbf{k}$, $\mathbf{w} = 3\mathbf{i} + 2\mathbf{k}$.

6. $\mathbf{u} = \begin{bmatrix} -1 & 2 & 0 \end{bmatrix}$, $\mathbf{v} = \begin{bmatrix} 5 & 5 & -1 \end{bmatrix}$, $\mathbf{w} = \begin{bmatrix} 2 & 0 & 1 \end{bmatrix}$.

7. $\mathbf{u} = \begin{bmatrix} 1 & 3 & 1 \end{bmatrix}$, $\mathbf{v} = \begin{bmatrix} -2 & 1 & 4 \end{bmatrix}$, $\mathbf{w} = -2\mathbf{v}$.

8. Use properties of the determinant to show that Equation (8) is valid.

**9.** Show that if **p** is in span{**u**, **v**}, then **n** = **u** × **v** is orthogonal to **p**.

*In Exercises 10–13, find the normal to the plane determined by* **u** *and* **v**.

**10.** **u** = 3**i** − 2**j**, **v** = **i** + 2**j** − 2**k**.

**11.** **u** = **i**, **v** = **k**.

**12.** **u** = **j**, **v** = **i**.

**13.** **u** = $\begin{bmatrix} 1 & 1 & -1 \end{bmatrix}$, **v** = $\begin{bmatrix} 2 & -2 & 3 \end{bmatrix}$.

*In Exercises 14–16, compute the area of the parallelogram determined by* **u** *and* **v**.

**14.** **u** = 2**i** − **j** + 3**k**, **v** = 2**j** − **k**.

**15.** **u** = $\begin{bmatrix} 0 & 2 & -3 \end{bmatrix}$, **v** = $\begin{bmatrix} 2 & 1 & 0 \end{bmatrix}$.

**16.** **u** = $\begin{bmatrix} 4 & 2 & -1 \end{bmatrix}$, **v** = $\begin{bmatrix} 0 & 1 & -3 \end{bmatrix}$.

**17.** Use (7) to show that (**u** × **v**)·**w** = **u**·(**v** × **w**).

**18.** Show that **u** × (**v** × **w**) = (**u**·**w**)**v** − (**u**·**v**)**w**.

**19.** Give an alternate development for the expression in Equation (9) by starting with

$$\|\mathbf{u} \times \mathbf{v}\|^2 = (\mathbf{u} \times \mathbf{v}) \cdot (\mathbf{u} \times \mathbf{v})$$

and using the results of Exercises 17 and 18.

**20.** Determine the volume of the parallelepiped determined by **u**, **v**, and **w**.

(a) **u** = $\begin{bmatrix} 1 & 2 & -1 \end{bmatrix}$, **v** = $\begin{bmatrix} 0 & 1 & 3 \end{bmatrix}$, **w** = $\begin{bmatrix} 5 & 1 & -2 \end{bmatrix}$.

(b) **u** = 3**i** − 2**j** + **k**, **v** = **i** + 2**j** − 2**k**, **w** = 2**i** + 3**j** + 5**k**.

**21.** Use properties of the determinant to show that

$$\left| \det \left( \begin{bmatrix} u_1 & u_2 & u_3 \\ v_1 & v_2 & v_3 \\ w_1 & w_2 & w_3 \end{bmatrix} \right) \right|$$

$$= \left| \det \left( \begin{bmatrix} v_1 & v_2 & v_3 \\ u_1 & u_2 & u_3 \\ w_1 & w_2 & w_3 \end{bmatrix} \right) \right|$$

$$= \left| \det \left( \begin{bmatrix} w_1 & w_2 & w_3 \\ v_1 & v_2 & v_3 \\ u_1 & u_2 & u_3 \end{bmatrix} \right) \right|$$

$$= |(\mathbf{u} \times \mathbf{v}) \cdot \mathbf{w}|.$$

**22.** Show that **u** × (**v** + **w**) = (**u** × **v**) + (**u** × **w**).

**23.** Show that for any scalar $t$, $(t\mathbf{u}) \times \mathbf{v} = t(\mathbf{u} \times \mathbf{v})$.

**24.** Determine the area of the triangle whose vertices are $P(3, -2, 4)$, $Q(1, 3 - 2)$, and $R(-2, 4, 1)$.

*In Exercises 25–27, find the vector that corresponds to each given line segment.*

**25.** Segment from $P(3, 1, 0)$ to $Q(5, -1, 3)$.

**26.** Segment from $P(2, -1, 4)$ to $R(4, 5, 6)$.

**27.** Segment from $O(0, 0, 0)$ to $S(3, -6, 4)$.

*In Exercises 28 and 29, determine an equation of the plane that contains P, Q, and R.*

**28.** $P(3, 1, 0)$, $Q(5, -1, 3)$, $R(4, 5, 6)$.

**29.** $P(0, 0, 0)$, $Q(6, 6, 2)$, $R(0, 5, 4)$.

*In Exercises 30 and 31, determine the acute angle in degrees between each given pair of planes.*

**30.** $4x - 3y + 6z = 10$, $2x - 4y + z = 5$.

**31.** $9x + 6y - 3z = 12$, $x + y - 5z = 2$.

**32.** In Example 7, the projections of the vertices of the triangle onto the $xy$-plane are $(0, 0, 0)$, $(2, 4, 0)$, and $(3, -2, 0)$. Compute the area of this triangle directly without the use of (14).

**33.** Let $R$ be the triangle determined by the points $(1, 2, 1)$, $(-3, -2, 4)$, and $(5, -1, 2)$; see Figure 10.

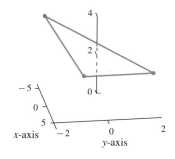

FIGURE 10

(a) Find area$(\text{proj}_{xy} R)$ using (14).

(b) Find area$(\text{proj}_{xy} R)$ directly using the coordinates of the projected vertices.

**34.** Let area$(\text{proj}_{yz} R)$ denote the projection of planar region $R$ onto the $yz$-plane. For the triangle $R$ in Example 7, compute area$(\text{proj}_{yz} R)$ using (14).

## True/False Review Questions

*Determine whether each of the following statements is true or false.*

**1.** If **u** = 4**i** + 2**j** − 3**k** and **v** = **i** + **j** + **k**, then **u** × **v** = 5**i** − 7**j** + 2**k**.

**2.** $\mathbf{u} \times \mathbf{v} = \mathbf{v} \times \mathbf{u}$.

**3.** If $\mathbf{u} \times \mathbf{v} = \mathbf{0}$, then $\mathbf{u}$ and $\mathbf{v}$ are parallel.

**4.** Vector $\mathbf{u} \times \mathbf{v}$ is orthogonal to $a\mathbf{u} + b\mathbf{v}$ for any scalars $a$ and $b$.

**5.** The angle between a pair of planes in $R^3$ is the angle between their normal vectors.

## Terminology

| Cross product | Normal vector |
| --- | --- |

Orthogonal vectors will play important roles in succeeding topics. Here we focus on obtaining orthogonal vectors in $R^3$ and use them to describe planes in 3-space.

- What does the cross product of a pair of vectors produce?
- Describe how to find conveniently the cross product of a pair of vectors.
- What is a normal vector to a plane? How is it related to a cross product?
- Given the equation of a plane, how do we find a normal to the plane purely by inspection?
- How do we find the angle of intersection between a pair of intersecting planes?
- If the cross product of a pair of nonzero vectors is the zero vector, what is the relationship between the pair of vectors? Explain the reasoning you used to formulate your answer.

## CHAPTER TEST

**1.** $T$ is a matrix transformation from $R^2$ to $R^2$ whose magnification factor is 3. Determine an associated matrix $\mathbf{A}$ for $T$, which has no zero entries.

**2.** A triangle has vertices $A(2, 1)$, $B(5, 4)$, and $C(3, 7)$. Find the area of triangle $ABC$.

**3.** Explain how to determine the area of the polygon shown below.

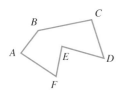

**4.** Determine all values of $k$ so that $\mathbf{A} = \begin{bmatrix} 1 & k & 4 \\ 2 & 3 & 1 \\ 1 & 0 & -1 \end{bmatrix}$ is singular.

**5.** Compute $\det(\mathbf{A})$ where $\mathbf{A} = \begin{bmatrix} 2 & 3 & -1 & 0 \\ 11 & 8 & -4 & 6 \\ 1 & 2 & -6 & -10 \\ 2 & 0 & 1 & 4 \end{bmatrix}$.

**6.** Use expansion about the third column to compute $\det(\mathbf{A})$ where $\mathbf{A} = \begin{bmatrix} 2 & -1 & 0 & 2 \\ 0 & 1 & 5 & 6 \\ 1 & 2 & 1 & 7 \\ 3 & 4 & 0 & -3 \end{bmatrix}$.

**7.** Compute the inverse of $\mathbf{A} = \begin{bmatrix} 1 & 1 & -1 \\ 0 & 1 & -1 \\ 1 & 1 & 1 \end{bmatrix}$ using adj($\mathbf{A}$).

**8.** Use Cramer's rule to determine $x_2$ for the linear system

$$\begin{aligned} 4x_1 + 2x_2 + x_3 &= 9 \\ 3x_1 \phantom{+ 2x_2} + x_3 &= 10 \\ x_1 + 2x_2 - x_3 &= -2. \end{aligned}$$

**9.** Points $A(1, 2, 1)$ and $B(3, -4, 2)$ lie in a plane through the origin in $R^3$. Find an equation of the plane.

**10.** Find the cosine of the angle between planes $5x + 2y + 3z = 1$ and $-x + y - 2z = 4$.

**11.** True or false:

    (a) $\mathbf{A}$ is singular if and only if $\det(\mathbf{A}) = 0$.

    (b) $\det(\mathbf{A}^T) = \det(\mathbf{A})$.

    (c) $\det(\mathbf{ABC}^{-1}) = \dfrac{\det(\mathbf{A})\,\det(\mathbf{B})}{\det(\mathbf{C})}$.

    (d) If $\mathbf{A}$ is nonsingular, then $\mathrm{adj}(\mathbf{A})$ is nonsingular.

    (e) $\mathbf{u} \times \mathbf{v} = \mathbf{v} \times \mathbf{u}$.

# C H A P T E R

# 4

# *EIGEN INFORMATION*

In Section 1.5 we investigated matrix transformations; that is, a function $f$ from $R^m$ to $R^n$ determined by an $m \times n$ matrix $\mathbf{A}$ so that $f(\mathbf{p}) = \mathbf{Ap}$, for $\mathbf{p}$ in $R^m$. In this chapter we only consider the case where $m = n$; that is, square matrices. We begin with $n = 2$ or 3 so that we can generate a geometric model that will aid in visualizing the concepts to be developed. We saw that such transformations did not preserve length in the sense that $\|\mathbf{Ap}\|$ need not be the same as $\|\mathbf{p}\|$. This was clearly evident in the case $n = 2$ for

$$\mathbf{A} = \begin{bmatrix} a & b \\ c & d \end{bmatrix}$$

when we showed that the image of the unit square was a parallelogram and the image of the unit circle was an ellipse. (See Example 8 and Exercise 29 in Section 1.5.) Since a vector $\mathbf{p}$ in $R^n$ has two basic properties, length and direction, it is reasonable to investigate whether $\mathbf{p}$ and $\mathbf{Ap}$ are ever parallel[1] for a given matrix $\mathbf{A}$. We note that $\mathbf{A0} = \mathbf{0}$; hence we need consider only nonzero vectors $\mathbf{p}$. We investigate this notion geometrically and algebraically in order to gain more information about a matrix.

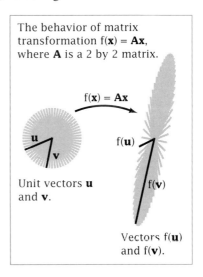

The behavior of matrix transformation f(**x**) = **Ax**, where **A** is a 2 by 2 matrix.

f(**x**) = **Ax**

**u**

**v**

Unit vectors **u** and **v**.

f(**u**)

f(**v**)

Vectors f(**u**) and f(**v**).

Image $f(\mathbf{u})$ is parallel to $\mathbf{u}$ and $f(\mathbf{v})$ is parallel to $\mathbf{v}$. See MATLAB routine, **mapcirc**.

[1] Two vectors in $R^n$ are parallel if they are in the same direction or in exactly opposite directions.

## 4.1 ■ EIGENVALUES AND EIGENVECTORS

For a given $2 \times 2$ matrix **A** we proceed to investigate pictorially parallel inputs and outputs of the matrix transformation $f(\mathbf{p}) = \mathbf{Ap}$. We choose an input vector **p** with tip on the unit circle and then sketch the resulting output vector **Ap**. (See Figure 1.)

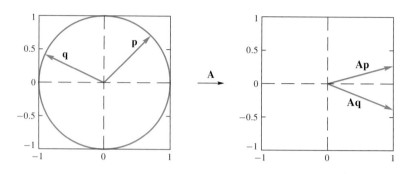

FIGURE 1

By choosing inputs from around the unit circle we can have every direction available for input. From such sketches it is difficult to check accurately for parallel inputs and outputs. In Figure 1, we might suspect that **q** and **Aq** are parallel, but we then need a careful algebraic check. Such a graphical search requires a lot of effort, without any guarantee of results. Thus we turn to the algebraic formulation of parallel, which says that one of the vectors must be a scalar multiple of the other:

$$\mathbf{Ap} = scalar * \mathbf{p}. \tag{1}$$

This formulation is much simpler to deal with computationally. In Example 1 we illustrate the use of (1) with several matrices that lead to simple geometric and algebraic investigations.

EXAMPLE 1    Determine vectors **p** so that **Ap** is parallel to **p**.

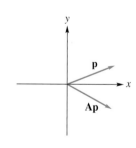

FIGURE 2

(a) Let $\mathbf{A} = \begin{bmatrix} 1 & 0 \\ 0 & -1 \end{bmatrix}$. The matrix transformation determined by **A** causes a reflection of a vector $\mathbf{p} = \begin{bmatrix} x \\ y \end{bmatrix}$ about the $x$-axis; see Figure 2. Computing the matrix transformation we get

$$\mathbf{Ap} = \begin{bmatrix} x \\ -y \end{bmatrix}.$$

Let $t$ be a scalar; then

$$\mathbf{Ap} = t\mathbf{p} \Leftrightarrow \begin{bmatrix} x \\ -y \end{bmatrix} = \begin{bmatrix} tx \\ ty \end{bmatrix} \Leftrightarrow x = tx \quad \text{and} \quad y = -ty.$$

If both $x$ and $y$ are nonzero, then $x = tx$ and $-y = ty$ cannot both be valid. Since $\mathbf{p} \neq \mathbf{0}$, we have two cases: $x = 0$ or $y = 0$.

1. Case $y = 0$: for $\mathbf{p} = \begin{bmatrix} x \\ 0 \end{bmatrix}$ with $x \neq 0$, $\mathbf{Ap} = t\mathbf{p}$ provided that $t = 1$.

2. Case $x = 0$: for $\mathbf{p} = \begin{bmatrix} 0 \\ y \end{bmatrix}$ with $y \neq 0$, $\mathbf{Ap} = t\mathbf{p}$ provided that $t = -1$.

Hence for any vector **p** along the $x$-axis or along the $y$-axis, **Ap** will be parallel to **p**.

(b) Let $\mathbf{A} = \begin{bmatrix} 2 & -1 \\ 0 & 3 \end{bmatrix}$. The matrix transformation that is determined by **A** acting on the unit square gives the parallelogram shown in Figure 3. For $\mathbf{p} = \begin{bmatrix} x \\ y \end{bmatrix}$. computing the matrix transformation we get

$$\mathbf{Ap} = \begin{bmatrix} 2x - y \\ 3y \end{bmatrix}.$$

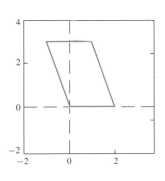

FIGURE 3

Let $t$ be a scalar; then

$$\mathbf{Ap} = t\mathbf{p} \quad \Leftrightarrow \quad \begin{bmatrix} 2x - y \\ 3y \end{bmatrix} = \begin{bmatrix} tx \\ ty \end{bmatrix}$$

$$\Leftrightarrow \quad \begin{aligned} 2x - \ y &= tx \\ 3y &= ty \end{aligned}$$

$$\Leftrightarrow \quad \begin{aligned} (2 - t)x &= y \\ (3 - t)y &= 0. \end{aligned}$$

From the requirement that $(3 - t)y = 0$, we have two cases: $y = 0$ or $y \neq 0$. Thus we see we have two cases:

1. Case $y = 0$: for $\mathbf{p} = \begin{bmatrix} x \\ 0 \end{bmatrix}$ with $x \neq 0$, we must have $(2 - t)x = 0$. Hence $t = 2$ and so $\mathbf{Ap} = 2\mathbf{p}$.
2. Case $y \neq 0$: Thus $(3 - t)y = 0$ implies $t = 3$. Then $(2 - t)x = y$ implies $y = -x$. Hence for $\mathbf{p} = \begin{bmatrix} x \\ -x \end{bmatrix}$ with $x \neq 0$, $\mathbf{Ap} = 3\mathbf{p}$.

Thus we have that for any vector **p** along the $x$-axis (see Case 1.) or any vector of the form $\begin{bmatrix} x \\ -x \end{bmatrix}$ (see Case 2.), **Ap** is parallel to **p**.

(c) Let

$$\mathbf{A} = \begin{bmatrix} \cos(\theta) & -\sin(\theta) \\ \sin(\theta) & \cos(\theta) \end{bmatrix}$$

with $0 \leq \theta < 2\pi$. The matrix transformation determined by **A** causes a rotation of a vector $\mathbf{p} = \begin{bmatrix} x \\ y \end{bmatrix}$ through an angle $\theta$ (see Figure 4). It follows that if $\theta \neq 0$ and $\theta \neq \pi$, then for every vector $\mathbf{p} \neq \mathbf{0}$, **Ap** is in a direction different from that of **p** so **p** and **Ap** are never parallel. If $\theta = 0$, then $\mathbf{A} = \mathbf{I}_2$ and hence $\mathbf{Ap} = \mathbf{p}$ so **p** and **Ap** are in the same direction. If $\theta = \pi$, then $\mathbf{A} = -\mathbf{I}_2$ and hence $\mathbf{Ap} = -\mathbf{p}$ so **p** and **Ap** are oppositely directed.   ■

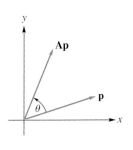

FIGURE 4

Example 1 showed that for a given matrix **A** there is quite a variety of possible answers to the parallel input and output question as given in (1). It is important to have a thorough understanding of this idea, hence we make the following definition and then review the contents of Example 1.

**Definition**    Let $\mathbf{A}$ be an $n \times n$ matrix. The scalar $\lambda$ is called an **eigenvalue**[2] of matrix $\mathbf{A}$ if there exists an $n \times 1$ vector $\mathbf{p}$, $\mathbf{p} \neq \mathbf{0}$, such that

$$\mathbf{Ap} = \lambda \mathbf{p}. \tag{2}$$

Every nonzero vector $\mathbf{p}$ satisfying (2) is called an eigenvector of $\mathbf{A}$ associated with eigenvalue $\lambda$ or an eigenvector corresponding to eigenvalue $\lambda$.

We will use the term **eigenpair** to mean an eigenvalue and an associated eigenvector. This emphasizes that for a given matrix $\mathbf{A}$ a solution of (2) requires both a scalar, an eigenvalue $\lambda$, and a nonzero vector, an eigenvector $\mathbf{p}$. Equation (2) is commonly called the **eigen equation**, and its solution is the goal of this chapter.

The results in Example 1 can be summarized as follows.

(a) For $\mathbf{A} = \begin{bmatrix} 1 & 0 \\ 0 & -1 \end{bmatrix}$ we found two eigenpairs:

$$\lambda_1 = 1, \mathbf{p}_1 = \begin{bmatrix} x \\ 0 \end{bmatrix}, x \neq 0 \quad \text{and} \quad \lambda_2 = -1, \mathbf{p}_2 = \begin{bmatrix} 0 \\ y \end{bmatrix}, y \neq 0.$$

Note that there are many eigenvectors associated with a particular eigenvalue.

(b) For $\mathbf{A} = \begin{bmatrix} 2 & -1 \\ 0 & 3 \end{bmatrix}$, we found two eigenpairs:

$$\lambda_1 = 2, \mathbf{p}_1 = \begin{bmatrix} x \\ 0 \end{bmatrix}, x \neq 0 \quad \text{and} \quad \lambda_2 = 3, \mathbf{p}_2 = \begin{bmatrix} x \\ -x \end{bmatrix}, x \neq 0.$$

Note that there are many eigenvectors associated with a particular eigenvalue.

(c) For $\mathbf{A} = \begin{bmatrix} \cos(\theta) & -\sin(\theta) \\ \sin(\theta) & \cos(\theta) \end{bmatrix}$ with $0 \leq \theta < 2\pi$ we had various results that depended upon the choice of $\theta$.

> Case $\theta = 0$: $\mathbf{A} = \mathbf{I}_2$, one eigenpair $\lambda = 1$, $\mathbf{p} \neq \mathbf{0}$.
>
> Case $\theta = \pi$: $\mathbf{A} = -\mathbf{I}_2$, one eigenpair $\lambda = -1$, $\mathbf{p} \neq \mathbf{0}$.
>
> Case $\theta \neq 0$ and $\theta \neq \pi$: seemingly no eigenpairs. This is not really true; it is just that the geometric approach we used was not good enough. We deal with this in the discussion that follows.

In the eigen equation in (2) there is no restriction on the entries of matrix $\mathbf{A}$, the type of scalar $\lambda$, or the entries of vector $\mathbf{p}$, other than that they cannot all be zero. It is possible that an eigenpair of a matrix with real entries can involve complex values. In such a case the geometric approach of parallel input and output fails to lead us to an appropriate solution. (We will show in Section 4.2 that the eigen equation always has solutions, some of which may be complex.) For $\theta = \pi/4$ the matrix in Example 1(c) is

$$\mathbf{A} = \frac{1}{2} \begin{bmatrix} \sqrt{2} & -\sqrt{2} \\ \sqrt{2} & \sqrt{2} \end{bmatrix}.$$

An eigen pair is

$$\lambda = \frac{\sqrt{2}}{2} + i \frac{\sqrt{2}}{2}, \mathbf{p} = \begin{bmatrix} i \\ 1 \end{bmatrix} \text{ where } i = \sqrt{-1}, \text{ the complex unit. (Verify.)}$$

---

[2]The term *eigenvalue* is a hybrid, where *eigen*, from German means, "proper." Eigenvalues are also called proper values or characteristic values; eigenvectors are also called proper vectors or characteristic vectors.

In order to obtain solutions of the eigen problem we will use an algebraic approach rather than rely on geometry. With this in mind we can make some important observations about eigenvalues and eigenvectors directly from (2).

---

### Eigen Properties

**Property 1.** If $(\lambda, \mathbf{p})$ is an eigenpair of $\mathbf{A}$, then so is $(\lambda, k\mathbf{p})$ for any scalar $k \neq 0$.

We see this as follows: $(\lambda, \mathbf{p})$ is an eigenpair of $\mathbf{A} \Rightarrow \mathbf{Ap} = \lambda\mathbf{p}$. $\mathbf{p} \neq \mathbf{0}$ and $k \neq 0 \Rightarrow k\mathbf{p} \neq \mathbf{0}$. Thus

$$\mathbf{A}(k\mathbf{p}) = k(\mathbf{Ap}) = k(\lambda\mathbf{p}) = \lambda(k\mathbf{p}),$$

which shows that $(\lambda, k\mathbf{p})$ is an eigenpair of $\mathbf{A}$.

**Property 2.** If $\mathbf{p}$ and $\mathbf{q}$ are eigenvectors corresponding to eigenvalue $\lambda$ of $\mathbf{A}$, then so is $\mathbf{p} + \mathbf{q}$ (assuming $\mathbf{p} + \mathbf{q} \neq \mathbf{0}$).[3]

We see this as follows: $(\lambda, \mathbf{p})$ is an eigenpair of $\mathbf{A} \Rightarrow \mathbf{Ap} = \lambda\mathbf{p}$. $(\lambda, \mathbf{q})$ is an eigenpair of $\mathbf{A} \Rightarrow \mathbf{Aq} = \lambda\mathbf{q}$. Thus,

$$\mathbf{A}(\mathbf{p} + \mathbf{q}) = \mathbf{Ap} + \mathbf{Aq} = \lambda\mathbf{p} + \lambda\mathbf{q} = \lambda(\mathbf{p} + \mathbf{q})$$

which shows that $(\lambda, \mathbf{p} + \mathbf{q})$ is an eigenpair of $\mathbf{A}$.

**Property 3.** If $(\lambda, \mathbf{p})$ is an eigenpair of $\mathbf{A}$, then for any positive integer $r$, $(\lambda^r, \mathbf{p})$ is an eigenpair of $\mathbf{A}^r$.

We see this as follows: $(\lambda, \mathbf{p})$ is an eigenpair of $\mathbf{A} \Rightarrow \mathbf{Ap} = \lambda\mathbf{p}$. Thus

$$\mathbf{A}^2\mathbf{p} = \mathbf{A}(\mathbf{Ap}) = \mathbf{A}(\lambda\mathbf{p}) = \lambda(\mathbf{Ap}) = \lambda(\lambda\mathbf{p}) = \lambda^2\mathbf{p},$$

$$\mathbf{A}^3\mathbf{p} = \mathbf{A}(\mathbf{A}^2\mathbf{p}) = \mathbf{A}(\lambda^2\mathbf{p}) = \lambda^2(\mathbf{Ap}) = \lambda^2(\lambda\mathbf{p}) = \lambda^3\mathbf{p},$$

$$\vdots$$

etc.

Hence $(\lambda^r, \mathbf{p})$ is an eigenpair of $\mathbf{A}^r$.

**Property 4.** If $(\lambda, \mathbf{p})$ and $(\mu, \mathbf{q})$ are eigenpairs of $\mathbf{A}$ with $\lambda \neq \mu$, then $\mathbf{p}$ and $\mathbf{q}$ are linearly independent. (See Section 2.3.)

We see this as follows: $(\lambda, \mathbf{p})$ is an eigenpair of $\mathbf{A} \Rightarrow \mathbf{Ap} = \lambda\mathbf{p}$. $(\mu, \mathbf{q})$ is an eigenpair of $\mathbf{A} \Rightarrow \mathbf{Aq} = \mu\mathbf{q}$. Let $s$ and $t$ be scalars so that $t\mathbf{p} + s\mathbf{q} = \mathbf{0}$, or equivalently $t\mathbf{p} = -s\mathbf{q}$. Multiplying this linear combination by $\mathbf{A}$ we get

$$\mathbf{A}(t\mathbf{p} + s\mathbf{q}) = t(\mathbf{Ap}) + s(\mathbf{Aq}) = t(\lambda\mathbf{p}) + s(\mu\mathbf{q}) = \lambda(t\mathbf{p}) + \mu(s\mathbf{q}) = \mathbf{0}.$$

Substituting for $t\mathbf{p}$ we get

$$\lambda(-s\mathbf{q}) + \mu(s\mathbf{q}) = \mathbf{0} \Leftrightarrow s(\lambda - \mu)\mathbf{q} = \mathbf{0}.$$

Since $\mathbf{q} \neq \mathbf{0}$ and $\lambda \neq \mu$, we have $s = 0$, but then

$$t\mathbf{p} = -0\mathbf{q} = \mathbf{0} \Leftrightarrow t = 0.$$

Hence the only way $t\mathbf{p} + s\mathbf{q} = \mathbf{0}$ is when $t = s = 0$, so $\mathbf{p}$ and $\mathbf{q}$ are linearly independent.

---

[3]If we modify Property 1 to permit the scalar $k$ to be zero and permit $\mathbf{p} + \mathbf{q} = \mathbf{0}$ in Property 2, then Properties 1 and 2 tell us that the set of eigenvectors associated with a particular eigenvalue form a closed set. See also Exercise 9.

Property 4 is often stated as follows: *Eigenvectors corresponding to distinct eigenvalues are linearly independent.* This property is particularly important and will be used repeatedly in upcoming topics.

Note that Properties 1 through 4 are valid for an eigenvalue $\lambda = 0$. Other eigen properties are developed in the exercises. In Sections 4.1 through 4.3 we list properties of eigenvalues and eigenvectors and number them consecutively 1 through 13. At the end of Section 4.3 we provide a summary of these properties.

Next we look at two applications involving eigenvalues and eigenvectors. The first is a Markov process or Markov chain, which was introduced in Section 1.3. The basic idea is that we have a matrix transformation $f(\mathbf{x}) = \mathbf{Ax}$, where $\mathbf{A}$ is a square probability matrix; that is, $0 \le a_{ij} \le 1$ and the sum of the entries in each column is 1. (We called $\mathbf{A}$ a transition matrix. See Exercises 69 and 70 in Section 1.3.) In a Markov process the matrix transformation is used to form successive compositions as follows: Let $\mathbf{s}_0$ be an initial vector. Then

$$\mathbf{s}_1 = f(\mathbf{s}_0) = \mathbf{As}_0$$
$$\mathbf{s}_2 = f(\mathbf{s}_1) = f(f(\mathbf{s}_0)) = \mathbf{As}_1 = \mathbf{A}^2\mathbf{s}_0$$
$$\vdots$$
$$\mathbf{s}_{n+1} = f(\mathbf{s}_n) = \mathbf{As}_n = \mathbf{A}^{n+1}\mathbf{s}_0$$
$$\vdots$$

(3)

If the sequence of vectors $\mathbf{s}_0, \mathbf{s}_1, \mathbf{s}_2, \ldots$ converges to a limiting vector $\mathbf{p}$, then we call $\mathbf{p}$ the steady state of the Markov process. (See Example 7 in Section 1.3.) If the process has a steady state $\mathbf{p}$, then

$$f(\mathbf{p}) = \mathbf{p} \quad \Leftrightarrow \quad \mathbf{Ap} = \mathbf{p}.$$

That is, the steady state is an eigenvector corresponding to an eigenvalue $\lambda = 1$ of the transition matrix $\mathbf{A}$.

EXAMPLE 2    Let

$$\mathbf{A} = \begin{bmatrix} \frac{1}{3} & \frac{2}{5} \\ \frac{2}{3} & \frac{3}{5} \end{bmatrix}$$

be the transition matrix of the Markov process of Example 7 in Section 1.3. By computations we displayed in Example 7, we observed that the process had a steady state. Hence $\lambda = 1$ is an eigenvalue of $\mathbf{A}$. To find a corresponding eigenvector, we proceed as in Example 1. We seek a vector $\mathbf{p} = \begin{bmatrix} x \\ y \end{bmatrix}$ so that

$$\mathbf{Ap} = \mathbf{p} \quad \Leftrightarrow \quad \begin{bmatrix} \frac{1}{3} & \frac{2}{5} \\ \frac{2}{3} & \frac{3}{5} \end{bmatrix}\begin{bmatrix} x \\ y \end{bmatrix} = \begin{bmatrix} x \\ y \end{bmatrix} \quad \Leftrightarrow \quad \begin{bmatrix} \frac{1}{3}x + \frac{2}{5}y \\ \frac{2}{3}x + \frac{3}{5}y \end{bmatrix} = \begin{bmatrix} x \\ y \end{bmatrix}$$

$$\Leftrightarrow \quad \begin{array}{l} \frac{1}{3}x + \frac{2}{5}y = x \\ \frac{2}{3}x + \frac{3}{5}y = y \end{array} \quad \Leftrightarrow \quad \begin{array}{l} -\frac{2}{3}x + \frac{2}{5}y = 0 \\ \frac{2}{3}x - \frac{2}{5}y = 0 \end{array}$$

$$\Leftrightarrow \quad \begin{bmatrix} -\frac{2}{3} & \frac{2}{5} \\ \frac{2}{3} & -\frac{2}{5} \end{bmatrix}\begin{bmatrix} x \\ y \end{bmatrix} = \begin{bmatrix} 0 \\ 0 \end{bmatrix}.$$

The solution to this homogeneous linear system gives $y = \frac{5}{3}x$, so we have eigenvector $\mathbf{p} = \begin{bmatrix} x \\ \frac{5}{3}x \end{bmatrix}$, $x \neq 0$, corresponding to eigenvalue $\lambda = 1$. If we let $x = 4500$, then $y = 7500$ so $\mathbf{p} = \begin{bmatrix} 4500 \\ 7500 \end{bmatrix}$ is the steady state vector. This is the same vector determined by successive computations in Example 7 of Section 1.3, where we started with the initial stock vector $\mathbf{s}_0 = \begin{bmatrix} 3000 \\ 9000 \end{bmatrix}$. Another interpretation of the steady state vector $\mathbf{p}$ is obtained by choosing $x$ so that

$$x + y = x + \tfrac{5}{3}x = 1, \quad x > 0.$$

Take $x = \frac{3}{8}$; then

$$\mathbf{p} = \begin{bmatrix} \frac{3}{8} \\ \frac{5}{8} \end{bmatrix}$$

represents the distribution of stock between rooms R1 and R2. We interpret this to say that regardless of the initial stock vector, with at least one positive entry, ultimately the "transition" of stock, as dictated by the information in $\mathbf{A}$, is that $\frac{3}{8} = 37.5\%$ will be in R1 and $\frac{5}{8} = 62.5\%$ will be in R2. Thus we see that the eigen equation (2) provides a means to solve a Markov process provided that a steady state exists or, equivalently, provided that $\lambda = 1$ is an eigenvalue of $\mathbf{A}$. (See Property 5 in Section 4.2 and the discussion following it.) ■

The matrix in Example 2 has another eigenpair,

$$\mu = -\frac{1}{15}, \quad \mathbf{q} = \begin{bmatrix} -1 \\ 1 \end{bmatrix}. \quad \text{(Verify.)}$$

Since the eigenvalues are distinct, vectors $\mathbf{p}$ and $\mathbf{q}$ are linearly independent by Property 4. It follows from Example 4 in Section 2.3 that $S = \{\mathbf{p}, \mathbf{q}\}$ is a basis for $R^2$. Hence any initial vector $\mathbf{s}_0$ can be expressed as linear combination of $\mathbf{p}$ and $\mathbf{q}$. Let $\mathbf{s}_0 = c_1\mathbf{p} + c_2\mathbf{q}$. Then the Markov process as described in (3) behaves as follows:

$$\mathbf{s}_1 = \mathbf{A}\mathbf{s}_0 = \mathbf{A}(c_1\mathbf{p} + c_2\mathbf{q}) = c_1(\lambda\mathbf{p}) + c_2(\mu\mathbf{q})$$

$$\mathbf{s}_2 = \mathbf{A}\mathbf{s}_1 = \mathbf{A}(c_1(\lambda\mathbf{p}) + c_2(\mu\mathbf{q})) = c_1(\lambda^2\mathbf{p}) + c_2(\mu^2\mathbf{q})$$

$$\vdots$$

$$\mathbf{s}_{n+1} = \mathbf{A}\mathbf{s}_n = c_1(\lambda^{n+1}\mathbf{p}) + c_2(\mu^{n+1}\mathbf{q})$$

$$\vdots$$

Since $\lambda = 1$ and $\mu = -\frac{1}{15}$,

$$\mathbf{s}_{n+1} = \mathbf{A}\mathbf{s}_n = c_1(\mathbf{p}) + c_2\left(\left(-\tfrac{1}{15}\right)^{n+1}\mathbf{q}\right)$$

and so

$$\lim_{n \to \infty} \mathbf{s}_{n+1} = c_1\mathbf{p}.$$

Thus the long-term behavior of the Markov process is a multiple of the steady state distribution vector $\mathbf{p} = \begin{bmatrix} \frac{3}{8} \\ \frac{5}{8} \end{bmatrix}$. This was verified computationally in Example 2.

The important observation to make at this point is as follows:

A basis consisting of eigenvectors of the transition matrix **A** provides a way to analyze the behavior of the process *without* computation of the sequence of vectors $\mathbf{s}_1, \mathbf{s}_2, \ldots$.

Hence we have a case where information imbedded in the matrix provides a means to determine the behavior of a process that uses the matrix.

As a second application of the use of eigen information in the analysis of the behavior of a process, we introduce a population model. Many of the ideas used with the Markov process apply in the population model, except that we expect populations to change, possibly drastically over long periods of time. Thus we will not be interested in a steady state, but rather in characterizing change in behavior of the population.

## Application:   **Leslie Population Model**

A Leslie population model[4] uses data collected about birth rates and survival rates of populations which are broken into age groups or categories. Field studies or laboratory studies must be conducted to provide the information on the rates. Once the rate information is collected, a model for predicting the changes of the populations within the age groups is constructed where the rates are entries of a matrix **A**. The basic equation is

$$(\text{new population vector}) = \mathbf{A} * (\text{old population vector}).$$

EXAMPLE 3    A laboratory study of a blood parasite has revealed that there are three age categories that determine its effect on the host. We refer to these categories as juvenile (J), adult (A), and senior (S). The parasite is capable of reproduction in each of these stages, and only a certain fraction of the organism survive to move on to the next stage. The birth and survival rates are shown in Table 1.

TABLE 1

| Rates | Categories | | |
|---|---|---|---|
| | J | A | S |
| Birth Rate | 0.1 | 0.8 | 0.4 |
| Survival Rate | 0.9 | 0.7 | 0 |

We interpret this information as follows.

For category J:
{ On the average, 10% of the current juvenile population will produce offspring.
On the average, 90% of the current juvenile population will survive to become adults. }

For category A:
{ On the average, 80% of the current adult population will produce offspring.
On the average, 70% of the current adult population will survive to become seniors. }

For category S:
{ On the average, 40% of the current senior population will produce offspring.
0% of the current senior population will survive. }

[4]See P. H. Leslie, "On the Use of Matrices in Certain Population Mathematics," *Biometrika* 33, 1945.

The new population of juveniles is determined by the total births from each of the categories. The information regarding the changes in the population is summarized in three equations. Let $x_1$ be the current population of juveniles, $x_2$ the current population of adults, and $x_3$ the current population of seniors. We have

$$
\begin{aligned}
0.1x_1 + 0.8x_2 + 0.4x_3 &= \text{ new juvenile population (birth information)} \\
0.9x_1 \qquad\qquad\quad &= \text{ new adult population (survival information)} \\
0.7x_2 \quad &= \text{new senior population (survival information)}
\end{aligned}
$$

Let

$$
\mathbf{A} = \begin{bmatrix} 0.1 & 0.8 & 0.4 \\ 0.9 & 0 & 0 \\ 0 & 0.7 & 0 \end{bmatrix}.
$$

Then the population changes are governed by the matrix equation

$$
\mathbf{A} * (\text{current population vector}) = (\text{new population vector})
$$

and $\mathbf{A}$ is called a **Leslie matrix**.

Let $\mathbf{s}_0$ be an initial population vector with components of juvenile, adult, and senior, respectively. Then future populations are given by

$$
\begin{aligned}
\mathbf{s}_1 &= \mathbf{A}\mathbf{s}_0 \\
\mathbf{s}_2 &= \mathbf{A}\mathbf{s}_1 = \mathbf{A}^2\mathbf{s}_0 \\
&\vdots \\
\mathbf{s}_{n+1} &= \mathbf{A}\mathbf{s}_n = \mathbf{A}^{n+1}\mathbf{s}_0 \\
&\vdots
\end{aligned}
$$

The matrix $\mathbf{A}$ has three distinct eigenvalues, so the corresponding eigenvectors are linearly independent. (See Property 4.) By Exercise 49 in Section 2.3, they form a basis for $R^3$; hence any initial vector $\mathbf{s}_0$ can be expressed as a linear combination of these eigenvectors. Let $(\lambda_i, \mathbf{p}_i)$, $i = 1, 2, 3$ denote the three eigenpairs. The eigenvalue that is largest in absolute value is called the **dominant eigenvalue**. Suppose that

$$
|\lambda_1| > |\lambda_2| \geq |\lambda_3|
$$

Then it follows that

$$
\begin{aligned}
\mathbf{s}_{n+1} = \mathbf{A}\mathbf{s}_n &= \mathbf{A}^{n+1}\mathbf{s}_0 \\
&= \mathbf{A}^{n+1}(c_1\mathbf{p}_1 + c_2\mathbf{p}_2 + c_3\mathbf{p}_3) \\
&= c_1(\lambda_1)^{n+1}\mathbf{p}_1 + c_2(\lambda_2)^{n+1}\mathbf{p}_2 + c_3(\lambda_3)^{n+1}\mathbf{p}_3.
\end{aligned}
$$

The ultimate behavior of the populations is controlled by the term containing the dominant eigenvalue:

$$
\lim_{n\to\infty} \mathbf{s}_{n+1} \approx \lim_{n\to\infty} c_1\lambda_1^{n+1}\mathbf{p}_1.
$$

If $|\lambda_1| < 1$, then the population of the organism is decreasing and will eventually die out. However, in our case the dominant eigenvalue is $|\lambda_1| \approx 1.0331$ so the population is increasing at a rate of about 3.3% each time we determine a new population. (We will verify the computation of $\lambda_1$ in Section 4.2.)   ■

## EXERCISES 4.1

**1.** Let $\mathbf{A} = \begin{bmatrix} 0 & 1 \\ 1 & 0 \end{bmatrix}$. The matrix transformation determined

by $\mathbf{A}$ causes a reflection of a vector $\mathbf{p} = \begin{bmatrix} x \\ y \end{bmatrix}$ about the

line $y = x$; that is, the $45°$ line. (See Figure 5.) We have

$$\mathbf{Ap} = \mathbf{A} \begin{bmatrix} x \\ y \end{bmatrix} = \begin{bmatrix} y \\ x \end{bmatrix}.$$

Let $t$ be a scalar. Use algebra as in Example 1 to determine two distinct eigenpairs of $\mathbf{A}$ so that $\mathbf{Ap} = t\mathbf{p}$.

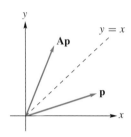

FIGURE 5

**2.** Let $\mathbf{A} = \begin{bmatrix} 1 & 2 \\ 0 & 1 \end{bmatrix}$. The matrix transformation determined
by $\mathbf{A}$ causes a shear in the $x$-direction by a factor of 2.
Figure 6 shows the image of the unit square that is the result of the matrix transformation determined by $\mathbf{A}$. For a

vector $\mathbf{p} = \begin{bmatrix} x \\ y \end{bmatrix}$ we have

$$\mathbf{Ap} = \mathbf{A} \begin{bmatrix} x \\ y \end{bmatrix} = \begin{bmatrix} x + 2y \\ y \end{bmatrix}.$$

Let $t$ be a scalar. Use algebra as in Example 1 to show that there is only one eigenpair of $\mathbf{A}$ so that $\mathbf{Ap} = t\mathbf{p}$.

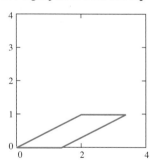

FIGURE 6

**3.** Let $\mathbf{A} = \begin{bmatrix} 1 & 1 \\ -2 & 4 \end{bmatrix}$.

(a) Show that

$$\lambda = 2 \quad \text{and} \quad \mathbf{p} = \begin{bmatrix} r \\ r \end{bmatrix}, r \neq 0,$$

is an eigenpair of $\mathbf{A}$.

(b) Show that

$$\mu = 3 \quad \text{and} \quad \mathbf{q} = \begin{bmatrix} s \\ 2s \end{bmatrix}, s \neq 0,$$

is an eigenpair of $\mathbf{A}$.

(c) Explain why $\mathbf{p}$ and $\mathbf{q}$ are linearly independent.

**4.** Let $\mathbf{A} = \begin{bmatrix} 1 & 4 \\ 1 & -2 \end{bmatrix}$.

(a) Show that

$$\lambda = 2 \quad \text{and} \quad \mathbf{p} = \begin{bmatrix} 4r \\ r \end{bmatrix}, r \neq 0,$$

is an eigenpair of $\mathbf{A}$.

(b) Show that

$$\mu = -3 \quad \text{and} \quad \mathbf{q} = \begin{bmatrix} -s \\ s \end{bmatrix}, s \neq 0,$$

is an eigenpair of $\mathbf{A}$.

(c) Explain why $\mathbf{p}$ and $\mathbf{q}$ are linearly independent.

**5.** Let $\mathbf{A} = \begin{bmatrix} 1 & 1 \\ 1 & 1 \end{bmatrix}$. There are two distinct eigenvalues for
$\mathbf{A}$, $\lambda = 0$ and $\mu = 2$.

(a) Use the eigen equation (2) to determine an eigenvector associated with $\lambda = 0$.

(b) Use the eigen equation (2) to determine an eigenvector associated with $\mu = 2$.

**6.** Let $\mathbf{A} = \begin{bmatrix} 1 & 2 & -1 \\ 1 & 0 & 1 \\ 4 & -4 & 5 \end{bmatrix}$.

(a) Show that

$$\lambda_1 = 1 \quad \text{and} \quad \mathbf{p}_1 = \begin{bmatrix} -1 \\ 1 \\ 2 \end{bmatrix}$$

is an eigenpair of $\mathbf{A}$.

(b) Show that

$$\lambda_2 = 2 \quad \text{and} \quad \mathbf{p}_2 = \begin{bmatrix} -2 \\ 1 \\ 4 \end{bmatrix}$$

is an eigenpair of $\mathbf{A}$.

(c) Show that

$$\lambda_3 = 3 \quad \text{and} \quad \mathbf{p}_3 = \begin{bmatrix} -1 \\ 1 \\ 4 \end{bmatrix}$$

is an eigenpair of $\mathbf{A}$.

(d) Explain why the set $\{\mathbf{p}_1, \mathbf{p}_2, \mathbf{p}_3\}$ is linearly independent.

**7.** A boat rental firm has two offices on a very scenic lake. The offices, denoted R and S, are near opposite ends of the lake. The offices are also on the bus route that has tours around the lake. Some customers drop off their boat rentals at the other office and return via the bus. Each day 20% of the available boats at office R are dropped off at office S, while only 15% of the available boats at office S are dropped off at office R. Let $r_n$ be the number of boats available at office R on day $n$ and $s_n$ be the number of boats available at office S on day $n$. Then equations

$$r_{n+1} = 0.80r_n + 0.15s_n$$
$$s_{n+1} = 0.20r_n + 0.85s_n$$

provide a model for the number of boats at the offices on the next day. Let

$$\mathbf{b}_n = \begin{bmatrix} r_n \\ s_n \end{bmatrix}$$

be the "boat vector" for day $n$ and

$$\mathbf{A} = \begin{bmatrix} 0.80 & 0.15 \\ 0.20 & 0.85 \end{bmatrix}.$$

Then the preceding linear system has the matrix form $\mathbf{b}_{n+1} = \mathbf{A}\mathbf{b}_n$.

(a) Let $\mathbf{b}_0 = \begin{bmatrix} 40 \\ 60 \end{bmatrix}$. Compute $\mathbf{b}_1$ through $\mathbf{b}_5$.

(b) Rounding to the nearest whole number, make a conjecture for the "long-term boat vector"; that is,

$$\lim_{n \to \infty} \mathbf{b}_{n+1}.$$

(c) The eigenvalues of $\mathbf{A}$ are $\lambda_1 = 1$ and $\lambda_2 = 0.65$. If we denote the corresponding eigenvectors as $\mathbf{p}_1$ and $\mathbf{p}_2$, respectively, then

$$\lim_{n \to \infty} \mathbf{b}_{n+1} = \lim_{n \to \infty} \mathbf{A}^{n+1}\mathbf{b}_0$$
$$= \lim_{n \to \infty} \mathbf{A}^{n+1}(c_1\lambda_1\mathbf{p}_1 + c_2\lambda_2\mathbf{p}_2).$$

Substitute for $\lambda_1$ and $\lambda_2$, perform the matrix multiplication by $\mathbf{A}^{n+1}$, and simplify the expression as much as possible.

(d) Complete the following statement: The long-term boat vector is a multiple of _____ .

**8.** Let $(\lambda, \mathbf{p})$ be an eigenpair of matrix $\mathbf{A}$.

(a) For what values of $\lambda$ will $\mathbf{p}$ and $\mathbf{A}\mathbf{p}$ be in the same direction with $\|\mathbf{A}\mathbf{p}\| > \|\mathbf{p}\|$?

(b) For what values of $\lambda$ will $\mathbf{p}$ and $\mathbf{A}\mathbf{p}$ be in the same direction with $\|\mathbf{A}\mathbf{p}\| < \|\mathbf{p}\|$?

(c) For what values of $\lambda$ will $\mathbf{p}$ and $\mathbf{A}\mathbf{p}$ be in the opposite directions?

**9.** Let $\mathbf{A}$ be an $n \times n$ matrix with eigenvalue $\lambda$. Let $S_\lambda$ denote the set of all eigenvectors of $\mathbf{A}$ associated with $\lambda$ together with the zero vector. Show that $S_\lambda$ is a subspace of $R^n$. We call $S_\lambda$ the **eigenspace associated with** $\lambda$. (*Hint*: See Properties 1 and 2.)

**10.** Why did we need to explicitly say that the zero vector was in set $S_\lambda$ in Exercise 9?

**11.** Let $\mathbf{A}$ be an $n \times n$ matrix with eigenvalues $\lambda_1$ and $\lambda_2$ where $\lambda_1 \neq \lambda_2$. Let $S_1$ and $S_2$ be the eigenspaces associated with $\lambda_1$ and $\lambda_2$, respectively. Explain why the zero vector is the only vector that is in both $S_1$ and $S_2$.

**12.** Let $(\lambda, \mathbf{p})$ be an eigenpair of $n \times n$ matrix $\mathbf{A}$. Show that $(\lambda + k, \mathbf{p})$ is an eigenpair of matrix $\mathbf{A} + k\mathbf{I}_n$. (This property says that adding a scalar multiple of the identity matrix to $\mathbf{A}$ just shifts the eigenvalues by the scalar multiplier.)

**13.** Let $\mathbf{A}$ be as in Exercise 3. Then $S_2$, the eigenspace associated with $\lambda = 2$, consists of all vectors of the form

$$t \begin{bmatrix} 1 \\ 1 \end{bmatrix}$$

and $S_3$, the eigenspace associated with $\mu = 3$, consists of all vectors of the form

$$s \begin{bmatrix} 1 \\ 2 \end{bmatrix}.$$

(a) Show that

$$\left\{ \begin{bmatrix} 1 \\ 1 \end{bmatrix}, \begin{bmatrix} 1 \\ 2 \end{bmatrix} \right\}$$

is a basis for $R^2$.

(b) Explain why we can say that every vector in $R^2$ is a linear combination of a vector from $S_2$ and a vector from $S_3$.

(c) Explain why it is reasonable to write $R^2 = S_2 + S_3$.

(d) Let

$$\mathbf{P} = \begin{bmatrix} 1 & 1 \\ 1 & 2 \end{bmatrix},$$

the matrix whose columns are eigenvectors of $\mathbf{A}$. Compute $\mathbf{P}^{-1}\mathbf{A}\mathbf{P}$.

**14.** Let $\mathbf{A}$ be as in Exercise 6 and $\mathbf{P} = \begin{bmatrix} \mathbf{p}_1 & \mathbf{p}_2 & \mathbf{p}_3 \end{bmatrix}$ the matrix whose columns are eigenvectors of $\mathbf{A}$. Compute $\mathbf{P}^{-1}\mathbf{A}\mathbf{P}$.

**15.** Let $\mathbf{A}$ be a square matrix.

(a) Suppose that the homogeneous system $\mathbf{A}\mathbf{x} = \mathbf{0}$ has a nontrivial solution $\mathbf{x} = \mathbf{p}$. Show that $\mathbf{p}$ is an eigenvector of $\mathbf{A}$ and determine the corresponding eigenvalue.

(b) Suppose that $(0, \mathbf{q})$ is an eigenpair of $\mathbf{A}$. Show that the homogeneous system $\mathbf{A}\mathbf{x} = \mathbf{0}$ has a nontrivial solution.

(c) From parts (a) and (b), state an eigen property of any singular matrix.

(d) From parts (a), (b) and (c), state an eigen property of any nonsingular matrix.

**16.** Let $f(\mathbf{x}) = \mathbf{A}\mathbf{x}$ be the matrix transformation determined by the $n \times n$ matrix $\mathbf{A}$. A subspace $W$ of $R^n$ is said to be **invariant** under the transformation $f$ if for any vector $\mathbf{w}$ in $W$, $f(\mathbf{w}) = \mathbf{A}\mathbf{w}$ is also in $W$. Explain why an eigenspace of $\mathbf{A}$ is invariant under $f(\mathbf{x}) = \mathbf{A}\mathbf{x}$.

**17.** A certain beetle can live to a maximum age of 3 years. Let

$x_1 = $ the population of beetles with age up to 1 year,

$x_2 = $ the population of beetles with age between 1 year and 2 years,

$x_3 = $ the population of beetles with age greater than 2 years.

A field study revealed that the Leslie matrix for this species is

$$\mathbf{A} = \begin{bmatrix} 0 & 0.7 & 0.8 \\ 0.8 & 0 & 0 \\ 0 & 0.8 & 0 \end{bmatrix}.$$

(a) Fill in the following table of information. (See Example 3.)

| Rates | Age Categories | | |
|---|---|---|---|
| | $x_1$ | $x_2$ | $x_3$ |
| Birth Rate | | | |
| Survival Rate | | | |

(b) Matrix $\mathbf{A}$ has three distinct eigenvalues. The dominant eigenvalue is $\lambda_1 \approx 1.0285$. Explain the long-term growth of the beetle population.

(c) Suppose we had a total population of 1000 beetles, which were distributed so that $x_1 = 400$, $x_2 = 400$, and $x_3 = 200$. Let

$$\mathbf{s}_0 = \begin{bmatrix} 400 \\ 400 \\ 200 \end{bmatrix}$$

be the initial population vector and $\mathbf{s}_n = \mathbf{A}_n \mathbf{s}_0$ be the population vector after $n$ years. The following data have been computed (and chopped to the nearest whole number):

$$\mathbf{s}_{19} = \begin{bmatrix} 738 \\ 574 \\ 446 \end{bmatrix} \quad \text{and} \quad \mathbf{s}_{20} = \begin{bmatrix} 759 \\ 590 \\ 459 \end{bmatrix}$$

Show that between the nineteenth and twentieth year the total population increased by about 2.85%.

**18.** If the Leslie matrix in Exercise 17 were

$$\mathbf{A} = \begin{bmatrix} 0 & 0.5 & 0.8 \\ 0.8 & 0 & 0 \\ 0 & 0.6 & 0 \end{bmatrix},$$

the dominant eigenvalue is $\lambda_1 \approx 0.9073$. Explain the long-term growth of the beetle population in this case.

## In MATLAB

*The concepts of eigenvalue and eigenvector are of such fundamental importance that it is advisable to have both a geometric and an algebraic feel for them. Here we use MATLAB to provide a geometric look at these ideas by developing visual models involving $2 \times 2$ matrices.*

**ML. 1.** In MATLAB type **help matvec**. Carefully read the description displayed. This routine lets you choose input vectors $\mathbf{x}$ for the matrix product $\mathbf{A}\mathbf{x}$, and then computes the output vector $\mathbf{y} = \mathbf{A}\mathbf{x}$. Both input and output are shown graphically. For a $2 \times 2$ matrix $\mathbf{A}$ we can experiment to approximate eigenvectors of $\mathbf{A}$ by trying to find vectors such that the output is parallel to the input. Start the routine in MATLAB by typing **matvec**; then choose to use the built-in demo. Your screen should look like the following.

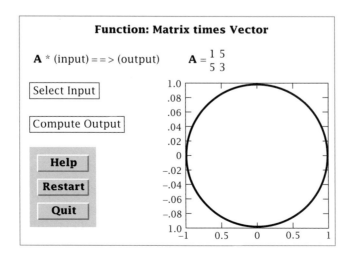

Use your mouse to click on the button **Select Input**. A message will appear directing you to click on the circumference of the circle to select an input vector **x**. After you make your selection, the coordinates of vector **x** are displayed and the vector is drawn from the center of the unit circle. Next click on the button **Compute Output**. The coordinates of $\mathbf{y} = \mathbf{A}\mathbf{x}$ are shown and the output vector is made into a unit vector so we can show it within the unit circle. A **More** button will appear. Click on it in order to make other choices for input vectors, and then repeat the process just described. Try to determine an input vector **x** so that the output $\mathbf{y} = \mathbf{A}\mathbf{x}$ is (nearly) parallel to **x**. Such a vector will be (close to) an eigenvector of matrix **A**. Once you have one eigenvector, search for another nonparallel one in a similar manner. (*Hint*: There is an eigenvector in the first quadrant.)

**ML. 2.** Use the technique with **matvec** described in Exercise ML.1 to approximate eigenvectors of each of the following matrices.

(a) $\begin{bmatrix} 4 & 0 \\ 1 & -3 \end{bmatrix}$.

(b) $\begin{bmatrix} 1 & 5 \\ 1 & 1 \end{bmatrix}$.

**ML. 3.** Using Equation (2) and the geometric approach from **matvec**, explain how you would recognize that the eigenvalue associated with an eigenvector was negative.

**ML. 4.** Searching for an eigenvector of a $2 \times 2$ matrix **A** using the procedure described for **matvec** can be automated by selecting vectors that encompass directions around the unit circle and then checking to see if the input and output are parallel. In MATLAB type **help evecsrch** and read the directions carefully. To execute this routine type **evecsrch** and then choose the built-in demo. The search starts by selecting an input at a randomly chosen point on the unit circle and then graphing the corresponding radius. Next the output is computed, scaled to a unit vector, and displayed in a contrasting color. If the input and output are parallel, the images are retained, otherwise both are erased.

When an eigenvector is found, its components are displayed.

**ML. 5.** Check the approximations you obtained in Exercise ML.2 by using the matrix in the routine **evecsrch**.

**ML. 6.** In routine **matvec** we selected input vectors **x** from the unit circle and displayed their image $\mathbf{y} = \mathbf{A}\mathbf{x}$ scaled to the unit circle. Using routine **evecsrch** we searched for inputs around the unit circle so that the images would be parallel to the input vector. Here we look at the entire set of images of the unit circle. In MATLAB type **help mapcirc** and read the description carefully. To use this routine type **mapcirc**, then select the built-in demo. From the second menu choose not to see the eigenvector information. Give a geometric description of the image of the unit circle in this case.

**ML. 7.** Use each of the following matrices **A** in routine **mapcirc** and write a brief description of the image.

(a) $\mathbf{A} = \begin{bmatrix} 1 & 0 \\ 0 & 4 \end{bmatrix}$.

(b) $\mathbf{A} = \begin{bmatrix} 3 & 0 \\ 0 & -2 \end{bmatrix}$.

(c) $\mathbf{A} = \begin{bmatrix} 1 & 4 \\ 0 & 1 \end{bmatrix}$.

(d) $\mathbf{A} = \begin{bmatrix} 1 & 0 \\ 5 & 1 \end{bmatrix}$.

(e) $\mathbf{A} = \begin{bmatrix} 1 & 3 \\ 1 & 4 \end{bmatrix}$.

(f) Complete the following. *Conjecture*: The image of the unit circle by these matrices is _____.

**ML. 8.** Repeat the procedure of ML.7 for each of the following matrices.

(a) $\mathbf{A} = \begin{bmatrix} 1 & 2 \\ 2 & 4 \end{bmatrix}$.

(b) $\mathbf{A} = \begin{bmatrix} 3 & 7 \\ 0 & 0 \end{bmatrix}$.

(c) $\mathbf{A} = \begin{bmatrix} 1 & 0 \\ -2 & 0 \end{bmatrix}$.

(d) Complete the following. *Conjecture*: The image of the unit circle by these matrices is _____.

(e) What property do these three matrices have in common (besides the fact that they are the same size)?

(f) Using your answer to part (e), explain why you should expect the answer for part (d).

## True/False Review Questions

*Determine whether each of the following statements is true or false.*

1. If **p** is an eigenvector of **A**, then so is $k\mathbf{p}$ for any scalar $k$.

2. An eigenvector **q** of **A** satisfies the property that $\mathbf{A}\mathbf{q}$ is parallel to **q**.

3. If 4 is an eigenvalue of **A**, then 64 is an eigenvalue of $\mathbf{A}^3$.

4. Eigenvectors corresponding to different eigenvalues of matrix **A** are linearly independent.

5. $\left( -3, \begin{bmatrix} -4 \\ 1 \end{bmatrix} \right)$ is an eigenpair of $\mathbf{A} = \begin{bmatrix} -2 & 4 \\ 1 & 1 \end{bmatrix}$.

## Terminology

| | |
|---|---|
| Eigenvalue | Eigenvector |
| Eigenpair | Eigen equation |
| Distinct eigenvalues | Dominant eigenvalue |
| Eigenspace | Invariant subspace |

- Explain how the notion of parallel vectors is related to eigenvectors.
- If $(\lambda, \mathbf{p})$ is an eigenpair of matrix $\mathbf{A}$, then state a (matrix) algebraic relation that must be true.
- Explain why we require that an eigenvector not be the zero vector.
- Explain why we can claim that linear combinations of eigenvectors for a matrix $\mathbf{A}$ are guaranteed to generate another eigenvector of $\mathbf{A}$. (Are there any [technical] restrictions to this claim?)
- Let $\mathbf{A}$ be a $2 \times 2$ matrix with a pair of distinct eigenvalues. Explain why we can say that eigenvectors corresponding to these eigenvalues form a basis for $R^2$.
- Let $\mathbf{A}$ be the transition matrix of a Markov chain. How are eigenvalues and corresponding eigenvectors related to the long-term behavior of the process?
- Let $\mathbf{A}$ be a Leslie matrix. How are eigenvalues and corresponding eigenvectors related to the long-term behavior of the populations?
- What is an eigenspace?

## 4.2 ■ COMPUTING EIGEN INFORMATION FOR SMALL MATRICES

In Section 4.1 we used an algebraic approach to compute the eigenvalues and eigenvectors of a $2 \times 2$ matrix. In this section we use the eigen equation $\mathbf{Ap} = \lambda \mathbf{p}$ together with determinants and the solution of homogeneous linear systems of equations to find eigenpairs. This approach is valid for $n \times n$ matrices of any size. However, some of the computational steps are quite tedious for $n > 4$ except for matrices with special structures. We indicate the potential difficulties, give further properties of eigenvalues, and present a method to approximate the dominant eigenvalue.

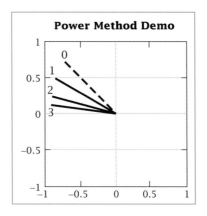

MATLAB routine, **pmdemo.m**

The eigen equation can be rearranged as follows:

$$\mathbf{Ap} = \lambda\mathbf{p} \Leftrightarrow \mathbf{Ap} = \lambda\mathbf{I}_n\mathbf{p} \Leftrightarrow \mathbf{Ap} - \lambda\mathbf{I}_n\mathbf{p} = \mathbf{0} \Leftrightarrow (\mathbf{A} - \lambda\mathbf{I}_n)\mathbf{p} = \mathbf{0}. \qquad (1)$$

The matrix equation $(\mathbf{A} - \lambda\mathbf{I}_n)\mathbf{p} = \mathbf{0}$ is a homogeneous linear system with coefficient matrix $\mathbf{A} - \lambda\mathbf{I}_n$. Since an eigenvector $\mathbf{p}$ cannot be the zero vector, this means we seek a nontrivial solution to the linear system $(\mathbf{A} - \lambda\mathbf{I}_n)\mathbf{p} = \mathbf{0}$. Thus $\mathbf{ns}(\mathbf{A} - \lambda\mathbf{I}_n) \neq \mathbf{0}$ or, equivalently, $\mathbf{rref}(\mathbf{A} - \lambda\mathbf{I}_n)$ must contain a zero row. It follows that matrix $\mathbf{A} - \lambda\mathbf{I}_n$ must be singular, so from Chapter 2,

$$\det(\mathbf{A} - \lambda\mathbf{I}_n) = 0. \qquad (2)$$

Equation (2) is called the **characteristic equation** of matrix $\mathbf{A}$ and solving it for $\lambda$ gives us the eigenvalues of $\mathbf{A}$. Because the determinant is a linear combination of particular products of entries of the matrix, the characteristic equation is really a polynomial equation of degree $n$. We call

$$c(\lambda) = \det(\mathbf{A} - \lambda\mathbf{I}_n) \qquad (3)$$

the **characteristic polynomial** of matrix $\mathbf{A}$. The eigenvalues are the solutions of (2) or, equivalently, the roots of the characteristic polynomial (3). Once we have the $n$ eigenvalues of $\mathbf{A}$, $\lambda_1, \lambda_2, \ldots, \lambda_n$, the corresponding eigenvectors are nontrivial solutions of the homogeneous linear systems

$$(\mathbf{A} - \lambda_i\mathbf{I}_n)\mathbf{p} = \mathbf{0} \quad \text{for } i = 1, 2, \ldots, n \qquad (4)$$

We summarize the computational approach for determining eigenpairs $(\lambda, \mathbf{p})$ as a two-step procedure:

**Step 1.** To find the eigenvalues of $\mathbf{A}$ compute the roots of the characteristic equation $\det(\mathbf{A} - \lambda\mathbf{I}_n) = 0$.

**Step 2.** To find an eigenvector corresponding to an eigenvalue $\mu$, compute a nontrivial solution to the homogeneous linear system $(\mathbf{A} - \mu\mathbf{I}_n)\mathbf{p} = \mathbf{0}$.

EXAMPLE 1    Let $\mathbf{A} = \begin{bmatrix} 1 & 1 \\ -2 & 4 \end{bmatrix}$. To find eigenpairs of $\mathbf{A}$ we follow the two-step procedure just given.

**Step 1.** Find the eigenvalues.

$$c(\lambda) = \det(\mathbf{A} - \lambda\mathbf{I}_2) = \det\left(\begin{bmatrix} 1 - \lambda & 1 \\ -2 & 4 - \lambda \end{bmatrix}\right)$$
$$= (1 - \lambda)(4 - \lambda) + 2 = \lambda^2 - 5\lambda + 6.$$

Thus the characteristic polynomial is a quadratic and the eigenvalues are the solutions of $\lambda^2 - 5\lambda + 6 = 0$. We factor the quadratic to get $(\lambda - 3)(\lambda - 2) = 0$, so the eigenvalues are $\lambda_1 = 3$ and $\lambda_2 = 2$.

**Step 2.** To find corresponding eigenvectors, we solve Equation (4). Case $\lambda_1 = 3$: We have that $(\mathbf{A} - 3\mathbf{I}_2)\mathbf{p} = \mathbf{0}$ has augmented matrix

$$\begin{bmatrix} -2 & 1 & | & 0 \\ -2 & 1 & | & 0 \end{bmatrix}$$

and its rref is

$$\begin{bmatrix} 1 & -\frac{1}{2} & | & 0 \\ 0 & 0 & | & 0 \end{bmatrix}. \quad \text{(Verify.)}$$

Thus if $\mathbf{p} = \begin{bmatrix} x_1 \\ x_2 \end{bmatrix}$ we have $x_1 = \frac{1}{2}x_2$ so

$$\mathbf{p} = \begin{bmatrix} \frac{1}{2}x_2 \\ x_2 \end{bmatrix}, \quad x_2 \neq 0.$$

Choosing $x_2 = 2$, to conveniently get integer entries, gives eigenpair

$$(\lambda_1, \mathbf{p}_1) = \left(3, \begin{bmatrix} 1 \\ 2 \end{bmatrix}\right).$$

Case $\lambda_2 = 2$: We have that $(\mathbf{A} - 2\mathbf{I}_2)\mathbf{p} = \mathbf{0}$ has augmented matrix

$$\begin{bmatrix} -1 & 1 & 0 \\ -2 & 2 & 0 \end{bmatrix}$$

and its rref is

$$\begin{bmatrix} 1 & -1 & 0 \\ 0 & 0 & 0 \end{bmatrix}. \quad \text{(Verify.)}$$

Thus if $\mathbf{p} = \begin{bmatrix} x_1 \\ x_2 \end{bmatrix}$ we have $x_1 = x_2$ so

$$\mathbf{p} = \begin{bmatrix} x_2 \\ x_2 \end{bmatrix} = x_2 \begin{bmatrix} 1 \\ 1 \end{bmatrix}, \quad x_2 \neq 0.$$

Choosing $x_2 = 1$ gives eigenpair

$$(\lambda_2, \mathbf{p}_2) = \left(2, \begin{bmatrix} 1 \\ 1 \end{bmatrix}\right). \qquad ■$$

Given that $\lambda$ is an eigenvalue of $\mathbf{A}$, then we know that matrix $\mathbf{A} - \lambda\mathbf{I}_n$ is singular and hence $\mathbf{rref}(\mathbf{A} - \lambda\mathbf{I}_n)$ will have at least one zero row. A homogeneous linear system, as in (4), whose coefficient matrix has rref with at least one zero row will have a solution set with at least one free variable. The free variables can be chosen to have any value as long as the resulting solution is not the zero vector. In Example 1 for eigenvalue $\lambda_1 = 3$, the general solution of (4) was

$$x_2 \begin{bmatrix} \frac{1}{2} \\ 1 \end{bmatrix}.$$

The free variable in this case could be any nonzero value. We chose $x_2 = 2$ to avoid fractions, but this is not required. If we chose $x_2 = \frac{1}{7}$, then

$$(\lambda_1, \mathbf{p}_1) = \left(3, \begin{bmatrix} \frac{1}{14} \\ \frac{1}{7} \end{bmatrix}\right)$$

is a valid eigenpair. (See Property 1 in Section 4.1.)

EXAMPLE 2    In Example 1(c) of Section 4.1, for an angle of rotation $\pi/4$ radians, the matrix of the transformation is

$$\mathbf{A} = \frac{1}{2}\begin{bmatrix} \sqrt{2} & -\sqrt{2} \\ \sqrt{2} & \sqrt{2} \end{bmatrix}.$$

To find its eigenpairs we follow our two-step procedure.

**Step 1.**   Find the eigenvalues.

$$c(\lambda) = \det(\mathbf{A} - \lambda \mathbf{I}_2) = \det\left(\begin{bmatrix} \frac{\sqrt{2}}{2} - \lambda & -\frac{\sqrt{2}}{2} \\ \frac{\sqrt{2}}{2} & \frac{\sqrt{2}}{2} - \lambda \end{bmatrix}\right)$$

$$= \left(\frac{\sqrt{2}}{2} - \lambda\right)^2 + \frac{1}{2} = \lambda^2 - \sqrt{2}\,\lambda + 1.$$

Applying the quadratic formula to find the roots of $c(\lambda) = 0$, we get

$$\lambda = \frac{\sqrt{2} \pm i\sqrt{2}}{2}.$$

Thus we have a pair of complex eigenvalues

$$\lambda_1 = \frac{\sqrt{2} + i\sqrt{2}}{2} \quad \text{and} \quad \lambda_2 = \frac{\sqrt{2} - i\sqrt{2}}{2}$$

even though all the entries in $\mathbf{A}$ are real.

**Step 2.**   To find corresponding eigenvectors, we solve Equation (4).

Case $\lambda_1 = \dfrac{\sqrt{2} + i\sqrt{2}}{2}$: We have that

$$\left(\mathbf{A} - \frac{\sqrt{2} + i\sqrt{2}}{2}\mathbf{I}_2\right)\mathbf{p} = \mathbf{0}$$

has augmented matrix

$$\begin{bmatrix} -i\frac{\sqrt{2}}{2} & -\frac{\sqrt{2}}{2} & \bigg| & 0 \\ \frac{\sqrt{2}}{2} & -i\frac{\sqrt{2}}{2} & \bigg| & 0 \end{bmatrix}$$

and its rref is

$$\begin{bmatrix} 1 & -i & \big| & 0 \\ 0 & 0 & \big| & 0 \end{bmatrix}. \quad \text{(Verify.)}$$

Thus if $\mathbf{p} = \begin{bmatrix} x_1 \\ x_2 \end{bmatrix}$ we have $x_1 = ix_2$ so

$$\mathbf{p} = \begin{bmatrix} ix_2 \\ x_2 \end{bmatrix} = x_2 \begin{bmatrix} i \\ 1 \end{bmatrix}, \quad x_2 \neq 0.$$

Choosing $x_2 = 1$ gives eigenpair

$$(\lambda_1, \mathbf{p}_1) = \left(\frac{\sqrt{2} + i\sqrt{2}}{2}, \begin{bmatrix} i \\ 1 \end{bmatrix}\right).$$

Case $\lambda_2 = \dfrac{\sqrt{2} - i\sqrt{2}}{2}$: By calculations similar to the previous case, we get eigenpair

$$(\lambda_2, \mathbf{p}_2) = \left(\frac{\sqrt{2} - i\sqrt{2}}{2}, \begin{bmatrix} -i \\ 1 \end{bmatrix}\right). \quad \text{(Verify.)}$$

The eigenvalues are the roots of the characteristic polynomial; thus it is possible that a numerical value can be a root more than once. For example,

$$\lambda^3 - 6\lambda^2 + 9\lambda = \lambda(\lambda^2 - 6\lambda + 9) = \lambda(\lambda - 3)^2$$

has roots 0, 3, 3. We say 3 is a **multiple root** or **repeated root**. The number of times the root is repeated is called its **multiplicity**. Thus it follows that a matrix can have repeated eigenvalues.

EXAMPLE 3   Let

$$A = \begin{bmatrix} 0 & -2 & 1 \\ 1 & 3 & -1 \\ 0 & 0 & 1 \end{bmatrix}.$$

To find eigenpairs of **A** we follow the two-step procedure given previously.

**Step 1.**   Find the eigenvalues.

$$c(\lambda) = \det(A - \lambda I_3) = \det\left(\begin{bmatrix} -\lambda & -2 & 1 \\ 1 & 3 - \lambda & -1 \\ 0 & 0 & 1 - \lambda \end{bmatrix}\right)$$

$$= -\lambda(3 - \lambda)(1 - \lambda) + 2(1 - \lambda)$$

$$= (1 - \lambda)(\lambda^2 - 3\lambda + 2)$$

$$= (1 - \lambda)(\lambda - 2)(\lambda - 1).$$

Thus the eigenvalues are $\lambda_1 = \lambda_2 = 1$ and $\lambda_3 = 2$. We have a repeated eigenvalue of multiplicity 2.

**Step 2.**   To find corresponding eigenvectors, we solve Equation (4). Case $\lambda_1 = \lambda_2 = 1$: We have that $(A - 1I_3)p = 0$ has augmented matrix

$$\begin{bmatrix} -1 & -2 & 1 & | & 0 \\ 1 & 2 & -1 & | & 0 \\ 0 & 0 & 0 & | & 0 \end{bmatrix}$$

and its rref is

$$\begin{bmatrix} 1 & 2 & -1 & | & 0 \\ 0 & 0 & 0 & | & 0 \\ 0 & 0 & 0 & | & 0 \end{bmatrix}. \quad \text{(Verify.)}$$

Thus if $p = \begin{bmatrix} x_1 \\ x_2 \\ x_3 \end{bmatrix}$ we have $x_1 = -2x_2 + x_3$ so

$$p = \begin{bmatrix} -2x_2 + x_3 \\ x_2 \\ x_3 \end{bmatrix} = x_2 \begin{bmatrix} -2 \\ 1 \\ 0 \end{bmatrix} + x_3 \begin{bmatrix} 1 \\ 0 \\ 1 \end{bmatrix}.$$

As long as we choose free variables $x_2$ and $x_3$ so that they are not both zero, we have an eigenvector corresponding to eigenvalue 1. From Chapter 1 we know that vectors

$$p_1 = \begin{bmatrix} -2 \\ 1 \\ 0 \end{bmatrix} \quad \text{and} \quad p_2 = \begin{bmatrix} 1 \\ 0 \\ 1 \end{bmatrix}$$

are linearly independent (explain why this is true) and hence we have two linearly independent eigenvectors corresponding to eigenvalue 1. These vectors form a basis for the eigenspace of eigenvalue 1. (See Exercise 9 in Section 4.1.)

Case $\lambda_3 = 2$: We have that $(\mathbf{A} - 2\mathbf{I}_3)\mathbf{p} = \mathbf{0}$ has augmented matrix

$$\left[\begin{array}{ccc|c} -2 & -2 & 1 & 0 \\ 1 & 1 & -1 & 0 \\ 0 & 0 & -1 & 0 \end{array}\right]$$

and its rref is

$$\left[\begin{array}{ccc|c} 1 & 1 & 0 & 0 \\ 0 & 0 & 1 & 0 \\ 0 & 0 & 0 & 0 \end{array}\right]. \quad \text{(Verify.)}$$

Thus if $\mathbf{p} = \begin{bmatrix} x_1 \\ x_2 \\ x_3 \end{bmatrix}$ we have $x_1 = -x_2$ and $x_3 = 0$ so

$$\mathbf{p} = \begin{bmatrix} -x_2 \\ x_2 \\ 0 \end{bmatrix} = x_2 \begin{bmatrix} -1 \\ 1 \\ 0 \end{bmatrix}, \quad x_2 \neq 0.$$

Choosing $x_2 = 1$ gives the eigenpair

$$(\lambda_3, \mathbf{p}_3) = \left( 2, \begin{bmatrix} -1 \\ 1 \\ 0 \end{bmatrix} \right).$$

■

Determining eigenpairs using our two-step procedure is not always easy even for $3 \times 3$ matrices. (See Example 4.) The solution of the characteristic equation requires that we find all the roots of a polynomial of degree 3. In some cases it is not simple to factor the cubic polynomial. While there is a formula for determining the roots of a cubic equation, known as Cardano's formula (similar to the quadratic formula) it is cumbersome to use. For $n \times n$ matrices, $n > 3$ the task of finding all the roots of the characteristic polynomial is even harder.

EXAMPLE 4　The Leslie matrix from Example 3 in Section 4.1 is

$$\mathbf{A} = \begin{bmatrix} 0.1 & 0.8 & 0.4 \\ 0.9 & 0 & 0 \\ 0 & 0.7 & 0 \end{bmatrix}.$$

Its characteristic equation is (verify)

$$c(\lambda) = \det(\mathbf{A} - \lambda \mathbf{I}_3) = -\lambda^3 + \tfrac{1}{10}\lambda^2 + \tfrac{18}{25}\lambda + \tfrac{63}{250}$$
$$= -\lambda^3 + 0.1\lambda^2 + 0.72\lambda + 0.252 = 0.$$

It is highly unlikely that we can factor this equation by hand. However, using an approximation procedure we can show that its roots, listed to four decimal places, are

$$\lambda_1 \approx 1.0331, \quad \lambda_2 \approx -0.4665 + 0.1621i, \quad \lambda_3 \approx -0.4665 - 0.1621i.$$

Unfortunately, we cannot use approximate eigenvalues to determine corresponding eigenvectors from (4). If $\lambda$ is not exactly an eigenvalue of $\mathbf{A}$, then $\mathbf{A} - \lambda \mathbf{I}_n$ will be nonsingular and (4) will have only the zero solution.

■

As we saw for Markov processes and Leslie population models in Section 4.1, knowing a particular eigenvalue and possibly an associated eigenvector can provide information about the behavior of the model. In the case of a Markov process we wanted to determine a stable state that corresponds to an eigenvector associated with eigenvalue $\lambda = 1$. A natural question is,

Does the transition matrix of a Markov process always have eigenvalue 1?

To show that the answer is yes, we need another property of eigenvalues. In Section 4.1 we developed Properties 1 through 4, so we continue the list here with Property 5.

**Property 5.** $\mathbf{A}$ and $\mathbf{A}^T$ have the same eigenvalues.
We see this by showing that $\mathbf{A}$ and $\mathbf{A}^T$ have the same characteristic polynomial:

$$\det(\mathbf{A}^T - \lambda \mathbf{I}_n) = \det(\mathbf{A}^T - \lambda \mathbf{I}_n^T) \quad \text{(since } \mathbf{I}_n \text{ is symmetric)}$$
$$= \det((\mathbf{A} - \lambda \mathbf{I}_n)^T) \quad \text{(by properties of the transpose)}$$
$$= \det(\mathbf{A} - \lambda \mathbf{I}_n). \quad \text{(by properties of determinants)}$$

Thus the eigenvalues must be the same since they are the roots of the characteristic polynomial.

A transition matrix $\mathbf{A}$ of a Markov process has entries $0 \le a_{ij} \le 1$ with the sum of the entries in each column equal to 1. Thus the sum of the entries in each row of $\mathbf{A}^T$ is 1 and we have

$$\mathbf{A}^T \begin{bmatrix} 1 \\ 1 \\ \vdots \\ 1 \end{bmatrix} = \begin{bmatrix} 1 \\ 1 \\ \vdots \\ 1 \end{bmatrix}.$$

Hence 1 is an eigenvalue of $\mathbf{A}^T$ and also of transition matrix $\mathbf{A}$.
    The characteristic equation is derived from a determinant:

$$c(\lambda) = \det(\mathbf{A} - \lambda \mathbf{I}_n) = 0.$$

Thus matrices for which the determinant calculation is especially simple may yield their eigenvalues with little work. We have the following property.

**Property 6.** If $\mathbf{A}$ is diagonal, upper triangular, or lower triangular, then its eigenvalues are its diagonal entries.

You are asked to verify this result in Exercises 42 and 43.
    Symmetric matrices arise frequently in applications. The next two properties provide information about their eigenvalues and eigenvectors.

**Property 7.** The eigenvalues of a symmetric matrix are real.

**Property 8.** Eigenvectors corresponding to distinct eigenvalues of a symmetric matrix are orthogonal.

To verify these properties, see Exercises 45 and 46, respectively.

For a square matrix $\mathbf{A}$ we have the following important questions:

Is $\mathbf{A}$ nonsingular or singular? (See Chapter 2.)

Is $\det(\mathbf{A})$ nonzero or zero? (See Chapter 3.)

Next we connect these to properties of the eigenvalues.

**Property 9.** $|\det(\mathbf{A})|$ is the product of the absolute values of the eigenvalues of $\mathbf{A}$. We see this as follows: If $\lambda_1, \lambda_2, \ldots, \lambda_n$ are the eigenvalues of $\mathbf{A}$, then they are the roots of its characteristic polynomial. Hence

$$c(\lambda) = \det(\mathbf{A} - \lambda \mathbf{I}_n) = (\lambda - \lambda_1)(\lambda - \lambda_2) \cdots (\lambda - \lambda_n).$$

Evaluating this polynomial at $\lambda = 0$ gives

$$\det(\mathbf{A}) = (-1)^n \lambda_1 \lambda_2 \cdots \lambda_n$$

so $|\det(\mathbf{A})| = |\lambda_1| \, |\lambda_2| \cdots |\lambda_n|$.

**Property 10.** $\mathbf{A}$ is nonsingular if and only if 0 is not an eigenvalue of $\mathbf{A}$, or equivalently, $\mathbf{A}$ is singular if and only if 0 is an eigenvalue of $\mathbf{A}$.
We see this as follows: $\mathbf{A}$ is nonsingular if and only if $\det(\mathbf{A}) \neq 0$. Since

$$|\det(\mathbf{A})| = |\lambda_1| \, |\lambda_2| \cdots |\lambda_n|,$$

$\det(\mathbf{A}) \neq 0$ if and only if zero is not an eigenvalue of $\mathbf{A}$. See also Exercise 15 in Section 4.1 for an alternate approach.

### Application:   The Power Method (Optional)

The power method is a way to approximate the dominant eigenvalue; that is, the eigenvalue of largest absolute value. We do this indirectly by first approximating an eigenvector associated with the dominant eigenvalue. Our development uses a number of properties of eigenvalues and information about bases and subspaces.

Let $\mathbf{A}$ be a real $n \times n$ matrix with eigenpairs $\{(\lambda_i, \mathbf{p}_i) \mid i = 1, 2, \ldots, n\}$ where we have numbered eigenpairs so that $|\lambda_1| > |\lambda_i|$, $i = 2, \ldots, n$, and $\{\mathbf{p}_1, \mathbf{p}_2, \ldots, \mathbf{p}_n\}$ are linearly independent; that is, they form a basis for $R^n$. To approximate the dominant eigenvalue $\lambda_1$ we proceed as follows. Let $\mathbf{u}_0$ be an initial guess for an eigenvector associated with $\lambda_1$. We assume that

$$\mathbf{u}_0 = \sum_{i=1}^{n} c_i \mathbf{p}_i \quad \text{where } c_1 \neq 0. \tag{5}$$

The power method computes a sequence of vectors $\{\mathbf{u}_k\}$ given by

$$\begin{aligned}
\mathbf{u}_1 &= \mathbf{A}\mathbf{u}_0, \\
\mathbf{u}_2 &= \mathbf{A}\mathbf{u}_1 = \mathbf{A}^2 \mathbf{u}_0, \\
\mathbf{u}_3 &= \mathbf{A}\mathbf{u}_2 = \mathbf{A}^3 \mathbf{u}_0, \\
&\vdots \\
\mathbf{u}_k &= \mathbf{A}^k \mathbf{u}_0, \\
&\vdots
\end{aligned} \tag{6}$$

By Property 3 of Section 4.1, the eigenvalues of $\mathbf{A}^k$ are the powers of those of $\mathbf{A}$. Our goal is to show how the sequence $\{\mathbf{u}_k\}$ can be used to approximate the dominant eigenvalue $\lambda_1$. We substitute (5) into the general expression in (6) to get

$$\mathbf{u}_k = \mathbf{A}^k \sum_{i=1}^{n} c_i \mathbf{p}_i = \sum_{i=1}^{n} c_i \mathbf{A}^k \mathbf{p}_i = \sum_{i=1}^{n} c_i (\lambda_i)^k \mathbf{p}_i$$

$$= c_1 (\lambda_1)^k \mathbf{p}_1 + c_2 (\lambda_2)^k \mathbf{p}_2 + \cdots + c_n (\lambda_n)^k \mathbf{p}_n.$$

Next we factor $(\lambda_1)^k$ out of each term to give

$$\mathbf{u}_k = (\lambda_1)^k \left( c_1 \mathbf{p}_1 + c_2 \frac{(\lambda_2)^k}{(\lambda_1)^k} \mathbf{p}_2 + \cdots + c_n \frac{(\lambda_n)^k}{(\lambda_1)^k} \mathbf{p}_n \right). \tag{7}$$

Since $|\lambda_i / \lambda_1| < 1$ for $i \geq 2$, as $k \to \infty$ we conclude that $\mathbf{u}_k$ approaches a multiple of eigenvector $\mathbf{p}_1$; that is, $\mathbf{u}_k \approx (\lambda_1)^k c_1 \mathbf{p}_1$. But also $\mathbf{u}_{k+1} \approx (\lambda_1)^{k+1} c_1 \mathbf{p}_1$. If we take the ratio of corresponding entries of vectors $\mathbf{u}_k$ and $\mathbf{u}_{k+1}$ we have

$$\lambda_1 \approx \frac{\text{entry}_j(\mathbf{u}_{k+1})}{\text{entry}_j(\mathbf{u}_k)}$$

assuming the denominator is not zero. We demonstrate this graphically in Example 5 and then from a numerical point of view in Example 6.

### EXAMPLE 5    Let

$$\mathbf{A} = \begin{bmatrix} 5 & 1 \\ 1 & -2 \end{bmatrix} \quad \text{and} \quad \mathbf{u}_0 = \begin{bmatrix} 1 \\ 1 \end{bmatrix}.$$

Figure 1 shows the sequence of vectors

$$\mathbf{u}_0, \quad \mathbf{u}_1 = \mathbf{A}\mathbf{u}_0, \quad \mathbf{u}_2 = \mathbf{A}\mathbf{u}_1, \quad \mathbf{u}_3 = \mathbf{A}\mathbf{u}_2, \quad \ldots .$$

For graphical display purposes we have scaled each of the vectors in the figure so that they are unit vectors. This preserves the direction of the vectors so the convergence of the approximations is not changed. We see that $\mathbf{u}_5$ and $\mathbf{u}_6$ are very nearly parallel. The numerical steps for approximating the dominant eigenvalue of a $3 \times 3$ matrix are shown in detail in Example 6.  ∎

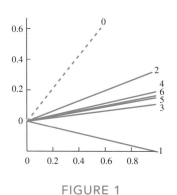

FIGURE 1

### EXAMPLE 6    Let

$$\mathbf{A} = \begin{bmatrix} 7 & 4 & 5 \\ -2 & 2 & -2 \\ 1 & 0 & 3 \end{bmatrix}.$$

To approximate the dominant eigenvalue of $\mathbf{A}$ we start with

$$\mathbf{u}_0 = \begin{bmatrix} 1 \\ 1 \\ 1 \end{bmatrix}$$

as an initial guess for a corresponding eigenvector. Next we compute vectors $\mathbf{u}_1 = \mathbf{A}\mathbf{u}_0, \mathbf{u}_2 = \mathbf{A}\mathbf{u}_1, \mathbf{u}_3 = \mathbf{A}\mathbf{u}_2, \ldots$. Then we form the ratio of corresponding components. As shown in the following table, the size of the entries in the vectors can grow quite rapidly and we may need the number $k$ of vectors to be large to get an accurate approximation. (We have only included enough steps to indicate the behavior of the method.)

It appears that the ratios of corresponding entries are converging to 6. Hence we conjecture that the dominant eigenvalue is 6.  ∎

| $k$ | $\mathbf{u}_k$ | | | entries ($\mathbf{u}_{k+1}/\mathbf{u}_k$) | | |
|---|---|---|---|---|---|---|
| 0 | 1 | 1 | 1 | | | |
| 1 | 16 | $-2$ | 4 | 16 | $-2$ | 4 |
| 2 | 124 | $-44$ | 28 | 7.7500 | 22.0000 | 7.0000 |
| 3 | 832 | $-392$ | 208 | 6.7097 | 8.9091 | 7.4286 |
| 4 | 5296 | $-2864$ | 1456 | 6.3654 | 7.3061 | 7.0000 |
| 5 | 32896 | $-19232$ | 9664 | 6.2115 | 6.7151 | 6.6374 |
| 6 | 201664 | $-123584$ | 61888 | 6.1304 | 6.4260 | 6.4040 |
| 10 | 269999104 | $-177204224$ | 88603648 | 6.0235 | 6.0719 | 6.0716 |
| 15 | $2.1137 * 10^{12}$ | $-1.4063 * 10^{12}$ | $0.7031 * 10^{12}$ | 6.0030 | 6.0092 | 6.0092 |

There is a modification of the power method, called the **scaled power method**, which inhibits the growth of the size of the entries of the vectors $\mathbf{u}_k$ and is recommended for use in computational work. We will not pursue this here.

# EXERCISES 4.2

*In Exercises 1–6, determine the characteristic polynomial of each given matrix. Express your answer as a polynomial in standard form.*

**1.** $\begin{bmatrix} 4 & 2 \\ 3 & 3 \end{bmatrix}$.

**2.** $\begin{bmatrix} 5 & -1 \\ 2 & 3 \end{bmatrix}$.

**3.** $\begin{bmatrix} 5 & -1 \\ 2 & 2 \end{bmatrix}$.

**4.** $\begin{bmatrix} 5 & -2 & 0 \\ 0 & -1 & 2 \\ 0 & 0 & 6 \end{bmatrix}$.

**5.** $\begin{bmatrix} 1 & 2 & 1 \\ 0 & 1 & 2 \\ -1 & 3 & 2 \end{bmatrix}$.

**6.** $\begin{bmatrix} 1+i & 2 \\ -i & i \end{bmatrix}$.

*In Exercises 7–15, find all the eigenvalues of each given matrix.*

**7.** $\begin{bmatrix} 3 & 2 \\ 6 & 2 \end{bmatrix}$.

**8.** $\begin{bmatrix} 3 & 2 \\ 6 & 4 \end{bmatrix}$.

**9.** $\begin{bmatrix} 5 & 2 \\ -1 & 3 \end{bmatrix}$.

**10.** $\begin{bmatrix} 2 & -2 & 3 \\ 0 & 3 & -2 \\ 0 & -1 & 2 \end{bmatrix}$.

**11.** $\begin{bmatrix} 2 & 1 & 2 \\ 2 & 2 & -2 \\ 3 & 1 & 1 \end{bmatrix}$.

**12.** $\begin{bmatrix} -2 & -4 & -8 \\ 1 & 0 & 0 \\ 0 & 1 & 0 \end{bmatrix}$.

**13.** $\begin{bmatrix} 0 & 1 \\ -1 & 0 \end{bmatrix}$.

**14.** $\begin{bmatrix} -1 & -1+i \\ 1 & 0 \end{bmatrix}$.

**15.** $\begin{bmatrix} 2-i & 2i & 0 \\ 1 & 0 & 0 \\ 0 & 1 & 0 \end{bmatrix}$.

*In Exercises 16–21, find an eigenvector corresponding to the eigenvalue λ.*

**16.** $\begin{bmatrix} 4 & 2 \\ 3 & 3 \end{bmatrix}, \lambda = 6$.

**17.** $\begin{bmatrix} 5 & 2 \\ -1 & 2 \end{bmatrix}, \lambda = 4$.

**18.** $\begin{bmatrix} 5 & 2 \\ -1 & 3 \end{bmatrix}, \lambda = 4+i$.

**19.** $\begin{bmatrix} 4 & 2 & 1 \\ 0 & 3 & 2 \\ 0 & 0 & 3 \end{bmatrix}, \lambda = 3$.

**20.** $\begin{bmatrix} -9 & -3 & 16 \\ 13 & 7 & 16 \\ 3 & 3 & 10 \end{bmatrix}, \lambda = -6$.

**21.** $\begin{bmatrix} 0 & 0 & 0 \\ 0 & 1 & 0 \\ 1 & 0 & 1 \end{bmatrix}, \lambda = 1$.

*In Exercises 22–28, find all the eigenvalues and a corresponding eigenvector for each given matrix.*

**22.** $\begin{bmatrix} 1 & 4 \\ 1 & -2 \end{bmatrix}$.

**23.** $\begin{bmatrix} 0 & -9 \\ 1 & 0 \end{bmatrix}$.

**24.** $\begin{bmatrix} -3 & 6 & 8 \\ 1 & 0 & 0 \\ 0 & 1 & 0 \end{bmatrix}$.

**25.** $\begin{bmatrix} 0 & 0 & 3 \\ 1 & 0 & -1 \\ 0 & 1 & 3 \end{bmatrix}$.

**26.** $\begin{bmatrix} 4 & 2 & -4 \\ 1 & 5 & -4 \\ 0 & 0 & 6 \end{bmatrix}$.

**27.** $\begin{bmatrix} 0 & 1 \\ -1 & 0 \end{bmatrix}$.

28. $\begin{bmatrix} 0 & -1 & 0 \\ 1 & 0 & 0 \\ 0 & 1 & 0 \end{bmatrix}$.

*In Exercises 29–33, determine a basis for the eigenspace corresponding to the eigenvalue λ.*

29. $\mathbf{A} = \begin{bmatrix} 0 & 0 & 1 \\ 0 & 1 & 0 \\ 1 & 0 & 0 \end{bmatrix}$, $\lambda = 1$.

30. $\mathbf{A} = \begin{bmatrix} 2 & 1 & 0 \\ 1 & 2 & 1 \\ 0 & 1 & 2 \end{bmatrix}$, $\lambda = 2$.

31. $\mathbf{A} = \begin{bmatrix} 3 & 0 & 0 \\ -2 & 3 & -2 \\ 2 & 0 & 5 \end{bmatrix}$, $\lambda = 3$.

32. $\mathbf{A} = \begin{bmatrix} 0 & -4 & 0 \\ 1 & 0 & 0 \\ 0 & 1 & 0 \end{bmatrix}$, $\lambda = 2i$.

33. $\mathbf{A} = \begin{bmatrix} 4 & 2 & 0 & 0 \\ 3 & 3 & 0 & 0 \\ 0 & 0 & 2 & 5 \\ 0 & 0 & 0 & 2 \end{bmatrix}$, $\lambda = 2$.

*A matrix $\mathbf{A}$ is called **defective** if $\mathbf{A}$ has an eigenvalue $\lambda$ of multiplicity $m > 1$ for which the corresponding eigenspace has a basis of fewer than $m$ vectors; that is, the dimension of the eigenspace corresponding to $\lambda$ is less than $m$. In Exercises 34–38, use the indicated eigenvalue of the given matrix to determine whether the matrix is defective.*

34. $\mathbf{A} = \begin{bmatrix} 8 & 7 \\ 0 & 8 \end{bmatrix}$, $\lambda = 8, 8$.

35. $\mathbf{A} = \begin{bmatrix} 3 & 0 & 0 \\ -2 & 3 & -2 \\ 2 & 0 & 5 \end{bmatrix}$, $\lambda = 3, 3, 5$.

36. $\mathbf{A} = \begin{bmatrix} 3 & 3 & 3 \\ 3 & 3 & 3 \\ -3 & -3 & -3 \end{bmatrix}$, $\lambda = 0, 0, 3$.

37. $\mathbf{A} = \begin{bmatrix} 0 & 0 & 1 & 0 \\ 0 & 0 & 0 & -1 \\ 1 & 0 & 0 & 0 \\ 0 & -1 & 0 & 0 \end{bmatrix}$, $\lambda = 1, 1, -1, -1$.

38. $\mathbf{A} = \begin{bmatrix} 4 & 2 & 0 & 0 \\ 3 & 3 & 0 & 0 \\ 0 & 0 & 2 & 5 \\ 0 & 0 & 0 & 2 \end{bmatrix}$, $\lambda = 1, 2, 2, 6$.

39. Let $\mathbf{A}$ be $n \times n$ with $n$ distinct eigenvalues. Explain why $\mathbf{A}$ is not defective.

40. Let $\mathbf{A}$ be the matrix in Example 1 and $\mathbf{P} = \begin{bmatrix} \mathbf{p}_1 & \mathbf{p}_2 \end{bmatrix}$ the matrix whose columns are eigenvectors corresponding to $\lambda_1$ and $\lambda_2$, respectively.

(a) Explain why $\mathbf{P}$ is nonsingular.

(b) Compute $\mathbf{P}^{-1}$.

(c) Compute $\mathbf{P}^{-1}\mathbf{A}\mathbf{P}$ and describe it in terms of the eigen information for $\mathbf{A}$.

41. Let $\mathbf{A}$ be the matrix in Example 3 and $\mathbf{P} = \begin{bmatrix} \mathbf{p}_1 & \mathbf{p}_2 & \mathbf{p}_3 \end{bmatrix}$ the matrix whose columns are eigenvectors corresponding to $\lambda_1$, $\lambda_2$, and $\lambda_3$, respectively.

(a) Explain why $\mathbf{P}$ is nonsingular.

(b) Compute $\mathbf{P}^{-1}$.

(c) Compute $\mathbf{P}^{-1}\mathbf{A}\mathbf{P}$ and describe it in terms of the eigen information for $\mathbf{A}$.

42. Explain why the eigenvalues of a diagonal matrix are its diagonal entries.

43. Explain why the eigenvalues of a upper (lower) triangular matrix are its diagonal entries.

44. Let $c$ be a complex number with real part $a$ and imaginary part $b$, so that $c = a+bi$, where $i = \sqrt{-1}$. (See Appendix 1.) The conjugate of $c$, denoted $\bar{c}$, is $\bar{c} = a - bi$. The conjugate of a matrix $\mathbf{A}$ is denoted by $\overline{\mathbf{A}}$ and $\overline{\mathbf{A}} = \begin{bmatrix} \overline{a_{jk}} \end{bmatrix}$; that is, we take the conjugate of each entry. Verify the following properties of the conjugate.

(a) $\overline{\overline{\mathbf{A}}} = \mathbf{A}$.

(b) $\overline{\mathbf{A} + \mathbf{B}} = \overline{\mathbf{A}} + \overline{\mathbf{B}}$.

(c) $\overline{\mathbf{A}\mathbf{B}} = \overline{\mathbf{A}}\,\overline{\mathbf{B}}$.

(d) $\overline{k\mathbf{A}} = k\overline{\mathbf{A}}$, for $k$ a real scalar.

(e) $\overline{k\mathbf{A}} = \bar{k}\,\overline{\mathbf{A}}$, for $k$ a complex scalar.

(f) $\overline{\mathbf{A}}^T = \overline{\mathbf{A}^T}$.

45. If $\mathbf{A}$ is an $n \times n$ symmetric matrix with eigenpair $(\lambda, \mathbf{p})$ then $\lambda$ is a real scalar. We verify this in the following steps by showing that $\lambda = \bar{\lambda}$. (See Exercise 44 and Appendix 1.)

(a) Explain why $\mathbf{A}\mathbf{p} = \lambda\mathbf{p}$.

(b) Explain why $\overline{\mathbf{A}\mathbf{p}} = \mathbf{A}\overline{\mathbf{p}}$.

(c) Explain why $\overline{\lambda\mathbf{p}} = \bar{\lambda}\overline{\mathbf{p}}$.

(d) Explain why $\overline{\mathbf{A}\mathbf{p}} = \overline{\lambda\mathbf{p}}$ and so $\mathbf{A}\overline{\mathbf{p}} = \bar{\lambda}\overline{\mathbf{p}}$.

(e) Explain why $(\mathbf{A}\overline{\mathbf{p}})^T = (\bar{\lambda}\overline{\mathbf{p}})^T$ implies that $\overline{\mathbf{p}}^T\mathbf{A} = \bar{\lambda}\overline{\mathbf{p}}^T$.

(f) In part (e) we can multiply both sides on the right by $\mathbf{p}$ to get $\overline{\mathbf{p}}^T\mathbf{A}\mathbf{p} = \bar{\lambda}\,\overline{\mathbf{p}}^T\mathbf{p}$. Explain why this is the same as $\lambda\overline{\mathbf{p}}^T\mathbf{p} = \bar{\lambda}\,\overline{\mathbf{p}}^T\mathbf{p}$.

(g) Explain why $\mathbf{p} \neq \mathbf{0}$.

(h) Explain why $\overline{\mathbf{p}}^T\mathbf{p} \neq \mathbf{0}$.

(i) Using the results of parts (f) through (h), explain why we can say that $\lambda = \bar{\lambda}$.

46. Show that eigenvectors corresponding to distinct eigenvalues of a symmetric matrix $\mathbf{A}$ are orthogonal. That is, if $(\lambda, \mathbf{p})$ and $(\mu, \mathbf{q})$ are eigenpairs of symmetric matrix $\mathbf{A}$ with $\lambda \neq \mu$, then $\mathbf{p} \cdot \mathbf{q} = \mathbf{p}^T\mathbf{q} = 0$.

(a) Explain why $\mathbf{Ap} = \lambda\mathbf{p}$ and $\mathbf{Aq} = \mu\mathbf{q}$.

(b) Multiply $\mathbf{Aq} = \mu\mathbf{q}$ on the left by $\mathbf{p}^T$ to get $\mathbf{p}^T\mathbf{Aq} = \mu\mathbf{p}^T\mathbf{q}$. Explain why we can replace $\mathbf{A}$ by $\mathbf{A}^T$ to get $\mathbf{p}^T\mathbf{A}^T\mathbf{q} = \mu\mathbf{p}^T\mathbf{q}$.

(c) Explain why $\mathbf{p}^T\mathbf{A}^T\mathbf{q} = \mu\mathbf{p}^T\mathbf{q}$ is the same as $(\mathbf{Ap})^T\mathbf{q} = \mu\mathbf{p}^T\mathbf{q}$ and why we next can write $\lambda\mathbf{p}^T\mathbf{q} = \mu\mathbf{p}^T\mathbf{q}$.

(d) Using $\lambda\mathbf{p}^T\mathbf{q} = \mu\mathbf{p}^T\mathbf{q}$, explain why we can conclude $\mathbf{p}^T\mathbf{q} = 0$.

47. If $\text{col}_j(\mathbf{A}) = k\text{col}_j(\mathbf{I}_n)$, then show that $k$ is an eigenvalue of $\mathbf{A}$.

48. If $\text{row}_j(\mathbf{A}) = k\text{row}_j(\mathbf{I}_n)$, then show that $k$ is an eigenvalue of $\mathbf{A}$.

49. Let $\mathbf{A} = \begin{bmatrix} 3 & -2 \\ 1 & 0 \end{bmatrix}$ and $\mathbf{u}_0 = \begin{bmatrix} 1 \\ 2 \end{bmatrix}$. When eight steps of the power method, $\mathbf{u}_k = \mathbf{A}^k\mathbf{u}_0$, $k = 1, 2, \ldots, 8$, are used the following vectors are obtained.

| $\mathbf{u}_0$ | $\mathbf{u}_1$ | $\mathbf{u}_2$ | $\mathbf{u}_3$ | $\mathbf{u}_4$ | $\mathbf{u}_5$ | $\mathbf{u}_6$ | $\mathbf{u}_7$ | $\mathbf{u}_8$ |
|---|---|---|---|---|---|---|---|---|
| 1 | -1 | -5 | -13 | -29 | -61 | -125 | -253 | -509 |
| 2 | 1 | -1 | 5 | -13 | -29 | -61 | -125 | -253 |

(a) Compute the ratios $\text{ent}_j(\mathbf{u}_{k+1})/\text{ent}_j(\mathbf{u}_k)$, $j = 1, 2$, $k = 0, 1, \ldots, 7$ and record this in the following table.

| $j$ ↓ \ $k \to$ | 0 | 1 | 2 | 3 | 4 | 5 | 6 | 7 |
|---|---|---|---|---|---|---|---|---|
| 1 | | | | | | | | |
| 2 | | | | | | | | |

(b) Conjecture a value for the dominant eigenvalue of $\mathbf{A}$. (*Hint:* It is an integer.)

(c) As a check, compute $\mathbf{A}^{20}\mathbf{u}_0$ and $\mathbf{A}^{21}\mathbf{u}_0$, then find the ratio of corresponding components.

50. Let $\mathbf{A} = \begin{bmatrix} 21 & 19 & 1 \\ 13 & -11 & -1 \\ -7 & -7 & 1 \end{bmatrix}$ and $\mathbf{u}_0 = \begin{bmatrix} 1 \\ 2 \\ 1 \end{bmatrix}$.

(a) Compute six steps of the power method, $\mathbf{u}_k = \mathbf{A}^k\mathbf{u}_0$, $k = 1, 2, \ldots, 6$.

(b) Form the ratios of just the first entries, $\text{ent}_1(\mathbf{u}_{k+1})/\text{ent}_1(\mathbf{u}_k)$, $k = 0, 1, \ldots, 6$ to estimate the dominant eigenvalue of $\mathbf{A}$. (*Hint:* It is an integer.)

(c) Find a corresponding eigenvector.

51. In Equation (7) assume that $|\lambda_1| > |\lambda_2| \geq |\lambda_3| \geq \cdots |\lambda_n|$. Form a conjecture for the relationship between $|\lambda_1|$ and $|\lambda_2|$ so that $\mathbf{u}_k \approx (\lambda_1)^k c_1\mathbf{p}_1$ for a small value of $k$.

52. In Equation (7) assume that $\lambda_1 = \lambda_2 > |\lambda_3| \geq \cdots |\lambda_n|$. Form a conjecture as to the behavior of the vectors $\mathbf{u}_k$ as $k$ gets large.

### In MATLAB

*This section showed how to compute eigenpairs of a matrix using a two-step process: (1) Find the eigenvalues; (2) Find associated eigenvectors. In MATLAB we can perform these steps with a sequence of commands that mirror the algebraic process of hand calculations. We describe both a numeric and symbolic approach together with a shortcut alternative in the following exercises.*

**ML. 1.** Let $\mathbf{A}$ be an $n \times n$ numeric matrix in MATLAB. The following two commands can be used to find the eigenvalues of $\mathbf{A}$.

**poly(A)** produces a vector of $n + 1$ coefficients for the characteristic polynomial of $\mathbf{A}$. The entries start with the coefficient of the highest degree term.

For example, $\mathbf{v} = [8 -5\ 3]$ corresponds to the polynomial $8t^2 - 5t + 3$.

**roots(v)** produces a list of the roots of the polynomial whose coefficients are in $\mathbf{v}$. The form of $\mathbf{v}$ is

$$v_1 t^n + v_2 t^{n-1} + \cdots + v_n t + v_{n+1}.$$

(a) Use **poly** to determine the coefficients of the characteristic polynomial of each of the following matrices. Write the characteristic polynomial in terms of powers of $t$.

$$\begin{bmatrix} 4 & 2 \\ 3 & 3 \end{bmatrix}, \begin{bmatrix} 5 & -1 \\ 2 & 2 \end{bmatrix},$$

$$\begin{bmatrix} 1 & 2 & 1 \\ 0 & 1 & 2 \\ -1 & 3 & 2 \end{bmatrix}, \begin{bmatrix} 2 & -1 & 0 & 0 \\ -1 & 2 & -1 & 0 \\ 0 & -1 & 2 & -1 \\ 0 & 0 & -1 & 2 \end{bmatrix}.$$

(b) Use **roots** to find the eigenvalues of each of the matrices in part (a).

(c) Let $\mathbf{A} = \begin{bmatrix} 5 & -8 & -1 \\ 4 & -7 & -4 \\ 0 & 0 & 4 \end{bmatrix}$. Describe the output of the command **roots(poly(A))**.

**ML. 2.** Once we have the exact eigenvalues of a matrix, associated eigenvectors can be determined using the **rref** command. This follows because an eigenvector is a nontrivial solution of the homogeneous linear system $(\mathbf{A} - \lambda\mathbf{I})\mathbf{p} = \mathbf{0}$.

(a) Matrix $\mathbf{A} = \begin{bmatrix} 5 & -8 & -1 \\ 4 & -7 & -4 \\ 0 & 0 & 4 \end{bmatrix}$ has eigenvalue $\lambda = -3$. Use **rref** to find a corresponding eigenvector.

(b) Matrix $\mathbf{A} = \begin{bmatrix} 1 & 2 & -1 \\ 1 & 0 & 1 \\ 4 & -4 & 5 \end{bmatrix}$ has eigenvalue $\lambda = 2$. Use **rref** to find a basis for the associated eigenspace.

(c) Matrix $\mathbf{A} = \begin{bmatrix} -2 & -7 & 0 \\ 0 & 5 & 0 \\ 7 & 7 & 5 \end{bmatrix}$ has eigenvalue $\lambda = 5$. Use **rref** to find a basis for the associated eigenspace.

(d) Matrix $\mathbf{A} = \begin{bmatrix} 5 & -8 & -1 \\ 4 & -7 & -4 \\ 0 & 0 & 4 \end{bmatrix}$ has an approximate eigenvalue $\lambda = 3.99$. Use **rref** to find a corresponding eigenvector. Were you successful? Explain.

**ML. 3.** The combination of commands **poly**, **roots**, and **rref** described in Exercises ML.1 and ML.2 require that we obtain the exact value for an eigenvalue in order to find a corresponding eigenvector. If we have the slightest error due to rounding or approximating quantities like square roots, then the eigenvector computation using **rref** will fail to give results. A computational alternative in MATLAB is to use the **eig** command, which will compute highly accurate approximations to both eigenvalues and eigenvectors. (See Example 4.) This command uses a different strategy to make successive approximations that eventually converge to the true values for most matrices.

> **eig(A)** displays a vector containing the eigenvalues of **A**.
>
> **[V,D] = eig(A)** displays eigenvectors of **A** as columns of the matrix **V** and the corresponding eigenvalues as the diagonal entries of matrix **D**.

(a) Let $\mathbf{A} = \begin{bmatrix} -6 & -3 & -5 \\ 5 & 2 & 5 \\ 3 & 3 & 2 \end{bmatrix}$. Use **eig** to find the eigenvalues and corresponding eigenvectors of **A**. Explain why eigenvectors of **A** form a basis for $R^3$.

(b) Let $\mathbf{A} = \begin{bmatrix} 5 & -8 & -1 \\ 4 & -7 & -4 \\ 0 & 0 & 4 \end{bmatrix}$. Compute the eigenpairs of **A** using **eig**. One of the eigenvalues is $\lambda = -3$. Compare the result from **eig** with the eigenvector you obtained in Exercise ML.2(a). Explain any differences between the vectors.

(c) Use **eig** to show that eigenvectors of

$$\mathbf{A} = \begin{bmatrix} 1 & 1 & -2 \\ 2 & 2 & 2 \\ 1 & -1 & 4 \end{bmatrix}$$

do not form a basis for $R^3$. Explain your steps.

**ML. 4.** (Optional: Requires the Symbolic Toolbox within MATLAB.) MATLAB can also determine eigen information of matrices that are symbolic. (See Chapter 8 for an introduction to symbolic quantities in MATLAB.) Commands **poly**, **rref**, and **eig** work in the same way as described previously. However, the **roots** command is replaced by the command **solve**. For example,

enter the following commands:

$$\mathbf{S = sym('[5,t;3,1]')}$$
$$\mathbf{poly(S)}$$
$$\mathbf{solve(poly(S))}$$
$$\mathbf{eig(S)}$$

The command **poly(S)** gives the characteristic polynomial in terms of the unknown $t$; note that the coefficients depend upon the parameter $t$. Both **solve(poly(S))** and **eig(S)** produce the eigenvalues of **S** in terms of $t$. We could use **rref** as we did in ML.2 to find corresponding eigenvectors, but it is simpler to use **eig**. Command **[V,D] = eig(S)** gives eigenvectors as columns of matrix **V**.

(a) What values of $t$ will ensure that the eigenvalues of **S** are real numbers?

(b) Determine a value of $t$ that will ensure that the eigenvalues of **S** are integers.

(c) Determine a value of $t$ that will ensure that the eigenvalues of **S** are the same value.

(d) Determine a value of $t$ that will ensure that 1 is an eigenvalue of $S$.

**ML. 5.** Matrix $\mathbf{A} = \begin{bmatrix} t & -1 & 0 \\ -1 & t & -1 \\ 0 & -1 & t \end{bmatrix}$ arose in the design of an electrical component. For the behavior of this component to fall within acceptable limits it is required that the eigenvalues of **A** all be smaller than 1. For what values of $t$, if any, is this possible?

**ML. 6.** In MATLAB type **help pmdemo**. Carefully read the description that appears. Start the routine by typing **pmdemo**.

(a) Choose the $2 \times 2$ matrix option. Read the description of the geometric aspects of the routine. Choose the built-in demo, select demo #2, and execute the power method. Follow the directions in the 'Message Window'. Click the 'MORE' button to see another set of approximations. Verify that the last vector shown is an approximate eigenvector of **A** as follows. Multiply this last vector by **A** and verify that the result is (approximately) a multiple of the vector. What, approximately, is the dominant eigenvalue?

(b) Repeat part (a) for the $3 \times 3$ matrix **A** of Example 6 with the specified initial vector.

**ML. 7.** Use **pmdemo** for $2 \times 2$ matrices with demo #4.

(a) From the geometric displays that are shown, describe the behavior of the sequence of approximate eigenvectors.

(b) Describe the matrix transformation defined by the matrix **A**.

(c) Briefly explain how the results of parts (a) and (b) are related.

**ML. 8.** Use **pmdemo** for $2 \times 2$ matrices with demo #3.

    (a) From the geometric displays that are shown, describe the behavior of the sequence of approximate eigenvectors.

    (b) Compute the eigenpairs of the matrix **A** using **eig**.

    (c) Briefly explain how the results of parts (a) and (b) are related.

**ML. 9.** Use **pmdemo** with $3 \times 3$ matrices. Execute each of the built-in demos. Briefly describe the behavior you observe for each demo. Select the 'MORE' button as needed. Use **eig**, as needed, to aid in formulating a reason for the behavior you observe.

## True/False Review Questions

*Determine whether each of the following statements is true or false.*

1. The characteristic equation of $\mathbf{A} = \begin{bmatrix} -2 & 4 \\ 1 & 1 \end{bmatrix}$ is $\lambda^2 + \lambda + 6$.

2. If 7 is an eigenvalue of $\mathbf{A}$, then $\mathbf{A} - 7\mathbf{I}_n$ is singular.

3. $\mathbf{A}$ and $\mathbf{A}^T$ have the same eigenpairs.

4. The eigenvalues of a matrix are its diagonal entries.

5. If $\lambda = 0$ is an eigenvalue of $\mathbf{A}$, then $\mathbf{A}$ is nonsingular.

6. If $\mathbf{A}$ has real entries and $\mathbf{A} = \mathbf{A}^T$, then all the eigenvalues of $\mathbf{A}$ are real numbers.

7. Any matrix that has repeated eigenvalues is defective.

8. The eigenvalues of $\mathbf{A} = \begin{bmatrix} 0 & -1 \\ 1 & 0 \end{bmatrix}$ are $\pm i$.

9. If $\mathbf{A}$ is symmetric with eigenpairs $(\lambda, \mathbf{p})$ and $(\mu, \mathbf{q})$ where $\lambda \neq \mu$, then $\mathbf{p} \cdot \mathbf{q} = 0$.

10. If $\mathbf{A}$ is $3 \times 3$ with eigenvalues $\lambda$, $\mu$, and $\alpha$, then $|\det(\mathbf{A})| = |\lambda \mu \alpha|$.

## Terminology

| | |
|---|---|
| Characteristic equation and polynomial | Repeated eigenvalues |
| Defective matrix | Power method |

From a computational point of view this section focused on determining eigenvalues and eigenvectors using the algebraic tools of a determinant and the solution of a homogeneous linear system. However, Properties 5 through 10 set the stage for further developments that utilize eigenvalues and eigenvectors in upcoming sections. It is these properties that let us extract pertinent information about a matrix based on its eigenpairs.

- What is the characteristic equation of a matrix? How many unknowns are there in the characteristic equation? Describe them.
- Describe the two-step procedure we developed for determining eigenpairs.
- Why is the two-step procedure not recommended for large matrices?
- What do we mean by a repeated eigenvalue?
- If a $3 \times 3$ matrix has eigenvalues $-2$, 6, and 6, how many linearly independent eigenvectors can it have? If there is more than one possible answer, explain each case.
- What is a defective matrix?

It is important to be able to use the special structure of certain matrices so that we can infer information regarding their eigenvalues and eigenvectors without actually performing computations. Use this notion to respond to the following questions and statements.

- How are the eigenvalues of a matrix and its transpose related?
- How can we identify the eigenvalues of a triangular matrix? Of a diagonal matrix?
- What property do all the eigenvalues of a symmetric matrix share?
- If $(\lambda, \mathbf{p})$ and $(\mu, \mathbf{q})$ are eigenpairs of a symmetric matrix with $\lambda \neq \mu$, then how are vectors $\mathbf{p}$ and $\mathbf{q}$ related?
- How are the eigenvalues of a matrix related to its determinant?
- If we know all the eigenvalues of a matrix, how can we determine whether it is singular?
- What does the power method approximate?
- Describe the power method.

## 4.3 ■ DIAGONALIZATION AND SIMILAR MATRICES

In Section 4.2 we showed how to compute eigenpairs $(\lambda, \mathbf{p})$ of a matrix $\mathbf{A}$ by determining the roots of the characteristic polynomial $c(\lambda) = \det(\mathbf{A} - \lambda \mathbf{I}_n)$ and then using the values of $\lambda$ to solve the associated homogeneous linear systems $(\mathbf{A} - \lambda \mathbf{I}_n)\mathbf{p} = \mathbf{0}$. For $2 \times 2$ and $3 \times 3$ matrices this two-step procedure could be carried out by hand-calculations. If we needed an approximation to the dominant eigenvalue of $\mathbf{A}$ an alternative was to use the power method. The two-step method is called a **direct method** since we get the eigenpairs as a result of a specific set of calculations. The power method is called an **iterative method** since we generate a sequence of approximations that is terminated when we are satisfied with the agreement of values between successive stages of the calculations.

For large matrices that appear in applications for which we need all the eigenvalues or more than just the dominant eigenvalue, iterative methods are used to obtain the approximations. Such methods use a computer to perform the arithmetic steps involved but rely on properties of matrices and eigenpairs to develop the requisite steps. In this section we develop further properties of eigenvalues and eigenvectors that are used to design iterative methods for eigen computations. In Section 4.6 we briefly investigate a prototype of one of the common iterative methods for eigen computation.

In our study of linear systems in Chapter 2, we used elementary row operations to produce a new linear system that was equivalent to the original linear system; that is, the two linear systems had the same solution set. In this section we show how to obtain an **equivalent eigenproblem**; that is, given a matrix $\mathbf{A}$ we show how to find a matrix $\mathbf{B}$ which has the same eigenvalues as $\mathbf{A}$. If we design our procedure efficiently, then the eigenvalues of $\mathbf{B}$ will be easy to obtain and hence we will have the eigenvalues of the original matrix $\mathbf{A}$. We say that a procedure that yields an equivalent eigenproblem **preserves eigenvalues**. By carefully recording the steps of such procedures we will be able to generate the corresponding eigenvectors.

We begin by defining an operation on a matrix that preserves eigenvalues.

**Definition**  A **similarity transformation** of an $n \times n$ matrix $\mathbf{A}$ is a function $f(\mathbf{A}) = \mathbf{P}^{-1}\mathbf{A}\mathbf{P}$, where $\mathbf{P}$ is an $n \times n$ nonsingular matrix. If $\mathbf{B} = \mathbf{P}^{-1}\mathbf{A}\mathbf{P}$, then we say $\mathbf{A}$ is **similar** to $\mathbf{B}$.

We can show (see Exercise 5) that if $\mathbf{A}$ is similar to $\mathbf{B}$, then $\mathbf{B}$ is similar to $\mathbf{A}$. Hence we say $\mathbf{A}$ and $\mathbf{B}$ are **similar matrices**.

Thus we have

> $\mathbf{A}$ and $\mathbf{B}$ are similar if and only if there exists a nonsingular matrix $\mathbf{P}$ such that $\mathbf{B} = \mathbf{P}^{-1}\mathbf{A}\mathbf{P}$.

In Sections 4.1 and 4.2 we listed properties of eigenvalues. We continue that list here with Property 11.

---

**Property 11.** Similar matrices have the same eigenvalues.
To see this we show that if $\mathbf{A}$ and $\mathbf{B}$ are similar, that is, $\mathbf{B} = \mathbf{P}^{-1}\mathbf{A}\mathbf{P}$, then they have the same characteristic polynomial:

$$
\begin{aligned}
\det(\mathbf{B} - \lambda\mathbf{I}_n) &= \det(\mathbf{P}^{-1}\mathbf{A}\mathbf{P} - \lambda\mathbf{I}_n) && \text{[since } \mathbf{B} = \mathbf{P}^{-1}\mathbf{A}\mathbf{P}] \\
&= \det(\mathbf{P}^{-1}\mathbf{A}\mathbf{P} - \lambda\mathbf{P}^{-1}\mathbf{I}_n\mathbf{P}) && \text{[since } \mathbf{P}^{-1}\mathbf{I}_n\mathbf{P} = \mathbf{I}_n] \\
&= \det(\mathbf{P}^{-1}\mathbf{A}\mathbf{P} - \mathbf{P}^{-1}\lambda\mathbf{I}_n\mathbf{P}) && \text{[property of scalar multiplication]} \\
&= \det(\mathbf{P}^{-1}(\mathbf{A} - \lambda\mathbf{I}_n)\mathbf{P}) && \text{[associativity of multiplication]} \\
&= \det(\mathbf{P}^{-1})\det(\mathbf{A} - \lambda\mathbf{I}_n)\det(\mathbf{P}) && \text{[property of determinants]} \\
&= \det(\mathbf{A} - \lambda\mathbf{I}_n). && \text{[since } \det(\mathbf{P}^{-1}) = 1/\det(\mathbf{P})]
\end{aligned}
$$

---

Thus it follows that similarity transformations preserve eigenvalues. If $\mathbf{A}$ and $\mathbf{B}$ are similar, with eigenpairs $(\lambda, \mathbf{p})$ and $(\lambda, \mathbf{q})$, respectively, then it is natural to ask about the relationship between the eigenvectors $\mathbf{p}$ and $\mathbf{q}$. We have $\mathbf{B} = \mathbf{P}^{-1}\mathbf{A}\mathbf{P}$ or, equivalently, $\mathbf{A} = \mathbf{P}\mathbf{B}\mathbf{P}^{-1}$; thus

$$
\begin{aligned}
\mathbf{A}\mathbf{p} = \lambda\mathbf{p} &\Leftrightarrow \mathbf{P}\mathbf{B}\mathbf{P}^{-1}\mathbf{p} = \lambda\mathbf{p} \\
&\Leftrightarrow \mathbf{P}^{-1}(\mathbf{P}\mathbf{B}\mathbf{P}^{-1}\mathbf{p}) = \mathbf{P}^{-1}(\lambda\mathbf{p}) \\
&\Leftrightarrow \mathbf{B}(\mathbf{P}^{-1}\mathbf{p}) = \lambda(\mathbf{P}^{-1}\mathbf{p}).
\end{aligned}
$$

This says that if $\mathbf{p}$ is an eigenvector of $\mathbf{A}$ corresponding to eigenvalue $\lambda$, then $\mathbf{q} = \mathbf{P}^{-1}\mathbf{p}$ is an eigenvector of $\mathbf{B}$ corresponding to $\lambda$. To summarize,

> If $\mathbf{B} = \mathbf{P}^{-1}\mathbf{A}\mathbf{P}$ and $(\lambda, \mathbf{p})$ is an eigenpair of $\mathbf{A}$, then $(\lambda, \mathbf{P}^{-1}\mathbf{p})$ is an eigenpair of $\mathbf{B}$.

EXAMPLE 1  Let

$$
\mathbf{A} = \begin{bmatrix} -6 & -7 & -9 \\ 3 & 4 & 4 \\ 7 & 7 & 9 \end{bmatrix} \quad \text{and} \quad \mathbf{P} = \begin{bmatrix} 2 & -2 & 3 \\ -1 & 1 & -1 \\ -1 & 2 & -3 \end{bmatrix}.
$$

An eigenpair of $\mathbf{A}$ is $\left(2, \begin{bmatrix} -2 \\ 1 \\ 1 \end{bmatrix}\right)$ (verify) and $\mathbf{P}$ is nonsingular with

$$\mathbf{P}^{-1} = \begin{bmatrix} 1 & 0 & 1 \\ 2 & 3 & 1 \\ 1 & 2 & 0 \end{bmatrix} \quad \text{(verify).}$$

Then

$$\mathbf{B} = \mathbf{P}^{-1}\mathbf{AP} = \begin{bmatrix} 2 & -2 & 3 \\ 0 & 3 & -2 \\ 0 & -1 & 2 \end{bmatrix}$$

is similar to $\mathbf{A}$ and

$$\mathbf{B}\left(\mathbf{P}^{-1}\begin{bmatrix} -2 \\ 1 \\ 1 \end{bmatrix}\right) = \mathbf{B}\begin{bmatrix} -1 \\ 0 \\ 0 \end{bmatrix} = \begin{bmatrix} -2 \\ 0 \\ 0 \end{bmatrix}$$

$$= 2\left(\begin{bmatrix} -1 \\ 0 \\ 0 \end{bmatrix}\right) = 2\left(\mathbf{P}^{-1}\begin{bmatrix} -2 \\ 1 \\ 1 \end{bmatrix}\right).$$

Thus $\left(2, \mathbf{P}^{-1}\begin{bmatrix} -2 \\ 1 \\ 1 \end{bmatrix}\right)$ is an eigenpair of $\mathbf{B}$.    ■

From Section 4.2 we know that the eigenvalues of a diagonal matrix, an upper triangular matrix, or a lower triangular matrix are the diagonal entries. With this in mind we pose the following similarity questions for an $n \times n$ matrix $\mathbf{A}$.

### Similarity Questions

1. Is $\mathbf{A}$ similar to a diagonal matrix? That is, does there exist a nonsingular matrix $\mathbf{P}$ so that $\mathbf{B} = \mathbf{P}^{-1}\mathbf{AP}$ is diagonal?
2. If $\mathbf{A}$ is not similar to a diagonal matrix, does there exist a nonsingular matrix $\mathbf{P}$ so that $\mathbf{B} = \mathbf{P}^{-1}\mathbf{AP}$ is upper (lower) triangular; that is, is $\mathbf{A}$ similar to a triangular matrix?
3. If $\mathbf{A}$ is similar to a diagonal or triangular matrix, how do we find the matrix $\mathbf{P}$ that performs the similarity transformation?

Questions 1 and 2 are **existence questions**; that is, is there some property of matrix $\mathbf{A}$ that will guarantee that such a transformation is possible? However, question 3 is a **construction question**; that is, construct the appropriate matrix $\mathbf{P}$. We will answer question 1 here and question 2 in Section 4.4. We will give a partial answer to question 3 in Section 4.4 and a more general answer in Section 4.5. In Section 4.6 we give an explicit iterative procedure for constructing $\mathbf{P}$.

We begin by deriving information about $\mathbf{A}$, if it is similar to a diagonal matrix. So let us assume that $\mathbf{P}^{-1}\mathbf{AP} = \mathbf{D}$, a diagonal matrix. Then by Property 10 the eigenvalues of $\mathbf{A}$ are the diagonal entries $d_{jj}$. We have

$$\mathbf{P}^{-1}\mathbf{AP} = \mathbf{D} \Leftrightarrow \mathbf{AP} = \mathbf{PD}.$$

Now express $\mathbf{P}$ as a matrix partitioned into its columns: $\mathbf{P} = \begin{bmatrix} \mathbf{p}_1 & \mathbf{p}_2 & \cdots & \mathbf{p}_n \end{bmatrix}$. Then we have

$$\mathbf{A}\begin{bmatrix} \mathbf{p}_1 & \mathbf{p}_2 & \cdots & \mathbf{p}_n \end{bmatrix} = \begin{bmatrix} \mathbf{p}_1 & \mathbf{p}_2 & \cdots & \mathbf{p}_n \end{bmatrix}\mathbf{D}.$$

Perform the multiplications to get (verify)

$$\begin{bmatrix} \mathbf{Ap}_1 & \mathbf{Ap}_2 & \cdots & \mathbf{Ap}_n \end{bmatrix} = \begin{bmatrix} d_{11}\mathbf{p}_1 & d_{22}\mathbf{p}_2 & \cdots & d_{nn}\mathbf{p}_n \end{bmatrix}$$

(The product **PD** requires careful inspection.) Equating corresponding columns gives

$$\mathbf{Ap}_1 = d_{11}\mathbf{p}_1, \quad \mathbf{Ap}_2 = d_{22}\mathbf{p}_2, \quad \ldots, \quad \mathbf{Ap}_n = d_{nn}\mathbf{p}_n$$

which implies that $\text{col}_j(\mathbf{P})$ is an eigenvector of $\mathbf{A}$ corresponding to eigenvalue $d_{jj}$. Since **P** is nonsingular, its columns are linearly independent and guaranteed not to be **0**. The steps here are reversible; that is, if **A** has $n$ linearly independent eigenvectors, then it is similar to a diagonal matrix. We summarize this important result as follows:

**Property 12.** An $n \times n$ matrix **A** is similar to a diagonal matrix if and only if **A** has $n$ linearly independent eigenvectors.

An immediate application of Property 12 is that there exist matrices with the same eigenvalues that are not similar. That is, the converse of Property 11 is not true. The following example provides an illustration of this statement.

EXAMPLE 2   Let $\mathbf{A} = \mathbf{I}_2$ and $\mathbf{B} = \begin{bmatrix} 1 & 1 \\ 0 & 1 \end{bmatrix}$. Both **A** and **B** have eigenvalue 1 with multiplicity 2. (Verify.) **A** is diagonal, so it is certainly similar to a diagonal matrix. However, matrix **B** is not similar to a diagonal matrix since it has only one linearly independent eigenvector. (Verify.) Thus **A** and **B** cannot be similar. (Explain.) ∎

From the point of view of using similarity transformations to find eigenvalues, Property 12 says that we need eigenvectors in order to find eigenvalues. This is just the reverse of the two-step method we developed in Section 4.2. Thus while Property 12 specifically tells us which matrices are similar to a diagonal matrix, it really is not useful for directly computing the eigenvalues of a matrix. However, it does lead us to some criteria that will enable us to recognize that certain matrices are similar to a diagonal matrix. At this point it is convenient to make the following definition.

**Definition**   A matrix **A** is called **diagonalizable** provided that it is similar to a diagonal matrix.

EXAMPLE 3   Let $\mathbf{A} = \begin{bmatrix} 1 & 1 \\ -2 & 4 \end{bmatrix}$, which is the matrix in Example 1 of Section 4.2. We showed that **A** has eigenpairs

$$(\lambda_1, \mathbf{p}_1) = \left(3, \begin{bmatrix} 1 \\ 2 \end{bmatrix}\right) \quad \text{and} \quad (\lambda_2, \mathbf{p}_2) = \left(2, \begin{bmatrix} 1 \\ 1 \end{bmatrix}\right).$$

By Property 4 in Section 4.1, $\{\mathbf{p}_1, \mathbf{p}_2\}$ is a linearly independent set. Hence **A** is similar to a diagonal matrix. If we let $\mathbf{P} = \begin{bmatrix} \mathbf{p}_1 & \mathbf{p}_2 \end{bmatrix}$, then the diagonal matrix is $\mathbf{D} = \begin{bmatrix} 3 & 0 \\ 0 & 2 \end{bmatrix}$. If we let $\mathbf{P} = \begin{bmatrix} \mathbf{p}_2 & \mathbf{p}_1 \end{bmatrix}$, then the diagonal matrix is $\mathbf{D} = \begin{bmatrix} 2 & 0 \\ 0 & 3 \end{bmatrix}$. Thus we see that *the diagonal matrix is not unique*, its diagonal entries can be reordered and they must correspond to the order in which the eigenvectors appear as columns in **P**. In addition, *the matrix* **P** *is not unique*, since any nonzero multiple of an eigenvector is another eigenvector corresponding to the same eigenvalue. (See Property 1 in Section 4.1.)                                                                    ■

   Property 4 in Section 4.1 gives us a way to recognize diagonalizable matrices. (It does not help determine eigenvalues but plays an important role in other respects.) We record this is in Property 13.

> **Property 13.** If an $n \times n$ matrix **A** has $n$ distinct eigenvalues, then **A** is diagonalizable.

**WARNING:**    If **A** has an eigenvalue with multiplicity greater than 1, then **A** may or may not be diagonalizable. We must check to see if each eigenvalue has as many linearly independent eigenvectors as its multiplicity. That is, in the language of the statement preceding Exercise 34 in Section 4.2, we must check to see if the matrix is defective or not. *A defective matrix is not diagonalizable.* (Explain.)

EXAMPLE 4    Let

$$A = \begin{bmatrix} 0 & -2 & 1 \\ 1 & 3 & -1 \\ 0 & 0 & 1 \end{bmatrix},$$

which is the matrix in Example 3 of Section 4.2. We showed that $\lambda = 1$ was an eigenvalue of multiplicity 2 and that the corresponding eigenspace had a basis of two eigenvectors. Since the remaining eigenvalue had a value other than 1, **A** is diagonalizable. (Explain.)                                                              ■

EXAMPLE 5    Let

$$A = \begin{bmatrix} 1 & 0 & 0 \\ 2 & 1 & 0 \\ 0 & 0 & 3 \end{bmatrix}.$$

Since **A** is lower triangular, its eigenvalues are 1, 1, and 3. Since it has an eigenvalue of multiplicity $> 1$, there is a chance it is not diagonalizable; that is, it may be defective. To check this, we determine the number of linearly independent eigenvectors associated with eigenvalue $\lambda = 1$; that is, the dimension of the eigenspace associated with $\lambda = 1$. To do this, we find **ns** $(A - 1I_3)$. We compute **rref**$(A - 1I_3)$ and obtain (verify)

$$\begin{bmatrix} 1 & 0 & 0 & | & 0 \\ 0 & 0 & 1 & | & 0 \\ 0 & 0 & 0 & | & 0 \end{bmatrix}.$$

Thus $x_1 = x_3 = 0$ and $x_2$ is a free variable. Since there is only one free variable $\dim(\mathbf{ns}\,(A - 1I_3)) = 1$, and so there is just one linearly independent eigenvector corresponding to $\lambda = 1$. Hence **A** is defective and is not diagonalizable.    ■

   Thus we have answered Similarity Question 1. Questions 2 and 3 are investigated in Section 4.4 through 4.6.
   In Sections 4.1 through 4.3 we developed properties of eigenvalues and eigenvectors and numbered them 1 through 13. Following is a complete list of these

properties. It provides a good review and is a convenient reference for the major concepts we have developed.

---

### Eigen Properties: Summary

**Property 1.** If $(\lambda, \mathbf{p})$ is an eigenpair of $\mathbf{A}$, then so is $(\lambda, k\mathbf{p})$ for any scalar $k \neq 0$.

**Property 2.** If $\mathbf{p}$ and $\mathbf{q}$ are eigenvectors corresponding to eigenvalue $\lambda$ of $\mathbf{A}$, then so is $\mathbf{p} + \mathbf{q}$ (assuming $\mathbf{p} + \mathbf{q} \neq \mathbf{0}$).

Note that 1 and 2 imply that the set of eigenvectors associated with an eigenvalue $\lambda$ form a subspace, provided that we include the zero vector.

**Property 3.** If $(\lambda, \mathbf{p})$ is an eigenpair of $\mathbf{A}$, then for any positive integer $r$, $(\lambda^r, \mathbf{p})$ is an eigenpair of $\mathbf{A}^r$.

**Property 4.** If $(\lambda, \mathbf{p})$ and $(\mu, \mathbf{q})$ are eigenpairs of $\mathbf{A}$ with $\lambda \neq \mu$, then $\mathbf{p}$ and $\mathbf{q}$ are linearly independent. That is, *eigenvectors corresponding to distinct eigenvalues are linearly independent.*

**Property 5.** $\mathbf{A}$ and $\mathbf{A}^T$ have the same eigenvalues.

**Property 6.** If $\mathbf{A}$ is diagonal, upper triangular, or lower triangular, then its eigenvalues are its diagonal entries.

**Property 7.** The eigenvalues of a symmetric matrix are real.

**Property 8.** Eigenvectors corresponding to distinct eigenvalues of a symmetric matrix are orthogonal.

**Property 9.** $|\det(\mathbf{A})|$ is the product of the absolute values of the eigenvalues of $\mathbf{A}$.

**Property 10.** $\mathbf{A}$ is nonsingular if and only if 0 is not an eigenvalue of $\mathbf{A}$, or, equivalently, $\mathbf{A}$ is singular if and only if 0 is an eigenvalue of $\mathbf{A}$.

**Property 11.** Similar matrices have the same eigenvalues.

**Property 12.** An $n \times n$ matrix $\mathbf{A}$ is similar to a diagonal matrix if and only if $\mathbf{A}$ has $n$ linearly independent eigenvectors.

**Property 13.** If an $n \times n$ matrix $\mathbf{A}$ has $n$ distinct eigenvalues, then $\mathbf{A}$ is diagonalizable.

---

## EXERCISES 4.3

*In Exercises 1–4, for each given matrix find two matrices that are similar to it.*

**1.** $\begin{bmatrix} 2 & 3 \\ 1 & 0 \end{bmatrix}$.

**2.** $\begin{bmatrix} 1 & -1 \\ 0 & 0 \end{bmatrix}$.

**3.** $\begin{bmatrix} 2 & 1 & 0 \\ 1 & 2 & 1 \\ 0 & 1 & 2 \end{bmatrix}$.

**4.** $\begin{bmatrix} 4 & 3 & -1 \\ 2 & 0 & 1 \\ 6 & 3 & 0 \end{bmatrix}$.

**5.** If $\mathbf{A}$ is similar to $\mathbf{B}$, then show that $\mathbf{B}$ is similar to $\mathbf{A}$.

**6.** If $\mathbf{A}$ is similar to $\mathbf{B}$ and $\mathbf{B}$ is similar to $\mathbf{C}$, then show that $\mathbf{A}$ is similar to $\mathbf{C}$. (*Hint*: There exist matrices $\mathbf{P}$ and $\mathbf{Q}$ so that $\mathbf{B} = \mathbf{P}^{-1}\mathbf{AP}$ and $\mathbf{C} = \mathbf{Q}^{-1}\mathbf{BQ}$. Find a matrix $\mathbf{R}$ such that $\mathbf{R}^{-1}\mathbf{AR} = \mathbf{C}$.)

**7.** If $\mathbf{A}$ is similar to $\mathbf{B}$, then show that $\mathbf{A}^2$ is similar to $\mathbf{B}^2$.

**8.** Explain why similar matrices have the same determinant.

**9.** Let $\mathbf{A} = \begin{bmatrix} -4 & 6 \\ -9 & 11 \end{bmatrix}$ and $\mathbf{P} = \begin{bmatrix} 1 & 2 \\ 1 & 3 \end{bmatrix}$.

  (a) Compute $\mathbf{B} = \mathbf{P}^{-1}\mathbf{AP}$.

  (b) Find two eigenpairs of $\mathbf{A}$, one for each eigenvalue.

  (c) Find two eigenpairs of $\mathbf{B}$, one for each eigenvalue.

**10.** Let $\mathbf{A} = \begin{bmatrix} 4 & -1 \\ 0 & 3 \end{bmatrix}$ and $\mathbf{P} = \begin{bmatrix} 2 & 1 \\ 5 & 3 \end{bmatrix}$.

  (a) Compute $\mathbf{B} = \mathbf{P}^{-1}\mathbf{AP}$.

  (b) Find two eigenpairs of $\mathbf{A}$, one for each eigenvalue.

  (c) Find two eigenpairs of $\mathbf{B}$, one for each eigenvalue.

**11.** Let

$$\mathbf{A} = \begin{bmatrix} -4 & 6 \\ -9 & 11 \end{bmatrix} \quad \text{and} \quad \mathbf{B} = \begin{bmatrix} 2 & 0 \\ 0 & 5 \end{bmatrix}.$$

The results of Exercise 9 show that **A** and **B** are similar. Here we investigate the geometric properties of the matrix transformation $f$ determined by **A** and the matrix transformation $g$ determined by **B**. Let

$$\mathbf{S} = \begin{bmatrix} 0 & 1 & 1 & 0 & 0 \\ 0 & 0 & 1 & 1 & 0 \end{bmatrix}$$

be the matrix of coordinates of the vertices of the unit square.

(a) Compute the vertices of the image of the unit square by the matrix transformation $f$; that is, $f(\mathbf{S}) = \mathbf{A} * \mathbf{S}$. Sketch the vertices of the image and connect them to form a parallelogram.

(b) Compute the vertices of the image of the unit square by the matrix transformation $g$; that is, $g(\mathbf{S}) = \mathbf{B} * \mathbf{S}$. Sketch the vertices of the image and connect them to form a parallelogram.

(c) Are the sketches from parts (a) and (b) identical? Explain.

(d) What geometric property of the images determined in parts (a) and (b) is the same? (*Hint*: Recall the magnification factor discussion in Section 3.1 and see Exercise 8.)

**12.** Let $\mathbf{A} = \begin{bmatrix} 0 & -2 & 1 \\ 1 & 3 & -1 \\ 0 & 0 & 1 \end{bmatrix}$ and $\mathbf{P} = \begin{bmatrix} 1 & 1 & 0 \\ 0 & 1 & 1 \\ 1 & 0 & 1 \end{bmatrix}$.

(a) Compute $\mathbf{B} = \mathbf{P}^{-1}\mathbf{AP}$.

(b) Verify that $\left( 2, \begin{bmatrix} -1 \\ 1 \\ 0 \end{bmatrix} \right)$ is an eigenpair of **A**.

(c) Determine vector **q** so that $(2, \mathbf{q})$ is an eigenpair of **B**.

*In Exercises 13–16, show that each given matrix is diagonalizable and find a diagonal matrix similar to the given matrix.*

**13.** $\begin{bmatrix} 4 & 2 \\ 3 & 3 \end{bmatrix}$.

**14.** $\begin{bmatrix} 3 & 2 \\ 6 & 4 \end{bmatrix}$.

**15.** $\begin{bmatrix} 2 & -2 & 3 \\ 0 & 3 & -2 \\ 0 & -1 & 2 \end{bmatrix}$.

**16.** $\begin{bmatrix} 0 & -2 & 1 \\ 1 & 3 & -1 \\ 0 & 0 & 1 \end{bmatrix}$.

*A matrix is not diagonalizable if and only if it is defective. In Exercises 17–19, show that each given matrix is defective.*

**17.** $\begin{bmatrix} 1 & 1 \\ 0 & 1 \end{bmatrix}$.

**18.** $\begin{bmatrix} 2 & 0 & 0 \\ 3 & 2 & 0 \\ 0 & 0 & 5 \end{bmatrix}$.

**19.** $\begin{bmatrix} 10 & 11 & 3 \\ -3 & -4 & -3 \\ -8 & -8 & -1 \end{bmatrix}$.

**20.** Find a $2 \times 2$ nondiagonal matrix that has eigenpairs

$$\left( 2, \begin{bmatrix} -1 \\ 2 \end{bmatrix} \right) \quad \text{and} \quad \left( -3, \begin{bmatrix} 1 \\ 1 \end{bmatrix} \right).$$

**21.** Find a $3 \times 3$ nondiagonal matrix that has eigenpairs

$$\left( -2, \begin{bmatrix} 1 \\ 0 \\ 1 \end{bmatrix} \right), \quad \left( -2, \begin{bmatrix} 0 \\ 1 \\ 1 \end{bmatrix} \right), \quad \text{and} \quad \left( 3, \begin{bmatrix} 1 \\ 1 \\ 1 \end{bmatrix} \right).$$

**22.** Let $\mathbf{C} = \begin{bmatrix} k & t \\ 0 & k \end{bmatrix}$ where $k$ is any real number and $t$ is any nonzero real number. Show that **C** is not diagonalizable.

**23.** Let $g$ be the function given by $g(\mathbf{A}) = \mathbf{P}^{-1}\mathbf{AP}$, where **A** is any $n \times n$ matrix and **P** is a fixed $n \times n$ matrix. Show that each of the following properties hold. (In Chapter 6 we call such functions linear transformations.)

(a) $g(\mathbf{A} + \mathbf{B}) = g(\mathbf{A}) + g(\mathbf{B})$, where **A** and **B** are any $n \times n$ matrices.

(b) $g(k\mathbf{A}) = kg(\mathbf{A})$, where **A** is any $n \times n$ matrix and $k$ is a scalar.

**24.** Explain why the function in Exercise 23 preserves eigenvalues; that is, **A** and $g(\mathbf{A})$ have the same eigenvalues.

## In MATLAB

*In Section 4.2, Exercise ML.3, we introduced the **eig** command. Review the features of the **eig** command by typing **help eig** or carefully read the description in Exercise ML.3 in Section 4.2.*

**ML. 1.** Use the **eig** command to determine which of the following matrices is diagonalizable.

(a) $\begin{bmatrix} 4 & 2 \\ 0 & 4 \end{bmatrix}$.

(b) $\begin{bmatrix} 4 & 0 & 0 \\ 0 & 1 & 5 \\ 0 & 0 & 1 \end{bmatrix}$.

(c) $\begin{bmatrix} 1 & 1 & 1 \\ 1 & 1 & 1 \\ 1 & 1 & 1 \end{bmatrix}$.

(d) $\begin{bmatrix} 3 & 0 & 0 \\ -2 & 3 & -2 \\ 2 & 0 & 5 \end{bmatrix}$.

(e) $\begin{bmatrix} 4 & 2 & 0 & 0 \\ 3 & 3 & 0 & 0 \\ 0 & 0 & 2 & 5 \\ 0 & 0 & 0 & 2 \end{bmatrix}$.

(f) $\begin{bmatrix} 0 & 0 & 5 & 0 \\ 0 & -5 & 0 & 0 \\ 5 & 0 & 0 & 0 \\ 0 & 0 & 0 & -5 \end{bmatrix}$.

**ML. 2.** Let $k$ be a nonzero real number, $\mathbf{A} = \begin{bmatrix} k & 0 \\ 0 & k \end{bmatrix}$, and

$\mathbf{B} = \begin{bmatrix} k & 0 \\ k & k \end{bmatrix}$.

(a) Explain why $\mathbf{A}$ is diagonalizable.
(b) For arbitrary $k$, show that $\mathbf{B}$ is not diagonalizable. (Do not try to use **eig**.)
(c) Choose a value for $k$, then use MATLAB to determine which, if any, of the following partitioned matrices is diagonalizable. ($\mathbf{O}_2$ is the $2 \times 2$ zero matrix; in MAT-LAB use command **zeros(2)**.)

$$\mathbf{C} = \begin{bmatrix} \mathbf{A} & \mathbf{O}_2 \\ \mathbf{O}_2 & \mathbf{B} \end{bmatrix}, \quad \mathbf{D} = \begin{bmatrix} \mathbf{A} & \mathbf{B} \\ \mathbf{O}_2 & \mathbf{B} \end{bmatrix},$$

$$\mathbf{E} = \begin{bmatrix} \mathbf{B} & \mathbf{O}_2 \\ \mathbf{O}_2 & \mathbf{B} \end{bmatrix}, \quad \mathbf{F} = \begin{bmatrix} \mathbf{B} & \mathbf{B} \\ \mathbf{B} & \mathbf{B} \end{bmatrix},$$

$$\mathbf{G} = \begin{bmatrix} \mathbf{A} & \mathbf{O}_2 \\ \mathbf{O}_2 & \mathbf{A} \end{bmatrix}.$$

## True/False Review Questions

*Determine whether each of the following statements is true or false.*

1. If $\mathbf{A}$ is $3 \times 3$ with eigenvalues $\lambda = 0, 2, -3$, then $\mathbf{A}$ is diagonalizable.

2. Every defective matrix is diagonalizable.

3. Similar matrices have the same eigenvalues.

4. If $\mathbf{A}$ is $4 \times 4$ and diagonalizable, then there exists a basis for $R^4$ consisting of eigenvectors of $\mathbf{A}$.

5. If 4 is an eigenvalue of $\mathbf{A}$, then corresponding eigenvectors belong to $\mathbf{ns}(\mathbf{A} - 4\mathbf{I}_n)$.

6. An $n \times n$ matrix $\mathbf{A}$ is diagonalizable if and only if $\mathbf{A}$ has $n$ linearly independent eigenvectors.

7. If $\mathbf{A}$ and $\mathbf{B}$ have the same eigenvalues, then they are similar.

8. If $\mathbf{P}^{-1}\mathbf{AP} = \mathbf{Q}^{-1}\mathbf{BQ}$, then $\mathbf{A}$ and $\mathbf{B}$ are similar.

9. If $\lambda$ is an eigenvalue of $\mathbf{A}$ and $\mathbf{B}$, then the corresponding eigenspaces are the same.

10. Every matrix is similar to a diagonal matrix.

## Terminology

| Similar matrices | Diagonalizable |
| --- | --- |

We have seen in previous sections that the eigenvalues of a matrix supply information about behavior of certain applications like Markov chains and Leslie population models. But eigenvalues also play a correspondingly important role in determining the behavior of matrix transformations, as we will see in upcoming sections. Much, but not necessarily all, of the behavior that results from a matrix transformation $f(\mathbf{x}) = \mathbf{Ax}$ from $R^n$ to $R^n$ can be accomplished by using a matrix similar to $\mathbf{A}$. If $\mathbf{A}$ is diagonalizable, then this similar diagonal matrix can be used to aid in analyzing the behavior that results from the matrix transformation. Thus the notions in this section can help us gain information about matrix $\mathbf{A}$ and hence about the behavior of the matrix transformation $f$.

- What does it mean to say that two matrices are similar?
- What eigen feature do similar matrices share?
- What does it mean to say that a matrix is diagonalizable?
- If matrix $\mathbf{A}$ is diagonalizable, then how are eigenvectors of $\mathbf{A}$ involved?
- Can a singular matrix be diagonalizable? Explain.
- Can a defective matrix be diagonalizable? Explain why or why not.

- If matrices **A** and **B** are $3 \times 3$ with eigenvalues 0, 3, and $-2$, name at least two distinct properties that **A** and **B** have in common.
- If matrices **A** and **B** are $3 \times 3$ with eigenvalues 3, 3, and $-2$, are they guaranteed to be similar? Explain.
- Let $\mathbf{D} = \begin{bmatrix} 3 & 0 \\ 0 & 7 \end{bmatrix}$. Explain how to find *all* the matrices **A** that are similar to **D**.
- Let $g$ be the function defined on the set of $2 \times 2$ matrices by the expression $g(\mathbf{C}) = \mathbf{P}^{-1}\mathbf{C}\mathbf{P}$ where **C** is any $2 \times 2$ matrix and **P** is a $2 \times 2$ matrix that is fixed. What properties do **C** and the image $g(\mathbf{C})$ share?

Complete the following statements.

- A $5 \times 5$ matrix is diagonalizable if and only if _____.
- A matrix whose eigenvalues are all different from one another is _____.

## 4.4 ■ ORTHOGONAL MATRICES AND SPECTRAL REPRESENTATION

In Section 4.3 we saw that an $n \times n$ matrix **A** was similar to a diagonal matrix if and only if it had $n$ linearly independent eigenvectors. (See Property 12 in Section 4.3.) This answered the first of the similarity questions posed in Section 4.3. In order to answer similarity question 2 posed in Section 4.3, "When is a matrix similar to a triangular matrix?", we need additional concepts. We will see that the answer to this question includes both real and complex square matrices, whether or not they are diagonalizable. In addition, we will get a valuable result for symmetric matrices that we use in two applications.

In Section 1.3 we defined the dot product of a pair of vectors **x** and **y** in $R^n$ as $\mathbf{x}{\cdot}\mathbf{y} = x_1 y_1 + x_2 y_2 + \cdots + x_n y_n$ and the complex dot product of a pair of vectors **x** and **y** in $C^n$ as $\overline{\mathbf{x}}{\cdot}\mathbf{y} = \overline{x_1} y_1 + \overline{x_2} y_2 + \cdots + \overline{x_n} y_n$. (See Exercise 21 in Section 1.3.) We note that for **x** and **y** in $R^n$, the complex dot product is the same as the dot product in $R^n$. Also in Section 1.3 we defined the length of a vector to be the square root of the dot product of the vector with itself; $\|\mathbf{x}\| = \sqrt{\mathbf{x}{\cdot}\mathbf{x}}$ for **x** in $R^n$ and $\|\mathbf{x}\| = \sqrt{\overline{\mathbf{x}}{\cdot}\mathbf{x}}$ for **x** in $C^n$. The length of **x** is also called the **norm of x**. If we think of vectors **x** and **y** as columns, then both the dot product and the norm can be expressed in terms of a row-by-column product. (*In order to incorporate both the real and complex cases it is convenient to use the conjugate transpose of a matrix, which was introduced in the Exercises in Section* 1.1.) The **conjugate transpose**[1] of an $n \times 1$ vector **x** is denoted $\mathbf{x}^*$ and is the $1 \times n$ vector given by

$$\mathbf{x}^* = \begin{bmatrix} \overline{x_1} & \overline{x_2} & \cdots & \overline{x_n} \end{bmatrix}.$$

It follows that the dot product of **x** and **y** is given by the row-by-column product $\mathbf{x}^*\mathbf{y}$ and the norm of a vector **x** in $R^n$ or $C^n$ can be expressed as $\|\mathbf{x}\| = \sqrt{\mathbf{x}^*\mathbf{x}}$.

---

**Definition**    A set $S$ of $n \times 1$ vectors in $R^n$ or $C^n$ is called an **orthogonal set** provided that none of the vectors is the zero vector and each pair of distinct vectors in $S$ is orthogonal; that is, for vectors $\mathbf{x} \neq \mathbf{y}$ in $S$, $\mathbf{x}^*\mathbf{y} = 0$.

---

[1]Recall that $\mathbf{A}^*$ is the conjugate transpose of **A**.

EXAMPLE 1   Each of the following sets is an orthogonal set. (Verify.)

(a) $S_1 = \left\{ \begin{bmatrix} 4 \\ 0 \end{bmatrix}, \begin{bmatrix} 0 \\ -1 \end{bmatrix} \right\}.$

(b) $S_2 = \left\{ \begin{bmatrix} 2 \\ 1 \\ 0 \end{bmatrix}, \begin{bmatrix} -1 \\ 2 \\ 3 \end{bmatrix}, \begin{bmatrix} 6 \\ -12 \\ 10 \end{bmatrix} \right\}.$

(c) $S_3 = \left\{ \begin{bmatrix} 1+i \\ 1-3i \end{bmatrix}, \begin{bmatrix} 1-2i \\ 1 \end{bmatrix} \right\}.$

(d) $S_4 = \left\{ \begin{bmatrix} -\frac{1}{\sqrt{5}} \\ 0 \\ \frac{2}{\sqrt{5}} \end{bmatrix}, \begin{bmatrix} \frac{2}{\sqrt{5}} \\ 0 \\ \frac{1}{\sqrt{5}} \end{bmatrix} \right\}.$

(e) $S_5 = \left\{ \begin{bmatrix} 2-i \\ 0 \\ 1+i \end{bmatrix}, \begin{bmatrix} 3-i \\ 4+2i \\ -3-4i \end{bmatrix}, \begin{bmatrix} -2+6i \\ 14-7i \\ 10 \end{bmatrix} \right\}.$   (f) $S_6 =$ columns of $I_n$.   ■

---

**Definition**   An orthogonal set of vectors $S$ is called an **orthonormal set** provided that each of its vectors has length or norm 1; that is, $\mathbf{x}^*\mathbf{x} = 1$, for each $\mathbf{x}$ in $S$.

---

EXAMPLE 2   Referring to Example 1, only $S_4$ and $S_6$ are orthonormal sets. (Verify.) Each of the others can be turned into an orthonormal set by replacing each vector $\mathbf{x}$ by its corresponding unit vector $\dfrac{\mathbf{x}}{\|\mathbf{x}\|}$. (See Exercise 15 in Section 1.4.) For $S_2$, the corresponding orthonormal set is

$$\left\{ \frac{1}{\sqrt{5}} \begin{bmatrix} 2 \\ 1 \\ 0 \end{bmatrix}, \frac{1}{\sqrt{14}} \begin{bmatrix} -1 \\ 2 \\ 3 \end{bmatrix}, \frac{1}{\sqrt{280}} \begin{bmatrix} 6 \\ -12 \\ 10 \end{bmatrix} \right\}.$$   ■

Orthogonal and orthonormal sets of vectors have an important property.

**An orthogonal or orthonormal set of vectors is linearly independent.**

To see this we proceed as follows: Let $S = \{\mathbf{v}_1, \mathbf{v}_2, \ldots, \mathbf{v}_k\}$ be an orthogonal set of $n$-vectors. If

$$\sum_{j=1}^{k} c_j \mathbf{v}_j = \mathbf{0},$$

then we will show that $c_j = 0$, $j = 1, 2, \ldots, k$. We have:

$$\mathbf{v}_i^* \sum_{j=1}^{k} c_j \mathbf{v}_j = \mathbf{v}_i^* \mathbf{0} \qquad \text{(multiplying each side on the left by } \mathbf{v}_i^*)$$

$$\Leftrightarrow \quad \sum_{j=1}^{k} c_j \mathbf{v}_i^* \mathbf{v}_j = \mathbf{v}_i^* \mathbf{0} = 0 \quad \text{(by properties of matrix multiplication)}$$

$$\Leftrightarrow \quad c_i \mathbf{v}_i^* \mathbf{v}_i = 0 \qquad \begin{array}{l} \text{(since } S \text{ is an orthogonal set } \mathbf{v}_i^* \mathbf{v}_j = 0 \\ \text{when } i \neq j) \end{array}$$

$$\Leftrightarrow \quad c_i = 0 \qquad \text{(since } \mathbf{v}_i^* \mathbf{v}_i \neq 0)$$

This is true for $i = 1, 2, \ldots, k$ so $S$ is a linearly independent set.

Next we define an important class of matrices that have rows and columns that form an orthonormal set. (See Exercise 16.)

**Definition**   A real (complex) $n \times n$ matrix **Q** whose columns form an orthonormal set is called an **orthogonal (unitary) matrix**.

To combine the real and complex cases, we will use the term unitary since every real matrix is a complex matrix whose entries have imaginary parts equal to zero. Unitary matrices have a variety of properties, which we list in the following box. Verification of these properties is left to the exercises. (See Exercises 16, 18, 19, and 20, respectively.)

---

### Properties of Unitary Matrices

1.  **Q** is unitary if and only if $\mathbf{Q}^*\mathbf{Q} = \mathbf{I}_n$.
    (*Note*: Unitary matrices are nonsingular and $\mathbf{Q}^{-1} = \mathbf{Q}^*$.)
2.  If **Q** is unitary and **x** and **y** are any $n$-vectors, then $\mathbf{x} \cdot \mathbf{y} = \mathbf{Qx} \cdot \mathbf{Qy}$.
3.  If **Q** is unitary and **x** is any $n$-vector, then $\|\mathbf{x}\| = \|\mathbf{Qx}\|$.
4.  If **Q** is unitary, then so is $\mathbf{Q}^*$.

---

A similarity transformation $f(\mathbf{A}) = \mathbf{P}^{-1}\mathbf{AP} = \mathbf{B}$ in which matrix **P** is unitary is called a **unitary similarity transformation**. If **A** is a real matrix and **P** is real, then the term **orthogonal similarity transformation** is often used. We say that **A** and **B** are **unitarily similar** or **orthogonally similar**, respectively.

Recall that one goal of this chapter is to determine the eigenvalues of a matrix **A**. As we stated in Section 4.3, one way to do this is to find a similarity transformation that results in a diagonal or triangular matrix so we can in effect "read off" the eigenvalues as the diagonal entries. It happens that unitary similarity transformations are just the right tool, as given in the following important result, which is known as **Schur's Theorem**.

$$\text{An } n \times n \text{ complex matrix A is unitarily similar} \tag{1}$$
$$\text{to an upper triangular matrix.}$$

We will not verify (1), but it does tell us that we can determine all the eigenvalues of **A** by finding an appropriate unitary matrix **P** so that $\mathbf{P}^{-1}\mathbf{AP} = \mathbf{P}^*\mathbf{AP}$ is upper triangular. The following statement is a special case that is useful in a variety of applications, two of which we investigate later in this section.

$$\text{Every Hermitian (symmetric) matrix is}$$
$$\text{unitarily (orthogonally) similar to a diagonal matrix.} \tag{2}$$

We indicate how to verify this statement in Exercise 24. An interpretation of this result is

**Every symmetric matrix is diagonalizable.**

In fact, we know even more. If **A** is symmetric, then all of its eigenvalues are real numbers, by Property 7 in Section 4.2, and we can find a corresponding set of eigenvectors that form an orthonormal set by (2). Combining these ideas, we can show that eigenvalues and eigenvectors are fundamental building blocks of a symmetric matrix. To illustrate this we first look at the image of the unit circle by a matrix transformation whose associated matrix is symmetric. We will see geometrically the role that the eigenvalues and eigenvectors play in determining the image. We follow this application with an algebraic view of the construction of a symmetric matrix from its eigenpairs and then again look at the idea in terms of geometric approximations of images.

Application:    **The Image of the Unit Circle by a Symmetric Matrix**[2]

In Example 8(b) of Section 1.5 we showed that the image of the unit circle by a matrix transformation whose associated matrix is diagonal was an ellipse centered at the origin with major and minor axes parallel to the coordinate axes; that is, in standard position. Here we investigate the image of the unit circle by a matrix transformation whose associated matrix $\mathbf{A}$ is symmetric. We will show the fundamental role of the eigenpairs of $\mathbf{A}$ in determining both the size and orientation of the image.

Let $\mathbf{A}$ be a $2 \times 2$ symmetric matrix. Then we know that $\mathbf{A}$ is orthogonally diagonalizable.[3] That is, there exists a $2 \times 2$ orthogonal matrix $\mathbf{P} = \begin{bmatrix} \mathbf{p}_1 & \mathbf{p}_2 \end{bmatrix}$ such that

$$\mathbf{P}^T \mathbf{A} \mathbf{P} = \mathbf{D} = \begin{bmatrix} \lambda_1 & 0 \\ 0 & \lambda_2 \end{bmatrix}.$$

Hence $\mathbf{A} = \mathbf{P}\mathbf{D}\mathbf{P}^T$ and the eigenpairs of $\mathbf{A}$ are $(\lambda_1, \mathbf{p}_1)$ and $(\lambda_2, \mathbf{p}_2)$ with $\{\mathbf{p}_1, \mathbf{p}_2\}$ an orthonormal set.

Any point on the unit circle in $R^2$ is represented by a vector (or point)

$$\mathbf{c} = \begin{bmatrix} \cos(t) \\ \sin(t) \end{bmatrix}, \quad 0 \le t < 2\pi.$$

Hence the image of the unit circle consists of all vectors (or points) $\mathbf{A}\mathbf{c} = (\mathbf{P}\mathbf{D}\mathbf{P}^T)\mathbf{c}$. It is convenient to view the image points as those obtained from a composition of matrix transformations; that is,

$$\mathbf{A}\mathbf{c} = \mathbf{P}(\mathbf{D}(\mathbf{P}^T\mathbf{c})). \tag{3}$$

This corresponds to three successive matrix transformations.

To determine the form or shape of this image, we use the following fact (see Exercise 13 in Section 4.5):

If $\mathbf{Q}$ is any $2 \times 2$ orthogonal matrix, then there is a real number $\varphi$ such that

$$\mathbf{Q} = \begin{bmatrix} \cos(\varphi) & -\sin(\varphi) \\ \sin(\varphi) & \cos(\varphi) \end{bmatrix} \quad \text{or} \quad \mathbf{Q} = \begin{bmatrix} \cos(\varphi) & \sin(\varphi) \\ \sin(\varphi) & -\cos(\varphi) \end{bmatrix}.$$

In the first case $\mathbf{Q}$ is a rotation matrix; see Section 1.5. In the second case $\mathbf{Q}$ performs a reflection (about the $x$-axis, the $y$-axis, the line $y = x$, or the line $y = -x$) or a rotation followed by a reflection through the $x$-axis.

With this information and the result of Example 8(b) in Section 1.5, we can determine the geometric form of the image of the unit circle. Using the composite form displayed in (3), we have the following actions (see Exercise 25):

$\mathbf{P}^T$ takes the unit circle to another unit circle,

$\mathbf{D}$ takes this unit circle to an ellipse in standard position, and    (4)

$\mathbf{P}$ rotates or reflects the ellipse.

Thus the image of the unit circle by a symmetric matrix is an ellipse with center at the origin, but possibly with its axes not parallel to the coordinate axes (see Figure 1).

---

[2]Since a matrix transformation $f(\mathbf{c}) = \mathbf{A}\mathbf{c}$ is performed by multiplication by $\mathbf{A}$, the following language is often used in place of direct reference to the function $f$: "The image by the matrix $\mathbf{A}$" or an equivalent wording.

[3]We are using only real matrices in this application; hence we will use the term *orthogonal* in place of *unitary* and *transpose* in place of *conjugate transpose*.

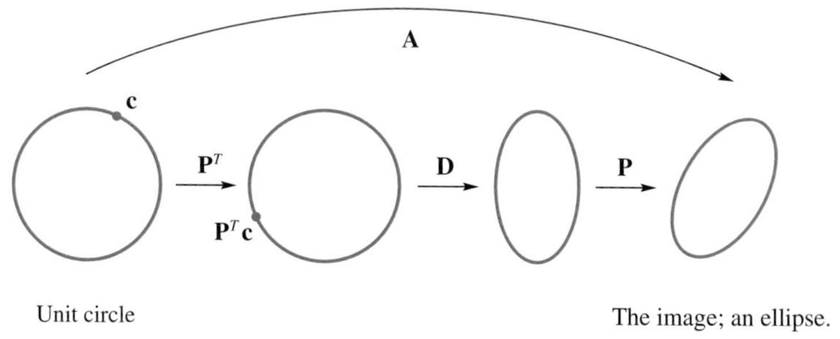

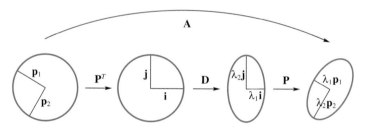

Unit circle

The image; an ellipse.

FIGURE 1

Next we show how the eigenpairs of $\mathbf{A}$ completely determine the elliptical image. The unit vectors $\mathbf{i} = \begin{bmatrix} 1 \\ 0 \end{bmatrix}$ and $\mathbf{j} = \begin{bmatrix} 0 \\ 1 \end{bmatrix}$ are on the unit circle. Then $\mathbf{P} * \mathbf{i} = \mathbf{p}_1$ and $\mathbf{P} * \mathbf{j} = \mathbf{p}_2$. (Verify.) But $\mathbf{p}_1$ and $\mathbf{p}_2$ are also on the unit circle since $\mathbf{P}$ is an orthogonal matrix. Hence

$$\mathbf{P}^{-1}\mathbf{p}_1 = \mathbf{P}^T\mathbf{p}_1 = \mathbf{i} \quad \text{and} \quad \mathbf{P}^{-1}\mathbf{p}_2 = \mathbf{P}^T\mathbf{p}_2 = \mathbf{j}.$$

In (3) let $\mathbf{c} = \mathbf{p}_1$; then we have

$$\mathbf{A}\mathbf{p}_1 = \mathbf{P}(\mathbf{D}(\mathbf{P}^T\mathbf{p}_1)) = \mathbf{P}(\mathbf{D}(\mathbf{i})) = \mathbf{P}(\lambda_1\mathbf{i}) = \lambda_1\mathbf{P}\mathbf{i} = \lambda_1\mathbf{p}_1 \qquad (5)$$

and also $\mathbf{A}\mathbf{p}_2 = \lambda_2\mathbf{p}_2$. (See Exercise 26.) But of course we knew this because of the eigen relationships. However, this sequence of steps shows that eigenvectors of $\mathbf{A}$ on the original unit circle become multiples of themselves in the elliptical image. Moreover, since $\mathbf{p}_1$ and $\mathbf{p}_2$ are orthogonal, so are $\mathbf{A}\mathbf{p}_1$ and $\mathbf{A}\mathbf{p}_2$ (Why?) and these are the axes of the elliptical image. We display this graphically in Figure 2. It follows that the elliptical image is completely determined by the eigenpairs of matrix $\mathbf{A}$.

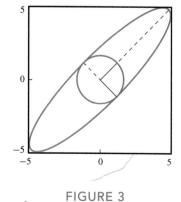

FIGURE 3

Unit Circle

The image; an ellipse.

FIGURE 2

**EXAMPLE 3** Let $\mathbf{A} = \begin{bmatrix} 3 & 4 \\ 4 & 3 \end{bmatrix}$. Then the eigenpairs with orthonormal eigenvectors are

$$(\lambda_1, \mathbf{p}_1) = \left( -1, \begin{bmatrix} \frac{1}{\sqrt{2}} \\ -\frac{1}{\sqrt{2}} \end{bmatrix} \right) \quad \text{and} \quad (\lambda_2, \mathbf{p}_2) = \left( 7, \begin{bmatrix} \frac{1}{\sqrt{2}} \\ \frac{1}{\sqrt{2}} \end{bmatrix} \right).$$

(Verify.) Figure 3 shows the unit circle with $\mathbf{p}_1$ and $\mathbf{p}_2$ displayed as solid line segments and the elliptical image of the unit circle with axes displayed as dashed line segments. ■

The preceding results and example show that

1. The eigenvalues determine the stretching of axes.
2. Eigenvectors determine the orientation of the images of the axes.

So indeed the eigenpairs completely determine the image. These results generalize to $n \times n$ symmetric matrices. The image of the unit $n$-ball is an $n$-dimensional ellipse. For $n = 3$, the image is an ellipsoid.

### Application:   Symmetric Images (Optional)

The previous application showed the role that the eigenpairs play in determining the image of the unit circle. However, there is another way to use the eigenpairs of a symmetric matrix that reveals the information contribution made by each individual eigenpair in algebraically constructing the original matrix. This construction has a number of implications, but we discuss it in terms of approximating digitized images in order to provide a foundation for a more general setting that is discussed in Section 4.7.

The following is called the **spectral representation of a symmetric matrix** and we state it without verification. For uniformity with subsequent material in Section 4.7 we use the conjugate transpose notation *, even though the vectors and matrices have all real entries here.

If $n \times n$ matrix $\mathbf{A}$ is symmetric with eigenvalues $\lambda_1, \lambda_2, \ldots, \lambda_n$ and the corresponding eigenvectors $\mathbf{p}_1, \mathbf{p}_2, \ldots, \mathbf{p}_n$ are chosen to form an orthonormal set, then

$$\mathbf{A} = \lambda_1 \mathbf{p}_1 \mathbf{p}_1^* + \lambda_2 \mathbf{p}_2 \mathbf{p}_2^* + \cdots + \lambda_n \mathbf{p}_n \mathbf{p}_n^*. \tag{6}$$

We note that (6) expresses the symmetric matrix $\mathbf{A}$ as a linear combination of matrices, $\mathbf{p}_j \mathbf{p}_j^*$, which are $n \times n$ since $\mathbf{p}_j$ is $n \times 1$ and $\mathbf{p}_j^*$ is $1 \times n$. The matrix $\mathbf{p}_j \mathbf{p}_j^*$ has a very simple construction; schematically (see Figure 4).

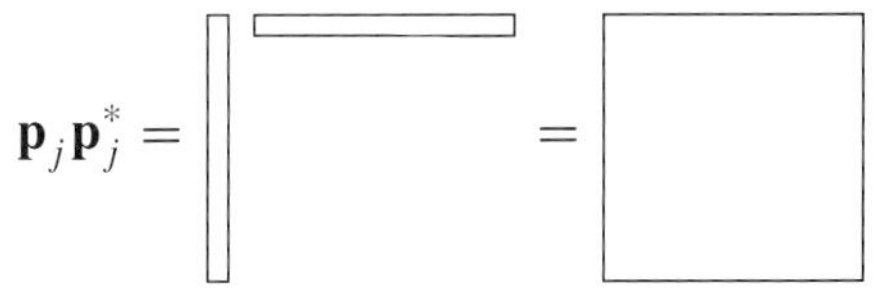

FIGURE 4

In fact $\mathbf{p}_j \mathbf{p}_j^*$ is an **outer product** (see Exercise 65 in Section 1.3.) so each row is a multiple of $\mathbf{p}_j^*$. Hence $\mathbf{rref}(\mathbf{p}_j \mathbf{p}_j^*)$ has one nonzero row and thus has rank 1. This means it contributes just one piece of information to the construction of matrix $\mathbf{A}$. So the spectral representation certainly reveals very basic information about matrix $\mathbf{A}$. In addition, if we let

$$\mathbf{P} = \begin{bmatrix} \mathbf{p}_1 & \mathbf{p}_2 & \cdots & \mathbf{p}_n \end{bmatrix} \quad \text{and} \quad \mathbf{D} = \begin{bmatrix} \lambda_1 & 0 & 0 & 0 \\ 0 & \lambda_2 & 0 & 0 \\ \vdots & \vdots & \ddots & \vdots \\ 0 & 0 & 0 & \lambda_n \end{bmatrix}$$

then $\mathbf{P}$ is unitary and (6) is equivalent to the matrix product

$$\mathbf{A} = \mathbf{PDP}^* \tag{6}$$

which is called the **spectral decomposition** of symmetric matrix $\mathbf{A}$.

A formal verification of the spectral representation is quite lengthy, but we can demonstrate the process in the case $n = 2$. This will reveal the pattern we could use in general and shows the importance of the orthonormal set of eigenvectors. Let $\mathbf{A}$ be a $2 \times 2$ symmetric matrix with eigenvalues $\lambda_1$ and $\lambda_2$ and corresponding orthonormal eigenvectors $\mathbf{p}_1$ and $\mathbf{p}_2$. Let $\mathbf{P} = \begin{bmatrix} \mathbf{p}_1 & \mathbf{p}_2 \end{bmatrix}$ and to make the manipulations simpler to see, let $\mathbf{p}_1 = \begin{bmatrix} a \\ b \end{bmatrix}$ and $\mathbf{p}_2 = \begin{bmatrix} c \\ d \end{bmatrix}$. Since a symmetric matrix is orthogonally diagonalizable, we obtain the following set of matrix equations:

$$\mathbf{P}^*\mathbf{A}\mathbf{P} = \mathbf{D} = \begin{bmatrix} \lambda_1 & 0 \\ 0 & \lambda_2 \end{bmatrix}$$

and, equivalently

$$
\begin{aligned}
\mathbf{A} = \mathbf{P} \begin{bmatrix} \lambda_1 & 0 \\ 0 & \lambda_2 \end{bmatrix} \mathbf{P}^* &= \begin{bmatrix} a & c \\ b & d \end{bmatrix} \begin{bmatrix} \lambda_1 a & \lambda_1 b \\ \lambda_2 c & \lambda_2 d \end{bmatrix} \\
&= \begin{bmatrix} \lambda_1 a^2 + \lambda_2 c^2 & \lambda_1 ab + \lambda_2 cd \\ \lambda_1 ab + \lambda_2 cd & \lambda_1 b^2 + \lambda_2 d^2 \end{bmatrix} \\
&= \begin{bmatrix} \lambda_1 a^2 & \lambda_1 ab \\ \lambda_1 ab & \lambda_1 b^2 \end{bmatrix} + \begin{bmatrix} \lambda_2 c^2 & \lambda_2 cd \\ \lambda_2 cd & \lambda_2 d^2 \end{bmatrix} \\
&= \lambda_1 \begin{bmatrix} a^2 & ab \\ ab & b^2 \end{bmatrix} + \lambda_2 \begin{bmatrix} c^2 & cd \\ cd & d^2 \end{bmatrix} \\
&= \lambda_1 \begin{bmatrix} a\begin{bmatrix} a & b \end{bmatrix} \\ b\begin{bmatrix} a & b \end{bmatrix} \end{bmatrix} + \lambda_2 \begin{bmatrix} c\begin{bmatrix} c & d \end{bmatrix} \\ d\begin{bmatrix} c & d \end{bmatrix} \end{bmatrix} \\
&= \lambda_1 \begin{bmatrix} a \\ b \end{bmatrix} \begin{bmatrix} a & b \end{bmatrix} + \lambda_2 \begin{bmatrix} c \\ d \end{bmatrix} \begin{bmatrix} c & d \end{bmatrix} \\
&= \lambda_1 \mathbf{p}_1 \mathbf{p}_1^* + \lambda_2 \mathbf{p}_2 \mathbf{p}_2^*.
\end{aligned}
$$

EXAMPLE 4    Let

$$\mathbf{A} = \begin{bmatrix} 4 & 1 & 0 \\ 1 & 4 & 0 \\ 0 & 0 & -1 \end{bmatrix}.$$

To determine the spectral representation of $\mathbf{A}$ we first compute its eigenpairs. We find it has three distinct eigenvalues and has the following eigenpairs (verify):

$$\left( 5, \begin{bmatrix} 1 \\ 1 \\ 0 \end{bmatrix} \right), \quad \left( 3, \begin{bmatrix} 1 \\ -1 \\ 0 \end{bmatrix} \right), \quad \text{and} \quad \left( -1, \begin{bmatrix} 0 \\ 0 \\ 1 \end{bmatrix} \right).$$

Since the eigenvalues are distinct, we are assured that the corresponding eigenvectors form an orthogonal set. (See Exercise 46 in Section 4.2.) Scaling them to be unit vectors gives us eigenpairs whose eigenvectors form an orthonormal set:

$$(\lambda_1, \mathbf{p}_1) = \left( 5, \frac{1}{\sqrt{2}} \begin{bmatrix} 1 \\ 1 \\ 0 \end{bmatrix} \right),$$

$$(\lambda_2, \mathbf{p}_2) = \left( 3, \frac{1}{\sqrt{2}} \begin{bmatrix} 1 \\ -1 \\ 0 \end{bmatrix} \right),$$

and

$$(\lambda_3, \mathbf{p}_3) = \left(-1, \begin{bmatrix} 0 \\ 0 \\ 1 \end{bmatrix}\right).$$

Then the spectral representation of $\mathbf{A}$ is

$$\mathbf{A} = \lambda_1 \mathbf{p}_1 \mathbf{p}_1^* + \lambda_2 \mathbf{p}_2 \mathbf{p}_2^* + \lambda_3 \mathbf{p}_3 \mathbf{p}_3^*$$

$$= 5 \left(\frac{1}{\sqrt{2}}\right)^2 \begin{bmatrix} 1 \\ 1 \\ 0 \end{bmatrix} \begin{bmatrix} 1 & 1 & 0 \end{bmatrix} + 3 \left(\frac{1}{\sqrt{2}}\right)^2 \begin{bmatrix} 1 \\ -1 \\ 0 \end{bmatrix} \begin{bmatrix} 1 & -1 & 0 \end{bmatrix}$$

$$+ (-1) \begin{bmatrix} 0 \\ 0 \\ 1 \end{bmatrix} \begin{bmatrix} 0 & 0 & 1 \end{bmatrix}$$

$$= \frac{5}{2} \begin{bmatrix} 1 & 1 & 0 \\ 1 & 1 & 0 \\ 0 & 0 & 0 \end{bmatrix} + \frac{3}{2} \begin{bmatrix} 1 & -1 & 0 \\ 1 & -1 & 0 \\ 0 & 0 & 0 \end{bmatrix} + (-1) \begin{bmatrix} 0 & 0 & 0 \\ 0 & 0 & 0 \\ 0 & 0 & 1 \end{bmatrix}.$$

In Example 4 the eigenvalues of $\mathbf{A}$ were distinct so the corresponding eigenvectors formed an orthogonal set. If a symmetric matrix has an eigenvalue of multiplicity greater than 1, then the linearly independent eigenvectors corresponding to the repeated eigenvalue are not necessarily orthogonal to one another. However, it is possible to "manufacture" orthogonal eigenvectors from those that are just linearly independent. The procedure for doing this is known as the **Gram-Schmidt process** and is discussed in Section 4.5. Given any set $S$ of linearly independent vectors that are a basis for a subspace $W$, the Gram-Schmidt process manufactures from $S$ an orthogonal basis $T$ for the same subspace $W$. In fact, by forcing each vector in $T$ to be a unit vector, we produce an orthonormal basis for subspace $W$.

In a spectral representation each of the outer products $\mathbf{p}_j \mathbf{p}_j^*$ contributes one new piece of information to the construction of the matrix $\mathbf{A}$. The eigenvalue $\lambda_j$ is the "weight" given to that single piece of information. The larger $|\lambda_j|$ the more important the information in $\mathbf{p}_j \mathbf{p}_j^*$. (We illustrate this by the following.)

Suppose that we have a large symmetric matrix $\mathbf{A}$ of information that must be transmitted quickly, but we really only need "most" of the information content. This might be the case if the symmetric matrix represented information from a pattern or a photo. Let's also assume that we can compute eigenpairs of the matrix with relative ease. Using the spectral representation of the matrix, we can approximate the image using **partial sums**[4] rather than all the eigen information. To do this we consider the eigenvalues as weights for the information contained in the eigenvectors and label things so that the eigenvalues are ordered with

$$|\lambda_1| \geq |\lambda_2| \geq \cdots \geq |\lambda_n|.$$

Then $\mathbf{A} \approx \lambda_1 \mathbf{p}_1 \mathbf{p}_1^* + \lambda_2 \mathbf{p}_2 \mathbf{p}_2^* + \cdots + \lambda_k \mathbf{p}_k \mathbf{p}_k^*$, where $k \leq n$. This scheme uses the information associated with the larger eigenvalues first and then adjoins that associated with the smaller eigenvalues.

---

[4]When fewer than the $n$ terms in (6) are combined, we call this a partial sum of the spectral representation.

EXAMPLE 5    For the geometric pattern shown in Figure 5, we digitize it using a 1 for a black block and 0 for a white block. Let **A** be the $9 \times 9$ matrix of corresponding zeros and ones in Figure 6.

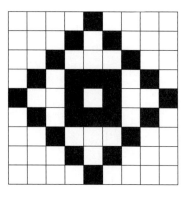

| 0 | 0 | 0 | 0 | 1 | 0 | 0 | 0 | 0 |
|---|---|---|---|---|---|---|---|---|
| 0 | 0 | 0 | 1 | 0 | 1 | 0 | 0 | 0 |
| 0 | 0 | 1 | 0 | 0 | 0 | 1 | 0 | 0 |
| 0 | 1 | 0 | 1 | 1 | 1 | 0 | 1 | 0 |
| 1 | 0 | 0 | 1 | 0 | 1 | 0 | 0 | 1 |
| 0 | 1 | 0 | 1 | 1 | 1 | 0 | 1 | 0 |
| 0 | 0 | 1 | 0 | 0 | 0 | 1 | 0 | 0 |
| 0 | 0 | 0 | 1 | 0 | 1 | 0 | 0 | 0 |
| 0 | 0 | 0 | 0 | 1 | 0 | 0 | 0 | 0 |

FIGURE 5                              FIGURE 6

We can approximate the geometric pattern using partial sums of the spectral representation. Using MATLAB we can show that the eigenvalues are, in order of magnitude to tenths, 3.7, 2, $-2$, 1.1, $-0.9$, 0, 0, 0, 0. Rather than display the approximations digitally, we show a pictorial representation where an entry will be a black square if the corresponding numerical entry is greater than or equal to one half in absolute value, otherwise it will be shown as a white square. We show the first two partial sums in Figures 7 and 8, respectively. The third partial sum reveals Figure 8 again and the fourth partial sum is the original pattern of Figure 5. If we had to transmit enough information to build a black-white approximation to the pattern, then we could just send the first four eigenvalues and their corresponding eigenvectors; that is, $4 + (4)9 = 40$ numerical values. This is a significant saving compared to transmitting all of matrix **A**, which requires 81 values.    ■

$$\mathbf{A} \approx 3.7\mathbf{p}_1\mathbf{p}_1^* \quad \Rightarrow$$

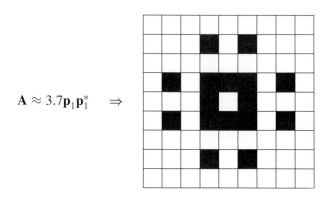

FIGURE 7

Naturally not many photo images are symmetric, and we need a large matrix to represent them. One technique is to imbed a nonsymmetric pattern or photo in a larger symmetric one and proceed as in Example 5. We illustrate this in Exercise ML.4. However, we can generalize these ideas to nonsymmetric and nonsquare matrices. The generalizations behave in much the same way as the use of the spectral representation illustrated here. We briefly introduce such concepts in Section 4.7.

$$\mathbf{A} \approx 3.7\mathbf{p}_1\mathbf{p}_1^* + 2\mathbf{p}_2\mathbf{p}_2^* \quad \Rightarrow$$

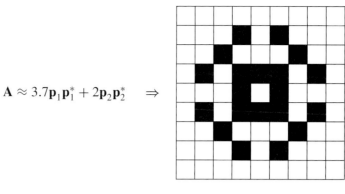

FIGURE 8

## EXERCISES 4.4

*In Exercises 1–4, show that each given set is an orthogonal set.*

**1.** $\left\{ \begin{bmatrix} 1 \\ 2 \end{bmatrix}, \begin{bmatrix} -2 \\ 1 \end{bmatrix} \right\}$.

**2.** $\left\{ \begin{bmatrix} 1 \\ 0 \\ 1 \end{bmatrix}, \begin{bmatrix} -3 \\ 4 \\ 3 \end{bmatrix} \right\}$.

**3.** $\left\{ \begin{bmatrix} 1 \\ 2 \\ 1 \end{bmatrix}, \begin{bmatrix} -1 \\ 1 \\ -1 \end{bmatrix}, \begin{bmatrix} 3 \\ 0 \\ -3 \end{bmatrix} \right\}$.

**4.** $\left\{ \begin{bmatrix} 1 \\ 2 \\ 0 \\ -1 \end{bmatrix}, \begin{bmatrix} -1 \\ 0 \\ 3 \\ -1 \end{bmatrix}, \begin{bmatrix} -2 \\ 2 \\ 0 \\ 2 \end{bmatrix} \right\}$.

**5.** Why is it impossible to have three vectors in $R^2$ that form an orthogonal set?

**6.** Find a (nonzero) vector $\mathbf{v}$ in $R^3$ that is orthogonal to each of the vectors in Exercise 2. (*Hint*: Let $\mathbf{v} = \begin{bmatrix} x_1 & x_2 & x_3 \end{bmatrix}^T$ and compute the dot product of each vector with $\mathbf{v}$ to obtain a system of equations to solve for the $x_i$.)

**7.** Find a (nonzero) vector $\mathbf{v}$ in $R^4$ that is orthogonal to each of the vectors in Exercise 4.

**8.** Convert each of the orthogonal sets in Exercises 1–4 to an orthonormal set.

*In Exercises 9–11, show that each given set of complex vectors is an orthogonal set.*

**9.** $\left\{ \begin{bmatrix} 1 + i \\ 2 - i \end{bmatrix}, \begin{bmatrix} 1 + 3i \\ -2 \end{bmatrix} \right\}$.

**10.** $\left\{ \begin{bmatrix} 2 \\ 2 + i \\ 1 + 2i \end{bmatrix}, \begin{bmatrix} 1 - i \\ 0 \\ -1.2 - 0.4i \end{bmatrix} \right\}$.

**11.** $\left\{ \begin{bmatrix} i \\ 2i \\ 1 + i \end{bmatrix}, \begin{bmatrix} 2 \\ -1 \\ 0 \end{bmatrix}, \begin{bmatrix} 1 \\ 2 \\ -2.5 + 2.5i \end{bmatrix} \right\}$.

**12.** Convert each of the orthogonal sets in Exercises 9–11 to an orthonormal set.

**13.** Let $\mathbf{v}_1 = \begin{bmatrix} 1 \\ 0 \\ -1 \end{bmatrix}$ and $\mathbf{v}_2 = \begin{bmatrix} 2 \\ 1 \\ 2 \end{bmatrix}$.

(a) Show that $\mathbf{v}_1$ and $\mathbf{v}_2$ are orthogonal vectors.

(b) Let $\mathbf{v}_3 = \begin{bmatrix} x_1 \\ x_2 \\ x_3 \end{bmatrix}$. Determine the entries of $\mathbf{v}_3$ so that $\{\mathbf{v}_1, \mathbf{v}_2, \mathbf{v}_3\}$ is an orthogonal set. (*Hint*: Use $\mathbf{v}_1 \cdot \mathbf{v}_3 = 0$ and $\mathbf{v}_2 \cdot \mathbf{v}_3 = 0$ to obtain a system of equations.)

(c) Convert $\{\mathbf{v}_1, \mathbf{v}_2, \mathbf{v}_3\}$ to an orthonormal set $\{\mathbf{u}_1, \mathbf{u}_2, \mathbf{u}_3\}$.

(d) Show that the matrix $\mathbf{U} = \begin{bmatrix} \mathbf{u}_1 & \mathbf{u}_2 & \mathbf{u}_3 \end{bmatrix}$ is unitary.

**14.** Let $\mathbf{v}_1 = \begin{bmatrix} 1 + i \\ i \\ i \end{bmatrix}$ and $\mathbf{v}_2 = \begin{bmatrix} 0 \\ 1 \\ -1 \end{bmatrix}$.

(a) Show that $\mathbf{v}_1$ and $\mathbf{v}_2$ are orthogonal vectors.

(b) Let $\mathbf{v}_3 = \begin{bmatrix} a + bi \\ 1 \\ 1 \end{bmatrix}$. Determine $a$ and $b$ so that $\{\mathbf{v}_1, \mathbf{v}_2, \mathbf{v}_3\}$ is an orthogonal set. (*Hint*: Use $\mathbf{v}_1 \cdot \mathbf{v}_3 = 0$ and $\mathbf{v}_2 \cdot \mathbf{v}_3 = 0$ to obtain a system of equations.)

(c) Convert $\{\mathbf{v}_1, \mathbf{v}_2, \mathbf{v}_3\}$ to an orthonormal set $\{\mathbf{u}_1, \mathbf{u}_2, \mathbf{u}_3\}$.

(d) Show that the matrix $\mathbf{U} = \begin{bmatrix} \mathbf{u}_1 & \mathbf{u}_2 & \mathbf{u}_3 \end{bmatrix}$ is unitary.

**15.** Let $\mathbf{A} = \begin{bmatrix} 1 & -1 & -1 \\ 0 & 1 & -2 \\ 1 & 1 & 1 \end{bmatrix}$.

(a) Compute $\mathbf{A}^*\mathbf{A}$. Is $\mathbf{A}$ a unitary matrix?

(b) If $\mathbf{A}$ is not a unitary matrix, then explain how to construct a unitary matrix from $\mathbf{A}$.

**16.** Let $\mathbf{Q} = \begin{bmatrix} \mathbf{q}_1 & \mathbf{q}_2 & \mathbf{q}_3 \end{bmatrix}$ be a $3 \times 3$ unitary matrix. Then
$$\mathbf{Q}^* = \begin{bmatrix} \mathbf{q}_1^* \\ \mathbf{q}_2^* \\ \mathbf{q}_3^* \end{bmatrix}.$$

(a) Use row-by-column products to show that $\mathbf{Q}^*\mathbf{Q} = \mathbf{I}_3$. (Note that this argument generalizes to $n \times n$ matrices.)

(b) Explain why we can say the columns of a unitary matrix form a linearly independent set.

(c) Explain why we can say the rows of a unitary matrix form a linearly independent set.

**17.** Explain why we can obtain the inverse of a unitary matrix at virtually 'no cost'.

**18.** If $\mathbf{Q}$ is unitary and $\mathbf{x}$ and $\mathbf{y}$ are any $n$-vectors, then show that $\mathbf{x} \cdot \mathbf{y} = \mathbf{Q}\mathbf{x} \cdot \mathbf{Q}\mathbf{y}$. (*Hint:* $\mathbf{x} \cdot \mathbf{y} = \mathbf{x}^*\mathbf{y}$.)

**19.** Let $\mathbf{Q}$ be unitary and $\mathbf{x}$ any $n$-vector.

(a) Show that $\|\mathbf{x}\| = \|\mathbf{Q}\mathbf{x}\|$. That is, a matrix transformation whose associated matrix is unitary preserves the length of a vector. (Hint: $\|\mathbf{x}\| = \sqrt{\mathbf{x}^*\mathbf{x}}$.)

(b) Use part (a) together with Exercise 18 to show that the angle between vectors $\mathbf{x}$ and $\mathbf{y}$ is the same as between $\mathbf{Q}\mathbf{x}$ and $\mathbf{Q}\mathbf{y}$. (That is, a matrix transformation whose associated matrix is unitary preserves angles between vectors.)

**20.** Show that if $\mathbf{Q}$ is unitary, then so is $\mathbf{Q}^*$.

**21.** Let $\mathbf{Q}$ be an orthogonal matrix.

(a) Why is $\det(\mathbf{Q}) \neq 0$?

(b) Show that $|\det(\mathbf{Q})| = 1$. (*Hint:* Use the fact that $\mathbf{Q}^T\mathbf{Q} = \mathbf{I}$.)

(c) Explain why we can say that a matrix transformation determined by an orthogonal matrix is "area invariant."

**22.** Show that the matrix
$$\mathbf{A} = \begin{bmatrix} \cos(\theta) & -\sin(\theta) \\ \sin(\theta) & \cos(\theta) \end{bmatrix},$$

which determines the matrix transformation that performs a rotation of a vector in $R^2$ by an angle $\theta$, is a unitary matrix.

**23.** Explain why we can say that matrix transformations determined by unitary matrices are "distortion free."

**24.** Let $\mathbf{A}$ be a symmetric (or Hermitian) matrix. Use Schur's Theorem to show that $\mathbf{A}$ is unitarily similar to a diagonal matrix. (*Hint:* If $\mathbf{U}$ is upper triangular, then $\mathbf{U}^*$ is lower triangular.)

**25.** Verify the three assertions made in (4).

**26.** Give a reason for each of the equalities used in (5).

**27.** Let $\mathbf{A} = \begin{bmatrix} 5.5 & 4.5 \\ 4.5 & 5.5 \end{bmatrix}$. There are two distinct eigenvalues and eigenpairs
$$(\lambda_1, \mathbf{x}_1) = \left(10, \begin{bmatrix} 1 \\ 1 \end{bmatrix}\right), \quad (\lambda_2, \mathbf{x}_2) = \left(1, \begin{bmatrix} 1 \\ -1 \end{bmatrix}\right).$$

(a) Determine an orthonormal set of eigenvectors $\mathbf{p}_1$ and $\mathbf{p}_2$ for $\lambda_1$ and $\lambda_2$, respectively.

(b) Construct the partial sums $\mathbf{S}_i = \lambda_i \mathbf{p}_i \mathbf{p}_i^*$, $i = 1, 2$ of the spectral representation of $\mathbf{A}$.

(c) Verify that $\mathbf{S}_1 + \mathbf{S}_2$ gives $\mathbf{A}$.

(d) We see that $\lambda_1$ is quite a bit larger than $\lambda_2$, so we should expect that $\mathbf{S}_1 = \lambda_1 \mathbf{p}_1 \mathbf{p}_1^*$ is quite close to $\mathbf{A}$. One way to measure the closeness is to compute the matrix $\mathbf{A} - \mathbf{S}_1$ and then compute the square root of the sum of the squares of its entries. Show that we get 1.

**28.** Let
$$\mathbf{A} = \left(\tfrac{1}{6}\right) \begin{bmatrix} 41 & 31 & 2 \\ 31 & 41 & -2 \\ 2 & -2 & 8 \end{bmatrix}$$
$$\approx \begin{bmatrix} 6.8333 & 5.1667 & 0.3333 \\ 5.1667 & 6.8333 & -0.3333 \\ 0.3333 & -0.3333 & 1.3333 \end{bmatrix}.$$

There are three distinct eigenvalues and eigenpairs
$$(\lambda_1, \mathbf{x}_1) = \left(12, \begin{bmatrix} 1 \\ 1 \\ 0 \end{bmatrix}\right), \quad (\lambda_2, \mathbf{x}_2) = \left(2, \begin{bmatrix} 1 \\ -1 \\ 1 \end{bmatrix}\right),$$

and
$$(\lambda_3, \mathbf{x}_3) = \left(1, \begin{bmatrix} 1 \\ -1 \\ -2 \end{bmatrix}\right).$$

(a) Determine an orthonormal set of eigenvectors $\mathbf{p}_1$, $\mathbf{p}_2$, and $\mathbf{p}_3$ for $\lambda_1$, $\lambda_2$, and $\lambda_3$, respectively.

(b) Construct the partial sums $\mathbf{S}_i = \lambda_i \mathbf{p}_i \mathbf{p}_i^*$, $i = 1, 2, 3$, of the spectral representation of $\mathbf{A}$.

(c) Verify that $\mathbf{S}_1 + \mathbf{S}_2 + \mathbf{S}_3$ gives $\mathbf{A}$.

(d) We see that $\lambda_1$ is quite a bit larger than $\lambda_2$ or $\lambda_3$, so we should expect that $\mathbf{S}_1 = \lambda_1 \mathbf{p}_1 \mathbf{p}_1^*$ is quite close to $\mathbf{A}$. One way to measure the closeness is to compute the matrix $\mathbf{A} - \mathbf{S}_1$ and then compute the square root of the sum of the squares of its entries. Show that we get approximately 2.23.

**29.** Let
$$\mathbf{A} = \left(\tfrac{1}{12}\right) \begin{bmatrix} 125 & 2 & 115 \\ 2 & 8 & -2 \\ 115 & -2 & 125 \end{bmatrix}$$
$$\approx \begin{bmatrix} 10.4167 & 0.1667 & 9.5833 \\ 0.1667 & 0.6667 & -0.1667 \\ 9.5833 & -0.1667 & 10.4167 \end{bmatrix}.$$

There are three distinct eigenvalues and eigenpairs

$$(\lambda_1, \mathbf{x}_1) = \left(20, \begin{bmatrix} 1 \\ 0 \\ 1 \end{bmatrix}\right), \quad (\lambda_2, \mathbf{x}_2) = \left(1, \begin{bmatrix} -1 \\ -1 \\ 1 \end{bmatrix}\right),$$

and

$$(\lambda_3, \mathbf{x}_3) = \left(0.5, \begin{bmatrix} -1 \\ 2 \\ 1 \end{bmatrix}\right).$$

(a) Determine an orthonormal set of eigenvectors $\mathbf{p}_1$, $\mathbf{p}_2$, and $\mathbf{p}_3$ for $\lambda_1$, $\lambda_2$, and $\lambda_3$, respectively.

(b) Construct the partial sums $\mathbf{S}_i = \lambda_i \mathbf{p}_i \mathbf{p}_i^*$, $i = 1, 2, 3$, of the spectral representation of $\mathbf{A}$.

(c) Verify that $\mathbf{S}_1 + \mathbf{S}_2 + \mathbf{S}_3$ gives $\mathbf{A}$.

(d) We see that $\lambda_1$ is quite a bit larger than $\lambda_2$ or $\lambda_3$, so we should expect that $\mathbf{S}_1 = \lambda_1 \mathbf{p}_1 \mathbf{p}_1^*$ is quite close to $\mathbf{A}$. One way to measure the closeness is to compute the matrix $\mathbf{A} - \mathbf{S}_1$ and then compute the square root of the sum of the squares of its entries. Show that we get approximately 1.118.

### In MATLAB

*We use* MATLAB *to illustrate geometric properties of matrix transformations defined by symmetric matrices. We show how the eigenpairs reveal the structure of the symmetric matrix and how they determine the properties of the image.*

**ML. 1.** Following the procedures outlined in Example 5, we digitize the pattern shown in Figure 9 to zeros and ones as shown in Figure 10. Enter the $9 \times 9$ matrix from Figure 10 into MATLAB and name it $\mathbf{A}$. We use the m-files **scan** and **specgen** to illustrate the approximations of this pattern using the spectral representation of matrix $\mathbf{A}$. First enter command **scan(A)** and check that you get a pattern identical to Figure 9. Next enter command **specgen(A)** to see the partial sum approximations from the spectral representation of $\mathbf{A}$; follow the screen directions. How many eigenpairs of information are required before we obtain the full image as shown in Figure 9?

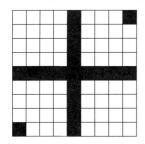

| 0 | 0 | 0 | 0 | 1 | 0 | 0 | 0 | 1 |
|---|---|---|---|---|---|---|---|---|
| 0 | 0 | 0 | 0 | 1 | 0 | 0 | 0 | 0 |
| 0 | 0 | 0 | 0 | 1 | 0 | 0 | 0 | 0 |
| 0 | 0 | 0 | 0 | 1 | 0 | 0 | 0 | 0 |
| 1 | 1 | 1 | 1 | 1 | 1 | 1 | 1 | 1 |
| 0 | 0 | 0 | 0 | 1 | 0 | 0 | 0 | 0 |
| 0 | 0 | 0 | 0 | 1 | 0 | 0 | 0 | 0 |
| 0 | 0 | 0 | 0 | 1 | 0 | 0 | 0 | 0 |
| 1 | 0 | 0 | 0 | 1 | 0 | 0 | 0 | 0 |

FIGURE 9                    FIGURE 10

**ML. 2.** Follow the directions in Exercise ML.1 for the image in Figure 11. In this case you will need to construct an $11 \times 11$ matrix of zeros and ones. Determine the number of eigenpairs of information required so that the image of a partial sum is easily recognizable as that in Figure 11.

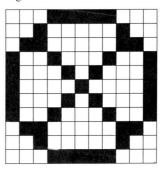

FIGURE 11

**ML. 3.** Follow the directions in Exercise ML.1 for the image of a smiley face in Figure 12. (To see the face, tilt your head to look up the main diagonal from the lower right corner.) In this case you will need to construct an $11 \times 11$ matrix of zeros and ones. Determine the number of eigenpairs of information that are required so that the image of a partial sum is easily recognizable as that in Figure 12.

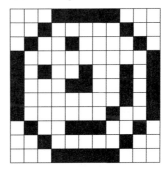

FIGURE 12

**ML. 4.** As we commented near the end of the section, a way to use the techniques of this section for a nonsymmetric pattern is to imbed it in a symmetric pattern. The trade-off is that we increase the size of the matrix for which we compute the spectral decomposition. Suppose that we have a square black and white pattern as in Example 5, but it is not symmetric. Let matrix $\mathbf{B}$ be the corresponding matrix of zeros and ones. To apply partial sums of the spectral representation, we construct the symmetric matrix

$$\mathbf{A} = \begin{bmatrix} \mathbf{Z} & \mathbf{B}^T \\ \mathbf{B} & \mathbf{Z} \end{bmatrix}$$

where $\mathbf{Z}$ is a square matrix of all zeros the same size as $\mathbf{B}$. We proceed as in Example 4 and Exercise ML.1.

(a) To demonstrate this procedure, use the following MATLAB commands. Choose the update image option in **specgen** enough times until you can clearly see the image reproduced.

> **load figtu**
> **scan(TU)**
> **Z=zeros(size(TU));**
> **A=[Z TU';TU Z];**
> **specgen(A)**

(b) Compute the eigenvalues of the matrix **A** of part (a). Make a conjecture about a pattern of the numerical values. Change the format to **long e** and inspect the display again. Does the pattern seem to still be valid?

(c) To pursue the notion in part (b) further, conduct five experiments like the following. Select an integer $n$ between 1 and 10. Use the following MATLAB commands.

> **n=** _____ ← enter your selection
> **B=fix(10*rand(n));**
> **Z=zeros(size(B));**
> **A=[Z B';B Z];**
> **eig(A)**

Does it seem that your conjecture is still valid? Explain. Formalize your conjecture into a mathematical statement.

## True/False Review Questions

*Determine whether each of the following statements is true or false.*

1. $\left\{ \begin{bmatrix} 1 \\ 0 \\ 2 \end{bmatrix}, \begin{bmatrix} -2 \\ 1 \\ 1 \end{bmatrix}, \begin{bmatrix} -2 \\ -5 \\ 1 \end{bmatrix} \right\}$ is an orthogonal set.

2. If $\{\mathbf{u}, \mathbf{v}, \mathbf{w}\}$ is an orthonormal set then $\mathbf{u} \cdot \mathbf{v} = \mathbf{u} \cdot \mathbf{w} = \mathbf{v} \cdot \mathbf{w} = 0$ and $\mathbf{u} \cdot \mathbf{u} = \mathbf{v} \cdot \mathbf{v} = \mathbf{w} \cdot \mathbf{w} = 1$.

3. There are sets of orthogonal vectors which are linearly dependent.

4. A unitary matrix is nonsingular.

5. If matrix transformation $T(\mathbf{x}) = \mathbf{A}\mathbf{x}$ from $R^n$ to $R^n$ is such that $\mathbf{A}$ is unitary, then $\|\mathbf{A}\mathbf{x}\| = \|\mathbf{x}\|$.

6. If $\mathbf{A} = \mathbf{A}^T$, then $\mathbf{A}$ is guaranteed to be similar to a diagonal matrix.

7. The rows of a unitary matrix are an orthogonal set.

8. The spectral decomposition of a symmetric matrix expresses it as a linear combination of matrices with rank 1.

9. If $\mathbf{P}$ is an $n \times n$ unitary matrix, then for any pair of $n$-vectors $\mathbf{x}$ and $\mathbf{y}$, $\mathbf{P}\mathbf{x} \cdot \mathbf{P}\mathbf{y} = \mathbf{x} \cdot \mathbf{y}$.

10. $\begin{bmatrix} 2 + 3i & 5 \\ -4i & 1 - i \end{bmatrix}^* = \begin{bmatrix} 2 - 3i & 5 \\ 4i & 1 + i \end{bmatrix}.$

## Terminology

| | |
|---|---|
| Norm | Orthogonal set |
| Orthonormal set | Orthogonal (unitary) matrix |
| Orthogonal (unitary) similarity transformation | Schur's Theorem |
| Symmetric image of the unit circle | Spectral representation of a symmetric matrix |
| Outer product | Spectral decomposition |

This section answers the question, What matrices are similar to a triangular matrix? (This is in contrast to the previous section, which focused on diagonalizable matrices.) The answer to this question has important consequences both theoretically and computationally. We touched on several theoretical implications for the class of symmetric matrices. It is the case that symmetric matrices arise in a variety of important applications so these results are widely used. We specifically showed how the image of the unit circle by a symmetric matrix transformation was completely determined by the eigenpairs of the matrix. We concluded the section by showing that any symmetric matrix can be built from the information in its eigenpairs and hence the matrix could be approximated by using a subset of the eigenpairs.

- What do orthogonal sets correspond to geometrically?
- Can an orthogonal set contain redundant information? Explain.
- How can any orthogonal set be converted to an orthonormal set?
- In $R^3$, what is the simplest orthonormal set?
- Suppose that the columns of an $n \times n$ matrix $\mathbf{A}$ form an orthonormal set. Explain why we are guaranteed that $\mathbf{A}$ is nonsingular.
- If $\mathbf{A}$ is a $3 \times 2$ matrix with orthogonal columns, then describe $\mathbf{A}^T \mathbf{A}$. If the columns were orthonormal, what is $\mathbf{A}^T \mathbf{A}$?
- What does it mean to say that two matrices are orthogonally (or unitarily) similar?
- Describe the set of all matrices that are similar to a triangular matrix.
- Suppose you have the set of matrices that satisfy the preceding statement and you remove all the diagonalizable matrices. What set remains? Explain.
- What special property of symmetric matrices is developed in this section?
- Describe the spectral decomposition of a symmetric matrix.
- How can every symmetric matrix be expressed as a linear combination involving its eigenpairs?
- Describe the procedure we developed for approximating a symmetric matrix by using only part of its eigenpairs.
- Describe the role played by the eigenpairs in determining the image of the unit circle when we use a symmetric matrix transformation.

## 4.5 ■ THE GRAM-SCHMIDT PROCESS

In Section 4.4 we showed that every symmetric (or Hermitian) matrix is unitarily similar to a diagonal matrix. (See Exercise 24, Section 4.4.) Thus if $\mathbf{A} = \mathbf{A}^*$, then there exists a unitary matrix $\mathbf{P}$ such that $\mathbf{P}^*\mathbf{AP} = \mathbf{D}$, a diagonal matrix.[1] It follows that the diagonal entries of $\mathbf{D}$ are the eigenvalues of $\mathbf{A}$ and that the columns of $\mathbf{P}$ form an orthonormal set of eigenvectors. If symmetric matrix $\mathbf{A}$ has a repeated eigenvalue $\mu$ with multiplicity $m > 1$, then there are guaranteed to exist $m$ linearly independent eigenvectors $\mathbf{q}_1, \mathbf{q}_2, \ldots, \mathbf{q}_m$ associated with $\mu$. However, there is no guarantee that the $m$ eigenvectors associated with $\mu$ are mutually orthogonal; that is, $\mathbf{q}_j^* \mathbf{q}_k$ with $j \neq k$ need not be zero. Thus we cannot use these eigenvectors as some of the columns of the unitary matrix $\mathbf{P}$ even if we scaled them to be unit vectors.

[1] We continue the practice of using the conjugate transpose $*$ so that both the real and complex cases are handled at the same time.

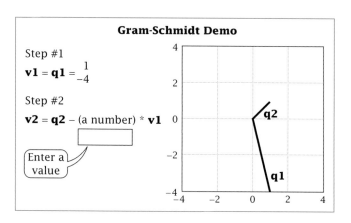

In MATLAB routine, **gsprac2.m**

EXAMPLE 1    Let $\mathbf{A} = \begin{bmatrix} 9 & -1 & -2 \\ -1 & 9 & -2 \\ -2 & -2 & 6 \end{bmatrix}$. Then the characteristic polynomial is

$$c(\lambda) = \lambda^3 - 24\lambda^2 + 180\lambda - 400 = (\lambda - 4)(\lambda - 10)^2$$

and it follows that we have eigenpairs (verify)

$$\left( 4, \begin{bmatrix} 1 \\ 1 \\ 2 \end{bmatrix} \right), \quad \left( 10, \begin{bmatrix} -1 \\ 1 \\ 0 \end{bmatrix} \right), \quad \text{and} \quad \left( 10, \begin{bmatrix} -2 \\ 0 \\ 1 \end{bmatrix} \right).$$

The set of eigenvectors given for the repeated eigenvalue 10 are linearly independent, but not orthogonal. (Verify.) Hence we cannot use them to construct the unitary matrix $\mathbf{P}$ so that $\mathbf{P}^*\mathbf{AP}$ is diagonal.    ■

Looking at things from another point of view, let $S = \{\mathbf{q}_1, \mathbf{q}_2, \ldots, \mathbf{q}_m\}$ be a basis for the eigenspace $W$ corresponding to $\mu$. We have seen that a subspace can have many different bases, so there *may* be a basis $T = \{\mathbf{u}_1, \mathbf{u}_2, \ldots, \mathbf{u}_m\}$ for $W$ that is an orthonormal set; that is, an **orthonormal basis**. An orthonormal basis can make certain computations simpler. For example, if $\mathbf{w}$ is a vector in subspace $W$, then it can be expressed as a linear combination of the basis vectors for $W$. If $W$ has orthonormal basis $T = \{\mathbf{u}_1, \mathbf{u}_2, \ldots, \mathbf{u}_m\}$, then the coefficients for the linear combination are easy to obtain. Let

$$\mathbf{w} = c_1\mathbf{u}_1 + c_2\mathbf{u}_2 + \cdots + c_m\mathbf{u}_m.$$

Then

$$\mathbf{u}_j^*\mathbf{w} = c_1\mathbf{u}_j^*\mathbf{u}_1 + \cdots + c_{j-1}\mathbf{u}_j^*\mathbf{u}_{j-1} + c_j\mathbf{u}_j^*\mathbf{u}_j + c_{j+1}\mathbf{u}_j^*\mathbf{u}_{j+1} + \cdots + c_m\mathbf{u}_j^*\mathbf{u}_m$$
$$= c_j$$

since $\mathbf{u}_j^*\mathbf{u}_k = 0$, $j \neq k$, and $\mathbf{u}_j^*\mathbf{u}_j = 1$. If we had used a basis that is not orthonormal, we would have had to solve a linear system of equations for the coefficients $c_1, c_2, \ldots, c_m$.

The Gram-Schmidt process is a computational algorithm that converts a basis for a subspace $W$ of $R^n$ or $C^n$ into an orthonormal basis for $W$. (We develop this in a more general setting in Section 6.3.) Whether we are seeking an orthonormal basis for an eigenspace or some other subspace $W$ of $R^n$ or $C^n$ has no bearing on the process that we describe in this section. The basic steps of the process use dot products, ideas about span and bases, and simple equations. We will use the row-by-column product notation $\mathbf{q}_j^*\mathbf{q}_k$ in place of the dot product notation $\mathbf{q}_j \cdot \mathbf{q}_k$ so that we can handle both real and complex $n$-vectors within the same algorithmic process. We start with any basis for $W$ and manufacture an orthogonal basis for $W$. Then we scale each of the orthogonal basis vectors to be a unit vector to obtain an orthonormal basis for $W$.

The actual steps indicated are algebraic, but there is a geometric interpretation to the steps that depends on projections. In Section 1.5 we developed the (orthogonal) projection of one vector onto another in $R^2$. For nonparallel column vectors $\mathbf{c}$ and $\mathbf{b}$ in $R^2$, the projection $\mathbf{p}$ of $\mathbf{c}$ onto $\mathbf{b}$ is given by

$$\mathbf{p} = \frac{(\mathbf{b} \cdot \mathbf{c})}{(\mathbf{b} \cdot \mathbf{b})}\mathbf{b} = \frac{\mathbf{b}^*\mathbf{c}}{\mathbf{b}^*\mathbf{b}}\mathbf{b}.$$

(It is convenient to use the following notation for the projection $\mathbf{p}$; $\mathbf{p} = \text{proj}_\mathbf{b}\mathbf{c}$.) This was depicted geometrically in Figures 13 and 14 of Section 1.5. We showed that if $\mathbf{c}$ and $\mathbf{b}$ were not parallel then vector $\mathbf{n} = \mathbf{c} - \mathbf{p}$ was orthogonal to $\mathbf{b}$ and vectors $\mathbf{b}$ and $\mathbf{n}$ were an orthogonal spanning set for $R^2$. We can say even more now:

- $\{\mathbf{c}, \mathbf{b}\}$ is a basis for $R^2$. (Explain why.)
- $\{\mathbf{b}, \mathbf{n}\}$ is an orthogonal basis for $R^2$. (Explain why.)

In effect, we constructed an orthogonal basis from the original basis $\{\mathbf{c}, \mathbf{b}\}$. The notion of projections can be extended to include the projection of a vector onto a subspace. We develop this in detail in Section 6.3. However, for our purposes now we use the following:

Let $\mathbf{c}$ be a vector in $R^n$ (or $C^n$) and $V$ a $k$-dimensional subspace with orthogonal basis $\{\mathbf{v}_1, \mathbf{v}_2, \ldots, \mathbf{v}_k\}$. The projection of $\mathbf{c}$ onto $V$, denoted $\text{proj}_V\mathbf{c}$, is the sum of the projections of $\mathbf{c}$ onto to each of the basis vectors $\mathbf{v}_j$. That is,

$$\text{proj}_V\mathbf{c} = \text{proj}_{\mathbf{v}_1}\mathbf{c} + \text{proj}_{\mathbf{v}_2}\mathbf{c} + \cdots + \text{proj}_{\mathbf{v}_k}\mathbf{c}$$

$$= \frac{\mathbf{v}_1^*\mathbf{c}}{\mathbf{v}_1^*\mathbf{v}_1}\mathbf{v}_1 + \frac{\mathbf{v}_2^*\mathbf{c}}{\mathbf{v}_2^*\mathbf{v}_2}\mathbf{v}_2 + \cdots + \frac{\mathbf{v}_k^*\mathbf{c}}{\mathbf{v}_k^*\mathbf{v}_k}\mathbf{v}_k.$$

We illustrate this for $R^3$ in Figure 1, where vector $\mathbf{c}$ is projected onto a plane $V$ determined by a pair of orthogonal vectors $\mathbf{v}_1$ and $\mathbf{v}_2$.

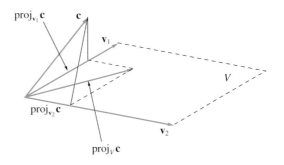

FIGURE 1

## The Gram-Schmidt Process

Let $S = \{\mathbf{q}_1, \mathbf{q}_2, \ldots, \mathbf{q}_m\}$ be a basis for subspace $W$. We construct an orthogonal basis $\{\mathbf{v}_1, \mathbf{v}_2, \ldots, \mathbf{v}_m\}$ for $W$ as follows:

1. Set $\mathbf{v}_1 = \mathbf{q}_1$.

2. Compute $\mathbf{v}_2 = \mathbf{q}_2 - \left(\dfrac{\mathbf{v}_1^* \mathbf{q}_2}{\mathbf{v}_1^* \mathbf{v}_1}\right)\mathbf{v}_1 = \mathbf{q}_2 - \text{proj}_{\mathbf{v}_1}\mathbf{q}_2$.

3. Now we show that $\mathbf{v}_1$ and $\mathbf{v}_2$ are orthogonal.

$$\mathbf{v}_1^* \mathbf{v}_2 = \mathbf{v}_1^*\left(\mathbf{q}_2 - \left(\frac{\mathbf{v}_1^* \mathbf{q}_2}{\mathbf{v}_1^* \mathbf{v}_1}\right)\mathbf{v}_1\right)$$

$$= \mathbf{q}_1^*\left(\mathbf{q}_2 - \left(\frac{\mathbf{q}_1^* \mathbf{q}_2}{\mathbf{q}_1^* \mathbf{q}_1}\right)\mathbf{q}_1\right) \quad \text{(since } \mathbf{v}_1 = \mathbf{q}_1\text{)}$$

$$= \mathbf{q}_1^* \mathbf{q}_2 - \left(\frac{\mathbf{q}_1^* \mathbf{q}_2}{\mathbf{q}_1^* \mathbf{q}_1}\right)\mathbf{q}_1^* \mathbf{q}_1 = 0$$

4. Next we observe that span$\{\mathbf{q}_1, \mathbf{q}_2\} = $ span$\{\mathbf{v}_1, \mathbf{v}_2\}$ since $\mathbf{v}_1 = \mathbf{q}_1$ and $\mathbf{v}_2$ is a linear combination of $\mathbf{q}_1$ and $\mathbf{q}_2$.

5. Compute

$$\mathbf{v}_3 = \mathbf{q}_3 - \left(\frac{\mathbf{v}_1^* \mathbf{q}_3}{\mathbf{v}_1^* \mathbf{v}_1}\right)\mathbf{v}_1 - \left(\frac{\mathbf{v}_2^* \mathbf{q}_3}{\mathbf{v}_2^* \mathbf{v}_2}\right)\mathbf{v}_2$$

$$= \mathbf{q}_3 - \text{proj}_{\mathbf{v}_1}\mathbf{q}_3 - \text{proj}_{\mathbf{v}_2}\mathbf{q}_3$$

6. Now we have that $\mathbf{v}_3$ is orthogonal to both $\mathbf{v}_1$ and $\mathbf{v}_2$. (Verify.) Observe that

$$\text{span}\{\mathbf{q}_1, \mathbf{q}_2, \mathbf{q}_3\} = \text{span}\{\mathbf{v}_1, \mathbf{v}_2, \mathbf{q}_3\} = \text{span}\{\mathbf{v}_1, \mathbf{v}_2, \mathbf{v}_3\}.$$

This follows since span$\{\mathbf{q}_1, \mathbf{q}_2\} = $ span$\{\mathbf{v}_1, \mathbf{v}_2\}$ and $\mathbf{v}_3$ is a linear combination of $\mathbf{v}_1$, $\mathbf{v}_2$, and $\mathbf{q}_3$.

7. Continuing in this way, we get

$$\mathbf{v}_j = \mathbf{q}_j - \sum_{k=1}^{j-1}\left(\frac{\mathbf{v}_k^* \mathbf{q}_j}{\mathbf{v}_k^* \mathbf{v}_k}\right)\mathbf{v}_k = \mathbf{q}_j - \sum_{k=1}^{j-1}\text{proj}_{\mathbf{v}_k}\mathbf{q}_j, \quad j = 2, \ldots, m.$$

We show the orthogonality and spanning properties in a similar manner. By construction, $\{\mathbf{v}_1, \mathbf{v}_2, \ldots, \mathbf{v}_m\}$ is an orthogonal basis for $W$. We scale each of these vectors to be a unit vector so that

$$\left\{\frac{\mathbf{v}_1}{\|\mathbf{v}_1\|}, \frac{\mathbf{v}_2}{\|\mathbf{v}_2\|}, \ldots, \frac{\mathbf{v}_m}{\|\mathbf{v}_m\|}\right\}$$

is an orthonormal basis for $W$.

EXAMPLE 2   Let

$$\mathbf{A} = \begin{bmatrix} 1 & 1 & 0 \\ 1 & 0 & 1 \\ 1 & 2 & 2 \end{bmatrix}.$$

Then $\mathbf{A}$ is nonsingular (verify) so its columns form a basis for $R^3$. To determine an orthogonal basis from

$$\{\mathbf{q}_1, \mathbf{q}_2, \mathbf{q}_3\} = \left\{ \begin{bmatrix} 1 \\ 1 \\ 1 \end{bmatrix}, \begin{bmatrix} 1 \\ 0 \\ 2 \end{bmatrix}, \begin{bmatrix} 0 \\ 1 \\ 2 \end{bmatrix} \right\}$$

we use the Gram-Schmidt process. We have

$$\mathbf{v}_1 = \mathbf{q}_1 = \begin{bmatrix} 1 \\ 1 \\ 1 \end{bmatrix}$$

$$\mathbf{v}_2 = \mathbf{q}_2 - \left( \frac{\mathbf{v}_1^*\mathbf{q}_2}{\mathbf{v}_1^*\mathbf{v}_1} \right) \mathbf{v}_1 = \begin{bmatrix} 1 \\ 0 \\ 2 \end{bmatrix} - \frac{3}{3} \begin{bmatrix} 1 \\ 1 \\ 1 \end{bmatrix} = \begin{bmatrix} 0 \\ -1 \\ 1 \end{bmatrix}$$

$$\mathbf{v}_3 = \mathbf{q}_3 - \left( \frac{\mathbf{v}_1^*\mathbf{q}_3}{\mathbf{v}_1^*\mathbf{v}_1} \right) \mathbf{v}_1 - \left( \frac{\mathbf{v}_2^*\mathbf{q}_3}{\mathbf{v}_2^*\mathbf{v}_2} \right) \mathbf{v}_2$$

$$= \begin{bmatrix} 0 \\ 1 \\ 2 \end{bmatrix} - \frac{3}{3} \begin{bmatrix} 1 \\ 1 \\ 1 \end{bmatrix} - \frac{1}{2} \begin{bmatrix} 0 \\ -1 \\ 1 \end{bmatrix} = \begin{bmatrix} -1 \\ 0.5 \\ 0.5 \end{bmatrix}.$$

Scaling each of the vectors $\mathbf{v}_1$, $\mathbf{v}_2$, and $\mathbf{v}_3$ to be a unit vector gives the orthonormal basis

$$\left\{ \frac{1}{\sqrt{3}} \begin{bmatrix} 1 \\ 1 \\ 1 \end{bmatrix}, \frac{1}{\sqrt{2}} \begin{bmatrix} 0 \\ -1 \\ 1 \end{bmatrix}, \frac{1}{\sqrt{1.5}} \begin{bmatrix} -1 \\ 0.5 \\ 0.5 \end{bmatrix} \right\}.$$

To verify our computations we can construct a matrix $\mathbf{P}$ whose columns are these three vectors and show that $\mathbf{P}$ is a unitary matrix.   ■

**EXAMPLE 3**   In Example 1, the eigenvalue $\lambda = 10$ had multiplicity 2. The corresponding eigenvectors

$$\mathbf{q}_1 = \begin{bmatrix} -1 \\ 1 \\ 0 \end{bmatrix} \quad \text{and} \quad \mathbf{q}_2 = \begin{bmatrix} -2 \\ 0 \\ 1 \end{bmatrix}$$

are linearly independent, but not orthogonal. Using the Gram-Schmidt process, we can determine a pair of orthogonal eigenvectors for eigenvalue $\lambda = 10$. We get

$$\mathbf{v}_1 = \mathbf{q}_1 = \begin{bmatrix} -1 \\ 1 \\ 0 \end{bmatrix}$$

$$\mathbf{v}_2 = \mathbf{q}_2 - \left( \frac{\mathbf{v}_1^*\mathbf{q}_2}{\mathbf{v}_1^*\mathbf{v}_1} \right) \mathbf{v}_1 = \begin{bmatrix} -2 \\ 0 \\ 1 \end{bmatrix} - \frac{2}{2} \begin{bmatrix} -1 \\ 1 \\ 0 \end{bmatrix} = \begin{bmatrix} -1 \\ -1 \\ 1 \end{bmatrix}$$

It follows that

$$\left\{ \begin{bmatrix} -1 \\ 1 \\ 0 \end{bmatrix}, \begin{bmatrix} -1 \\ -1 \\ 1 \end{bmatrix}, \begin{bmatrix} 1 \\ 1 \\ 2 \end{bmatrix} \right\}$$

is a set of orthogonal eigenvectors for matrix **A**. Scaling these vectors to be unit vectors and making them the columns of matrix **P**, we get

$$\mathbf{P} = \begin{bmatrix} \frac{-1}{\sqrt{2}} & \frac{-1}{\sqrt{3}} & \frac{1}{\sqrt{6}} \\ \frac{1}{\sqrt{2}} & \frac{-1}{\sqrt{3}} & \frac{1}{\sqrt{6}} \\ 0 & \frac{1}{\sqrt{3}} & \frac{2}{\sqrt{6}} \end{bmatrix}.$$

We have that

$$\mathbf{P}^*\mathbf{AP} = \begin{bmatrix} 10 & 0 & 0 \\ 0 & 10 & 0 \\ 0 & 0 & 4 \end{bmatrix}. \quad \text{(Verify.)}$$

If we had chosen to place the orthonormal vectors into matrix **P** in a different order, then **P**\***AP** would still be diagonal, but the diagonal entries, the eigenvalues, could appear in a different order. Hence we see that the unitary matrix that performs the similarity transformation of a symmetric (or Hermitian) matrix to diagonal form is not unique. ■

## EXERCISES 4.5

*Vectors*

$$\mathbf{v}_1 = \left(\frac{1}{\sqrt{2}}\right)\begin{bmatrix} 1 \\ 1 \\ 0 \end{bmatrix}, \quad \mathbf{v}_2 = \left(\frac{1}{\sqrt{3}}\right)\begin{bmatrix} 1 \\ -1 \\ 1 \end{bmatrix},$$

*and*

$$\mathbf{v}_3 = \left(\frac{1}{\sqrt{6}}\right)\begin{bmatrix} 1 \\ -1 \\ -2 \end{bmatrix}$$

*are an orthonormal basis for $R^3$. In Exercises 1–3, express each given vector as a linear combination of these basis vectors.*

**1.** $\mathbf{w} = \begin{bmatrix} 1 \\ 0 \\ 0 \end{bmatrix}.$   **2.** $\mathbf{w} = \begin{bmatrix} 1 \\ 2 \\ 1 \end{bmatrix}.$   **3.** $\mathbf{w} = \begin{bmatrix} 1 \\ 0 \\ 1 \end{bmatrix}.$

**4.** $S = \left\{\begin{bmatrix} 1 \\ 1 \end{bmatrix}, \begin{bmatrix} 1 \\ 2 \end{bmatrix}\right\}$ is a basis for $R^2$.

(a) Express $\mathbf{w} = \begin{bmatrix} 7 \\ 9 \end{bmatrix}$ in terms of the basis vectors in $S$.

(b) Use the Gram-Schmidt process to transform the $S$-basis into an orthonormal basis for $R^2$. Then express $\mathbf{w}$ in terms of this basis.

**5.** A subspace $W$ of $R^3$ has basis

$$\left\{\begin{bmatrix} 2 \\ 1 \\ 0 \end{bmatrix}, \begin{bmatrix} 1 \\ 0 \\ 1 \end{bmatrix}\right\}.$$

Use the Gram-Schmidt process to find an orthonormal basis for $W$.

**6.** A subspace $W$ of $R^4$ has basis

$$\left\{\begin{bmatrix} 1 \\ 0 \\ 2 \\ -1 \end{bmatrix}, \begin{bmatrix} 0 \\ 1 \\ 1 \\ 2 \end{bmatrix}, \begin{bmatrix} 1 \\ 0 \\ 0 \\ 1 \end{bmatrix}\right\}.$$

Use the Gram-Schmidt process to find an orthonormal basis for $W$.

**7.** Use the Gram-Schmidt process to transform the basis

$$\left\{\begin{bmatrix} 0 \\ 1 \\ 1 \end{bmatrix}, \begin{bmatrix} 1 \\ 0 \\ 2 \end{bmatrix}, \begin{bmatrix} 2 \\ 0 \\ 1 \end{bmatrix}\right\}$$

into an orthonormal basis for $R^3$.

**8.** Use the Gram-Schmidt process to transform the basis

$$\left\{\begin{bmatrix} 1 \\ 0 \\ 1 \end{bmatrix}, \begin{bmatrix} 1 \\ 0 \\ -2 \end{bmatrix}, \begin{bmatrix} -2 \\ 2 \\ 1 \end{bmatrix}\right\}$$

into an orthonormal basis for $R^3$.

**9.** Can the set of vectors

$$\left\{\begin{bmatrix} 1 \\ 0 \\ 1 \end{bmatrix}, \begin{bmatrix} 1 \\ 0 \\ -2 \end{bmatrix}, \begin{bmatrix} -2 \\ 0 \\ 1 \end{bmatrix}\right\}$$

be transformed into an orthonormal basis for $R^3$? Explain your answer.

**10.** Find an orthonormal basis for the null space of the homogeneous linear system

$$x_1 + x_2 - x_3 = 0$$
$$2x_1 + x_2 + 2x_3 = 0.$$

**11.** Find an orthonormal basis for the null space of the homogeneous linear system

$$2x_1 \quad - x_3 + x_4 = 0$$
$$x_2 + 2x_3 - x_4 = 0.$$

**12.** Let $A = \begin{bmatrix} 1 & 2 & -1 \\ 1 & 0 & 1 \\ 4 & -4 & 5 \end{bmatrix}$.

(a) Find a basis for $R^3$ consisting of eigenvectors of $A$.

(b) Apply the Gram-Schmidt process to the basis you found in part (a). Are the resulting vectors eigenvectors of $A$? Explain.

**13.** Let $Q = \begin{bmatrix} a & c \\ b & d \end{bmatrix}$ be any $2 \times 2$ orthogonal matrix.

(a) Compute $QQ^T$.

(b) Use the fact that $Q$ is orthogonal and the results of part (a) to get the three equations

$$a^2 + b^2 = 1 \tag{1}$$
$$c^2 + d^2 = 1 \tag{2}$$
$$ac + bd = 0. \tag{3}$$

(c) Using (1), explain why there exists a real number $\varphi_1$ such that $a = \cos(\varphi_1)$ and $b = \sin(\varphi_1)$.

(d) Using (2), explain why there exists a real number $\varphi_2$ such that $c = \cos(\varphi_2)$ and $d = \sin(\varphi_2)$.

(e) Use (3) and a trigonometric identity for $\cos(\varphi_2 - \varphi_1)$ to show that $\cos(\varphi_2 - \varphi_1) = 0$.

(f) Use Exercise 21(b) in Section 4.4 to show that

$$ad - bc = \pm 1. \tag{4}$$

(g) Use (4) and a trigonometric identity for $\sin(\varphi_2 - \varphi_1)$ to show that $\sin(\varphi_2 - \varphi_1) = \pm 1$.

(h) Use parts (e) and (g) to explain why $\varphi_2 - \varphi_1 = \pm \dfrac{\pi}{2}$.

(i) Show that either

$$Q = \begin{bmatrix} \cos(\varphi) & -\sin(\varphi) \\ \sin(\varphi) & \cos(\varphi) \end{bmatrix}$$

or

$$Q = \begin{bmatrix} \cos(\varphi) & \sin(\varphi) \\ \sin(\varphi) & -\cos(\varphi) \end{bmatrix}.$$

(j) In the second case in part (i) explain why $Q$ could be a reflection about the $x$-axis, the $y$-axis, the line $y = x$, or the line $y = -x$.

## In MATLAB

*We use* MATLAB *to provide a step-by-step demonstration of the situation illustrated in Figure 1. In addition, we also use* MATLAB *as a smart tutor to show the intimate connection between the algebraic steps and geometric principles of the Gram-Schmidt process in $R^2$ and $R^3$.*

**ML. 1.** In MATLAB type **help projxy** and read the description that appears.

(a) Execute routine **projxy** using input vector $\begin{bmatrix} 5 & 9 & 15 \end{bmatrix}$. Write a formula for the 'orthogonal projector' $S$ that appears in the final figure.

(b) Experiment with several other input vectors for this routine.

**ML. 2.** The Gram-Schmidt process manufactures an orthonormal set of vectors from a linearly independent set. The computational process is algebraic, but there are geometric analogs of the steps involved. In $R^2$, we trade in a pair of vectors $q_1$ and $q_2$ that form a parallelogram for a pair of vectors $v_1$ and $v_2$ that form a rectangle in such a way that span$\{q_1, q_2\}$ = span$\{v_1, v_2\}$. At the final step we trade in $v_1$ and $v_2$ for a pair of orthogonal unit vectors that form a square. We illustrate this process in MATLAB using routine **gsprac2**. To get an overview of the routine type **help gsprac2** and carefully read the description. To start the routine type **gsprac2**. As you use this routine follow the directions that appear in the Message Window.

(a) Use **gsprac2**, choosing demo #1. Compute the number to type into the yellow box by hand. Check that the final pair of vectors $u_1$ and $u_2$ are orthogonal by computing their dot product by hand. (Your result should be quite small but may not be exactly zero since not all the decimal places are displayed.)

(b) Use **gsprac2**, choosing demo #3. Compute the number to type into the yellow box by hand. Check that the final pair of vectors $u_1$ and $u_2$ are orthogonal by computing their dot product by hand. (Your result should be quite small but may not be exactly zero since not all the decimal places are displayed.)

(c) Choose a pair of linearly independent 2-vectors. Plot them on graph paper. Form a conjecture as to what the Gram-Schmidt process will construct for the orthonormal vectors and sketch them on your diagram. Now use **gsprac2** to check your conjecture.

**ML. 3.** MATLAB routine **gsprac3** provides a view of the Gram-Schmidt process in $R^3$ and follows closely the steps used in Exercise ML.2. In this routine you are asked to type in the three required coefficients. We recommend that you carefully do the computations by hand or using a calculator before entering the values. You can enter fractions or decimal approximations that are accurate to four digits. The routine automatically checks the values, requiring a correct or sufficiently close approximation before you can continue to the next step.

(a) Use **gsprac3**, choosing demo #1. Check that the final set of vectors $\mathbf{u}_1$, $\mathbf{u}_2$, and $\mathbf{u}_3$ are orthonormal by computing the appropriate dot products by hand. (The computations for checking orthogonality should result in values that are quite small but may not be exactly zero since not all the decimal places are displayed.)

(b) Use **gsprac3**, choosing demo #4. Explain why two of the coefficients are extremely easy to compute.

(c) Use **gsprac3** to find an orthonormal basis for the row space of matrix

$$\begin{bmatrix} 1 & 0 & 0 \\ 1 & 2 & 0 \\ 1 & 2 & 3 \end{bmatrix}.$$

## True/False Review Questions

*Determine whether each of the following statements is true or false.*

1. If the Gram-Schmidt process applied to vectors $\{\mathbf{v}_1, \mathbf{v}_2\}$ yields $\{\mathbf{u}_1, \mathbf{u}_2\}$, then span$\{\mathbf{v}_1, \mathbf{v}_2\}$ = span$\{\mathbf{u}_1, \mathbf{u}_2\}$.

2. The Gram-Schmidt process converts a basis into an orthonormal basis.

3. If the Gram-Schmidt process is applied to $\left\{ \begin{bmatrix} 1 \\ 1 \\ 1 \end{bmatrix}, \begin{bmatrix} 1 \\ 0 \\ 2 \end{bmatrix} \right\}$, the corresponding output as an orthogonal set is $\left\{ \begin{bmatrix} 1 \\ 1 \\ 1 \end{bmatrix}, \begin{bmatrix} 1 \\ 0 \\ -1 \end{bmatrix} \right\}$.

4. If $\mathbf{A}$ is $3 \times 3$ and defective, then the Gram-Schmidt process can determine an orthonormal basis of eigenvectors of $\mathbf{A}$ for $C^3$.

5. If $\mathbf{A}$ is $4 \times 4$ and symmetric with distinct eigenvalues, then we do not need to use the Gram-Schmidt process to find an orthonormal basis of eigenvectors of $\mathbf{A}$ for $R^4$.

## Terminology

| | |
|---|---|
| Orthonormal basis | Projection |
| Gram-Schmidt process | |

This section develops a process or algorithm that takes a set $S$ of linearly independent vectors and produces a set $T$ of orthonormal vectors such that span$(S)$ = span$(T)$.

- Sketch the simplest orthonormal basis for $R^2$. For $R^3$.
- What is the advantage of having an orthonormal basis instead of an ordinary basis? Explain.
- Describe geometrically the projection of one vector onto another.
- Carefully describe the Gram-Schmidt process step by step. Do this in a narrative form without equations. Carefully indicate the role of projections.
- How can the Gram-Schmidt process be used as part of the construction of a unitary matrix $\mathbf{P}$ such that $\mathbf{P}^*\mathbf{AP}$ is diagonal where $\mathbf{A}$ is symmetric?
- If $\mathbf{A}$ is symmetric with distinct eigenvalues, explain why we do not need to use the Gram-Schmidt process to obtain a unitary matrix $\mathbf{P}$ such that $\mathbf{P}^*\mathbf{AP}$ is diagonal.

## 4.6 ■ THE QR-FACTORIZATION (OPTIONAL)

In this section we introduce a factorization of a matrix that has become very important in revealing the inherent information contained in a matrix. Rather than present a detailed justification of these topics, we will use the development of the Gram-Schmidt process as given in Section 4.5 to provide a foundation for this topic. More comprehensive treatments are available in the references listed at the end of this section. We will use MATLAB to illustrate some of the computations involved in this section.

Let $\mathbf{A}$ be an $m \times n$ matrix with $m \geq n$. Denote the columns of $\mathbf{A}$ by $\mathbf{a}_1, \mathbf{a}_2, \ldots, \mathbf{a}_n$, respectively, and assume that they are linearly independent. Then the Gram-Schmidt process[1] as shown in Section 4.5 will produce an orthonormal set of $n$ vectors, $\mathbf{q}_1, \mathbf{q}_2, \ldots, \mathbf{q}_n$ in $R^m$, such that

$$\text{span}\{\mathbf{a}_1, \mathbf{a}_2, \ldots, \mathbf{a}_j\} = \text{span}\{\mathbf{q}_1, \mathbf{q}_2, \ldots, \mathbf{q}_j\}$$

for $j = 1, 2, \ldots, n$. Table 1 gives the notation and the linear combinations involved.

TABLE 1  **Gram-Schmidt Computations[+]**

| Step # | Orthogonal Vectors | Orthonormal Vectors |
|:---:|:---:|:---:|
| 1 | $\mathbf{v}_1 = \mathbf{a}_1$ | $\mathbf{q}_1 = \dfrac{\mathbf{v}_1}{\|\mathbf{v}_1\|}$ |
| 2 | $\mathbf{v}_2 = \mathbf{a}_2 - \dfrac{\mathbf{v}_1^* \mathbf{a}_2}{\mathbf{v}_1^* \mathbf{v}_1}\mathbf{v}_1$ | $\mathbf{q}_2 = \dfrac{\mathbf{v}_2}{\|\mathbf{v}_2\|}$ |
| 3 | $\mathbf{v}_3 = \mathbf{a}_3 - \dfrac{\mathbf{v}_1^* \mathbf{a}_3}{\mathbf{v}_1^* \mathbf{v}_1}\mathbf{v}_1 - \dfrac{\mathbf{v}_2^* \mathbf{a}_3}{\mathbf{v}_2^* \mathbf{v}_2}\mathbf{v}_2$ | $\mathbf{q}_3 = \dfrac{\mathbf{v}_3}{\|\mathbf{v}_3\|}$ |
| ⋮ | ⋮ | ⋮ |
| $n$ | $\mathbf{v}_n = \mathbf{a}_n - \displaystyle\sum_{j=1}^{n-1} \dfrac{\mathbf{v}_j^* \mathbf{a}_n}{\mathbf{v}_j^* \mathbf{v}_j}\mathbf{v}_j$ | $\mathbf{q}_n = \dfrac{\mathbf{v}_n}{\|\mathbf{v}_n\|}$ |

[+]We continue the practice of using the conjugate transpose * so that both the real and complex cases are handled at the same time.

Each step of the Gram-Schmidt process can be rearranged to express a column of $\mathbf{A}$ as a linear combination of the orthonormal columns $\mathbf{q}_1, \mathbf{q}_2, \ldots, \mathbf{q}_n$ as shown in Table 2. To make the connections between Tables 1 and 2, it is helpful to recall that

$$\mathbf{q}_j = \frac{\mathbf{v}_j}{\|\mathbf{v}_j\|} \quad \text{and} \quad \|\mathbf{v}_j\|^2 = \mathbf{v}_j^* \mathbf{v}_j.$$

[1]The notation for the vectors used in the Gram-Schmidt process is different in this section so that we can use names that naturally correspond to the standard notation used for the techniques we develop.

TABLE 2  **Columns of A in terms of the Orthonormal Vectors**

| Step # | |
|---|---|
| 1 | $\mathbf{a}_1 = \mathbf{v}_1 = \|\mathbf{v}_1\|\mathbf{q}_1$ |
| 2 | $\mathbf{a}_2 = \dfrac{\mathbf{v}_1^* \mathbf{a}_2}{\mathbf{v}_1^* \mathbf{v}_1}\mathbf{v}_1 + \mathbf{v}_2 = \dfrac{\mathbf{v}_1^* \mathbf{a}_2}{\|\mathbf{v}_1\|}\mathbf{q}_1 + \|\mathbf{v}_2\|\mathbf{q}_2$ |
| 3 | $\mathbf{a}_3 = \dfrac{\mathbf{v}_1^* \mathbf{a}_3}{\mathbf{v}_1^* \mathbf{v}_1}\mathbf{v}_1 + \dfrac{\mathbf{v}_2^* \mathbf{a}_3}{\mathbf{v}_2^* \mathbf{v}_2}\mathbf{v}_2 + \mathbf{v}_3 = \dfrac{\mathbf{v}_1^* \mathbf{a}_3}{\|\mathbf{v}_1\|}\mathbf{q}_1 + \dfrac{\mathbf{v}_2^* \mathbf{a}_3}{\|\mathbf{v}_2\|}\mathbf{q}_2 + \|\mathbf{v}_3\|\mathbf{q}_3$ |
| $\vdots$ | $\vdots$ |
| $n$ | $\mathbf{a}_n = \displaystyle\sum_{j=1}^{n-1} \dfrac{\mathbf{v}_j^* \mathbf{a}_n}{\mathbf{v}_j^* \mathbf{v}_j}\mathbf{v}_j + \mathbf{v}_n = \displaystyle\sum_{j=1}^{n-1} \dfrac{\mathbf{v}_j^* \mathbf{a}_n}{\|\mathbf{v}_j\|}\mathbf{q}_j + \|\mathbf{v}_n\|\mathbf{q}_n$ |

Next define $\mathbf{R}$ to be the $n \times n$ matrix of coefficients of the $\mathbf{q}$'s in Table 2; that is,

$$\text{col}_1(\mathbf{R}) = \text{coefficient from Step 1} = \begin{bmatrix} \|\mathbf{v}_1\| \\ 0 \\ 0 \\ \vdots \\ 0 \end{bmatrix},$$

$$\text{col}_2(\mathbf{R}) = \text{coefficient from Step 2} = \begin{bmatrix} (\mathbf{v}_1^*\mathbf{a}_2)/\|\mathbf{v}_1\| \\ \|\mathbf{v}_2\| \\ 0 \\ \vdots \\ 0 \end{bmatrix},$$

$$\text{col}_3(\mathbf{R}) = \text{coefficient from Step 3} = \begin{bmatrix} (\mathbf{v}_1^*\mathbf{a}_3)/\|\mathbf{v}_1\| \\ (\mathbf{v}_2^*\mathbf{a}_3)/\|\mathbf{v}_2\| \\ \|\mathbf{v}_3\| \\ \vdots \\ 0 \end{bmatrix},$$

and so on until

$$\text{col}_n(\mathbf{R}) = \text{coefficient from Step } n = \begin{bmatrix} (\mathbf{v}_1^*\mathbf{a}_n)/\|\mathbf{v}_1\| \\ (\mathbf{v}_2^*\mathbf{a}_n)/\|\mathbf{v}_2\| \\ \vdots \\ (\mathbf{v}_{n-1}^*\mathbf{a}_n)/\|\mathbf{v}_{n-1}\| \\ \|\mathbf{v}_n\| \end{bmatrix}.$$

Hence $\mathbf{R}$ is a nonsingular upper triangular matrix with

$$\mathbf{R} = \begin{bmatrix} \|\mathbf{v}_1\| & (\mathbf{v}_1^*\mathbf{a}_2)/\|\mathbf{v}_1\| & (\mathbf{v}_1^*\mathbf{a}_3)/\|\mathbf{v}_1\| & \cdots & (\mathbf{v}_1^*\mathbf{a}_n)/\|\mathbf{v}_1\| \\ 0 & \|\mathbf{v}_2\| & (\mathbf{v}_2^*\mathbf{a}_3)/\|\mathbf{v}_2\| & \cdots & (\mathbf{v}_2^*\mathbf{a}_n)/\|\mathbf{v}_2\| \\ 0 & 0 & \|\mathbf{v}_3\| & \cdots & (\mathbf{v}_3^*\mathbf{a}_n)/\|\mathbf{v}_3\| \\ \vdots & \vdots & \vdots & \ddots & \vdots \\ 0 & 0 & 0 & 0 & \|\mathbf{v}_n\| \end{bmatrix}. \tag{1}$$

(See Exercise 6.)

Let $\mathbf{Q} = \begin{bmatrix} \mathbf{q}_1 & \mathbf{q}_2 & \cdots & \mathbf{q}_n \end{bmatrix}$. Then we can express the set of linear combinations in Table 2 as the matrix product

$$\mathbf{A} = \mathbf{QR} \tag{2}$$

since $\text{col}_j(\mathbf{A}) = \mathbf{Q} * \text{col}_j(\mathbf{R})$. (See Section 1.3.) We summarize this result in the following statement.

> If matrix $\mathbf{A}$ has linearly independent columns, then there is a matrix $\mathbf{Q}$ with orthonormal columns and a nonsingular upper triangular matrix $\mathbf{R}$ such that $\mathbf{A} = \mathbf{QR}$. (3)

We call (2) the **QR-factorization** of matrix $\mathbf{A}$. We have the following special case.

> Any nonsingular matrix $\mathbf{A}$ can be expressed as the product of a unitary matrix $\mathbf{Q}$ and a nonsingular upper triangular matrix $\mathbf{R}$; that is, $\mathbf{A} = \mathbf{QR}$. (4)

(See Exercises 4, 5, and 6.)

EXAMPLE 1   Let

$$\mathbf{A} = \begin{bmatrix} 1 & 1 & 0 \\ 1 & 0 & 1 \\ 1 & 2 & 2 \end{bmatrix}.$$

In Example 1 in Section 4.5 we applied the Gram-Schmidt process to the columns of $\mathbf{A}$. Using the notation established in Tables 1 and 2 we find the following:

$$\mathbf{v}_1 = \mathbf{a}_1 = \begin{bmatrix} 1 \\ 1 \\ 1 \end{bmatrix} \qquad \text{or} \qquad \mathbf{a}_1 = \sqrt{3}\,\mathbf{q}_1,$$

$$\mathbf{q}_1 = \frac{1}{\sqrt{3}} \begin{bmatrix} 1 \\ 1 \\ 1 \end{bmatrix}$$

$$\mathbf{v}_2 = \mathbf{a}_2 - 1\mathbf{v}_1 = \begin{bmatrix} 0 \\ -1 \\ 1 \end{bmatrix} \qquad \text{or} \qquad \mathbf{a}_2 = \frac{3}{\sqrt{3}}\,\mathbf{q}_1 + \sqrt{2}\,\mathbf{q}_2,$$

$$\mathbf{q}_2 = \frac{1}{\sqrt{2}} \begin{bmatrix} 0 \\ -1 \\ 1 \end{bmatrix}$$

$$\mathbf{v}_3 = \mathbf{a}_3 - 1\mathbf{v}_1 - 0.5\mathbf{v}_2 = \begin{bmatrix} -1 \\ 0.5 \\ 0.5 \end{bmatrix} \qquad \text{or} \qquad \mathbf{a}_3 = \frac{3}{\sqrt{3}}\,\mathbf{q}_1 + \frac{1}{\sqrt{2}}\,\mathbf{q}_2 + \sqrt{1.5}\,\mathbf{q}_3.$$

$$\mathbf{q}_3 = \frac{1}{\sqrt{1.5}} \begin{bmatrix} -1 \\ 0.5 \\ 0.5 \end{bmatrix}$$

Thus we have

$$\mathbf{Q} = \begin{bmatrix} \frac{1}{\sqrt{3}} & 0 & -\frac{1}{\sqrt{1.5}} \\ \frac{1}{\sqrt{3}} & -\frac{1}{\sqrt{2}} & \frac{0.5}{\sqrt{1.5}} \\ \frac{1}{\sqrt{3}} & \frac{1}{\sqrt{2}} & \frac{0.5}{\sqrt{1.5}} \end{bmatrix}$$

and

$$\mathbf{R} = \begin{bmatrix} \sqrt{3} & \frac{3}{\sqrt{3}} & \frac{3}{\sqrt{3}} \\ 0 & \sqrt{2} & \frac{1}{\sqrt{2}} \\ 0 & 0 & \sqrt{1.5} \end{bmatrix} = \begin{bmatrix} \sqrt{3} & \sqrt{3} & \sqrt{3} \\ 0 & \sqrt{2} & \frac{1}{\sqrt{2}} \\ 0 & 0 & \sqrt{1.5} \end{bmatrix}.$$

So

$$\mathbf{A} = \mathbf{QR} = \begin{bmatrix} \frac{1}{\sqrt{3}} & 0 & -\frac{1}{\sqrt{1.5}} \\ \frac{1}{\sqrt{3}} & -\frac{1}{\sqrt{2}} & \frac{0.5}{\sqrt{1.5}} \\ \frac{1}{\sqrt{3}} & \frac{1}{\sqrt{2}} & \frac{0.5}{\sqrt{1.5}} \end{bmatrix} \begin{bmatrix} \sqrt{3} & \sqrt{3} & \sqrt{3} \\ 0 & \sqrt{2} & \frac{1}{\sqrt{2}} \\ 0 & 0 & \sqrt{1.5} \end{bmatrix}.$$

Verify that $\mathbf{QR}$ gives $\mathbf{A}$. ■

If $\mathbf{A}$ is nonsingular, then the QR-factorization provides us with another method for solving the linear system $\mathbf{Ax} = \mathbf{b}$. We have equivalent matrix equations

$$\mathbf{Ax} = \mathbf{b}, \quad \mathbf{QRx} = \mathbf{b}, \quad \mathbf{Q}^*\mathbf{QRx} = \mathbf{Q}^*\mathbf{b}, \quad \mathbf{Rx} = \mathbf{Q}^*\mathbf{b} \quad \text{(since } \mathbf{Q} \text{ is unitary).} \quad (5)$$

Since $\mathbf{R}$ is nonsingular and upper triangular, we can be assured that back substitution will succeed in solving for $\mathbf{x}$.

EXAMPLE 2   Use the QR-factorization of

$$\mathbf{A} = \begin{bmatrix} 1 & 1 & 0 \\ 1 & 0 & 1 \\ 1 & 2 & 2 \end{bmatrix}$$

to solve the linear system $\mathbf{Ax} = \mathbf{b}$ where $\mathbf{b} = \begin{bmatrix} 1 \\ 4 \\ 7 \end{bmatrix}$. We have from Example 1 and the relationships in (5) that

$$\mathbf{Ax} = \mathbf{b} \Leftrightarrow \mathbf{Rx} = \mathbf{Q}^*\mathbf{b}$$

$$\Leftrightarrow \mathbf{Rx} = \begin{bmatrix} \frac{1}{\sqrt{3}} & \frac{1}{\sqrt{3}} & \frac{1}{\sqrt{3}} \\ 0 & -\frac{1}{\sqrt{2}} & \frac{1}{\sqrt{2}} \\ -\frac{1}{\sqrt{1.5}} & \frac{0.5}{\sqrt{1.5}} & \frac{0.5}{\sqrt{1.5}} \end{bmatrix} \begin{bmatrix} 1 \\ 4 \\ 7 \end{bmatrix} = \begin{bmatrix} \frac{12}{\sqrt{3}} \\ \frac{3}{\sqrt{2}} \\ \frac{4.5}{\sqrt{1.5}} \end{bmatrix}$$

$$\Leftrightarrow \begin{bmatrix} \sqrt{3} & \sqrt{3} & \sqrt{3} \\ 0 & \sqrt{2} & \frac{1}{\sqrt{2}} \\ 0 & 0 & \sqrt{1.5} \end{bmatrix} \begin{bmatrix} x_1 \\ x_2 \\ x_3 \end{bmatrix} = \begin{bmatrix} \frac{12}{\sqrt{3}} \\ \frac{3}{\sqrt{2}} \\ \frac{4.5}{\sqrt{1.5}} \end{bmatrix}.$$

Back substitution gives $x_3 = 3$, $x_2 = 0$, $x_1 = 1$. (Verify.) ■

The QR-factorization of a coefficient matrix of a linear system can also be used when $\mathbf{A}$ is not square as long as the columns of $\mathbf{A}$ are linearly independent. This technique is often used in computing least squares models for data sets. In Section 2.6 we showed how to compute the line of best fit in two ways. The QR-factorization gives us a third technique and is often used in computer algorithms. From Equation

(4) of Section 2.6, the overdetermined linear system for computing the slope $m$ and $y$-intercept $b$ of the line of best fit is given by

$$\mathbf{Ax} = \mathbf{y} \Leftrightarrow \begin{bmatrix} x_1 & 1 \\ x_2 & 1 \\ \vdots & \vdots \\ \vdots & \vdots \\ x_n & 1 \end{bmatrix} \begin{bmatrix} m \\ b \end{bmatrix} = \begin{bmatrix} y_1 \\ y_2 \\ \vdots \\ \vdots \\ y_n \end{bmatrix}. \tag{6}$$

The coefficient matrix for a set of distinct $x$-values has linearly independent columns; hence we can obtain its QR-factorization. We illustrate this in Example 3.

**EXAMPLE 3**   For the altitude-temperature data given in Table 1 in Section 2.6, the linear system of the form in (6) is

$$\begin{bmatrix} 1 & 1 \\ 5 & 1 \\ 10 & 1 \\ 15 & 1 \\ 20 & 1 \\ 30 & 1 \\ 36 & 1 \end{bmatrix} \begin{bmatrix} m \\ b \end{bmatrix} = \begin{bmatrix} 56 \\ 41 \\ 23 \\ 5 \\ -15 \\ -47 \\ -69 \end{bmatrix}.$$

The QR-factorization of the coefficient matrix, recording only four decimal places and rounding, is

$$\mathbf{QR} = \begin{bmatrix} 0.0184 & 0.6258 \\ 0.0921 & 0.5223 \\ 0.1842 & 0.3929 \\ 0.2763 & 0.2636 \\ 0.3684 & 0.1342 \\ 0.5526 & -0.1245 \\ 0.6632 & -0.2797 \end{bmatrix} \begin{bmatrix} 54.2863 & 2.1552 \\ 0 & 1.5346 \end{bmatrix}.$$

Applying the procedure in (5) gives

$$\mathbf{R} \begin{bmatrix} m \\ b \end{bmatrix} = \mathbf{Q}^* \mathbf{y} = \begin{bmatrix} -66.8309 \\ 89.9507 \end{bmatrix}$$

whose solution is $m = -3.5582$, $b = 58.6159$ (verify), the same as in (3) in Section 2.6.   ■

The Gram-Schmidt process is quite sensitive to small errors that are introduced when numbers are rounded or chopped, when we use decimal approximations to values like $\sqrt{2}$ or $\sqrt{3}$, or generally where there is any loss of accuracy due to using approximations. Hence alternative methods are used in computer algorithms for the computation of the QR-factorization. (See Exercise 9 for a brief look at how MATLAB handles QR-factorization.)

The QR-factorization is not restricted to matrices with linearly independent columns. We state the general result next without justification, which can be found in the references given at the end of this section.

An $m \times n$ complex matrix A can be written as $A = QR$ where Q is an $m \times n$ matrix with orthonormal columns and R is an $n \times n$ upper triangular matrix.

In the case that **A** is square, $m = n$, then matrix **Q** is a unitary matrix.

The QR-factorization of a matrix can be used to build an iterative procedure to approximate all the eigenvalues of a matrix. Such methods have been the focus of many groups whose goal has been to develop fast, highly accurate, and widely applicable algorithms. We give an outline of the fundamentals of these methods and call our procedure the **basic QR-method** for eigen approximation.

The basic QR-method for approximating the eigenvalues of an $n \times n$ complex matrix **A** generates a sequence of $n \times n$ matrices that are unitarily similar and approach either an upper triangular matrix or a matrix that is nearly upper triangular. Since each matrix of the sequence is similar to the original matrix, the eigenvalues are preserved (see Exercise 10). The limit of the sequence either reveals the eigenvalues as the diagonal entries of an upper triangular matrix or the eigenvalues can easily be determined from the nearly triangular form that appears. The key to the basic QR-method is to be able to obtain QR-factorizations quickly and accurately. We next describe the deceptively simple basic QR-method.

## The Basic QR-Method

1. Let $\mathbf{A}_0 = \mathbf{A}$. Obtain the QR-factorization and denote it as $\mathbf{A}_0 = \mathbf{Q}_0 \mathbf{R}_0$.
2. Define $\mathbf{A}_1 = \mathbf{R}_0 \mathbf{Q}_0$. Obtain the QR-factorization and denote it as

$$\mathbf{A}_1 = \mathbf{Q}_1 \mathbf{R}_1.$$

3. Define $\mathbf{A}_2 = \mathbf{R}_1 \mathbf{Q}_1$. Obtain the QR-factorization and denote it as

$$\mathbf{A}_2 = \mathbf{Q}_2 \mathbf{R}_2.$$

$$\vdots$$

In general, obtain the QR-factorization of $\mathbf{A}_{k-1}$ and denote it as

$$\mathbf{A}_{k-1} = \mathbf{Q}_{k-1} \mathbf{R}_{k-1}.$$

Define $\mathbf{A}_k = \mathbf{R}_{k-1} \mathbf{Q}_{k-1}, k \geq 1$.

To form $\mathbf{A}_k$, we take the product of the QR-factors from the previous step in reverse order. Thus we have a sequence of matrices $\mathbf{A}_0, \mathbf{A}_1, \mathbf{A}_2, \ldots,$. We summarize the behavior of this sequence as follows:

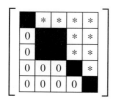

* represents a possible nonzero value, there are three $1 \times 1$ blocks that will contain eigenvalues and a $2 \times 2$ block that corresponds to a complex eigenvalue and its conjugate.

FIGURE 1

1. If **A** has all real entries and all its eigenvalues are real numbers, then $\lim_{k \to \infty} \mathbf{A}_k =$ upper triangular matrix. Thus the eigenvalues of **A** are the diagonal entries of the upper triangular matrix.
2. If **A** has all real entries but some of its eigenvalues are not real numbers, then $\lim_{k \to \infty} \mathbf{A}_k =$ nearly upper triangular matrix. By *nearly upper triangular* we mean a matrix that can have $2 \times 2$ diagonal blocks. (See Figure 1.) Thus the eigenvalues of **A** are the diagonal entries that are $1 \times 1$ blocks or can easily be obtained from the $2 \times 2$ diagonal blocks.
3. If **A** is complex and we use complex arithmetic throughout the computations, then $\lim_{k \to \infty} \mathbf{A}_k =$ upper triangular matrix. Thus the eigenvalues of **A** are the diagonal entries of the upper triangular matrix.

It is beyond the scope of our development to provide a detailed justification of the preceding statements; see the references at the end of the section. However, we illustrate the computations involved in the following examples. We have used MATLAB to perform the QR-factorizations. See also Exercise 11.

EXAMPLE 4   Let

$$A = \begin{bmatrix} 2 & 3 \\ 3 & 2 \end{bmatrix}.$$

Since $A$ is symmetric, all its eigenvalues are real so we expect that the basic QR-method will produce an upper triangular matrix that will reveal the eigenvalues of $A$ as the diagonal entries. The QR-factorizations were obtained using MATLAB. We have the following results (entries are recorded to only four decimal places):

$$A = A_0 = Q_0 R_0 = \begin{bmatrix} -0.5547 & -0.8321 \\ -0.8321 & 0.5547 \end{bmatrix} \begin{bmatrix} -3.6056 & -3.3282 \\ 0 & -1.3868 \end{bmatrix}$$

$$A_1 = R_0 Q_0 = \begin{bmatrix} 4.7692 & 1.1538 \\ 1.1538 & -0.7692 \end{bmatrix}$$

$$= Q_1 R_1 = \begin{bmatrix} -0.9720 & -0.2352 \\ -0.2352 & 0.9720 \end{bmatrix} \begin{bmatrix} -4.9068 & -0.9406 \\ 0 & -1.0190 \end{bmatrix}$$

$$A_2 = R_1 Q_1 = \begin{bmatrix} 4.9904 & 0.2396 \\ 0.2396 & -0.9904 \end{bmatrix}$$

$$= Q_2 R_2 = \begin{bmatrix} -0.9988 & -0.0480 \\ -0.0480 & 0.9988 \end{bmatrix} \begin{bmatrix} -4.9962 & -0.1918 \\ 0 & -1.0008 \end{bmatrix}$$

$$A_3 = R_2 Q_2 = \begin{bmatrix} 4.9996 & 0.0480 \\ 0.0480 & -0.9996 \end{bmatrix}$$

$$\vdots$$

$$A_{10} = R_9 Q_9 = \begin{bmatrix} 5.0000 & 0.0000 \\ 0.0000 & -1.0000 \end{bmatrix}.$$

It appears that

$$\lim_{k \to \infty} A_k = \begin{bmatrix} 5 & 0 \\ 0 & -1 \end{bmatrix}$$

and hence the eigenvalues of $A$ are $\lambda = 5, -1$. Verify that this is the case by computing the characteristic polynomial of $A$ and finding its roots.   ■

EXAMPLE 5   Let $A = \begin{bmatrix} 3 & 3 & 4 \\ 1 & 0 & 0 \\ 0 & 1 & 0 \end{bmatrix}$. Applying the basic QR-method, we obtain the following results:

$$\mathbf{A} = \mathbf{A}_0 = \mathbf{Q}_0\mathbf{R}_0$$

$$= \begin{bmatrix} -0.9487 & 0.2176 & -0.2294 \\ -0.3162 & -0.6529 & 0.6882 \\ 0 & 0.7255 & 0.6882 \end{bmatrix} \begin{bmatrix} -3.1623 & -2.8460 & -3.7947 \\ 0 & 1.3784 & 0.8706 \\ 0 & 0 & -0.9177 \end{bmatrix}$$

$$\mathbf{A}_1 = \mathbf{R}_0\mathbf{Q}_0$$

$$= \begin{bmatrix} 3.9000 & -1.5830 & -3.8450 \\ -0.4359 & -0.2684 & 1.5479 \\ 0 & -0.6657 & -0.6316 \end{bmatrix}$$

$$= \mathbf{Q}_1\mathbf{R}_1$$

$$= \begin{bmatrix} -0.9938 & -0.0615 & -0.0925 \\ 0.1111 & -0.5502 & -0.8276 \\ 0 & -0.8328 & 0.5536 \end{bmatrix} \begin{bmatrix} -3.9243 & 1.5734 & 3.9932 \\ 0 & 0.7994 & -0.0892 \\ 0 & 0 & -1.2750 \end{bmatrix}$$

$$\mathbf{A}_2 = \mathbf{R}_1\mathbf{Q}_1$$

$$= \begin{bmatrix} 4.0714 & -3.9332 & 1.2964 \\ 0.0888 & -0.3655 & -0.7110 \\ 0 & 1.0618 & -0.7059 \end{bmatrix}$$

$$\mathbf{A}_3 = \begin{bmatrix} 3.9855 & -2.2644 & 3.5690 \\ -0.0239 & -0.7576 & 0.9356 \\ 0 & -0.8650 & -0.2279 \end{bmatrix}$$

$$\vdots$$

$$\mathbf{A}_{10} = \begin{bmatrix} 4.0000 & 1.1651 & -4.0430 \\ 0.0000 & -0.4246 & -1.1695 \\ 0 & 0.6461 & -0.5754 \end{bmatrix}.$$

The eigenvalues of $\mathbf{A}$ are 4 and the eigenvalues of the $2 \times 2$ diagonal block

$$\mathbf{B} = \begin{bmatrix} -0.4246 & -1.1695 \\ 0.6461 & -0.5754 \end{bmatrix}.$$

Forming the characteristic polynomial of $\mathbf{B}$ and finding its roots we get,

$$c(\mathbf{B}) \approx \lambda^2 + 0.9999\lambda + 1$$

so the roots are approximately $-0.4999 \pm 0.8660i$. The exact eigenvalues of $\mathbf{A}$ are 4 and $-0.5 \pm \left(\frac{\sqrt{3}}{2}\right)i$ (verify), so we see that our 10 steps of the basic QR-method gave good approximations. ■

---

## ■ REFERENCES FOR SECTION 4.6

GOLUB, G. and C. VAN LOAN, *Matrix Computations*, 3rd edition, Baltimore, Maryland: The Johns Hopkins University Press, 1996.

HILL, D., *Experiments in Computational Matrix Algebra*. New York: Random House, 1988 (distributed by McGraw-Hill).

KAHANER, D., C. MOLER, AND S. NASH, *Numerical Methods and Software*. Englewood Cliffs, N.J.: Prentice Hall, 1989.

STEWART, G., *Introduction to Matrix Computations*, New York: Academic Press, 1973.

# EXERCISES 4.6

**1.** Let $\mathbf{A} = \begin{bmatrix} \mathbf{a}_1 & \mathbf{a}_2 & \mathbf{a}_3 \end{bmatrix} = \begin{bmatrix} 1 & 2 & 1 \\ 0 & 0 & 1 \\ 2 & 1 & 0 \end{bmatrix}$. The Gram-Schmidt process produces the orthogonal set of vectors

$$\{\mathbf{v}_1, \mathbf{v}_2, \mathbf{v}_3\} = \left\{ \begin{bmatrix} 1 \\ 0 \\ 2 \end{bmatrix}, \begin{bmatrix} 1.2 \\ 0 \\ -0.6 \end{bmatrix}, \begin{bmatrix} 0 \\ 1 \\ 0 \end{bmatrix} \right\}.$$

  (a) Determine the corresponding orthonormal set $\{\mathbf{q}_1, \mathbf{q}_2, \mathbf{q}_3\}$.
  (b) Use Equation (1) to find the matrix $\mathbf{R}$.
  (c) Find the QR-factorization of $\mathbf{A}$.
  (d) Verify that $\mathbf{QR} = \mathbf{A}$.

**2.** Let $\mathbf{A} = \begin{bmatrix} \mathbf{a}_1 & \mathbf{a}_2 & \mathbf{a}_3 \end{bmatrix} = \begin{bmatrix} 1 & 0 & 1 \\ 0 & 1 & -1 \\ 0 & 1 & 1 \\ 1 & 0 & -1 \end{bmatrix}$.

  (a) Show that if the Gram-Schmidt process is applied to the columns of $\mathbf{A}$ then the orthogonal basis produced is $\mathbf{v}_j = \mathbf{a}_j$, $j = 1, 2, 3$.
  (b) Determine the corresponding orthonormal set $\{\mathbf{q}_1, \mathbf{q}_2, \mathbf{q}_3\}$.
  (c) Use Equation (1) to find the matrix $\mathbf{R}$.
  (d) Find the QR-factorization of $\mathbf{A}$.
  (e) Verify that $\mathbf{QR} = \mathbf{A}$.

**3.** Determine the QR-factorization of

$$\mathbf{A} = \begin{bmatrix} 1 & 0 & 2 \\ 0 & 1 & 0 \\ 2 & 1 & 1 \\ 1 & 2 & 1 \end{bmatrix}.$$

Use the techniques given in Tables 1 and 2.

**4.** In (3) the matrix $\mathbf{Q}$ is said to have orthonormal columns.
  (a) Explain why this is true. (*Hint*: See Table 1.)
  (b) If $\mathbf{Q}$ is $m \times n$, then find $\mathbf{Q}^*\mathbf{Q}$.

**5.** In (4) the matrix $\mathbf{Q}$ is said to be unitary. Explain why this is true. (*Hint*: See Table 1.)

**6.** In (1), (3), and (4) the upper triangular matrix $\mathbf{R}$ is said to be nonsingular. Explain why this is true. [*Hint*: See (1).]

**7.** The matrix $\mathbf{A}$ has QR-factorization given by

$$\mathbf{Q} = \begin{bmatrix} -\frac{1}{\sqrt{3}} & \frac{1}{\sqrt{2}} & -\frac{1}{\sqrt{6}} \\ -\frac{1}{\sqrt{3}} & 0 & \frac{2}{\sqrt{6}} \\ \frac{1}{\sqrt{3}} & \frac{1}{\sqrt{2}} & \frac{1}{\sqrt{6}} \end{bmatrix}$$

and

$$\mathbf{R} = \begin{bmatrix} -\sqrt{3} & 0 & 0 \\ 0 & \sqrt{2} & \frac{1}{\sqrt{2}} \\ 0 & 0 & \frac{1}{\sqrt{1.5}} \end{bmatrix}.$$

Solve the linear system $\mathbf{Ax} = \begin{bmatrix} 3 \\ 3 \\ 0 \end{bmatrix}$.

**8.** The matrix $\mathbf{A}$ has QR-factorization given by

$$\mathbf{Q} = \begin{bmatrix} -\frac{1}{\sqrt{3}} & \frac{1}{\sqrt{2}} & \frac{1}{\sqrt{6}} & 0 \\ -\frac{1}{\sqrt{3}} & 0 & -\frac{2}{\sqrt{6}} & 0 \\ 0 & 0 & 0 & 1 \\ -\frac{1}{\sqrt{3}} & -\frac{1}{\sqrt{2}} & \frac{1}{\sqrt{6}} & 0 \end{bmatrix}$$

and

$$\mathbf{R} = \begin{bmatrix} -\sqrt{3} & 0 & -\frac{1}{\sqrt{3}} & -\frac{2}{\sqrt{3}} \\ 0 & \sqrt{2} & 0 & \frac{1}{\sqrt{2}} \\ 0 & 0 & -\frac{2}{\sqrt{6}} & -\frac{1}{\sqrt{3}} \\ 0 & 0 & 0 & 1 \end{bmatrix}.$$

Solve the linear system $\mathbf{Ax} = \begin{bmatrix} 2 \\ 2 \\ 1 \\ 1 \end{bmatrix}$.

**9.** MATLAB has a command that produces the QR-factorization of any matrix $\mathbf{A}$. The command is **qr**, and further information is available using **help qr**. The technique used by MATLAB is not the Gram-Schmidt process. However, if $\mathbf{A}$ is nonsingular, then the output from command **qr(A)** is the same as we would get using the Gram-Schmidt process as shown in Tables 1 and 2. If $\mathbf{A}$ is $m \times n$, $m \neq n$, or singular, then the **qr** command produces an $m \times m$ unitary matrix $\mathbf{Q}$ and an $m \times n$ matrix $\mathbf{R}$ with zeros below its diagonal entries so that $\mathbf{A} = \mathbf{QR}$. ($\mathbf{R}$ is essentially upper triangular but need not be square.) Use the **qr** command to obtain QR-factorizations of each of the following matrices.

  (a) $\begin{bmatrix} 2 & -1 & 0 \\ 1 & 2 & 1 \\ 0 & 1 & 0 \end{bmatrix}$.

  (b) $\begin{bmatrix} 1 & 1 & 0 \\ 0 & 1 & 1 \\ 1 & 0 & 2 \\ 1 & 0 & 1 \end{bmatrix}$.

(c) $\begin{bmatrix} 1 & 1 & 1 \\ 0 & 1 & 0 \\ 2 & 0 & 0 \\ 1 & 1 & 0 \\ -1 & 1 & 1 \end{bmatrix}$.

(a) $\begin{bmatrix} 3 & -1 \\ 3 & -1 \end{bmatrix}$, $k = 2$.

(b) $\begin{bmatrix} 5 & 2 \\ 8 & 5 \end{bmatrix}$, $k = 5$.

**10.** Let $\mathbf{A}$ be a square matrix and $\mathbf{A}_0 = \mathbf{A}$ as in the basic QR-method.

   (a) Show that $\mathbf{A}_1$ is unitarily similar to $\mathbf{A}_0$.
   (b) Show that $\mathbf{A}_2$ is unitarily similar to $\mathbf{A}_1$.
   (c) Explain why $\mathbf{A}_k$ has the same eigenvalues as $\mathbf{A}$.

(c) $\begin{bmatrix} 2 & -2 & 1 \\ 0 & 3 & 0 \\ 1 & 2 & 2 \end{bmatrix}$, $k = 10$.

(d) $\begin{bmatrix} 3 & 12 & -10 \\ 1 & -1 & 5 \\ -2 & -8 & 5 \end{bmatrix}$, $k = 10$.

**11.** There is a MATLAB routine to perform the basic QR-method. Use **help basicqr** for details. Use this routine to find the eigenvalues of each of the following matrices using $k$ iterations. If a $2 \times 2$ diagonal block occurs, determine its eigenvalues. Compare the results with those obtained using the **eig** command.

(e) $\begin{bmatrix} -1 & -2 & 3 & -1 \\ -6 & 2 & 0 & 6 \\ -8 & -4 & 4 & 8 \\ -9 & -2 & 3 & 7 \end{bmatrix}$, $k = 20$.

## True/False Review Questions

*Determine whether each of the following statements is true or false.*

**1.** $\begin{bmatrix} -1 & -1 \\ -1 & 1 \end{bmatrix} \begin{bmatrix} -2 & -3 \\ 0 & 1 \end{bmatrix}$ is a QR-factorization of $\begin{bmatrix} 2 & 2 \\ 2 & 4 \end{bmatrix}$.

**2.** If $\mathbf{A}$ is any nonsingular matrix, then we can find a unitary matrix $\mathbf{Q}$ and upper triangular matrix $\mathbf{R}$ so that $\mathbf{A} = \mathbf{QR}$.

**3.** If we have a QR-factorization of $n \times n$ matrix $\mathbf{A}$, then we can solve $\mathbf{Ax} = \mathbf{b}$ without row reduction.

**4.** The basic QR-method for eigen approximation is based on similarity transformations.

**5.** If $\mathbf{A}$ is nonsingular with QR-factorization $\mathbf{A} = \mathbf{QR}$, then $|\det(\mathbf{A})| = |\det(\mathbf{R})|$.

## Terminology

| QR-factorization | Basic QR-method |
| --- | --- |

This section shows how to rewrite a matrix into a product of two matrices (sometimes referred to as a decomposition of the matrix) that have a particularly easy and useful structure. We then used this decomposition to obtain alternative procedures for computing both the solution of a nonsingular linear system and for determining the eigenvalues of a matrix, quite a remarkable set of results.

• What do we mean by the QR-factorization of a matrix?
• If the columns of $\mathbf{A}$ are linearly independent, then we showed how to get the QR-factorization using a technique studied previously. What technique did we use? Describe in narrative form how we got the matrices $\mathbf{Q}$ and $\mathbf{R}$.
• Describe the alternative based on a QR-factorization for solving the linear system $\mathbf{Ax} = \mathbf{b}$ when $\mathbf{A}$ is nonsingular.
• Describe in narrative form the way a QR-factorization is used to estimate the eigenvalues of a matrix.

## 4.7 ■ SINGULAR VALUE DECOMPOSITION (OPTIONAL)

In Section 4.4 we developed the spectral representation of a symmetric matrix; see (5) in Section 4.4. The spectral representation showed how to build a symmetric matrix as a linear combination of the outer products of its eigenvectors using the corresponding eigenvalues as coefficients. Two natural questions arise.

1. If $\mathbf{A}$ is square but not symmetric, is there a way to express $\mathbf{A}$ as a linear combination analogous to the spectral representation?
2. If $\mathbf{A}$ is rectangular, is there a way to express $\mathbf{A}$ as a linear combination analogous to the spectral representation?

Fortunately, we can answer both of these questions affirmatively and, in fact, the same result can be applied to both cases. We state the result without justification. (See the references listed at the end of this section.)

---

**The Singular Value Decomposition of a Matrix**

Let $\mathbf{A}$ be an $m \times n$ complex matrix. Then there exist unitary matrices $\mathbf{U}$ of size $m \times m$ and $\mathbf{V}$ of size $n \times n$ such that

$$\mathbf{A} = \mathbf{U}\mathbf{S}\mathbf{V}^* \tag{1}$$

where $\mathbf{S}$ is an $m \times n$ matrix with nondiagonal entries all zero and $s_{11} \geq s_{22} \geq \cdots \geq s_{pp} \geq 0$ where $p = \min\{m, n\}$.

---

The diagonal entries of $\mathbf{S}$ are called the **singular values** of $\mathbf{A}$, the columns of $\mathbf{U}$ are called the **left singular vectors** of $\mathbf{A}$; and the columns of $\mathbf{V}$ are called the **right singular vectors** of $\mathbf{A}$. The singular value decomposition of $\mathbf{A}$ in (1) can be expressed as the following linear combination:

$$\mathbf{A} = \text{col}_1(\mathbf{U})s_{11}\text{col}_1(\mathbf{V})^* + \text{col}_2(\mathbf{U})s_{22}\text{col}_2(\mathbf{V})^* + \cdots + \text{col}_p(\mathbf{U})s_{pp}\text{col}_p(\mathbf{V})^*. \tag{2}$$

which has the same form as the spectral representation of a symmetric matrix as given in Equation (5) in Section 4.4.

To determine the matrices $\mathbf{U}$, $\mathbf{S}$, and $\mathbf{V}$ in the singular value decomposition given in (1) we start as follows. An $n \times n$ Hermitian matrix (symmetric if $\mathbf{A}$ is real) related to $\mathbf{A}$ is $\mathbf{A}^*\mathbf{A}$. By results from Section 4.5 there exists a unitary $n \times n$ matrix $\mathbf{V}$ such that

$$\mathbf{V}^*(\mathbf{A}^*\mathbf{A})\mathbf{V} = \mathbf{D}$$

where $\mathbf{D}$ is a diagonal matrix whose diagonal entries $\lambda_1, \lambda_2, \ldots, \lambda_n$ are the eigenvalues of $\mathbf{A}^*\mathbf{A}$. Let $\mathbf{v}_j$ denote column $j$ of $\mathbf{V}$; then $(\mathbf{A}^*\mathbf{A})\mathbf{v}_j = \lambda_j\mathbf{v}_j$. Multiply both sides of this expression on the left by $\mathbf{v}_j^*$ and then rearrange the expression as

$$\mathbf{v}_j^*(\mathbf{A}^*\mathbf{A})\mathbf{v}_j = \lambda_j\mathbf{v}_j^*\mathbf{v}_j \Leftrightarrow (\mathbf{A}\mathbf{v}_j)^*(\mathbf{A}\mathbf{v}_j) = \lambda_j\mathbf{v}_j^*\mathbf{v}_j \Leftrightarrow \|\mathbf{A}\mathbf{v}_j\|^2 = \lambda_j\|\mathbf{v}_j\|^2.$$

Since the length of a vector is nonnegative, the last expression implies that $\lambda_j \geq 0$. Hence each eigenvalue of $\mathbf{A}^*\mathbf{A}$ is nonnegative. If necessary, renumber the eigenvalues of $\mathbf{A}^*\mathbf{A}$ so that $\lambda_1 \geq \lambda_2 \geq \cdots \geq \lambda_n$; then define $s_{jj} = \sqrt{\lambda_j}$. We note that since $\mathbf{V}$ is a unitary matrix, each of its columns is a unit vector; that is, $\|\mathbf{v}_j\| = 1$. Hence $s_{jj} = \|\mathbf{A}\mathbf{v}_j\|$. (Verify.) Thus we have

the singular values of $\mathbf{A}$ are the square roots of the eigenvalues of $\mathbf{A}^*\mathbf{A}$.

Finally, we determine the $m \times m$ unitary matrix $\mathbf{U}$. If we believe the matrix equation in (1), let's see what the columns of $\mathbf{U}$ should look like.

- Since $\mathbf{U}$ is to be unitary, its columns must be an orthonormal set, and hence are linearly independent $m \times 1$ vectors.
- The matrix $\mathbf{S}$ is block diagonal and has the form

$$\mathbf{S} = \left[ \begin{array}{cccc|c} s_{11} & 0 & \cdots & 0 & \\ 0 & s_{22} & \ddots & 0 & \mathbf{O}_{p,\,n-p} \\ \vdots & \ddots & \ddots & 0 & \\ 0 & \cdots & 0 & s_{pp} & \\ \hline & \multicolumn{3}{c|}{\mathbf{O}_{m-p,\,p}} & \mathbf{O}_{m-p,\,n-p} \end{array} \right].$$

- From (1), $\mathbf{AV} = \mathbf{US}$, so

$$\mathbf{A} \begin{bmatrix} \mathbf{v}_1 & \mathbf{v}_2 & \cdots & \mathbf{v}_n \end{bmatrix}$$

$$= \begin{bmatrix} \mathbf{u}_1 & \mathbf{u}_2 & \cdots & \mathbf{u}_m \end{bmatrix} \left[ \begin{array}{cccc|c} s_{11} & 0 & \cdots & 0 & \\ 0 & s_{22} & \ddots & 0 & \mathbf{O}_{p,\,n-p} \\ \vdots & \ddots & \ddots & 0 & \\ 0 & \cdots & 0 & s_{pp} & \\ \hline & \multicolumn{3}{c|}{\mathbf{O}_{m-p,\,p}} & \mathbf{O}_{m-p,\,n-p} \end{array} \right].$$

This implies that we need to require that $\mathbf{Av}_j = s_{jj}\mathbf{u}_j$ for $j = 1, 2, \ldots, p$. (Verify.)

- However, $\mathbf{U}$ must have $m$ orthonormal columns and $m \geq p$. In order to construct $\mathbf{U}$ we need an orthonormal basis for $R^m$ whose first $p$ vectors are $\mathbf{u}_j = \dfrac{1}{s_{jj}}\mathbf{Av}_j$. In Section 2.3 we showed how to extend any linearly independent subset of a subspace to a basis. We use that technique here to obtain the remaining $m - p$ columns of $\mathbf{U}$. (This is necessary only if $m > p$.) Since these $m - p$ columns are not unique, matrix $\mathbf{U}$ is not unique. (Neither is $\mathbf{V}$ if any of the eigenvalues of $\mathbf{A}^*\mathbf{A}$ are repeated.)

It can be shown that the preceding construction gives matrices $\mathbf{U}$, $\mathbf{S}$, and $\mathbf{V}$ so that $\mathbf{A} = \mathbf{USV}^*$. We illustrate the process in Example 1.

EXAMPLE 1    To find the singular value decomposition of

$$\mathbf{A} = \begin{bmatrix} 2 & -4 \\ 2 & 2 \\ -4 & 0 \\ 1 & 4 \end{bmatrix}$$

we follow the steps outlined previously. First we compute $\mathbf{A}^*\mathbf{A}$ and then compute its eigenvalues and eigenvectors. We obtain

$$\mathbf{A}^*\mathbf{A} = \begin{bmatrix} 25 & 0 \\ 0 & 36 \end{bmatrix}$$

and since it is diagonal we know that its eigenvalues are its diagonal entries. It follows that $\begin{bmatrix} 1 \\ 0 \end{bmatrix}$ is an eigenvector for eigenvalue 25 and that $\begin{bmatrix} 0 \\ 1 \end{bmatrix}$ is an eigenvector for eigenvalue 36. (Verify.) We label the eigenvalues in decreasing magnitude as $\lambda_1 = 36$ and $\lambda_2 = 25$ with corresponding eigenvectors $\mathbf{v}_1 = \begin{bmatrix} 0 \\ 1 \end{bmatrix}$ and $\mathbf{v}_2 = \begin{bmatrix} 1 \\ 0 \end{bmatrix}$. Hence

$$\mathbf{V} = \begin{bmatrix} \mathbf{v}_1 & \mathbf{v}_2 \end{bmatrix} = \begin{bmatrix} 0 & 1 \\ 1 & 0 \end{bmatrix}$$

and $s_{11} = 6$ and $s_{22} = 5$. It follows that

$$S = \left[ \begin{array}{cc} 6 & 0 \\ 0 & 5 \\ \hline \mathbf{O}_2 \end{array} \right].$$

Next we determine the matrix $\mathbf{U}$ starting with the first two columns

$$\mathbf{u}_1 = \frac{1}{s_{11}} \mathbf{A} \mathbf{v}_1 = \frac{1}{6} \begin{bmatrix} -4 \\ 2 \\ 0 \\ 4 \end{bmatrix} \quad \text{and} \quad \mathbf{u}_2 = \frac{1}{s_{22}} \mathbf{A} \mathbf{v}_2 = \frac{1}{5} \begin{bmatrix} 2 \\ 2 \\ -4 \\ 1 \end{bmatrix}.$$

The remaining two columns are found by extending the linearly independent vectors in $\{\mathbf{u}_1, \mathbf{u}_2\}$ to a basis for $R^4$ and then applying the Gram-Schmidt process. We proceed as follows. We first compute

$$\mathbf{rref}\left( \begin{bmatrix} \mathbf{u}_1 & \mathbf{u}_2 & \mathbf{I}_4 \end{bmatrix} \right) = \begin{bmatrix} 1 & 0 & 0 & 0 & \frac{3}{8} & \frac{3}{2} \\ 0 & 1 & 0 & 0 & -\frac{5}{4} & 0 \\ 0 & 0 & 1 & 0 & \frac{3}{4} & 1 \\ 0 & 0 & 0 & 1 & \frac{3}{8} & -\frac{1}{2} \end{bmatrix}.$$

This tells us that the set $\{\mathbf{u}_1, \mathbf{u}_2, \mathbf{e}_1, \mathbf{e}_2\}$ is a basis for $R^4$. (Explain why.) Next we apply the Gram-Schmidt process to this set to find the unitary matrix $\mathbf{U}$. Recording the results to six decimals, we obtain

$$\mathbf{U} = \begin{bmatrix} \mathbf{u}_1 & \mathbf{u}_2 & \begin{matrix} 0.628932 & 0 \\ 0.098933 & 0.847998 \\ 0.508798 & 0.317999 \\ 0.579465 & -0.423999 \end{matrix} \end{bmatrix}.$$

Thus we have the singular value decomposition of $\mathbf{A}$ with

$$\mathbf{A} = \mathbf{U}\mathbf{S}\mathbf{V}^* = \begin{bmatrix} \mathbf{u}_1 & \mathbf{u}_2 & \begin{matrix} 0.628932 & 0 \\ 0.098933 & 0.847998 \\ 0.508798 & 0.317999 \\ 0.579465 & -0.423999 \end{matrix} \end{bmatrix} \left[ \begin{array}{cc} 6 & 0 \\ 0 & 5 \\ \hline \mathbf{O}_2 \end{array} \right] \begin{bmatrix} \mathbf{v}_1 & \mathbf{v}_2 \end{bmatrix}^*$$

$$= \begin{bmatrix} -\frac{2}{3} & \frac{2}{5} & 0.628932 & 0 \\ \frac{1}{3} & \frac{2}{5} & 0.098933 & 0.847998 \\ 0 & -\frac{4}{5} & 0.508798 & 0.317999 \\ \frac{2}{3} & \frac{1}{5} & 0.579465 & -0.423999 \end{bmatrix} \begin{bmatrix} 6 & 0 \\ 0 & 5 \\ 0 & 0 \\ 0 & 0 \end{bmatrix} \begin{bmatrix} 0 & 1 \\ 1 & 0 \end{bmatrix}^*.$$   ■

The singular decomposition of a matrix $\mathbf{A}$ has been called one of the most useful tools in terms of the information that it reveals. Its use in an ever-increasing number of applications is a testament to its ability to reveal a variety of information for science and engineering. For example, previously in Section 2.3 we indicated that we could compute rank($\mathbf{A}$) by determining the number of nonzero rows in the reduced row echelon form of $\mathbf{A}$. An implicit assumption in this statement is that all the computational steps in the row operations would use exact arithmetic. Unfortunately, in most computing environments, when we perform row operations exact arithmetic is not used. Rather floating point arithmetic, which is a model of exact arithmetic, is used. In doing so we may lose accuracy because of the accumulation of small errors in the arithmetic steps. In some cases this loss of accuracy is enough to introduce doubt into the computation of rank. The following list of properties, which we state without proof, indicate how the singular value decomposition of a matrix can be used to compute its rank.

---

**Properties of Singular Values and Vectors**

1. Let $\mathbf{A}$ be an $m \times n$ matrix and let $\mathbf{B}$ and $\mathbf{C}$ be nonsingular matrices of size $m \times m$ and $n \times n$, respectively. Then

$$\text{rank}(\mathbf{B}\mathbf{A}) = \text{rank}(\mathbf{A}) = \text{rank}(\mathbf{A}\mathbf{C}).$$

2. The rank of $\mathbf{A}$ is the number of nonzero singular values of $\mathbf{A}$. (Multiple singular values are counted according to their multiplicity.)

---

Because matrices $\mathbf{U}$ and $\mathbf{V}$ of a singular value decomposition are unitary, it can be shown that most of the errors due to using the floating-point arithmetic occur in the computation of the singular values. The size of the matrix $\mathbf{A}$ and characteristics of the floating point arithmetic are often used to determine a threshold value below which singular values are considered zero. It has been argued that singular values and singular value decomposition give us a computationally reliable way to compute the rank of $\mathbf{A}$.

In addition to determining rank, the singular value decomposition of a matrix also provides orthonormal bases for the fundamental subspaces associated with a linear system of equations. We state without proof the following.

---

3. Let $\mathbf{A}$ be an $m \times n$ real matrix of rank $r$ with singular value decomposition $\mathbf{U}\mathbf{S}\mathbf{V}^*$. Then

   i. The first $r$ columns of $\mathbf{U}$ are an orthonormal basis for the column space of $\mathbf{A}$.
   ii. The first $r$ columns of $\mathbf{V}$ are an orthonormal basis for the row space of $\mathbf{A}$.
   iii. The last $n - r$ columns of $\mathbf{V}$ are an orthonormal basis for the null space of $\mathbf{A}$.

---

Next we see that (1) has the same form as the spectral decomposition of a symmetric matrix as given in (7) in Section 4.4. MATLAB has a command for computing the singular value decomposition of a matrix (see Exercises 8 through 10). We illustrate (2) in Example 2.

EXAMPLE 2    Let

$$\mathbf{A} = \begin{bmatrix} 1 & -1 \\ 1 & 0 \\ 1 & 1 \end{bmatrix}.$$

Then using MATLAB we find that the singular value decomposition $\mathbf{A} = \mathbf{USV}^*$ is given by

$$\mathbf{U} = \begin{bmatrix} 0.5774 & 0.7071 & 0.4082 \\ 0.5774 & 0.0000 & -0.8165 \\ 0.5774 & -0.7071 & 0.4082 \end{bmatrix}, \quad \mathbf{S} = \begin{bmatrix} 1.7321 & 0 \\ 0 & 1.4142 \\ 0 & 0 \end{bmatrix},$$

and

$$\mathbf{V} = \begin{bmatrix} 1 & 0 \\ 0 & -1 \end{bmatrix}.$$

Thus the singular values are $1.7321 \approx \sqrt{3}$ and $1.4142 \approx \sqrt{2}$. We have that

$$\mathrm{col}_1(\mathbf{U})\, s_{11}\, \mathrm{col}_1(\mathbf{V})^* + \mathrm{col}_2(\mathbf{U})\, s_{22}\, \mathrm{col}_2(\mathbf{V})^*$$

$$= \begin{bmatrix} 0.5774 \\ 0.5774 \\ 0.5774 \end{bmatrix} 1.7321 \begin{bmatrix} 1 \\ 0 \end{bmatrix}^* + \begin{bmatrix} 0.7071 \\ 0.0000 \\ -0.7071 \end{bmatrix} 1.4142 \begin{bmatrix} 0 \\ -1 \end{bmatrix}^*$$

$$= \begin{bmatrix} 1.0001 & -1.0000 \\ 1.0001 & 0 \\ 1.0001 & 1.0000 \end{bmatrix} \approx \mathbf{A}.$$

This computation was performed with just the decimal places displayed. Had we used all the decimal places available in MATLAB, the preceding linear combination would be even closer to $\mathbf{A}$. ■

Using singular value decomposition we can generalize the symmetric image application of Section 4.4. We no longer require that the image be "digitized" into a square symmetric matrix. The singular values of a matrix are all nonnegative and are arranged as the diagonal entries of $\mathbf{S}$ in decreasing size. Hence we can use the following partial sum expression derived from (2) to generate approximations to images, just as we did in Section 4.4.

$$\sum_{j=1}^{k} \mathrm{col}_j(\mathbf{U}) s_{jj} \mathrm{col}_j(\mathbf{V})^*, \quad k \leq p.$$

In fact, the collection of singular values and vectors used to form the partial sum is a form of **data compression** of the image. For a large image, it is significantly less costly to transmit the singular values and vectors that generate the approximation than the entire rectangular image. For a small image, like a fingerprint, which is digitized into a large matrix, the same thing is true.[1] As an introduction to digital compression, we illustrate these ideas in Example 3 with a small, crudely digitized example.

EXAMPLE 3    A word is encoded as a $5 \times 23$ matrix $\mathbf{A}$ using zeros and ones. (See Example 5 in Section 4.4 .) To show the word graphically we assign a white square to a zero and a black square to a 1. We use partial sums of the singular value decomposition, $\mathbf{A} = \mathbf{USV}^*$, to generate approximations to the word. We use the same display scheme as in Example 5 of Section 4.4: an entry will be a black square if the corresponding numerical entry is greater than or equal to one half in absolute value, otherwise it will be shown as a white square. Treating this as a recognition

---

[1] Satellite pictures and fingerprints are digitized and the data is compressed using schemes more robust than singular value decomposition. However, the principles mentioned here still apply.

game, we want to identify the word using a partial sum as small as possible. Figure 1 contains the graphical representation of the first partial sum, $\mathrm{col}_1(\mathbf{U})s_{11}\mathrm{col}_1(\mathbf{V})^*$. One might suspect that there are four letters in the word, but it is difficult to guess at the letters.

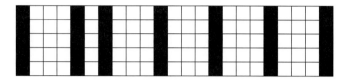

FIGURE 1

Figure 2 contains the graphical representation of the second partial sum,

$$\mathrm{col}_1(\mathbf{U})\, s_{11} \,\mathrm{col}_1(\mathbf{V})^* + \mathrm{col}_2(\mathbf{U})\, s_{22} \,\mathrm{col}_2(\mathbf{V})^*.$$

More information is revealed in this approximation, and we can guess that the image is the word MATH, written in block letters.

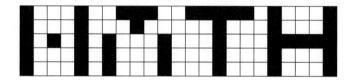

FIGURE 2

There are five singular values of the $5 \times 23$ matrix $\mathbf{A}$ containing zeros and ones for this word. They are 5.7628, 2.4092, 2.1698, 1.5145, and 0.9919. To generate the approximation in Figure 2 we used only the first two singular values and their corresponding singular vectors. (The singular values were computed using MATLAB. See the statements preceding Exercise 8 for more information on using MATLAB.)

■

In Section 4.4 we looked at the image of the unit circle by a symmetric matrix. The key to analyzing the form of the image was the spectral decomposition. Next we generalize the matrix transformation to any $2 \times 2$ matrix and use the singular value decomposition to determine the form of the image.

### Application:   The Image of the Unit Circle by a 2 × 2 Matrix

Let $\mathbf{A}$ be any $2 \times 2$ real matrix. Then we know that $\mathbf{A}$ has a singular value decomposition $\mathbf{A} = \mathbf{U}\mathbf{S}\mathbf{V}^T$ where $\mathbf{U} = \begin{bmatrix} \mathbf{u}_1 & \mathbf{u}_2 \end{bmatrix}$ and $\mathbf{V} = \begin{bmatrix} \mathbf{v}_1 & \mathbf{v}_2 \end{bmatrix}$ are $2 \times 2$ orthogonal matrices. The matrix $\mathbf{S}$ is a $2 \times 2$ diagonal matrix with the singular values of $\mathbf{A}$ as its diagonal entries. The analysis of the form of the image of the unit circle in this case is nearly the same as in Section 4.4, except that we use singular vectors of $\mathbf{A}$ in place of eigenvectors. The image is again an ellipse. We have the following important relations:

$$\mathbf{A} = \mathbf{U}\mathbf{S}\mathbf{V}^T \Leftrightarrow \mathbf{A}\mathbf{V} = \mathbf{U}\mathbf{S}$$

$$\Leftrightarrow \mathbf{A}\begin{bmatrix} \mathbf{v}_1 & \mathbf{v}_2 \end{bmatrix} = \begin{bmatrix} \mathbf{u}_1 & \mathbf{u}_2 \end{bmatrix} \begin{bmatrix} s_{11} & 0 \\ 0 & s_{22} \end{bmatrix}$$

$$\Leftrightarrow \mathbf{A}\mathbf{v}_1 = s_{11}\mathbf{u}_1 \text{ and } \mathbf{A}\mathbf{v}_2 = s_{22}\mathbf{u}_2.$$

Thus the image of the columns of $\mathbf{V}$ are scalar multiples of the columns of $\mathbf{U}$. This is reminiscent of the eigen relationships. Since $\mathbf{V}$ is orthogonal its columns are orthonormal vectors and $\mathbf{V}^T = \mathbf{V}^{-1}$, so $\mathbf{V}^T\mathbf{v}_1 = \mathbf{i}$ and $\mathbf{V}^T\mathbf{v}_2 = \mathbf{j}$. Figure 3 is analogous to Figure 7 in Section 4.3, with eigenvectors replaced by singular vectors.

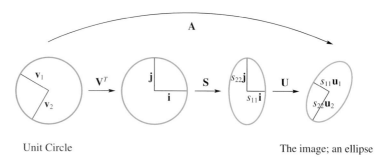

Unit Circle                                                    The image; an ellipse

FIGURE 3

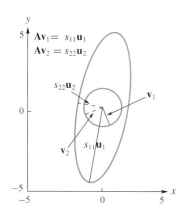

FIGURE 4

EXAMPLE 4   Let $\mathbf{A} = \begin{bmatrix} 1 & 1 \\ -2 & 4 \end{bmatrix}$. Using MATLAB we find that the singular value decomposition is $\mathbf{A} = \mathbf{U}\mathbf{S}\mathbf{V}^T$, where

$$\mathbf{U} = \begin{bmatrix} -0.1091 & -0.9940 \\ -0.9940 & 0.1091 \end{bmatrix}, \quad \mathbf{S} = \begin{bmatrix} 4.4966 & 0 \\ 0 & 1.3343 \end{bmatrix},$$

and

$$\mathbf{V} = \begin{bmatrix} 0.4179 & -0.9085 \\ -0.9085 & -0.4179 \end{bmatrix}.$$

Figure 4 shows the unit circle with vectors $\mathbf{v}_1$ and $\mathbf{v}_2$ displayed and the elliptical image with its major and minor axes displayed as $s_{11}\mathbf{u}_1$ and $s_{22}\mathbf{u}_2$, respectively.   ■

These results generalize to $n \times n$ matrices. The image of the unit $n$-ball is an $n$-dimensional ellipse. For $n = 3$, the image is an ellipsoid (see Figure 5).

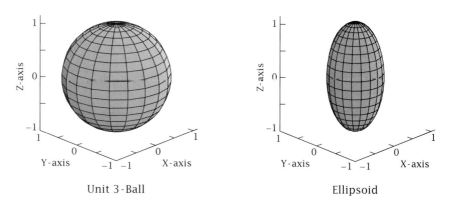

Unit 3-Ball                                              Ellipsoid

FIGURE 5

## ■ REFERENCES FOR SECTION 4.7

GOLUB, G. and C. VAN LOAN, *Matrix Computations*, 3rd Ed. Baltimore, Maryland: The Johns Hopkins University Press, 1996.

HERN, T. AND C. LONG, "Viewing Some Concepts and Applications in Linear Algebra," in *Visualization in Teaching and Learning Mathematics*, *MAA Notes* 19, pp. 173–190, The Mathematical Association of America, 1991.

HILL, D., *Experiments in Computational Matrix Algebra*, New York: Random House, 1988 (distributed by McGraw-Hill).

HILL, D., "Nuggets of Information," in *Using MATLAB in the Classroom*, pp.77–93. Englewood Cliffs, N.J.: Prentice-Hall, 1993.

KAHANER, D., C. MOLER, AND S. NASH, *Numerical Methods and Software*, Englewood Cliffs, N.J.: Prentice Hall, 1989.

STEWART, G., *Introduction to Matrix Computations*, New York: Academic Press, 1973.

## EXERCISES 4.7

*In Exercises 1–4, find the singular values of each given matrix.*

**1.** $A = \begin{bmatrix} 5 & 0 \\ 0 & 0 \\ 0 & -1 \end{bmatrix}$.    **2.** $A = \begin{bmatrix} 1 & 1 \\ 1 & 1 \end{bmatrix}$.

**3.** $A = \begin{bmatrix} 1 & 2 \\ 0 & 1 \\ -2 & 1 \end{bmatrix}$.    **4.** $A = \begin{bmatrix} 1 & 0 & 1 & -1 \\ 0 & 1 & -1 & 1 \end{bmatrix}$.

**5.** Determine the singular value decomposition of

$$A = \begin{bmatrix} 1 & -4 \\ -2 & 2 \\ 2 & 4 \end{bmatrix}.$$

**6.** Determine the singular value decomposition of

$$A = \begin{bmatrix} 1 & 1 \\ -1 & 1 \end{bmatrix}.$$

**7.** Determine the singular value decomposition of

$$A = \begin{bmatrix} 1 & 0 & 1 \\ 1 & 1 & -1 \\ -1 & 1 & 0 \\ 0 & 1 & 1 \end{bmatrix}.$$

*There is a MATLAB command, **svd**, that computes the singular value decomposition of a matrix **A**. Used in the form **svd(A)**, the output is a list of the singular values of **A**. In the form **[U,S,V] = svd(A)** we get matrices **U**, **S**, and **V** such that **A** = **USV***. In Exercises 8–10, use the **svd** command to determine the singular values of each given matrix.*

**8.** $\begin{bmatrix} 5 & 2 \\ 8 & 5 \end{bmatrix}$.    **9.** $\begin{bmatrix} 3 & 12 & -10 \\ 1 & -1 & 5 \\ -2 & -8 & 5 \end{bmatrix}$.

**10.** $\begin{bmatrix} 1 & 5 & 9 \\ 2 & 6 & 10 \\ 3 & 7 & 11 \\ 4 & 8 & 12 \end{bmatrix}$.

*An $m \times n$ matrix **A** is said to have full rank if rank$(A) = \min\{m, n\}$. The singular value decomposition lets us measure how close **A** is to not having full rank. If any singular value is zero, then **A** does not have full rank. If $s_{\min}$ is the smallest singular value of **A** and $s_{\min} \neq 0$, then the distance from the set of matrices with rank $r = \min\{m, n\} - 1$ is $s_{\min}$. In Exercises 11–13, determine the distance from each given matrix to the matrix of the same size with rank $\min\{m, n\} - 1$. (Use MATLAB to find the singular values.)*

**11.** $\begin{bmatrix} 1 & 3 & 2 \\ 4 & 1 & 0 \\ 2 & -5 & -4 \\ 1 & 2 & 1 \end{bmatrix}$.    **12.** $\begin{bmatrix} 1 & 2 & 3 \\ 1 & 0 & 1 \\ 3 & 6 & 9 \end{bmatrix}$.

**13.** $\begin{bmatrix} 1 & 0 & 0 & -1 & 0 \\ -1 & 1 & 1 & 0 & 0 \\ 0 & 0 & 1 & -2 & -1 \\ 0 & 1 & 0 & 1 & 1 \end{bmatrix}$.

**14.** Prove Property 2 (see page 296).

**15.** Verify the following property of singular values.

If **A** is symmetric, then the singular values of **A** are the absolute values of the eigenvalues of **A**.

**16.** Another property of singular values is the following.

Let $k \leq$ rank$(A)$. Then the matrix of rank $k$ that is "closest"[2] to **A** is

$$\sum_{j=1}^{k} \mathrm{col}_j(U)\, s_{jj}\, \mathrm{col}_j(V)^*.$$

---

[2] We have not developed the necessary concepts to verify that the singular value decomposition can be used to construct closest matrices of a given rank. Please refer to G. W. Stewart, *Introduction to Matrix Computations*, ( New York: Academic Press, 1973) or G. Golub and C. Van Loan, *Matrix Computations*, 3rd ed. (Baltimore: The Johns Hopkins University Press, 1996).

Use this statement to find the matrix of rank 2 closest to

$$A = \begin{bmatrix} 1 & 2 & 3 \\ 4 & 5 & 6 \\ 7 & 8 & 0 \end{bmatrix}.$$

17. Using Figure 3, explain the form of the image if $s_{22} = 0$.

18. Define matrix

$$A = \begin{bmatrix} 2 & -1 \\ 1 & 0 \end{bmatrix}$$

in MATLAB. Then enter the following set of commands to see the unit circle and its elliptical image under the matrix transformation determined by $A$. Print out the figure generated.

```
close all
t=0:.05:2*pi+.2;
x=cos(t);y=sin(t);
c=[x;y];
Ac=A*c;
axis([-3 3 -3 3]),axis(axis),
hold on,axis('square')
figure(gcf),plot(x,y,'k-',Ac(1,:),Ac(2,:),'b-')
hold off
```

(a) Compute the singular value decomposition of $A$.
(b) In the unit circle draw the vectors $v_1$ and $v_2$.
(c) Sketch in the major and minor axes for the ellipse.
(d) In the ellipse draw the vectors $Av_1$ and $Av_2$.

## True/False Review Questions

*Determine whether each of the following statements is true or false.*

1. If $A$ is square with singular value decomposition $A = USV^*$, then $A$ is nonsingular provided every diagonal entry of $S$ is not zero.

2. The singular value decomposition of a matrix $A$ expresses it as a linear combination of matrices of rank 1.

3. If $A$ is symmetric, then its singular values are its eigenvalues.

4. Rank($A$) is the number of nonzero singular values of $A$.

5. Every matrix has a singular value decomposition.

## Terminology

| Singular value decomposition | Singular values and vectors |
| --- | --- |

The singular value decomposition of a matrix is a factorization into a product of matrices that can be used to reveal the basic information and building blocks of the matrix. We indicated how the factors are constructed from the original matrix and how orthonormal sets were involved. We then stated properties that could be derived from the factorization. We showed that this factorization generalized the spectral decomposition discussed in Section 4.5 and how it could be used as a form of data compression. We also analyzed the image of the unit circle for an arbitrary matrix transformation using the information in the singular value decomposition.

• Describe the factors in the singular value decomposition of a matrix. Explicitly state the form and properties of each term.
• How are the singular values of a matrix related to the eigenvalues of a related matrix? Explain.
• Explain how the singular value decomposition generalizes the spectral decomposition of a symmetric matrix.
• How does the singular value aid in the computation of rank?
• If $A$ is symmetric, how are its singular values related to its eigenvalues?
• What role do the singular values and singular vectors play in forming the geometric image of the unit circle?

# CHAPTER TEST (For Sections 4.1–4.5.)

**1.** Determine the characteristic polynomial of $\mathbf{A} = \begin{bmatrix} 1 & 1 & 4 \\ 2 & 0 & -4 \\ -1 & 1 & 5 \end{bmatrix}$.

**2.** Determine the eigenpairs of $\mathbf{A} = \begin{bmatrix} 0 & 0 & 1 \\ 0 & 1 & 0 \\ 0 & 0 & 1 \end{bmatrix}$.

**3.** Show that $\mathbf{A} = \begin{bmatrix} 3 & 1 & 0 \\ 0 & 3 & 1 \\ 0 & 0 & -2 \end{bmatrix}$ is defective.

**4.** Complete the entries of the following table with the correct responses of YES, NO, or INSUFFICIENT INFORMATION TO DECIDE (abbreviated IITD.)

|  | diagonalizable | nonsingular |
|---|---|---|
| A is $3 \times 3$ with eigenvalues $0, 2, -1$ |  |  |
| A is $4 \times 4$ with eigenvalues $2, 2, 4, -3$ |  |  |
| A is $4 \times 4$ symmetric with eigenvalues $-1, 0, 0, 2$ |  |  |

**5.** Apply the Gram-Schmidt process to the set $\left\{ \begin{bmatrix} -1 \\ 1 \\ 0 \end{bmatrix}, \begin{bmatrix} -1 \\ 0 \\ 1 \end{bmatrix}, \begin{bmatrix} 1 \\ 1 \\ 1 \end{bmatrix} \right\}$.

**6.** Find a basis for the eigenspace corresponding to eigenvalue $\lambda = 3$ of $\mathbf{A} = \begin{bmatrix} 3 & 0 & 0 & 0 \\ 0 & 3 & 1 & 0 \\ 0 & 0 & 3 & 0 \\ 0 & 0 & 0 & 4 \end{bmatrix}$.

**7.** Show that if $\mathbf{A}$ is nonsingular and diagonalizable then $\mathbf{A}^{-1}$ is diagonalizable.

**8.** $\mathbf{A}$ is symmetric with eigenpairs $(1, \mathbf{v})$, $(3, \mathbf{u})$, and $(-2, \mathbf{w})$ in which the eigenvectors form an orthonormal set. Construct the spectral decomposition of $\mathbf{A}$.

**9.** Let $\mathbf{A} = \begin{bmatrix} 3 & 0 & 1 \\ 0 & 1 & 0 \\ -1 & 0 & 3 \end{bmatrix}$. Is $\mathbf{A}$ a unitary matrix? If not, construct a unitary matrix $\mathbf{Q}$ closely related to $\mathbf{A}$.

**10.** Answer the following "True" or "False."

    (a) An orthogonal matrix is nonsingular.

    (b) If $\mathbf{A}$ is diagonalizable, then it has distinct eigenvalues.

    (c) If $\mathbf{A}$ is singular, then one of its eigenvalues is 0.

    (d) Similar matrices have the same eigenvalues.

    (e) If $(\lambda, \mathbf{p})$ and $(\mu, \mathbf{q})$ are eigenpairs of $\mathbf{A}$ with $\lambda \neq \mu$, then $(\lambda + \mu, \mathbf{p} + \mathbf{q})$ is another eigenpair of $\mathbf{A}$.

# C H A P T E R

# 5

# *VECTOR SPACES*

Mathematics has been called the science of patterns. The identification of patterns and common features in seemingly diverse situations provides us with opportunities to unify information. This approach can lead to the development of classification schemes and structures, which can be studied independently of a particular setting or application. Thus we can develop ideas that apply to each and every member of the class based upon properties that they have in common. The result is called an abstract model or abstract structure. In many ways this can help us work with more difficult and comprehensive ideas.

In Chapters 1 through 4 we focused on a matrix. We saw that its structure, the arrangement of rows and columns of information, was applicable to a variety of situations. We showed how to manipulate matrices to obtain information about the solution set of a linear system of equations and how to obtain information about the behavior of a matrix transformation. In both of these situations subspaces played an important role: row space, column space, null space, and eigenspace. In effect, subspace structures provided a window into information of various kinds contained within the matrix. Here we study subspaces in more detail and introduce a mathematical structure, the **vector space**, which includes subspaces.

Linear algebra uses extensively a structure called a vector space. The properties of a vector space give the study of matrices a more geometrical foundation. The corresponding geometrical viewpoint on topics aids both in understanding as well as in suggesting new approaches.

## 5.1 ■ THE VECTOR SPACE STRUCTURE

A vector space is an abstract mathematical structure that consists of a set $V$ of objects together with two operations on these objects that satisfy a certain set of rules. When we identity a particular example as a vector space, we will automatically be able to attribute to it certain properties that hold for all vector spaces. In a sense these properties are a legacy of the fact that we have a vector space structure.

In this section we define an abstract vector space and discuss several examples which are familiar from our work in Chapters 1 through 4. We then introduce new examples that will be important in applications we meet later.

---

**Definition**    Let $V$ be a set with two operations $\oplus$, and $\odot$. V is called a real **vector space**, denoted $(V, \oplus, \odot)$, if the following properties or axioms hold:

1. If $\mathbf{v}$ and $\mathbf{w}$ belong to V, then so does $\mathbf{v} \oplus \mathbf{w}$. (We say $V$ is closed with respect to operation $\oplus$.)

    (a) $\mathbf{v} \oplus \mathbf{w} = \mathbf{w} \oplus \mathbf{v}$, for all $\mathbf{v}$ and $\mathbf{w}$ in $V$. (Commutativity of $\oplus$.)
    (b) $(\mathbf{v} \oplus \mathbf{w}) \oplus \mathbf{u} = \mathbf{v} \oplus (\mathbf{w} \oplus \mathbf{u})$ for all $\mathbf{u}$, $\mathbf{v}$, and $\mathbf{w}$ in $V$. (Associativity of $\oplus$.)
    (c) There exists a unique member $\mathbf{z}$ in $V$, such that $\mathbf{v} \oplus \mathbf{z} = \mathbf{z} \oplus \mathbf{v} = \mathbf{v}$, for any $\mathbf{v}$ in $V$. ($\mathbf{z}$ is called the *identity* or *zero vector* for $\oplus$.)
    (d) For each $\mathbf{v}$ in $V$ there exists a unique member $\mathbf{w}$ in $V$ such that $\mathbf{v} \oplus \mathbf{w} = \mathbf{w} \oplus \mathbf{v} = \mathbf{z}$. ($\mathbf{w}$ is denoted $-\mathbf{v}$ and called the negative of $\mathbf{v}$.)

2. If $\mathbf{v}$ is any member of $V$ and $k$ is any real number, then $k \odot \mathbf{v}$ is a member of $V$. (We say $V$ is closed with respect to $\odot$.)

    (e) $k \odot (\mathbf{v} \oplus \mathbf{w}) = (k \odot \mathbf{v}) \oplus (k \odot \mathbf{w})$ for any $\mathbf{v}$ and $\mathbf{w}$ in $V$ and any real number $k$. (Distributivity of $\odot$ over $\oplus$.)
    (f) $(k + j) \odot \mathbf{v} = (k \odot \mathbf{v}) \oplus (j \odot \mathbf{v})$ for any $\mathbf{v}$ in $V$ and any real numbers $j$ and $k$. (Distributivity of addition of real numbers over $\odot$.)
    (g) $k \odot (j \odot \mathbf{v}) = (kj) \odot \mathbf{v}$, for any $\mathbf{v}$ in $V$ and any real numbers $j$ and $k$.
    (h) $1 \odot \mathbf{v} = \mathbf{v}$, for any $\mathbf{v}$ in $V$.

---

The following terminology is used.

- The members of $V$ are called **vectors** and need not be directed line segments or row or column vectors. Consider the term *vector* as a generic name for a member of a vector space.
- Real numbers are called **scalars**.
- The operation $\oplus$ is called **vector addition**. The actual definition of this operation need not resemble ordinary addition. The term *vector addition* is a generic name for an operation that combines **two** vectors from $V$ and produces another vector of $V$.
- The operation $\odot$ is called **scalar multiplication**. It combines a real scalar and a vector from $V$ to produce another vector in $V$.
- If this definition holds with real numbers replaced by complex numbers, then the structure $(V, \oplus, \odot)$ is called a **complex vector space**. In this case the complex numbers are called scalars.

The definition of a vector space looks complicated because $V$, $\oplus$, and $\odot$ have not been given specific meanings. In a particular example the members of set $V$ must be explicitly specified and the action of the operations $\oplus$ and $\odot$ must be stated. Hence we can check properties 1 and 2 and properties a through h directly. If any of these properties is not satisfied, then set $V$ with operations $\oplus$ and $\odot$ is not a vector space.

The name *vector space* conjures up the image of a directed line segment in $R^2$ or $R^3$ as we introduced in Section 1.1. (See Figures 2 and 3 in Section 1.1.) This is, of course, where the name is derived from. Matrices and $n$-vectors give us examples of vector spaces with which we are already familiar.

EXAMPLE 1   Let $V = R^n$, $\oplus$ be ordinary addition of $n$-vectors, and $\odot$ be ordinary scalar multiplication of $n$-vectors. (See Section 1.2.) To verify that $R^n$ is a vector space with these operations we must explicitly show that properties 1, 2, and a through h in the definition of a vector space are valid. The closures have been verified in Exercise 37 in Section 1.2. Properties a, b, e, f, and g are cited at the beginning of Section 1.4. To complete the verification that $R^n$ is a vector space with these operations, we now show properties c, d, and h.

Property c.   Certainly a candidate for the identity for vector addition is the $n$-vector $\mathbf{z} = \begin{bmatrix} 0 & 0 & \cdots & 0 \end{bmatrix}$. This follows because for

$$\mathbf{v} = \begin{bmatrix} v_1 & v_2 & \cdots & v_n \end{bmatrix}$$

we have

$$\mathbf{v} \oplus \mathbf{z} = \begin{bmatrix} v_1 + 0 & v_2 + 0 & \cdots & v_n + 0 \end{bmatrix}$$
$$= \begin{bmatrix} v_1 & v_2 & \cdots & v_n \end{bmatrix} = \mathbf{v}$$

and

$$\mathbf{z} \oplus \mathbf{v} = \begin{bmatrix} 0 + v_1 & 0 + v_2 & \cdots & 0 + v_n \end{bmatrix}$$
$$= \begin{bmatrix} v_1 & v_2 & \cdots & v_n \end{bmatrix} = \mathbf{v}.$$

Thus $\mathbf{v} \oplus \mathbf{z} = \mathbf{z} \oplus \mathbf{v} = \mathbf{v}$. It remains to show that $\mathbf{z} = \begin{bmatrix} 0 & 0 & \cdots & 0 \end{bmatrix}$ is unique; that is, $\mathbf{z} = \begin{bmatrix} 0 & 0 & \cdots & 0 \end{bmatrix}$ is the only member of $R^n$ for which it is true that $\mathbf{v} \oplus \mathbf{z} = \mathbf{z} \oplus \mathbf{v} = \mathbf{v}$ for every vector $\mathbf{v}$ in $R^n$. Suppose there is another vector $\mathbf{w}$ in $R^n$ for which $\mathbf{v} \oplus \mathbf{w} = \mathbf{w} \oplus \mathbf{v} = \mathbf{v}$ for every vector $\mathbf{v}$ in $R^n$. Then it must be that $\mathbf{z} \oplus \mathbf{w} = \mathbf{w}$ and $\mathbf{w} \oplus \mathbf{z} = \mathbf{z}$. But vector addition is commutative; that is, $\mathbf{z} \oplus \mathbf{w} = \mathbf{w} \oplus \mathbf{z}$. Hence $\mathbf{z} = \mathbf{w}$ and so there is a unique identity for operation $\oplus$. (Previously we denoted $\begin{bmatrix} 0 & 0 & \cdots & 0 \end{bmatrix}$ as $\mathbf{0}$.)

Property d.   For $\mathbf{v} = \begin{bmatrix} v_1 & v_2 & \cdots & v_n \end{bmatrix}$, a candidate for $-\mathbf{v}$ is

$$\begin{bmatrix} -v_1 & -v_2 & \cdots & -v_n \end{bmatrix}.$$

We have

$$\mathbf{v} \oplus \begin{bmatrix} -v_1 & -v_2 & \cdots & -v_n \end{bmatrix}$$
$$= \begin{bmatrix} v_1 - v_1 & v_2 - v_2 & \cdots & v_n - v_n \end{bmatrix}$$
$$= \begin{bmatrix} 0 & 0 & \cdots & 0 \end{bmatrix} = \mathbf{0}$$

and

$$\begin{bmatrix} -v_1 & -v_2 & \cdots & -v_n \end{bmatrix} \oplus \mathbf{v}$$
$$= \begin{bmatrix} -v_1 + v_1 & -v_2 + v_2 & \cdots & -v_n + v_n \end{bmatrix}$$
$$= \begin{bmatrix} 0 & 0 & \cdots & 0 \end{bmatrix} = \mathbf{0}$$

so there exists a negative for every vector in $R^n$. It remains to show that it is unique. Let

$$\mathbf{w} = \begin{bmatrix} -v_1 & -v_2 & \cdots & -v_n \end{bmatrix}.$$

Let $\mathbf{u} = \begin{bmatrix} u_1 & u_2 & \cdots & u_n \end{bmatrix}$ be another vector such that $\mathbf{v} \oplus \mathbf{u} = \mathbf{u} \oplus \mathbf{v} = \mathbf{0}$. Then

$$\begin{aligned}
\mathbf{w} &= \mathbf{w} \oplus \mathbf{0} && \text{(since } \mathbf{0} \text{ is the identity)} \\
&= \mathbf{w} \oplus (\mathbf{v} \oplus \mathbf{u}) && \text{(since } \mathbf{0} = \mathbf{v} \oplus \mathbf{u}) \\
&= (\mathbf{w} \oplus \mathbf{v}) \oplus \mathbf{u} && \text{(by the associativity of } \oplus \text{ property)} \\
&= \mathbf{0} \oplus \mathbf{u} && \text{(since } \mathbf{w} \oplus \mathbf{v} = \mathbf{0}) \\
&= \mathbf{u}. && \text{(since } \mathbf{0} \text{ is the identity)}
\end{aligned}$$

Hence there is a unique negative for every vector in $R^n$.

Property h. For $\mathbf{v} = \begin{bmatrix} v_1 & v_2 & \cdots & v_n \end{bmatrix}$ we have

$$1 \odot \mathbf{v} = \begin{bmatrix} 1v_2 & 1v_2 & \cdots & 1v_n \end{bmatrix} = \begin{bmatrix} v_1 & v_2 & \cdots & v_n \end{bmatrix} = \mathbf{v}.$$

It follows that $R^n$ is a real vector space since all ten properties that define a vector space are satisfied. ■

In fact, Example 1 has shown that we have a **family of real vector spaces** $(R^n, \oplus, \odot)$. Thus, in particular $R^1$ (the real numbers), $R^2$ (2-vectors), and $R^3$ (3-vectors) are real vector spaces. In a similar fashion, we can verify that $(C^n, \oplus, \odot)$ is a **family of complex vector spaces**. (The operations $\oplus$ and $\odot$ are the standard operations of matrix addition and scalar multiplication.)

**EXAMPLE 2**    Let $V$ be $R_{m \times n}$, the set of $m \times n$ matrices with real entries; let $\oplus$ be ordinary matrix addition; and let $\odot$ be ordinary scalar multiplication. To verify that $V$ is a vector space with these operations we must explicitly show that properties 1, 2, and a through h in the definition of a vector space are valid. The closures have been verified in Exercise 37 in Section 1.2. Properties a, b, e, f, and g are developed in Section 1.2 and Exercises 52 and 53 in that section. To complete the verification that $V$ is a vector space with these operations, we need to show properties c, d, and h. The identical proofs that were used in Example 1 are applicable here with the $n$-vectors replaced by $m \times n$ matrices with real entries and the identity is $\mathbf{0}$, the $m \times n$ matrix of all zeros . Again we have a family of real vector spaces, one for each value of $m$ and $n$. (Of course, Example 1 is a particular case of $m \times n$ matrices. We chose to do it first because the verification of the properties is simpler to write out than those for general $m \times n$ matrices.) If $V$ is $C_{m \times n}$, the set of $m \times n$ complex matrices, then $(V, \oplus, \odot)$ is a family of complex vector spaces. ■

The vector spaces $R^n$, $C^n$, $R_{m \times n}$, and $C_{m \times n}$ are the standard ones that we should expect to have available based upon our study of linear algebra in Chapters 1 through 4. There are many other vector spaces that are useful in mathematics, and we introduce several of these in Examples 3 through 5.

**EXAMPLE 3**    Let $P_n$ be the set of all polynomials of degree $n$ or less with real coefficients. (A polynomial is said to be of degree $k$ if its highest power term with a nonzero coefficient is $x^k$.) Define vector addition $\oplus$ to be the sum of polynomials obtained by adding coefficients of like power terms and scalar multiplication $\odot$ to be the multiplication of a polynomial by a constant obtained by multiplying the coefficient of each term by the scalar. If $\mathbf{p} = a_n x^n + a_{n-1} x^{n-1} + \cdots + a_1 x + a_0$ and $\mathbf{q} = b_n x^n + b_{n-1} x^{n-1} + \cdots + b_1 x + b_0$ are vectors in $V$, then

$$\mathbf{p} \oplus \mathbf{q} = (a_n + b_n)x^n + (a_{n-1} + b_{n-1})x^{n-1} + \cdots + (a_1 + b_1)x + (a_0 + b_0)$$

and for any real scalar $k$

$$k \odot \mathbf{p} = ka_n x^n + ka_{n-1} x^{n-1} + \cdots + ka_1 x + ka_0.$$

Since each of the coefficients $(a_n + b_n)$, $(a_{n-1} + b_{n-1})$, ..., $(a_1 + b_1)$, and $(a_0 + b_0)$ is a sum of real numbers, the result is a real number. Thus $\mathbf{p} \oplus \mathbf{q}$ is a polynomial of degree $n$ or less, hence a vector in $P_n$. Therefore, $P_n$ is closed under vector addition; that is, Property 1 in the definition of a vector space is satisfied. Similarly, $k \odot \mathbf{p}$ is a polynomial of degree $n$ or less, hence a vector in $P_n$. So $P_n$ is closed under scalar multiplication. The remaining eight properties of the definition of a vector space can be demonstrated to be true. [See Exercise 11. In Exercise 12 you are asked to use the ideas developed in Section 1.2 for associating a polynomials with an $(n + 1)$-vector to obtain the verification in an indirect manner.] Thus $P_n$ is a vector space. We note that the zero element is $\mathbf{z} = 0x^n + 0x^{n-1} + \cdots + 0x + 0 = 0$; that is, the function that is zero at each value of $x$. We call this the **zero polynomial**. (We agree to say that the zero polynomial has no degree.) ■

Another vector space related to $P_n$ is $P$, the **vector space of all polynomials**. The verification of properties 1, 2, and a through h is similar to those in Example 3 and Exercise 11. We will refer to the vector space of all polynomials later in this chapter and in succeeding chapters.

EXAMPLE 4   Let $V$ be the set of all real valued functions that are defined on the interval $[a, b]$. Then $\mathbf{f}$ is a member of $V$ provided that, for each value $t$ in the interval $[a, b]$, $\mathbf{f}$ evaluated at $t$, $\mathbf{f}(t)$, is defined and is a real number. If $\mathbf{f}$ and $\mathbf{g}$ are in $V$, we define vector addition by

$$(\mathbf{f} \oplus \mathbf{g})(t) = \mathbf{f}(t) + \mathbf{g}(t).$$

Note that the $+$ sign on the right side means addition of real numbers. Since $\mathbf{f}$ and $\mathbf{g}$ are in $V$, $\mathbf{f}(t)$ and $\mathbf{g}(t)$ are real numbers and the sum of real numbers is another real number. Thus $V$ is closed under $\oplus$. If $\mathbf{f}$ is in $V$ and $k$ is any real scalar, we define scalar multiplication by

$$(k \odot \mathbf{f})(t) = k\mathbf{f}(t).$$

Note that the right side is a product of real numbers and hence another real number. Thus $V$ is closed under $\odot$. Properties a through h of the definition of a vector space can be verified (see Exercise 13) by relying on the corresponding properties of real numbers. It follows that the set of all real valued functions on $[a, b]$ is a vector space. Similarly, the set of all real valued functions on $(-\infty, \infty)$ is a vector space. ■

EXAMPLE 5   (Requires Calculus)   Let $V$ be $C(a, b)$, the set of all real valued continuous functions that are defined on the interval $(a, b)$. Then $\mathbf{f}$ is a member of $V$ provided that it is continuous on $(a, b)$ [that is, intuitively, we can draw its graph over $(a, b)$ without removing our pencil from the paper] and for each value $t$ in the interval $(a, b)$, $\mathbf{f}$ evaluated at $t$, $\mathbf{f}(t)$, is defined and is a real number. If $\mathbf{f}$ and $\mathbf{g}$ are in $V$, we define vector addition by

$$(\mathbf{f} \oplus \mathbf{g})(t) = \mathbf{f}(t) + \mathbf{g}(t).$$

Note that the + sign on the right side means addition of real numbers. Since **f** and **g** are in $V$, $\mathbf{f}(t)$ and $\mathbf{g}(t)$ are real numbers and the sum of real numbers is another real number. From calculus we recall that the sum of functions that are continuous at a point produces another function continuous at the same point. Thus $V$ is closed under $\oplus$. Using the fact from calculus that a constant multiple of a continuous function yields a continuous function and the same definition for $\odot$ as in Example 4, we can say that $V$ is closed under $\odot$. It can be verified that properties a through h in the definition of a vector space are satisfied, hence $C(a, b)$ is a vector space. (As in $P_n$, the zero vector is the function that is zero at each point.) In a similar manner, it follows that $C[a, b]$ and $C(-\infty, \infty)$ are vector spaces. ■

If $(V, \oplus, \odot)$ is a vector space, then there are certain properties that hold in addition to the ten specified in the definition of a vector space. These properties are valid regardless of the type of vectors in $V$ or the definition of the operations of $\oplus$ and $\odot$. Although the properties in the following list seem reasonable and natural, they require verification, especially given the variety of vector spaces that appear in Examples 1 through 5.

**Vector Space Properties**

Let $(V, \oplus, \odot)$ be a real (or complex) vector space, with vectors **u**, **v**, and **w** in $V$, scalars $c$ and $k$, and identity vector **z**.

i. $0 \odot \mathbf{u} = \mathbf{z}$   (zero times any vector is the identity vector)
ii. $k \odot \mathbf{z} = \mathbf{z}$   (any scalar multiple of the identity vector is the identity vector)
iii. If $k \odot \mathbf{u} = \mathbf{z}$, then either $k = 0$ or $\mathbf{u} = \mathbf{z}$.
iv. $(-1) \odot \mathbf{u} = -\mathbf{u}$   (negative 1 times a vector is the negative of the vector)
v. $-(-\mathbf{u}) = \mathbf{u}$   (the negative of vector $-\mathbf{u}$ is **u**)
vi. If $\mathbf{u} \oplus \mathbf{v} = \mathbf{u} \oplus \mathbf{w}$, then $\mathbf{v} = \mathbf{w}$.
vii. If $\mathbf{u} \neq \mathbf{z}$ and $k \odot \mathbf{u} = c \odot \mathbf{u}$, then $k = c$.

In Examples 6 and 7 we show the verification of Properties i and v, respectively, and leave the remaining properties as exercises. We provide a reason for each step of the process, as you should do when working out exercises of this type.

EXAMPLE 6   Verification of Property i, zero times any vector is the identity vector: $0 \odot \mathbf{u} = \mathbf{z}$.

$0 \odot \mathbf{u} = (0 + 0) \odot \mathbf{u}$   (by properties of real or complex numbers)
$\quad\quad = (0 \odot \mathbf{u}) \oplus (0 \odot \mathbf{u})$   (by f in the definition of a vector space)

$(0 \odot \mathbf{u})$ is a vector in $V$ so it has a negative, $-(0 \odot \mathbf{u})$. Add $-(0 \odot \mathbf{u})$ to both sides and we have

$$(0 \odot \mathbf{u}) \oplus (-(0 \odot \mathbf{u})) = (0 \odot \mathbf{u}) \oplus (0 \odot \mathbf{u}) \oplus (-(0 \odot \mathbf{u})).$$

Now we use the fact that a vector added to its negative gives the identity vector to rewrite this expression as
$$\mathbf{z} = (0 \odot \mathbf{u}) \oplus \mathbf{z}$$

which gives $\mathbf{z} = (0 \odot \mathbf{u})$. (Explain why this final expression follows from the preceding expression.) ■

**EXAMPLE 7** Verification of Property v, the negative of vector $-\mathbf{u}$ is $\mathbf{u}$:

$$-(-\mathbf{u}) = \mathbf{u}.$$

One way to proceed is as follows.

$$
\begin{aligned}
-(-\mathbf{u}) &= -((-1) \odot \mathbf{u}) && \text{(by iv)}\\
&= (-1) \odot ((-1) \odot \mathbf{u}) && \text{(by iv again)}\\
&= (-1)(-1) \odot \mathbf{u} && \text{(by g in the definition of a vector space)}\\
&= 1 \odot \mathbf{u} && \text{(by properties of real numbers)}\\
&= \mathbf{u}. && \text{(by h in the definition of a vector space)} \qquad ■
\end{aligned}
$$

In Examples 1 through 5 we used sets of vectors familiar from Chapters 1 through 4 or from calculus together with the "usual" operations on the vectors in these sets. If we change the set $V$ or the definition of either operation $\oplus$ or $\odot$, then we must carefully check the ten conditions to have $(V, \oplus, \odot)$ qualify to be called a vector space. We illustrate this next.

**EXAMPLE 8** Let $V$ be the set of all $2 \times 1$ real matrices with both entries positive. For vectors $\mathbf{v} = \begin{bmatrix} v_1 \\ v_2 \end{bmatrix}$ and $\mathbf{w} = \begin{bmatrix} w_1 \\ w_2 \end{bmatrix}$ in $V$, we define vector addition as

$$\mathbf{v} \oplus \mathbf{w} = \begin{bmatrix} v_1 w_1 \\ v_2 w_2 \end{bmatrix}$$

and scalar multiplication by a real scalar $k$ is defined as

$$k \odot \mathbf{v} = \begin{bmatrix} v_1^k \\ v_2^k \end{bmatrix}.$$

We claim that $(V, \oplus, \odot)$ is a real vector space. We verify properties 1, c, and e for a vector space and leave the remaining properties as exercises.

Verification of 1: Since $\mathbf{v}$ and $\mathbf{w}$ both have positive entries, and the product of positive numbers is positive, we have that

$$\mathbf{v} \oplus \mathbf{w} = \begin{bmatrix} v_1 w_1 \\ v_2 w_2 \end{bmatrix}$$

is a $2 \times 1$ matrix with positive entries, so $V$ is closed under vector addition.

Verification of c: Let $\mathbf{z} = \begin{bmatrix} 1 \\ 1 \end{bmatrix}$. We can easily show that $\mathbf{v} \oplus \mathbf{z} = \mathbf{z} \oplus \mathbf{v} = \mathbf{v}$. However, we must show that this is the only vector for which this is true for all $\mathbf{v}$ in $V$. We proceed as follows: Suppose there is another vector $\mathbf{u}$ so that $\mathbf{v} \oplus \mathbf{u} = \mathbf{u} \oplus \mathbf{v} = \mathbf{v}$ for all $\mathbf{v}$ in $V$. Then since $\mathbf{z}$ is in $V$, we have $\mathbf{u} = \mathbf{z} \oplus \mathbf{u} = \mathbf{z}$. Hence we have a unique identity. (Note that we used the same reasoning to show uniqueness in Example 1.)

Verification of e: We compute both sides of the relation and show that they are the same. We have

$$k \odot (\mathbf{v} \oplus \mathbf{w}) = k \odot \begin{bmatrix} v_1 w_1 \\ v_2 w_2 \end{bmatrix} = \begin{bmatrix} (v_1 w_1)^k \\ (v_2 w_2)^k \end{bmatrix}$$

and

$$(k \odot \mathbf{v}) \oplus (k \odot \mathbf{w}) = \begin{bmatrix} (v_1)^k \\ (v_2)^k \end{bmatrix} \oplus \begin{bmatrix} (w_1)^k \\ (w_2)^k \end{bmatrix}$$

$$= \begin{bmatrix} (v_1)^k (w_1)^k \\ (v_2)^k (w_2)^k \end{bmatrix} = \begin{bmatrix} (v_1 w_1)^k \\ (v_2 w_2)^k \end{bmatrix}. \qquad ■$$

**Strategy:**    When dealing with a set $V$ and operations that are not familiar, checking the ten vector properties necessary for a vector space can be quite laborious. It is recommended that we *first check properties 1 and 2*, the closures, *and then property* c, the existence of an identity or zero vector. If these are satisfied, we then check the remaining seven properties. If any property fails to hold, then we can immediately stop and state that $(V, \oplus, \odot)$ is *not* a vector space.

EXAMPLE 9    Let V be the set of all $2 \times 1$ matrices with real entries. For $\mathbf{v}$ and $\mathbf{w}$ in $V$, define vector addition as

$$\mathbf{v} \oplus \mathbf{w} = \begin{bmatrix} v_1 \\ v_2 \end{bmatrix} \oplus \begin{bmatrix} w_1 \\ w_2 \end{bmatrix} = \begin{bmatrix} \max\{v_1, w_1\} \\ v_2 + w_2 \end{bmatrix}$$

and scalar multiplication as ordinary real scalar multiplication; that is, for a real scalar $k$

$$k \odot \mathbf{v} = k \odot \begin{bmatrix} v_1 \\ v_2 \end{bmatrix} = \begin{bmatrix} k v_1 \\ k v_2 \end{bmatrix}.$$

Since the result of both $\oplus$ and $\odot$ is another $2 \times 1$ matrix with real entries, properties 1 and 2 are valid. That is, $V$ is closed with respect to both $\oplus$ and $\odot$. To verify property c, the existence of a unique identity vector, we must find a vector $\mathbf{z} = \begin{bmatrix} z_1 \\ z_2 \end{bmatrix}$ such that

$$\mathbf{v} \oplus \mathbf{z} = \begin{bmatrix} \max\{v_1, z_1\} \\ v_2 + z_2 \end{bmatrix} = \begin{bmatrix} v_1 \\ v_2 \end{bmatrix} = \mathbf{v}.$$

A natural candidate for $z_2$ is zero. However, $z_1$ must be chosen so that $\max\{v_1, z_1\} = v_1$ for all real numbers. If $z_1 = 0$, then this fails to hold when $v_1 < 0$. If $z_1 > 0$ or $z_1 < 0$, then $\max\{v_1, z_1\} \neq v_1$ for $v_1 < z_1$. Thus no matter what value is specified for $z_1$, there exists a vector $\mathbf{v}$ in $V$ such that $\mathbf{v} \oplus \mathbf{z} \neq \mathbf{v}$. Hence there is no identity vector and so $(V, \oplus, \odot)$ is not a vector space. ■

We conclude this section with two examples that provide further information about the nature of vector spaces.

EXAMPLE 10    Let $(V, \oplus, \odot)$ be a real (or complex) vector space. Suppose that $V$ consists of a single vector $\mathbf{u}$. Then by the closure properties we have $\mathbf{u} \oplus \mathbf{u} = \mathbf{u}$ and $k \odot \mathbf{u} = \mathbf{u}$ for any scalar $k$. It follows that $\mathbf{u}$ must be the identity vector. ■

EXAMPLE 11    Let $(V, \oplus, \odot)$ be a real (or complex) vector space. Suppose that $V$ contains a vector $\mathbf{u}$ that is not the identity vector $\mathbf{z}$. Then $\mathbf{u} \oplus \mathbf{u}$ is in $V$ but is not $\mathbf{z}$. Similarly, successive sums of $\mathbf{u}$ with itself are in $V$, but are not $\mathbf{z}$. In addition, for $k \neq 0$, $k \odot \mathbf{u}$ is in $V$ but the result is not $\mathbf{z}$. It follows that $V$ must contain infinitely many distinct vectors in addition to $\mathbf{z}$. ■

Examples 10 and 11 tell us that the only vector space that has finitely many vectors is the vector space that consists of a single identity vector. This vector space is called the **zero vector space**.

## EXERCISES 5.1

*Identification of properties common to sets of objects is one way to begin the investigation of an abstract model or structure. In Exercises 1–5, make a list of features common to the set described.*

1. All triangles that are similar to one another.

2. All triangles that are congruent to one another.

3. All $2 \times 2$ nonsingular matrices.

4. All $2 \times 2$ singular matrices.

5. Lines in the plane that go through the origin.

*An important aspect of a vector space is the closure properties. As we indicated, these should be checked before trying to verify the other eight properties. In Exercises 6–10, determine whether the set $V$ with the specified operations $\oplus$ and $\odot$ satisfies the closure properties.*

6. $V$ is the set of all ordered pairs of real numbers $(x, y)$ with both $x$ and $y$ positive. Vector addition $\oplus$ is defined as $(x, y) \oplus (r, s) = (x+r, y+s)$ and scalar multiplication with real numbers is defined as $k \odot (x, y) = (kx, ky)$.

7. $V$ is the set of all ordered triples of real numbers of the form $(0, x, y)$. Vector addition $\oplus$ is defined as $(0, x, y) \oplus (0, r, s) = (0, x+r, y+s)$ and scalar multiplication with real numbers is defined as $k \odot (0, x, y) = (0, 0, ky)$.

8. $V$ is the set of all $2 \times 2$ nonsingular matrices with real entries. Vector addition $\oplus$ is the standard addition of matrices and scalar multiplication with real numbers is the standard scalar multiplication of matrices.

9. $V$ is the set of all $2 \times 2$ matrices $\mathbf{A}$ with real entries such that $\mathbf{Ax} = \mathbf{0}$ where $\mathbf{x} = \begin{bmatrix} 1 \\ 2 \end{bmatrix}$. Vector addition $\oplus$ is the standard addition of matrices, and scalar multiplication with real numbers is the standard scalar multiplication of matrices.

10. $V$ is the set of all $2 \times 2$ matrices $\mathbf{A}$ with real entries such that $\mathbf{x} = \begin{bmatrix} 1 \\ 5 \end{bmatrix}$ is an eigenvector of $\mathbf{A}$. Vector addition $\oplus$ is the standard addition of matrices, and scalar multiplication with real numbers is the standard scalar multiplication of matrices.

11. For $P_n$ defined in Example 3, verify properties a through h in the definition of a vector space.

12. In Section 1.2 we associated an $(n + 1)$-vector with each member of $P_n$. Explain how we can infer from this association that $P_n$ is a vector space.

13. For the set of all real valued functions defined on $[a, b]$ discussed in Example 4, verify properties a through h in the definition of a vector space.

14. For $C(a, b)$ defined in Example 5, verify properties a through h in the definition of a vector space.

*Let $(V, \oplus, \odot)$ be a real (or complex) vector space, with vectors $\mathbf{u}$, $\mathbf{v}$, and $\mathbf{w}$ in $V$, scalars $c$ and $k$, and zero vector $\mathbf{z}$. In Exercises 15–19, verify each of the given vector space properties.*

15. $k \odot \mathbf{z} = \mathbf{z}$. [*Hint*: $k \odot \mathbf{z} = k \odot (\mathbf{z} \oplus \mathbf{z})$; expand, add $-(k \odot \mathbf{z})$ to each side and simplify.]

16. If $k \odot \mathbf{u} = \mathbf{z}$, then either $k = 0$ or $\mathbf{u} = \mathbf{z}$. (*Hint*: Break into cases $k = 0$ and $k \neq 0$; in the latter case multiply by scalar $1/k$.)

17. $(-1) \odot \mathbf{u} = -\mathbf{u}$. [*Hint*: Use the fact that there is a unique negative; add $\mathbf{u}$ to $((-1) \odot \mathbf{u})$; use properties to rearrange and show that the result is $\mathbf{z}$.]

18. If $\mathbf{u} \oplus \mathbf{v} = \mathbf{u} \oplus \mathbf{w}$, then $\mathbf{v} = \mathbf{w}$. (*Hint*: Add $-\mathbf{u}$ to both sides and simplify.)

19. If $\mathbf{u} \neq \mathbf{z}$ and $k \odot \mathbf{u} = c \odot \mathbf{u}$, then $k = c$. [*Hint*: Add $-(c \odot \mathbf{u})$ to both sides, simplify, and use Exercise 16.]

20. Let $(V, \oplus, \odot)$ be a real (or complex) vector space with identity vector $\mathbf{z}$. Show that the negative of $\mathbf{z}$ is $\mathbf{z}$.

21. For the set $V$ and operations defined in Example 8, verify properties a, b, d, 2, f, g, and h in the definition of a vector space.

22. Let $V$ be the set of all real numbers with vector addition defined as $\mathbf{u} \oplus \mathbf{v} = u - v$ and scalar multiplication by real numbers defined as $k \odot \mathbf{u} = k\mathbf{u}$. Determine whether or not $(V, \oplus, \odot)$ is a real vector space.

23. Let $V$ be the set of all $2 \times 2$ matrices with real entries. For $\mathbf{A}$ and $\mathbf{B}$ in $V$, define vector addition by

$$\mathbf{A} \oplus \mathbf{B} = \begin{bmatrix} a_{11} & a_{12} \\ a_{21} & a_{22} \end{bmatrix} \oplus \begin{bmatrix} b_{11} & b_{12} \\ b_{21} & b_{22} \end{bmatrix}$$

$$= \begin{bmatrix} a_{11}b_{11} & a_{12}b_{12} \\ a_{21}b_{21} & a_{22}b_{22} \end{bmatrix}$$

and define scalar multiplication by real numbers by

$$k \odot \mathbf{A} = k \odot \begin{bmatrix} a_{11} & a_{12} \\ a_{21} & a_{22} \end{bmatrix} = \begin{bmatrix} k + a_{11} & k + a_{12} \\ k + a_{21} & k + a_{22} \end{bmatrix}.$$

(a) Describe the operation $\oplus$ in words.

(b) Describe the operation $\odot$ in words.

(c) Determine whether or not $(V, \oplus, \odot)$ is a real vector space. If it is not a vector space, list all the properties that fail to hold.

## True/False Review Questions

*Determine whether each of the following statements is true or false.*

1. A vector space can have more than one zero vector.

2. If $\mathbf{u}$, $\mathbf{v}$, and $\mathbf{w}$ are vectors in a vector space $V$, then $5\mathbf{u} - 4\mathbf{v} + 7\mathbf{w}$ is also a vector in $V$.

3. $\dfrac{1}{x}$ belongs to vector space $P_2$.

4. Any scalar multiple of the (additive) identity of a vector space gives the (additive) identity back.

5. In a vector space scalar multiplication distributes over vector addition.

6. If $\mathbf{w}$ is in a vector space $V$, then $((-1) \odot (-\mathbf{w})) = \mathbf{w}$.

7. In a vector space, the scalar zero times any vector produces the zero vector.

8. To check to see if a set is a vector space, we need only show closure using $\oplus$ and closure using $\odot$.

9. In $R^3$ any plane is a vector space.

10. A vector space can consist entirely of just 10 vectors.

## Terminology

| | |
|---|---|
| Vector space | Closed with respect to $\oplus$ and $\odot$ |
| Vectors | Identity or zero vector |
| $P_n$; $P$ | $C(a, b)$; $C[a, b]$; $C(-\infty, \infty)$ |
| The zero vector space | |

This section dealt with a unifying structure called a vector space. We use the term *unifying* since any set with a pair of operations that satisfies the ten axioms of a vector space automatically satisfies a list of other properties. In many cases these properties seem to be natural extensions of our familiar operations from Chapters 1 through 4. However, even if the operations are not something we have seen before, the mere fact that we are working in a vector space guarantees that certain properties hold and we need not verify them. Hence we are able to determine information about a vector space without specifying its members or the explicit operations involved. Of course, the matrices and $n$-vectors we dealt with in Chapters 1 through 4 are particular cases of vector spaces. In succeeding sections we will see how matrices play an important role in developing further information about the nature and construction of abstract vector spaces.

Answer the following to review your understanding of the material in this section.

- What does it mean to say that a vector space $V$ is closed with respect to the operation $\oplus$?
- What does it mean to say that a vector space $V$ is closed with respect to the operation $\odot$?
- What distinguishes a real vector space from a complex vector space?
- What do we mean when we say that a zero vector in a vector space need not have zeros involved with it?

- Describe the vectors in $P_4$.
- Describe the vectors in $C(a, b)$.
- Describe the vectors in $P$.
- Every vector **w** in a vector space $V$ has a negative. What is the result of vector addition of **w** and its negative?
- When trying to determine whether or not $(V, \oplus, \odot)$ is a vector space, what strategy did we recommend for checking the ten axioms?

Given that we are within a vector space, complete each of the following.

- Any scalar times the zero vector yields _____.
- If **u** and **w** are both negatives of vector **v**, then _____.
- The zero vector added to a vector **v** yields _____.
- Negative one times any vector **v** yields _____.
- If $\mathbf{u} \oplus \mathbf{v} = \mathbf{u} \oplus \mathbf{w}$, then we can prove that $\mathbf{v} = \mathbf{w}$ by _____.
- Zero times any vector **v** yields _____.

## 5.2 ■ SUBSPACES AND SPAN

In much of our work in Chapters 1 through 4 we were concerned not with an entire vector space like $R^n$, $C^n$, $R_{m \times n}$, or $C_{m \times n}$, but just a portion of it. In a number of important instances we needed to form linear combinations and be sure that these new vectors (or matrices) retained a particular property. Here we formally develop corresponding ideas for abstract vector spaces and investigate their construction.

Let $(V, \oplus, \otimes)$ be a vector space with $W$ a subset of the vectors in $V$. If $W$, with the same operations $\oplus$ and $\otimes$, is a vector space in its own right, then we call $(W, \oplus, \otimes)$ a **subspace** of vector space $(V, \oplus, \otimes)$. (We will often just use the shorter phrase "$W$ is a subspace of $V$.") Since all the vectors in $W$ are also in $V$, properties a, b, e, f, g, and h of the definition of a vector space given in Section 5.1 are valid. (Informally, we say these properties are inherited.) However, the closure properties 1 and 2, the identity (property c), and the negative (property d) are different. Namely, if **r** and **s** are vectors in $W$, then both $\mathbf{r} \oplus \mathbf{s}$ and $k \otimes \mathbf{r}$, for any scalar $k$, must be in $W$, not just in $V$. We must also show that the identity vector **z** is in $W$ and the negative of each vector in $W$ must be in $W$. A careful inspection of these requirements leads to the following result. This result provides a shortcut for determining whether a subset $W$ of vector space $V$ is a subspace.

### Subspace Criterion

Let $(V, \oplus, \otimes)$ be a real (or complex) vector space. A subset $W$ of $V$ is a subspace if and only if

  i. $W$ is closed with respect to vector addition; that is, for **r** and **s** in $W$, $\mathbf{r} \oplus \mathbf{s}$ is in $W$;

and

  ii. $W$ is closed with respect to scalar multiplication; that is, for **r** in $W$ and $k$ any scalar, $k \otimes \mathbf{r}$ is in $W$.

The verification of the subspace criterion requires that we show two properties:

1. If $W$ is a subspace of $V$, then i and ii are valid. (See Exercise 1.)
2. If i and ii are valid, then $W$ is a subspace of $V$. (See Exercise 2.)

In summary, the subspace criterion says that $W$ is a subspace of vector space $V$ provided that $W$ is closed for both vector addition and scalar multiplication.

Every vector space $V$ has at least two subspaces $W$. The whole space $W = V$ and $W = \{\mathbf{z}\}$, where $\mathbf{z}$ is the identity vector. (See Example 10 in Section 5.1.) These subspaces are called the **trivial subspaces**.

In the remainder of this section and throughout the rest of this book, we adopt the following conventions:

- Vector addition $\mathbf{u} \oplus \mathbf{v}$ will be denoted $\mathbf{u} + \mathbf{v}$.
- Scalar multiplication $k \odot \mathbf{u}$ will be denoted $k\mathbf{u}$.
- The context of the example or discussion will tell us the particular definitions of vector addition and scalar multiplication.
- The identity vector $\mathbf{z}$ will be called the **zero vector**.[1] In $R^n$, $C^n$, $R_{m \times n}$, and $C_{m \times n}$ we will denote the zero vector as $\mathbf{0}$.
- The subspace $W = \{\mathbf{z}\}$ will be called the **zero subspace**.

In Chapters 1 through 4 we developed certain subspaces of $R^n$, $C^n$, $R_{m \times n}$, and $C_{m \times n}$, which played important roles in the information that we were able to obtain from a matrix. In Example 1 we review these important subspaces.

EXAMPLE 1    In each of the following the operation of vector addition is the standard addition of matrices, and scalar multiplication is the scalar multiple that we defined in Section 1.2.

(a) Let $\mathbf{A}$ be a real $m \times n$ matrix. Then the row space of $\mathbf{A}$, $\mathbf{row}(\mathbf{A})$, is a subspace of $(R^n, +, \cdot)$. (See Section 2.3.)

(b) Let $\mathbf{A}$ be a real $m \times n$ matrix. Then the column space of $\mathbf{A}$, $\mathbf{col}(\mathbf{A})$, is a subspace of $(R^m, +, \cdot)$. (See Section 2.3.)

(c) Let $\mathbf{A}$ be a real $m \times n$ matrix. Then the null space of $\mathbf{A}$, $\mathbf{ns}(\mathbf{A})$, is a subspace of $(R^n, +, \cdot)$. (See Section 2.3.)

(d) Let $\mathbf{A}$ be a complex $n \times n$ matrix, $\lambda$ an eigenvalue of $\mathbf{A}$, and $W$ the set of all eigenvectors corresponding to $\lambda$ together with the zero vector. Then $W$ is a subspace of $(C^n, +, \cdot)$, called the **eigenspace** associated with eigenvalue $\lambda$. (See Exercise 9 in Section 4.1.)

Parts (a), (b), and (c) remain valid when *real* is replaced by *complex*, and part (d) already applies to real square matrices.    ■

EXAMPLE 2    In Section 1.4 we showed that every line in the plane through the origin is of the form $k\mathbf{b}$ where $\mathbf{b}$ is a nonzero vector in $R^2$. Let $\mathbf{b}$ be a fixed nonzero vector in $R^2$ and let $W_\mathbf{b}$ be the set of all real scalar multiples of $\mathbf{b}$. (Geometrically $W_\mathbf{b}$ is a line in the direction of vector $\mathbf{b}$ through the origin.) We claim that $W_\mathbf{b}$ is a subspace of $R^2$. To verify this claim we use the Subspace Criterion.

[1] The identity vector or zero vector need not contain any zero entries or be zero. Its content depends upon the definitions of vector addition and scalar multiplication. See Example 8 in Section 5.1.

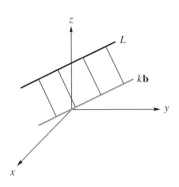

FIGURE 1

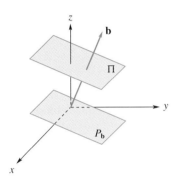

FIGURE 2

Let $\mathbf{r}$ and $\mathbf{s}$ be in $W_{\mathbf{b}}$ and let $k$ be any real scalar. Then $\mathbf{r} = c_1\mathbf{b}$ and $\mathbf{s} = c_2\mathbf{b}$ for some real scalars $c_1$ and $c_2$. We have

$$\mathbf{r} + \mathbf{s} = c_1\mathbf{b} + c_2\mathbf{b} = (c_1 + c_2)\mathbf{b} \quad \text{and} \quad k\mathbf{r} = k(c_1\mathbf{b}) = (kc_1)\mathbf{b}.$$

The result of both vector addition and scalar multiplication is in $W_{\mathbf{b}}$, so $W_{\mathbf{b}}$ is closed with respect to both operations. Thus $W_{\mathbf{b}}$ is a subspace of $R^2$.

It follows from Section 1.4 that any line $L$ not through the origin is a translation of a subspace of $R^2$.    ■

EXAMPLE 3    Following the ideas in Section 1.4 we can say that a line through the origin in 3-space has the form $k\mathbf{b}$ where $\mathbf{b}$ is a nonzero vector in $R^3$. Using the same approach as in Example 2, we can show that $W_{\mathbf{b}}$, the set of all real scalar multiples of vector $\mathbf{b}$, is a subspace of $R^3$. In addition, any line $L$ in $R^3$ not through the origin is a translation of a subspace of $R^3$. (See Figure 1.)

A plane in 3-space is characterized by its normal vector $\mathbf{b}$ (a nonzero vector in $R^3$) and a point $p = (p_1, p_2, p_3)$ lying in the plane. That is, the plane consists of all line segments through point $p$ which are orthogonal to vector $\mathbf{b}$. Let $P_{\mathbf{b}}$ be the set of all vectors in $R^3$ that are orthogonal to $\mathbf{b}$. We claim that $P_{\mathbf{b}}$ is a plane containing the origin and is a subspace of $R^3$. Since a vector is a directed line segment connecting the origin (the initial point) to the terminal point of the vector, all the members of $P_{\mathbf{b}}$ contain the origin. Since every member of $P_{\mathbf{b}}$ is orthogonal to $\mathbf{b}$, $P_{\mathbf{b}}$ satisfies the criteria necessary to be a plane. To show it is a subspace, we use the Subspace Criterion.

Let $\mathbf{r}$ and $\mathbf{s}$ be in $P_{\mathbf{b}}$ and let $k$ be any real scalar. Then $\mathbf{r} \cdot \mathbf{b} = 0$ and $\mathbf{s} \cdot \mathbf{b} = 0$. We have

$$(\mathbf{r} + \mathbf{s}) \cdot \mathbf{b} = \mathbf{r} \cdot \mathbf{b} + \mathbf{s} \cdot \mathbf{b} = 0 + 0 = 0 \quad \text{and} \quad (k\mathbf{r} \cdot \mathbf{b}) = k(\mathbf{r} \cdot \mathbf{b}) = k0 = 0.$$

The result of both vector addition and scalar multiplication is in $P_{\mathbf{b}}$; hence $P_{\mathbf{b}}$ is closed with respect to both operations. Thus $P_{\mathbf{b}}$ is a subspace of $R^3$.

In analogy with lines, any plane $\Pi$ in $R^3$ not through the origin is a translation of some subspace $P_{\mathbf{b}}$. (See Figure 2.)    ■

EXAMPLE 4    Let $W$ be the set of all nonsingular matrices in vector space $(R_{n \times n}, +, \cdot)$. Is $W$ a subspace of $R_{n \times n}$? We use the Subspace Criterion and check the closures of $W$. If $\mathbf{A}$ is in $W$, then $0\mathbf{A} = \mathbf{0}$, the zero matrix. But the zero matrix is singular. (Why?) Hence $W$ is not closed under scalar multiplication and thus is not a subspace.    ■

EXAMPLE 5    Let $P_3$ be the vector space of polynomials of degree 3 or less as defined in Example 3 in Section 5.1. Let $W$ be the set of all polynomials of degree 2 or less. Using the development in Example 3 of Section 5.1, $W = P_2$, so it is a vector space in its own right, hence a subspace of $P_3$. In an analogous fashion $U$, the set of all polynomials of degree 1 or less, is really $P_1$, and hence a subspace of both $P_2$ and $P_3$. In general, if $k$ and $n$ are nonnegative integers with $k \le n$, $P_k$ is a subspace of $P_n$.    ■

The nesting of the subspaces of polynomials as described in Example 5 can be used to make an observation about least squares approximations, which we introduced in Section 2.6. We showed that the procedure for obtaining the "least squares line" found the polynomial in $P_1$ that minimized the sum of the squares of the vertical deviations. In Exercise 8 in Section 2.6 we introduced the least squares quadratic, which can be shown to be the polynomial in $P_2$ that minimizes the sum of the squares of the vertical deviations. Since $P_1$ is a subspace of $P_2$, the search for the least squares quadratic considers all the possible polynomials in $P_1$ as potential solutions. Hence the actual sum of the deviations for the least squares quadratic must be less than or equal to the sum of the deviations for the least squares line.

Because of the closure properties of vector spaces and subspaces, we know that sums and scalar multiples of vectors "remain in the space." The other properties of vector addition and scalar multiplication permit us to combine addition and scalar multiplication of vectors in a variety of ways. We summarize these possible manipulations by formally defining a linear combination, which we used extensively in Chapters 1 through 4.

---

**Definition**   Let $\mathbf{v}_1, \mathbf{v}_2, \ldots, \mathbf{v}_k$ be vectors in a vector space $V$ and $c_1, c_2, \ldots, c_k$ be scalars. The expression

$$c_1\mathbf{v}_1 + c_2\mathbf{v}_2 + \cdots + c_k\mathbf{v}_k = \sum_{j=1}^{k} c_j\mathbf{v}_j$$

is a **linear combination** of vectors $\mathbf{v}_1, \mathbf{v}_2, \ldots, \mathbf{v}_k$.

---

It follows by the closure properties *that a linear combination of vectors "remains in the space."* The Subspace Criterion can be restated as follows:

> $W$ is a subspace of a vector space $V$ provided that $W$ is closed with respect to linear combinations.

Forming linear combinations of vectors in a vector space $V$ produces more vectors in the space. Hence linear combinations provide a building or construction process. As with many such processes, several natural questions, which deal with capabilities and efficiencies, arise. Here we focus on capabilities, and we treat efficiencies in Section 5.3.

### Questions about Linear Combinations

1. Is there a set $S$ of vectors in $V$ such that every vector in $V$ can be written as a linear combination of the members of set $S$?

   If there is such a set $S$, then $S$ is called a **set of generators** or a **spanning set** for $V$. (See Section 1.2.) We use the notation **span(S)** to denote the set of all linear combinations of the members of a set $S$. If $\text{span}(S) = V$, then $S$ is a set of generators (a spanning set) for $V$. We say $S$ **spans** $V$.

2. For a given set of vectors $S$, how do we determine if a particular vector is in $\text{span}(S)$?

In Chapters 1 through 4 we answered these questions for certain subspaces of $R^n$, $C^n$, $R_{m \times n}$, or $C_{m \times n}$. What is a bit surprising is that the techniques we used there apply in general to abstract vector spaces. The link between the particular cases involving row space, column space, and null space and the abstract case is the commonality of expressions, which are linear combinations. In fact, the definitions we gave and the results we developed in Chapters 1 through 4 for the concepts stated in the preceding questions are almost identical to the corresponding idea for abstract spaces. For completeness we provide formal definitions and statements of the corresponding results in the following discussion. At times we leave the step-by-step verifications as exercises, whose solution is very close to developments given in Chapters 1 through 4.

**The Structure of span(S)**

If $S$ is a set of vectors in a vector space $V$, then span($S$) is a subspace of $V$.

To verify this statement, see Exercise 53 in Section 2.3 and Exercise 20 at the end of this section.

## When is a set $S$ a spanning set for a vector space $V$?

We first look at the situation in $R^n$ (or $C^n$) and its subspaces. Let $S = \{\mathbf{v}_1, \mathbf{v}_2, \dots, \mathbf{v}_k\}$ be a set of $n$-vectors that you suspect is a spanning set for your space. Let $\mathbf{w}$ be *any* vector in your space. Then $S$ is a spanning set provided there exist scalars $c_1, c_2, \dots, c_k$ so that

$$\sum_{j-1}^{k} c_j \mathbf{v}_j = \mathbf{w}.$$

That is, every vector in the space can be constructed from the members of $S$. Using properties of $n$-vectors (for convenience, assume they are columns) and the meaning of equality of $n$-vectors, the preceding linear combination can be expressed as an associated linear system of equations $\mathbf{Ac} = \mathbf{w}$ where

$$\mathbf{A} = \begin{bmatrix} \mathbf{v}_1 & \mathbf{v}_2 & \cdots & \mathbf{v}_k \end{bmatrix} \quad \text{and} \quad \mathbf{c} = \begin{bmatrix} c_1 \\ c_2 \\ \vdots \\ c_k \end{bmatrix}.$$

It follows that $S$ is a spanning set if and only if the linear system $\mathbf{Ac} = \mathbf{w}$ is consistent for all vectors $\mathbf{w}$. In the language of Section 2.3, *S is a spanning set only if every vector $\mathbf{w}$ in the space is in the column space of matrix $\mathbf{A}$.*

Thus for $R^n$ or $C^n$ and their subspaces, the spanning question has been reduced to investigating the consistency of a linear system of equations. We illustrate this in Example 6.

EXAMPLE 6    Determine if

$$S = \{\mathbf{v}_1, \mathbf{v}_2, \mathbf{v}_3, \mathbf{v}_4\} = \left\{ \begin{bmatrix} 2 \\ 1 \\ 3 \end{bmatrix}, \begin{bmatrix} 1 \\ 2 \\ 1 \end{bmatrix}, \begin{bmatrix} 3 \\ 3 \\ 4 \end{bmatrix}, \begin{bmatrix} 3 \\ 0 \\ 5 \end{bmatrix} \right\}$$

spans $R^3$. Let $\mathbf{w} = \begin{bmatrix} r \\ s \\ t \end{bmatrix}$ be an arbitrary vector in $R^3$. Using the preceding discussion, the linear combination

$$\sum_{j=1}^{4} c_j \mathbf{v}_j = \mathbf{w}$$

corresponds to the linear system $\mathbf{Ac} = \mathbf{w}$, where

$$\mathbf{A} = \begin{bmatrix} 2 & 1 & 3 & 3 \\ 1 & 2 & 3 & 0 \\ 3 & 1 & 4 & 5 \end{bmatrix}.$$

To investigate the consistency of $\mathbf{Ac} = \mathbf{w}$, we proceed to compute $\mathbf{rref}\left(\begin{bmatrix} \mathbf{A} & \mathbf{w} \end{bmatrix}\right)$. We obtain

$$
\begin{bmatrix} 2 & 1 & 3 & 3 & r \\ 1 & 2 & 3 & 0 & s \\ 3 & 1 & 4 & 5 & t \end{bmatrix}_{R_2 \leftrightarrow R_1} \Rightarrow \begin{bmatrix} 1 & 2 & 3 & 0 & s \\ 2 & 1 & 3 & 3 & r \\ 3 & 1 & 4 & 5 & t \end{bmatrix}_{\substack{-2R_1 + R_2 \\ -3R_1 + R_3}}
$$

$$
\Rightarrow \begin{bmatrix} 1 & 2 & 3 & 0 & s \\ 0 & -3 & -3 & 3 & -2s + r \\ 0 & -5 & -5 & 5 & -3s + t \end{bmatrix}_{\substack{-\frac{1}{3}R_2 \\ -\frac{1}{5}R_3}}
$$

$$
\Rightarrow \begin{bmatrix} 1 & 2 & 3 & 0 & s \\ 0 & 1 & 1 & -1 & \frac{2}{3}s - \frac{1}{3}r \\ 0 & 1 & 1 & -1 & \frac{3}{5}s - \frac{1}{5}t \end{bmatrix}_{-R_2 + R_3}
$$

$$
\Rightarrow \begin{bmatrix} 1 & 2 & 3 & 0 & s \\ 0 & 1 & 1 & -1 & \frac{2}{3}s - \frac{1}{3}r \\ 0 & 0 & 0 & 0 & \frac{1}{3}r - \frac{1}{15}s - \frac{1}{5}t \end{bmatrix}.
$$

At this point we stop and make the following observation. Since $r$, $s$, and $t$ are arbitrary real numbers, the last row is of the form $\begin{bmatrix} 0 & 0 & 0 & 0 & * \end{bmatrix}$, where $*$ need not be zero. Hence for an arbitrary vector $\mathbf{w}$ the linear system $\mathbf{Ac} = \mathbf{w}$ may be inconsistent and so $\mathrm{span}(S) \neq R^3$. However, we can use the final augmented matrix to precisely describe the subspace $\mathrm{span}(S)$. Span $(S)$ consists of all vectors $\begin{bmatrix} r \\ s \\ t \end{bmatrix}$ such that $\frac{1}{3}r - \frac{1}{15}s - \frac{1}{5}t = 0$ since only in those cases is the linear system consistent. To get further information about $\mathrm{span}(S)$, multiply this equation by 15 to clear the fractions, obtaining $5r - s - 3t = 0$. Solve this equation for $s$ to avoid fractions, obtaining $s = 5r - 3t$. Then vectors in $\mathrm{span}(S)$ are of the form

$$
\begin{bmatrix} r \\ s \\ t \end{bmatrix} = \begin{bmatrix} r \\ 5r - 3t \\ t \end{bmatrix} = \begin{bmatrix} r \\ 5r \\ 0 \end{bmatrix} + \begin{bmatrix} 0 \\ -3t \\ t \end{bmatrix} = r \begin{bmatrix} 1 \\ 5 \\ 0 \end{bmatrix} + t \begin{bmatrix} 0 \\ -3 \\ 1 \end{bmatrix}.
$$

It follows that every vector in $\mathrm{span}(S)$ is a linear combination of $\begin{bmatrix} 1 \\ 5 \\ 0 \end{bmatrix}$ and $\begin{bmatrix} 0 \\ -3 \\ 1 \end{bmatrix}$.

Thus

$$
\left\{ \begin{bmatrix} 1 \\ 5 \\ 0 \end{bmatrix}, \begin{bmatrix} 0 \\ -3 \\ 1 \end{bmatrix} \right\}
$$

is a set of generators for $\mathrm{span}(S)$.  ■

**General Strategy to determine if span(S) = vector space V.**
For an abstract vector space or subspace $V$, the common properties of linear combinations imply that we can use the steps that worked in $R^n$ (or $C^n$) to determine if a set of vectors $S = \{\mathbf{v}_1, \mathbf{v}_2, \ldots, \mathbf{v}_k\}$ in $V$ spans the space $V$. Proceed as follows:

- Form a linear combination of the members of set $S$ with unknown coefficients $c_1, c_2, \ldots, c_k$.
- Set the linear combination equal to an arbitrary vector $\mathbf{w}$ of $V$.
- Use the particular meaning of the operations of addition and scalar multiplication in the space to construct an associated linear system of equations, where the unknowns are the scalars $c_1, c_2, \ldots, c_k$.
- Determine if the system is consistent for all vectors $\mathbf{w}$. If it is, then set $S$ spans $V$; otherwise span$(S)$ will be some subspace of $V$.

EXAMPLE 7   Let $V$ be the set of all $2 \times 1$ matrices with both entries positive real numbers. If

$$\mathbf{v} = \begin{bmatrix} v_1 \\ v_2 \end{bmatrix} \quad \text{and} \quad \mathbf{w} = \begin{bmatrix} w_1 \\ w_2 \end{bmatrix}$$

are in $V$, we define vector addition and scalar multiplication by a real scalar $k$, respectively by

$$\mathbf{v} + \mathbf{w} = \begin{bmatrix} v_1 w_1 \\ v_2 w_2 \end{bmatrix} \quad \text{and} \quad k\mathbf{v} = \begin{bmatrix} v_1^k \\ v_2^k \end{bmatrix}.$$

(In Example 8 in Section 5.1 we showed that $V$, with these operations, is a real vector space.) Let

$$S = \{\mathbf{v}_1, \mathbf{v}_2\} = \left\{ \begin{bmatrix} 4 \\ 6 \end{bmatrix}, \begin{bmatrix} 3 \\ 2 \end{bmatrix} \right\}.$$

Does span$(S) = V$? Following the outline given above for $\mathbf{w} = \begin{bmatrix} r \\ s \end{bmatrix}$ where $r > 0$ and $s > 0$, we have

$$c_1 \mathbf{v}_1 + c_2 \mathbf{v}_2 = \begin{bmatrix} 4^{c_1} 3^{c_2} \\ 6^{c_1} 2^{c_2} \end{bmatrix} = \begin{bmatrix} r \\ s \end{bmatrix}.$$

Equating corresponding matrix entries leads to the (nonlinear) system of equations

$$4^{c_1} 3^{c_2} = r$$
$$6^{c_1} 2^{c_2} = s.$$

Taking the natural logarithm of both sides of each equation and using properties of logarithms leads to the linear system of equations

$$c_1 \ln(4) + c_2 \ln(3) = \ln(r)$$
$$c_1 \ln(6) + c_2 \ln(2) = \ln(s).$$

This linear system is consistent for all positive values of $r$ and $s$ since the coefficient matrix is nonsingular. (Verify.) Hence span$(S) = V$.   ■

EXAMPLE 8   Let $W$ be the set of all $2 \times 2$ matrices of the form $\begin{bmatrix} a & b \\ c & 0 \end{bmatrix}$ where $a, b,$ and $c$ are arbitrary real numbers. We can show that $W$ is a subspace of $R_{2 \times 2}$.

Determine if span($S$) = $W$, where

$$S = \left\{ \begin{bmatrix} 2 & 1 \\ 3 & 0 \end{bmatrix}, \begin{bmatrix} 1 & 2 \\ 1 & 0 \end{bmatrix}, \begin{bmatrix} 3 & 0 \\ 5 & 0 \end{bmatrix}, \begin{bmatrix} -2 & 2 \\ -4 & 0 \end{bmatrix} \right\}.$$

If not, determine the set of all $2 \times 2$ matrices $\begin{bmatrix} r & t \\ s & 0 \end{bmatrix}$ that belong to span($S$).

We can work with the $2 \times 2$ matrices directly or we can use the following association of $2 \times 2$ matrices with vectors in $R^4$, which is often easier. Corresponding to each $2 \times 2$ matrix is a 4-vector obtained by "stringing together" its columns. Using this approach, $S$ becomes the set of 4-vectors

$$\mathbf{v}_1 = \begin{bmatrix} 2 \\ 3 \\ 1 \\ 0 \end{bmatrix}, \quad \mathbf{v}_2 = \begin{bmatrix} 1 \\ 1 \\ 2 \\ 0 \end{bmatrix}, \quad \mathbf{v}_3 = \begin{bmatrix} 3 \\ 5 \\ 0 \\ 0 \end{bmatrix}, \quad \mathbf{v}_4 = \begin{bmatrix} -2 \\ -4 \\ 2 \\ 0 \end{bmatrix}.$$

Following the strategy given previously, we form the linear combination $\sum_{k=1}^{4} c_k \mathbf{v}_k$ and set it equal to an arbitrary vector

$$\mathbf{w} = \begin{bmatrix} r \\ s \\ t \\ 0 \end{bmatrix}.$$

We form the corresponding linear system whose augmented matrix is

$$\begin{bmatrix} 2 & 1 & 3 & -2 & | & r \\ 3 & 1 & 5 & -4 & | & s \\ 1 & 2 & 0 & 2 & | & t \\ 0 & 0 & 0 & 0 & | & 0 \end{bmatrix}.$$

Computing the reduced row echelon form, we get (verify)

$$\begin{bmatrix} 1 & 0 & 2 & -2 & | & -r + s \\ 0 & 1 & -1 & 2 & | & 3r - 2s \\ 0 & 0 & 0 & 0 & | & -5r + 3s + t \\ 0 & 0 & 0 & 0 & | & 0 \end{bmatrix}.$$

We see that for an arbitrary vector $\mathbf{w}$ this system is inconsistent since, $-5r + 3s + t$ need not be zero. Hence $S$ is not a spanning set of subspace $W$. It follows that span($S$) consists of all $2 \times 2$ matrices such that $-5r + 3s + t = 0$. Solving this expression for $t$, we have $t = -3s + 5r$; hence the matrices in span($S$) are of the form

$$\begin{bmatrix} r & -3s + 5r \\ s & 0 \end{bmatrix} = r \begin{bmatrix} 1 & 5 \\ 0 & 0 \end{bmatrix} + s \begin{bmatrix} 0 & -3 \\ 1 & 0 \end{bmatrix}$$

where $r$ and $s$ are arbitrary real numbers. (We return to this example in Section 5.3.)

■

## Application:   Trail Mix Varieties

Trail mix is a blend of cereal, dried fruit, chocolate chips, nuts, and marshmallows. It is used by athletes, hikers, and campers as a snack food and as a dietary supplement. By varying the amount of each ingredient, different types of mix can be created. For example, the amount of chocolate affects the calorie content, the amount and type of cereal affects the fiber content, and the amount of nuts affects the fat content. Since different groups want trail mix with different characteristics, merchants want to be able to produce custom mixes.

A wholesaler prepares three types of trail mix that is sold to retailers, who can then use it to blend special orders for customers. The wholesaler's basic mixes come in 80-oz (5-lb) bags and have the following composition. (The weight of each ingredient is in ounces.)

|  | Trail Mix Varieties | | |
|  | Energy Booster | High Fiber | Low Fat |
| --- | --- | --- | --- |
| Wheat Cereal | 10 | 20 | 30 |
| Rice Cereal | 10 | 20 | 30 |
| Dried Fruit | 15 | 15 | 10 |
| Chocolate Chips | 20 | 10 | 0 |
| Nuts | 15 | 10 | 5 |
| Marshmallows | 10 | 5 | 5 |

For example, a scout troop might want a blend made from 10 bags of Energy Boost (**E**), 5 bags of High Fiber (**H**), and 5 bags of Low Fat (**L**); that is

$$\text{Scout Mix} = 10\mathbf{E} + 5\mathbf{H} + 5\mathbf{L}.$$

The basic types of trail mix are represented by a 6-vector (see the preceding table), and custom blends are 6-vectors that are in span($\{\mathbf{E}, \mathbf{H}, \mathbf{L}\}$).

Some customers specify a total amount of each type of ingredient that they want in their blend. In such cases the retailer must determine the linear combination of the basic mixes that will match the specifications. If necessary, the retailer will use partial bags with the understanding that the customer will be charged for the number of bags opened to generate the desired mix. For example, a camping group wants 70 lb of mix to contain 190 oz of wheat cereal, 200 oz of rice cereal, 210 oz of dried fruit, 220 oz of chocolate chips, 180 oz of nuts, and 120 oz of marshmallows. Thus we want to determine if

$$\mathbf{b} = \begin{bmatrix} 190 \\ 200 \\ 210 \\ 220 \\ 180 \\ 120 \end{bmatrix}$$

is in span($\{\mathbf{E}, \mathbf{H}, \mathbf{L}\}$). Following the preceding strategy, we determine if the linear system associated with the linear combination $c_1\mathbf{E} + c_2\mathbf{H} + c_3\mathbf{L} = \mathbf{b}$ is consistent or not. Forming the augmented matrix, we compute its rref and obtain (verify)

$$\begin{bmatrix} 1 & 0 & 0 & 10 \\ 0 & 1 & 0 & 2 \\ 0 & 0 & 1 & 2 \\ 0 & 0 & 0 & 0 \\ 0 & 0 & 0 & 0 \\ 0 & 0 & 0 & 0 \end{bmatrix}.$$

It follows that the linear system is consistent and that we use 10 bags of Energy Boost, 2 bags of High Fiber, and 2 bags of Low Fat to fill this special order. (Explain why anyone allergic to nuts would not buy any variety of the trail mix in span($\{\mathbf{E}, \mathbf{H}, \mathbf{L}\}$).)

# EXERCISES 5.2

1. Explain why if $W$ is a subspace of vector space $V$, then $W$ is closed with respect to addition and scalar multiplication.

2. Here we want to verify that if a subset $W$ of a vector space $V$ is closed with respect to both addition and scalar multiplication then $W$ is a subspace of $V$. Let $\mathbf{r}$ be in $W$. Supply a reason for each of the following.

   (a) $(-1)\mathbf{r}$ is in $W$ because _____.

   (b) The negative of $\mathbf{r}$ is in $W$ because _____.

   (c) $\mathbf{r} + (-1)\mathbf{r}$ is in $W$ because _____.

   (d) $\mathbf{r} + (-1)\mathbf{r} = \mathbf{z}$, the identity because _____.

   (e) $\mathbf{z}$ is in $W$ because _____.

   (f) Properties a, b, e, f, g, and h of the definition of a vector space are valid for $W$ because _____.

3. Let $W$ be the set of all singular matrices in vector space $(R_{n \times n}, +, \cdot)$. Is $W$ a subspace of $R_{n \times n}$?

4. Let $W$ be the set of all symmetric matrices in vector space $(R_{n \times n}, +, \cdot)$. Is $W$ a subspace of $R_{n \times n}$?

5. Let $W$ be the set of all diagonal matrices in vector space $(R_{n \times n}, +, \cdot)$. Is $W$ a subspace of $R_{n \times n}$?

6. Let $W$ be the set of all polynomials in vector space $(P_2, +, \cdot)$ that have the form $ax^2 + bx$, where $a$ and $b$ are any real numbers. Is $W$ a subspace of $P_2$?

7. Let $W$ be the set of all polynomials in vector space $(P_3, +, \cdot)$ that have the form $ax^3 + bx + 1$, where $a$ and $b$ are any real numbers. Is $W$ a subspace of $P_3$?

8. Let $W$ be the set of all polynomials in vector space $(P_2, +, \cdot)$ that have the form $ax^2 + bx + c$, where $a$, $b$, and $c$ are real numbers such that $a + b + c = 0$. Is $W$ a subspace of $P_2$?

9. Let $W$ be the set of all vectors $f$ in vector space $C(-\infty, \infty)$ such that $f(5) = 0$. Is $W$ a subspace of $C(-\infty, \infty)$?

10. Let $W$ be the set of all vectors $\mathbf{f}$ in vector space $C(-\infty, \infty)$ such that $f(0) = 3$. Is $W$ a subspace of $C(-\infty, \infty)$ ?

11. Let $W$ be the set of all vectors $\mathbf{f}$ in vector space $C(-\infty, \infty)$ that are differentiable functions. Is $W$ a subspace of $C(-\infty, \infty)$?

12. Let $W$ be the set of all vectors $(a, b, c)$ in vector space $R^3$ of the form $c = 2a - b$. Is W a subspace of $R^3$?

13. Let $W$ be the set of all vectors $(a, b, c)$ in vector space $R^3$ of the form $c = 7$. Is W a subspace of $R^3$?

14. Let $W$ be the set of all vectors $(a, b, c)$ in vector space $R^3$ that are orthogonal to $\mathbf{v} = (1, 2, -3)$. Is $W$ a subspace of $R^3$?

15. Let $W$ be the set of all vectors $(a, b, c)$ in vector space $R^3$ that are orthogonal to both $\mathbf{v} = (1, 2, -3)$ and $\mathbf{u} = (3, 0, 4)$. Is $W$ a subspace of $R^3$?

16. Let $W$ be the set of all matrices of the form $\begin{bmatrix} a & b & 0 \\ c & 0 & d \end{bmatrix}$ in vector space $(R_{2 \times 3}, +, \cdot)$, where $b = a + c$. Is $W$ a subspace of $R_{2 \times 3}$?

17. Let $W$ be the set of all matrices of the form $\begin{bmatrix} a & b & c \\ 0 & 0 & d \end{bmatrix}$ in vector space $(R_{2 \times 3}, +, \cdot)$, where $b > 0$. Is $W$ a subspace of $R_{2 \times 3}$?

18. Let $W$ be the set of all matrices $\mathbf{A}$ in vector space $(R_{2 \times 3}, +, \cdot)$ such that $\mathbf{x} = \begin{bmatrix} 1 \\ -1 \\ 0 \end{bmatrix}$ is in $\mathbf{ns}(\mathbf{A})$. Is $W$ a subspace of $R_{2 \times 3}$?

19. $P_k$ is the vector space of all polynomials of degree $k$ or less. (See Example 5.)

   (a) Describe $P_0$.

   (b) Let the vector space $P_k$ be represented by the closed polygonal region in Figure 3. On Figure 3 draw a figure to represent $P_{k-1}$ and another to show $P_{k+1}$.

FIGURE 3

20. Verify that if $S$ is a set of vectors in vector space $(V, +, \cdot)$, then span$(S)$ is a subspace of $V$.

*In Exercises 21–24, determine whether the given vector $\mathbf{v}$ belongs to* span$(\{\mathbf{v}_1, \mathbf{v}_2, \mathbf{v}_3\})$ *where*

$$\mathbf{v}_1 = \begin{bmatrix} 1 \\ 0 \\ 1 \end{bmatrix}, \quad \mathbf{v}_2 = \begin{bmatrix} 1 \\ -1 \\ 0 \end{bmatrix}, \quad and \quad \mathbf{v}_3 = \begin{bmatrix} 1 \\ 1 \\ 2 \end{bmatrix}.$$

21. $\mathbf{u} = \begin{bmatrix} 2 \\ 1 \\ 3 \end{bmatrix}$.

22. $\mathbf{u} = \begin{bmatrix} 1 \\ 1 \\ 1 \end{bmatrix}$.

23. $\mathbf{u} = \begin{bmatrix} 5 \\ -1 \\ 4 \end{bmatrix}$.

24. $\mathbf{u} = \begin{bmatrix} 3 \\ 8 \\ 10 \end{bmatrix}$.

*In Exercises 25–28, determine whether the vector $\mathbf{u}$ belongs to* span$(\{\mathbf{v}_1, \mathbf{v}_2, \mathbf{v}_3\})$ *where*

$$\mathbf{v}_1 = \begin{bmatrix} 1 & -1 \\ 0 & 3 \end{bmatrix}, \quad \mathbf{v}_2 = \begin{bmatrix} 1 & 1 \\ 0 & 2 \end{bmatrix},$$

*and*

$$\mathbf{v}_3 = \begin{bmatrix} 2 & 2 \\ -1 & 1 \end{bmatrix}.$$

**25.** $\mathbf{u} = \begin{bmatrix} 2 & 1 \\ -1 & 9 \end{bmatrix}.$     **26.** $\mathbf{u} = \begin{bmatrix} -3 & -1 \\ 3 & 2 \end{bmatrix}.$

**27.** $\mathbf{u} = \begin{bmatrix} 3 & -2 \\ 3 & 2 \end{bmatrix}.$     **28.** $\mathbf{u} = \begin{bmatrix} 1 & 0 \\ 2 & 1 \end{bmatrix}.$

*In Exercises 29–31, refer to Example 7.*

**29.** Let $S = \left\{ \begin{bmatrix} 4 \\ 1 \end{bmatrix}, \begin{bmatrix} 6 \\ 1 \end{bmatrix} \right\}$. Does span$(S) = V$?

**30.** Let $S = \left\{ \begin{bmatrix} 5 \\ 1 \end{bmatrix}, \begin{bmatrix} 1 \\ 5 \end{bmatrix} \right\}$. Does span$(S) = V$?

**31.** Let $S = \left\{ \begin{bmatrix} p \\ q \end{bmatrix}, \begin{bmatrix} e \\ e \end{bmatrix} \right\}$ where $p > 0, q > 0$, and $e$ is the natural logarithm base. Determine all values of $p$ and $q$ such that span$(S) = V$.

**32.** It is known that for

$$S = \left\{ \begin{bmatrix} 2 \\ 3 \\ -1 \end{bmatrix}, \begin{bmatrix} -4 \\ 1 \\ 0 \end{bmatrix}, \begin{bmatrix} 0 \\ 7 \\ -2 \end{bmatrix}, \begin{bmatrix} 10 \\ 1 \\ -1 \end{bmatrix} \right\},$$

span$(S) \neq R^3$. Determine the set of all vectors $\mathbf{w} = \begin{bmatrix} r \\ s \\ t \end{bmatrix}$ that lie in span$(S)$.

**33.** It is known that for

$$S = \left\{ 2x^2 + 5x - 3, -x^2 + 2x + 1, 6x^2 + 6x - 8 \right\},$$

span$(S) \neq P_2$. Determine the set of all polynomials $rx^2 + sx + t$ that lie in span$(S)$.

**34.** Determine if

$$S = \left\{ \begin{bmatrix} 1 \\ 3 \\ -1 \end{bmatrix}, \begin{bmatrix} 2 \\ 1 \\ 1 \end{bmatrix}, \begin{bmatrix} 0 \\ 5 \\ -3 \end{bmatrix}, \begin{bmatrix} 1 \\ 1 \\ 1 \end{bmatrix} \right\}$$

spans $R^3$. If not, find the set of all vectors $\mathbf{w} = \begin{bmatrix} r \\ s \\ t \end{bmatrix}$ that lie in span$(S)$.

**35.** Answer the following about the application on trail mix varieties.

(a) Can the retailer construct trail mix that corresponds to the 6-vector

$$\begin{bmatrix} 225 \\ 240 \\ 165 \\ 90 \\ 100 \\ 60 \end{bmatrix}?$$

If he can, how many bags of each type of the three basic mixes must be used?

(b) Suppose that the retailer received a special order for trail mix with equal amounts of each of the ingredients. Can the order be filled using bags of three basic mixes? If it can, how many bags of each type of the three basic mixes must be used?

(c) The wholesaler has introduced a fourth mix, called the Regular Mix (**R**), which is described by the 6-vector

$$\mathbf{R} = \begin{bmatrix} 20 \\ 22 \\ 15 \\ 8 \\ 9 \\ 10 \end{bmatrix}.$$

Must the retailer stock this variety in order to provide as many custom blends as possible? Explain your answer.

## True/False Review Questions

*Determine whether each of the following statements is true or false.*

**1.** If $\mathbf{A}$ is any real $m \times n$ matrix, then $\mathbf{ns}(\mathbf{A})$ is a subspace of $R^n$.

**2.** If $\lambda$ is an eigenvalue of an $n \times n$ matrix $\mathbf{A}$, then the set of all eigenvectors of $\mathbf{A}$ corresponding to $\lambda$ is a subspace of $R^n$.

**3.** Every vector space has at least two subspaces.

**4.** If $S$ is a subset of vector space $V$, then to show $S$ is not a subspace we can show that one of the closure properties fails to hold.

**5.** In $R^3$ any plane is a subspace.

**6.** $P_2$ is a subspace of $P_3$.

**7.** A subset $S$ of a vector space $V$ is a subspace if every linear combination of vectors from $S$ is also a vector in $S$.

**8.** If $S$ is any subset of a vector space $V$, then span($S$) is a subspace of $V$.

**9.** If $\{\mathbf{w}_1, \mathbf{w}_2, \mathbf{w}_3\}$ is a set of generators for vector space $V$, then span$\{\mathbf{w}_1, \mathbf{w}_2, \mathbf{w}_3\} = V$.

**10.** If $U$ and $W$ are subspaces of vector space $V$, then there is at least one vector that is in both $U$ and $W$.

## Terminology

| | |
|---|---|
| Subspace; subspace criterion | Trivial subspaces |
| Subspaces connected with linear systems of equations | Linear combination |
| Span of a set | Determining if the span of a set is the whole space |

The focus of this section is as follows:

> Given that we have a vector space $V$ and a set $S$ of vectors in $V$, can the members of $S$ be used to obtain every vector in $V$?

If the answer is yes, then we have a way to build the entire space from the set $S$. If the answer is no, then we want to be able to determine what portion of $V$ we can construct from the set $S$.

The ideas in this section are the foundation for developments in subsequent sections and chapters. Respond to the following to review and link together the developments in this section.

- What does it mean to say that $W$ is a subspace of vector space $V$? (Do not tell how to test to see if $W$ is a subspace.)
- If you suspect that $W$ is a subspace of vector space $V$, how do you check to see if your suspicion is correct? (You should be able to answer this in two ways; one way requires a two-step test and the other is stated in one step.)
- If $V$ is a vector space, we are guaranteed that it has at least two subspaces. Name these two subspaces. When are these two subspaces identical? Different?
- An $m \times n$ matrix has three subspaces associated with it. Name them and describe them.
- A homogeneous linear system $\mathbf{Ax} = \mathbf{0}$ has a subspace associated with it. Name it and describe it.
- For a square matrix $\mathbf{A}$, describe the eigenspaces associated with $\mathbf{A}$.
- Let $S$ be a set of vectors in vector space $V$. What is span($S$)? Is span($S$) a subspace of $V$? Explain.
- What does it mean to say that a set $S$ spans vector space $V$?
- If $V$ is a vector space and we select any set of vectors $S$ from $V$, is it always true that $S$ is a spanning set for $V$? Explain.
- Let $S$ be a set of vectors in vector space $V$. Explain the process we use to determine whether or not $S$ is a spanning set for $V$. (As part of your explanation, include the role played by a system of equations; Chapters 1 through 4 focused on systems of equations, and this is the link between abstract vector spaces and linear systems.)
- Suppose that span($S$) $\neq$ vector space $V$. Explain how we determine the subspace of $V$ that is equal to span($S$).

## 5.3 ■ BASES AND DIMENSION

In Section 5.2 we developed the structure of vector spaces and subspaces by considering sets of vectors that generated all the members of the space using linear combinations. We called such sets spanning sets. We also investigated the set of all linear combinations of a set $S$ of vectors; that is, the span of set $S$, span($S$). We showed that span($S$) was a subspace. In this section we take a closer look at spanning sets and determine conditions under which a spanning set is as efficient as possible (that is, spanning sets that contain as little information as possible but still generate the space).

We begin by posing several additional questions about linear combinations within a vector space $V$. (Questions 1 and 2 were stated and addressed in Section 5.2.) The developments in this section provide a complete answer to Question 3 and a partial answer to 4. We answer Question 5 in Chapter 6.

### Questions about Linear Combinations (continued)

3. Is there a smallest spanning set for a vector space? If so, how do we find it?
4. Are there some spanning sets that have advantages over other spanning sets? If so, what are the advantages and how can we exploit them?
5. If vector **u** belongs to $V$, but not to a subspace $W$ of $V$, is there a member of $W$ that is closest to **u**? If so, how do we find it?

In Chapters 1 through 4 we answered these questions for certain subspaces of $R^n$, $C^n$, $R_{m \times n}$ or $C_{m \times n}$. Here we will develop the corresponding concepts to obtain answers for abstract vector spaces. At times we leave the verification of certain properties as exercises. In most cases the verification is similar to that discussed in Chapters 1 through 4.

In order to answer Question 3 we use the following idea, which provides an important property about a relationship among vectors in a vector space.

---

**Definition**   The vectors $\mathbf{v}_1, \mathbf{v}_2, \ldots, \mathbf{v}_k$ in a vector space $V$ are called **linearly dependent** if there exist scalars $c_1, c_2, \ldots, c_k$, not all zero, such that linear combination

$$c_1 \mathbf{v}_1 + c_2 \mathbf{v}_2 + \cdots + c_k \mathbf{v}_k = \sum_{j=1}^{k} c_j \mathbf{v}_j = \text{the zero vector.} \tag{1}$$

Otherwise the vectors $\mathbf{v}_1, \mathbf{v}_2, \ldots, \mathbf{v}_k$ are called **linearly independent**. That is, $\mathbf{v}_1, \mathbf{v}_2, \ldots, \mathbf{v}_k$ are linearly independent provided that the *only* linear combination that results in the zero vector is the one where all the scalars used are zero.

---

If $S = \{\mathbf{v}_1, \mathbf{v}_2, \ldots, \mathbf{v}_k\}$ is a set of vectors in vector space $V$, then set $S$ is either linearly independent or linearly dependent. We emphasize that for $\mathbf{v}_1, \mathbf{v}_2, \ldots, \mathbf{v}_k$ in $V$, the linear combination $0\mathbf{v}_1 + 0\mathbf{v}_2 + \cdots + 0\mathbf{v}_k$ always gives the zero vector. The important point in the preceding definition is whether or not it is possible to have Equation (1) satisfied with at least one of the scalars different from zero.

If $V = R^n$, $C^n$, $R_{m \times n}$, or $C_{m \times n}$ with the usual operations of addition and scalar multiplication, then to determine whether vectors $\mathbf{v}_1, \mathbf{v}_2, \ldots, \mathbf{v}_k$ are linearly independent or linearly dependent, we investigated the null space of a homogeneous linear system constructed from Equation (1). If the null space consists only of the zero vector, then $\mathbf{v}_1, \mathbf{v}_2, \ldots, \mathbf{v}_k$ are linearly independent. (See the discussions and

Example 1 in Section 2.3.) The following example emphasizes our use of the words zero vector in Equation (1).

EXAMPLE 1    Let $V$ be the vector space of all $2 \times 1$ matrices with both entries positive real numbers. If

$$\mathbf{v} = \begin{bmatrix} v_1 \\ v_2 \end{bmatrix} \quad \text{and} \quad \mathbf{w} = \begin{bmatrix} w_1 \\ w_2 \end{bmatrix}$$

are in $V$, we define vector addition and scalar multiplication by a real scalar $k$, respectively, by

$$\mathbf{v} + \mathbf{w} = \begin{bmatrix} v_1 w_1 \\ v_2 w_2 \end{bmatrix} \quad k\mathbf{v} = \begin{bmatrix} v_1^k \\ v_2^k \end{bmatrix}.$$

(See Example 8 in Section 5.1.) Determine whether $\mathbf{v} = \begin{bmatrix} 2 \\ 3 \end{bmatrix}$ and $\mathbf{w} = \begin{bmatrix} 4 \\ 9 \end{bmatrix}$ are linearly independent or linearly dependent. Forming the expression in Equation (1), we have

$$c_1 \mathbf{v} + c_2 \mathbf{w} = c_1 \begin{bmatrix} 2 \\ 3 \end{bmatrix} + c_2 \begin{bmatrix} 4 \\ 9 \end{bmatrix} = \text{zero vector} = \begin{bmatrix} 1 \\ 1 \end{bmatrix}.$$

Using the particular definitions for addition and scalar multiplication for this vector space, we compute the linear combination as

$$c_1 \begin{bmatrix} 2 \\ 3 \end{bmatrix} + c_2 \begin{bmatrix} 4 \\ 9 \end{bmatrix} = \begin{bmatrix} 2^{c_1} \\ 3^{c_1} \end{bmatrix} + \begin{bmatrix} 4^{c_2} \\ 9^{c_2} \end{bmatrix} = \begin{bmatrix} 2^{c_1} 4^{c_2} \\ 3^{c_1} 9^{c_2} \end{bmatrix} = \begin{bmatrix} 1 \\ 1 \end{bmatrix}.$$

Equating corresponding entries, we have

$$2^{c_1} 4^{c_2} = 1 \quad \text{and} \quad 3^{c_1} 9^{c_2} = 1.$$

Then

$$2^{c_1} 2^{2c_2} = 2^{c_1 + 2c_2} = 1 \quad \text{and} \quad 3^{c_1} 3^{2c_2} = 3^{c_1 + 2c_2} = 1.$$

From properties of exponents, it follows that we have to solve the homogeneous linear system

$$c_1 + 2c_2 = 0$$
$$c_1 + 2c_2 = 0$$

We find (verify) that $c_1 = -2c_2$, so there are infinitely many nontrivial solutions. Hence $\mathbf{v}$ and $\mathbf{w}$ are linearly dependent. (Note that $\mathbf{w} = 2\mathbf{v}$; that is, $\mathbf{w}$ is a scalar multiple of $\mathbf{v}$ for the scalar multiplication operation defined in this example.)    ■

We present the following facts about linearly independent and linearly dependent sets of vectors from a vector space $V$.[1] (In order to illustrate the verification of properties involving linearly independent or dependent sets in an abstract vector space, we show the development of three of the properties in Examples 2 through 4. The other are left as exercises.)

---

[1]Corresponding properties for $R^n$ or $C^n$ were developed in Section 2.3.

**Properties of Linear Independent/Dependent Sets**

1. Any set containing the zero vector is linearly dependent. (See Example 2.)
2. A set in which one vector is a linear combination of other vectors in the set is linearly dependent. (See Exercise 20.)
3. If $S$ is a linearly dependent set, then at least one vector is a linear combination of the other vectors in $S$. (See Example 3.)
4. The span of a linearly dependent set $S$ can be generated without using all the vectors in $S$. (See Example 4.)
5. The span of a linearly independent set $T$ can only be formed by using all the vectors in $T$. (See Exercise 23.)

EXAMPLE 2   Verify:

Any set containing the zero vector is linearly dependent (Property 1).

From the definition of linear dependence, we must show that we can find a linear combination of the vectors that yields the zero vector and not all the coefficients are zero.

We proceed as follows: Let $\mathbf{z}$ be the zero vector in the vector space $V$ and $S = \{\mathbf{v}_1, \mathbf{v}_2, \ldots, \mathbf{v}_k, \mathbf{z}\}$. Using (1), we form the linear combination

$$c_1\mathbf{v}_1 + c_2\mathbf{v}_2 + \cdots + c_k\mathbf{v}_k + c_{k+1}\mathbf{z} = \mathbf{z}. \tag{2}$$

Using the properties of vector spaces from Section 5.1, we know that any scalar times the zero vector gives the zero vector. Thus we set coefficients $c_j = 0$, for $j = 1, 2, \ldots, k$ and $c_{k+1} \neq 0$. Then (2) is true and we have a linear combination of the vectors in $S$ that gives the zero vector and not all the coefficients are zero. Hence $S$ is a linearly dependent set.   ■

EXAMPLE 3   Verify:

If $S$ is a linearly dependent set, then at least one vector is a linear combination of the other vectors in $S$ (Property 3).

From the definition of linear dependence, we know that there is a linear combination of the vectors in $S$ that yields the zero vector and not all the coefficients are zero. We start with such an expression and manipulate it to get the result.

We proceed as follows: Let $S = \{\mathbf{v}_1, \mathbf{v}_2, \ldots, \mathbf{v}_k\}$ be a linearly dependent set and $\mathbf{z}$ denote the zero vector in $V$. Then there exist scalars $c_1, c_2, \ldots, c_k$, not all zero, such that

$$c_1\mathbf{v}_1 + c_2\mathbf{v}_2 + \cdots + c_k\mathbf{v}_k = \mathbf{z}. \tag{3}$$

Suppose that $c_1 \neq 0$. (If $c_1 = 0$, reorder the vectors in $S$ so that the one named $\mathbf{v}_1$ does not have a zero coefficient; we know that there is at least one such vector.) Since $c_1 \neq 0$, we can multiply both sides of (3) by its reciprocal and then rearrange the resulting expression to get

$$\mathbf{v}_1 = -\frac{1}{c_1}(c_2\mathbf{v}_2 + \cdots + c_k\mathbf{v}_k) + \frac{1}{c_1}\mathbf{z}. \tag{4}$$

But a scalar multiple of the zero vector gives the zero vector and any vector added to the zero vector does not change. Hence (4) can be simplified to give

$$\mathbf{v}_1 = -\frac{1}{c_1}(c_2\mathbf{v}_2 + \cdots + c_k\mathbf{v}_k). \tag{5}$$

Thus (5) shows that one of the vectors in $S$ is a linear combination of the other vectors in $S$.    ■

EXAMPLE 4    Verify:

The span of a linearly dependent set $S$ can be generated without using all the vectors in $S$ (Property 4).

The span of set of vectors is the subspace of all linear combinations of the vectors. Thus given any linear combination of the vectors in $S$, we must show that it can be expressed without using all the vectors of $S$. We will use the result developed in Example 3.

We proceed as follows: Let $S = \{\mathbf{v}_1, \mathbf{v}_2, \ldots, \mathbf{v}_k\}$ be a linearly dependent set. Then, as shown in Example 3, (at least) one vector in $S$ is a linear combination of the others. As in Example 3, we will assume it is $\mathbf{v}_1$; that is,

$$\mathbf{v}_1 = b_2\mathbf{v}_2 + b_3\mathbf{v}_3 + \cdots + b_k\mathbf{v}_k = \sum_{j=2}^{k} b_j\mathbf{v}_j. \tag{6}$$

If $\mathbf{w}$ is any vector in span$(S)$, then there exist scalars $c_1, c_2, \ldots, c_k$ so that

$$\mathbf{w} = c_1\mathbf{v}_1 + c_2\mathbf{v}_2 + \cdots + c_k\mathbf{v}_k.$$

Replace $\mathbf{v}_1$ in the preceeding expression by the linear combination from (6) and collect terms to obtain

$$\mathbf{w} = c_1 \left( \sum_{j=2}^{k} b_j\mathbf{v}_j \right) + c_2\mathbf{v}_2 + \cdots + c_k\mathbf{v}_k \tag{7}$$

$$= (c_1b_2 + c_2)\mathbf{v}_2 + (c_1b_3 + c_3)\mathbf{v}_3 + \cdots + (c_1b_k + c_k)\mathbf{v}_k.$$

Observe that (7) expresses an arbitrary vector in span$(S)$ as a linear combination of $\mathbf{v}_2, \mathbf{v}_3, \ldots, \mathbf{v}_k$. Hence not all the vectors in $S$ are required to generate span$(S)$.    ■

Property 4 of the preceding list of properties implies that a linearly dependent set contains redundant[2] information, since fewer vectors can span the same space. Correspondingly, property 5 implies that a linearly independent set contains the minimal amount of information needed to span the space. Hence we make the following definition.

---

**Definition**    The vectors in a set $S$ from a vector space $V$ are said to form a **basis** for $V$ if span$(S) = V$ and $S$ is a linearly independent set.

---

Using property 5, we can say that a basis for a vector space is a smallest spanning set for a vector space since we need all of its information content in order to construct all the vectors of the space.

EXAMPLE 5    In Example 6 in Section 5.2 we showed that

$$\text{span}(S) = \text{span}\left\{ \begin{bmatrix} 2 \\ 1 \\ 3 \end{bmatrix}, \begin{bmatrix} 1 \\ 2 \\ 1 \end{bmatrix}, \begin{bmatrix} 3 \\ 3 \\ 4 \end{bmatrix}, \begin{bmatrix} 3 \\ 0 \\ 5 \end{bmatrix} \right\} = \text{span}\left\{ \begin{bmatrix} 1 \\ 5 \\ 0 \end{bmatrix}, \begin{bmatrix} 0 \\ -3 \\ 1 \end{bmatrix} \right\}.$$

[2]By *redundant information* we mean that some of the information content is available from more than just one source in the set. That is, certain vectors are not needed or are superfluous.

In fact, we can say more: $\begin{bmatrix} 1 \\ 5 \\ 0 \end{bmatrix}$ and $\begin{bmatrix} 0 \\ -3 \\ 1 \end{bmatrix}$ form a basis for span($S$). To see that this pair of vectors is linearly independent, set a linear combination of them equal to the zero vector and verify that the only solution is the zero solution. (This is analogous to the development we gave in Section 2.3 for finding a basis for the null space. See Examples 1 and 2 in Section 2.3.) ■

### If $S$ is a spanning set for a vector space, then how do we determine a basis for the space?

Our approach for any vector space or subspace V is as follows: Let $S = \{\mathbf{v}_1, \mathbf{v}_2, \ldots, \mathbf{v}_k\}$ be vectors so that span($S$) $= V$, your space. Our procedure is to cast out of set $S$ any vectors that depend on the other vectors in $S$. We do this by checking whether successively larger subsets are linearly independent. If they are, then we keep adjoining vectors and checking. If at any stage the set is linearly dependent, then we drop the last vector that we adjoined and continue to check the remaining vectors.

- Delete from $S$ any zero vectors. Suppose there are $p \leq k$ vectors remaining. Order these vectors and denote them as $\mathbf{w}_1, \mathbf{w}_2, \ldots, \mathbf{w}_p$. Then span($\{\mathbf{w}_1, \mathbf{w}_2, \ldots, \mathbf{w}_p\}$) $= V$. (Why?)
- Let $T = \{\mathbf{w}_1\}$. $T$ is a linearly independent set. (Why?)
- Next let $T = \{\mathbf{w}_1, \mathbf{w}_2\}$. Determine if $T$ is linearly dependent. If it is, discard vector $\mathbf{w}_2$ so that $T = \{\mathbf{w}_1\}$.
- Repeat the preceding step after adjoining vector $\mathbf{w}_3$ to $T$. Continue until you exhaust the vectors available. The vectors remaining in $T$ will span $V$ and be a linearly independent set. That is, you will have a basis that is a subset of the original spanning set.

EXAMPLE 6　Let

$$S = \{\mathbf{v}_1, \mathbf{v}_2, \mathbf{v}_3, \mathbf{v}_4, \mathbf{v}_5\} = \left\{ \begin{bmatrix} 1 \\ 2 \\ 1 \end{bmatrix}, \begin{bmatrix} 0 \\ 0 \\ 0 \end{bmatrix}, \begin{bmatrix} 2 \\ 4 \\ 2 \end{bmatrix}, \begin{bmatrix} 1 \\ 1 \\ 0 \end{bmatrix}, \begin{bmatrix} 2 \\ 3 \\ 1 \end{bmatrix} \right\}$$

be vectors so that span($S$) $= V$. Find a basis for $V$ that is a subset of $S$. Using the steps outlined previously, we cast out the zero vector and rename the vectors as

$$\{\mathbf{w}_1, \mathbf{w}_2, \mathbf{w}_3, \mathbf{w}_4\} = \left\{ \begin{bmatrix} 1 \\ 2 \\ 1 \end{bmatrix}, \begin{bmatrix} 2 \\ 4 \\ 2 \end{bmatrix}, \begin{bmatrix} 1 \\ 1 \\ 0 \end{bmatrix}, \begin{bmatrix} 2 \\ 3 \\ 1 \end{bmatrix} \right\}.$$

Then $T = \{\mathbf{w}_1\}$ is linearly independent since it contains one nonzero vector. Let $T = \{\mathbf{w}_1, \mathbf{w}_2\}$. Checking to see if $T$ is linearly independent, we compute

$$\mathbf{rref}\left(\begin{bmatrix} \mathbf{w}_1 & \mathbf{w}_2 \mid \mathbf{0} \end{bmatrix}\right) = \begin{bmatrix} 1 & 2 \mid 0 \\ 0 & 0 \mid 0 \\ 0 & 0 \mid 0 \end{bmatrix}.$$

Since (the number of columns of the coefficient matrix) $-$ (number of leading ones) $= (2) - (1) > 0$, there is an arbitrary constant in the general solution, so $T$ is a linearly dependent set. (Alternatively, note $\mathbf{w}_2 = 2\mathbf{w}_1$.) Thus we cast out $\mathbf{w}_2$. Next we set $T = \{\mathbf{w}_1, \mathbf{w}_3\}$. We compute

$$\mathbf{rref}\left(\begin{bmatrix} \mathbf{w}_1 & \mathbf{w}_3 & | & \mathbf{0} \end{bmatrix}\right) = \left[\begin{array}{cc|c} 1 & 0 & 0 \\ 0 & 1 & 0 \\ 0 & 0 & 0 \end{array}\right].$$

Since (the number of columns of the coefficient matrix) − (number of leading ones) = 0, the only solution is the zero solution so $T$ is linearly independent. Finally, we set $T = \{\mathbf{w}_1, \mathbf{w}_3, \mathbf{w}_4\}$. We compute

$$\mathbf{rref}\left(\begin{bmatrix} \mathbf{w}_1 & \mathbf{w}_3 & \mathbf{w}_4 & | & \mathbf{0} \end{bmatrix}\right) = \left[\begin{array}{ccc|c} 1 & 0 & 1 & 0 \\ 0 & 1 & 1 & 0 \\ 0 & 0 & 0 & 0 \end{array}\right].$$

Since (the number of columns of the coefficient matrix) − (number of leading ones) = (3) − (2) > 0, there is an arbitrary constant in the general solution, so $T$ is a linearly dependent set. Thus we cast out $\mathbf{w}_4$. Hence $T = \{\mathbf{w}_1, \mathbf{w}_3\}$ is a basis for $V$. ■

The casting out of vectors that depend upon preceding vectors is quite tedious. However, at the end of this section we develop a more efficient procedure for determining a linearly independent subset of a set of vectors from $R^n$ or $C^n$.

## Showing that a set is a basis

If we need to show that a particular set $S = \{\mathbf{v}_1, \mathbf{v}_2, \ldots, \mathbf{v}_k\}$ is a basis for a space, then we must show that $S$ both spans the space and is linearly independent. This requires that we investigate to see if there exist scalars $c_1, c_2, \ldots, c_k$ so that

$$\sum_{j=1}^{k} c_j \mathbf{v}_j = \mathbf{w},$$

for an arbitrary vector $\mathbf{w}$ in the space, and if there exist nonzero scalars $c_1, c_2, \ldots, c_k$ so that

$$\sum_{j=1}^{k} c_j \mathbf{v}_j = \text{zero vector.}$$

In many situations a linear system of equations is associated with these expressions. In such cases the coefficient matrix is the same, while the right side of the system is different. We can solve these linear systems at the same time by adjoining two augmented columns to the coefficient matrix and then computing the rref. We illustrate this in the next example.

EXAMPLE 7    Show that $S = \{\mathbf{v}_1, \mathbf{v}_2, \mathbf{v}_3\} = \{3x^2 + 2x + 1, x^2 + x + 1, x^2 + 1\}$ is a basis for $P_2$, the vector space of all polynomials of degree 2 or less. Let $\mathbf{w} = ax^2 + bx + c$ be an arbitrary vector of $P_2$. Then to verify span$(S) = P_2$, we set

$$c_1(3x^2 + 2x + 1) + c_2(x^2 + x + 1) + c_3(x^2 + 1) = ax^2 + bx + c.$$

Expanding and collecting like terms on the left side gives

$$(3c_1 + c_2 + c_3)x^2 + (2c_1 + c_2)x + (c_1 + c_2 + c_3)1 = ax^2 + bx + c.$$

Equating coefficients of like terms from each side leads to the linear system of equations

$$\begin{aligned}3c_1 + c_2 + c_3 &= a\\ 2c_1 + c_2 &= b\\ c_1 + c_2 + c_3 &= c\end{aligned} \quad \text{or} \quad \begin{bmatrix} 3 & 1 & 1 \\ 2 & 1 & 0 \\ 1 & 1 & 1 \end{bmatrix}\begin{bmatrix} c_1 \\ c_2 \\ c_3 \end{bmatrix} = \begin{bmatrix} a \\ b \\ c \end{bmatrix}.$$

To combine the span part with the independence/dependence part, we form the partitioned matrix

$$\left[\begin{array}{ccc|c|c} 3 & 1 & 1 & a & 0 \\ 2 & 1 & 0 & b & 0 \\ 1 & 1 & 1 & c & 0 \end{array}\right]$$

and compute its reduced row echelon form. We obtain (verify)

$$\left[\begin{array}{ccc|c|c} 1 & 0 & 0 & \frac{1}{2}a - \frac{1}{2}b & 0 \\ 0 & 1 & 0 & -a+b+c & 0 \\ 0 & 0 & 1 & \frac{1}{2}a - b + \frac{1}{2}c & 0 \end{array}\right]$$

which has (the number of columns of the coefficient matrix) − (number of leading ones) = 0. Hence there is a unique solution to both linear systems and it follows that $S$ spans $P_2$ and is linearly independent. So $S$ is a basis for $P_2$. ■

To address Question 4 in the questions about linear combinations (page 325), we introduce the bases discussed in Example 8 that follows. We present only a partial answer to Question 4. We address the question further in Chapter 6.

EXAMPLE 8   In some vector spaces there are bases that are easy to use and arise in a natural fashion because of the type of vector in the space. Here we record such bases for a number of important vector spaces. Each of the bases listed here is called the **natural basis** or **standard basis** for the vector space. With these bases it is very easy to form linear combinations. As we saw in Chapter 4, in analyzing the behavior of matrix transformations it is easiest to have a basis for $R^n$ consisting of eigenvectors of the matrix associated with the transformation. Thus the choice of a basis can play an important role in applications.

(a) The vectors $e_1 = \begin{bmatrix} 1 \\ 0 \end{bmatrix}$ and $e_2 = \begin{bmatrix} 0 \\ 1 \end{bmatrix}$ form the natural basis for $R^2$ and $C^2$.

(b) The vectors $e_1 = \begin{bmatrix} 1 \\ 0 \\ 0 \end{bmatrix}$, $e_2 = \begin{bmatrix} 0 \\ 1 \\ 0 \end{bmatrix}$, and $e_3 = \begin{bmatrix} 0 \\ 0 \\ 1 \end{bmatrix}$ form the natural basis for $R^3$ and $C^3$.

(c) Let $e_j = \text{col}_j(I_n)$, $j = 1, 2, \ldots, n$, then $\{e_1, e_2, \ldots, e_n\}$ is the natural basis for $R^n$ and $C^n$.

(d) The set of polynomials $\{x^n, x^{n-1}, \ldots, x, 1\}$ is the natural basis for $P_n$.

(e) The set of $2 \times 2$ matrices

$$\left\{\begin{bmatrix} 1 & 0 \\ 0 & 0 \end{bmatrix}, \begin{bmatrix} 0 & 1 \\ 0 & 0 \end{bmatrix}, \begin{bmatrix} 0 & 0 \\ 1 & 0 \end{bmatrix}, \begin{bmatrix} 0 & 0 \\ 0 & 1 \end{bmatrix}\right\}$$

is the natural basis for $R_{2\times2}$ and $C_{2\times2}$.

(f) Let $E_{ij}$ be the $m \times n$ matrix all of whose entries are zero except the $(i, j)$-entry, which is 1. Then the $mn$ vectors $E_{ij}$ are the natural basis for $R_{m\times n}$ and $C_{m\times n}$. ■

Next we develop two fundamental results about the number of vectors in different bases for the same vector space.

1. Observe that if $\{\mathbf{v}_1, \mathbf{v}_2, \ldots, \mathbf{v}_k\}$ is a basis for a space and $c$ is a nonzero scalar, then $\{c\mathbf{v}_1, \mathbf{v}_2, \ldots, \mathbf{v}_k\}$ is also a basis. (See Exercise 24.) Thus a vector space always has infinitely many bases.

2. Observe that any subset of a linearly independent set is also linearly independent. (See Exercise 25.)

### The number of vectors in a linearly independent set

If $S = \{\mathbf{v}_1, \mathbf{v}_2, \ldots, \mathbf{v}_k\}$ is a basis for a vector space $V$ and $T = \{\mathbf{w}_1, \mathbf{w}_2, \ldots, \mathbf{w}_r\}$ is a linearly independent subset of vectors in $V$, then $r \leq k$.

The verification of this result is a clever casting out of linearly dependent vectors. Let $T_1 = \{\mathbf{w}_1, \mathbf{v}_1, \mathbf{v}_2, \ldots, \mathbf{v}_k\}$. Since $S$ is a basis, $\text{span}(T_1) = \text{span}(S)$ and $\mathbf{w}_1$ is a linear combination of the vectors in $S$. Hence $T_1$ is a linearly dependent set and there exist scalars $c_1, c_2, \ldots, c_k$ such that

$$\mathbf{w}_1 = c_1\mathbf{v}_1 + c_2\mathbf{v}_2 + \cdots + c_k\mathbf{v}_k. \tag{8}$$

Since $\mathbf{w}_1$ is a member of a linearly independent set $T$, $\mathbf{w}_1 \neq \mathbf{0}$, so at least one of the coefficients $c_1, c_2, \ldots, c_k$ is not zero. Suppose that $c_j \neq 0$, then we have

$$c_j\mathbf{v}_j = \mathbf{w}_1 - (c_1\mathbf{v}_1 + c_2\mathbf{v}_2 + \cdots + c_{j-1}\mathbf{v}_{j-1} + c_{j+1}\mathbf{v}_{j+1} + \cdots + c_k\mathbf{v}_k)$$

or, equivalently,

$$\mathbf{v}_j = \frac{1}{c_j}\left[\mathbf{w}_1 - (c_1\mathbf{v}_1 + c_2\mathbf{v}_2 + \cdots + c_{j-1}\mathbf{v}_{j-1} + c_{j+1}\mathbf{v}_{j+1} + \cdots + c_k\mathbf{v}_k)\right].$$

So $\mathbf{v}_j$ is a linear combination of the other vectors in $T_1$. It follows that vectors $\mathbf{w}_1, \mathbf{v}_1, \mathbf{v}_2, \ldots, \mathbf{v}_{j-1}, \mathbf{v}_{j+1}, \ldots, \mathbf{v}_k$ still span $V$ and in fact are linearly independent. (See Exercise 33.) Hence $S_1 = \{\mathbf{w}_1, \mathbf{v}_1, \mathbf{v}_2, \ldots, \mathbf{v}_{j-1}, \mathbf{v}_{j+1}, \ldots, \mathbf{v}_k\}$ is a basis for $V$. Next let

$$T_2 = \{\mathbf{w}_2, \mathbf{w}_1, \mathbf{v}_1, \mathbf{v}_2, \ldots, \mathbf{v}_{j-1}, \mathbf{v}_{j+1}, \ldots, \mathbf{v}_k\}.$$

Since $S_1$ is a basis for $V$, $T_2$ is linearly dependent. We now use the fact that some vector in $T_2$ is a linear combination of the preceding vectors in $T_2$.[3] Since $T$ is linearly independent, this vector cannot be $\mathbf{w}_1$, so it is $\mathbf{v}_i$, $i \neq j$. Repeat this process over and over. If the $\mathbf{v}$ vectors are all cast out before we run out of $\mathbf{w}$ vectors, then one of the $\mathbf{w}$ vectors is a linear combination of the others, which implies that $T$ is linearly dependent. Thus we would have a contradiction of the statement that $T$ is linearly independent. Hence we must conclude that the number $r$ of $\mathbf{w}$ vectors is less than or equal to the number $k$ of $\mathbf{v}$ vectors; $r \leq k$.

If $S = \{\mathbf{v}_1, \mathbf{v}_2, \ldots, \mathbf{v}_k\}$ and $T = \{\mathbf{w}_1, \mathbf{w}_2, \ldots, \mathbf{w}_r\}$ are bases for the same vector space, then $k = r$.

To verify the preceding statement, a major result, we note that the first result implies that $r \leq k$. Since $T$ is also a basis, we can reverse the roles of sets $S$ and $T$ and get that $k \leq r$. Hence $k = r$.

Thus, although a vector space has many bases, we have just shown that *for a particular vector space $V$, all bases have the same number of vectors*. We can then make the following definition.

---

[3]See p. 121 in *Elementary Linear Algebra*, 7th Edition, by B. Kolman and D. Hill, Prentice Hall, Upper Saddle River, NJ, 2000. We have omitted the verification of this result since it is used only in this section.

**Definition**    The **dimension** of a nonzero vector space $V$ is the number of vectors in a basis for $V$. We often write dim($V$) for the dimension of $V$. Since the zero vector space has no linearly independent subset, it is natural to say that the zero vector space, $\{\mathbf{z}\}$, has dimension zero.

Referring to Example 8, we have

$$\dim(R^2) = \dim(C^2) = 2$$
$$\dim(R^3) = \dim(C^3) = 3$$
$$\dim(R^n) = \dim(C^n) = n$$
$$\dim(P_n) = n + 1$$
$$\dim(R_{m \times n}) = \dim(C_{m \times n}) = mn.$$

From Section 2.3 for an $m \times n$ matrix $\mathbf{A}$ we have

$$\dim(\mathbf{row}(\mathbf{A})) = \text{number of leading ones in } \mathbf{rref}(\mathbf{A})$$
$$\dim(\mathbf{col}(\mathbf{A})) = \text{number of leading ones in } \mathbf{rref}(\mathbf{A}^T)$$
$$\dim(\mathbf{ns}(\mathbf{A})) = \text{number of free variables.}$$

Most of the vector spaces that we use in this book are finite dimensional; that is, their dimension if a finite number. However, there are many important vector spaces whose dimension is infinite. In particular, $C(a, b)$, $C[a, b]$, $C(-\infty, \infty)$, and $P$, the vector space of all polynomials, are not finite dimensional. (See Exercise 27.)

The idea of dimension, of course, applies to subspaces. In Example 2 of Section 5.2 the dimension of a line through the origin is one, while in Example 3 of Section 5.2 the dimension of a plane through the origin is two. In Example 5, dim(span($S$)) is two.

## Linearly independent sets and bases

We next consider the following statement.

> If $S$ is a linearly independent subset of vectors in a finite dimensional vector space $V$, then there is a basis for $V$, which contains $S$.

This is sometimes stated as follows: *Any linearly independent set can be extended to form a basis.* Hence we can build bases containing particular linearly independent sets and thus construct bases for particular needs. The result is verified by casting out dependent vectors. (See Exercise 29.)

From the definition of a basis, a set of $k$ vectors in a vector space $V$ is a basis for $V$ if it spans $V$ and is linearly independent. However, if we are given the additional information that the dimension of vector space $V$ is $k$, $k$ *finite*, we need only verify one of the two conditions. This is the content of the parts (a) and (b) in the following set of statements.

> Let $V$ be a vector space with dim($V$) $= k$, and $S = \{\mathbf{v}_1, \mathbf{v}_2, \ldots, \mathbf{v}_k\}$ a set of $k$ vectors in $V$.
> (a)  If $S$ is linearly independent, then it is a basis for $V$.
> (b)  If span($S$) $= V$, then $S$ is a basis for $V$.
> (c)  Any set of more than $k$ vectors from $V$ is linearly dependent.
> (d)  Any set of fewer than $k$ vectors cannot be a basis for $V$.

In order to illustrate further the verification of properties of linearly independent sets we show the development of properties (a) and (c) above in Examples 9 and 10, respectively. The other properties are left as exercises. (See Exercise 30.)

EXAMPLE 9    Let $V$ be a vector space with $\dim(V) = k$, and $S = \{\mathbf{v}_1, \mathbf{v}_2, \ldots, \mathbf{v}_k\}$ a set of $k$ vectors in $V$. Verify:

If $S$ is linearly independent, then it is a basis for $V$.

Since $S$ is a linearly independent set, it can be extended to a basis for $V$. But adjoining any vectors to $S$ to obtain a basis would produce a basis with more than $k$ vectors. This is not possible since $\dim(V) = k$. Hence $S$ must itself be a basis for $V$.    ■

EXAMPLE 10    Let $V$ be a vector space with $\dim(V) = k$, and $S = \{\mathbf{v}_1, \mathbf{v}_2, \ldots, \mathbf{v}_k\}$ a set of $k$ vectors in $V$. Verify:

Any set of more than $k$ vectors from $V$ is linearly dependent.

Suppose that $T$ is a set of $n$ vectors from $V$ where $n > k$. If $T$ were a linearly independent set, then it could be extended to a basis for $V$. Hence we would have a basis with more than $k$ vectors, which is impossible since $\dim(V) = k$. Thus $T$ must be a linearly dependent set.    ■

## Expressing a vector in terms of a basis

The preceding results deal with counting the number of vectors in a basis. Here we consider the way a vector can be written in terms of a basis. We have the following major result.

If $S = \{\mathbf{v}_1, \mathbf{v}_2, \ldots, \mathbf{v}_k\}$ is a basis for a vector space $V$, then every vector in $V$ can be written in one and only one way as a linear combination of the vectors in $S$.

We say that **a vector is uniquely expressible in terms of a basis**. (We have ignored changing the order of the vectors in $S$ to make this statement.) We leave the verification of this result to Exercise 31.

We can also show (Exercise 32) that if $S = \{\mathbf{v}_1, \mathbf{v}_2, \ldots, \mathbf{v}_k\}$ is a set of nonzero vectors in a vector space $V$ such that every vector in $V$ can be written uniquely as a linear combination of the vectors in $S$, then $S$ is a basis for $V$.

We postpone the discussion of Question 5 in the Questions about Linear Combinations (see page 325) until Chapter 6.

Application:    **Determining a basis from a spanning set of vectors from $R^n$ or $C^n$**

The procedure for obtaining a basis from a spanning set by successively casting out dependent vectors requires that we investigate the solution of a homogeneous linear system at each stage. Here we present an alternate strategy for sets of vectors in $R^n$ or $C^n$.

Let $S = \{\mathbf{v}_1, \mathbf{v}_2, \ldots, \mathbf{v}_k\}$ be a set of $k$ vectors in $R^n$ or $C^n$ and $W = \text{span}(S)$. To find a basis for subspace $W$ that is a subset of $S$, we use the following steps.

**Step 1.**    Form the linear combination $\displaystyle\sum_{j=1}^{k} c_j \mathbf{v}_j = \mathbf{0}$.

**Step 2.**   Construct the augmented matrix of the homogeneous linear system that is associated with this linear combination and determine its reduced row echelon form.

**Step 3.**   The numbers of the columns in which the leading 1's appear, correspond to the subscripts of the members of $S$ that form a basis for $W$. In effect, the leading 1's "point" to the subset of vectors of $S$ that form a basis for $W = \text{span}(S)$.

We omit the proof of this result, but we note that the columns with leading 1's correspond to the unknowns that are not free variables in the general solution to the homogeneous linear system. We illustrate this procedure in the next example.

EXAMPLE 11   Let $S$ be the set of 3-vectors in Example 6. Then the augmented matrix associated with the linear combination

$$c_1\mathbf{v}_1 + c_2\mathbf{v}_2 + \cdots + c_5\mathbf{v}_5 = \sum_{j=1}^{5} c_j\mathbf{v}_j = \mathbf{0}$$

is

$$\begin{bmatrix} 1 & 0 & 2 & 1 & 2 & | & 0 \\ 2 & 0 & 4 & 1 & 3 & | & 0 \\ 1 & 0 & 2 & 0 & 1 & | & 0 \end{bmatrix}.$$

Forming its rref, we obtain

$$\begin{bmatrix} 1 & 0 & 2 & 0 & 1 & | & 0 \\ 0 & 0 & 0 & 1 & 1 & | & 0 \\ 0 & 0 & 0 & 0 & 0 & | & 0 \end{bmatrix}.$$

(Verify.)   The leading 1's point to vectors $\mathbf{v}_1$ and $\mathbf{v}_4$. Thus $\{\mathbf{v}_1, \mathbf{v}_4\}$ is a basis for span($S$). This is the same pair of vectors as determined in Example 6.   ■

If the vectors in set $S$ are reordered, then the columns of the corresponding homogeneous system are reordered in the same fashion. Hence the leading 1's in the rref could point to another subset that is a basis for span(S). The order of the vectors in the original spanning set determines which basis for span($S$) is obtained. Referring to Example 6, if

$$S = \{\mathbf{v}_3, \mathbf{v}_2, \mathbf{v}_1, \mathbf{v}_5, \mathbf{v}_4\} = \left\{ \begin{bmatrix} 2 \\ 4 \\ 2 \end{bmatrix}, \begin{bmatrix} 0 \\ 0 \\ 0 \end{bmatrix}, \begin{bmatrix} 1 \\ 2 \\ 1 \end{bmatrix}, \begin{bmatrix} 2 \\ 3 \\ 1 \end{bmatrix}, \begin{bmatrix} 1 \\ 1 \\ 0 \end{bmatrix} \right\},$$

then the rref of the corresponding homogeneous linear system is

$$\begin{bmatrix} 1 & 0 & \frac{1}{2} & 0 & -\frac{1}{2} & | & 0 \\ 0 & 0 & 0 & 1 & 1 & | & 0 \\ 0 & 0 & 0 & 0 & 0 & | & 0 \end{bmatrix}.$$

Hence $\{\mathbf{v}_3, \mathbf{v}_5\}$ is a basis for span($S$).

## EXERCISES 5.3

*In Exercises 1–3, determine whether the given set of vectors is linearly independent in $R^3$.*

1. $S = \left\{ \begin{bmatrix} 1 \\ 2 \\ 5 \end{bmatrix}, \begin{bmatrix} 2 \\ 2 \\ -1 \end{bmatrix} \right\}$.

2. $S = \left\{ \begin{bmatrix} 1 \\ 2 \\ -3 \end{bmatrix}, \begin{bmatrix} 4 \\ 2 \\ 1 \end{bmatrix}, \begin{bmatrix} 2 \\ 6 \\ -5 \end{bmatrix} \right\}$.

3. $S = \left\{ \begin{bmatrix} 1 \\ 2 \\ 5 \end{bmatrix}, \begin{bmatrix} 2 \\ 2 \\ -1 \end{bmatrix}, \begin{bmatrix} 0 \\ 1 \\ 0 \end{bmatrix}, \begin{bmatrix} 0 \\ 0 \\ 2 \end{bmatrix} \right\}$.

*In Exercises 4–8, determine whether the given set of vectors is linearly independent in $P_2$.*

4. $S = \{x^2 + 1, x - 2, x + 3\}$.

5. $S = \{2x^2 + x + 2, x^2 - 3x + 1, x^2 + 11x + 1\}$.

6. $S = \{x^2 - 2x - 1, 2x^2 + 3x, 7x + 2\}$.

7. $S = \{x^2, x + 2, x - 3, x^2 + 2x\}$.

8. $S = \{x^2, x^2 + x\}$.

*In Exercises 9–11, determine which of the sets is a basis for $R^3$.*

9. $S = \left\{ \begin{bmatrix} 1 \\ 2 \\ 5 \end{bmatrix}, \begin{bmatrix} 2 \\ 2 \\ -1 \end{bmatrix} \right\}$.

10. $S = \left\{ \begin{bmatrix} 1 \\ 2 \\ -3 \end{bmatrix}, \begin{bmatrix} 4 \\ 2 \\ 1 \end{bmatrix}, \begin{bmatrix} 2 \\ 6 \\ -5 \end{bmatrix} \right\}$.

11. $S = \left\{ \begin{bmatrix} 1 \\ 2 \\ 1 \end{bmatrix}, \begin{bmatrix} 2 \\ 1 \\ 2 \end{bmatrix}, \begin{bmatrix} 3 \\ 0 \\ 3 \end{bmatrix} \right\}$.

*In Exercises 12–14, determine which of the sets is a basis for $P_3$.*

12. $S = \{x^3 - x, x^3 + x - 1, x - 2\}$.

13. $S = \{x^3, x^2 - x, x + 1, 1\}$.

14. $S = \{x^3 + x^2 + 1, x^2 + x + 1, x^3 + 1, x^2 + x\}$.

15. Let $S = \left\{ \begin{bmatrix} 1 \\ -2 \\ 1 \end{bmatrix}, \begin{bmatrix} -2 \\ 4 \\ -2 \end{bmatrix}, \begin{bmatrix} 0 \\ 0 \\ 0 \end{bmatrix}, \begin{bmatrix} 1 \\ 1 \\ 0 \end{bmatrix} \right\}$.

   (a) Determine a subset $T$ of $S$ that is a basis for span($S$).
   (b) Is span($T$) = $R^3$? Explain your answer.

16. Let $S = \left\{ \begin{bmatrix} 1 \\ 0 \\ 0 \\ -1 \end{bmatrix}, \begin{bmatrix} 2 \\ 1 \\ 2 \\ 0 \end{bmatrix}, \begin{bmatrix} 1 \\ 1 \\ 2 \\ 1 \end{bmatrix}, \begin{bmatrix} 3 \\ 2 \\ 4 \\ 0 \end{bmatrix}, \begin{bmatrix} 0 \\ 1 \\ 2 \\ 0 \end{bmatrix} \right\}$.

   (a) Determine a subset $T$ of $S$ that is a basis for span($S$).
   (b) Is span($T$) = $R^4$? Explain your answer.

17. Let $S = \left\{ \begin{bmatrix} 1 & 1 \\ 0 & 0 \end{bmatrix}, \begin{bmatrix} 1 & 2 \\ 1 & 0 \end{bmatrix}, \begin{bmatrix} 3 & 4 \\ 1 & 0 \end{bmatrix}, \begin{bmatrix} 0 & 1 \\ 1 & 0 \end{bmatrix}, \begin{bmatrix} 1 & 2 \\ -1 & 0 \end{bmatrix}, \begin{bmatrix} 3 & 5 \\ 1 & 1 \end{bmatrix} \right\}$.

   (a) Determine a subset $T$ of $S$ that is a basis for span($S$).
   (b) Is span($T$) = $R_{2 \times 2}$? Explain your answer.

18. The Sweep-Clean company makes products for industry that aid in absorbing spills of oil, water, and other industrial fluids that can accumulate in the workplace. Over the years their development team has found that they can supply effective absorbents by combining five basic mixtures of five different ingredients. Each of the basic mixtures and their composition is shown by the 5-vectors in the following table. The entries in the table are pounds of the corresponding material. We will designate the five basic mixtures as $\mathbf{v}_1$ through $\mathbf{v}_5$.

| Sweep-Clean Basic Products | | | | | |
|---|---|---|---|---|---|
| | $\mathbf{v}_1$ | $\mathbf{v}_2$ | $\mathbf{v}_3$ | $\mathbf{v}_4$ | $\mathbf{v}_5$ |
| Diatomaceous Earth | 2 | 0 | 1 | 0 | 1 |
| Corn Cob | 2 | 3 | 1 | 1 | 2 |
| Sand | 0 | 1 | 0 | 0 | 0 |
| Recycled Rag Fiber | 0 | 0 | 0 | 2 | 0 |
| Shredded Cardboard | 0 | 0 | 2 | 1 | 0 |

   (a) Show that Sweep-Clean can produce any type of absorbent that contains the five ingredients listed in the preceding table using their basic mixtures.

   (b) The development team has noticed that there are a lot of requests for a mixture described by the 5-vector

$$\mathbf{u} = \begin{bmatrix} 0.5 \\ 2 \\ 0.5 \\ 0 \\ 1 \end{bmatrix}.$$

   They want to add it to their basic mixtures but still maintain only five basic mixtures. Which of $\mathbf{v}_1$ through $\mathbf{v}_5$ should be replaced by $\mathbf{u}$? Explain how you arrived at your decision.

19. For an abstract vector space $(V, +, \cdot)$ show that any set of vectors containing the zero vector is linearly dependent.

20. For an abstract vector space $(V, +, \cdot)$ show that any set of vectors in which one vector is a linear combination of other vectors in the set is linearly dependent.

21. For an abstract vector space $(V, +, \cdot)$ show that any linearly dependent set of vectors $S$ must contain one vector that is a linear combination of the other vectors of $S$.

22. For an abstract vector space $(V, +, \cdot)$ show that for a linearly dependent set of vectors $S$, span$(S)$ can be generated without using all the vectors in $S$.

23. For an abstract vector space $(V, +, \cdot)$ show that for a linearly independent set of vectors $S$, span$(S)$ cannot be generated without using all the vectors in $S$.

24. For an abstract vector space $(V, +, \cdot)$ show that if $\{\mathbf{v}_1, \mathbf{v}_2, \ldots, \mathbf{v}_k\}$ is a basis for $V$ and $c$ is a nonzero scalar then $\{c\mathbf{v}_1, \mathbf{v}_2, \ldots, \mathbf{v}_k\}$ is also a basis for V.

25. For an abstract vector space $(V, +, \cdot)$ show that any subset of a linearly independent set $S$ is also linearly independent.

26. Show by example in $R^3$ that a subset of a linearly dependent set may be either linearly dependent or linearly independent.

27. Describe a set $S$ of vectors that forms a basis for the vector space $P$ of all polynomials. How many vectors are in $S$? Explain why we say $P$ is not finite dimensional.

28. Use Exercise 25 to explain why we can say that vector spaces $C(a, b)$, $C[a, b]$, and $C(-\infty, \infty)$ are not finite dimensional.

29. Let $S$ be a linearly independent subset of a finite dimensional abstract vector space $(V, +, \cdot)$. Show that there is a basis for $V$ that includes the vectors in set $S$. [*Hint*: Let $S = \{\mathbf{v}_1, \mathbf{v}_2, \ldots, \mathbf{v}_k\}$, $\dim(V) = n \geq k$, and $\{\mathbf{w}_1, \mathbf{w}_2, \ldots, \mathbf{w}_n\}$ be a basis for $V$. Note that the set $\{\mathbf{v}_1, \mathbf{v}_2, \ldots, \mathbf{v}_k, \mathbf{w}_1, \mathbf{w}_2, \ldots, \mathbf{w}_n\}$ is linearly dependent. Argue that we can cast out $n - k$ of the $\mathbf{v}$'s.]

30. Let $(V, +, \cdot)$ be an abstract vector space with $\dim(V) = k$, and $S = \{\mathbf{v}_1, \mathbf{v}_2, \ldots, \mathbf{v}_k\}$ a set of $k$ vectors in $V$. Show each of the following.
   (a) If span$(S) = V$, then $S$ is a basis for $V$.
   (b) Any set of fewer than $k$ vectors cannot be a basis for $V$.

31. Show that if $S = \{\mathbf{v}_1, \mathbf{v}_2, \ldots, \mathbf{v}_k\}$ is a basis for a vector space $V$, then every vector in $V$ can be written in one and only one way as a linear combination of the vectors in $S$. (*Hint*: Let vector $\mathbf{v}$ in $V$ be expressed in two ways in terms of the basis $S$;

$$\mathbf{v} = \sum_{j=1}^{k} a_j \mathbf{v}_j \quad \text{and} \quad \mathbf{v} = \sum_{j=1}^{k} b_j \mathbf{v}_j.$$

Then show that $a_j = b_j$.)

32. Let $S = \{\mathbf{v}_1, \mathbf{v}_2, \ldots, \mathbf{v}_k\}$ be a set of nonzero vectors in a vector space $V$ such that every vector in $V$ can be written uniquely as a linear combination of the vectors in $S$. Show that $S$ is a basis for $V$. [*Hint*: Show that span$(S) = V$ and that $S$ is a linearly independent set.]

33. In our verification of the statement
   "If $S = \{\mathbf{v}_1, \mathbf{v}_2, \ldots, \mathbf{v}_k\}$ is a basis for a vector space $V$ and $T = \{\mathbf{w}_1, \mathbf{w}_2, \ldots, \mathbf{w}_r\}$ is a linearly independent subset of vectors in $V$, then $r \leq k$."
   We claimed that $\{\mathbf{w}_1, \mathbf{v}_1, \mathbf{v}_2, \ldots, \mathbf{v}_{j-1}, \mathbf{v}_{j+1}, \ldots, \mathbf{v}_k\}$ was a linearly independent set. To verify this, proceed as follows. Form a linear combination of vectors $\mathbf{w}_1, \mathbf{v}_1, \mathbf{v}_2, \ldots, \mathbf{v}_{j-1}, \mathbf{v}_{j+1}, \ldots, \mathbf{v}_k$, with unknown coefficients $a_i$, $i = 1, 2, \ldots, k$ and set it equal to the zero vector. Replace $\mathbf{w}_1$ by the expression in (8) and show that each $a_i$ is zero.

## True/False Review Questions

*Determine whether each of the following statements is true or false.*

1. If span$(S) = V$, a vector space, then the vectors in $S$ are also a basis for $V$.

2. If $B = \{\mathbf{v}_1, \mathbf{v}_2, \ldots, \mathbf{v}_k\}$ is a basis for vector space $V$, then every vector in $V$ can be expressed as a linear combination of the vectors in $B$.

3. Any set containing the zero vector cannot be a basis for a vector space.

4. If $\dim(V) = 4$ and $W$ is a subspace of $V$ with basis $\{\mathbf{w}_1, \mathbf{w}_2\}$, then we can find vectors $\mathbf{v}_1$ and $\mathbf{v}_2$ so that $\{\mathbf{w}_1, \mathbf{w}_2, \mathbf{v}_1, \mathbf{v}_2\}$ is a basis for $V$.

5. If $S$ is any linearly dependent set in vector space $V$, then span$(S)$ cannot be the same as $V$.

6. If $\dim(V) = n$ and $S = \{\mathbf{v}_1, \mathbf{v}_2, \ldots, \mathbf{v}_k\}$ is any linearly independent set in $V$, then $k \leq n$.

7. If $S$ is any linearly dependent set in vector space $V$, then at least one vector in $S$ is a linear combination of the other vectors in $S$.

8. If $\dim(V) = n$ and $S = \{\mathbf{v}_1, \mathbf{v}_2, \ldots, \mathbf{v}_k\}$ is any linearly dependent set in $V$, then $k \leq n$.

9. If $\dim(V) = n$, then every basis for $V$ must have exactly $n$ vectors.

10. If $\mathbf{A}$ is $m \times n$, then $\dim(\mathbf{row}(\mathbf{A})) \leq m$.

## Terminology

| | |
|---|---|
| Linearly dependent/independent sets | Basis |
| Natural basis or standard basis | Dimension of a vector space |

This section focused on bases for a vector space or subspace. There were results that indicated how to build bases from other sets and others that in effect told us that the number of vectors in a basis for a particular space did not change. We further characterized the way in which members of a vector space could be expressed in terms of a basis. All the ideas in this section point to the fact that *when we know a basis for a vector space, we in effect know everything about the space.* However, since there can be many bases for a vector space, a particular type of basis may have an advantage over another basis. We saw that this was the case in Chapter 4, and we will see this again in Chapter 6.

Respond to the following to review the ideas developed in this section. (We indicate the zero vector of a vector space $V$ by $\mathbf{z}$.)

- Why is $\{\mathbf{v}, \mathbf{w}, \mathbf{z}\}$ linearly dependent?
- Let $S = \{\mathbf{v}, \mathbf{w}\}$ be a subset of $V$. Given that $\mathbf{v} \neq \mathbf{w}$, explain why we are unable to determine if $S$ is linearly independent or linearly dependent.
- Let $S = \{\mathbf{v}, \mathbf{w}, \mathbf{v} + \mathbf{w}\}$ be a subset of $V$. Why can we say that $S$ is linearly dependent?
- If $S$ is a linearly dependent set, explain why we can say that it includes redundant information.
- If $T$ is a linearly dependent set and S is a subset of $T$, is it possible that span($S$) = span($T$)? Explain.
- If $T$ is a linearly independent set and $S$ is a subset of $T$, is it possible that span($S$) = span($T$)? Explain.
- What is the difference between a spanning set of a vector space and a basis?
- If we know $S$ is a linearly dependent set, explain how we determine a basis for span($S$).
- Explain how to determine if a set is a basis.
- If $V$ is not $R^n$, $C^n$, $R_{m \times n}$, or $C_{m \times n}$ what major idea from Chapters 1 through 4 is used as part of the process to determine if a set $S$ is a basis for $V$?
- What does dimension count?
- Why we can we say that the dimension of a vector space is invariant?
- If $S$ is a linearly independent set in $V$, how can we construct a basis for $V$ that contains $S$?
- If $T$ is a basis for $V$, then in how many ways can vector $\mathbf{v}$ in $V$ be expressed in terms of the vectors in $V$?

## CHAPTER TEST

1. Let $V$ be the set of all $2 \times 2$ real matrices with operations $\oplus$ and $\odot$ defined as follows:

$$\begin{bmatrix} a_1 & a_2 \\ a_3 & a_4 \end{bmatrix} \oplus \begin{bmatrix} b_1 & b_2 \\ b_3 & b_4 \end{bmatrix} = \begin{bmatrix} a_1 & b_2 \\ b_3 & a_4 \end{bmatrix}, \quad k \odot \begin{bmatrix} a_1 & a_2 \\ a_3 & a_4 \end{bmatrix} = \begin{bmatrix} ka_1 & ka_2 \\ ka_3 & ka_4 \end{bmatrix}.$$

Determine if $(V, \oplus, \odot)$ is a vector space.

**2.** Let $V = P_2$, the vector space of all polynomials of degree 2 or less. Let $W$ be all polynomials of the form $kx^2$, where $k$ is any (real) scalar. Determine if $W$ is a subspace of $V$.

**3.** Let $V = R^2$ and $W$ be all 2-vectors of the form $\begin{bmatrix} t \\ mt + b \end{bmatrix}$, $t$ any real number. For what values of $m$ and $b$ will $W$ be a subspace of $V$?

**4.** Explain why all subspaces of a vector space $V$ have at least one vector in common.

**5.** Let $V = R_{2 \times 2}$ and $W$ be the subspace of $V$ of all vectors of the form $\begin{bmatrix} a & a+b \\ 0 & b \end{bmatrix}$. Find a basis for $W$.

**6.** Let $V = C[0, 1]$ and $W$ be the set of all functions $f(t)$ in $V$ such that $\int_0^1 f(t)\, dt = 0$. Show that $W$ is a subspace of $V$.

**7.** Let $V = R_{2 \times 2}$ and $W = \text{span} \left\{ \begin{bmatrix} 1 & 0 \\ 0 & 0 \end{bmatrix}, \begin{bmatrix} 1 & 0 \\ 0 & 1 \end{bmatrix} \right\}$. Determine a basis for $V$ that contains $\begin{bmatrix} 1 & 0 \\ 0 & 0 \end{bmatrix}$ and $\begin{bmatrix} 1 & 0 \\ 0 & 1 \end{bmatrix}$.

**8.** Let $V = R_{2 \times 2}$ and $W$ be the subspace of all vectors $\mathbf{A}$ in $V$ such that $\mathbf{ns(A)}$ contains $\begin{bmatrix} 1 \\ 1 \end{bmatrix}$. Find a basis for $W$.

**9.** If $\dim(V) = k$, explain why any set of $k + 1$ (or more) vectors from $V$ is linearly dependent.

**10.** True or False:

(a) Every spanning set for $R^3$ contains at least 3 vectors.

(b) Every linearly independent set in $R^3$ contains at least 3 vectors.

(c) Every line in $R^2$ is a subspace.

(d) Every subspace of a vector space contains the zero vector.

(e) If $W$ is a subspace of $n$-dimensional space $V$ with $\dim(W) = n$, then $W = V$.

(f) $P$ is a subspace of $C(-\infty, \infty)$.

(g) Every vector space contains infinitely many vectors.

(h) If $U$ is a subspace of $W$, which is a subspace of $V$, then $U$ is also a subspace of $V$.

(i) If $\text{span}(S) = V$, then $S$ contains a basis for $V$.

(j) The set of all diagonal $2 \times 2$ matrices is a subspace of $R_{2 \times 2}$.

# 6

# *INNER PRODUCT SPACES*

So far our primary use of vector spaces and subspaces has been as algebraic structures. For instance, the algebra of linear combinations of vectors was used to
generate solutions of a homogeneous linear system $\mathbf{Ax} = \mathbf{0}$. We also used bases
of eigenvectors to develop approximations and compress information contained in
a matrix. However, using the dot product in $R^n$ we saw that we could determine the
length of a vector and, at least for $R^2$ and $R^3$, we developed the notion of an angle
between vectors. These concepts hinted at geometric aspects of vector spaces and
subspaces.

We begin with a brief review of the dot product in vector spaces $R^n$ and $C^n$. We
discuss orthogonal vectors, the length of a vector, and the distance between vectors.
In succeeding sections dot products and their properties are extended to abstract
vector spaces; in particular, function spaces—that is, vector spaces where the vectors are functions such as in $P_n$ and $C(-\infty, \infty)$. We also develop a general result on
how to define norms and the angle between vectors. We use these developments to
construct the projection of a vector onto a subspace. Finally, we revisit least squares
approximations in this more general setting.

## 6.1 ■ THE DOT PRODUCT IN $R^n$ AND $C^n$

The dot product of a pair of vectors $\mathbf{x}$ and $\mathbf{y}$ in $R^n$ was defined in Section 1.3 as

$$\mathbf{x} \cdot \mathbf{y} = \begin{bmatrix} x_1 & x_2 & \cdots & x_n \end{bmatrix} \cdot \begin{bmatrix} y_1 & y_2 & \cdots & y_n \end{bmatrix}$$

$$= x_1 y_1 + x_2 y_2 + \cdots + x_n y_n = \sum_{j=1}^{n} x_j y_j. \tag{1}$$

Recall that we write vectors from $R^n$ as either rows, as in this case, or columns. We frequently computed dot products of vectors from $R^n$ in Chapters 1 through 4. (See Examples 1 and 2 in Section 1.3.) [We refer to (1) as the **standard dot product** on $R^n$.]

The complex dot product of a pair of vectors $\mathbf{x}$ and $\mathbf{y}$ in $C^n$ was defined in Exercise 21 in Section 1.3 as

$$\overline{\mathbf{x}} \cdot \mathbf{y} = \sum_{j=1}^{n} \overline{x}_j y_j \tag{2}$$

where the bar indicates the conjugate. (See Appendix 1.) Since every real number is a complex number with imaginary part equal to zero, we could use (2) as the definition of the dot product in both $R^n$ and $C^n$. However, it is convenient at this time to introduce a notation for dot products that can be used for $R^n$ or $C^n$, where we let the nature of the vectors determine whether to use the computational steps in expressions (1) or (2). For vectors $\mathbf{x}$ and $\mathbf{y}$ let $(\mathbf{x}, \mathbf{y})$ denote the dot product. If $\mathbf{x}$ and $\mathbf{y}$ are in $R^n$, then we use expression (1), and if $\mathbf{x}$ and $\mathbf{y}$ are in $C^n$, then we use expression (2).

### EXAMPLE 1

(a)  Let $\mathbf{x} = \begin{bmatrix} 3 \\ 2 \\ -1 \\ 0 \end{bmatrix}$ and $\mathbf{y} = \begin{bmatrix} 2 \\ -2 \\ 5 \\ 8 \end{bmatrix}$. Then from (1) we have

$$(\mathbf{x}, \mathbf{y}) = (3)(2) + (2)(-2) + (-1)(5) + (0)(8) = 6 - 4 - 5 + 0 = -3.$$

(b)  Let $\mathbf{v} = \begin{bmatrix} 2 - i \\ 2i \\ 4 + 3i \end{bmatrix}$ and $\mathbf{w} = \begin{bmatrix} 1 + 2i \\ 3 - 2i \\ 2 \end{bmatrix}$. Then from (2) we have

$$(\mathbf{v}, \mathbf{w}) = (\overline{2 - i})(1 + 2i) + (\overline{2i})(3 - 2i) + (\overline{4 + 3i})(2) = 4 - 7i. \qquad ■$$

The **length** of a vector $\mathbf{x}$ in $R^n$ or $C^n$ is denoted by $\|\mathbf{x}\|$ and is determined from the expression

$$\|\mathbf{x}\| = \sqrt{(\mathbf{x}, \mathbf{x})} = \sqrt{\sum_{j=1}^{n} |x_j|^2}.$$

This is a generalization of the standard definition of length in a plane. (See Section 1.4.) The symbol $\|\mathbf{x}\|$ is read "the **norm** of $\mathbf{x}$." For $\mathbf{x} \neq \mathbf{0}$,

$$\mathbf{u} = \frac{1}{\|\mathbf{x}\|} \mathbf{x}$$

is called a **unit vector** and $\|\mathbf{u}\| = 1$.

EXAMPLE 2

(a) For $\mathbf{x} = \begin{bmatrix} 2 \\ -3 \\ 1 \end{bmatrix}$ we have unit vector

$$\frac{1}{\|\mathbf{x}\|}\mathbf{x} = \frac{1}{\sqrt{(\mathbf{x}, \mathbf{x})}}\begin{bmatrix} 2 \\ -3 \\ 1 \end{bmatrix} = \frac{1}{\sqrt{14}}\begin{bmatrix} 2 \\ -3 \\ 1 \end{bmatrix}.$$

(b) For $\mathbf{v} = \begin{bmatrix} 2 - i \\ 2i \\ 4 + 3i \end{bmatrix}$ we have unit vector

$$\frac{1}{\|\mathbf{v}\|}\mathbf{v} = \frac{1}{\sqrt{(\mathbf{v}, \mathbf{v})}}\begin{bmatrix} 2-i \\ 2i \\ 4+3i \end{bmatrix} = \frac{1}{\sqrt{\mathbf{v}\cdot\mathbf{v}}}\begin{bmatrix} 2-i \\ 2i \\ 4+3i \end{bmatrix} = \frac{1}{\sqrt{34}}\begin{bmatrix} 2-i \\ 2i \\ 4+3i \end{bmatrix}.$$

See also Exercise 15 in Section 1.4.    ■

We say that a pair of vectors $\mathbf{x}$ and $\mathbf{y}$ are **orthogonal** provided that $(\mathbf{x}, \mathbf{y}) = 0$; that is, their dot product is zero. In Chapter 4 we saw that orthogonal basis vectors play a fundamental role in a variety of situations.

EXAMPLE 3    If $\mathbf{x} = \begin{bmatrix} 3 \\ 2 \\ -1 \\ 4 \end{bmatrix}$, then vector $\mathbf{r} = \begin{bmatrix} 1 \\ s \\ 2 \\ t \end{bmatrix}$ is orthogonal to $\mathbf{x}$ for any scalars $s$ and $t$ that satisfy $2s + 4t = -1$. (Verify.) Thus there are infinitely many vectors that are orthogonal to $\mathbf{x}$. Also note that if $\mathbf{x}$ is any vector in $R^n$ with vector $\mathbf{r}$ orthogonal to $\mathbf{x}$, then $k\mathbf{r}$ is orthogonal to $\mathbf{x}$ for any real scalar $k$. (Verify.)    ■

The dot product gives us a tool with which we can define the distance between vectors. For a pair of vectors $\mathbf{x}$ and $\mathbf{y}$ in $R^n$ or $C^n$, the **distance** between the two vectors is given by

$$D(\mathbf{x}, \mathbf{y}) = \|\mathbf{x} - \mathbf{y}\|.$$

Thus we see that the distance between a pair of vectors is the norm of their difference.

EXAMPLE 4

(a) Let $\mathbf{x} = \begin{bmatrix} 3 \\ 2 \\ -1 \\ 0 \end{bmatrix}$ and $\mathbf{y} = \begin{bmatrix} 2 \\ -2 \\ 5 \\ 8 \end{bmatrix}$. Then

$$D(\mathbf{x}, \mathbf{y}) = \|\mathbf{x} - \mathbf{y}\| = \left\| \begin{bmatrix} 3 \\ 2 \\ -1 \\ 0 \end{bmatrix} - \begin{bmatrix} 2 \\ -2 \\ 5 \\ 8 \end{bmatrix} \right\| = \left\| \begin{bmatrix} 1 \\ 4 \\ -6 \\ -8 \end{bmatrix} \right\| = \sqrt{117}. \quad \text{(Verify).}$$

(b) Let $\mathbf{v} = \begin{bmatrix} 2-i \\ 2i \\ 4+3i \end{bmatrix}$ and $\mathbf{w} = \begin{bmatrix} 1+2i \\ 3-2i \\ 2 \end{bmatrix}$. Then

$$D(\mathbf{x} - \mathbf{y}) = \|\mathbf{x} - \mathbf{y}\| = \left\| \begin{bmatrix} 2-i \\ 2i \\ 4+3i \end{bmatrix} - \begin{bmatrix} 1+2i \\ 3-2i \\ 2 \end{bmatrix} \right\|$$

$$= \left\| \begin{bmatrix} 1-3i \\ -3+4i \\ 2+3i \end{bmatrix} \right\| = \sqrt{48}. \quad \text{(Verify.)}$$

■

In Section 6.2 we investigate vector spaces and dot products in a more general setting. There we obtain results that are independent of the nature of the vectors in the space.

## EXERCISES 6.1

**1.** Vectors $\mathbf{x} = \begin{bmatrix} 2 \\ 1 \\ 0 \end{bmatrix}$, $\mathbf{y} = \begin{bmatrix} 3 \\ 2 \\ -1 \end{bmatrix}$, and $\mathbf{s} = \begin{bmatrix} -1 \\ 4 \\ 2 \end{bmatrix}$ are in $R^3$.

(a) Compute the unit vector $\dfrac{1}{\|\mathbf{x}\|}\mathbf{x}$.

(b) Compute $(\mathbf{x}, \mathbf{y})$, $(\mathbf{x}, \mathbf{s})$, and $(\mathbf{x}, 2\mathbf{y} - 3\mathbf{s})$.

(c) Compute $D(\mathbf{x}, \mathbf{y})$ and $D(\mathbf{x}, \mathbf{s})$. Is $\mathbf{y}$ closer to $\mathbf{x}$ than to $\mathbf{s}$? Explain.

(d) Find a nonzero vector $\mathbf{v}$ in $R^3$ that is orthogonal to $\mathbf{s}$.

(e) Find a nonzero vector $\mathbf{w}$ in $R^3$ that is orthogonal to both $\mathbf{x}$ and $\mathbf{y}$.

**2.** Vectors $\mathbf{x} = \begin{bmatrix} 4-i \\ 2+2i \end{bmatrix}$, $\mathbf{y} = \begin{bmatrix} 1+i \\ 3i \end{bmatrix}$, and $\mathbf{s} = \begin{bmatrix} 1+i \\ 2 \end{bmatrix}$ are in $C^2$.

(a) Compute the unit vector $\dfrac{1}{\|\mathbf{x}\|}\mathbf{x}$.

(b) Compute $(\mathbf{x}, \mathbf{y})$, $(\mathbf{x}, \mathbf{s})$, and $(\mathbf{x}, \mathbf{y} + \mathbf{s})$.

(c) Compute $D(\mathbf{x}, \mathbf{y})$ and $D(\mathbf{x}, \mathbf{s})$. Is $\mathbf{x}$ closer to $\mathbf{y}$ than to $\mathbf{s}$? Explain.

**3.** Vectors $\mathbf{x} = \begin{bmatrix} 1 \\ 0 \\ -1 \end{bmatrix}$, $\mathbf{y} = \begin{bmatrix} 1 \\ 1 \\ 2 \end{bmatrix}$, and $\mathbf{s} = \begin{bmatrix} 1 \\ -3 \\ 1 \end{bmatrix}$ are in $R^3$.

(a) Show that both $\mathbf{x}$ and $\mathbf{y}$ are orthogonal to $\mathbf{s}$.

(b) Show that every vector in span($\{\mathbf{x}, \mathbf{y}\}$) is orthogonal to $\mathbf{s}$.

(c) Is $\mathbf{s}$ in span($\{\mathbf{x}, \mathbf{y}\}$)? Explain.

**4.** Vectors $\mathbf{x} = \begin{bmatrix} 2 \\ 1 \\ 2 \\ 1 \end{bmatrix}$, $\mathbf{y} = \begin{bmatrix} 0 \\ 1 \\ -1 \\ 0 \end{bmatrix}$, and $\mathbf{s} = \begin{bmatrix} 1 \\ -3 \\ -3 \\ 7 \end{bmatrix}$ are in $R^4$.

(a) Show that both $\mathbf{x}$ and $\mathbf{y}$ are orthogonal to $\mathbf{s}$.

(b) Show that every vector in span($\{\mathbf{x}, \mathbf{y}\}$) is orthogonal to $\mathbf{s}$.

(c) Is $\mathbf{s}$ in span($\{\mathbf{x}, \mathbf{y}\}$)? Explain.

**5.** Let $V$ be the vector space $R^3$. Let $W$ be the set of all vectors in $V$ that are orthogonal to $\mathbf{w} = \begin{bmatrix} 1 \\ 1 \\ -2 \end{bmatrix}$.

(a) Show that $W$ is a subspace of $V$.

(b) Find a basis for $W$.

(c) What is the dimension of $W$?

(d) For any vector $\mathbf{x}$ in $R^3$, show that we can express $\mathbf{x}$ as a linear combination of $\mathbf{w}$ and the basis found in part (b).

**6.** Let $V$ be the vector space $R^4$. Let $W$ be the set of all vectors in $V$ that are orthogonal to $\mathbf{w} = \begin{bmatrix} 1 \\ -1 \\ 0 \\ 1 \end{bmatrix}$.

(a) Show that $W$ is a subspace of $V$.

(b) Find a basis for $W$.

(c) What is the dimension of $W$?

(d) For any vector $\mathbf{x}$ in $R^4$, show that we can express $\mathbf{x}$ as a linear combination of $\mathbf{w}$ and the basis found in part (b).

**7.** Let $\mathbf{x} = \begin{bmatrix} 2 \\ t \end{bmatrix}$ and $\mathbf{y} = \begin{bmatrix} 1 \\ -2 \end{bmatrix}$. Determine all the values of $t$ so that $D(\mathbf{x}, \mathbf{y}) = 2$.

**8.** Let $\mathbf{x} = \begin{bmatrix} -1 \\ 2 \\ 3 \end{bmatrix}$ and $\mathbf{y} = \begin{bmatrix} t \\ -2 \\ 3 \end{bmatrix}$. Determine all the values of $t$ so that $D(\mathbf{x}, \mathbf{y}) = 5$.

**9.** Let $\mathbf{x} = \begin{bmatrix} 1 \\ 2 \\ t \end{bmatrix}$ and $\mathbf{y} = \begin{bmatrix} r \\ 1 \\ 3 \end{bmatrix}$.

(a) Make a sketch of all the ordered pairs $(t, r)$ such that $(\mathbf{x}, \mathbf{y}) = 0$.

(b) Make a sketch of all the ordered pairs $(t, r)$ such that $D(\mathbf{x}, \mathbf{y}) = 2$.

### In MATLAB

*In Section 1.3 we defined the dot product of any two n-vectors* $\mathbf{x}$ *and* $\mathbf{y}$. *As we stated at the end of Section 1.3, in* MATLAB *this computation is most easily performed when both vectors are columns, that is, $n \times 1$ matrices. Whether the vectors have real or complex entries, the dot product is given by* $\mathbf{x}' * \mathbf{y}$. *In the real case this just performs a row* ($\mathbf{x}'$) *by column* ($\mathbf{y}$) *product. In the complex case* $\mathbf{x}'$ *is a row of conjugates of the entries of column* $\mathbf{x}$ *so the result is our complex dot product.*

*(See Exercise* ML.1. *in Section 1.3.)*

### True/False Review Questions

*Determine whether each of the following statements is true or false.*

1. The length of vector $\mathbf{x}$ from $R^n$ is $(\mathbf{x}, \mathbf{x})$.

2. If $\mathbf{x}$ is any vector in $R^n$, then we can find a unit vector parallel to $\mathbf{x}$.

3. For $\mathbf{x}$ and $\mathbf{y}$ in $R^n$, the distance between them is $\sqrt{(\mathbf{x} - \mathbf{y}, \mathbf{x} - \mathbf{y})}$.

4. If $\mathbf{x}$, $\mathbf{y}$, and $\mathbf{w}$ are in $R^n$ with $(\mathbf{x}, \mathbf{w}) = (\mathbf{y}, \mathbf{w}) = 0$, then for any vector $\mathbf{z}$ in span$\{\mathbf{x}, \mathbf{y}\}$, $(\mathbf{z}, \mathbf{w}) = 0$.

5. If $\mathbf{x}$, $\mathbf{y}$ and $\mathbf{z}$ are in $R^3$ with $(\mathbf{x}, \mathbf{y}) = 0$ and $(\mathbf{y}, \mathbf{z}) = 0$, then $(\mathbf{x}, \mathbf{z}) = 0$.

### Terminology

| | |
|---|---|
| Dot product in $R^n$ and $C^n$ | The length or norm of a vector |
| Unit vector | Orthogonal vectors |
| The distance between vectors | |

The material in this section is a review of ideas covered in Sections 1.3 and 1.4. We have included these ideas again to lay a foundation for generalizations of the topics to abstract vector spaces in Section 6.2. A thorough understanding of the terminology and ideas of this section will make it easier to master the material in upcoming sections.

- How are dot products and row-by-column products related?
- In terms of the dot product, how do we compute the norm of a vector?
- Explain how to construct a unit vector from a nonzero vector $\mathbf{v}$ of $R^n$ or $C^n$.
- Under what condition are two vectors orthogonal?
- Sketch vector $\mathbf{v} = \begin{bmatrix} 1 \\ 1 \end{bmatrix}$ in $R^2$. Draw three different vectors that are orthogonal to $\mathbf{v}$.
- Give a geometric description of all vectors in $R^3$ orthogonal to $\mathbf{v} = \begin{bmatrix} 1 \\ 2 \\ 1 \end{bmatrix}$.
- How do we compute the distance between a pair of vectors in $R^n$ or $C^n$?
- What is the usual name for the distance between a pair of vectors in $R^1$?

## 6.2 ■ INNER PRODUCTS AND NORMS

In this section we develop generalizations of a dot product, called an **inner product**, for various vector spaces. In addition, we show how to develop **norms**, measures of length, for vectors in these spaces. In some cases norms are based on inner products, like in Section 6.1 for $R^n$. However, there are several popular norms that can be developed differently. We focus on real vector spaces and treat complex vector spaces in the exercises.

---

**Definition**   Let V be a real vector space. An **inner product** on $V$ is a function that assigns to each ordered pair of vectors **u** and **v** in $V$ a real number, denoted by $(\mathbf{u}, \mathbf{v})$, satisfying the following:
(a) $(\mathbf{u}, \mathbf{u}) > 0$ for $\mathbf{u} \neq \mathbf{z}$, the zero vector of $V$, and $(\mathbf{u}, \mathbf{u}) = 0$ if and only if $\mathbf{u} = \mathbf{z}$.
(b) $(\mathbf{u}, \mathbf{v}) = (\mathbf{v}, \mathbf{u})$ for any vectors **u** and **v** in $V$.
(c) $(\mathbf{u}, \mathbf{v} + \mathbf{w}) = (\mathbf{u}, \mathbf{v}) + (\mathbf{u}, \mathbf{w})$ for any vectors **u**, **v**, and **w** in $V$.
(d) $(k\mathbf{u}, \mathbf{v}) = k(\mathbf{u}, \mathbf{v})$ for any vectors **u** and **v** in $V$ and any real scalar $k$.

---

From the preceding definition it follows that the following properties are also valid:
(e) $(\mathbf{v} + \mathbf{w}, \mathbf{u}) = (\mathbf{v}, \mathbf{u}) + (\mathbf{w}, \mathbf{u})$ for any vectors **u**, **v**, and **w** in $V$.
(f) $(\mathbf{u}, k\mathbf{v}) = k(\mathbf{u}, \mathbf{v})$ for any vectors **u** and **v** in $V$ and any real scalar $k$.
(g) $\left( \mathbf{u}, \displaystyle\sum_{j=1}^{m} c_j \mathbf{v}_j \right) = \displaystyle\sum_{j=1}^{m} c_j (\mathbf{u}, \mathbf{v}_j)$ for any vectors **u** and $\mathbf{v}_1, \mathbf{v}_2, \ldots, \mathbf{v}_m$ in $V$ and any real scalars $c_1, c_2, \ldots, c_m$.

We leave the verification of properties (e), (f) and (g) to the exercises. (See Exercises 1 through 3.)

EXAMPLE 1    The standard dot product on $R^n$ (see Section 6.1) is an inner product on $R^n$. The verification of properties (c) and (d) is Exercise 15 of Section 1.3. To verify property (a), we proceed as follows:

$$(\mathbf{u}, \mathbf{u}) = \mathbf{u} \cdot \mathbf{u} = \sum_{j=1}^{n} u_j u_j = \sum_{j=1}^{n} (u_j)^2.$$

Thus for $\mathbf{u} \neq \mathbf{0}$, some $u_j \neq 0$ so $(\mathbf{u}, \mathbf{u}) > 0$. If $(\mathbf{u}, \mathbf{u}) = 0$, then $\displaystyle\sum_{j=1}^{n} (u_j)^2 = 0$. Hence each $u_j = 0$, so $\mathbf{u} = \mathbf{0}$. If $\mathbf{u} = \mathbf{0}$, then $u_j = 0$ for each $j$ and we have

$$(\mathbf{u}, \mathbf{u}) = \sum_{j=1}^{n} (u_j)^2 = 0.$$

To verify property (b), we proceed as follows:

$$(\mathbf{u}, \mathbf{v}) = \mathbf{u} \cdot \mathbf{v} = \sum_{j=1}^{n} u_j v_j = \sum_{j=1}^{n} v_j u_j = \mathbf{v} \cdot \mathbf{u} = (\mathbf{v}, \mathbf{u}).$$

We refer to the standard dot product on $R^n$ as the **standard inner product** on $R^n$. ■

EXAMPLE 2    Let $V = C[0, 1]$, the vector space of all continuous real valued functions on $[0, 1]$. (See Example 5 in Section 5.1.) For functions **f** and **g** in $C[0, 1]$, define

$$(\mathbf{f}, \mathbf{g}) = \int_0^1 \mathbf{f}(t)\mathbf{g}(t)\, dt.$$

We now verify that this is an inner product on $C[0, 1]$.

For Property (a): Using results from calculus, we have for $\mathbf{f} \neq \mathbf{0}$, the zero function,

$$(\mathbf{f}, \mathbf{f}) = \int_0^1 \mathbf{f}(t)\mathbf{f}(t)\, dt = \int_0^1 (\mathbf{f}(t))^2\, dt > 0.$$

Moreover, if $(\mathbf{f}, \mathbf{f}) = 0$, then $\mathbf{f} = \mathbf{0}$. Conversely, if $\mathbf{f} = \mathbf{0}$, then

$$(\mathbf{f}, \mathbf{f}) = \int_0^1 \mathbf{f}(t)\mathbf{f}(t)\, dt = \int_0^1 (\mathbf{0})^2\, dt = 0.$$

For Property (b): From properties for functions, we have

$$(\mathbf{f}, \mathbf{g}) = \int_0^1 \mathbf{f}(t)\mathbf{g}(t)\, dt = \int_0^1 \mathbf{g}(t)\mathbf{f}(t)\, dt = (\mathbf{g}, \mathbf{f}).$$

For Property (c): From properties of integrals, we have

$$(\mathbf{f}, \mathbf{g} + \mathbf{h}) = \int_0^1 \mathbf{f}(t)(\mathbf{g}(t) + \mathbf{h}(t))\, dt$$

$$= \int_0^1 \mathbf{f}(t)\mathbf{g}(t)\, dt + \int_0^1 \mathbf{f}(t)\mathbf{h}(t)\, dt$$

$$= (\mathbf{f}, \mathbf{g}) + (\mathbf{f}, \mathbf{h}).$$

For Property (d): From properties of integrals, we have

$$(k\mathbf{f}, \mathbf{g}) = \int_0^1 k\mathbf{f}(t)\mathbf{g}(t)\, dt = k \int_0^1 \mathbf{f}(t)\mathbf{g}(t)\, dt = k(\mathbf{f}, \mathbf{g}). \qquad ■$$

From Example 2 we have for $\mathbf{f}(t) = t^2$ and $\mathbf{g}(t) = t + 3$ that

$$(\mathbf{f}, \mathbf{g}) = \int_0^1 t^2(t + 3)\, dt = \int_0^1 (t^3 + 3t^2)\, dt = \left[\frac{t^4}{4} + t^3\right]_0^1 = \frac{5}{4}.$$

**EXAMPLE 3**    Let $V = P$, the vector space of all polynomials over interval $[0, 1]$. (See the comment immediately after Example 3 in Section 5.1.) If $\mathbf{p}$ and $\mathbf{q}$ are polynomials, we define

$$(\mathbf{p}, \mathbf{q}) = \int_0^1 \mathbf{p}(t)\mathbf{q}(t)\, dt.$$

Since this vector space is a subspace of vector space $C[0, 1]$ and the expression for $(\mathbf{p}, \mathbf{q})$ corresponds to that given in Example 2, this is an inner product for the vector space of all polynomials over the interval $[0, 1]$. As an example, we have for $\mathbf{p}(t) = t$ and $\mathbf{q}(t) = t^2 - 0.75t$ that

$$(\mathbf{p}, \mathbf{q}) = \int_0^1 t(t^2 - 0.75t)\, dt = \left[0.25t^4 - 0.25t^3\right]_0^1 = 0. \qquad ■$$

In Examples 2 and 3 the interval $[0, 1]$ can be replaced by any finite interval $[a, b]$. It follows that $(\mathbf{f}, \mathbf{g}) = \int_a^b \mathbf{f}(t)\mathbf{g}(t)\, dt$ is an inner product for both $C[a, b]$ and the vector space of all polynomials over interval $[a, b]$. In Example 3, if the interval were changed from $[0, 1]$ to $[-1, 1]$, then the value of the inner product can change. We have that

$$(\mathbf{p}, \mathbf{q}) = \int_{-1}^1 \mathbf{p}(t)\mathbf{q}(t)\, dt = \int_{-1}^1 t(t^2 - 0.75t)\, dt = \left[0.25t^4 - 0.25t^3\right]_{-1}^1 = -0.5.$$

Our next example shows how to construct an inner product on any finite-dimensional vector space. We use facts established about bases in Section 5.3.

**EXAMPLE 4**    Let $V$ be any finite-dimensional real vector space and let $S = \{\mathbf{u}_1, \mathbf{u}_2, \dots, \mathbf{u}_n\}$ be an **ordered basis**[1] for $V$. Then any vector in $V$ can be written as a linear combination of the vectors in $S$ in one and only one way. (See Section 5.3.) If $\mathbf{v}$ and $\mathbf{w}$ are in $V$ with

$$\mathbf{v} = \sum_{j=1}^{n} a_j \mathbf{u}_j = a_1 \mathbf{u}_1 + a_2 \mathbf{u}_2 + \dots + a_n \mathbf{u}_n$$

and

$$\mathbf{w} = \sum_{j=1}^{n} b_j \mathbf{u}_j = b_1 \mathbf{u}_1 + b_2 \mathbf{u}_2 + \dots + b_n \mathbf{u}_n$$

we define

$$(\mathbf{v}, \mathbf{w}) = a_1 b_1 + a_2 b_2 + \dots + a_n b_n.$$

The verification that this function is an inner product closely follows the procedure for verifying that the standard dot product on $R^n$ is an inner product. (See Example 1.) In this instance, we must keep in mind that the zero vector $\mathbf{z}$ of $V$ is the linear combination of the vectors in $S$ in which each coefficient is zero.    ■

**EXAMPLE 5**    Let $V = R^2$. For $\mathbf{u} = \begin{bmatrix} u_1 \\ u_2 \end{bmatrix}$ and $\mathbf{v} = \begin{bmatrix} v_1 \\ v_2 \end{bmatrix}$ define function

$$(\mathbf{u}, \mathbf{v}) = u_1 v_1 - u_2 v_1 - u_1 v_2 + 3 u_2 v_2.$$

We claim that this function is an inner product on $R^2$, of course, not the standard inner product as in Example 1. We verify Property (a) for an inner product and relegate the verification of the remaining three properties to Exercise 10.

For Property (a): 
$$\begin{aligned}
(\mathbf{u}, \mathbf{u}) &= u_1^2 - 2 u_1 u_2 + 3 u_2^2 \\
&= u_1^2 - 2 u_1 u_2 + u_2^2 + 2 u_2^2 \\
&= (u_1 - u_2)^2 + 2 u_2^2.
\end{aligned}$$

If $\mathbf{u} \neq \mathbf{0}$, then $(\mathbf{u}, \mathbf{u}) > 0$. Moreover, if $(\mathbf{u}, \mathbf{u}) = 0$, then $u_1 = u_2$ and $u_2 = 0$, so $\mathbf{u} = \mathbf{0}$. Conversely, if $\mathbf{u} = \mathbf{0}$, then $(\mathbf{u}, \mathbf{u}) = 0$.    ■

**EXAMPLE 6**    Let $V = R^3$. For $\mathbf{u} = \begin{bmatrix} u_1 \\ u_2 \\ u_3 \end{bmatrix}$ and $\mathbf{v} = \begin{bmatrix} v_1 \\ v_2 \\ v_3 \end{bmatrix}$ define function

$$(\mathbf{u}, \mathbf{v}) = \max \{|u_1 v_3|, |u_2 v_2|, |u_3 v_1|\}.$$

Is this function an inner product on $V$? In this case we see that for $\mathbf{u} = \begin{bmatrix} 1 \\ 0 \\ 0 \end{bmatrix}$, $(\mathbf{u}, \mathbf{u}) = 0$. (Verify.) Hence Property (a) is not satisfied, and so this function is not an inner product.    ■

---

[1] By an ordered basis, we mean a basis in which we maintain the order in which the vectors are listed. If $S = \{\mathbf{u}_1, \mathbf{u}_2, \dots, \mathbf{u}_n\}$ is an ordered basis for $V$, then $T = \{\mathbf{u}_2, \mathbf{u}_1, \dots, \mathbf{u}_n\}$ is a different ordered basis for $V$.

We make the following definition to provide an easy reference to real vector spaces that have an inner product.

---

**Definition**   A real vector space that has an inner product defined on it is called an **inner product space**. If the space is finite dimensional, it is called a **Euclidean space**.

---

As we have seen, an inner product space can have more than one inner product defined on it. (See Examples 1 and 5.) In certain situations it is advantageous to define an inner product that reflects aspects of the problem under study. This is done in order to take advantage of known facts about the particular space and to develop appropriate mechanisms to derive further information about a space. Such constructions are beyond the scope of our current development.

In a real inner product space $V$ we can use the inner product to develop a measure of the size of vectors and the distance between vectors. For $V = R^n$ with the standard inner product, the length or norm of a vector $\mathbf{x}$ is given by

$$\|\mathbf{x}\| = \sqrt{(\mathbf{x}, \mathbf{x})} = \sqrt{\sum_{j=1}^{n} |x_j|^2}.$$

(See Section 6.1.) For a general inner product space we make the following definition.

---

**Definition**   In an inner product space $V$, the (inner product) **norm** of a vector $\mathbf{v}$ is denoted $\|\mathbf{v}\|$ and computed by the expression

$$\|\mathbf{v}\| = \sqrt{(\mathbf{v}, \mathbf{v})}.$$

---

Thus the (inner product) norm of a vector $\mathbf{v}$ is the square root of the inner product of vector $\mathbf{v}$ with itself. Since $(\mathbf{v}, \mathbf{v}) \geq 0$, for any vector $\mathbf{v}$ in $V$, this is always defined. (We will generally use just the term *norm*.)

EXAMPLE 7

(a) For $R^2$ with the standard inner product, we have

$$\left\| \begin{bmatrix} 1 \\ -2 \end{bmatrix} \right\| = \sqrt{(1)(1) + (-2)(-2)} = \sqrt{5}.$$

(b) For the inner product space of polynomials $P$ in Example 3, we have

$$\|t\| = \sqrt{(t, t)} = \sqrt{\int_0^1 t^2\, dt} = \sqrt{\frac{1}{3}}.$$

(c) For the inner product space in Example 5, we have

$$\left\| \begin{bmatrix} 1 \\ -2 \end{bmatrix} \right\| = \sqrt{(1)^2 - 2(1)(-2) + 3(-2)^2} = \sqrt{17}.$$

■

The notation of double bars for a norm is no accident. It is used to indicate that norms are generalizations of absolute value, which provides a measure of the size of real numbers and a measure of the distance between numbers on the real line. In this regard, norms have two properties that are generalizations of properties of absolute value. We state these properties next and then discuss their verification.

If $V$ is an inner product space with norm given by $\|\mathbf{v}\| = \sqrt{(\mathbf{v}, \mathbf{v})}$, then

i. For any real scalar $k$ and any vector $\mathbf{v}$ in $V$, $\|k\mathbf{v}\| = |k|\,\|\mathbf{v}\|$.

ii. For any vectors $\mathbf{u}$ and $\mathbf{v}$ in $V$, $\|\mathbf{u} + \mathbf{v}\| \le \|\mathbf{u}\| + \|\mathbf{v}\|$.
   (This is called the **triangle inequality**.)

We leave the verification of property i as Exercise 43 since it follows directly from properties of the inner product. However, property ii, the triangle inequality, requires a more detailed investigation. As we shall see, we need a particular relationship between the dot product of a pair of vectors and the norms of the individual vectors. Let's begin with the following.

$$\|\mathbf{u} + \mathbf{v}\|^2 = (\mathbf{u} + \mathbf{v}, \mathbf{u} + \mathbf{v}) = (\mathbf{u}, \mathbf{u}) + 2(\mathbf{u}, \mathbf{v}) + (\mathbf{v}, \mathbf{v})$$
$$= \|\mathbf{u}\|^2 + 2(\mathbf{u}, \mathbf{v}) + \|\mathbf{v}\|^2.$$

We also have that

$$(\|\mathbf{u}\| + \|\mathbf{v}\|)^2 = \|\mathbf{u}\|^2 + 2\|\mathbf{u}\|\,\|\mathbf{v}\| + \|\mathbf{v}\|^2.$$

Comparing the right sides of these two expressions, we see that if $(\mathbf{u}, \mathbf{v}) \le \|\mathbf{u}\|\,\|\mathbf{v}\|$, then it follows that

$$\|\mathbf{u} + \mathbf{v}\|^2 = \|\mathbf{u}\|^2 + 2(\mathbf{u}, \mathbf{v}) + \|\mathbf{v}\|^2 \le \|\mathbf{u}\|^2 + 2\|\mathbf{u}\|\,\|\mathbf{v}\| + \|\mathbf{v}\|^2$$
$$= (\|\mathbf{u}\| + \|\mathbf{v}\|)^2$$

and upon taking the square root of both sides we get the triangle inequality. The following result, known as the **Cauchy-Schwarz-Bunyakovsky**[2] inequality, provides the appropriate condition. (We use the shortened term *Cauchy-Schwarz inequality*.)

### The Cauchy-Schwarz Inequality

If $\mathbf{u}$ and $\mathbf{v}$ are any two vectors in an inner product space $V$, then

$$|(\mathbf{u}, \mathbf{v})|^2 \le \|\mathbf{u}\|^2\,\|\mathbf{v}\|^2. \tag{1}$$

The verification of the Cauchy-Schwarz inequality is not difficult but requires a clever observation to start the process. In Exercise 44 we provide a step-by-step approach to establish this inequality. At the end of this section we provide an alternate approach that uses ideas that we developed earlier and introduces concepts that are useful in further generalizations of inner product spaces.

The Cauchy-Schwarz inequality appears in different forms depending upon the definition of the particular inner product in the space $V$. We illustrate this for several spaces in the next example.

---

[2]The result we refer to provides a good example of how nationalistic feelings make their way into science. In Russia this result is generally known as Bunyakovsky's inequality (after Viktor Yakovlevich Bunyakovsky, (1804–1889)). His proof of the Cauchy-Schwarz inequality appeared in one of his monographs in 1859, 25 years before Schwarz published his proof of the inequality. In France it is often referred to as Cauchy's inequality and in Germany it is frequently called Schwarz's inequality. In an attempt to distribute credit for the result among all three contenders, a minority of authors refer to the result as the CBS inequality. We adopt the term *Cauchy-Schwarz inequality*.

EXAMPLE 8

(a) In $R^n$ with the standard inner product, the Cauchy-Schwarz inequality implies that

$$(\mathbf{u}, \mathbf{v})^2 = \left(\sum_{j=1}^{n} u_j v_j\right)^2 \leq \left(\sum_{j=1}^{n} u_j\right)^2 \left(\sum_{j=1}^{n} v_j\right)^2 = \|\mathbf{u}\|^2 \|\mathbf{v}\|^2.$$

(b) In the inner product space of polynomials $P$ in Example 3, the Cauchy-Schwarz inequality implies that

$$(\mathbf{p}(t), \mathbf{q}(t))^2 = \left(\int_0^1 p(t)q(t)\, dt\right)^2$$

$$\leq \left(\int_0^1 p(t)^2\, dt\right) \left(\int_0^1 q(t)^2\, dt\right) = \|\mathbf{p}(t)\|^2 \|\mathbf{q}(t)\|^2. \quad \blacksquare$$

---

**Definition**    If $V$ is an inner product space, then the **distance between** two vectors $\mathbf{u}$ and $\mathbf{v}$ in $V$ is given by $D(\mathbf{u}, \mathbf{v}) = \|\mathbf{u} - \mathbf{v}\|$.

---

Thus the distance between vectors $\mathbf{u}$ and $\mathbf{v}$ is the square root of the inner product $\mathbf{u} - \mathbf{v}$ with itself: $D(\mathbf{u}, \mathbf{v}) = \|\mathbf{u} - \mathbf{v}\| = \sqrt{(\mathbf{u} - \mathbf{v}, \mathbf{u} - \mathbf{v})}$. This is a generalization of the distance between points in the plane since for $P(x_1, y_1)$ and $Q(x_2, y_2)$

$$D(P, Q) = \sqrt{(x_2 - x_1)^2 + (y_2 - y_1)^2} = \left\| \begin{bmatrix} x_1 \\ y_1 \end{bmatrix} - \begin{bmatrix} x_2 \\ y_2 \end{bmatrix} \right\|.$$

EXAMPLE 9    In the inner product space of polynomials $P$ in Example 3, to determine the distance between vectors $\mathbf{p}(t) = t^2$ and $\mathbf{q}(t) = 1 - t$, compute

$$D(\mathbf{p}(t), \mathbf{q}(t)) = \|\mathbf{p}(t) - \mathbf{q}(t)\| = \|t^2 - (1 - t)\| = \sqrt{\int_0^1 (t^2 - 1 + t)^2\, dt}$$

$$= \sqrt{\int_0^1 (t^4 + 2t^3 - t^2 - 2t + 1)\, dt} = \sqrt{\frac{11}{30}}. \quad \blacksquare$$

In our development so far, norms and hence the distance between vectors required that we have an inner product defined on a real vector space. That is, we were working in an inner product space. However, there are norms, and hence distance measures, that are used frequently that are not obtained from inner products. We introduce such concepts by first defining a norm and then giving several examples. In such situations we are working in a vector space $V$ but need not have an inner product defined on the space.

---

**Definition**    A **norm** on a real vector space $V$ is a function $\mathbf{N}: V \to R$, the real numbers, such that the following properties hold:

1. For any vector $\mathbf{v}$ in $V$ that is not the identity vector, $\mathbf{N}(\mathbf{v}) > 0$, and $\mathbf{N}(\mathbf{v}) = 0$ if and only if $\mathbf{v}$ is the identity vector.
2. $\mathbf{N}(k\mathbf{v}) = |k|\mathbf{N}(\mathbf{v})$, for any real scalar $k$ and any vector $\mathbf{v}$.
3. $\mathbf{N}(\mathbf{u} + \mathbf{v}) \leq \mathbf{N}(\mathbf{u}) + \mathbf{N}(\mathbf{v})$, for any vectors $\mathbf{u}$ and $\mathbf{v}$.

---

We showed previously that an inner product norm given by $\|\mathbf{v}\| = \sqrt{(\mathbf{v}, \mathbf{v})}$ satisfies these three properties. The following example introduces two norms on $R^n$ that are not derived from an inner product.

EXAMPLE 10   Let $V = R^n$.

(a) The **1-norm**, denoted $\|\mathbf{v}\|_1$, is defined by

$$N(\mathbf{v}) = \|\mathbf{v}\|_1 = |v_1| + |v_2| + \cdots + |v_n| = \sum_{j=1}^{n} |v_j|.$$

We leave the verification of three properties in the definition of a norm to Exercise 32.

(b) The **infinity norm**, denoted $\|\mathbf{v}\|_\infty$, is defined by

$$N(\mathbf{v}) = \|\mathbf{v}\|_\infty = \max\{|v_1|, |v_2|, \ldots, |v_n|\}.$$

We leave the verification of three properties in the definition of a norm to Exercise 33.   ■

The norm based on the standard inner product in $R^n$ is called the **2-norm**, following the notation of Example 10:

$$N(\mathbf{v}) = \|\mathbf{v}\|_2 = \sqrt{|v_1|^2 + |v_2|^2 + \cdots + |v_n|^2} = \sqrt{\sum_{j=1}^{n} |v_j|^2}.$$

A norm often used in approximation theory where we want to measure the distance from a function to a polynomial approximation in the vector space $C[a, b]$ (see Example 2) is called the **maximum norm** or **infinity norm** and is defined by

$$\|\mathbf{f}\|_\infty = \max_{[a,b]} |f(x)|.$$

The verification of the three properties in the definition of a norm is quite straightforward. In fact, the steps we use involve basic properties of absolute values. (See Exercise 34.)

## Verification of the Cauchy-Schwarz Inequality (Optional)

Let $V$ be an inner product space with $\mathbf{u}$ and $\mathbf{v}$ vectors in $V$, neither of which is the identity vector. Note that if either $\mathbf{u}$ or $\mathbf{v}$ is the identity vector of inner product space $V$, then the Cauchy-Schwarz inequality given in (1) is true since both sides would have numerical value zero. (Verify.) Define $2 \times 2$ matrix $\mathbf{G}$ as

$$\mathbf{G} = \begin{bmatrix} (\mathbf{u}, \mathbf{u}) & (\mathbf{u}, \mathbf{v}) \\ (\mathbf{v}, \mathbf{u}) & (\mathbf{v}, \mathbf{v}) \end{bmatrix}.$$

We note that $\mathbf{G}$ is symmetric since $(\mathbf{u}, \mathbf{v}) = (\mathbf{v}, \mathbf{u})$ in a real vector space. (Matrix $\mathbf{G}$ is called the **Grammian** of vectors $\mathbf{u}$ and $\mathbf{v}$.) We have the following properties of matrix $\mathbf{G}$.

1. For any vector $\mathbf{c} = \begin{bmatrix} c_1 \\ c_2 \end{bmatrix}$ in $R^2$, $\mathbf{c}^T \mathbf{G} \mathbf{c} \geq 0$. To verify this result we compute $\mathbf{c}^T \mathbf{G} \mathbf{c}$:

$$\mathbf{c}^T \mathbf{G} \mathbf{c} = \begin{bmatrix} c_1 & c_2 \end{bmatrix} \begin{bmatrix} (\mathbf{u}, \mathbf{u}) & (\mathbf{u}, \mathbf{v}) \\ (\mathbf{u}, \mathbf{v}) & (\mathbf{v}, \mathbf{v}) \end{bmatrix} \begin{bmatrix} c_1 \\ c_2 \end{bmatrix}$$

$$= \begin{bmatrix} c_1(\mathbf{u}, \mathbf{u}) + c_2(\mathbf{u}, \mathbf{v}) \\ c_1(\mathbf{u}, \mathbf{v}) + c_2(\mathbf{v}, \mathbf{v}) \end{bmatrix} \begin{bmatrix} c_1 \\ c_2 \end{bmatrix}$$

$$= c_1^2(\mathbf{u}, \mathbf{u}) + 2c_1 c_2(\mathbf{u}, \mathbf{v}) + c_2^2(\mathbf{v}, \mathbf{v})$$

$$= (c_1\mathbf{u}, c_1\mathbf{u}) + (c_1\mathbf{u}, c_2\mathbf{v}) + (c_2\mathbf{v}, c_1\mathbf{u}) + (c_2\mathbf{v}, c_2\mathbf{v})$$

$$= [(c_1\mathbf{u}, c_1\mathbf{u}) + (c_1\mathbf{u}, c_2\mathbf{v})] + [(c_2\mathbf{v}, c_1\mathbf{u}) + (c_2\mathbf{v}, c_2\mathbf{v})]$$

$$= (c_1\mathbf{u}, c_1\mathbf{u} + c_2\mathbf{v}) + (c_2\mathbf{v}, c_1\mathbf{u} + c_2\mathbf{v})$$

$$= (c_1\mathbf{u} + c_2\mathbf{v}, c_1\mathbf{u} + c_2\mathbf{v})$$

$$= \|c_1\mathbf{u} + c_2\mathbf{v}\|^2 \geq 0.$$

2. All the eigenvalues of $\mathbf{G}$ are $\geq 0$. To verify this result, let $(\lambda, \mathbf{y})$ be an eigenpair of $\mathbf{G}$. Then $\mathbf{G}\mathbf{y} = \lambda\mathbf{y}$. Multiplying by $\mathbf{y}^T$ on the left gives $\mathbf{y}^T \mathbf{G}\mathbf{y} = \lambda\mathbf{y}^T\mathbf{y}$. Since $\mathbf{y}$ is an eigenvector it is not the zero vector, so $\mathbf{y}^T\mathbf{y} \neq 0$. Hence we can solve for $\lambda$:

$$\lambda = \frac{\mathbf{y}^T\mathbf{G}\mathbf{y}}{\mathbf{y}^T\mathbf{y}}.$$

By this result the numerator is $\geq 0$ and the denominator is positive; hence $\lambda \geq 0$. It follows that all the eigenvalues of $\mathbf{G}$ are $\geq 0$.

3. $\det(\mathbf{G}) \geq 0$. To verify this, recall from Chapter 4 that a symmetric matrix is orthogonally similar to a diagonal matrix $\mathbf{D}$ whose diagonal entries are its eigenvalues. Hence there exists an orthogonal matrix $\mathbf{P}$ (recall $\mathbf{P}^T = \mathbf{P}^{-1}$) such that $\mathbf{G} = \mathbf{P}\mathbf{D}\mathbf{P}^T$. Hence

$$\det(\mathbf{G}) = \det(\mathbf{P}) \det(\mathbf{D}) \det(\mathbf{P}^T)$$

$$= \det(\mathbf{D}) \qquad \text{(verify)}$$

$$= \text{product of the eigenvalues of } \mathbf{G} \geq 0.$$

The Cauchy-Schwarz inequality $(\mathbf{u}, \mathbf{v})^2 \leq \|\mathbf{u}\|^2 \|\mathbf{v}\|^2$ follows from the preceding results since $\det(\mathbf{G}) = (\mathbf{u}, \mathbf{u})(\mathbf{v}, \mathbf{v}) - |(\mathbf{u}, \mathbf{v})|^2 \geq 0$.

## EXERCISES 6.2

1. In an inner product space $V$, verify that $(\mathbf{v} + \mathbf{w}, \mathbf{u}) = (\mathbf{v}, \mathbf{u}) + (\mathbf{w}, \mathbf{u})$ for any vectors $\mathbf{u}$, $\mathbf{v}$, and $\mathbf{w}$ in $V$.

2. In an inner product space $V$, verify that $(\mathbf{u}, k\mathbf{v}) = k(\mathbf{u}, \mathbf{v})$ for any vectors $\mathbf{u}$ and $\mathbf{v}$ in $V$ and any real scalar $k$.

3. In an inner product space $V$, verify that

$$\left( \mathbf{u}, \sum_{j=1}^{m} c_j \mathbf{v}_j \right) = \sum_{j=1}^{m} c_j (\mathbf{u}, \mathbf{v}_j)$$

for any vectors $\mathbf{u}$ and $\mathbf{v}_1, \mathbf{v}_2, \ldots, \mathbf{v}_m$ in $V$ and any real scalars $c_1, c_2, \ldots, c_m$.

*Let $\mathbf{u}$, $\mathbf{v}$, and $\mathbf{w}$ be vectors in the inner product space $R^3$ with the standard inner product. In Exercises 4–6, compute each*

*indicated expression given that*

$$\mathbf{u} = \begin{bmatrix} 1 \\ 2 \\ -1 \end{bmatrix}, \quad \mathbf{v} = \begin{bmatrix} 3 \\ 0 \\ 4 \end{bmatrix}, \quad and \quad \mathbf{w} = \begin{bmatrix} -2 \\ 1 \\ 5 \end{bmatrix}.$$

4. $(\mathbf{u}, 3\mathbf{v} + 2\mathbf{w})$.

5. $(4\mathbf{u}, -3\mathbf{v})$.

6. $\dfrac{(\mathbf{u}, \mathbf{v})}{(\mathbf{u}, \mathbf{u})}\mathbf{v} + \dfrac{(\mathbf{u}, \mathbf{w})}{(\mathbf{u}, \mathbf{u})}\mathbf{w}$.

*For the inner product space in Example 2, compute the expressions in Exercises 7–9.*

7. $(t, t)$.

8. $(t, \sin(2\pi t))$.

9. $\left( \sqrt{t^2 + 1}, t \right)$.

**10.** Verify properties (b) through (d) of an inner product for the function defined in Example 5.

*Let $V = R^2$, $\mathbf{u} = \begin{bmatrix} u_1 \\ u_2 \end{bmatrix}$, and $\mathbf{v} = \begin{bmatrix} v_1 \\ v_2 \end{bmatrix}$. Determine whether each given function in Exercises 11–13 is an inner product.*

**11.** $(\mathbf{u}, \mathbf{v}) = u_1v_1 + 4u_2v_2$.

**12.** $(\mathbf{u}, \mathbf{v}) = (u_1 - v_1)^2 + (u_2 + v_2)^2$.

**13.** $(\mathbf{u}, \mathbf{v}) = u_1v_1 + u_2v_1 + u_1v_2 + u_2v_2$.

**14.** Let $V = R^3$ and $\mathbf{u} = \begin{bmatrix} u_1 \\ u_2 \\ u_3 \end{bmatrix}$ and $\mathbf{v} = \begin{bmatrix} v_1 \\ v_2 \\ v_3 \end{bmatrix}$. Determine whether $(\mathbf{u}, \mathbf{v}) = u_1v_3 + u_3v_1$ is an inner product.

**15.** Let $V$ be the vector space of all real $2 \times 2$ diagonal matrices. (See Exercise 5 in Section 5.2.) For vectors

$$\mathbf{A} = \begin{bmatrix} a_{11} & 0 \\ 0 & a_{22} \end{bmatrix} \quad \text{and} \quad \mathbf{B} = \begin{bmatrix} b_{11} & 0 \\ 0 & b_{22} \end{bmatrix}$$

in $V$, define the function $(\mathbf{A}, \mathbf{B}) = a_{11}b_{11} + a_{22}b_{22}$. Determine whether this function is an inner product.

**16.** Let $V$ be the vector space of all real $2 \times 2$ matrices. For vectors $\mathbf{A}$ and $\mathbf{B}$ in $V$, define the function $(\mathbf{A}, \mathbf{B}) = \det(\mathbf{AB})$. Determine whether this function is an inner product.

*Let $V$ be a real inner product space with identity vector $\mathbf{z}$. Verify each statement in Exercises 17–20.*

**17.** $(\mathbf{z}, \mathbf{z}) = 0$.

**18.** $(\mathbf{v}, \mathbf{z}) = (\mathbf{z}, \mathbf{v}) = 0$ for any vector $\mathbf{v}$ in $V$.

**19.** If $(\mathbf{u}, \mathbf{v}) = 0$ for all vectors $\mathbf{v}$ in $V$, then $\mathbf{u} = \mathbf{z}$.

**20.** If $(\mathbf{u}, \mathbf{w}) = (\mathbf{v}, \mathbf{w})$ for all $\mathbf{w}$ in $V$, then $\mathbf{u} = \mathbf{v}$.

**21.** Compute the length of vectors $\mathbf{u}$, $\mathbf{v}$, and $\mathbf{w}$ given in the statement preceding Exercise 4.

*For the inner product space in Example 2, compute the given expression in Exercises 22–25.*

**22.** $\|t\|$.

**23.** $\|e^t\|$.

**24.** $\left\|\sqrt{t^2 + 1}\right\|$.

**25.** $\|\sin(2\pi t)\|$.

**26.** Let $\mathbf{u}$ and $\mathbf{v}$ be any vectors in an inner product space. Verify the parallelogram law

$$\|\mathbf{u} + \mathbf{v}\|^2 + \|\mathbf{u} - \mathbf{v}\|^2 = 2\|\mathbf{u}\|^2 + 2\|\mathbf{v}\|^2.$$

**27.** Let $V$ be a real inner product space with the distance between vectors in $\mathbf{u}$ and $\mathbf{v}$ given by $D(\mathbf{u}, \mathbf{v}) = \|\mathbf{u} - \mathbf{v}\| = \sqrt{(\mathbf{u} - \mathbf{v}, \mathbf{u} - \mathbf{v})}$. Verify that $D(\mathbf{u}, \mathbf{v}) = D(\mathbf{v}, \mathbf{u})$.

*For the inner product space in Example 2, determine the distance between each pair of vectors in Exercises 28–31.*

**28.** $t, t^2$.

**29.** $t^2 + 1, t^2 - 1$.

**30.** $\sin(t), \cos(t)$.

**31.** $2t + 1, 1$.

**32.** Verify that $\mathbf{N}(\mathbf{v}) = \|\mathbf{v}\|_1$ in Example 10(a) is a norm.

**33.** Verify that $\mathbf{N}(\mathbf{v}) = \|\mathbf{v}\|_\infty$ in Example 10(b) is a norm.

**34.** Verify that $\mathbf{N}(\mathbf{f}) = \|\mathbf{f}\|_\infty = \max_{[0,1]} |f(x)|$ is a norm for the vector space $C[0, 1]$.

*For the vectors $\mathbf{u}$, $\mathbf{v}$, and $\mathbf{w}$ in $R^3$ given in statement preceding Exercise 4, compute the following.*

**35.** $\|\mathbf{u}\|_1, \|\mathbf{u}\|_\infty$.

**36.** $\|\mathbf{v}\|_1, \|\mathbf{v}\|_\infty$.

**37.** $\|\mathbf{w}\|_1, \|\mathbf{w}\|_\infty$.

**38.** For $\mathbf{v}$ in $R^n$, verify that $\|\mathbf{v}\|_\infty \leq \|\mathbf{v}\|_1$ and $\|\mathbf{v}\|_1 \leq n\|\mathbf{v}\|_\infty$.

**39.** Draw a picture of the set of vectors $\mathbf{v}$ in $R^2$ such that $\|\mathbf{v}\|_1 = 1$.

**40.** Draw a picture of the set of vectors $\mathbf{v}$ in $R^2$ such that $\|\mathbf{v}\|_2 = 1$.

**41.** Draw a picture of the set of vectors $\mathbf{v}$ in $R^2$ such that $\|\mathbf{v}\|_\infty = 1$.

**42.** Let $V = R^n$ with the standard inner product. Let $\mathbf{P}$ be an $n \times n$ orthogonal matrix. (See Section 4.4.) Verify that $\|\mathbf{Pv}\| = \|\mathbf{v}\|$ for any vector $\mathbf{v}$ in $V$. (This implies that a matrix transformation using an orthogonal matrix preserves lengths of vectors.)

**43.** If $V$ is a real inner product space with the norm given by $\|\mathbf{v}\| = \sqrt{(\mathbf{v}, \mathbf{v})}$, then show that for any real scalar $k$ and any vector $\mathbf{v}$ in $V$, $\|k\mathbf{v}\| = |k|\,\|\mathbf{v}\|$.

**44.** If $\mathbf{u}$ or $\mathbf{v}$ is the identity vector of an inner product space $V$, then $(\mathbf{u}, \mathbf{v})^2 = 0$ and $\|\mathbf{u}\|^2\,\|\mathbf{v}\|^2 = 0$, so the Cauchy-Schwarz inequality holds. Let $\mathbf{u}$ be a vector in $V$ which is not the identity vector.

(a) For a real scalar $r$, show that $(r\mathbf{u} + \mathbf{v}, r\mathbf{u} + \mathbf{v})$ can be written in the form $ar^2 + 2br + c$ where $a = (\mathbf{u}, \mathbf{u})$, $b = (\mathbf{u}, \mathbf{v})$, and $c = (\mathbf{v}, \mathbf{v})$.

(b) Let $p(r) = ar^2 + 2br + c$ as in part (a). Explain why $p(r) \geq 0$ for all values of $r$.

(c) Use the fact that $p(r) \geq 0$ for all values of $r$ to explain why the quadratic $p(r)$ has at most one real root.

(d) Given that $p(r)$ has at most one real root, explain why $4b^2 - 4ac \leq 0$.

(e) Given that $4b^2 - 4ac \leq 0$, verify the Cauchy-Schwarz inequality.

**45.** Let **u** and **v** be vectors in inner product space $V$, neither of which is the identity vector. Show that

$$-1 \leq \frac{(\mathbf{u}, \mathbf{v})}{\|\mathbf{u}\| \|\mathbf{v}\|} \leq 1.$$

**Definition**   Let $W$ be a complex vector space. A **complex inner product** on $W$ is a function that assigns to each ordered pair of vectors **u** and **v** in $V$ a complex number, denoted by $(\mathbf{u}, \mathbf{v})$, satisfying the following:

(a) $(\mathbf{u}, \mathbf{u}) > 0$ for $\mathbf{u} \neq \mathbf{z}$, the zero vector of $W$, and $(\mathbf{u}, \mathbf{u}) = 0$ if and only if $\mathbf{u} = \mathbf{z}$.

(b) $(\mathbf{u}, \mathbf{v}) = \overline{(\mathbf{v}, \mathbf{u})}$ for any vectors **u** and **v** in $W$.

(c) $(\mathbf{v} + \mathbf{w}, \mathbf{u}) = (\mathbf{v}, \mathbf{u}) + (\mathbf{w}, \mathbf{u})$ for any vectors **u**, **v**, and **w** in $W$.

(d) $(k\mathbf{u}, \mathbf{v}) = \overline{k}(\mathbf{u}, \mathbf{v})$ for any vectors **u** and **v** in $W$ and any complex scalar $k$.

**Definition**   A complex vector space that has an **inner product** defined on it is called a **complex inner product space** or a **unitary space**.

**46.** Let $W = C^n$ be the vector space of $n$-vectors that are columns with complex entries. Define $(\mathbf{u}, \mathbf{v}) = \overline{\mathbf{u}}^T \mathbf{v} = \mathbf{u}^* \mathbf{v}$. (See Section 1.1 and Exercise 21 in Section 1.3.) Verify that this function is a complex inner product and hence $C^n$ is a unitary space.

**47.** If **u** and **v** are vectors in a unitary space, then verify that

$$\det\left(\begin{bmatrix} (\mathbf{u}, \mathbf{u}) & (\mathbf{u}, \mathbf{v}) \\ (\mathbf{v}, \mathbf{u}) & (\mathbf{v}, \mathbf{v}) \end{bmatrix}\right) = (\mathbf{u}, \mathbf{u})(\mathbf{v}, \mathbf{v}) - |(\mathbf{u}, \mathbf{v})|^2$$

(It follows that the Cauchy-Schwarz inequality is valid in a unitary space.)

## True/False Review Questions

*Determine whether each of the following statements is true or false.*

**1.** In an inner product space, we can define the length of a vector as the square root of the inner product of the vector with itself.

**2.** The triangle inequality is valid in every inner product space.

**3.** In an inner product space, the distance between two vectors is the norm of their difference.

**4.** The Cauchy-Schwarz Inequality is valid in every inner product space.

**5.** In vector space $C[a, b]$, $\|\mathbf{f}\|_\infty = 0$ if and only if $f(x) = 0$ for all $x$ in $[a, b]$.

## Terminology

| | |
|---|---|
| Inner Product | Standard inner product on $R^n$ |
| Ordered basis | Real inner product space |
| Euclidean space | Norm |
| Triangle inequality | Cauchy-Schwarz inequality |
| Distance between vectors | 1-, 2-, infinity norm |
| Complex inner product space | Unitary space |

The focus of this section is real vector spaces on which an inner product has been defined. We developed properties of such spaces that are analogous to those in $R^n$ with the standard inner product. We further showed how to determine the length of a vector and the distance between vectors. To review the notions developed in this section, respond to the following statements and questions.

● In order for a function defined on a pair of vectors from a real vector space to be called an inner product, it must satisfy three properties. List these properties and discuss their meaning.

- Let $V$ be a real vector space with an inner product and let each of the following be a linear combination of vectors in $V$:

$$\mathbf{u} = \sum_{i=1}^{3} c_i \mathbf{u}_i, \quad \mathbf{v} = \sum_{j=1}^{2} c_j \mathbf{v}_j.$$

  Display the inner product of $\mathbf{u}$ and $\mathbf{v}$ in as simple terms as possible.
- Explain why ordered bases are needed in Example 4.
- What is an inner product space? How are a Euclidean space and an inner product space related?
- What is a unitary space? How is it different from an inner product space?
- Explain how we use an inner product to define a norm.
- What physical interpretation do we give to the norm of a vector?
- In this section the Cauchy-Schwarz inequality is used to establish the triangle inequality. Explain the algebraic steps involved.
- How is the distance between two vectors defined in an inner product space?
- A norm can be defined on a vector space without an inner product. Explain how this is done.
- In words, explain each of the following, where $\mathbf{v}$ is in $R^n$: $\|\mathbf{v}\|_1$, $\|\mathbf{v}\|_2$, and $\|\mathbf{v}\|_\infty$.

## 6.3 ■ ORTHOGONAL BASES AND PROJECTIONS

In this section we continue our investigation of inner product spaces. Here we focus on geometrically oriented relationships between vectors. Using the Cauchy-Schwarz inequality from Section 6.2, we develop a measure of angles between vectors. Next we show how to obtain bases for inner product spaces that consist of vectors that are mutually "perpendicular." Then we show how properties of such bases are used to easily express a vector as a linear combination of the basis vectors. In addition, we use these bases to develop approximations to vectors in the space in terms of vectors from specified subspaces. Our developments generalize corresponding ideas for $R^n$ that appeared in Sections 1.4, 1.5, 4.2, 4.5, and 4.6.

In Section 6.2 we developed the Cauchy-Schwarz inequality, which says that for a pair of vectors $\mathbf{u}$ and $\mathbf{v}$ in an inner product space (or unitary space) $V$

$$|(\mathbf{u}, \mathbf{v})|^2 \leq \|\mathbf{u}\|^2 \|\mathbf{v}\|^2$$

where $\|\mathbf{u}\| = \sqrt{(\mathbf{u}, \mathbf{u})}$. For inner product spaces, Exercise 45 in Section 6.2 tells us that

$$-1 \leq \frac{(\mathbf{u}, \mathbf{v})}{\|\mathbf{u}\| \|\mathbf{v}\|} \leq 1$$

for any pair of nonzero[1] vectors $\mathbf{u}$ and $\mathbf{v}$. It follows that there is a unique angle $\theta$ between $0$ and $\pi$ such that

$$\cos(\theta) = \frac{(\mathbf{u}, \mathbf{v})}{\|\mathbf{u}\| \|\mathbf{v}\|}. \tag{1}$$

---

[1] By *nonzero* we mean that neither $\mathbf{u}$ nor $\mathbf{v}$ is the identity vector $\mathbf{z}$ of the space.

**Definition**    The **angle** $\theta$ between a pair of nonzero vectors **u** and **v** in an inner product space is given by

$$\theta = \arccos\left(\frac{(\mathbf{u}, \mathbf{v})}{\|\mathbf{u}\|\,\|\mathbf{v}\|}\right).$$

Equivalently, we say that the angle $\theta$ satisfies Equation (1) with $0 \leq \theta \leq \pi$.

EXAMPLE 1    Let $V = R^2$ and $\mathbf{u} = \begin{bmatrix} 1 \\ -3 \end{bmatrix}$ and $\mathbf{v} = \begin{bmatrix} 6 \\ 2 \end{bmatrix}$.

(a) For the standard inner product we have that the angle $\theta$ between **u** and **v** satisfies

$$\cos(\theta) = \frac{(\mathbf{u}, \mathbf{v})}{\|\mathbf{u}\|\,\|\mathbf{v}\|} = \frac{0}{\sqrt{10}\,\sqrt{40}} = 0. \qquad \text{(Verify.)}$$

Hence $\theta = \pi/2$ radians.

(b) For the inner product of Example 5 in Section 6.2 we have that the angle $\theta$ between **u** and **v** satisfies

$$\cos(\theta) = \frac{(\mathbf{u}, \mathbf{v})}{\|\mathbf{u}\|\,\|\mathbf{v}\|} = \frac{4}{\sqrt{34}\,\sqrt{24}}. \qquad \text{(Verify.)}$$

Hence $\theta$ is approximately 1.43 radians; $\theta \approx 1.43$ radians.    ■

From Example 1 we see that different inner products generally produce different angle measures between vectors. In fact, the concept of an angle between vectors in an inner product space often does not agree with geometric concepts that we may attach to 'vectors' when the objects are used outside the structure of an inner product space. We illustrate this in Example 2. It is convenient to introduce the following terminology since in much of our work we will now use inner product spaces like that in Example 2. An inner product space whose vectors are functions, as opposed to $n$-vectors, will be called a **function space**; $C[a, b]$ and $P$ are function spaces.

EXAMPLE 2    Let $V$ be the function space $C[0, 1]$ with inner product

$$(\mathbf{f}, \mathbf{g}) = \int_0^1 f(t)g(t)\,dt.$$

The angle $\theta$ between $f(t) = t$ and $g(t) = t + 1$ satisfies

$$\cos(\theta) = \frac{(\mathbf{f}, \mathbf{g})}{\|\mathbf{f}\|\,\|\mathbf{g}\|} = \frac{\int_0^1 t(t+1)\,dt}{\sqrt{\int_0^1 t^2\,dt}\,\sqrt{\int_0^1 (t+1)^2\,dt}} = \frac{\frac{5}{6}}{\sqrt{\frac{1}{3}}\,\sqrt{\frac{7}{3}}} \approx 0.9449.$$

(Verify.) It follows that $\theta \approx 0.3335$ radians $\approx 19.1066°$. Since geometrically $y = f(t)$ and $y = g(t)$ are parallel lines, the concept of the angle between them is not based on their graphs but solely relies on the definition of the inner product.    ■

The trigonometric functions sine and cosine play prominent roles in a variety of applications. A number of applications employ these functions and powers of these functions in the function space $C[-\pi, \pi]$. Example 3 shows an important relationship between sine and cosine.

EXAMPLE 3   Let $V$ be the function space $C[-\pi.\pi]$ with inner product

$$(\mathbf{f}, \mathbf{g}) = \int_{-\pi}^{\pi} f(t)g(t)\,dt.$$

The angle between $f(t) = \cos(t)$ and $g(t) = \sin(t)$ satisfies

$$\cos(\theta) = \frac{(\mathbf{f}, \mathbf{g})}{\|\mathbf{f}\|\,\|\mathbf{g}\|} = \frac{\displaystyle\int_{-\pi}^{\pi} \cos(t)\sin(t)\,dt}{\sqrt{\displaystyle\int_{-\pi}^{\pi} \cos^2(t)\,dt}\ \sqrt{\displaystyle\int_{-\pi}^{\pi} \sin^2(t)\,dt}} = \frac{0}{\sqrt{\pi}\,\sqrt{\pi}} = 0.$$

It follows that $\theta = \pi/2$ radians.   ■

In Sections 4.3 through 4.7 we saw that mutually perpendicular $n$-vectors could be used to gain information about matrices. The same is true in inner product spaces. Thus we have the following.

---

**Definition**   Vectors $\mathbf{u}$ and $\mathbf{v}$ in an inner product space $V$ are called **orthogonal** provided $(\mathbf{u}, \mathbf{v}) = 0$.

---

The vectors in Example 1(a) are orthogonal and the vectors in Example 3 are orthogonal. Again from Sections 4.3 through 4.7 we saw that sets of orthogonal vectors in $R^n$ were used to reveal information or restructure information. Thus we have the following.

---

**Definition**   Let $V$ be an inner product space. A set $S$ of nonzero vectors in $V$ is called an **orthogonal set** if any two distinct vectors in $S$ are orthogonal. If, in addition, each vector in $S$ is of unit length, then $S$ is called an **orthonormal set**.

---

We note that if $\mathbf{v}$ is a nonzero vector of an inner product space, then we can always find a vector $\mathbf{u}$ of unit length, called a **unit vector**, that makes the same angle as $\mathbf{v}$ with any other vector[2]. If we let

$$\mathbf{u} = \frac{1}{\|\mathbf{v}\|}\mathbf{v} = \frac{\mathbf{v}}{\|\mathbf{v}\|},$$

we have

$$\|\mathbf{u}\| = \sqrt{(\mathbf{u}, \mathbf{u})} = \sqrt{\left(\frac{1}{\|\mathbf{v}\|}\mathbf{v}, \frac{1}{\|\mathbf{v}\|}\mathbf{v}\right)} = \sqrt{\frac{(\mathbf{v}, \mathbf{v})}{\|\mathbf{v}\|^2}} = \sqrt{\frac{\|\mathbf{v}\|^2}{\|\mathbf{v}\|^2}} = 1.$$

EXAMPLE 4   Let $V$ be the function space $C[-\pi, \pi]$ with inner product

$$(\mathbf{f}, \mathbf{g}) = \int_{-\pi}^{\pi} f(t)g(t)\,dt.$$

---

[2] Instead of saying "makes the same angle as $\mathbf{v}$ with any other vector," we often say "in the same direction as $\mathbf{v}$."

The set $S = \{1, \cos(t), \sin(t)\}$ is an orthogonal set in $V$. To verify this we compute the inner products $(1, \cos(t))$, $(1, \sin(t))$, and $(\cos(t), \sin(t))$. We have

$$(1, \cos(t)) = \int_{-\pi}^{\pi} \cos(t)\, dt = \sin(t)]_{-\pi}^{\pi} = 0,$$

$$(1, \sin(t)) = \int_{-\pi}^{\pi} \sin(t)\, dt = -\cos(t)]_{-\pi}^{\pi} = 0,$$

$$(\cos(t), \sin(t)) = 0 \quad \text{(see Example 3).}$$

The set

$$T = \left\{ \frac{1}{\|1\|}, \frac{\cos(t)}{\|\cos(t)\|}, \frac{\sin(t)}{\|\sin(t)\|} \right\} = \left\{ \frac{1}{\sqrt{2\pi}}, \frac{\cos(t)}{\sqrt{\pi}}, \frac{\sin(t)}{\sqrt{\pi}} \right\}$$

is an orthonormal set in $V$. (Verify.)    ■

## Orthogonal Sets and Linear Independence

An important result about orthogonal (and orthonormal) sets of vectors in an inner product space is the following.

Let $S = \{\mathbf{u}_1, \mathbf{u}_2, \ldots, \mathbf{u}_n\}$ be a finite orthogonal set of vectors in an inner product space $V$. Then $S$ is a linearly independent set.

The verification of this result follows by using the same steps that we used in Section 4.4 in the special case of $C^n$. We merely need to use the inner product notation $(\mathbf{u}, \mathbf{v})$ in place of the matrix multiplication form of the standard inner product. (See Exercise 8.) It follows that **an orthonormal set is also linearly independent**.

EXAMPLE 5    Let $V$ be the function space $C[-\pi, \pi]$ with inner product

$$(\mathbf{f}, \mathbf{g}) = \int_{-\pi}^{\pi} f(t)g(t)\, dt.$$

Let $S = \{1, \cos(t), \sin(t), \cos(2t), \sin(2t), \ldots, \cos(nt), \sin(nt), \ldots\}$, where $n$ is a positive integer. We see that $S$ is an infinite set. The relationships

$$(1, \cos(nt)) = \int_{-\pi}^{\pi} \cos(nt)\, dt = \frac{1}{n} \sin(nt) \Big]_{-\pi}^{\pi} = 0,$$

$$(1, \sin(nt)) = \int_{-\pi}^{\pi} \sin(nt)\, dt = -\frac{1}{n} \cos(nt) \Big]_{-\pi}^{\pi} = 0,$$

$$(\cos(nt), \sin(nt)) = \int_{-\pi}^{\pi} \cos(nt) \sin(nt)\, dt = 0,$$

$$(\cos(mt), \cos(nt)) = \int_{-\pi}^{\pi} \cos(mt) \cos(nt)\, dt = 0 \quad \text{for } m \neq n,$$

$$(\sin(mt), \sin(nt)) = \int_{-\pi}^{\pi} \sin(mt) \sin(nt)\, dt = 0 \quad \text{for } m \neq n,$$

demonstrate that $(\mathbf{f}, \mathbf{g}) = 0$ whenever $\mathbf{f}$ and $\mathbf{g}$ are distinct members of $S$. Hence every finite subset of functions from $S$ is an orthogonal set of nonzero vectors. Thus the preceding result implies that any finite subset of functions from $S$ is linearly independent.  ∎

## Linear Combinations and Orthonormal Bases

As we saw in Sections in 4.3 through 4.7, orthonormal bases are highly efficient for certain tasks. An orthonormal basis for an inner product space also provides similar efficiencies. As an example we have the following result.

> If $S = \{\mathbf{u}_1, \mathbf{u}_2, \ldots, \mathbf{u}_n\}$ is an orthonormal basis for an inner product space $V$, then for any vector $\mathbf{v}$ in $V$ we have
>
> $$\mathbf{v} = (\mathbf{u}_1, \mathbf{v})\mathbf{u}_1 + (\mathbf{u}_2, \mathbf{v})\mathbf{u}_2 + \cdots + (\mathbf{u}_n, \mathbf{v})\mathbf{u}_n = \sum_{j=1}^{n}(\mathbf{u}_j, \mathbf{v})\mathbf{u}_j.$$

The verification of this result is a nice application of inner products and the properties of an orthonormal set. Since $S$ is a basis, we know $\mathbf{v}$ can be expressed as a linear combination of $\{\mathbf{u}_1, \mathbf{u}_2, \ldots, \mathbf{u}_n\}$. We need only show that the coefficients required are the inner products of $\mathbf{v}$ with the respective basis vectors $\mathbf{u}_j$. Thus let

$$\mathbf{v} = c_1\mathbf{u}_1 + c_2\mathbf{u}_2 + \cdots + c_n\mathbf{u}_n.$$

Compute the following inner products and simplify:

$$\begin{aligned}
(\mathbf{u}_j, \mathbf{v}) &= c_1(\mathbf{u}_j, \mathbf{u}_1) + c_2(\mathbf{u}_j, \mathbf{u}_2) + \cdots + c_n(\mathbf{u}_j, \mathbf{u}_n) \\
&= c_j(\mathbf{u}_j, \mathbf{u}_j) \quad \text{since } S \text{ is orthonormal, } (\mathbf{u}_j, \mathbf{u}_k) = 0 \text{ when } j \neq k \\
&= c_j \quad \text{since } \mathbf{u}_j \text{ is a unit vector}
\end{aligned}$$

It follows that

$$\mathbf{v} = \sum_{j=1}^{n}(\mathbf{u}_j, \mathbf{v})\mathbf{u}_j.$$

This result shows that to write a vector in $V$ as a linear combination of an orthonormal basis we need not solve a linear system as we had to do in a number of instances in Chapters 1 and 5. The necessary coefficients are obtained immediately from inner product computations. We note that if the basis is only orthogonal, then

$$c_j = \frac{(\mathbf{u}_j, \mathbf{v})}{(\mathbf{u}_j, \mathbf{u}_j)}$$

and

$$\mathbf{v} = \sum_{j=1}^{n}\frac{(\mathbf{u}_j, \mathbf{v})}{(\mathbf{u}_j, \mathbf{u}_j)}\mathbf{u}_j. \tag{2}$$

Before we develop further applications of orthonormal bases in function spaces, we need to be sure we can construct them. In Section 4.5 we showed how to use the Gram-Schmidt process to obtain an orthonormal basis from an arbitrary basis for $C^n$ or $R^n$ and any subspace of these spaces. Fortunately, the Gram-Schmidt process generalizes to inner product spaces and the necessary steps are identical to those given in Section 4.5 except we use the inner product of the space in which we are working rather than the standard inner product of $C^n$ or $R^n$. We present the first few steps of the Gram-Schmidt process for inner product spaces without the verifications we supplied in Section 4.5.

## The Gram-Schmidt Process of Orthogonalization

Let $S = \{\mathbf{q}_1, \mathbf{q}_2, \ldots, \mathbf{q}_m\}$ be a basis for subspace $W$ of an inner product space $V$. (Assume that $W$ is not the trivial subspace consisting of the identity vector alone.) We construct an orthogonal basis $\{\mathbf{v}_1, \mathbf{v}_2, \ldots, \mathbf{v}_m\}$ for $W$ as follows: Set $\mathbf{v}_1 = \mathbf{q}_1$. Compute

$$\mathbf{v}_2 = \mathbf{q}_2 - \frac{(\mathbf{v}_1, \mathbf{q}_2)}{(\mathbf{v}_1, \mathbf{v}_1)} \mathbf{v}_1. \tag{3}$$

We can show that $\mathbf{v}_1$ and $\mathbf{v}_2$ are orthogonal. (See Exercise 11(a).) It follows that $\text{span}\{\mathbf{q}_1, \mathbf{q}_2\} = \text{span}\{\mathbf{v}_1, \mathbf{v}_2\}$. (See Exercise 11(b).) Next compute

$$\mathbf{v}_3 = \mathbf{q}_3 - \frac{(\mathbf{v}_1, \mathbf{q}_3)}{(\mathbf{v}_1, \mathbf{v}_1)} \mathbf{v}_1 - \frac{(\mathbf{v}_2, \mathbf{q}_3)}{(\mathbf{v}_2, \mathbf{v}_2)} \mathbf{v}_2. \tag{4}$$

We can show that $\mathbf{v}_3$ is orthogonal to both $\mathbf{v}_1$ and $\mathbf{v}_2$. (See Exercise 11(c).) It follows that

$$\text{span}\{\mathbf{q}_1, \mathbf{q}_2, \mathbf{q}_3\} = \text{span}\{\mathbf{v}_1, \mathbf{v}_2, \mathbf{q}_3\} = \text{span}\{\mathbf{v}_1, \mathbf{v}_2, \mathbf{v}_3\}.$$

(See Exercise 11(d).) Continuing in this way, we get

$$\mathbf{v}_j = \mathbf{q}_j - \sum_{k=1}^{j-1} \frac{(\mathbf{v}_k, \mathbf{q}_j)}{(\mathbf{v}_k, \mathbf{v}_k)} \mathbf{v}_k, \qquad j = 1, 2, \ldots, m. \tag{5}$$

We show the orthogonality and spanning in a similar manner. By construction, $\{\mathbf{v}_1, \mathbf{v}_2, \ldots, \mathbf{v}_m\}$ is an orthogonal basis for $W$. We scale each of these vectors to be a unit vector so that

$$\left\{ \mathbf{u}_1 = \frac{\mathbf{v}_1}{\|\mathbf{v}_1\|}, \mathbf{u}_2 = \frac{\mathbf{v}_2}{\|\mathbf{v}_2\|}, \ldots, \mathbf{u}_m = \frac{\mathbf{v}_m}{\|\mathbf{v}_m\|} \right\}$$

is an orthonormal basis for $W$.

**EXAMPLE 6**    Let $V$ be the inner product space $C[0, 1]$ with the inner product

$$(\mathbf{f}, \mathbf{g}) = \int_0^1 f(t)g(t)\, dt.$$

Let $W$ be the subspace $\text{span}\{t, t^2\}$. The set $S = \{t, t^2\}$ is a basis for $W$. To find an orthonormal basis for $W$, we apply the Gram-Schmidt process. Let $\mathbf{q}_1 = t$ and $\mathbf{q}_2 = t^2$. Now let $\mathbf{v}_1 = \mathbf{q}_1 = t$. Then

$$\mathbf{v}_2 = \mathbf{q}_2 - \frac{(\mathbf{v}_1, \mathbf{q}_2)}{(\mathbf{v}_1, \mathbf{v}_1)} \mathbf{v}_1 = t^2 - \frac{\displaystyle\int_0^1 t t^2\, dt}{\displaystyle\int_0^1 t t\, dt} t = t^2 - \frac{\frac{1}{4}}{\frac{1}{3}} t = t^2 - \frac{3}{4} t.$$

To verify that $T = \{\mathbf{v}_1, \mathbf{v}_2\}$ is an orthonormal set, compute

$$(\mathbf{v}_1, \mathbf{v}_2) = \int_0^1 t \left( t^2 - \frac{3}{4} t \right) dt = \left( \frac{t^4}{4} - \frac{t^3}{4} \right)\Big]_0^1 = 0.$$

Next we compute

$$(\mathbf{v}_1, \mathbf{v}_1) = \sqrt{\frac{1}{3}} \quad \text{and} \quad (\mathbf{v}_2, \mathbf{v}_2) = \sqrt{\frac{1}{80}}$$

and use these values to normalize $\mathbf{v}_1$ and $\mathbf{v}_2$. We get

$$\mathbf{u}_1 = t\sqrt{3} \quad \text{and} \quad \mathbf{u}_2 = \sqrt{80}\left(t^2 - \frac{3}{4}t\right). \qquad \text{(Verify.)}$$

Then $\{\mathbf{u}_1, \mathbf{u}_2\}$ is an orthonormal basis for $W$.   ■

EXAMPLE 7   The set

$$S = \{1, \cos(t), \sin(t), \cos(2t), \sin(2t), \dots, \cos(nt), \sin(nt), \dots\}$$

from Example 5 is orthogonal in function space $C[-\pi, \pi]$. To obtain an orthonormal set, we need only convert each vector to a unit vector. Using results from Example 4 and the following relationships (verify)

$$(\cos(nt), \cos(nt)) = \int_{-\pi}^{\pi} \cos^2(nt)\, dt = \pi,$$

$$(\sin(nt), \sin(nt)) = \int_{-\pi}^{\pi} \sin^2(nt)\, dt = \pi,$$

where $n$ is a positive integer, we have that

$$T = \left\{ \frac{1}{\sqrt{2\pi}}, \frac{\cos(t)}{\sqrt{\pi}}, \frac{\sin(t)}{\sqrt{\pi}}, \frac{\cos(2t)}{\sqrt{\pi}}, \frac{\sin(2t)}{\sqrt{\pi}}, \dots, \frac{\cos(nt)}{\sqrt{\pi}}, \frac{\sin(nt)}{\sqrt{\pi}}, \dots \right\}$$

is such that any finite subset is an orthonormal set in function space $C[-\pi, \pi]$.   ■

## Approximations from Subspaces

We can now illustrate how the structure of an inner product space, in a particular function space, is used to determine approximations to vectors using members of a particular subspace. This type of application is widely used in a variety of settings.

Let $\mathbf{f}$ be a vector in inner product space $C[a, b]$ that has inner product

$$(\mathbf{h}, \mathbf{g}) = \int_a^b h(t)g(t)\, dt.$$

If $\mathbf{f}$ is a complicated function, we might want to approximate it by a simpler function, say a polynomial. For example, suppose we wish to construct an approximation to $\mathbf{f}$ from the subspace $W = \text{span}\{t, t^2\}$. Using Example 6, we have an orthonormal basis for $W$ given by

$$\{\mathbf{u}_1, \mathbf{u}_2\} = \left\{ t\sqrt{3}, \sqrt{80}\left(t^2 - \tfrac{3}{4}t\right) \right\}.$$

Rather than just any approximation from $W$, it is reasonable to ask for the "best approximation" to $\mathbf{f}$ from $W$. Since an inner product space has the structure that lets us measure the distance between vectors, we interpret *best* to mean the vector $\mathbf{p}^*$ in $W$ such that $\|\mathbf{f} - \mathbf{p}^*\|$ is as small as possible.[3] That is, $\|\mathbf{f} - \mathbf{p}^*\| \le \|\mathbf{f} - \mathbf{p}\|$ for all vectors $\mathbf{p}$ in $W$. Finding such a vector is equivalent to minimizing

$$\int_a^b (f(t) - p(t))^2\, dt$$

among all vectors $\mathbf{p}$ in $W$. We now develop a procedure for finding such a best approximation. We start by determining a relationship that the vector in $W$ that is closest must satisfy. We call the relationship a *characterization* of the closest vector.

---

[3] Here $*$ does not mean conjugate transpose, rather it is used to denote a particular vector in the space.

Let $V$ be an inner product space and $W$ be a subspace of $V$. Let $\mathbf{v}$ be a vector in $V$ and suppose $\mathbf{p}^*$ is a vector in $W$ such that

$$(\mathbf{v} - \mathbf{p}^*, \mathbf{w}) = 0 \qquad \text{for all vectors } \mathbf{w} \text{ in } W.$$

Then $\|\mathbf{v} - \mathbf{p}^*\| \le \|\mathbf{v} - \mathbf{w}\|$ for all vectors $\mathbf{w}$ in $W$ with equality holding only for $\mathbf{w} = \mathbf{p}^*$. (This is sometimes stated as follows: $\mathbf{v} - \mathbf{p}^*$ is **normal** to the subspace $W$.)

To verify this statement we observe that minimizing $\|\mathbf{v} - \mathbf{w}\|$ is the same as minimizing $\|\mathbf{v} - \mathbf{w}\|^2$. Working with $\|\mathbf{v} - \mathbf{w}\|^2$ simplifies expressions in the following. We have

$$\|\mathbf{v} - \mathbf{w}\|^2 = \|\mathbf{v} - \mathbf{p}^* + \mathbf{p}^* - \mathbf{w}\|^2 \quad \text{(adding zero in disguise to get } \mathbf{p}^* \text{ into the expression)}$$

$$= \|(\mathbf{v} - \mathbf{p}^*) - (\mathbf{w} - \mathbf{p}^*)\|^2 \quad \text{(grouping and rewriting the expression)}$$

$$= ((\mathbf{v} - \mathbf{p}^*) - (\mathbf{w} - \mathbf{p}^*), (\mathbf{v} - \mathbf{p}^*) - (\mathbf{w} - \mathbf{p}^*)) \quad \text{(writing as an inner product)}$$

$$= ((\mathbf{v} - \mathbf{p}^*), (\mathbf{v} - \mathbf{p}^*) - (\mathbf{w} - \mathbf{p}^*))$$
$$- ((\mathbf{w} - \mathbf{p}^*), (\mathbf{v} - \mathbf{p}^*) - (\mathbf{w} - \mathbf{p}^*)) \quad \text{(expanding)}$$

$$= ((\mathbf{v} - \mathbf{p}^*), (\mathbf{v} - \mathbf{p}^*)) - ((\mathbf{v} - \mathbf{p}^*), (\mathbf{w} - \mathbf{p}^*))$$
$$- ((\mathbf{w} - \mathbf{p}^*), (\mathbf{v} - \mathbf{p}^*)) + ((\mathbf{w} - \mathbf{p}^*), (\mathbf{w} - \mathbf{p}^*)) \quad \text{(expanding more)}$$

$$= \|\mathbf{v} - \mathbf{p}^*\|^2 - 2((\mathbf{v} - \mathbf{p}^*), (\mathbf{w} - \mathbf{p}^*)) + \|\mathbf{w} - \mathbf{p}^*\|^2. \quad \text{[collecting terms and writing in terms of norms]}$$

Since $\mathbf{w} - \mathbf{p}^*$ is in $W$, $((\mathbf{v} - \mathbf{p}^*), (\mathbf{w} - \mathbf{p}^*)) = 0$; hence we get

$$\|\mathbf{v} - \mathbf{w}\|^2 = \|\mathbf{v} - \mathbf{p}^*\|^2 + \|\mathbf{w} - \mathbf{p}^*\|^2.$$

Since $\|\mathbf{w} - \mathbf{p}^*\| \ge 0$ we have that $\|\mathbf{v} - \mathbf{w}\|^2 \ge \|\mathbf{v} - \mathbf{p}^*\|^2$. Then it follows that $\|\mathbf{v} - \mathbf{w}\|^2 > \|\mathbf{v} - \mathbf{p}^*\|^2$ unless $\mathbf{w} = \mathbf{p}^*$. Hence $\mathbf{p}^*$ is closer to $\mathbf{v}$ than to any other vector in $W$.

The preceding result gives a property that the closest vector must satisfy. Next we show how to construct the closest vector.

Let $V$ be an inner product space and $W$ be a subspace of $V$ with $\dim(W) = m$. If $\{\mathbf{w}_1, \mathbf{w}_2, \ldots, \mathbf{w}_m\}$ is an orthogonal basis for $W$, then the vector in $W$ closest to the vector $\mathbf{v}$ in $V$ is given by

$$\mathbf{p}^* = \frac{(\mathbf{v}, \mathbf{w}_1)}{(\mathbf{w}_1, \mathbf{w}_1)}\mathbf{w}_1 + \frac{(\mathbf{v}, \mathbf{w}_2)}{(\mathbf{w}_2, \mathbf{w}_2)}\mathbf{w}_2 + \cdots + \frac{(\mathbf{v}, \mathbf{w}_m)}{(\mathbf{w}_m, \mathbf{w}_m)}\mathbf{w}_m. \qquad (6)$$

That is, $\|\mathbf{v} - \mathbf{p}^*\| \le \|\mathbf{v} - \mathbf{w}\|$ for all vectors $\mathbf{w}$ in $W$.

To verify this statement we show that $\mathbf{v} - \mathbf{p}^*$ is orthogonal to every vector in $W$, where $\mathbf{p}^*$ is given by the expression in (6). Once we do this, we have verified the characterization of the closest vector given previously, and hence this result is valid.

**Strategy:**   We take advantage of the structure of an inner product space by noting that a vector is orthogonal to every vector in $W$ if and only if it is orthogonal to a basis for $W$. Thus rather than deal with all the vectors in $W$ we need only deal with the vectors in a basis.

Using this approach, we show that $\mathbf{v} - \mathbf{p}^*$ is orthogonal to every vector in the orthogonal basis $\{\mathbf{w}_1, \mathbf{w}_2, \ldots, \mathbf{w}_m\}$. To this end we compute $(\mathbf{w}_j, \mathbf{v} - \mathbf{p}^*)$ as follows:

$$(\mathbf{w}_j, \mathbf{v} - \mathbf{p}^*) = (\mathbf{w}_j, \mathbf{v}) - (\mathbf{w}_j, \mathbf{p}^*)$$

$$= (\mathbf{w}_j, \mathbf{v}) - \left( \mathbf{w}_j, \frac{(\mathbf{v}, \mathbf{w}_1)}{(\mathbf{w}_1, \mathbf{w}_1)}\mathbf{w}_1 + \frac{(\mathbf{v}, \mathbf{w}_2)}{(\mathbf{w}_2, \mathbf{w}_2)}\mathbf{w}_2 + \cdots + \frac{(\mathbf{v}, \mathbf{w}_m)}{(\mathbf{w}_m, \mathbf{w}_m)}\mathbf{w}_m \right)$$

$$= (\mathbf{w}_j, \mathbf{v}) - \frac{(\mathbf{v}, \mathbf{w}_1)}{(\mathbf{w}_1, \mathbf{w}_1)}(\mathbf{w}_j, \mathbf{w}_1) - \frac{(\mathbf{v}, \mathbf{w}_2)}{(\mathbf{w}_2, \mathbf{w}_2)}(\mathbf{w}_j, \mathbf{w}_2)$$

$$- \cdots - \frac{(\mathbf{v}, \mathbf{w}_j)}{(\mathbf{w}_j, \mathbf{w}_j)}(\mathbf{w}_j, \mathbf{w}_j) - \cdots - \frac{(\mathbf{v}, \mathbf{w}_m)}{(\mathbf{w}_m, \mathbf{w}_m)}(\mathbf{w}_j, \mathbf{w}_m).$$

Recall that $(\mathbf{w}_j, \mathbf{w}_k) = 0$ for $j \neq k$. Thus the last expression simplifies so that we have

$$(\mathbf{w}_j, \mathbf{v} - \mathbf{p}^*) = (\mathbf{w}_j, \mathbf{v}) - \frac{(\mathbf{v}, \mathbf{w}_j)}{(\mathbf{w}_j, \mathbf{w}_j)}(\mathbf{w}_j, \mathbf{w}_j) = (\mathbf{w}_j, \mathbf{v}) - (\mathbf{w}_j, \mathbf{v}) = 0.$$

This true for $j = 1, 2, \ldots, m$ so the result follows.

The preceding result leads us to the following definition.

**Definition**   Let $V$ be an inner product space and $W$ a subspace of $V$. For a specified vector in $V$, the vector $\mathbf{p}^*$ in $W$ that is closest to $\mathbf{v}$ is called the **projection** of $\mathbf{v}$ onto the subspace $W$.

The projection of $\mathbf{v}$ onto subspace $W$ is denoted $\text{proj}_W \mathbf{v}$. Thus from Equation (6) we have

$$\text{proj}_W \mathbf{v} = \frac{(\mathbf{v}, \mathbf{w}_1)}{(\mathbf{w}_1, \mathbf{w}_1)}\mathbf{w}_1 + \frac{(\mathbf{v}, \mathbf{w}_2)}{(\mathbf{w}_2, \mathbf{w}_2)}\mathbf{w}_2 + \cdots + \frac{(\mathbf{v}, \mathbf{w}_m)}{(\mathbf{w}_m, \mathbf{w}_m)}\mathbf{w}_m$$

in terms of the orthogonal basis $\{\mathbf{w}_1, \mathbf{w}_2, \ldots, \mathbf{w}_m\}$ for $W$. If the basis is orthonormal, then we have

$$\text{proj}_W \mathbf{v} = (\mathbf{v}, \mathbf{w}_1)\mathbf{w}_1 + (\mathbf{v}, \mathbf{w}_2)\mathbf{w}_2 + \cdots + (\mathbf{v}, \mathbf{w}_m)\mathbf{w}_m. \tag{7}$$

The projection of a vector onto a subspace is often called the **orthogonal projection**. This terminology follows from the fact that the formula for the projection is given in terms of an orthogonal or orthonormal basis. Recall that we can "manufacture" such bases from an ordinary basis using the Gram-Schmidt process. This development in inner product spaces generalizes the discussion of projections given for $R^n$ in Section 1.5. In Section 1.5 we limited ourselves to projections of one vector onto another; that is, onto a one-dimensional subspace.

## Projections and the Gram-Schmidt Process

Using projections, we can express the steps for the Gram-Schmidt process in an alternate form that reveals the geometric nature of the algorithm. As previously, let $V$ be an inner product space and $W$ a subspace with basis $S = \{\mathbf{q}_1, \mathbf{q}_2, \ldots, \mathbf{q}_m\}$. We set $\mathbf{v}_1 = \mathbf{q}_1$. Then from (3) we compute

$$\mathbf{v}_2 = \mathbf{q}_2 - \frac{(\mathbf{v}_1, \mathbf{q}_2)}{(\mathbf{v}_1, \mathbf{v}_1)}\mathbf{v}_1.$$

But from (6) we see that

$$\frac{(\mathbf{v}_1, \mathbf{q}_2)}{(\mathbf{v}_1, \mathbf{v}_1)}\mathbf{v}_1 = \text{proj}_{\mathbf{v}_1}\mathbf{q}_2$$

hence

$$\mathbf{v}_2 = \mathbf{q}_2 - \frac{(\mathbf{v}_1, \mathbf{q}_2)}{(\mathbf{v}_1, \mathbf{v}_1)}\mathbf{v}_1 = \mathbf{q}_2 - \text{proj}_{\mathbf{v}_1}\mathbf{q}_2.$$

As before, $\mathbf{v}_1$ and $\mathbf{v}_2$ are orthogonal and $\text{span}\{\mathbf{q}_1, \mathbf{q}_2\} = \text{span}\{\mathbf{v}_1, \mathbf{v}_2\}$. Next we compute

$$\mathbf{v}_3 = \mathbf{q}_3 - \frac{(\mathbf{v}_1, \mathbf{q}_3)}{(\mathbf{v}_1, \mathbf{v}_1)}\mathbf{v}_1 - \frac{(\mathbf{v}_2, \mathbf{q}_3)}{(\mathbf{v}_2, \mathbf{v}_2)}\mathbf{v}_2.$$

From (6) we have

$$\frac{(\mathbf{v}_1, \mathbf{q}_3)}{(\mathbf{v}_1, \mathbf{v}_1)}\mathbf{v}_1 + \frac{(\mathbf{v}_2, \mathbf{q}_3)}{(\mathbf{v}_2, \mathbf{v}_2)}\mathbf{v}_2 = \text{proj}_{\text{span}(\{\mathbf{v}_1, \mathbf{v}_2\})}\mathbf{q}_3.$$

Hence

$$\mathbf{v}_3 = \mathbf{q}_3 - \text{proj}_{\text{span}(\{\mathbf{v}_1, \mathbf{v}_2\})}\mathbf{q}_3.$$

We have that $\mathbf{v}_3$ is orthogonal to both $\mathbf{v}_1$ and $\mathbf{v}_2$ and

$$\text{span}\{\mathbf{q}_1, \mathbf{q}_2, \mathbf{q}_3\} = \text{span}\{\mathbf{v}_1, \mathbf{v}_2, \mathbf{q}_3\} = \text{span}\{\mathbf{v}_1, \mathbf{v}_2, \mathbf{v}_3\}.$$

Continuing in this manner it follows that

$$\mathbf{v}_j = \mathbf{q}_j - \sum_{k=1}^{j-1}\frac{(\mathbf{v}_k, \mathbf{q}_j)}{(\mathbf{v}_k, \mathbf{v}_k)}\mathbf{v}_k = \mathbf{q}_j - \text{proj}_{\text{span}(\{\mathbf{v}_1, \mathbf{v}_2, \ldots, \mathbf{v}_{j-1}\})}\mathbf{q}_j, \quad j = 1, 2, \ldots, m.$$

Thus the vectors $\{\mathbf{v}_1, \mathbf{v}_2, \ldots, \mathbf{v}_m\}$ form an orthogonal basis for subspace $W$. We see that each vector $\mathbf{v}_j$ is a linear combination of the corresponding vector $\mathbf{q}_j$ and the projection of $\mathbf{q}_j$ onto the subspace spanned by the orthogonal vectors generated so far. It is also the case that $\|\mathbf{v}_j\|$ is the distance from $\mathbf{q}_j$ to the subspace spanned by the orthogonal vectors generated so far. (Verify.)

## Distance from a Vector to a Subspace

In addition to computing the projection using Equation (6) or (7), we also have a measure of the distance from the vector $\mathbf{v}$ to the subspace $W$. We make this precise in the following definition.

---

**Definition**    In an inner product space $V$, the **distance from a vector v to a subspace $W$** is

$$\|\mathbf{v} - \text{proj}_W\mathbf{v}\| = \min_{\mathbf{w} \text{ in } W}\|\mathbf{v} - \mathbf{w}\|.$$

---

We illustrate the computation and the use of projections in the next two examples. In addition, we examine the important use of projections in the function space $C[-\pi, \pi]$ onto the subspace with basis

$$S = \{1, \cos(t), \sin(t), \cos(2t), \sin(2t), \ldots, \cos(nt), \sin(nt), \ldots\}$$

in an application at the end of the section. We investigate another important instance of projections in Section 6.4.

**EXAMPLE 8**   Let $V$ be the inner product space $R^3$ with the standard inner product and let $W$ be the two-dimensional subspace with orthogonal basis $\{\mathbf{w}_1, \mathbf{w}_2\}$ where

$$\mathbf{w}_1 = \begin{bmatrix} 2 \\ -1 \\ -2 \end{bmatrix} \quad \text{and} \quad \mathbf{w}_2 = \begin{bmatrix} 1 \\ 0 \\ 1 \end{bmatrix}.$$

To find $\text{proj}_W \mathbf{v}$ where $\mathbf{v} = \begin{bmatrix} 2 \\ 1 \\ 3 \end{bmatrix}$, we use Equation (6). We have (verify)

$$\text{proj}_W \mathbf{v} = \frac{(\mathbf{v}, \mathbf{w}_1)}{(\mathbf{w}_1, \mathbf{w}_1)}\mathbf{w}_1 + \frac{(\mathbf{v}, \mathbf{w}_2)}{(\mathbf{w}_2, \mathbf{w}_2)}\mathbf{w}_2 = -\frac{3}{9}\mathbf{w}_1 + \frac{5}{2}\mathbf{w}_2 = \begin{bmatrix} \frac{33}{18} \\ \frac{3}{9} \\ \frac{57}{18} \end{bmatrix}.$$

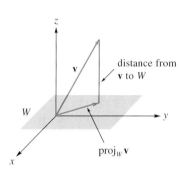

W

distance from
v to W

$\text{proj}_W \mathbf{v}$

FIGURE 1

Then the distance from $\mathbf{v}$ to subspace $W$ is $\|\mathbf{v} - \text{proj}_W \mathbf{v}\| = \sqrt{\frac{1}{2}} \approx 0.7071$. (Verify.) Since we are in $R^3$, the two-dimensional subspace $W$ is a plane and $\|\mathbf{v} - \text{proj}_W \mathbf{v}\|$ represents the perpendicular distance from the terminal point of vector $\mathbf{v}$ to plane $W$. See Figure 1 for a geometric look.   ■

**EXAMPLE 9**   Refer to Example 6. Let $\mathbf{v} = t^3$. Then, using Equation (7), we have

$$\text{proj}_W \mathbf{v} = (\mathbf{v}, \mathbf{u}_1)\mathbf{u}_1 + (\mathbf{v}, \mathbf{u}_2)\mathbf{u}_2$$
$$= \int_0^1 t^3 \sqrt{3}\, t\, dt\ \mathbf{u}_1 + \int_0^1 t^3 \sqrt{80}(t^2 - 0.75t)\, dt\ \mathbf{u}_2$$
$$= \frac{\sqrt{3}}{5}\mathbf{u}_1 + \frac{\sqrt{80}}{60}\mathbf{u}_2 \approx 0.3464\mathbf{u}_1 + 0.1491\mathbf{u}_2.$$

(Verify the computations of the coefficients in the preceding expression.) Substituting for the orthonormal basis vectors gives us

$$\text{proj}_W (t^3) \approx 0.3464 \left(\sqrt{3}\, t\right) + 0.1491 \left(\sqrt{80}\left(t^2 - 0.75t\right)\right)$$
$$\approx -0.4t + 1.3336t^2.$$

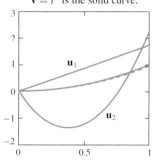

The projection is the
dashed curve.
$\mathbf{v} = t^3$ is the solid curve.

$\mathbf{u}_1$

$\mathbf{u}_2$

FIGURE 2

Figure 2 shows that the vector $\mathbf{v}$ and the projection of $\mathbf{v}$ onto subspace $W$ are graphically nearly indistinguishable. In fact, the distance from $\mathbf{v}$ to subspace $W$ is

$$\|\mathbf{v} - \text{proj}_W \mathbf{v}\| \approx 0.0252.$$   ■

In the preceding two examples we started, respectively, with an orthogonal and orthonormal basis for subspace W. In the event that we have just a standard basis, then we must first apply the Gram-Schmidt process and use the resulting orthogonal or orthonormal basis, whichever may be easier to work with in a particular case. (Especially for hand calculations, it is sometimes easier to work with an orthogonal basis.) While the computations[4] may entail quite a few manipulations, there is a definite procedure for computing the projection of a vector onto a subspace.

Application:    **Introduction to Fourier Analysis**

Many phenomena exhibit cyclical behavior. In nature there are seasonal fluctuations, daily fluctuations of tides, the repetitive structure of crystals, and even the weather pattern called El Niño. There is repetitive behavior exhibited in every branch of science including such areas as electrical signals and circuits, acoustics, heat flow, optics, and mechanical systems. A basic task of Fourier analysis is to represent often complicated, cyclical behavior with linear combinations of simpler functions involving sines and cosines. While we cannot delve too deeply into the riches of Fourier analysis, we can describe certain fundamentals in terms of inner product spaces and orthogonal families of functions. We have laid some foundations for this in Examples 3, 4, 5, and 7. Here we will present some of that information in a slightly different style so that it is easier to develop approximations.

As mentioned previously, Fourier analysis is typically used to build models of phenomena that exhibit cyclical or periodic behavior. In this brief introduction we construct approximations to continuous functions over a closed interval. The functions we use are chosen so that the computations are relatively easy and short. Hence the functions will not necessarily exhibit true periodic behavior. It is a testament to the versatility of Fourier analysis that in such cases we often generate approximations that are quite good.

We begin with a brief description of terminology and notation. We will be working in the function space $V = C[-\pi, \pi]$ with inner product

$$(\mathbf{f}, \mathbf{g}) = \frac{1}{\pi} \int_{-\pi}^{\pi} f(t)g(t)\, dt.$$

This a modification of the inner product used in Examples 3, 4, 5, and 7. (See Exercise 19.) The use of the factor $\dfrac{1}{\pi}$ will make a number of our computations easier. It follows that any subset of

$$S = \left\{ \frac{1}{\sqrt{2}}, \cos(t), \sin(t), \cos(2t), \sin(2t), \ldots \right\}$$

is an orthonormal set using this inner product. Let

$$S_n = \left\{ \frac{1}{\sqrt{2}}, \cos(t), \sin(t), \cos(2t), \sin(2t), \ldots, \cos(nt), \sin(nt) \right\}$$

and $T_n = \text{span}(S_n)$. Then $T_n$ is a $(2n+1)$-dimensional subspace of $V$ with orthonormal basis $S_n$. We call the vectors in $T_n$ **trigonometric polynomials** of order $n$ or less. The projection of a vector $\mathbf{f}$ in $V$ onto subspace $T_n$ is obtained using Equation (7). It is common to call the inner product of $\mathbf{f}$ with vectors in basis $S_n$ the **Fourier coefficients** of $\mathbf{f}$, and we adopt the following labeling pattern:

$$a_0 = \left( \mathbf{f}, \frac{1}{\sqrt{2}} \right), \quad a_j = (\mathbf{f}, \cos(jt)), \quad b_j = (\mathbf{f}, \sin(jt)) \qquad \text{for } j = 1, 2, \ldots, n.$$

---

[4]When working with function spaces, it is recommended that integrals that arise be computed using computer algebra systems. MATLAB has its symbolic toolbox that uses Maple as its symbolic engine. Most other high-level computational environments have access to similar symbolic software. The integrals in Examples 6 and 9 were computed using MATLAB's **int** command.

Then from Equation (7) we have

$$\text{proj}_{T_n} \mathbf{f} = a_0 \frac{1}{\sqrt{2}} + \sum_{j=1}^{n} a_j \cos(jt) + \sum_{j=1}^{n} b_j \sin(jt). \tag{8}$$

This expression is called the **$n$th-order Fourier approximation** of **f** or the **$n$th-order trigonometric polynomial approximation** of **f**.

EXAMPLE 10    Let $f(t) = |t|$. To determine the first-order Fourier approximation of **f**, we compute the Fourier coefficients $a_0$, $a_1$, and $b_1$ and then form the expression that corresponds to Equation (8) for this case. We have

$$a_0 = \frac{1}{\pi} \int_{\pi}^{\pi} |t| \frac{1}{\sqrt{2}} \, dt = \frac{1}{\pi \sqrt{2}} \int_{-\pi}^{0} (-t) \, dt + \frac{1}{\pi \sqrt{2}} \int_{0}^{\pi} (t) \, dt = \frac{\pi}{\sqrt{2}},$$

$$a_1 = \frac{1}{\pi} \int_{-\pi}^{\pi} |t| \cos(t) \, dt = \frac{1}{\pi} \int_{-\pi}^{0} (-t) \cos(t) \, dt + \frac{1}{\pi} \int_{0}^{\pi} (t) \cos(t) \, dt = \frac{-4}{\pi},$$

$$b_1 = \frac{1}{\pi} \int_{-\pi}^{\pi} |t| \sin(t) \, dt = \frac{1}{\pi} \int_{-\pi}^{0} (-t) \sin(t) \, dt + \frac{1}{\pi} \int_{0}^{\pi} (t) \sin(t) \, dt = 0.$$

Hence

$$\text{proj}_{T_1} |t| = a_0 \frac{1}{\sqrt{2}} + a_1 \cos(t) + b_1 \sin(t) = \frac{\pi}{2} - \frac{4}{\pi} \cos(t).$$

A graphical comparison is given in Figure 3.

To determine the second-order Fourier approximation, we need only compute $a_2 = (|t|, \cos(2t))$ and $b_2 = (|t|, \sin(2t))$. Then

$$\text{proj}_{T_2} |t| = \frac{\pi}{2} - \frac{4}{\pi} \cos(t) + a_2 \cos(2t) + b_2 \sin(2t).$$

(See Exercise 20.)    ■

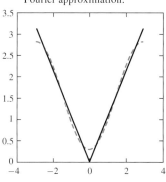

Solid curve is $|t|$.
Dashed curve is the first order Fourier approximation.

FIGURE 3

A natural question that arises from Example 10 is, which approximation, $\text{proj}_{T_1} |t|$ or $\text{proj}_{T_2} |t|$, is better. We can answer this using the criteria we developed for the distance from a vector to a subspace along with the observation that space $T_1$ is a subspace of $T_2$. We have

$$\left\| |t| - \text{proj}_{T_1} |t| \right\| = \min_{\mathbf{g} \text{ in } T_1} \| |t| - \mathbf{g} \| \geq \min_{\mathbf{h} \text{ in } T_2} \| |t| - \mathbf{h} \| = \| |t| - \text{proj}_{T_2} |t| \|.$$

Hence the second-order Fourier approximation over the "larger" subspace $T_2$ is closer to $|t|$ than the first-order Fourier approximation. (This result is true in general when one space is a subspace of the other.) It seems natural to compute successively higher Fourier approximations and hence generate a sequence of more accurate linear combinations that approximate a function. This is a preview of an approach in the theory of approximation of functions and, in particular, the study of Fourier series. Fourier series play a fundamental role in various area of mathematics, science, and engineering and are often studied in advanced courses.

## EXERCISES 6.3[5]

*Let $V = R^2$ or $R^3$, as appropriate, with the standard inner product. Determine the cosine of the angle between each given pair of vectors in Exercises 1–4.*

1. $\mathbf{u} = \begin{bmatrix} 1 \\ 1 \end{bmatrix}, \mathbf{v} = \begin{bmatrix} 1 \\ 3 \end{bmatrix}$.

2. $\mathbf{u} = \begin{bmatrix} -1 \\ -2 \end{bmatrix}, \mathbf{v} = \begin{bmatrix} 1 \\ 4 \end{bmatrix}$.

3. $\mathbf{u} = \begin{bmatrix} 2 \\ 1 \\ 2 \end{bmatrix}, \mathbf{v} = \begin{bmatrix} 1 \\ 0 \\ -1 \end{bmatrix}$.

4. $\mathbf{u} = \begin{bmatrix} 2 \\ 3 \\ -2 \end{bmatrix}, \mathbf{v} = \begin{bmatrix} 1 \\ 2 \\ 0 \end{bmatrix}$.

*Let $V$ be the inner product space in Example 2. Determine the cosine of the angle between each given pair of vectors in Exercises 5–7.*

5. $t, t^2$.        6. $t, e^t$.        7. $\sin(t), \cos(t)$.

8. Verify that a finite orthogonal set of vectors in an inner product space is linearly independent.

9. Express each of the following vectors in terms of the orthonormal basis determined in Example 6.
   (a) $5t - t^2$.        (b) $t + 4t^2$.

10. Let $V = C[-1, 1]$ be the inner product space with inner product
$$(\mathbf{f}, \mathbf{g}) = \int_{-1}^{1} f(t)g(t)\,dt.$$
Let $\mathbf{p}_0(t) = 1$, $\mathbf{p}_1(t) = t$, and $\mathbf{p}_2(t) = 0.5(3t^2 - 1)$.
   (a) Show that $S = \{\mathbf{p}_0(t), \mathbf{p}_1(t), \mathbf{p}_2(t)\}$ is an orthogonal set.
   (b) Determine an orthonormal basis for span($S$).
   (c) Express vector $\mathbf{f}(t) = t^2 + 2t$ in terms of the orthonormal basis determined in part (b).

11. Let $S = \{\mathbf{q}_1, \mathbf{q}_2, \dots, \mathbf{q}_m\}$ be a basis for subspace $W$ of an inner product space $V$. The Gram-Schmidt process is to be used to produce an orthogonal basis $\{\mathbf{v}_1, \mathbf{v}_2, \dots, \mathbf{v}_m\}$ for $W$. We have $\mathbf{v}_1 = \mathbf{q}_1$.
   (a) Verify that $\mathbf{v}_1$ and $\mathbf{v}_2 = \mathbf{q}_2 - \dfrac{(\mathbf{v}_1, \mathbf{q}_2)}{(\mathbf{v}_1, \mathbf{v}_1)}\mathbf{v}_1$ are orthogonal.
   (b) Explain why span$\{\mathbf{q}_1, \mathbf{q}_2\}$ = span$\{\mathbf{v}_1, \mathbf{v}_2\}$.
   (c) Verify that
$$\mathbf{v}_3 = \mathbf{q}_3 - \frac{(\mathbf{v}_1, \mathbf{q}_3)}{(\mathbf{v}_1, \mathbf{v}_1)}\mathbf{v}_1 - \frac{(\mathbf{v}_2, \mathbf{q}_3)}{(\mathbf{v}_2, \mathbf{v}_2)}\mathbf{v}_2$$
   is orthogonal to both $\mathbf{v}_1$ and $\mathbf{v}_2$.

(d) Explain why
$$\text{span}\{\mathbf{q}_1, \mathbf{q}_2, \mathbf{q}_3\} = \text{span}\{\mathbf{v}_1, \mathbf{v}_2, \mathbf{q}_3\}$$
$$= \text{span}\{\mathbf{v}_1, \mathbf{v}_2, \mathbf{v}_3\}.$$

*Let $V$ be the inner product space of Example 2. Determine an orthogonal basis for the subspaces with the bases specified in Exercises 12–14.*

12. $\{1, t, t^2\}$.        13. $\{1, t, e^t\}$.        14. $\{1, e^t, e^{-t}\}$.

15. Let $V$ be the inner product space of Example 2. Let $S = \{1, t^2, t^4\}$, then an orthonormal basis for span($S$) is
$$\{\mathbf{w}_1, \mathbf{w}_2, \mathbf{w}_3\}$$
$$= \left\{ 1, \frac{3\sqrt{5}}{2}\left(t^2 - \frac{1}{3}\right), \frac{3}{8}(35t^4 - 30t^2 + 3) \right\}.$$
Let
$$\mathbf{p} = \frac{1}{2}\mathbf{w}_1 + \frac{\sqrt{5}}{8}\mathbf{w}_2 - \frac{1}{16}\mathbf{w}_3.$$
Verify that $\mathbf{p} = \text{proj}_{\text{span}(S)}t$. [*Hint:* Verify $(t - \mathbf{p}, \mathbf{w}_j) = 0$.]

16. Let $V$ be the inner product space $R^3$ with the standard inner product and let $W$ be the two-dimensional subspace with orthogonal basis $\{\mathbf{w}_1, \mathbf{w}_2\}$ where
$$\mathbf{w}_1 = \begin{bmatrix} 3 \\ -1 \\ -2 \end{bmatrix} \quad \text{and} \quad \mathbf{w}_2 = \begin{bmatrix} 2 \\ 0 \\ 3 \end{bmatrix}.$$
Find $\text{proj}_W \mathbf{v}$ for each of the following vectors.
   (a) $\mathbf{v} = \begin{bmatrix} 1 \\ -1 \\ 2 \end{bmatrix}$.        (b) $\mathbf{v} = \begin{bmatrix} 1 \\ 1 \\ 8 \end{bmatrix}$.

17. Let $V = C[0, 1]$ with the inner product as given in Example 6 and let $W$ be the subspace with orthonormal basis
$$\{\mathbf{u}_1, \mathbf{u}_2\} = \left\{ t\sqrt{3}, \sqrt{80}\left(t^2 - \tfrac{3}{4}t\right) \right\}.$$
Compute $\text{proj}_W e^t$ and the distance from $e^t$ to $W$.

18. In Exercise 10 the set of three orthogonal polynomials given are called the Legendre polynomials of degrees, 0, 1, and 2, respectively. Let $S = \{\mathbf{p}_0(t), \mathbf{p}_1(t), \mathbf{p}_2(t)\}$. Compute the projection of each of the following functions onto span($S$).
   (a) $\sin(t)$.        (b) $e^t$.

19. Let $V$ be an inner product space with inner product given by $(\mathbf{u}, \mathbf{v})$ for vectors $\mathbf{u}$ and $\mathbf{v}$ in $V$. Let $r$ be any positive scalar. Show that $r(\mathbf{u}, \mathbf{v})$ is another inner product on $V$.

---

[5]In this set of exercises we recommend that a computer algebra system be used to compute the inner product of vectors in function spaces.

**20.** Referring to Example 10,

    (a) Compute the Fourier coefficients $a_2$ and $b_2$.

    (b) Find the third-order Fourier approximation.

**21.** For function space $V = C[-\pi, \pi]$ with inner product

$$(\mathbf{f}, \mathbf{g}) = \frac{1}{\pi} \int_{-\pi}^{\pi} f(t)g(t)\, dt,$$

determine the first-order Fourier approximation to each of the following.

    (a) $t$.         (b) $e^t$.         (c) $\sin(2t)$.

## True/False Review Questions

*Determine whether each of the following statements is true or false.*

**1.** The Cauchy-Schwarz inequality in an inner product space is used to define an angle between vectors.

**2.** In an inner product space, any linearly independent set of vectors is also an orthogonal set.

**3.** If $\{\mathbf{u}_1, \mathbf{u}_2, \mathbf{u}_3\}$ is an orthonormal basis for inner product space $V$, then any vector $\mathbf{v}$ in $V$ can be expressed as $\mathbf{v} = (\mathbf{u}_1, \mathbf{v})\mathbf{u}_1 + (\mathbf{u}_2, \mathbf{v})\mathbf{u}_2 + (\mathbf{u}_3, \mathbf{v})\mathbf{u}_3$.

**4.** In inner product space $V$, if $\mathbf{q} = \text{proj}_w \mathbf{v}$, then $(\mathbf{v} - \mathbf{q}, \mathbf{w}) = 0$.

**5.** If $W$ is a subspace of inner product space $V$, then for any vector $\mathbf{v}$ in $V$, the distance from $\mathbf{v}$ to $W$ is $\|\mathbf{v} - \text{proj}_W \mathbf{v}\|$.

## Terminology

| | |
|---|---|
| Angle between vectors | Orthogonal and orthonormal sets |
| Orthonormal basis | The Gram-Schmidt process |
| Projection of a vector onto a subspace | Distance from a vector to a subspace |
| Fourier approximation | |

This section continued our study of inner product spaces by developing relationships between vectors, vectors and bases, and vectors and subspaces. Orthonormal bases played a key role in providing relatively simple expressions for a number of the concepts. We also developed an algorithm for converting an ordinary basis into an orthonormal basis.

    The concepts in this section play a primary role in the important application discussed in Section 6.4. To review the material of this section, respond to the following questions and statements.

- What relationship between vectors $\mathbf{u}$ and $\mathbf{v}$ is the key to the definition of the angle between these vectors?
- Give an example of a function space.
- Why can we say that the orthogonality of vectors depends on the inner product? Discuss.
- How are orthonormal sets different from orthogonal sets?
- An orthonormal set is automatically an orthogonal set, but it is also a _____ set. Hence orthonormal sets are prime candidates for inclusion within bases.
- If we have an orthonormal basis for an inner product space $V$, why can we say it is easy to express any vector in the space in terms of such a basis? Explain.
- Describe how to manufacture an orthonormal basis from an ordinary basis for an inner product space.

- Let $W$ be a subspace of inner product space $V$ and let $\mathbf{v}$ be a vector in $V$. What do we mean when we say that we want to find the best approximation to $\mathbf{v}$ from $W$?

- In terms of the inner product, how do we characterize the best approximation of $\mathbf{v}$? That is, what inner product relation involving $\mathbf{v}$ and the vectors in $W$ must be satisfied? Express this relationship in terms of the (inner product) norm.

- What is the projection of $\mathbf{v}$ onto a subspace $W$? If you know a basis for $W$, how can you compute the projection?

- How is projection involved in computing the distance from a vector $\mathbf{v}$ to a subspace $W$?

- Describe a Fourier approximation. State the inner product space in which you work and any bases or subspaces involved.

## 6.4 ■ LEAST SQUARES

In Section 2.6 we discussed the line of best fit. Using a calculus approach we determined the line $y = mt + b$ that came closest to a data set $D = \{(t_i, y_i) \mid i = 1, 2, \ldots, n\}$ of distinct ordered pairs. The values of m and b were determined so that the sum of the squares of the deviations $(mt_i + b - y_i)^2$ would be as small as possible. At that time we indicated that an alternative approach that emphasized the use of linear algebra could be developed. Here we investigate such an approach. At the same time we generalize the type of function that can be used to develop a mathematical model for a data set. We also investigate least squares approximation from function spaces.

We begin by recasting the line of best fit into an inner product space approach so we can use the developments of the previous sections of this chapter. Consider the data set $D = \{(t_i, y_i) \mid i = 1, 2, \ldots, n\}$ as a sample of a process that exhibits a linear relationship between the independent value $t_i$ and the corresponding observed, or dependent, value $y_i$. That is, the data set $D$ is in some sense well approximated by a line $y = mt + b$. In an ideal situation we would have $mt_i + b = y_i$ for each data point of $D$. However, as is usually the case, the corresponding linear system

$$
\begin{array}{l}
mt_1 + b = y_1 \\
mt_2 + b = y_2 \\
\vdots \\
\vdots \\
mt_n + b = y_n
\end{array}
\quad \text{or} \quad
\begin{bmatrix} t_1 & 1 \\ t_2 & 1 \\ \vdots & \vdots \\ \vdots & \vdots \\ t_n & 1 \end{bmatrix}
\begin{bmatrix} m \\ b \end{bmatrix}
=
\begin{bmatrix} y_1 \\ y_2 \\ \vdots \\ \vdots \\ y_n \end{bmatrix}
\tag{1}
$$

is inconsistent. Writing (1) in matrix form $\mathbf{Ax} = \mathbf{y}$ where

$$
\mathbf{A} = \begin{bmatrix} t_1 & 1 \\ t_2 & 1 \\ \vdots & \vdots \\ \vdots & \vdots \\ t_n & 1 \end{bmatrix}, \quad
\mathbf{x} = \begin{bmatrix} m \\ b \end{bmatrix}, \quad \text{and} \quad
\mathbf{y} = \begin{bmatrix} y_1 \\ y_2 \\ \vdots \\ \vdots \\ y_n \end{bmatrix}
$$

we interpret the inconsistent system as saying that vector $\mathbf{y}$ does not belong to the column space of $\mathbf{A}$, $\mathbf{col(A)}$. Let $V = R^n$ be an inner product space with the standard inner product; then $W = \mathbf{col(A)}$ is a subspace of $V$. Since we consider linear system $\mathbf{Ax} = \mathbf{y}$ to be inconsistent, we change our point of view.

**Strategy:** Determine a vector $\hat{\mathbf{x}}$ in $R^2$ so that $\mathbf{A}\hat{\mathbf{x}}$ is as close to vector $\mathbf{y}$ as possible. Since $\mathbf{A}\hat{\mathbf{x}}$ is a linear combination of the columns of $\mathbf{A}$, it is in subspace $W = \mathbf{col}(\mathbf{A})$. By the requirement that $\mathbf{A}\hat{\mathbf{x}}$ be as close as possible to $\mathbf{y}$, it follows that $\mathbf{A}\hat{\mathbf{x}} = \text{proj}_W \mathbf{y}$. We use the results of Section 6.3 to find $\hat{\mathbf{x}}$.

To determine $\hat{\mathbf{x}}$ we use the fact that $\mathbf{y} - \text{proj}_W \mathbf{y} = \mathbf{y} - \mathbf{A}\hat{\mathbf{x}}$ must be orthogonal to every vector in $W$. In particular, $\mathbf{y} - \mathbf{A}\hat{\mathbf{x}}$ is orthogonal to a basis for $W$. The most convenient basis for $W$ consists of the columns of $\mathbf{A}$. (Explain why the columns of $\mathbf{A}$ are a basis for $\mathbf{col}(\mathbf{A})$.) Thus

$$(\mathbf{col}_1(\mathbf{A}), \mathbf{y} - \mathbf{A}\hat{\mathbf{x}}) = 0 \quad \text{and} \quad (\mathbf{col}_2(\mathbf{A}), \mathbf{y} - \mathbf{A}\hat{\mathbf{x}}) = 0. \tag{2}$$

Recalling that the standard inner product of a pair of column vectors $\mathbf{u}$ and $\mathbf{v}$ in $R^n$ can be expressed as $(\mathbf{u}, \mathbf{v}) = \mathbf{u}^T \mathbf{v}$, then (2) can be written in the form

$$\mathbf{col}_1(\mathbf{A})^T (\mathbf{y} - \mathbf{A}\hat{\mathbf{x}}) = 0 \quad \text{and} \quad \mathbf{col}_2(\mathbf{A})^T (\mathbf{y} - \mathbf{A}\hat{\mathbf{x}}) = 0. \tag{3}$$

From properties of matrix multiplication, we can combine the result in (3) into the matrix product $\mathbf{A}^T (\mathbf{y} - \mathbf{A}\hat{\mathbf{x}}) = \mathbf{0}$ or, equivalently,

$$\mathbf{A}^T (\mathbf{A}\hat{\mathbf{x}} - \mathbf{y}) = \mathbf{0}$$

and can express it as the linear system

$$\mathbf{A}^T \mathbf{A}\hat{\mathbf{x}} = \mathbf{A}^T \mathbf{y}. \tag{4}$$

In (4), $\mathbf{A}^T \mathbf{A}$ is a square matrix (here $2 \times 2$) and the right side $\mathbf{A}^T \mathbf{y}$ is a column (here $2 \times 1$). Any solution $\hat{\mathbf{x}}$ is called a **least squares solution** of the linear system $\mathbf{A}\mathbf{x} = \mathbf{y}$. This is an unfortunate choice of words, since in general $\mathbf{A}\hat{\mathbf{x}} \neq \mathbf{y}$, rather $\mathbf{A}\hat{\mathbf{x}}$ is the vector in $W = \mathbf{col}(\mathbf{A})$ as close as possible to $\mathbf{y}$. That is, from Section 6.3

$$\|\mathbf{A}\hat{\mathbf{x}} - \mathbf{y}\| = \min_{\mathbf{w} \text{ in } W} \|\mathbf{w} - \mathbf{y}\|$$

and

$$\|\mathbf{A}\hat{\mathbf{x}} - \mathbf{y}\|^2 = \min_{\mathbf{w} \text{ in } W} \|\mathbf{w} - \mathbf{y}\|^2 = \min_{\mathbf{w} \text{ in } W} \left( \sum_{j=1}^{n} (w_j - y_j)^2 \right).$$

Thus a solution to the linear system of equations in (4) produces a vector $\hat{\mathbf{x}} = \begin{bmatrix} \hat{m} \\ \hat{b} \end{bmatrix}$

so that line $y = \hat{m}t + \hat{b}$ is as close as possible (in the least squares sense) to the data in $D$. This is the same as the solution of the linear system in Equation (2) in Section 2.6, which produces the line of best fit.

Several comments are needed at this time.

- If a linear combination of functions other than 1 and $t$ (which generate lines) are used to build a model for the data set $D$, then a similar development is valid except that matrix $\mathbf{A}$ and the vector of unknowns $\mathbf{x}$ will change in content and possibly size.

- This procedure is often called **discrete least squares approximation** since we are constructing a model for the discrete data set $D$.

- The linear system in (4) is called the **normal system** of equations.

- A least squares solution $\mathbf{A}\hat{\mathbf{x}}$ is the vector in $\mathbf{col}(\mathbf{A})$ closest to $\mathbf{y}$. However, in many cases we do not use $\mathbf{A}\hat{\mathbf{x}}$ explicitly; rather the entries of $\hat{\mathbf{x}}$ become coefficients in a linear combination of functions, like $mx + b$. Geometrically, the graph of this expression comes close to the data in $D$ but need not go through any of the data points.

- Whether we start with a data set $D$ or just with an inconsistent linear system $\mathbf{Ax} = \mathbf{y}$, a solution $\hat{\mathbf{x}}$ of the normal system of equations generates a least squares solution $\mathbf{A}\hat{\mathbf{x}}$.

- This approach avoided the construction of an orthogonal or orthonormal basis for subspace $W$, which was needed in Section 6.3. However, it is known that the solution of the normal system using row operations can be quite sensitive to roundoff errors encountered when computer arithmetic is used in the solution process. An alternative procedure is discussed later in this section.

- The coefficient matrix of the normal system is not always guaranteed to be nonsingular in more general cases. Hence there can be more than one least squares solution. The treatment of this general case is left to advanced courses and we focus on the special case where matrix $\mathbf{A}^T\mathbf{A}$ is nonsingular. The result that follows addresses this important case.

If $\mathbf{A}$ is an $m \times n$ matrix with linearly independent columns, then $\mathbf{A}^T\mathbf{A}$ is nonsingular and the linear system $\mathbf{Ax} = \mathbf{y}$ has a unique least squares solution given by $\hat{\mathbf{x}} = (\mathbf{A}^T\mathbf{A})^{-1}\mathbf{A}^T\mathbf{y}$.

To verify this result we note that $\mathbf{A}^T\mathbf{A}$ is nonsingular provided that the homogeneous linear system $\mathbf{A}^T\mathbf{Ax} = \mathbf{0}$ has only the trivial solution $\mathbf{x} = \mathbf{0}$. Multiplying both sides of $\mathbf{A}^T\mathbf{Ax} = \mathbf{0}$ by $\mathbf{x}^T$ on the left gives

$$0 = \mathbf{x}^T\mathbf{A}^T\mathbf{Ax} = (\mathbf{Ax})^T(\mathbf{Ax}) = (\mathbf{Ax}, \mathbf{Ax})$$

using the standard inner product on $R^m$. Since the inner product of a vector with itself is zero if and only if the vector is the zero vector, we have that $\mathbf{Ax} = \mathbf{0}$. But this expression implies that we have a linear combination of the columns of $\mathbf{A}$ that gives the zero vector. Since the columns of $\mathbf{A}$ are linearly independent, the only way this can be true is if $\mathbf{x} = \mathbf{0}$. That is, $\mathbf{A}^T\mathbf{Ax} = \mathbf{0}$ has only the trivial solution, so $\mathbf{A}^T\mathbf{A}$ is nonsingular. Thus the normal system has a unique solution and it is $\hat{\mathbf{x}} = (\mathbf{A}^T\mathbf{A})^{-1}\mathbf{A}^T\mathbf{y}$.

In Examples 1 and 2 we show the construction of least squares approximations that utilize expressions other than a line. Following these examples, we summarize the general approach to the computation of discrete least squares models.

EXAMPLE 1    The following data show atmospheric pollutant $y_i$ (relative to an Environmental Protection Agency [EPA] standard) at half-hour intervals $t_i$.

| $t_i$ | 1 | 1.5 | 2 | 2.5 | 3 | 3.5 | 4 | 4.5 | 5 |
|-------|------|------|------|------|------|------|------|------|-------|
| $y_i$ | −0.15 | 0.24 | 0.68 | 1.04 | 1.21 | 1.15 | 0.86 | 0.41 | −0.08 |

A plot of these data, as shown in Figure 1, suggests that a quadratic polynomial

$$p(t) = a_2t^2 + a_1t + a_0$$

may produce a good model. From the expressions $p(t_i) = a_2t_i^2 + a_1t_i + a_0 = y_i$ we

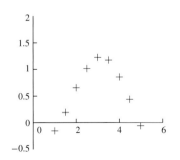

FIGURE 1

get the system of equations

$$
\begin{aligned}
a_2(1)^2 + \quad a_1(1) + a_0 &= -0.15 \\
a_2(1.5)^2 + a_1(1.5) + a_0 &= \phantom{-}0.24 \\
\vdots \qquad\quad \vdots \qquad \vdots \qquad &\vdots \\
a_2(5)^2 + \quad a_1(5) + a_0 &= -0.08
\end{aligned}
$$

which can be written in matrix form as $\mathbf{Ax} = \mathbf{y}$ where

$$
\mathbf{A} = \begin{bmatrix} 1 & 1 & 1 \\ 2.25 & 1.5 & 1 \\ 4 & 2 & 1 \\ 6.25 & 2.5 & 1 \\ 9 & 3 & 1 \\ 12.25 & 3.5 & 1 \\ 16 & 4 & 1 \\ 20.25 & 4.5 & 1 \\ 25 & 5 & 1 \end{bmatrix}, \quad \mathbf{x} = \begin{bmatrix} a_2 \\ a_1 \\ a_0 \end{bmatrix}, \quad \text{and} \quad \mathbf{y} = \begin{bmatrix} -0.15 \\ 0.24 \\ 0.68 \\ 1.04 \\ 1.21 \\ 1.15 \\ 0.86 \\ 0.41 \\ -0.05 \end{bmatrix}.
$$

Forming the normal system $\mathbf{A}^T\mathbf{A}\hat{\mathbf{x}} = \mathbf{A}^T\mathbf{y}$, we get

$$
\begin{bmatrix} 1583.25 & 378 & 96 \\ 378 & 96 & 27 \\ 96 & 27 & 9 \end{bmatrix} \begin{bmatrix} a_2 \\ a_1 \\ a_0 \end{bmatrix} = \begin{bmatrix} 55.4 \\ 16.86 \\ 5.39 \end{bmatrix}.
$$

Solving this system, we obtain (verify)

$$
\hat{\mathbf{x}} = \begin{bmatrix} a_2 \\ a_1 \\ a_0 \end{bmatrix} \approx \begin{bmatrix} -0.3238 \\ 1.9889 \\ -1.9137 \end{bmatrix}
$$

so the quadratic polynomial model is

$$
p(t) = -0.3238t^2 + 1.9889t - 1.9137.
$$

Figure 2 shows the data set indicated with $+$ and the graph of $p(t)$. We see that $p(t)$ is close to each data point but is not required to go through any of the data.  ■

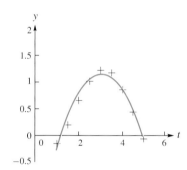

FIGURE 2

EXAMPLE 2  The following data are a sample from a process that contains cyclical behavior that seems to rise. (See Figure 3; the dots represent the ordered pairs of data and we have connected successive points with straight-line segments to get an impression of the behavior of this data set.)

FIGURE 3

| $t_i$ | 0.1 | 0.4 | 1.1 | 1.6 | 2.1 | 2.6 | 3.2 | 3.6 | 4.3 | 4.6 | 5.2 | 5.7 |
| --- | --- | --- | --- | --- | --- | --- | --- | --- | --- | --- | --- | --- |
| $y_i$ | 0.7 | 1.5 | 0.1 | 2.2 | 3.0 | 1.7 | 3.5 | 4.5 | 3.3 | 4.8 | 6.1 | 4.9 |

The construction of a least squares model for any data set requires that we specify a linear combination of functions that we feel will adequately capture the behavior of the data. In this case the cyclical behavior suggests that the use of a trigonometric function may be warranted. On the interval $[0, 6]$ there are about four complete oscillations, and the first four data points are connected in a form that suggests a sine function. Hence we will include $\sin(4t)$ in the linear combination. The "wave" seems to rise along a straight line; hence a polynomial in $\text{span}\{1, t\}$

will be included. This crude analysis suggests that we try to model that data with a function of the form

$$g(t) = a_2 \sin(4t) + a_1 t + a_0.$$

From the expressions $g(t_i) = a_2 \sin(4t_i) + a_1 t_i + a_0 = y_i$ we get the system of equations

$$a_2 \sin(0.4) + a_1(0.1) + a_0 = 0.7$$
$$a_2 \sin(1.6) + a_1(0.4) + a_0 = 1.5$$
$$\vdots \qquad \qquad \vdots \qquad \vdots \qquad \vdots$$
$$a_2 \sin(22.8) + a_1(5.7) + a_0 = 4.9$$

which can be written in matrix form as $\mathbf{Ax} = \mathbf{y}$ where

$$\mathbf{A} = \begin{bmatrix} 0.3894 & 0.1 & 1.0 \\ 0.9996 & 0.4 & 1.0 \\ -0.9516 & 1.1 & 1.0 \\ 0.1165 & 1.6 & 1.0 \\ 0.8546 & 2.1 & 1.0 \\ -0.8278 & 2.6 & 1.0 \\ 0.2315 & 3.2 & 1.0 \\ 0.9657 & 3.6 & 1.0 \\ -0.9969 & 4.3 & 1.0 \\ -0.4346 & 4.6 & 1.0 \\ 0.9288 & 5.2 & 1.0 \\ -0.7235 & 5.7 & 1.0 \end{bmatrix}, \quad \mathbf{x} = \begin{bmatrix} a_2 \\ a_1 \\ a_0 \end{bmatrix}, \quad \text{and} \quad \mathbf{y} = \begin{bmatrix} 0.7 \\ 1.5 \\ 0.1 \\ 2.2 \\ 3.0 \\ 1.7 \\ 3.5 \\ 4.5 \\ 3.3 \\ 4.8 \\ 6.1 \\ 4.9 \end{bmatrix}.$$

Forming the normal system $\mathbf{A}^T \mathbf{A}\hat{\mathbf{x}} = \mathbf{A}^T \mathbf{y}$, we get

$$\begin{bmatrix} 7.0404 & -2.1419 & 0.5517 \\ -2.1419 & 137.4900 & 34.5000 \\ 0.5517 & 34.5000 & 12.0000 \end{bmatrix} \begin{bmatrix} a_2 \\ a_1 \\ a_0 \end{bmatrix} = \begin{bmatrix} 4.9903 \\ 138.3400 \\ 36.3000 \end{bmatrix}.$$

Solving this system, we obtain (verify)

$$\hat{\mathbf{x}} = \begin{bmatrix} a_2 \\ a_1 \\ a_0 \end{bmatrix} \approx \begin{bmatrix} 0.9964 \\ 0.9841 \\ 0.1500 \end{bmatrix}$$

so the model for the data is

$$g(t) = 0.9964 \sin(4t) + 0.9841t + 0.1500.$$

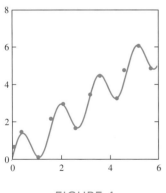

FIGURE 4

Figure 4 shows the data set with the graph of $g(t)$. We see that $g(t)$ is close to each data point but is not required to go through the data. (For convenience we have only recorded figures to four decimal places.) This data set was constructed by graphing $t + \sin(4t)$ over $[0, 6]$, sampling the graph at 12 points and then rounding the coordinates to tenths. This in effect introduced small errors into the data. The resulting model provides a good representation of the data.  ■

## General Approach for Discrete Least Squares Computations

Examples 1 and 2 illustrate a general form for discrete least squares approximation. Let $\{f_1(t), f_2(t), \ldots, f_k(t)\}$ be a basis for a subspace $W$ of a function space $V$. To construct a least squares approximation to a data set $D = \{(t_i, y_i) \mid i = 1, 2, \ldots, n\}$, where $n > k$, from the vectors in $W$ we form the expression

$$g(t) = a_1 f_1(t) + a_2 f_2(t) + \cdots + a_k f_k(t) = \sum_{j=1}^{k} a_j f_j(t).$$

The coefficients $a_1, a_2, \ldots, a_k$ are to be determined. This expression is called the **model equation**. Ideally, we would like to determine the coefficients (sometimes called **weights**) $a_1, a_2, \ldots, a_k$ such that

$$y_i = a_1 f_1(t_i) + a_2 f_2(t_i) + \cdots + a_k f_k(t_i)$$

for each data point $t_i$, $i = 1, 2, \ldots, n$. This set of equations leads to the linear system $\mathbf{Ax} = \mathbf{y}$, where

$$\mathbf{A} = \begin{bmatrix} f_1(t_1) & f_2(t_1) & \cdots & f_k(t_1) \\ f_1(t_2) & f_2(t_2) & \cdots & f_k(t_2) \\ \vdots & \vdots & \cdots & \vdots \\ \vdots & \vdots & \cdots & \vdots \\ f_1(t_n) & f_2(t_n) & \cdots & f_k(t_n) \end{bmatrix}, \quad \mathbf{x} = \begin{bmatrix} a_1 \\ a_2 \\ \vdots \\ \vdots \\ a_k \end{bmatrix}, \quad \text{and} \quad \mathbf{y} = \begin{bmatrix} y_1 \\ y_2 \\ \vdots \\ \vdots \\ y_k \end{bmatrix}.$$

As is often the case, the system $\mathbf{Ax} = \mathbf{y}$ is inconsistent, so we determine a least squares solution $\hat{\mathbf{x}}$ from the corresponding normal system of equations.

Denote the entries of $\hat{\mathbf{x}}$ as $\hat{a}_j$, $j = 1, 2, \ldots, k$, then the equation that models the data in the least squares sense is given by

$$\hat{g}(t) = \hat{a}_1 f_1(t) + \hat{a}_2 f_2(t) + \cdots + \hat{a}_k f_k(t).$$

In general, $\hat{g}(t_i) \neq y_i$ (they may be equal, but there is no guarantee). To measure the error incurred using the model $\hat{g}(t)$ for data set $D$, we let $e_i = y_i - \hat{g}(t_i)$ and define vector $\mathbf{e} = \begin{bmatrix} e_1 & e_2 & \cdots & e_n \end{bmatrix}^T$; then

$$\mathbf{e} = \mathbf{y} - \mathbf{A}\hat{\mathbf{x}}.$$

Recall that $\mathbf{A}\hat{\mathbf{x}} = \mathrm{proj}_{\mathrm{col}(\mathbf{A})}\mathbf{y}$, and by results from Section 6.3 we are guaranteed that $\|\mathbf{e}\| = \|\mathbf{y} - \mathbf{A}\hat{\mathbf{x}}\|$ is as small as possible. That is,

$$\|\mathbf{e}\|^2 = (\mathbf{e}, \mathbf{e}) = (\mathbf{y} - \mathbf{A}\hat{\mathbf{x}}, \mathbf{y} - \mathbf{A}\hat{\mathbf{x}}) = \sum_{i=1}^{n} \left[ y_i - \sum_{j=1}^{k} \hat{a}_j f_j(t_i) \right]^2$$

is minimized. We say that $\hat{\mathbf{x}}$, the least squares solution, minimizes the sum of the squares of the deviations between the observations $y_i$ and the values $\hat{g}(t_i)$ predicted by the model equation.

## Introduction to Continuous Least Squares Approximation

A discrete least squares approximation "fits" or comes as close as possible (in the least squares sense) to data in a set $D$. In many applications it is necessary to approximate a continuous function in terms of functions from some prescribed function space. We may be seeking a polynomial approximation from $P_n$, the inner product space of polynomials of degree $n$ or less, or an approximation in terms of trigonometric polynomials like Fourier approximation (see Section 6.3). We can use our development of projections from Section 6.3, as we show next.

Let $\mathbf{f}$ be a vector in $C[a, b]$ with inner product

$$(\mathbf{g}, \mathbf{h}) = \int_a^b g(t)h(t)\,dt.$$

Let $W$ be a subspace of $C[a, b]$ with basis vectors $\mathbf{w}_1, \mathbf{w}_2, \ldots, \mathbf{w}_k$. (These basis vectors are functions in $C[a, b]$.) Then the vector in $W$ closest to $\mathbf{f}$ is $\text{proj}_W \mathbf{f}$ and from Section 6.3 we know that $\mathbf{f} - \text{proj}_W \mathbf{f}$ is orthogonal to every member of $W$. It follows that $\mathbf{f} - \text{proj}_W \mathbf{f}$ is orthogonal to each of the basis vectors for $W$; hence we have

$$(\mathbf{w}_j, \mathbf{f} - \text{proj}_W \mathbf{f}) = 0 \quad j = 1, 2, \ldots, k. \tag{5}$$

[This is the identical approach used for the line of best fit in (2).] However, $\text{proj}_W \mathbf{f}$ is in $W$ so there exist scalars $c_1, c_2, \ldots, c_k$ such that

$$\text{proj}_W \mathbf{f} = \sum_{i=1}^{k} c_i \mathbf{w}_i.$$

From Equation (5) we obtain

$$
\begin{aligned}
0 &= \left( \mathbf{w}_j, \mathbf{f} - \sum_{i=1}^{k} c_i \mathbf{w}_i \right) = \int_a^b w_j(t) \left( f(t) - \sum_{i=1}^{k} c_i w_i(t) \right) dt \\
&= \int_a^b w_j(t) f(t) \, dt - \sum_{i=1}^{k} c_i \int_a^b w_j(t) w_i(t) \, dt \quad \text{for } j = 1, 2, \ldots, k.
\end{aligned}
\tag{6}
$$

The $k$ equations in (6) can be arranged into the following linear system of equations:

$$
c_1 \int_a^b w_1(t) w_1(t) \, dt + c_2 \int_a^b w_1(t) w_2(t) \, dt
$$
$$
+ \cdots + c_k \int_a^b w_1(t) w_k(t) \, dt = \int_a^b f(t) w_1(t) \, dt,
$$

$$
c_1 \int_a^b w_2(t) w_1(t) \, dt + c_2 \int_a^b w_2(t) w_2(t) \, dt
$$
$$
+ \cdots + c_k \int_a^b w_2(t) w_k(t) \, dt = \int_a^b f(t) w_2(t) \, dt, \tag{7}
$$

$$\vdots$$

$$
c_1 \int_a^b w_k(t) w_1(t) \, dt + c_2 \int_a^b w_k(t) w_2(t) \, dt
$$
$$
+ \cdots + c_k \int_a^b w_k(t) w_k(t) \, dt = \int_a^b f(t) w_k(t) \, dt.
$$

Since $\int_a^b w_j(t) w_i(t) \, dt = (\mathbf{w}_j, \mathbf{w}_i)$, (7) can be expressed in matrix form[1] as

$$
\begin{bmatrix}
(\mathbf{w}_1, \mathbf{w}_1) & (\mathbf{w}_1, \mathbf{w}_2) & \cdots & \cdots & (\mathbf{w}_1, \mathbf{w}_k) \\
(\mathbf{w}_2, \mathbf{w}_1) & (\mathbf{w}_2, \mathbf{w}_2) & \cdots & \cdots & (\mathbf{w}_2, \mathbf{w}_k) \\
\vdots & \vdots & \cdots & \cdots & \vdots \\
\vdots & \vdots & \cdots & \cdots & \vdots \\
(\mathbf{w}_k, \mathbf{w}_1) & (\mathbf{w}_k, \mathbf{w}_2) & \cdots & \cdots & (\mathbf{w}_k, \mathbf{w}_k)
\end{bmatrix}
\begin{bmatrix}
c_1 \\ c_2 \\ \vdots \\ c_k
\end{bmatrix}
=
\begin{bmatrix}
(\mathbf{w}_1, \mathbf{f}) \\ (\mathbf{w}_2, \mathbf{f}) \\ \vdots \\ \vdots \\ (\mathbf{w}_k, \mathbf{f})
\end{bmatrix}. \tag{8}
$$

[1]Note that the use of the inner product leads us to a linear system of equations whether we are solving a discrete or continuous least squares problem.

The coefficient matrix in (8) is called the **Grammian** of vectors $\{\mathbf{w}_1, \mathbf{w}_2, \ldots, \mathbf{w}_k\}$ and can be shown to be nonsingular since these vectors are linearly independent. (See also Section 6.2.) The linear system in (8) is called the **normal system of equations**, and this process is called **continuous least squares approximation**.

The Grammian matrix can be quite sensitive to roundoff errors incurred in computing the inner products and in the procedure used to solve the linear system when computer arithmetic is involved. Hence it is customary to use an orthogonal or orthonormal basis for $W$. In such cases the Grammian becomes diagonal (see Exercise 9) and the solution of the normal system is easy to compute.

To illustrate the continuous least squares process we compute a polynomial approximation to $\cos(\pi t)$ in Example 3. This example is for illustrative purposes only since it is unrealistic to approximate such a well-behaved function as $\cos(\pi t)$.

EXAMPLE 3    Let $V = C[0, 1]$ with its usual inner product. To find the least squares approximation to $f(t) = \cos(\pi t)$ from subspace $P_2$ we proceed as follows. Let $\mathbf{w}_1 = 1$, $\mathbf{w}_2 = t$ and $\mathbf{w}_3 = t^2$, then we know that $\{\mathbf{w}_1, \mathbf{w}_2, \mathbf{w}_3\}$ is a basis for $W$. Forming the normal system of equations as in (8), we have

$$\begin{bmatrix} \int_0^1 1\, dt & \int_0^1 t\, dt & \int_0^1 t^2\, dt \\ \int_0^1 t\, dt & \int_0^1 t^2\, dt & \int_0^1 t^3\, dt \\ \int_0^1 t^2\, dt & \int_0^1 t^3\, dt & \int_0^1 t^4\, dt \end{bmatrix} \begin{bmatrix} c_1 \\ c_2 \\ c_3 \end{bmatrix} = \begin{bmatrix} \int_0^1 \cos(\pi t)\, dt \\ \int_0^1 t \cos(\pi t)\, dt \\ \int_0^1 t^2 \cos(\pi t)\, dt \end{bmatrix}.$$

Performing the integrations in the preceding entries, we get the linear system

$$\begin{bmatrix} 1 & \frac{1}{2} & \frac{1}{3} \\ \frac{1}{2} & \frac{1}{3} & \frac{1}{4} \\ \frac{1}{3} & \frac{1}{4} & \frac{1}{5} \end{bmatrix} \begin{bmatrix} c_1 \\ c_2 \\ c_3 \end{bmatrix} = \begin{bmatrix} 0 \\ -\dfrac{2}{\pi^2} \\ -\dfrac{2}{\pi^2} \end{bmatrix}$$

whose solution is

$$c_1 = \frac{12}{\pi^2} \approx 1.2159,$$

$$c_2 = -\frac{24}{\pi^2} \approx -2.4317,$$

$$c_3 = 0.$$

Thus we have that the least squares approximation to $\cos(\pi t)$ from subspace $P_2$ is

$$\operatorname{proj}_{P_2}(\cos(\pi t)) = \frac{12}{\pi^2} - \frac{24}{\pi^2} t.$$

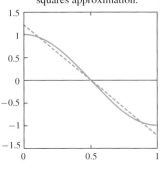

Solid curve is $\cos(\pi t)$. Dashed curve is the least squares approximation.

FIGURE 5

Figure 5 shows $\cos(\pi t)$ and this least squares approximation over $[0, 1]$. The error in this approximation is

$$\left\| \cos(\pi t) - \operatorname{proj}_{P_2}(\cos(\pi t)) \right\| = \sqrt{ \int_0^1 \left( \cos(\pi t) - \left( \frac{12}{\pi^2} - \frac{24}{\pi^2} t \right) \right)^2 dt }$$

$$= \sqrt{ \frac{1}{2} - \frac{48}{\pi^4} }$$

$$\approx \sqrt{0.0072328} \approx 0.085046. \qquad \blacksquare$$

## Least Squares Solutions Using QR-Factorization (Optional)

We have mentioned that the normal system of equations for both the discrete and continuous least squares computations can be quite sensitive to small errors introduced by computer arithmetic. Instead of solving the normal system directly using row reduction or LU-factorization (see Section 2.5), an alternative procedure is to use the QR-factorization (see Section 4.6).

For the discrete least squares problem that is characterized by an inconsistent linear system $\mathbf{Ax} = \mathbf{y}$, where $\mathbf{A}$ has linearly independent columns, we first obtain the QR-factorization of $\mathbf{A}$ and then the least squares solution as follows:

$$\mathbf{Ax} = \mathbf{y} \quad \text{or} \quad \mathbf{QRx} = \mathbf{y} \qquad (\text{since } \mathbf{A} = \mathbf{QR})$$
$$\mathbf{Q}^T\mathbf{QRx} = \mathbf{Q}^T\mathbf{y} \qquad (\text{multiply both sides on the left by } \mathbf{Q}^T) \quad (9)$$
$$\mathbf{I}_n\mathbf{Rx} = \mathbf{Q}^T\mathbf{y} \quad \text{or} \quad \mathbf{Rx} = \mathbf{Q}^T\mathbf{y} \qquad (\text{since } \mathbf{Q} \text{ is orthogonal, } \mathbf{Q}^T\mathbf{Q} = \mathbf{I}_n)$$

Finally, we solve $\mathbf{Rx} = \mathbf{Q}^T\mathbf{y}$ by back substitution since $\mathbf{R}$ is upper triangular. In terms of a formula, the least squares solution is $\hat{\mathbf{x}} = \mathbf{R}^{-1}\mathbf{Q}^T\mathbf{y}$, but we do not explicitly compute $\mathbf{R}^{-1}$; rather we perform the back substitution. Example 3 in Section 4.6 illustrates this procedure.

In the case of continuous least squares, we start with the normal system of equations whose coefficient matrix is the Grammian. We obtain its QR-factorization and follow the same procedure as in (9). We illustrate this in Example 4.

**EXAMPLE 4**   Using Example 3, we found that the normal system of equations is given by

$$\begin{bmatrix} 1 & \frac{1}{2} & \frac{1}{3} \\ \frac{1}{2} & \frac{1}{3} & \frac{1}{4} \\ \frac{1}{3} & \frac{1}{4} & \frac{1}{5} \end{bmatrix} \begin{bmatrix} c_1 \\ c_2 \\ c_3 \end{bmatrix} = \begin{bmatrix} 0 \\ -\dfrac{2}{\pi^2} \\ -\dfrac{2}{\pi^2} \end{bmatrix}.$$

Denote the coefficient matrix of the normal system by $\mathbf{A}$; then its QR-factorization is

$$\mathbf{A} = \mathbf{QR} = \begin{bmatrix} 0.8571 & -0.5016 & -0.1170 \\ 0.4286 & 0.5685 & 0.7022 \\ 0.2857 & 0.6521 & -0.7022 \end{bmatrix} \begin{bmatrix} 1.1667 & 0.6429 & 0.4500 \\ 0 & 0.1017 & 0.1053 \\ 0 & 0 & -0.0039 \end{bmatrix}.$$

Multiplying both sides on the left by $\mathbf{Q}^T$ gives the upper triangular system

$$\begin{bmatrix} 1.1667 & 0.6429 & 0.4500 \\ 0 & 0.1017 & 0.1053 \\ 0 & 0 & -0.0039 \end{bmatrix} \begin{bmatrix} c_1 \\ c_2 \\ c_3 \end{bmatrix} = \begin{bmatrix} -0.1447 \\ -0.2473 \\ 0 \end{bmatrix}$$

whose solution is

$$c_1 \approx 1.2159,$$
$$c_2 \approx -2.4317,$$
$$c_3 \approx 0,$$

just as in Example 3. (MATLAB was used to obtain the QR-factorization. We have only displayed the numerical values to 4 decimal places.)   ■

## El Niño; the Southern Oscillation (Optional)

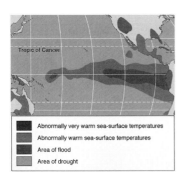

Tropic of Cancer

■ Abnormally very warm sea-surface temperatures
▨ Abnormally warm sea-surface temperatures
□ Area of flood
▨ Area of drought

© *1999 Prentice Hall*

El Niño is a cyclic phenomenon that occurs in the eastern Pacific and can have a profound influence on weather in vast regions around the world. El Niño is the result of cycles involving warming and cooling of the surface of the Pacific off the west coast of South America. These cycles are the result of winds, ocean currents, and the upwelling of colder water from ocean depths. The yearly effects are not always the same, but the year-to-year changes do follow a somewhat regular pattern that is called El Niño.

It has been found that the cycles of warming and cooling of these Pacific waters induce a distinctive change in sea level pressure. Scientists have recorded these changes in order to study and analyze the cycles. One particular set of data involves the pressure at Darwin, Australia compared to the pressure at Tahiti. The difference between the two pressures is used to generate an index number that has been recorded monthly for decades. Table 1 shows this index for the years 1993 through 1997. When the index is positive we have ocean cooling, and when it is negative we have ocean warming. An extreme positive index can indicate heavy rainfall in South America but severe droughts in eastern Australia. An extreme negative index often triggers the reverse situation.

TABLE 1

| Mon<br>Yr. | Jan | Feb | Mar | Apr | May | Jun | Jul | Aug | Sep | Oct | Nov | Dec |
|---|---|---|---|---|---|---|---|---|---|---|---|---|
| 1993 | −2.0 | −2.1 | −1.8 | −2.6 | −1.0 | −2.2 | −1.8 | −2.4 | −1.3 | −2.5 | −0.3 | 0.1 |
| 1994 | −.5 | −.1 | −2.2 | −2.9 | −1.7 | −1.5 | −2.9 | −3.0 | −3.0 | −2.6 | −1.2 | −2.6 |
| 1995 | −1.0 | −0.8 | 0.4 | −1.8 | −1.2 | −0.4 | 0.6 | −0.1 | 0.5 | −0.5 | −0.1 | −1.3 |
| 1996 | 1.7 | −0.2 | 1.2 | 1.1 | 0.2 | 1.6 | 1.0 | 0.7 | 1.0 | 0.7 | −0.3 | 1.3 |
| 1997 | 0.8 | 2.6 | −1.9 | −1.4 | −3.0 | −3.2 | −1.7 | −3.4 | −2.6 | −3.1 | −2.3 | −2.1 |

One way to investigate such cycles is to develop least squares approximations using trigonometric polynomials of various orders. (See Section 6.3.) For the 60 index values in Table 1 we construct least squares approximations of the form

$$a_0 + \sum_{k=1}^{n} (a_k \cos(kt) + b_k \sin(kt))$$

where $n$ is the order of the trigonometric polynomial. Figures 6 and 7 show the approximations of orders $n = 30$ and $n = 15$, respectively. The data from Table 1 are denoted by the plus signs in the figures, and the least squares approximations have been evaluated at extra (time) values in order to generate a continuous approximation. The approximation of order 30 actually goes through each of the data points and thus tracks all the details of the data. The approximation in Figure 7 is of order 15. It does a reasonable job of indicating the trends of El Niño over the five-year period, and we can see the major fluctuations due to the annual cycles. The lower the order of the trigonometric polynomial, the fewer computations that are required to generate the coefficients. The determination of the smallest order that accurately captures the trends of the cycles is beyond the scope of this course. (For more information on this and other aspects of least squares approximation using trigonometric polynomials, see *Numerical Methods and Software* by D. Kahner, C. Moler, and S. Nash, Prentice Hall, Inc., 1989. Good sources for more information on El Niño are available on the World Wide Web; just use a search engine with the keyword El Niño.)

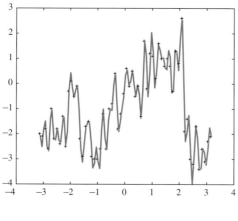

FIGURE 6                                        FIGURE 7

## EXERCISES 6.4[2]

**1.** Use the normal equations to determine the line of best fit to the following data that shows U.S. Health Expenditures[3] (in billions of dollars). From the model you develop, predict the expenditures in 1999.

| Year | Personal Health Care |
|------|----------------------|
| 1985 | 376.4 |
| 1990 | 614.7 |
| 1991 | 676.2 |
| 1992 | 739.8 |
| 1993 | 786.5 |
| 1994 | 831.7 |

**2.** Use the normal equations to determine the line of best fit to the following data which shows Public Debt of the United States[4]. From the model you develop, predict the debt in 2000.

| Year | Debt. per capita in $ |
|------|------------------------|
| 1970 | 1,814 |
| 1975 | 2,475 |
| 1980 | 3,985 |
| 1985 | 7,598 |
| 1990 | 13,000 |
| 1995 | 18,930 |

**3.** For the data in Exercise 2, determine the least squares

quadratic polynomial approximation using the normal system of equations. Compare this model with that obtained in Exercise 2 by computing the error in each case.

**4.** Determine a least squares model of the form $c_1 t + c_2 e^{-t}$ to the data in the following table.

| $x$ | $y$ | $x$ | $y$ |
|-----|-----|-----|-----|
| 0 | 1.0052 | 0.6 | 1.1487 |
| 0.1 | 1.0131 | 0.7 | 1.1928 |
| 0.2 | 1.0184 | 0.8 | 1.25 |
| 0.3 | 1.0403 | 0.9 | 1.3105 |
| 0.4 | 1.0756 | 1.0 | 1.3610 |
| 0.5 | 1.0998 | | |

**5.** Let $V = C[0, 1]$ with its usual inner product. Find the least squares approximation to $\mathbf{f}(t) = \sin(\pi t)$ from subspace $P_2$ with basis $\{\mathbf{w}_1, \mathbf{w}_2, \mathbf{w}_3\}$ where $\mathbf{w}_1 = 1$, $\mathbf{w}_2 = t$, $\mathbf{w}_3 = t^2$.

*In Exercises 6 and 7, let $V = C[0, 1]$ with its usual inner product and $\mathbf{f}(t) = e^t$.*

**6.** Find the least squares approximation to $\mathbf{f}(t)$ from subspace $P_1$ with basis $\{\mathbf{w}_1, \mathbf{w}_2\}$ where $\mathbf{w}_1 = 1$ and $\mathbf{w}_2 = t$.

**7.** Find the least squares approximation to $\mathbf{f}(t)$ from subspace $P_2$ with basis $\{\mathbf{w}_1, \mathbf{w}_2, \mathbf{w}_3\}$ where $\mathbf{w}_1 = 1$, $\mathbf{w}_2 = t$ and $\mathbf{w}_3 = t^2$.

[2]In this set of exercises we recommend that a computer algebra system be used to compute the inner product of vectors in function spaces and that the normal systems be determined and solved using a software package.
[3]*The World Almanac and Book of Facts 1997*, edited by Robert Famighetti, Mahwah, N.J.: K-III Reference Corporation, , 1996.
[4]ibid.

8. Let $V = C[-\pi, \pi]$ with inner product

$$(\mathbf{f}, \mathbf{g}) = \int_{-\pi}^{\pi} f(t)g(t)\, dt$$

Find the least squares approximation to $\mathbf{f}(t) = t\sin(t)$ from subspace $W = \text{span}\{\mathbf{w}_1, \mathbf{w}_2, \mathbf{w}_3\}$ where $\mathbf{w}_1 = 1$, $\mathbf{w}_2 = \cos(t)$ and $\mathbf{w}_3 = \sin(t)$.

9. Verify that the Grammian matrix, see (8), is diagonal if basis $\{\mathbf{w}_1, \mathbf{w}_2, \ldots, \mathbf{w}_k\}$ is an orthogonal set of vectors.

10. The Legendre polynomials (see Exercises 10 and 18 in Section 6.3) are an orthogonal basis for the subspace $P$ of polynomials in function space $C[-1, 1]$. The Legendre polynomials of degrees 0, 1, and 2 are, respectively, $p_0(t) = 1$, $p_1(t) = t$, and $p_2(t) = 0.5(3t^2 - 1)$. Determine the continuous least squares approximation by polynomials of degree 2 to

$$\mathbf{f}(t) = \frac{1}{1 + t^2}.$$

11. (**MATLAB required.**) Let $f(t) = |\sin(t)|^{\sin(t)}$. A sketch of $f$ over $[-\pi, \pi]$ is shown in Figure 8. We sample this curve at 32 equispaced points and investigate least squares trigonometric approximations.

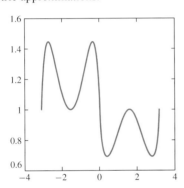

FIGURE 8

(a) To see the least squares approximation by trigonometric polynomials of order 16, use the following MATLAB commands.

```
f = 'abs(sin(t))^sin(t)';
t = -pi:pi/16:pi;
t = t(1:32);
y = eval(vectorize(f));
trigpoly(t,y,16);
```

Briefly describe the graph that is displayed in relation to the data set of 32 points, which is indicated by o's.

(b) Assuming that you executed the first MATLAB commands in part (a), now use command

```
trigpoly(t,y,8);
```

to see the least squares approximation by trigonometric polynomials of order 8. Briefly describe the graph

that is displayed by the routine **trigpoly** in comparison to the data set of 32 points, which is shown by o's in this routine.

(c) Describe how the graphs displayed in parts (a) and (b) differ.

12. Sometimes linear combinations of functions from an inner product space are inadequate to obtain an accurate model to a discrete data set $D$. There are other types of functions, beside linear combinations, that can be used. One approach is to define expressions that involve unknown constants that must be determined. If the data set $D$ is nearly steadily increasing or nearly steadily decreasing, then expressions of the form

$$y = \frac{1}{a + bt}, \quad y = a + \frac{b}{t}, \quad y = ae^{bt},$$

$$y = at^b, \quad y = \frac{t}{a + bt}, \quad \text{or} \quad y = a + b\ln(t)$$

have been used successfully for certain data sets. In such cases we must transform the expression into a linear combination in order to obtain a linear system of two equations in two unknowns so that we can determine parameters $a$ and $b$. We use this technique for the following situation.

Gauge is a measure of shotgun bore. Gauge numbers originally referred to the number of lead balls with the diameter equal to that of the gun barrel that could be made from a pound of lead. Thus a 16-gauge shotgun's bore was smaller than a 12-gauge shotgun's (see *The World Almanac and Book of Facts 1993*, New York, Pharos Books, 1992, page 290.) Today, an international agreement assigns millimeter measures to each gauge. The following table gives such information for popular gauges of shotguns.

| Gauge | Bore Diameter in mm |
|-------|---------------------|
| $t$   | $y$                 |
| 6     | 23.34               |
| 10    | 19.67               |
| 12    | 18.52               |
| 14    | 17.60               |
| 16    | 16.81               |
| 20    | 15.90               |

(a) Develop a model of the form $y = ae^{bt}$ for the gauge data in the preceding table. We first note that by taking the natural log of both sides of this expression, we obtain the linear combination $\ln(y) = \ln(a) + bt$. Let $\ln(a) = c_1$ and $b = c_2$. Upon substituting the data from the table for $t$ and $y$, respectively, we obtain the linear system

$$\ln(23.34) = c_1 + c_2(6)$$
$$\ln(19.67) = c_1 + c_2(10)$$
$$\ln(18.52) = c_1 + c_2(12)$$
$$\ln(17.60) = c_1 + c_2(14)$$
$$\ln(16.81) = c_1 + c_2(16)$$
$$\ln(15.90) = c_1 + c_2(20).$$

Verify that this system is inconsistent, find its least squares solution, determine $a$ and $b$, and then predict the bore diameter for an 18-gauge shotgun.

(b) Develop a model of the form

$$y = \frac{1}{a + bt}$$

for the gauge data in the preceding table. Determine a transformation of this expression that yields a linear combination of terms. Rename the coefficients $c_1$ and $c_2$ as in part (a). Construct the linear system obtained from substituting the data from the table for $t$ and $y$. Find its least squares solution, determine $a$ and $b$, and then predict the bore diameter for an 18-gauge shotgun.

## True/False Review Questions

*Determine whether each of the following statements is true or false.*

1. The least squares solution $\hat{\mathbf{x}}$ of system $\mathbf{Ax} = \mathbf{y}$ is the vector such that $\operatorname{proj}_{\operatorname{col}(\mathbf{A})}\mathbf{y} = \mathbf{A}\hat{\mathbf{x}}$.

2. If $\mathbf{A}$ has linearly independent columns, then the least squares solution of $\mathbf{Ax} = \mathbf{y}$ is unique.

3. If a linear system has a nonsingular coefficient matrix, then its solution is the same as the least squares solution of the linear system.

4. If $p(x)$ is the least squares quadratic polynomial to a set of data $\{(x_i, y_i) \mid i = 1, 2, \ldots, n\}$, then $p(x_i) = y_i$.

5. If $y = mx + b$ is the least squares line to a set of data $\{(x_i, y_i) \mid i = 1, 2, \ldots, n\}$, then $\sum_{i=1}^{n}(y_i - (mx_i + b))^2$ is minimized.

## Terminology

| | |
|---|---|
| Line of best fit | Least squares solution |
| Discrete least squares | Continuous least squares |
| Normal system of equations | Grammian |
| QR-factorization | El Niño |

This section reviewed the notion of the line of best fit to a data set. We then showed how to perform the computations that determine the coefficients of the equation of the line using projections. We saw that the approach could be generalized to obtain a technique for determining the coefficients for least squares computations using other function forms. In fact, the same steps are involved in the computation of continuous least squares approximations. It is the structure of an inner product space that makes this possible.

Least squares approximation are used in a wide variety of applications and in many different disciplines. Understanding the basic computational procedure and why it works is valuable knowledge for anyone who needs to use models for data. Respond to the following questions and statements to review the material in this section.

- What is minimized when we determine the line of best fit to a data set $D$?
- What do we mean by the least squares solution to a linear system $\mathbf{Ax} = \mathbf{y}$? Explain why the term *solution* in this phrase is misleading.
- Describe the role that projections play in determining the line of best fit.

- Under what circumstances do we use the term *discrete least squares*?
- Under what circumstances do we use the term *continuous least squares*?
- What is the normal system of equations in the discrete case? In the continuous case?
- Give a description of the computational steps for *discrete least squares*.
- Give a description of the computational steps for *continuous least squares*.
- How can the QR-factorization be used in the solution of a least squares problem?
- What is El Niño?

## 6.5 ■ ORTHOGONAL COMPLEMENTS (OPTIONAL)

In Section 6.3 we developed the notion of the projection $\mathbf{p}^*$ of a vector $\mathbf{v}$ onto a subspace $W$ and showed how to compute $\mathbf{p}^* = \text{proj}_W \mathbf{v}$ in terms of a basis for $W$. An important step was to show that $\mathbf{v} - \mathbf{p}^*$ was orthogonal to every vector in W. In Section 1.5 we developed the special case of the projection of one vector onto another vector in $R^2$. We displayed this geometrically in Figure 13 of Section 1.5 and made the same observation from a right triangle diagram. The feature from our discussions about projections that we want to emphasize is that the vector $\mathbf{v}$ can be expressed as a linear combination of a vector from subspace $W$ and a vector orthogonal to every vector in $W$:

$$\mathbf{v} = \mathbf{p}^* + (\mathbf{v} - \mathbf{p}^*).$$

This algebraic expression seems obvious, but it is the "nature" of the vectors $\mathbf{p}^*$ (in $W$) and $\mathbf{v} - \mathbf{p}^*$ (orthogonal to W) that is important.

In this section we investigate how to express an arbitrary vector from an inner product space as a linear combination of vectors from subspaces whose vectors are orthogonal to one another. We begin with several examples and then give definitions that make precise the relationships required for our development.

**EXAMPLE 1**   Let $L_1$ and $L_2$ be a pair of perpendicular lines through the origin in $R^3$. Then $L_1$ and $L_2$ are subspaces of $R^3$ that are orthogonal to one another. A special case of this is given by the $x$-axis, $k(1, 0, 0)$, and the $y$-axis, $t(0, 1, 0)$, as shown in Figure 1. (Here $k$ and $t$ are any real scalars.) It is not possible to express every vector in $R^3$ as a linear combination of vectors from subspaces $L_1$ and $L_2$. (Explain.)    ■

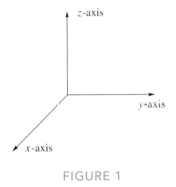

FIGURE 1

**EXAMPLE 2**   Let $U$ be the subspace of $R^3$ with basis $\{(1, 0, 0), (0, 1, 0)\}$ and $W$ be the subspace with basis $\{(0, 0, 1)\}$. Then every vector in $U$ is of the form $a(1, 0, 0) + b(0, 1, 0)$ and every vector in $W$ is of the form $c(0, 0, 1)$. It follows that every vector in $U$ is orthogonal to every vector in $W$. (Verify.) Geometrically $U$ is the $xy$-plane and $W$ is the $z$-axis; see Figure 1. In this case every vector in $R^3$ can be expressed as a linear combination of a vector from $U$ and a vector from $W$. This follows since the natural basis for $R^3$ is $\{(1, 0, 0), (0, 1, 0), (0, 0, 1)\}$.    ■

**EXAMPLE 3**   Let $U$ be the subspace of $R^3$ with basis $\{(1, 0, 0), (0, 1, 0)\}$ and $W$ the subspace with basis $\{(0, 1, 0), (0, 0, 1)\}$. Then every vector in $U$ is of the form $a(1, 0, 0) + b(0, 1, 0)$ and every vector in $W$ is of the form $c(0, 1, 0) + d(0, 0, 1)$. However, not every vector in $U$ is orthogonal to every vector in $W$; $(2, 1, 0) \cdot (0, 5, -1) = 5$. Geometrically $U$ is the $xy$-plane and $W$ is the $yz$-plane;

see Figure 1. While we see that the $xy$-plane and the $yz$-plane are perpendicular, the corresponding subspaces are not orthogonal to one another. That is, not every pair of vectors, one from $U$ and the other from $W$, are orthogonal.    ■

We need the following definition in order to clarify our notion of orthogonality between subspaces.

---

**Definition**   Let $U$ and $W$ be subspaces of an inner product space $V$. $U$ and $W$ are **orthogonal subspaces** provided that $(\mathbf{u}, \mathbf{w}) = 0$ for every vector $\mathbf{u}$ in $U$ and every vector $\mathbf{w}$ in $W$.

---

$L_1$ and $L_2$ are orthogonal subspaces in Example 1 and $U$ and $W$ are orthogonal subspaces in Example 2. However, $U$ and $W$ in Example 3 are not orthogonal subspaces.

In Section 4.5 we developed the Gram-Schmidt process for constructing an orthonormal basis for a subspace of $R^n$ and in Section 6.3 for subspaces of general inner product spaces. The span of distinct subsets of such bases provides us with orthogonal subspaces. However, there is an alternative procedure that has additional properties, and we develop it next.

---

**Definition**   Let $W$ be a subspace of an inner product space $V$. The set of all vectors in $V$ that are orthogonal to all the vectors in $W$ is called the **orthogonal complement** of $W$ in $V$ and is denoted by $W^\perp$ (read "$W$ perp").

---

EXAMPLE 4    In Section 3.6 we showed that a plane $P$ through the origin in $R^3$ is characterized by its normal vector $\mathbf{n} = (a, b, c)$; that is, a vector orthogonal to the plane. If $\mathbf{v} = (x, y, z)$ is any vector in $P$, then $\mathbf{n} \cdot \mathbf{v} = 0$ and the equation of plane $P$ is given by

$$\mathbf{n} \cdot \mathbf{v} = ax + by + cz = 0.$$

It follows that the orthogonal complement of $P$, $P^\perp$, is span$\{\mathbf{n}\}$, which is the line through the origin in $R^3$ with direction vector $\mathbf{n}$.    ■

EXAMPLE 5    Let $V$ be the inner product space of polynomials of degree 2 or less, $P_2$, over $[0, 1]$ with the inner product

$$(\mathbf{p}, \mathbf{q}) = \int_0^1 \mathbf{p}(t)\mathbf{q}(t)\, dt.$$

Let $W = \text{span}\{1, t\}$. Then to compute $W^\perp$ we proceed as follows. We determine all the vectors $\mathbf{v} = at^2 + bt + c$ in $V$ that are orthogonal to all the vectors in $W$. It suffices to determine all the vectors $\mathbf{v}$ that are orthogonal to the basis vectors 1 and $t$. (Explain.) Hence we require that

$$(\mathbf{v}, 1) = \int_0^1 (at^2 + bt + c)\, dt = \frac{a}{3} + \frac{b}{2} + c = 0,$$

$$(\mathbf{v}, t) = \int_0^1 (at^2 + bt + c)t\, dt = \frac{a}{4} + \frac{b}{3} + \frac{c}{2} = 0.$$

This gives us the homogeneous linear system

$$\left[\begin{array}{ccc|c} \frac{1}{3} & \frac{1}{2} & 1 & 0 \\ \frac{1}{4} & \frac{1}{3} & \frac{1}{2} & 0 \end{array}\right]$$

whose reduced row echelon form is

$$\left[\begin{array}{ccc|c} 1 & 0 & -6 & 0 \\ 0 & 1 & 6 & 0 \end{array}\right] \quad \text{(verify)}.$$

Then we have $a = 6c$ and $b = -6c$ and so $W^{\perp} = \{c(6t^2 - 6t + 1), c \text{ any scalar}\}$. It follows that $\{1, t, 6t^2 - 6t + 1\}$ is a basis for $P_2$ (verify) and hence every vector $\mathbf{u}$ in $P_2$ can be written in the form

$$\mathbf{u} = c_1(\text{a vector in } W) + c_2(\text{a vector in } W^{\perp}).$$  ■

The orthogonal complement of a subspace has particular properties, which we list next. We leave the verification of most of these to the exercises.

**Properties of the Orthogonal Complement**
Let $V$ be an inner product space with subspace $W$.

a. $W^{\perp}$ is a subspace of $V$.
b. The only vector in both $W$ and $W^{\perp}$ is the zero vector.
c. If $W$ is finite dimensional, then every vector in $V$ can be written uniquely as the sum of a vector from $W$ and another vector from $W^{\perp}$.
d. If $W$ is finite dimensional, then $(W^{\perp})^{\perp} = W$.
e. If $\dim(V) = n$, then $\dim(W) + \dim(W^{\perp}) = n$.

The verification of statement c is as follows: Let $\dim(W) = m$. Then $W$ has a basis consisting of $m$ vectors. By the Gram-Schmidt process, we can transform this basis to an orthonormal basis. Thus let $S = \{\mathbf{w}_1, \mathbf{w}_2, \ldots, \mathbf{w}_m\}$ be an orthonormal basis for $W$. If $\mathbf{v}$ is a vector in $V$, let

$$\mathbf{w} = (\mathbf{v}, \mathbf{w}_1)\mathbf{w}_1 + (\mathbf{v}, \mathbf{w}_2)\mathbf{w}_2 + \cdots + (\mathbf{v}, \mathbf{w}_m)\mathbf{w}_m$$

and

$$\mathbf{u} = \mathbf{v} - \mathbf{w}.$$

We see that $\mathbf{w}$ is in $W$ since it a linear combination of vectors that are a basis for $W$ and $\mathbf{u}$ is $V$. We next show that $\mathbf{u}$ is in $W^{\perp}$. We note that $\mathbf{w} = \text{proj}_W \mathbf{v}$ (see Equation 7 in Section 6.3) and so $\mathbf{u} = \mathbf{v} - \text{proj}_W \mathbf{v}$ which is orthogonal to every vector in $W$ (see Section 6.3). Thus $\mathbf{u}$ is in $W^{\perp}$. It follows that $\mathbf{v} = \mathbf{w} + \mathbf{u}$, and since the projection of $\mathbf{v}$ onto $W$ is unique so are $\mathbf{w}$ and $\mathbf{u}$.

We note that statement c is another way of viewing projections and their relationship to the vector that is projected. Statement c also motivates the following definition.

**Definition**   If $U$ and $W$ are subspaces of a vector space $V$ and each vector $\mathbf{v}$ in $V$ can be written uniquely as a sum $\mathbf{u} + \mathbf{w}$, where $\mathbf{u}$ is in $U$ and $\mathbf{w}$ is in $W$, then we say that $V$ is a **direct sum** of $U$ and $W$ and we write $V = U \oplus W$.

Certainly direct sums apply to inner product spaces as well, and from the preceding statements we have that for any finite-dimensional subspace $W$ of an inner product space $V$, $V = W \oplus W^{\perp}$. Such a decomposition gives us a way to think about inner product spaces that is reminiscent of a Cartesian plane.

EXAMPLE 6    Let $\mathbf{A}$ be an $m \times n$ real matrix. Then $\mathbf{ns}(\mathbf{A})$ is a subspace of $R^n$. If column vector $\mathbf{x}$ is in $\mathbf{ns}(\mathbf{A})$, then $\mathbf{x}$ is orthogonal to each row of $\mathbf{A}$ since $\mathbf{Ax} = \mathbf{0}$. We can interpret this as follows:

$\mathbf{x}$ in $\mathbf{ns}(\mathbf{A})$ is orthogonal to each column of $\mathbf{A}^T$; thus $\mathbf{x}$ is in $\mathbf{col}(\mathbf{A}^T)^{\perp}$.

Alternatively, if $\mathbf{x}$ is in $\mathbf{col}(\mathbf{A}^T)^{\perp}$, then $\mathbf{x}$ is orthogonal to every column in $\mathbf{A}^T$. Hence

$$0 = \mathrm{col}_j(\mathbf{A}^T) \cdot \mathbf{x} = \mathrm{row}_j(\mathbf{A}) \cdot \mathbf{x}$$

and we have $\mathbf{Ax} = \mathbf{0}$; that is, $\mathbf{x}$ is in $\mathbf{ns}(\mathbf{A})$. Thus it follows that $\mathbf{ns}(\mathbf{A}) = \mathbf{col}(\mathbf{A}^T)^{\perp}$. Hence

$$R^n = \mathbf{ns}(\mathbf{A}) \oplus \mathbf{ns}(\mathbf{A})^{\perp} = \mathbf{ns}(\mathbf{A}) \oplus \mathbf{col}(\mathbf{A}^T)$$

using the preceding statements and results. Notice that this implies that

$$\dim(R^n) = n = \dim(\mathbf{ns}(\mathbf{A})) + \dim(\mathbf{col}(\mathbf{A}^T))$$
$$= (\text{number of free variables in the solution of } \mathbf{Ax} = \mathbf{0}) + \mathrm{rank}(\mathbf{A}).$$

Recall that $\mathrm{rank}(\mathbf{A}) = $ number of linearly independent rows or columns of $\mathbf{A}$ (or $\mathbf{A}^T$). We saw this result previously in Section 2.2, where we wrote it in the form

degrees of freedom = number of unknowns − number of pivots.

We summarize the results of this example as follows:

$$\mathbf{ns}(\mathbf{A}) = \text{the orthogonal complement of } \mathbf{col}(\mathbf{A}^T) = \mathbf{col}(\mathbf{A}^T)^{\perp} \qquad (1)$$

or, equivalently, replacing $\mathbf{A}$ by $\mathbf{A}^T$, we have

$$\mathbf{ns}(\mathbf{A}^T) = \text{the orthogonal complement of } \mathbf{col}(\mathbf{A}) = \mathbf{col}(\mathbf{A})^T. \qquad (2)$$

■

We have used the following relation in Example 6. If $U$ is the orthogonal complement of $W$, then $W$ is the orthogonal complement of $U$. Hence we simply say that $U$ and $W$ are orthogonal complements. Hence from (1) we have the equivalent result

$$\mathbf{ns}(\mathbf{A})^{\perp} = \mathbf{col}(\mathbf{A}^T) \qquad (3)$$

and from (2) the equivalent result

$$\mathbf{ns}(\mathbf{A}^T)^{\perp} = \mathbf{col}(\mathbf{A}) \qquad (4)$$

We will use (1) through (4) in the next section when we investigate the fundamental subspaces associated with a linear system.

EXERCISES 6.5

1. Let $W$ be a subspace of inner product space $V$. Show that $W^{\perp}$ is also a subspace of $V$.

2. Let $W$ be a subspace of inner product space $V$. Show that the zero vector, $\mathbf{0}$, is the only vector in both $W$ and $W^{\perp}$.

3. Let $V$ be an inner product space. What is $V^{\perp}$? What is $\{\mathbf{0}\}^{\perp}$?

4. Let $W$ be a finite-dimensional subspace of inner product space $V$. Show that $(W^{\perp})^{\perp} = W$.

**5.** Let $W$ be a subspace of inner product space $V$ with $\dim(V) = n$. Show that $\dim(W) + \dim(W^\perp) = n$. (*Hint*: Let $\{\mathbf{w}_1, \mathbf{w}_2, \ldots, \mathbf{w}_r\}$ be a basis for $W$ and $\{\mathbf{u}_1, \mathbf{u}_2, \ldots, \mathbf{u}_s\}$ a basis for $W^\perp$ and show that $\{\mathbf{w}_1, \mathbf{w}_2, \ldots, \mathbf{w}_r, \mathbf{u}_1, \mathbf{u}_2, \ldots, \mathbf{u}_s\}$ is a basis for $V$.)

**6.** Let $V = C[0, 1]$ with inner product

$$(\mathbf{f}, \mathbf{g}) = \int_0^1 f(t)g(t)\, dt.$$

(See Example 2 in Section 6.3.) Let $W$ be the subspace $P$ of all polynomials. It can be shown that $W^\perp = \{\mathbf{0}\}$. What is $(W^\perp)^\perp$? Why doesn't this contradict the result in Exercise 4?

**7.** Refer to Example 5. But now let $W = \mathrm{span}\{1, t^2\}$. Compute $W^\perp$.

**8.** Refer to Example 5. But now let $W = \mathrm{span}\{t, t^2\}$. Compute $W^\perp$.

**9.** Let $V$ be the inner product space of polynomials of degree 3 or less, $P_3$, with the inner product

$$(\mathbf{p}, \mathbf{q}) = \int_0^1 \mathbf{p}(t)\mathbf{q}(t)\, dt.$$

Let $W = \mathrm{span}\{1, t\}$. Then determine $W^\perp$.

**10.** Let $V$ be the inner product space of polynomials of degree 3 or less, $P_3$, with the inner product

$$(\mathbf{p}, \mathbf{q}) = \int_0^1 \mathbf{p}(t)\mathbf{q}(t)\, dt.$$

Let $W = \mathrm{span}\{1, t^3\}$. Then determine $W^\perp$.

**11.** Let $W$ be the subspace of $R^3$ with orthonormal basis $\{\mathbf{w}_1, \mathbf{w}_2\}$ where

$$\mathbf{w}_1 = \begin{bmatrix} 0 \\ 1 \\ 0 \end{bmatrix} \quad \text{and} \quad \mathbf{w}_2 = \begin{bmatrix} \frac{1}{\sqrt{5}} \\ 0 \\ \frac{2}{\sqrt{5}} \end{bmatrix}.$$

Write vector $\mathbf{v} = \begin{bmatrix} 1 \\ 2 \\ -1 \end{bmatrix}$ as $\mathbf{w} + \mathbf{u}$ with $\mathbf{w}$ in $W$ and $\mathbf{u}$ in $W^\perp$. (See Example 5.)

**12.** Let $V$ be the inner product space $C[-\pi, \pi]$ with inner product

$$(\mathbf{f}, \mathbf{g}) = \frac{1}{\pi} \int_{-\pi}^{\pi} f(t)g(t)\, dt.$$

Let $W = \mathrm{span}\{1, \sin(t), \cos(t)\}$. Write the vector $\mathbf{v} = t - 1$ as $\mathbf{w} + \mathbf{u}$, with $\mathbf{w}$ in $W$ and $\mathbf{u}$ in $W^\perp$. (*Hint*: First determine an orthonormal basis for W.)

**13.** Let $A = \begin{bmatrix} 1 & 0 & 1 & 1 \\ 1 & 2 & 0 & 0 \end{bmatrix}$.

(a) Determine a basis for $\mathbf{ns}(A)$.
(b) Determine a basis for $\mathbf{ns}(A)^\perp$.
(c) Express vector $\mathbf{v} = \begin{bmatrix} 3 & 2 & 4 & 0 \end{bmatrix}^T$ in terms of the bases from parts (a) and (b).

**14.** Let $A = \begin{bmatrix} 0 & 1 & 2 & -1 \\ 1 & 2 & 0 & -1 \\ -2 & -1 & 6 & 5 \end{bmatrix}$.

(a) Determine a basis for $\mathbf{ns}(A)$.
(b) Determine a basis for $\mathbf{ns}(A)^\perp$.
(c) Express vector $\mathbf{v} = \begin{bmatrix} 8 & 0 & 3 & 1 \end{bmatrix}^T$ in terms of the bases from parts (a) and (b).

**15.** Let $\{\mathbf{v}_1, \mathbf{v}_2\}$ be an orthonormal basis for $R^2$. Explain why we can write $R^2 = \mathrm{span}\{\mathbf{v}_1\} \oplus \mathrm{span}\{\mathbf{v}_2\}$.

**16.** Let $\{\mathbf{v}_1, \mathbf{v}_2, \mathbf{v}_3\}$ be an orthonormal basis for $R^3$. Explain why we can write $R^3 = \mathrm{span}\{\mathbf{v}_1\} \oplus \mathrm{span}\{\mathbf{v}_2\} \oplus \mathrm{span}\{\mathbf{v}_3\}$.

**17.** Let $A$ be an $n \times n$ symmetric matrix with distinct eigenvalues $\lambda_1, \lambda_2, \ldots, \lambda_n$. Explain why we can say that $R^n$ is the direct sum of the corresponding eigenspaces.

**18.** Use Equation (4) to explain each of the following statements.

(a) If $A\mathbf{x} = \mathbf{b}$ is consistent, then every solution of $A^T\mathbf{y} = \mathbf{0}$ is orthogonal to $\mathbf{b}$.
(b) If $A\mathbf{x} = \mathbf{b}$ is inconsistent, then there is some solution of $A^T\mathbf{y} = \mathbf{0}$ that is not orthogonal to $\mathbf{b}$.

**19.** Let $A$ be a $3 \times 3$ matrix with $\dim(\mathbf{col}(A)) = 2$.

(a) Draw a plane that represents subspace $\mathbf{col}(A)$ of $R^3$. For part (a) of Exercise 18, draw vector $\mathbf{b}$ and $\mathbf{ns}(A^T)$ in relation to this plane.
(b) Draw a plane that represents subspace $\mathbf{col}(A)$ of $R^3$. For part (b) of Exercise 18, draw vector $\mathbf{b}$ and $\mathbf{ns}(A^T)$ in relation to this plane.

## True/False Review Questions

*Determine whether each of the following statements is true or false.*

**1.** If $V$ is an inner product space $V$, then $V^\perp = \{\mathbf{0}\}$.

**2.** To determine if a pair of subspaces $U$ and $W$ of an inner product space $V$ are orthogonal, we need only check that every basis vector for $U$ is orthogonal to every basis vector for $W$.

**3.** If $P$ is a plane through the origin in $R^3$, then $P^\perp$ is a line through the origin.

**4.** If $V$ is an inner product space with $\dim(V) = 5$ and $W$ is a 3-dimensional subspace of $V$, then $\dim(W^{\perp}) = 3$.

**5.** $\mathbf{ns}(\mathbf{A}) = \mathbf{col}(\mathbf{A}^T)^{\perp}$.

### Terminology

| | |
|---|---|
| Projections | Orthogonal subspaces |
| Orthogonal complement | Direct sum |

This section extended the idea that a single vector could be written as a linear combination of its projection into a subspace and a vector orthogonal to the subspace. The extension involved the splitting of a finite-dimensional inner product space $V$ into two subspaces that are orthogonal to one another such that every vector in $V$ could be written as a linear combination using a vector from each of the subspaces.

- Let $\mathbf{v}$ and $\mathbf{w}$ be nonparallel vectors in $R^2$. Draw a diagram that illustrates how $\mathbf{v}$ can be written as a linear combination of a vector in span$\{\mathbf{w}\}$ and a vector orthogonal to span$\{\mathbf{w}\}$. What do we call the vector in span$\{\mathbf{w}\}$?
- Let $\mathbf{v} = \begin{bmatrix} 1 & 2 & 3 \end{bmatrix}$. Draw a diagram that illustrates how $\mathbf{v}$ can be written as a linear combination of a vector in the $xy$-plane and a vector parallel to the $z$-axis in $R^3$. What do we call the vector in the $xy$-plane?
- Explain what it means to say that $U$ and $W$ are orthogonal subspaces of an inner product space $V$.
- In $R^3$, make a list of at least three pairs of orthogonal subspaces.
- If $U$ and $W$ are orthogonal subspaces of $R^3$, is it true that any vector in $R^3$ can be expressed as linear combination of a vector from $U$ and a vector from $W$? Explain.
- Let inner product space $V$ be represented by the rectangle in Figure 2 and let subspace $W$ correspond to the circle. Carefully shade the region of $V$ that corresponds to $W^{\perp}$.

FIGURE 2

- If vector space $V$ is the direct sum of subspaces $U$ and $W$, explain why we can say that $U$ and $W$ "fill" $V$ with only a very small overlap.
- If $V$ is a finite-dimensional inner product space and $U$ is any subspace of $V$, how can we obtain a direct sum for $V$ involving $U$? Explain.
- Explain why $\mathbf{ns}(\mathbf{A}) = \mathbf{col}(\mathbf{A}^T)^{\perp}$.
- Why is $\mathbf{ns}(\mathbf{A})^{\perp} = \mathbf{col}(\mathbf{A}^T)$.

---

## 6.6 ■ LINEAR SYSTEMS AND THEIR SUBSPACES (OPTIONAL)

The material in this section is not new but is collected from previous sections in this chapter and preceding chapters. We want to point out the intimate connection between linear systems and certain subspaces of $R^k$. To this end we will restate properties that have appeared earlier and indicate relationships between them. Roughly put, we develop the anatomy of linear systems from a vector space point of view.

Let **A** be an $m \times n$ real matrix. Then we have the following results concerning a linear system with coefficient matrix **A**.

1. **Ax** = **b** is consistent if and only if **b** is in **col(A)**. [Note that **col(A)** is a subspace of $R^m$.]
2. **Ay** = **0** if and only if **y** is in **ns(A)**. [Note that **ns(A)** is a subspace of $R^n$ and **ns(A)** = **col($A^T$)**$^\perp$; see (1) in Section 6.5.]
3. If the general solution of **Ax** = **b** contains a free variable, then any of its solutions can be written in the form **x** = $\mathbf{x}_p + \mathbf{x}_h$ where $\mathbf{x}_p$ satisfies $\mathbf{A}\mathbf{x}_p = \mathbf{b}$ and $\mathbf{x}_h$ satisfies $\mathbf{A}\mathbf{x}_h = \mathbf{0}$. Hence every member of the set of solutions is a linear combination of a member of **col(A)** and of **col($A^T$)**$^\perp$.

It follows that **col(A)** and **ns(A)** = **col($A^T$)**$^\perp$ are fundamentally important subspaces associated with a consistent linear system. To emphasize this importance further we have

$$n = \dim(R^n) = \dim(\mathbf{col(A)}) + \dim(\mathbf{ns(A)})$$

which we can express in computational terms as

$$n = \big(\text{number of leading 1's in } \mathbf{rref}\left(\left[\, \mathbf{A} \mid \mathbf{b} \,\right]\right)\big) + (\text{number of free variables}).$$

Next, let's change our point of view to the linear system with coefficient matrix $\mathbf{A}^T$. It follows that subspaces **col($A^T$)** and **ns($A^T$)** are fundamentally important. But from (3) in Section 6.5 we have **col($A^T$)** = **ns(A)**$^\perp$ and from (2) in Section 6.5 we have **ns($A^T$)** = **col(A)**$^\perp$. With these connections we have the summary given in Table 1 and we make additional observations about consistent linear systems with coefficient matrices **A** and $\mathbf{A}^T$.

TABLE 1

| | Coefficient Matrix A | | Coefficient Matrix $A^T$ | |
|---|---|---|---|---|
| Consistent Linear System | **Ax** = **b** | | $\mathbf{A}^T\mathbf{y} = \mathbf{c}$ | |
| General Solution | **x** = $\mathbf{x}_p$ + | $\mathbf{x}_h$ | **y** = $\mathbf{y}_p$ + | $\mathbf{y}_h$ |
| | ↓ | ↓ | ↓ | ↓ |
| | in **col(A)** | in **ns(A)** | in **col($A^T$)** | in **ns($A^T$)** |
| | | | = **ns(A)**$^\perp$ | = **col(A)**$^\perp$ |
| Vector Space Relations | $R^n$ = **ns(A)** $\oplus$ **ns(A)**$^\perp$ | | $R^m$ = **ns($A^T$)** $\oplus$ **ns($A^T$)**$^\perp$ | |
| | $R^m$ = **col(A)** $\oplus$ **col(A)**$^\perp$ | | $R^n$ = **col($A^T$)** $\oplus$ **col($A^T$)**$^\perp$ | |
| Orthogonal Vectors[a] | $(\mathbf{x}_h, \mathbf{y}_p) = 0$ | | $(\mathbf{x}_p, \mathbf{y}_h) = 0$ | |

[a]See Exercise 18 in Section 6.5

In Chapter 2 we focused on the computational aspects, and in preceding sections of this chapter we focused on subspace relations. Table 1 connects these ideas. The fundamental subspaces and the relationships among the "natural" parts of the general solution of a consistent linear system provide a glimpse of the intricate features of solution sets.

Next we consider the case of an inconsistent linear system $\mathbf{Ax} = \mathbf{b}$. As we saw in Section 6.4, we alter our notion of a solution and effectively determine the vector closest to $\mathbf{b}$ in $\mathbf{col}(\mathbf{A})$. That is, we determine the least squares solution, which is the algebraic solution of the associated linear system $\mathbf{A}^T\mathbf{Ax} = \mathbf{A}^T\mathbf{b}$. We considered only the case where $\mathbf{A}^T\mathbf{A}$ was nonsingular, so there was a unique least squares solution $\hat{\mathbf{x}}$. It followed that $\hat{\mathbf{x}} = (\mathbf{A}^T\mathbf{A})^{-1}\mathbf{A}^T\mathbf{y}$ and that the vector in $\mathbf{col}(\mathbf{A})$ closest to $\mathbf{b}$ is given by $\mathbf{A}\hat{\mathbf{x}}$. More advanced treatments of least squares consider the case where $\mathbf{A}^T\mathbf{A}$ is not nonsingular so there can be more than one least squares solution. We do not pursue that here but merely comment that vector space connections similar to that given in Table 1 can be developed and now involve $\mathbf{A}^T\mathbf{A}$ and $\mathbf{AA}^T$.

## CHAPTER TEST

Let $V$ be the inner product space $C[0, 1]$ with inner product $(\mathbf{f}, \mathbf{g}) = \int_0^1 f(t)g(t)\,dt$ and $\mathbf{f}(t) = 2t^2$ and $\mathbf{g}(t) = 3t$.

1. Compute $(\mathbf{f}, \mathbf{g})$.

2. Compute the distance from $\mathbf{f}$ to $\mathbf{g}$.

3. Compute the cosine of angle between $\mathbf{f}$ and $\mathbf{g}$.

4. Compute $\text{proj}_{\mathbf{f}}\mathbf{g}$.

5. Determine an orthogonal basis for span$\{\mathbf{f}, \mathbf{g}\}$.

Let $V$ be the inner product space $P_3$ with inner product $(\mathbf{p}, \mathbf{q}) = \int_0^1 p(t)q(t)\,dt$. Let $W$ be the subspace of $P_3$ with basis $\{1, t^3\}$. (*Note*: Solve 6 through 8 in order.)

6. Find a basis for $W^\perp$.

7. Express $5t^3 + 4t^2 + 3t$ as the sum of a vector from $W$ and a vector from $W^\perp$.

8. Find an orthogonal basis for $W$.

9. Compute $\text{proj}_W t^2$.

10. Sample the function $f(t) = t^3$ at $t = 1.0, 2.1, 3.3, 5.0, 5.6$. Determine the least squares quadratic approximation to the data set.

# 7

# *LINEAR TRANSFORMATIONS*

This chapter deals with the functions of linear algebra—that is, rules that define a relationship between two sets, generally called the domain and range, so that for each member of the domain the rule assigns exactly one member of the range. Here the domain and range will be vector spaces and hence their members are vectors. In preceding chapters we defined a number of functions, including the transpose of a matrix (Section 1.1), the length or norm of a vector (Sections 1.4 and 6.2), matrix transformations (Section 1.5), reduced row echelon form of a matrix (Section 2.2), rank of a matrix (Section 2.3), the determinant (Section 3.1), and the projection of a vector onto a subspace (Section 6.3).

We will separate the functions we encounter into two classes and focus our attention on the class of functions that behave in a natural way when they act on linear combinations of vectors. One goal is to provide appropriate structures to study the behavior of such special functions, and another is to determine a simple way to represent them. In our previous developments we have laid the foundation for many of the ideas of this chapter, and we will cite them as we need them.

## 7.1 ■ TRANSFORMATIONS

Two common synonyms for function are transformation and mapping. In linear algebra the term *transformation* is often used. For the transformations we study the domain and range sets are vector spaces. For clarity we make the following definition and introduce the particular terminology commonly used in linear algebra.

---

**Definition**   A **transformation** $T$ is a rule that assigns to each vector **v** in a vector space $V$ a single vector **w** in a vector space $W$.

---

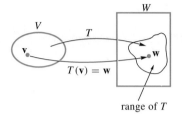

FIGURE 1

We use the notation $T: V \rightarrow W$ to indicate that $T$ is a transformation from vector space $V$ to vector space $W$. Then for **v** in $V$ we write $\mathbf{w} = T(\mathbf{v})$ when vector **w** in $W$ is assigned by the rule $T$ to vector **v**. We call $\mathbf{w} = T(\mathbf{v})$ the **image** of **v**. The **range** of $T$, denoted **range** $T$, is the set of all vectors in $W$ that are images of some vector from $V$. Sometimes the range of $T$ is called the **image of** $V$. See Figure 1.

**EXAMPLE 1**   Let $V$ and $W$ be any vector spaces. For **v** in $V$, define $T(\mathbf{v}) = \mathbf{0}$, the zero vector in $W$. We call $T$ the **zero** or **null transformation**. The range of $T$ consists of the single vector **0** in $W$ (a rather uninteresting transformation).   ■

**EXAMPLE 2**   Let $V$ be any vector space and $V = W$. Define $T(\mathbf{v}) = \mathbf{v}$ for **v** in $V$. We call $T$ the **identity transformation**. The range of $T$ is all of $V$. (The identity transformation is often denoted by $I$.)   ■

**EXAMPLE 3**   Let $V =$ vector space of all $2 \times 2$ real matrices and $W =$ the real numbers, $R^1$. For **A** in $V$, define $T(\mathbf{A}) = \det(\mathbf{A})$. The range of $T$ is all of $R^1$, since for any real number $x$,

$$\det\left(\begin{bmatrix} x & 0 \\ 0 & 1 \end{bmatrix}\right) = x.$$

As in Example 1, we see that there are many vectors in $V$ that can be assigned to the same real number $x$ in $W$, since

$$\det\left(\begin{bmatrix} x & 9 \\ 0 & 1 \end{bmatrix}\right) = \det\left(\begin{bmatrix} x & 0 \\ -4 & 1 \end{bmatrix}\right) = \det\left(\begin{bmatrix} 2x & 0 \\ 0 & 0.5 \end{bmatrix}\right) = x. \qquad ■$$

Linear combinations of vectors involve the operations of addition of vectors and scalar multiplication of vectors. Hence they are the fundamental way to manipulate vectors in a vector space. Next we define the class of transformations that preserve linear combinations in the sense that the image of a linear combination is a linear combination of the images of the individual vectors.

---

**Definition**   Let $V$ and $W$ be vector spaces. A transformation $T: V \rightarrow W$ is called a **linear transformation**[1] of $V$ into $W$ provided that

(a) $T(\mathbf{u} + \mathbf{v}) = T(\mathbf{u}) + T(\mathbf{v})$ for any vectors **u** and **v** in $V$.
(b) $T(k\mathbf{v}) = kT(\mathbf{v})$ for any vector **v** in $V$, and any scalar $k$.

---

[1] In Section 1.5 we discussed linear transformations from $R^n$ to $R^m$.

The two parts of the previous definition can be combined into one: $T$ is a linear transformation provided that $T(k_1\mathbf{u} + k_2\mathbf{v}) = k_1 T(\mathbf{u}) + k_2 T(\mathbf{v})$ for any vectors $\mathbf{u}$ and $\mathbf{v}$ in $V$ and any scalars $k_1$ and $k_2$. In fact, this behavior for linear transformations is true regardless of the number of vectors; that is,

$$T(k_1\mathbf{v}_1 + k_2\mathbf{v}_2 + \cdots + k_n\mathbf{v}_n) = k_1 T(\mathbf{v}_1) + k_2 T(\mathbf{v}_2) + \cdots + k_n T(\mathbf{v}_n).$$

Any transformation that is not a linear transformation (that is, that violates either property (a) or (b) or both in the preceding definition) is called a **nonlinear transformation**. Nonlinear transformations are important but tend not to have properties that are common to all such transformations. There is little in the way of a unifying theory for nonlinear transformations; hence we focus our development on linear transformations. Many of the transformations that measure a size or assign a number as the image tend to be nonlinear.

EXAMPLE 4   Each of the following transformations is nonlinear. We verify two of these and leave the others as exercises.

(a) Let $V = R^n$ and $W$ be $R^1$. Define

$$T(\mathbf{v}) = \|\mathbf{v}\| = \sqrt{\sum_{j=1}^{n}(v_j)^2}.$$

$T$ is a nonlinear transformation since $T(\mathbf{u}+\mathbf{v}) = \|\mathbf{u}+\mathbf{v}\| \neq \|\mathbf{u}\|+\|\mathbf{v}\| = T(\mathbf{u})+ T(\mathbf{v})$ for some vectors $\mathbf{u}$ and $\mathbf{v}$ in $V$. To see that this is the case, let $\mathbf{u} = \text{col}_1(\mathbf{I}_n)$ and $\mathbf{v} = \text{col}_2(\mathbf{I}_n)$. Then $T(\mathbf{u} + \mathbf{v}) = \sqrt{2}$ (verify) and $T(\mathbf{u}) + T(\mathbf{v}) = 2$ (verify).

(b) Let $V = R_{m \times n}$ and $W = R_{m \times n}$. Define $T(\mathbf{A}) = \mathbf{rref}(\mathbf{A})$. $T$ is a nonlinear transformation since it violates both properties (a) and (b) in the definition of a linear transformation. (See Exercise 3.)

(c) Let $V = R_{m \times n}$ and $W = R^1$. Define $T(\mathbf{A}) = \text{rank}(\mathbf{A})$. $T$ is a nonlinear transformation since it violates both properties (a) and (b) in the definition of a linear transformation. To see this, consider the case $m = 2$, $n = 3$, with scalar $k = 2$, and $\mathbf{A} = \begin{bmatrix} 1 & 1 & 1 \\ 1 & 1 & 1 \end{bmatrix}$. Then

$$T(k\mathbf{A}) = \text{rank}(2\mathbf{A}) = \text{rank}\left(\begin{bmatrix} 2 & 2 & 2 \\ 2 & 2 & 2 \end{bmatrix}\right)$$
$$= \text{the number of leading 1's in } \mathbf{rref}(2\mathbf{A})$$
$$= 1 \quad \text{(verify)}.$$

However, $kT(\mathbf{A}) = 2T(\mathbf{A}) = 2\,\text{rank}(\mathbf{A}) = 2 \cdot$ (the number of leading 1's in $\mathbf{rref}(\mathbf{A})$) $= 2$ (verify). Hence $T(k\mathbf{A}) \neq kT(\mathbf{A})$.

(d) Let $V = R_{n \times n}$ and $W = R^1$. Define $T(\mathbf{A}) = \det(\mathbf{A})$. $T$ is a nonlinear transformation since it violates both properties (a) and (b) in the definition of a linear transformation. (See Exercise 4.) ■

## Linear Transformations

We now turn our attention to linear transformations. The next example is one of the most important for our study of linear transformations. We introduced it in Section 1.5 and here we recast it in the language introduced in this section.

**EXAMPLE 5**   Let $V = R^n$ and $W = R^m$. For a given $m \times n$ matrix $\mathbf{A}$, define $T(\mathbf{v}) = \mathbf{Av}$ for $\mathbf{v}$ in $V$. In Section 1.5 we called this a **matrix transformation** and verified that it was a linear transformation using the properties of matrix multiplication, matrix addition, and scalar multiplication. Thus **every matrix transformation is a linear transformation**. Also from Section 1.5 we know that the range of a matrix transformation is the span of the columns of $\mathbf{A}$, that is, $\mathbf{col}(\mathbf{A})$. ■

As soon as we recognize a transformation as a matrix transformation, we automatically know that it is a linear transformation. We need not verify the definition for the particular transformation.

**EXAMPLE 6**   Let $V = R^3$ and $W = R^2$. Define $T : V \to W$ by

$$T(\mathbf{v}) = T\left(\begin{bmatrix} v_1 \\ v_2 \\ v_3 \end{bmatrix}\right) = \begin{bmatrix} v_1 \\ v_2 \end{bmatrix} = \mathbf{w}.$$

Let $\mathbf{A} = \begin{bmatrix} 1 & 0 & 0 \\ 0 & 1 & 0 \end{bmatrix}$; then note that $T(\mathbf{v}) = \mathbf{Av}$ (verify). Thus $T$ is a matrix transformation and hence a linear transformation. $T$ is called the **projection of $R^3$ onto $R^2$** or the **projection of $R^3$ onto the $xy$-plane**. See Example 6 and Figures 5 and 6 in Section 1.5. ■

For a particular transformation $T : V \to W$, it may not be easy to determine whether a matrix $\mathbf{A}$ exists so that $T(\mathbf{v}) = \mathbf{Av}$. In such a case we can attempt to verify the definition of a linear transformation directly. See Exercises 12–20.

**EXAMPLE 7**   Let $W$ be the vector space of all real valued continuous functions, $C(-\infty, \infty)$, and $V$ the subspace of $W$ of all differentiable functions, denoted $C^1(-\infty, \infty)$. Define transformation $T : V \to W$ by $T(f(t)) = f'(t)$, where $f'(t) = \dfrac{d}{dt} f(t)$ is the derivative of $f$. From calculus we know that

$$\frac{d}{dt}(f(t) + g(t)) = f'(t) + g'(t) \quad \text{and} \quad \frac{d}{dt}(kf(t)) = kf'(t).$$

Hence differentiation is a linear transformation between these function spaces. ■

By similar arguments, second derivatives, third derivatives, etc. are also linear transformations between appropriate vector spaces.

**EXAMPLE 8**   Let $V = C[a, b]$, the vector space of all real valued functions that are integrable over the interval $[a, b]$. Let $W = R^1$. Define transformation $T : V \to W$ by $T(f(t)) = \int_a^b f(t)\, dt$. It is easy to show that $T$ is a linear transformation. (See Exercise 24.)

**EXAMPLE 9**   Let $V$ be an inner product space with subspace $W$ where $\dim(W) = k$. Define transformation $T : V \to W$ by $T(\mathbf{v}) = \text{proj}_W \mathbf{v}$. We can show that $T$ is a linear transformation; see Exercise 25. Hence we can say that the projection of a linear combination is the linear combination of the individual projections. ■

## Properties of Linear Transformations

Next we investigate properties common to all linear transformations. Some of these parallel the properties of matrix transformations, and we introduce terminology traditionally used with linear transformations.

Let $T: V \rightarrow W$ be a linear transformation. Then

1. The image of the zero vector is the zero vector; that is, let $\mathbf{0}_V$ be the zero vector in vector space $V$ and $\mathbf{0}_W$ be the zero vector in vector space $W$, then $T(\mathbf{0}_V) = \mathbf{0}_W$. If $T$ were a matrix transformation, then this result would be immediate since a matrix times a zero vector yields a zero vector. For a general linear transformation $T$ we must use properties of a vector space and the fact that $T$ satisfies properties (a) and (b) in the definition of a linear transformation. See Exercise 26 for the details.

2. $T(\mathbf{u}-\mathbf{v}) = T(\mathbf{u})-T(\mathbf{v})$, for any vectors $\mathbf{u}$ and $\mathbf{v}$ in $V$. This follows immediately from properties of a vector space and the fact that $T$ is a linear transformation. Recall that $-\mathbf{v} = (-1)\mathbf{v}$; hence

$$\begin{aligned}
T(\mathbf{u} - \mathbf{v}) &= T(\mathbf{u} + (-1)\mathbf{v}) && [\text{since } -\mathbf{v} = (-1)\mathbf{v}] \\
&= T(\mathbf{u}) + T((-1)\mathbf{v}) && [\text{since } T \text{ is linear}] \\
&= T(\mathbf{u}) + (-1)T(\mathbf{v}) && [\text{since } T \text{ is linear}] \\
&= T(\mathbf{u}) - T(\mathbf{v}). && [\text{since } (-1)T(\mathbf{v}) = -T(\mathbf{v})]
\end{aligned}$$

3. If $\dim(V) = n$ and $S = \{\mathbf{v}_1, \mathbf{v}_2, \ldots, \mathbf{v}_n\}$ is a basis for $V$, then every image of a vector from $V$ is a linear combination of the images $\{T(\mathbf{v}_1), T(\mathbf{v}_2), \ldots, T(\mathbf{v}_n)\}$. (We say that a linear transformation is completely determined by its action on a basis.) We leave the verification of this property to Exercise 27.

## EXERCISES 7.1

1. Show that the transformation defined in Example 1 is a linear transformation.

2. Show that the transformation defined in Example 2 is a linear transformation.

3. Show that the transformation defined in Example 4(b) is nonlinear. (*Hint*: It suffices to use $2 \times 2$ matrices.)

4. Show that the transformation defined in Example 4(d) is nonlinear. (*Hint*: It suffices to use $2 \times 2$ matrices.)

5. Let $V = R_{m \times n}$ and $W = R_{n \times m}$. Define $T: V \rightarrow W$ by $T(\mathbf{A}) = \mathbf{A}^T$. Show that $T$ is a linear transformation. (*Hint*: See Properties of the Transpose in Section 1.2.)

*In Exercises 6–11, determine a $2 \times 2$ matrix $\mathbf{A}$ such that linear transformation $T: R^2 \rightarrow R^2$ can be expressed as $T(\mathbf{v}) = \mathbf{Av}$ and write a sentence that gives a geometric description of the transformation.*

6. $T\left(\begin{bmatrix} v_1 \\ v_2 \end{bmatrix}\right) = \begin{bmatrix} v_1 \\ -v_2 \end{bmatrix}$.

7. $T\left(\begin{bmatrix} v_1 \\ v_2 \end{bmatrix}\right) = \begin{bmatrix} v_2 \\ v_1 \end{bmatrix}$.

8. $T\left(\begin{bmatrix} v_1 \\ v_2 \end{bmatrix}\right) = \begin{bmatrix} v_1 + kv_2 \\ v_2 \end{bmatrix}$.

9. $T\left(\begin{bmatrix} v_1 \\ v_2 \end{bmatrix}\right) = \begin{bmatrix} v_1 \\ 0 \end{bmatrix}$.

10. $T\left(\begin{bmatrix} v_1 \\ v_2 \end{bmatrix}\right) = \begin{bmatrix} kv_1 \\ v_2 \end{bmatrix}$.

11. $T\left(\begin{bmatrix} v_1 \\ v_2 \end{bmatrix}\right) = \begin{bmatrix} 2v_1 \\ -5v_2 \end{bmatrix}$.

*In Exercises 12–15, either show that the transformation $T$ is nonlinear or determine a matrix $\mathbf{A}$ to express the transformation as a matrix transformation, $T(\mathbf{v}) = \mathbf{Av}$.*

12. $T: R^2 \rightarrow R^2$ defined by $T\left(\begin{bmatrix} v_1 \\ v_2 \end{bmatrix}\right) = \begin{bmatrix} 1 \\ v_2 \end{bmatrix}$.

13. $T: R^2 \rightarrow R^3$ defined by $T\left(\begin{bmatrix} v_1 \\ v_2 \end{bmatrix}\right) = \begin{bmatrix} v_1 + v_2 \\ v_2 \\ v_1 - v_2 \end{bmatrix}$.

14. $T: R^2 \rightarrow R^3$ defined by

$$T\left(\begin{bmatrix} v_1 \\ v_2 \end{bmatrix}\right) = \begin{bmatrix} v_1 v_2 \\ v_2 \\ v_1 + 3v_2 \end{bmatrix}.$$

**15.** $T : R^3 \rightarrow R^2$ defined by

$$T\left(\begin{bmatrix} v_1 \\ v_2 \\ v_3 \end{bmatrix}\right) = \begin{bmatrix} v_1 - v_2 + v_3 \\ (v_2)^2 \end{bmatrix}.$$

*In Exercises 16–20, determine whether each given transformation is a linear transformations.*

**16.** $T : R_{2 \times 2} \rightarrow R^1$ defined by $T\left(\begin{bmatrix} a & b \\ c & d \end{bmatrix}\right) = a + b + c + d.$

**17.** $T : R_{2 \times 2} \rightarrow R^1$ defined by $T\left(\begin{bmatrix} a & b \\ c & d \end{bmatrix}\right) = abcd.$

**18.** $T : R_{n \times n} \rightarrow R^1$ defined by $T(\mathbf{A}) = \sum_{j=1}^{n} a_{jj}$. (This transformation is called the **trace** of a matrix.)

**19.** $T : R_{n \times n} \rightarrow R_{n \times n}$ defined by

$$T(\mathbf{A}) = \begin{bmatrix} a_{11} & 0 & \cdots & \cdots & 0 \\ 0 & a_{22} & 0 & \cdots & 0 \\ 0 & 0 & \ddots & \ddots & 0 \\ 0 & \cdots & \ddots & \ddots & 0 \\ 0 & \cdots & \cdots & 0 & a_{nn} \end{bmatrix}.$$

(This transformation is often written $T(\mathbf{A}) = \text{diag}(\mathbf{A})$, meaning the image is a diagonal matrix whose diagonal entries are the diagonal entries of $\mathbf{A}$.)

**20.** $T : R_{n \times n} \rightarrow R_{n \times n}$ defined by $T(\mathbf{A}) = \mathbf{A}^T \mathbf{A}.$

*In Exercises 21–23, determine whether the given transformation is a linear transformation. [Here $P_k$ is the vector space of all polynomials of degree $k$ or less and $p'(t)$ denotes the derivative of $p(t)$ with respect to $t$.]*

**21.** $T : P_2 \rightarrow P_3$ defined by $T(p(t)) = t^3 p(0) + p(t).$

**22.** $T : P_1 \rightarrow P_2$ defined by $T(p(t)) = tp(t) + 5.$

**23.** $T : P_2 \rightarrow P_3$ defined by $T(p(t)) = t^3 p'(0) + t^2 p(0).$

**24.** Show that the transformation defined in Example 8 is a linear transformation. (*Hint*: Use properties of integrals from calculus.)

**25.** Refer to Example 9, where $T : V \rightarrow W$ by $T(\mathbf{v}) = \text{proj}_W \mathbf{v}$. We show that $T$ is a linear transformation by following the steps below.

(a) Explain why we can assume there exists an orthonormal basis $\{\mathbf{w}_1, \mathbf{w}_2, \ldots, \mathbf{w}_k\}$ for $W$.

(b) Let $\mathbf{u}$ and $\mathbf{v}$ be vectors in $V$. Determine expressions for $\text{proj}_W \mathbf{u}$ and $\text{proj}_W \mathbf{v}$.

(c) For $\mathbf{u}$ and $\mathbf{v}$ in part (b), determine the expressions for $\text{proj}_W (\mathbf{u} + \mathbf{v})$.

(d) Show that $\text{proj}_W \mathbf{u} + \text{proj}_W \mathbf{v} = \text{proj}_W (\mathbf{u} + \mathbf{v}).$

(e) Show that for scalar $k$, $\text{proj}_W (k\mathbf{v}) = k \cdot \text{proj}_W \mathbf{v}.$

(f) Explain why, having successfully verified parts (a) through (e), we can conclude that $T$ is a linear transformation.

**26.** Let $T : V \rightarrow W$ be a linear transformation. We show that $T(\mathbf{0}_V) = \mathbf{0}_W$.

(a) We have that $\mathbf{0}_V = \mathbf{0}_V + \mathbf{0}_V$. Explain why $T(\mathbf{0}_V) = T(\mathbf{0}_V) + T(\mathbf{0}_V)$.

(b) Express $T(\mathbf{0}_V) + T(\mathbf{0}_V)$ as a scalar multiple of a vector.

(c) Add $-T(\mathbf{0}_V)$ to both sides of the equation in part (a), and use part (b) to conclude that $T(\mathbf{0}_V) = \mathbf{0}_W$. Give a reason for each step as you simplify the expression.

**27.** Let $T : V \rightarrow W$ be a linear transformation. To verify the following statement, use the steps indicated.

If $\dim(V) = n$ and $S = \{\mathbf{v}_1, \mathbf{v}_2, \ldots, \mathbf{v}_n\}$ is a basis for $V$, then every image of a vector from $V$ is a linear combination of the images $\{T(\mathbf{v}_1), T(\mathbf{v}_2), \ldots, T(\mathbf{v}_n)\}$.

(a) Let $\mathbf{v}$ be a vector in $V$. Express $\mathbf{v}$ in terms of the vectors in $S$.

(b) Compute the image of $T(\mathbf{v})$ and show that it is in $\text{span}\{T(\mathbf{v}_1), T(\mathbf{v}_2), \ldots, T(\mathbf{v}_n)\}$.

## True/False Review Questions

*Determine whether each of the following statements is true or false. In the following, $V$ and $W$ are vector spaces.*

**1.** A transformation, $T : V \rightarrow W$ is a function that assigns a vector $\mathbf{w}$ in $W$ to a vector $\mathbf{v}$ from $V$.

**2.** The range of a transformation $T : V \rightarrow W$ is always all of $W$.

**3.** If $T : V \rightarrow W$ is linear, then the image of the zero vector in $V$ is the zero vector in $W$.

**4.** If $\{\mathbf{v}_1, \mathbf{v}_2, \mathbf{v}_3\}$ is a basis for $V$, then every image using linear transformation $T : V \rightarrow W$ is a linear combination of $T(\mathbf{v}_1)$, $T(\mathbf{v}_2)$, and $T(\mathbf{v}_3)$.

**5.** Let $V = W = R^1$ and transformation $T : V \rightarrow W$ be $T(\mathbf{v}) = |\mathbf{v}|$. Then $T$ is a linear transformation.

**6.** Let $V = W = R^3$ and $A$ be any $3 \times 3$ matrix. $T : V \rightarrow W$, where $T(\mathbf{v}) = \mathbf{A}^2 \mathbf{v}$ is a nonlinear transformation.

**7.** Let $V = R_{2 \times 2}$, $W = R^1$, and $T : V \rightarrow W$ be given by $T(\mathbf{A}) = \max\{|a_{11}|, |a_{12}|, |a_{21}|, |a_{22}|\}$. Then $T$ is a linear transformation.

**8.** Let $V = W = R_{2\times2}$ and $T : V \to W$ be given by $T(\mathbf{A}) = \begin{bmatrix} 0 & a_{12} \\ a_{21} & 0 \end{bmatrix}$. Then $T$ is a linear transformation.

**9.** Let $V = W = R_{3\times3}$ and $T : V \to W$ be given by $T(\mathbf{A}) = \mathbf{A}^3$. Then $T$ is a linear transformation.

**10.** Every matrix transformation is a linear transformation.

## Terminology

| Transformations | Image of a vector |
|---|---|
| Range $T$ | Linear transformation |
| Nonlinear transformation | A linear transformation is completely determined by its action on a basis |

A wide variety of mathematical applications involve rules, or functions. This section focused on the special functions between vectors spaces known as linear transformations. Linear transformations and their properties are fundamental to the study of mathematics. A firm understanding of the basic concepts presented here is assumed in succeeding sections.

- It is customary to use the term *transformation* rather than function for a rule that assigns to a vector $\mathbf{v}$ of a vector space $V$ a single vector $\mathbf{w}$ in a vector space $W$. Is such a transformation a function in the same sense as something like $f(x) = x^2$? Explain.
- What do we mean when we say that under the transformation $T$, $\mathbf{w}$ is the image of $\mathbf{v}$? Write this in the form of an equation.
- If $T : V \to W$, where is the range of $T$? Describe all vectors in range of $T$.
- If $T : V \to W$ is a linear transformation, why are we sure that the zero vector is in range of $T$?
- Explain the major differences between a nonlinear transformation and a linear transformation in terms of their behavior on linear combinations of vectors.
- Show that every matrix transformation is a linear transformation.
- Give two examples of transformations on matrices that are not linear.
- If $\{\mathbf{v}_1, \mathbf{v}_2, \mathbf{v}_3\}$ is a basis for $V$ and $T(\mathbf{v}_1) = \mathbf{w}_1$, $T(\mathbf{v}_2) = \mathbf{w}_2$, and $T(\mathbf{v}_3) = \mathbf{w}_3$, explain how to compute $T(\mathbf{v})$ where $\mathbf{v}$ is any vector in $V$.
- Suppose $T : V \to W$ such that for basis $\{\mathbf{v}_1, \mathbf{v}_2, \mathbf{v}_3\}$ of $V$ we have $T(\mathbf{v}_1) = \mathbf{w}_1$, $T(\mathbf{v}_2) = \mathbf{w}_2$, and $T(\mathbf{v}_3) = \mathbf{w}_3$. If $\{\mathbf{v}_1, \mathbf{v}_2, \mathbf{v}_3\}$ is a basis for $W$, describe the range of $T$. Explain your answer.

## 7.2 ■ ASSOCIATED SUBSPACES AND OPERATIONS INVOLVING LINEAR TRANSFORMATIONS

In the previous section we focused on identifying and investigating examples of linear transformations. We also cited several properties common to all linear transformations. In this section we develop subspaces associated with the domain and the set of images of a linear transformation. These concepts will lead us in a natural way to two important types of linear transformations. We will see how these special types of transformations are related to particular classes of matrices that we studied in Chapter 1. In addition, we investigate operations involving linear transformations that mimic matrix operations.

In Section 6.6 we summarized information about the fundamental subspaces associated with a linear system of equations. Here we develop parallel notions for a general linear transformation. We begin with two definitions and several examples and then develop further information about these concepts.

---

**Definition**    Let $T: V \rightarrow W$ be a linear transformation of vector space $V$ into vector space $W$. The **kernel** of $T$, denoted, **ker**$(T)$, is the subset of $V$ consisting of all vectors **v** such that $T(\mathbf{v}) = \mathbf{0}_W$.

---

The kernel is the set of vectors **v** in $V$ whose image is the zero vector in $W$. This is analogous to the set of all solutions of a homogeneous linear system $\mathbf{Ax} = \mathbf{0}$. In fact, if $T$ is a matrix transformation with associated matrix $\mathbf{A}$, then **ker**$(T)$ is **ns**$(\mathbf{A})$, the null space of $\mathbf{A}$.

EXAMPLE 1    Let $T: P_2 \rightarrow P_1$ be defined by $T(at^2+bt+c) = (a+b)t+(b-c)$. $T$ is a linear transformation. (Verify.) Then **ker**$(T)$ is the set of all polynomials in $P_2$ such that $T(at^2 + bt + c) = (a + b)t + (b - c) = 0$. It follows that $a + b = 0$ and $b - c = 0$. Hence we have the homogeneous linear system

$$
\begin{aligned}
a + b \quad\;\; &= 0 \\
b - c &= 0
\end{aligned}
$$

which in matrix form is given by

$$
\begin{bmatrix} 1 & 1 & 0 \\ 0 & 1 & -1 \end{bmatrix} \begin{bmatrix} a \\ b \\ c \end{bmatrix} = \begin{bmatrix} 0 \\ 0 \end{bmatrix}.
$$

The general solution of this system is $a = -c$, $b = c$. (Verify.) It follows that **ker**$(T)$ is the set of all quadratic polynomials of the form $-ct^2 + ct + c = c(-t^2 + t + 1)$, where $c$ is any real number.    ■

EXAMPLE 2    Let

$$
\mathbf{C} = \begin{bmatrix} 1 & 1 & 0 \\ 0 & 2 & 2 \end{bmatrix}
$$

and let $T: R_{2\times2} \rightarrow R_{2\times3}$ be given by $T(\mathbf{A}) = \mathbf{AC}$. Using properties of matrix algebra, we can show that $T$ is a linear transformation. (Verify.) However, $T$ is not a matrix transformation since it does not map $R^n$ into $R^m$. (See Section 1.5.) We have that **ker**$(T)$ is the set of all $2 \times 2$ matrices $\begin{bmatrix} a & b \\ c & d \end{bmatrix}$ such that

$$
\begin{bmatrix} a & b \\ c & d \end{bmatrix} \begin{bmatrix} 1 & 1 & 0 \\ 0 & 2 & 2 \end{bmatrix} = \begin{bmatrix} a & a + 2b & 2b \\ c & c + 2d & 2d \end{bmatrix} = \begin{bmatrix} 0 & 0 & 0 \\ 0 & 0 & 0 \end{bmatrix} = \mathbf{0}.
$$

It follows that $a = b = c = d = 0$ (verify), so **ker**$(T)$ consists of the single vector

$$
\begin{bmatrix} 0 & 0 \\ 0 & 0 \end{bmatrix} = \mathbf{0}.
$$

Thus if any entry of $\mathbf{A}$ is different from zero, then $T(\mathbf{A})$ is not $\mathbf{0}$.    ■

EXAMPLE 3   Let $T$ be the linear transformation given in Example 7 in Section 7.1; $T(f(t)) = f'(t)$, the derivative of $f$. In this case $W = C(-\infty, \infty)$ and $\mathbf{0}_W =$ the zero function; that is, the function which is zero for all values of $t$. Then $\mathbf{ker}(T)$ is the set of differentiable functions whose derivative is zero. From calculus we have that $\mathbf{ker}(T)$ is the set of all constant functions.   ■

Next we turn our attention to the collection of images of vectors when $T$ is a linear transformation.

**Definition**   Let $T: V \rightarrow W$ be a linear transformation mapping vector space $V$ into vector space $W$. The **range** or **image** of $T$, denoted, $\mathbf{R}(T)$, is the subset of $W$ consisting of all vectors $\mathbf{w}$ for which there exists a vector $\mathbf{v}$ in $V$ such that $T(\mathbf{v}) = \mathbf{w}$.

*The range of $T$ is the subset of $W$ consisting of elements that are images of vectors from $V$. If $T$ is a matrix transformation from $R^n \rightarrow R^m$ with associated matrix $\mathbf{A}$, then $\mathbf{R}(T)$ is the column space of $\mathbf{A}$, a subset of $R^m$.*

EXAMPLE 4   For the linear transformation in Example 1, $\mathbf{R}(T)$ is the set of all polynomials $k_1 t + k_2$ in $P_1$ for which there is a polynomial $at^2 + bt + c$ in $P_2$ such that

$$T(at^2 + bt + c) = (a + b)t + (b - c) = k_1 t + k_2.$$

It follows that $a + b = k_1$ and $b - c = k_2$. The corresponding linear system in matrix form is

$$\begin{bmatrix} 1 & 1 & 0 \\ 0 & 1 & -1 \end{bmatrix} \begin{bmatrix} a \\ b \\ c \end{bmatrix} = \begin{bmatrix} k_1 \\ k_2 \end{bmatrix}.$$

This system is consistent for all values of $k_1$ and $k_2$ (verify) so the range of $T$ is all of $W$; that is, $\mathbf{R}(T) = W = P_1$.   ■

EXAMPLE 5   For the linear transformation in Example 2, $\mathbf{R}(T)$ is the set of all $2 \times 3$ matrices that are an image of a $2 \times 2$ matrix; that is, all matrices of the form

$$\begin{bmatrix} a & b \\ c & d \end{bmatrix} \begin{bmatrix} 1 & 1 & 0 \\ 0 & 2 & 2 \end{bmatrix} = \begin{bmatrix} a & a + 2b & 2b \\ c & c + 2d & 2d \end{bmatrix}.$$

By inspecting the form of this image we can say that a $2 \times 3$ matrix is in $\mathbf{R}(T)$ provided that its second column is the sum of columns 1 and 3. Hence $\mathbf{R}(T) \neq R_{2 \times 3}$, but just a portion of the vector space of all $2 \times 3$ matrices.   ■

## The Structure of $\mathbf{ker}(T)$ and $\mathbf{R}(T)$

The following result tells us about the structure of both the kernel and the range of a linear transformation. With this result we will be able to delve further into the behavior of linear transformations.

Let $T: V \rightarrow W$ be a linear transformation of vector space $V$ into vector space $W$. Then
(a) $\mathbf{ker}(T)$ is a subspace of $V$.
(b) $\mathbf{R}(T)$ is a subspace of $W$.

To verify each of these statements we show that the set is closed with respect to addition and scalar multiplication. (See Exercises 2 and 3.)

Figure 1 shows these subspaces in relation to vector spaces $V$ and $W$, respectively. We note that not all vectors in $V$ are in $\mathbf{ker}(T)$, but of course all the images are in $\mathbf{R}(T)$. We also see that more than one vector from $\mathbf{v}$ can have the same image and that not every vector in $W$ need be in the range of $T$; for example, $\mathbf{w}_5$. We investigate these aspects of linear transformations later in this section.

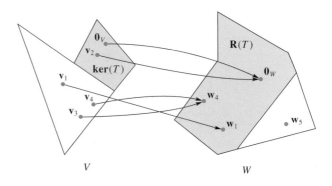

FIGURE 1

If $T$ is a matrix transformation with associated $m \times n$ matrix $\mathbf{A}$, then $\mathbf{ker}(T) = \mathbf{ns}(\mathbf{A})$ and $\mathbf{R}(T) = \mathbf{col}(\mathbf{A})$. Thus Figure 1 also provides a representation of $\mathbf{ns}(\mathbf{A})$ in $R^n$ and $\mathbf{col}(\mathbf{A})$ in $R^m$. Hence we have concepts for general linear transformations that are analogous to those we have developed for matrices.

The kernel of $T$ is a subspace of $V$ and the range of $T$ is a subspace of $W$. However, there is a relation between bases for these spaces that we develop next.

Let $\dim(V) = n$ and $T : V \to W$ be a linear transformation with $\{\mathbf{v}_1, \mathbf{v}_2, \ldots, \mathbf{v}_k\}$ a basis for $\mathbf{ker}(T)$. There exist vectors $\mathbf{v}_{k+1}, \mathbf{v}_{k+2}, \ldots, \mathbf{v}_n$ so that $\{\mathbf{v}_1, \ldots, \mathbf{v}_n\}$ is a basis for $V$ and $\{T(\mathbf{v}_{k+1}), T(\mathbf{v}_{k+2}), \ldots, T(\mathbf{v}_n)\}$ is a basis for $\mathbf{R}(T)$. To verify this result we must show that a basis $\{\mathbf{v}_1, \ldots, \mathbf{v}_k, \mathbf{v}_{k+1}, \mathbf{v}_{k+2}, \ldots, \mathbf{v}_n\}$ exists for $V$ and then that $\{T(\mathbf{v}_{k+1}), T(\mathbf{v}_{k+2}), \ldots, T(\mathbf{v}_n)\}$ spans $\mathbf{R}(T)$ and is linearly independent. To illustrate the use of vector space and linear transformation properties, we verify this result in Example 6. The steps we use provide a close look at the interplay between linear transformations and bases.

EXAMPLE 6    There is a relation between bases for $\mathbf{ker}(T)$ and $\mathbf{R}(T)$ given in the following statement.

Let $\dim(V) = n$ and $T : V \to W$ be a linear transformation with $\{\mathbf{v}_1, \mathbf{v}_2, \ldots, \mathbf{v}_k\}$ a basis for $\mathbf{ker}(T)$. There exist vectors $\mathbf{v}_{k+1}, \mathbf{v}_{k+2}, \ldots, \mathbf{v}_n$ so that $\{\mathbf{v}_1, \ldots, \mathbf{v}_n\}$ is a basis for $V$ and $\{T(\mathbf{v}_{k+1}), T(\mathbf{v}_{k+2}), \ldots, T(\mathbf{v}_n)\}$ is a basis for $\mathbf{R}(T)$.

To verify this statement we first recall that a basis for a subspace can be extended to a basis for the entire space. (See Exercise 29 in Section 5.3.) Here the subspace of $V$ is $\mathbf{ker}(T)$ so there exist vectors $\mathbf{v}_{k+1}, \mathbf{v}_{k+2}, \ldots, \mathbf{v}_n$ so that $\{\mathbf{v}_1, \ldots, \mathbf{v}_n\}$ is a basis for $V$. Next, if $\mathbf{v}$ is in $V$, then it can be expressed in terms of the basis $\{\mathbf{v}_1, \mathbf{v}_2, \ldots, \mathbf{v}_n\}$; we have

$$\mathbf{v} = \sum_{j=1}^{n} c_j \mathbf{v}_j.$$

It follows that

$$T(\mathbf{v}) = T\left(\sum_{j=1}^{n} c_j \mathbf{v}_j\right) = \sum_{j=1}^{n} c_j T(\mathbf{v}_j) = \sum_{j=k+1}^{n} c_j T(\mathbf{v}_j)$$

since $T$ is a linear transformation and $T(\mathbf{v}_1) = T(\mathbf{v}_2) = \cdots = T(\mathbf{v}_k) = \mathbf{0}_W$. Thus the image of any vector $\mathbf{v}$ from $V$ is in $\text{span}\{T(\mathbf{v}_{k+1}), T(\mathbf{v}_{k+2}), \ldots, T(\mathbf{v}_n)\}$ and hence $\{T(\mathbf{v}_{k+1}), T(\mathbf{v}_{k+2}), \ldots, T(\mathbf{v}_n)\}$ spans $\mathbf{R}(T)$. It remains to show that $\{T(\mathbf{v}_{k+1}), T(\mathbf{v}_{k+2}), \ldots, T(\mathbf{v}_n)\}$ is linearly independent. We proceed as follows. Suppose that

$$\sum_{j=k+1}^{n} b_j T(\mathbf{v}_j) = \mathbf{0}_W. \tag{1}$$

Since $T$ is a linear transformation, we have

$$\mathbf{0}_W = \sum_{j=k+1}^{n} b_j T(\mathbf{v}_j) = T\left(\sum_{j=k+1}^{n} b_j \mathbf{v}_j\right)$$

which says that $\sum_{j=k+1}^{n} b_j \mathbf{v}_j$ is in $\mathbf{ker}(T)$. But the vectors $\mathbf{v}_{k+1}, \mathbf{v}_{k+2}, \ldots, \mathbf{v}_n$ are not in $\mathbf{ker}(T)$; hence this linear combination of vectors must be the zero vector. Since $\{\mathbf{v}_{k+1}, \mathbf{v}_{k+2}, \ldots, \mathbf{v}_n\}$ is a linearly independent set, the only way

$$\sum_{j=k+1}^{n} b_j \mathbf{v}_j = \mathbf{0}_V$$

is when all the coefficients are zero; that is, $b_{k+1} = b_{k+2} = \cdots = b_n = 0$. Hence, $\{T(\mathbf{v}_{k+1}), T(\mathbf{v}_{k+2}), \ldots, T(\mathbf{v}_n)\}$ is a linearly independent set. Since

$$\{T(\mathbf{v}_{k+1}), T(\mathbf{v}_{k+2}), \ldots, T(\mathbf{v}_n)\}$$

both spans $\mathbf{R}(T)$ and is linearly independent, it is a basis.   ■

A careful inspection of the result in Example 6 gives the following relation:

$$\dim(V) = \dim(\mathbf{ker}(T)) + \dim(\mathbf{R}(T)).$$

This generalizes the matrix result that says

$$\text{number of columns of } \mathbf{A} = \dim(\mathbf{ns}(\mathbf{A})) + \dim(\mathbf{col}(\mathbf{A})).$$

See Sections 2.3, 6.5, and 6.6.

## One-to-One Linear Transformations

A linear transformation $T$ can map more than one vector from $V$ to the same image. In Examples 1 and 3, $\mathbf{ker}(T)$ consisted of more than one vector each of whose image was $\mathbf{0}_W$. In Figure 1, both $\mathbf{v}_3$ and $\mathbf{v}_4$ got mapped to $\mathbf{w}_4$. Linear transformations in which each vector from $V$ has a different image in $W$ exhibit important properties. With the following definition we make this notion precise and introduce the associated terminology.

---

**Definition**   A linear transformation $T: V \rightarrow W$ is called **one-to-one** if for $\mathbf{v}_1 \neq \mathbf{v}_2$ in $V$, then $T(\mathbf{v}_1) \neq T(\mathbf{v}_2)$ in $W$. An equivalent statement is that $T$ is one-to-one if $T(\mathbf{v}_1) = T(\mathbf{v}_2)$ implies that $\mathbf{v}_1 = \mathbf{v}_2$.

---

As you might suspect from the comments before the preceding definition, we can use information about the kernel of $T$ to determine if it is one-to-one. We have the following result.

> Let $T: V \to W$ be a linear transformation of vector space $V$ into vector space $W$. $T$ is one-to-one if and only if $\mathbf{ker}(T) = \{\mathbf{0}_V\}$.

To illustrate the use of concepts of one-to-one and the kernel of a linear transformation, we verify this result in Example 7.

EXAMPLE 7    To verify that a linear transformation $T$ of vector space $V$ into vector space $W$ is one-to-one if and only if $\mathbf{ker}(T) = \{\mathbf{0}_V\}$ we show two properties:

1. If $T$ is one-to-one, then $\mathbf{ker}(T) = \{\mathbf{0}_V\}$.
2. If $\mathbf{ker}(T) = \{\mathbf{0}_V\}$, then $T$ is one-to-one.

To verify (1) we proceed as follows. Let $\mathbf{v}$ be in $\mathbf{ker}(T)$. Then $T(\mathbf{v}) = \mathbf{0}_W$. But from Section 7.1 we know that the image of the zero vector is the zero vector; that is, $T(\mathbf{0}_V) = \mathbf{0}_W$. Since $T$ is one-to-one we must have that $\mathbf{v} = \mathbf{0}_V$. Hence $\mathbf{ker}(T)$ consists of the single vector $\mathbf{0}_V$.

To verify (2) we proceed as follows. Suppose that for $\mathbf{u}$ and $\mathbf{v}$ in $V$, $T(\mathbf{u}) = T(\mathbf{v})$. Thus $T(\mathbf{u}) - T(\mathbf{v}) = \mathbf{0}_W$. Since $T$ is a linear transformation this equation is equivalent to $T(\mathbf{u} - \mathbf{v}) = \mathbf{0}_W$. Hence vector $\mathbf{u} - \mathbf{v}$ is in $\mathbf{ker}(T)$. But $\mathbf{ker}(T)$ is exactly the zero vector of $V$ and so $\mathbf{u} - \mathbf{v} = \mathbf{0}_V$; hence $\mathbf{u} = \mathbf{v}$. This implies that if the images are the same, then the original vectors are identical; that is, $T$ is one-to-one.    ■

It follows that the linear transformations in Examples 1 and 3 are not one-to-one, but the linear transformation in Example 2 is one-to-one. If $T$ is a matrix transformation with associated matrix $\mathbf{A}$, then $T$ is one-to-one if and only if $\mathbf{Ax} = \mathbf{0}$ has only the trivial solution—that is, provided $\mathbf{ns}(\mathbf{A}) = \{\mathbf{0}\}$. For a square matrix $\mathbf{A}$, $T$ is one-to-one if and only if $\mathbf{A}$ is nonsingular. This follows from properties of nonsingular matrices developed in Section 2.4. (See Exercise 11.)

Note that we can also state the preceding result as follows: $T$ is one-to-one if and only if $\dim(\mathbf{ker}(T)) = 0$. (Recall that we agreed to say that the dimension of the zero subspace is zero. See Section 5.2.)

If a linear transformation $T: V \to W$ is one-to-one, then the only vector from $V$ whose image is $\mathbf{0}_W$ is $\mathbf{0}_V$. This follows from one of the properties of all linear transformations that we verified in Section 7.1: The image of the zero vector is the zero vector. For a linear transformation that is one-to-one, this observation can be extended to say that any linear combination of vectors in $V$ that is not the zero vector must have an image that is not the zero vector in $W$. The following result generalizes this statement even further.

> A linear transformation $T: V \to W$ is one-to-one if and only if the image of every linearly independent subset of vectors in $V$ is a linearly independent set of vectors in $W$.

To verify this result we show two properties:

1. If $T$ is one-to-one, then the image of a linearly independent set is a linearly independent set.
2. If the image of every linearly independent subset of vectors in $V$ is a linearly independent set of vectors in $W$, then $T$ is one-to-one.

We leave the verification of property 1 to Exercise 12 and show the verification of property 2 here.

Suppose that the image of every linearly independent subset of vectors in $V$ is a linearly independent set of vectors in $W$. Let $\mathbf{v}$ be a vector in $V$, where $\mathbf{v} \neq \mathbf{0}_V$. Then $\{\mathbf{v}\}$ is a linearly independent subset of $V$ and $\{T(\mathbf{v})\}$ is a linearly independent subset of $W$. Hence $T(\mathbf{v}) \neq \mathbf{0}_W$. This says that the image of every nonzero vector in $V$ is not $\mathbf{0}_W$, hence $\mathbf{ker}(T) = \{\mathbf{0}_W\}$ and so $T$ is one-to-one.

## Onto Linear Transformations

Another special type of linear transformation involves properties of the range. In Example 5 we noted that $\mathbf{R}(T) \neq W = R_{2 \times 3}$, and in Example 4 we saw that $\mathbf{R}(T) = W = P_1$. A linear transformation $T$ in which the range is the entire vector space $W$ in a sense "covers" $W$. That is, every vector in $W$ is the image of at least one vector from $V$. With the following definition we make this notion precise and introduce the associated terminology.

---

**Definition**   Let $T: V \rightarrow W$ be a linear transformation of vector space $V$ into vector space $W$. $T$ is called **onto** if $\mathbf{R}(T) = W$.

---

It follows that the linear transformation in Example 4 is onto, while that in Example 5 is not. If $T$ is a matrix transformation with associated $m \times n$ matrix $\mathbf{A}$, then $T$ is onto provided $\mathbf{A}\mathbf{x} = \mathbf{b}$ has a solution for each vector $\mathbf{b}$ in $R^m$. That is, $T$ is onto provided the columns of $\mathbf{A}$ span $R^n$. For a square matrix $\mathbf{A}$, $T$ is onto if and only if $\mathbf{A}$ is nonsingular. This follows from properties of nonsingular matrices developed in Section 2.4. (See Exercise 13.)

Some linear transformations are both one-to-one and onto. We investigate this important class of linear transformations next. We start with an example and then develop some properties.

EXAMPLE 8   Let $T: P_2 \rightarrow P_2$ be defined by $T(at^2 + bt + c) = ct^2 + bt + a$. $T$ is a linear transformation (verify) and $\mathbf{ker}(T)$ is the set of all polynomials in $P_2$ such that $T(at^2 + bt + c) = ct^2 + bt + a = 0$. It follows that $a = b = c = 0$; hence $\mathbf{ker}(T) = \{\mathbf{0}_V\} = \{0\}$ and so $T$ is one-to-one. To investigate $\mathbf{R}(T)$ we ask what vectors in $P_2$ can be images. Let $\mathbf{w} = k_2 t^2 + k_1 t + k_0$ be an arbitrary vector in $P_2$. Is there a vector $\mathbf{v} = at^2 + bt + c$ in $P_2$ such that $T(\mathbf{v}) = \mathbf{w}$? We have $T(\mathbf{v}) = ct^2 + bt + a = \mathbf{w} = k_2 t^2 + k_1 t + k_0$ provided that $c = k_2$, $b = k_1$, and $a = k_0$. Hence for any vector $\mathbf{w}$ in $P_2$, there is a vector $\mathbf{v}$ in $P_2$ such that $T(\mathbf{v}) = \mathbf{w}$ and so $T$ is onto. (See Exercise 14 for another approach to this example.)   ■

From our preceding remarks about a matrix transformation $T$ whose associated matrix is nonsingular, we see that such linear transformations are both one-to-one and onto. It is important to note that such transformations take vector space $V$ into vector space $W$ where $\dim(V) = \dim(W)$ since $V = W = R^n$. Two natural questions that arise for a general linear transformation $T: V \rightarrow W$ where $\dim(V) = \dim(W)$ are the following:

If $T$ is one-to-one, is it onto?

If $T$ is onto, is it one-to-one?

The answer to both of these questions is yes. (See Exercise 15.) Hence we have another parallel result between matrix transformation and linear transformations.

## Composition of Linear Transformations

To continue our development of properties of linear transformations, we next look at an operation involving linear transformations. See Exercises 16 and 17.

---

**Definition**   Let $V_1$ be an $n$-dimensional vector space, $V_2$ an $m$-dimensional vector space, and $V_3$ a $p$-dimensional vector space. Let $T_1\colon V_1 \to V_2$ and $T_2\colon V_2 \to V_3$ be linear transformations. The **composition** of $T_2$ with $T_1$, denoted $T_2 \circ T_1$, is the transformation defined by $(T_2 \circ T_1)(\mathbf{v}) = T_2(T_1(\mathbf{v}))$ for $\mathbf{v}$ in $V_1$.

---

It is easy to verify that $T_2 \circ T_1$ is a linear transformation from $V_1$ to $V_3$. (See Exercise 16.) If $T\colon V \to V$, then $T \circ T$ is written $T^2$. If $T_1$ and $T_2$ are matrix transformations with associated $m \times n$ matrix $\mathbf{A}$ and $p \times m$ matrix $\mathbf{B}$, respectively, then $T_2 \circ T_1$ is a matrix transformation with associated matrix $\mathbf{BA}$. (See Section 1.5, where we also noted that the row-by-column definition of matrix products is exactly the right procedure to determine the composition of matrix transformations.)

---

**Definition**   Let $T\colon V \to W$ be a linear transformation of vector space $V$ into vector space $W$. $T$ is called **invertible** if there exists a unique linear transformation $T^{-1}\colon W \to V$ such that $T^{-1} \circ T = I_V$ and $T \circ T^{-1} = I_W$, where $I_V =$ identity linear transformation on $V$ and $I_W =$ identity linear transformation on $W$. (We call $T^{-1}$ the **inverse** of $T$.)

---

If $T$ is a matrix transformation with associated nonsingular matrix $\mathbf{A}$, then it follows that $T$ is invertible since the matrix transformation with associated matrix $\mathbf{A}^{-1}$ would define the matrix transformation $T^{-1}$. To determine when a general linear transformation is invertible, we must rely on the properties that an invertible matrix transformation exhibits; namely, it is one-to-one and onto. Thus we are led to the following result, which we state without verification.

> Let $T\colon V \to W$ be a linear transformation of vector space $V$ into vector space $W$. $T$ is invertible if and only if $T$ is one-to-one and onto. Moreover, $T^{-1}$ is a linear transformation and the inverse of $T^{-1}$ is $T$; $(T^{-1})^{-1} = T$.

If we combine this result with two previous results, which we restate as

I. A linear transformation $T\colon V \to W$ is one-to-one if and only if the image of every linearly independent subset of vectors in $V$ is a linearly independent set of vectors in $W$ (see Exercise 12).

II. For a linear transformation $T\colon V \to W$ where $\dim(V) = \dim(W)$ we have the following: If $T$ is one-to-one, then $T$ is onto; and if $T$ is onto, then $T$ is one-to-one (see Exercise 15).

then we obtain the following

> Let $T\colon V \to W$ be a linear transformation and $\dim(V) = \dim(W)$. $T$ is invertible if and only if the image of a basis for $V$ is a basis for $W$.

To verify this result we show two properties:

1. If $T$ is invertible, then the image of a basis is a basis.
2. If the image of a basis is a basis, then $T$ is invertible.

To verify (1) we note that if $T$ is invertible it is one-to-one; hence result I implies that the image of a basis is a linearly independent set in $W$. But since $\dim(V) = \dim(W)$, this linearly independent set has the same number of vectors as a basis for $W$. Thus it must also be a basis for $W$.

To verify (2) we note that if the image of a basis for $V$ is a basis for $W$, then $\dim(\mathbf{ker}(T)) = 0$ so $T$ is one-to-one. It follows that result II implies that $T$ is also onto and hence $T$ must be invertible.

## Connections between Matrices and Linear Transformations

We conclude this section by summarizing some of the information about matrices, matrix transformations, and invertible transformations. This further emphasizes the close connection between two major concepts in linear algebra: matrices and linear transformations.

| $n \times n$ Matrix A | Matrix Transformation $T(\mathbf{x}) = \mathbf{Ax}$ | Linear Transformation $T : V \to W$ with $\dim(V) = \dim(W)$ |
|---|---|---|
| $\mathbf{A}$ nonsingular. | $T$ is invertible if and only if $\mathbf{A}$ nonsingular. | $T$ is invertible. |
| $\mathbf{ns(A)} = \mathbf{0}$ and $\mathbf{Ax} = \mathbf{b}$ is consistent for any $\mathbf{b}$. | $\mathbf{ker}(T) = \{\mathbf{0}\}$ implies $T$ is one-to-one and $\mathbf{R}(T) = R^n$ implies $T$ is onto. | $T$ is one-to-one and $T$ is onto. |

# EXERCISES 7.2

**1.** Let $T : R^4 \to R^2$ be the linear transformation defined by

$$T\left(\begin{bmatrix} v_1 \\ v_2 \\ v_3 \\ v_4 \end{bmatrix}\right) = \begin{bmatrix} v_1 + v_3 \\ v_2 + v_4 \end{bmatrix}.$$

(a) Is $\begin{bmatrix} 2 \\ 3 \\ -2 \\ 3 \end{bmatrix}$ in $\mathbf{ker}(T)$?

(b) Is $\begin{bmatrix} 4 \\ -2 \\ -4 \\ 2 \end{bmatrix}$ in $\mathbf{ker}(T)$?

(c) Is $\begin{bmatrix} 1 \\ 2 \end{bmatrix}$ in $\mathbf{R}(T)$?

(d) Describe the entries of all the vectors in $\mathbf{ker}(T)$.

(e) Describe the entries of all the vectors in $\mathbf{R}(T)$.

**2.** Let $T : V \to W$ be a linear transformation of vector space $V$ into vector space $W$. Show that $\mathbf{ker}(T)$ is a subspace of $V$ by verifying each of the following.

(a) Let $\mathbf{u}$ and $\mathbf{v}$ be in $\mathbf{ker}(T)$. Show that $\mathbf{u}+\mathbf{v}$ is in $\mathbf{ker}(T)$.

(b) Let $\mathbf{v}$ be in $\mathbf{ker}(T)$ and $k$ be any scalar. Show that $k\mathbf{v}$ is in $\mathbf{ker}(T)$.

**3.** Let $T : V \to W$ be a linear transformation of vector space $V$ into vector space $W$. Show that $\mathbf{R}(T)$ is a subspace of $W$ by verifying each of the following.

(a) Let $\mathbf{w}_1$ and $\mathbf{w}_2$ be in $\mathbf{R}(T)$. Show that $\mathbf{w}_1 + \mathbf{w}_2$ is in $\mathbf{R}(T)$.

(b) Let $\mathbf{w}$ be in $\mathbf{R}(T)$ and $k$ be any scalar. Show that $k\mathbf{w}$ is in $\mathbf{R}(T)$.

**4.** Let $V$ be the set of all symmetric matrices in vector space $R_{2\times 2}$.

(a) Show that $V$ is a subspace of $R_{2\times 2}$.

(b) Let $T : V \to R^1$ be a transformation defined by

$$T\left(\begin{bmatrix} a & b \\ b & c \end{bmatrix}\right) = b.$$

Show that $T$ is a linear transformation.

(c) Describe the vectors in $\mathbf{ker}(T)$. Is $T$ one-to-one?

(d) Determine $\mathbf{R}(T)$. Is $T$ onto?

5. Let $T: P_2 \rightarrow P_2$ be a transformation defined by

$$T(at^2 + bt + c) = (a - c)t^2 + 2(b + c).$$

(a) Show that $T$ is a linear transformation.

(b) Determine $\mathbf{ker}(T)$. Is $T$ one-to-one?

(c) Determine $\mathbf{R}(T)$. Is $T$ onto?

6. Let $\mathbf{C} = \begin{bmatrix} 1 & 2 \\ 0 & 1 \end{bmatrix}$ and let $T: R_{2 \times 2} \rightarrow R_{2 \times 2}$ be defined by $T(\mathbf{A}) = \mathbf{AC}$.

(a) Show that $T$ is a linear transformation.

(b) Determine $\mathbf{ker}(T)$. Is $T$ one-to-one?

(c) Determine $\mathbf{R}(T)$. Is $T$ onto? [*Hint*: Note that $\mathbf{C}$ is nonsingular. Then for any $\mathbf{B}$ in $R_{2 \times 2}$, $T(\mathbf{BC}^{-1}) = \mathbf{B}$.]

7. Let $T: R^4 \rightarrow R^3$ be the linear transformation defined by

$$T\left(\begin{bmatrix} v_1 \\ v_2 \\ v_3 \\ v_4 \end{bmatrix}\right) = \begin{bmatrix} v_1 + v_2 \\ v_3 + v_4 \\ v_1 + v_3 \end{bmatrix}.$$

(a) Show that $T$ is a matrix transformation by finding its associated matrix $\mathbf{A}$.

(b) Find a basis for $\mathbf{ker}(T)$ and $\dim(\mathbf{ker}(T))$.

(c) Find a basis for $\mathbf{R}(T)$ and $\dim(\mathbf{R}(T))$.

8. Let $T: P_2 \rightarrow P_3$ be the linear transformation given by $T(p(t)) = t^2 p'(t)$.

(a) Find a basis for $\mathbf{ker}(T)$ and $\dim(\mathbf{ker}(T))$.

(b) Find a basis for $\mathbf{R}(T)$ and $\dim(\mathbf{R}(T))$.

9. The relationship between bases for $\mathbf{ker}(T)$ and $\mathbf{R}(T)$ given in the following statement was verified in Example 6.

Let $\dim(V) = n$ and $T: V \rightarrow W$ be a linear transformation with $\{\mathbf{v}_1, \mathbf{v}_2, \dots, \mathbf{v}_k\}$ a basis for $\mathbf{ker}(T)$. There exist vectors $\mathbf{v}_{k+1}, \mathbf{v}_{k+2}, \dots, \mathbf{v}_n$ so that $\{\mathbf{v}_1, \dots, \mathbf{v}_n\}$ is a basis for $V$ and $\{T(\mathbf{v}_{k+1}), T(\mathbf{v}_{k+2}), \dots, T(\mathbf{v}_n)\}$ is a basis for $\mathbf{R}(T)$.

[To construct the bases for $\mathbf{ker}(T)$ and $\mathbf{R}(T)$ we will make use of a technique introduced in Section 5.3.[1] ]

Let $T: R^4 \rightarrow R^3$ be the matrix transformation whose associated matrix $\mathbf{A}$ is

$$\mathbf{A} = \begin{bmatrix} 2 & 1 & 0 & 1 \\ 3 & 2 & 1 & 1 \\ 0 & 1 & 1 & 0 \end{bmatrix}.$$

(a) Show that a basis for $\mathbf{ker}(T)$ is the single vector

$$\mathbf{v} = \begin{bmatrix} 0 \\ -1 \\ 1 \\ 1 \end{bmatrix}.$$

(b) We know that the columns of $\mathbf{I}_4$ are a basis for $R^4$. Call these columns $\mathbf{e}_1, \mathbf{e}_2, \mathbf{e}_3$, and $\mathbf{e}_4$, respectively. From the ordered set $S = \{\mathbf{v}, \mathbf{e}_1, \mathbf{e}_2, \mathbf{e}_3, \mathbf{e}_4\}$ use the technique cited in Section 5.3 to determine a subset of $S$ that is a basis for $R^4$ and includes $\mathbf{v}$. Call this basis $Q$ and denote its members as $\mathbf{v}_1 = \mathbf{v}, \mathbf{v}_2, \mathbf{v}_3, \mathbf{v}_4$.

(c) Determine a basis for $\mathbf{R}(T)$ using the $Q$ basis from part (b).

(d) In part (b) we chose the natural basis for $R^4$, but any other basis could have been used. Show that $\{\mathbf{w}_1, \mathbf{w}_2, \mathbf{w}_3, \mathbf{w}_4\}$ where

$$\mathbf{w}_1 = \begin{bmatrix} 3 \\ 1 \\ 0 \\ -2 \end{bmatrix}, \quad \mathbf{w}_2 = \begin{bmatrix} 1 \\ 2 \\ 0 \\ 1 \end{bmatrix},$$

$$\mathbf{w}_3 = \begin{bmatrix} 2 \\ -1 \\ 1 \\ 0 \end{bmatrix}, \quad \mathbf{w}_4 = \begin{bmatrix} 0 \\ 1 \\ 0 \\ 0 \end{bmatrix}$$

is a basis for $R^4$. Then repeat parts (b) and (c) to determine another basis for $\mathbf{R}(T)$.

10. Determine which of the following linear transformations is one-to-one.

(a) The null transformation from $R^2$ to $R^3$. (See Example 1 in Section 7.1.)

(b) The identity transformation from $R^3$ to $R^3$. (See Example 2 in Section 7.1.)

(c) $T: R^3 \rightarrow R^2$ given by

$$T(\mathbf{v}) = T\left(\begin{bmatrix} v_1 \\ v_2 \\ v_3 \end{bmatrix}\right) = \begin{bmatrix} v_1 \\ v_2 \end{bmatrix} = \mathbf{w}.$$

(See Example 6 in Section 7.1.)

(d) $T: R_{n \times n} \rightarrow R^1$ defined by

$$T(\mathbf{A}) = \sum_{j=1}^{n} a_{jj}.$$

(This transformation is called the **trace** of a matrix.)

(e) $T: R_{n \times n} \rightarrow R_{n \times n}$ defined by $T(\mathbf{A}) = \mathbf{CA}$, for $\mathbf{C}$ a fixed nonsingular $n \times n$ matrix.

11. Let $T$ be a matrix transformation with associated square matrix $\mathbf{A}$. Show that $T$ is one-to-one if and only if $\mathbf{A}$ is nonsingular. (*Hint*: Use properties of nonsingular matrices developed in Section 2.4.)

[1]See "Application: Determining a Basis from a Spanning Set of Vectors from $R^n$ or $C^n$" (page 334).

**12.** Let $T: V \rightarrow W$ be a linear transformation of vector space $V$ into vector space $W$. Show that if $T$ is one-to-one, then the image of a linearly independent set is a linearly independent set. [*Hint*: Let $\{\mathbf{v}_1, \mathbf{v}_2, \ldots, \mathbf{v}_k\}$ be a linearly independent set in $V$. Assume that a linear combination of the images of these vectors is equal to $\mathbf{0}_W$. Use the fact that $T(\mathbf{0}_V) = \mathbf{0}_W$ and that $T$ is one-to-one to show that all the coefficients must be zero.]

**13.** Let $T$ be a matrix transformation with associated square matrix $\mathbf{A}$. Show that $T$ is onto if and only if $\mathbf{A}$ is nonsingular. (*Hint*: Use properties of nonsingular matrices developed in Section 2.4.)

**14.** Let's agree to write a vector in $P_2$ in the form $at^2 + bt + c$. That is, the term involving $t^2$ is written first, the term involving $t$ second, and the constant term last. Then with each vector in $P_2$ we can associate a unique vector in $R^3$; the polynomial $at^2 + bt + c$ corresponds to the 3-vector

$$\begin{bmatrix} a \\ b \\ c \end{bmatrix}.$$

Referring to Example 8, we see that linear transformation $T$ can just as easily be expressed as $T: R^3 \rightarrow R^3$ where

$$T\left(\begin{bmatrix} a \\ b \\ c \end{bmatrix}\right) = \begin{bmatrix} c \\ b \\ a \end{bmatrix}.$$

(a) Show that $T: R^3 \rightarrow R^3$ is a matrix transformation and determine the associated matrix $\mathbf{A}$.

(b) Use properties of $\mathbf{A}$ to show that T is one-to-one and onto, hence invertible.

**15.** Let $T: V \rightarrow W$ be a linear transformation where $\dim(V) = \dim(W)$. Use the fact that $\dim(\mathbf{ker}(T)) + \dim(\mathbf{R}(T)) = \dim(V)$ for each of the following.

(a) Show that if $T$ is one-to-one, then $T$ is onto.

(b) Show that if $T$ is onto, then $T$ is one-to-one.

**16.** Let $V_1$ be an $n$-dimensional vector space, $V_2$ an $m$-dimensional vector space, and $V_3$ a $p$-dimensional vector space. Let $T_1: V_1 \rightarrow V_2$ and $T_2: V_2 \rightarrow V_3$ be linear transformations. Show that $T_2 \circ T_1$ is a linear transformation from $V_1$ to $V_3$.

**17.** Let $T_1: V \rightarrow W$ and $T_2: V \rightarrow W$ be linear transformations from vector space $V$ to vector space $W$.

(a) We define the sum of transformations $T_1$ and $T_2$ as the transformation $(T_1 + T_2): V \rightarrow W$ where $(T_1 + T_2)(\mathbf{v}) = T_1(\mathbf{v}) + T_2(\mathbf{v})$. Show that $(T_1 + T_2)$ is a linear transformation.

(b) Let $k$ be any scalar. We define the scalar multiple of transformation $T_1$ as the transformation $kT_1: V \rightarrow W$ where $kT_1(\mathbf{v}) = kT_1(\mathbf{v})$. Show that $kT_1$ is a linear transformation.

**18.** Let $T: R^4 \rightarrow R^6$ be a linear transformation.

(a) If $\dim(\mathbf{ker}(T)) = 2$, what is $\dim(\mathbf{R}(T))$?

(b) If $\dim(\mathbf{R}(T)) = 3$, what is $\dim(\mathbf{ker}(T))$?

**19.** Let $T: V \rightarrow R^5$ be a linear transformation.

(a) If $T$ is onto and $\dim(\mathbf{ker}(T)) = 2$, what is $\dim(V)$?

(b) If $T$ is one-to-one and onto, what is $\dim(V)$?

**20.** Let $T$ be a matrix transformation with associated $m \times n$ matrix $\mathbf{A}$. Show that $T$ is onto if and only if rank$(\mathbf{A}) = m$.

## True/False Review Questions

*Determine whether each of the following statements is true or false. In the following, $T$ is a linear transformation.*

**1.** If $T: V \rightarrow W$, then $\mathbf{R}(T)$ is a subset of $V$.

**2.** If $\mathbf{v}_1$ and $\mathbf{v}_2$ are in $\mathbf{ker}(T)$, then so is $\mathbf{v}_1 - \mathbf{v}_2$.

**3.** If $T$ is a matrix transformation with associated matrix $\mathbf{A}$, then $\mathbf{ker}(T) = \mathbf{ns}(\mathbf{A})$.

**4.** If $T: V \rightarrow W$, then $\dim(V) = \dim(\mathbf{ker}(T)) + \dim(\mathbf{R}(T))$.

**5.** If $T$ is a matrix transformation with associated $n \times n$ matrix $\mathbf{A}$, then $T$ is one-to-one if and only if $\mathbf{rref}(\mathbf{A}) = \mathbf{I}_n$.

**6.** If $T: V \rightarrow W$ is onto, then every vector in $W$ is the image of one or more vectors from $V$.

**7.** Let $V = R_{2\times2}$, $W = R^1$, and $T: V \rightarrow W$ be given by $T(\mathbf{A}) = a_{11} + a_{22}$. Then $\mathbf{ker}(T)$ is all $2 \times 2$ matrices of the form $\begin{bmatrix} 0 & b \\ c & 0 \end{bmatrix}$ where $b$ and $c$ are any real numbers.

**8.** Let $T: V \rightarrow W$ be one-to-one and $\{\mathbf{v}_1, \mathbf{v}_2, \mathbf{v}_3\}$ be a basis for $V$. It is possible that the set of images $\{T(\mathbf{v}_1), T(\mathbf{v}_2), T(\mathbf{v}_3)\}$ is linearly dependent.

**9.** $T$ is invertible if and only if $T$ is one-to-one and onto.

**10.** If $T: P_1 \rightarrow P_1$, where $T(p(x)) = p'(x)$, then $T$ is one-to-one.

## Terminology

| Kernel of $T$; $\mathbf{ker}(T)$ | Range of $T$; $\mathbf{R}(T)$ |
|---|---|
| $\dim(V) = \dim(\mathbf{ker}(T)) + \dim(\mathbf{R}(T))$ | One-to-one |
| Onto | An invertible linear transformation. |

The section deals with linear transformations and the subspaces associated with such transformations. We developed a characterization of the relationship between subspaces associated with a general linear transformation and compared them to the subspaces associated with a matrix or matrix transformation, which we studied earlier. It important that you carefully make the appropriate identifications between the subspaces associated with these closely related concepts. Advanced studies in linear algebra use linear transformation as a tool to study a variety of objects. The better we understand linear transformations, the easier it is to cope with advanced topics. The following questions and statements provide an opportunity to review these topics and relationships.

Let $T : V \rightarrow W$ be a linear transformation.

- Is $\mathbf{ker}(T)$ a subspace of $V$ or of $W$? Explain.
- Describe all the vectors in $\mathbf{ker}(T)$.
- Is $\mathbf{R}(T)$ a subspace of $V$ or of $W$? Explain.
- If $T$ is a matrix transformation with associated matrix $\mathbf{A}$, what is $\mathbf{ker}(T)$?
- Describe all the vectors in $\mathbf{R}(T)$.
- If $T$ is a matrix transformation with associated matrix $\mathbf{A}$, what is $\mathbf{R}(T)$?
- We showed that $\dim(V) = \dim(\mathbf{ker}(T)) + \dim(\mathbf{R}(T))$. Does this mean if we take a basis for $\mathbf{ker}(T)$ and a basis for $\mathbf{R}(T)$, define $S$ to be the set consisting of $\mathbf{ker}(T)$ and $\mathbf{R}(T)$, then $S$ is a basis for $V$? Explain.
- What does it mean to say that a linear transformation is one-to-one?
- Give an example of a linear transformation that is not one-to-one.
- How can we use the subspaces associated with a linear transformation to test if the transformation is one-to-one?
- What does it mean to say that a linear transformation is onto?
- Give an example of a linear transformation that is not onto.
- How can we use the subspaces associated with a linear transformation to test if the transformation is onto?
- Describe an important property of a linear transformation that is both one-to-one and onto.
- Suppose that $T_1 : V \rightarrow W$ and $T_2 : V \rightarrow W$. Describe what it means to form a linear combination of linear transformations $T_1$ and $T_2$.

## 7.3 ■ MATRIX REPRESENTATION

In Section 7.2 we saw that a general linear transformation had properties that paralleled the properties of matrix transformations and their associated matrices. Such similarities are not coincidental. In this section we shall show that if $T : V \rightarrow W$ is a linear transformation between $n$-dimensional vector space $V$ and $m$-dimensional vector space $W$, then $T$ can be "represented" as a matrix transformation between $R^n$ and $R^m$. That is, there exists a unique $m \times n$ matrix $\mathbf{A}$ such that the action that

$T$ performs on a vector $\mathbf{v}$ can be modeled by the matrix product $\mathbf{Ax}$. The vector $\mathbf{x}$ is in $R^n$ and is associated with vector $\mathbf{v}$ in a natural way. Thus, we can study linear transformations on finite-dimensional vector spaces by studying the matrix transformations that model them. We say that the matrix transformation **represents** or **acts in place of the linear transformation** $T$.

As we noted in Sections 2.3 and 5.3, if we know a basis for a vector space, then in effect we know the entire space. In Section 7.1 we noted that a linear transformation is completely determined by its action on a basis for its domain. It follows that the matrix representation of a linear transformation is highly dependent upon bases, not only for its domain $V$, but also for its range $\mathbf{R}(T)$. We can obtain different, but equivalent, matrix representations for $T$ when different bases are used. In this section we show how to use ordered bases to obtain a matrix representation of a linear transformation. In Section 7.4 we show how to modify a matrix representation if we change bases.

If $V$ is an $n$-dimensional vector space, then we know that $V$ has a basis with $n$ vectors in it. We will speak of an **ordered basis** $S = \{\mathbf{v}_1, \mathbf{v}_2, \ldots, \mathbf{v}_n\}$ for $V$; thus $S_1 = \{\mathbf{v}_2, \mathbf{v}_1, \ldots, \mathbf{v}_n\}$ or any other change in the order of the basis vectors is considered a different ordered basis for $V$.

If $S = \{\mathbf{v}_1, \mathbf{v}_2, \ldots, \mathbf{v}_n\}$ is an ordered basis for the $n$-dimensional vector space $V$, then every vector $\mathbf{v}$ in $V$ can be uniquely expressed as a linear combination of the vectors in $S$:

$$\mathbf{v} = c_1\mathbf{v}_1 + c_2\mathbf{v}_2 + \cdots + c_n\mathbf{v}_n = \sum_{j=1}^{n} c_j\mathbf{v}_j$$

where $c_1, c_2, \ldots, c_n$ are scalars. (See Section 5.3.) This motivates the following definition.

---

**Definition**  Let $S = \{\mathbf{v}_1, \mathbf{v}_2, \ldots, \mathbf{v}_n\}$ be an ordered basis for the $n$-dimensional vector space $V$ and let $\mathbf{v}$ be a vector in $V$. The **coordinates of $\mathbf{v}$ relative to the ordered basis** $S$, denoted $\mathbf{coord}_S(\mathbf{v})$, are the unique scalars $c_1, c_2, \ldots, c_n$ such that

$$\mathbf{v} = \sum_{j=1}^{n} c_j\mathbf{v}_j$$

and we write

$$\mathbf{coord}_S(\mathbf{v}) = \begin{bmatrix} c_1 \\ c_2 \\ \vdots \\ c_n \end{bmatrix}_S.$$

---

To find the coordinates of a vector relative to an ordered basis $S$, we must determine scalars so that

$$\mathbf{v} = c_1\mathbf{v}_1 + c_2\mathbf{v}_2 + \cdots + c_n\mathbf{v}_n = \sum_{j=1}^{n} c_j\mathbf{v}_j.$$

Upon substituting in the expressions for the vectors $\mathbf{v}, \mathbf{v}_1, \mathbf{v}_2, \ldots, \mathbf{v}_n$ we solve a linear system of equations for the coefficients $c_1, c_2, \ldots, c_n$. In Example 1 we show this process in $R^3$ for a particular ordered basis. In Example 2 we present a variety of coordinate computations, omitting the details of solving the linear systems involved.

EXAMPLE 1    Let $V = R^3$ and let $S = \{v_1, v_2, v_3\}$ be an ordered basis for $V$ where

$$v_1 = \begin{bmatrix} 1 \\ 1 \\ 0 \end{bmatrix}, \quad v_2 = \begin{bmatrix} 2 \\ 0 \\ 1 \end{bmatrix}, \quad \text{and} \quad v_3 = \begin{bmatrix} 0 \\ 1 \\ 2 \end{bmatrix}.$$

To find $\mathbf{coord}_S \left( \begin{bmatrix} 1 \\ 1 \\ -5 \end{bmatrix} \right)$ we determine scalars $c_1, c_2, c_3$ such that

$$c_1 \begin{bmatrix} 1 \\ 1 \\ 0 \end{bmatrix} + c_2 \begin{bmatrix} 2 \\ 0 \\ 1 \end{bmatrix} + c_3 \begin{bmatrix} 0 \\ 1 \\ 2 \end{bmatrix} = \begin{bmatrix} 1 \\ 1 \\ -5 \end{bmatrix}.$$

This expression leads to a linear system whose augmented matrix is (verify)

$$\begin{bmatrix} 1 & 2 & 0 & | & 1 \\ 1 & 0 & 1 & | & 1 \\ 0 & 1 & 2 & | & -5 \end{bmatrix}.$$

From the reduced row echelon form of this system we find that $c_1 = 3$, $c_2 = -1$, and $c_3 = -2$. (Verify.) Hence

$$\mathbf{coord}_S \left( \begin{bmatrix} 1 \\ 1 \\ -5 \end{bmatrix} \right) = \begin{bmatrix} 3 \\ -1 \\ -2 \end{bmatrix}_S.$$

■

EXAMPLE 2    Coordinate computations follow the same general pattern as illustrated in Example 1, regardless of the nature of the vector space $V$.

(a) Let $V = R^2$, and let $S = \left\{ \begin{bmatrix} 1 \\ 1 \end{bmatrix}, \begin{bmatrix} -1 \\ 2 \end{bmatrix} \right\}$ be an ordered basis for $V$. Then

$$\mathbf{coord}_S \left( \begin{bmatrix} 1 \\ 4 \end{bmatrix} \right) = \begin{bmatrix} 2 \\ 1 \end{bmatrix}_S. \quad \text{(Verify.)}$$

(b) Let $V = R^2$ and let $T = \left\{ \begin{bmatrix} 1 \\ -2 \end{bmatrix}, \begin{bmatrix} 4 \\ 1 \end{bmatrix} \right\}$ be an ordered basis for $V$. Then

$$\mathbf{coord}_T \left( \begin{bmatrix} 1 \\ 4 \end{bmatrix} \right) = \begin{bmatrix} -\frac{5}{3} \\ \frac{2}{3} \end{bmatrix}_T. \quad \text{(Verify.)}$$

(c) Let $V = P_2$ and let $S = \{t - 1, t + 1, t^2\}$ be an ordered basis for $V$. Then

$$\mathbf{coord}_S (4t^2 + 6t - 4) = \begin{bmatrix} 5 \\ 1 \\ 4 \end{bmatrix}_S. \quad \text{(Verify.)}$$

(d) Let $V = R_{2 \times 2}$ and let $S = \left\{ \begin{bmatrix} 1 & 1 \\ 0 & 0 \end{bmatrix}, \begin{bmatrix} 1 & 0 \\ 1 & 0 \end{bmatrix}, \begin{bmatrix} 1 & 0 \\ 0 & 1 \end{bmatrix}, \begin{bmatrix} 1 & 1 \\ 1 & 1 \end{bmatrix} \right\}$ be an ordered basis for $V$. Then

$$\mathbf{coord}_S \left( \begin{bmatrix} 2 & 3 \\ 0 & 1 \end{bmatrix} \right) = \begin{bmatrix} 2 \\ -1 \\ 0 \\ 1 \end{bmatrix}_S. \quad \text{(Verify.)}$$

■

In Example 2, note that in parts (a) and (b) the coordinates of the same vector relative to different ordered bases are not the same. To verify the results of Example 2 you need only form the linear combination of the basis vectors using the coordinates as coefficients and show that the result is the original vector.

The standard way we graph a vector in $R^2$ or $R^3$ is to use the natural bases and "step along" the coordinate axes according to the components of the vector. The same procedure is applied to graph a vector relative to any basis. The next example illustrates this in $R^2$ for a nonstandard basis.

EXAMPLE 3    Let

$$S = \{\mathbf{v}_1, \mathbf{v}_2\} = \left\{ \begin{bmatrix} 1 \\ 2 \end{bmatrix}, \begin{bmatrix} -2 \\ 3 \end{bmatrix} \right\}$$

be an ordered basis for $R^2$. (Verify.) To graph the vector $\mathbf{w} = \begin{bmatrix} 7 \\ 0 \end{bmatrix}$ relative to the $S$-basis we first find $\mathbf{coord}_S(\mathbf{w})$. Then we "step along" the axes determined by the $S$-basis using the components of the coordinate vector. We find that

$$\mathbf{coord}_S \left( \begin{bmatrix} 7 \\ 0 \end{bmatrix} \right) = \begin{bmatrix} 3 \\ -2 \end{bmatrix}_S . \quad \text{(Verify.)}$$

The $S$-basis is shown in Figure 1(a). Figure 1(b) shows 3 steps along the $\mathbf{v}_1$-axis and $-2$ steps along the $\mathbf{v}_2$-axis. We then use the parallelogram rule for addition of vectors to complete the formation of vector $\mathbf{w}$ in terms of the $S$-basis. This is depicted in Figure 1(c).    ■

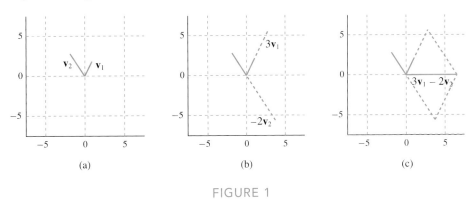

(a)                              (b)                              (c)

FIGURE 1

## A Matrix Transformation Corresponding to a Linear Transformation

The operation of determining coordinates relative to an ordered basis $S$ defines a transformation from $V$ to $R^n$; $\mathbf{coord}_S \colon V \rightarrow R^n$. We call this the **coordinate transformation** or **map**. It is easy to show that $\mathbf{coord}_S$ is a linear transformation that is both one-to-one and onto. (See Exercise 9.) Hence from Section 7.2 we have that $\mathbf{coord}_S$ is invertible. It follows that

Every vector in $V$ corresponds to a unique $n$-vector.

and

Every $n$-vector corresponds to a unique vector in $V$.

Thus, in a sense, the $n$-dimensional spaces $V$ and $R^n$ are "interchangeable"; the technical mathematical label we apply to such situations is that the spaces are **isomorphic**. (We investigate this in more detail at the end of this section.) Figure 2 indicates the manner in which we perform the interchange of information on vectors and the way that the computation of $T(\mathbf{v})$ is replaced with a matrix transformation.

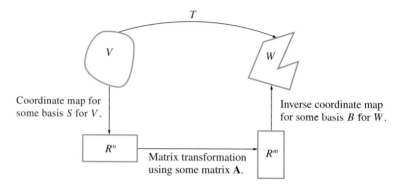

FIGURE 2

Referring to Figure 2, we have the following. (The order of operations follows the arrow pattern $\downarrow \rightarrow \uparrow$ to replace the direct use of T.)

1. Assume that $T$ is a linear transformation from $n$-dimensional space $V$ into $m$-dimensional space $W$.
2. Assume that $S = \{\mathbf{v}_1, \mathbf{v}_2, \ldots, \mathbf{v}_n\}$ is an ordered basis for $V$ and $B = \{\mathbf{w}_1, \mathbf{w}_2, \ldots, \mathbf{w}_m\}$ is an ordered basis for $W$.
3. Instead of going directly from $V$ to $W$ by the transformation $T$, we proceed as follows.
   (a) For $\mathbf{v}$ in $V$, compute its coordinates relative to the $S$-basis. (Step $\downarrow$)
   (b) Find an $m \times n$ matrix $\mathbf{A}$ that transforms the coordinates relative to $S$ in $R^n$ into coordinates relative to $B$ in $W$. (Step $\rightarrow$)
   (c) Use the inverse of the coordinate map from $W$ to $R^m$ to find the image of $\mathbf{v}$; that is $T(\mathbf{v})$. (Step $\uparrow$)

Once the ordered bases $S$ and $B$ are specified, we need only determine the matrix $\mathbf{A}$ that maps coordinates in $R^n$ relative to the $S$-basis to coordinates in $R^m$ relative to the $B$-basis. We construct the matrix $\mathbf{A}$ only once for the specified bases. We start by using the original linear transformation $T$ to find the image of each of the basis vectors in the $S$-basis, build the matrix $\mathbf{A}$, transform coordinates, and finally construct the image vector. The details for these steps are presented next.

**The $\downarrow$ step.**
Compute the coordinates of the vector $\mathbf{v}$ relative to the $S$-basis for $V$:

$$\mathbf{coord}_S(\mathbf{v}) = \begin{bmatrix} c_1 \\ c_2 \\ \vdots \\ c_n \end{bmatrix}_S.$$

This may involve the solution of a linear system of equations.

**The → step.**

This step requires that we transform $\mathbf{coord}_S(\mathbf{v})$ into $\mathbf{coord}_B(T(\mathbf{v}))$. The key is to construct a matrix so the transformation can be done with a matrix product. We first construct the matrix and then show that it will perform such a transformation.

- Compute $T(\mathbf{v}_j)$ for $j = 1, 2, \ldots, n$ directly; that is, use the definition of $T$ to compute these images.
- Express each image in terms of the $B$-basis for $W$. That is, find coefficients $a_{1j}, a_{2j}, \ldots, a_{mj}$ such that

$$T(\mathbf{v}_j) = a_{1j}\mathbf{w}_1 + a_{2j}\mathbf{w}_2 + \cdots + a_{mj}\mathbf{w}_m.$$

This is the computation of $\mathbf{coord}_B(T(\mathbf{v}_j))$ for $j = 1, 2, \ldots, n$. It requires the solution of a linear system for each basis vector in $S$.

- Define a matrix $\mathbf{A}$ containing the coordinates of the images of the $S$-basis vectors as follows: $\mathbf{col}_j(\mathbf{A}) = \mathbf{coord}_B(T(\mathbf{v}_j))$. We have

$$\mathbf{A} = \begin{bmatrix} \mathbf{coord}_B(T(\mathbf{v}_1)) & \mathbf{coord}_B(T(\mathbf{v}_2)) & \cdots & \cdots & \mathbf{coord}_B(T(\mathbf{v}_n)) \end{bmatrix}$$

$$= \begin{bmatrix} a_{11} & a_{12} & \cdots & \cdots & a_{1n} \\ a_{21} & a_{22} & \cdots & \cdots & a_{2n} \\ \vdots & \vdots & \cdots & \cdots & \vdots \\ \vdots & \vdots & \cdots & \cdots & \vdots \\ a_{m1} & a_{m2} & \cdots & \cdots & a_{mn} \end{bmatrix}.$$

- We now show that matrix $\mathbf{A}$ will transform the coordinates of $\mathbf{v}$ into the coordinates of $T(\mathbf{v})$.

$$\mathbf{coord}_B(T(\mathbf{v})) = \mathbf{coord}_B\left(T\left(\sum_{j=1}^{n} c_j\mathbf{v}_j\right)\right)$$

$$= \mathbf{coord}_B\left(\sum_{j=1}^{n} c_j T(\mathbf{v}_j)\right) \quad \text{(since } T \text{ is linear)}$$

$$= \sum_{j=1}^{n} c_j\, \mathbf{coord}_B(T(\mathbf{v}_j)) \quad \text{(since } \mathbf{coord}_B \text{ is linear)}$$

$$= \sum_{j=1}^{n} c_j \mathbf{col}_j(\mathbf{A})$$

$$= c_1 \begin{bmatrix} a_{11} \\ a_{21} \\ \vdots \\ \vdots \\ a_{m1} \end{bmatrix}_B + c_2 \begin{bmatrix} a_{12} \\ a_{22} \\ \vdots \\ \vdots \\ a_{m2} \end{bmatrix}_B + \cdots + c_n \begin{bmatrix} a_{1n} \\ a_{2n} \\ \vdots \\ \vdots \\ a_{mn} \end{bmatrix}_B$$

$$= \begin{bmatrix} a_{11} & a_{12} & \cdots & \cdots & a_{1n} \\ a_{21} & a_{22} & \cdots & \cdots & a_{2n} \\ \vdots & \vdots & \cdots & \cdots & \vdots \\ \vdots & \vdots & \cdots & \cdots & \vdots \\ a_{m1} & a_{m2} & \cdots & \cdots & a_{mn} \end{bmatrix} \begin{bmatrix} c_1 \\ c_2 \\ \vdots \\ c_n \end{bmatrix}_S = \mathbf{A} * \mathbf{coord}_S(\mathbf{v})$$

**The → step (continued)**

Thus the matrix product $\mathbf{A} * \mathbf{coord}_S(\mathbf{v})$ gives the coordinates of the image $T(\mathbf{v})$ relative to the $B$-basis in $W$.

**The ↑ step.**

Finally, we use the coordinates of $T(\mathbf{v})$ computed by the matrix transformation in the previous step to determine $T(\mathbf{v})$. If

$$\mathbf{coord}_B(T(\mathbf{v})) = \mathbf{A} * \mathbf{coord}_S(\mathbf{v}) = \begin{bmatrix} k_1 \\ k_2 \\ \vdots \\ \vdots \\ k_m \end{bmatrix}_B,$$

then we determine the image of $\mathbf{v}$ as

$$T(\mathbf{v}) = k_1\mathbf{w}_1 + k_2\mathbf{w}_2 + \cdots + k_m\mathbf{w}_m.$$

This step applies the inverse of the linear transformation $\mathbf{coord}_B$.

To summarize, the image $T(\mathbf{v})$ is, in effect, computed as the matrix product

$$\mathbf{A} * \mathbf{coord}_S(\mathbf{v}) \tag{1}$$

since this gives us the coordinates of $T(\mathbf{v})$ relative to the $B$-basis in $W$. The matrix $\mathbf{A}$ in (1), which consists of the coordinates of the images of the $S$-basis vectors, is called the **representation of $T$ relative to the ordered bases $S$ and $B$.** We also say that matrix $\mathbf{A}$ *represents $T$ with respect to $S$ and $B$.*

Figure 3 expresses the representation of $T$ in the language and notation that we have used in this section.

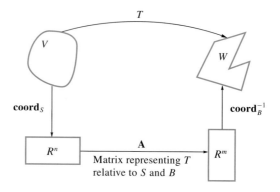

FIGURE 3

**EXAMPLE 4**    Let $T: P_2 \rightarrow P_1$ be defined by $T(p(t)) = p'(t)$ and consider ordered bases $S = \{t^2, t, 1\}$ and $B = \{t, 1\}$ for $P_2$ and $P_1$, respectively. To determine the matrix $\mathbf{A}$ representing $T$ relative to $S$ and $B$, we proceed as follows.

Compute the images of the vectors in the $S$-basis:

$$T(t^2) = 2t, \quad T(t) = 1, \quad T(1) = 0.$$

Next we find the coordinates of the images relative to the $B$-basis. (Usually we need to solve linear systems to find coordinates, but the bases here are the natural bases for the vector spaces so we can obtain the coordinates by inspection.) (Verify.)

$$\mathbf{coord}_B(T(t^2)) = \mathbf{coord}_B(2t) = \begin{bmatrix} 2 \\ 0 \end{bmatrix}_B$$

$$\mathbf{coord}_B(T(t)) = \mathbf{coord}_B(1) = \begin{bmatrix} 0 \\ 1 \end{bmatrix}_B$$

$$\mathbf{coord}_B(T(1)) = \mathbf{coord}_B(0) = \begin{bmatrix} 0 \\ 0 \end{bmatrix}_B.$$

Hence the matrix $\mathbf{A}$ representing $T$ relative to $S$ and $B$ is

$$\mathbf{A} = \begin{bmatrix} \mathbf{coord}_B(T(t^2)) & \mathbf{coord}_B(T(t)) & \mathbf{coord}_B(T(1)) \end{bmatrix} = \begin{bmatrix} 2 & 0 & 0 \\ 0 & 1 & 0 \end{bmatrix}.$$

Let $\mathbf{v} = 5t^2 - 3t + 2$. Then to compute the image $T(\mathbf{v})$ using the computational path $\downarrow \rightarrow \uparrow$ as in Figure 3, we proceed as follows.

Step $\downarrow$ :  $\mathbf{coord}_S(5t^2 - 3t + 2) = \begin{bmatrix} 5 \\ -3 \\ 2 \end{bmatrix}_S.$

Step $\rightarrow$ :  $\mathbf{A} * \begin{bmatrix} 5 \\ -3 \\ 2 \end{bmatrix}_S = \begin{bmatrix} 2 & 0 & 0 \\ 0 & 1 & 0 \end{bmatrix} \begin{bmatrix} 5 \\ -3 \\ 2 \end{bmatrix}_S = \begin{bmatrix} 10 \\ -3 \end{bmatrix}_B = \mathbf{coord}_B(T(\mathbf{v})).$

Step $\uparrow$ :  $T(\mathbf{v}) = 10t - 3.$

We can verify this result by direct computation of the derivative of $5t^2 - 3t + 2$. ■

EXAMPLE 5   Let $T$ be defined as in Example 4 and consider the ordered bases $S = \{1, t, t^2\}$ and $B = \{t, 1\}$ for $P_2$ and $P_1$, respectively. Following the procedure as illustrated in Example 4, we find that the matrix representing $T$ relative to these ordered bases $S$ and $B$ is (verify)

$$\begin{bmatrix} 0 & 0 & 2 \\ 0 & 1 & 0 \end{bmatrix}.$$

Since the only change was the order of the vectors in the basis for $P_2$, the matrix in this case has its rows reordered in the same manner. ■

EXAMPLE 6   Let $T$ be defined as in Example 4 and consider the ordered bases $S = \{t^2, 1, t\}$ and $B = \{t + 1, t - 1\}$ for $P_2$ and $P_1$, respectively. We follow the procedure as illustrated in Example 4. Compute the images of the vectors in the $S$-basis:

$$T(t^2) = 2t, \quad T(1) = 0, \quad T(t) = 1.$$

Next we find the coordinates of the image relative to the $B$-basis. Here we will need to solve linear systems to determine some of the coordinates.

To find $\mathbf{coord}_B(T(t^2)) = \mathbf{coord}_B(2t)$, we must find scalars $a_1$ and $a_2$ such that

$$a_1(t + 1) + a_2(t - 1) = 2t.$$

Expanding and equating coefficients of like powers of $t$, we obtain the linear system (verify)

$$a_1 + a_2 = 2$$
$$a_1 - a_2 = 0$$

whose solution is $a_1 = 1$ and $a_2 = 1$ (verify). Hence

$$\mathbf{coord}_B(T(t^2)) = \mathbf{coord}_B(2t) = \begin{bmatrix} 1 \\ 1 \end{bmatrix}_B.$$

It follows that

$$\mathbf{coord}_B(T(1)) = \mathbf{coord}_B(0) = \begin{bmatrix} 0 \\ 0 \end{bmatrix}_B.$$

To find $\mathbf{coord}_B(T(t)) = \mathbf{coord}_B(1)$, we must find scalars $a_1$ and $a_2$ such that

$$a_1(t+1) + a_2(t-1) = 1.$$

Expanding and equating coefficients of like powers of $t$, we obtain the linear system (verify)

$$a_1 + a_2 = 0$$
$$a_1 - a_2 = 1$$

whose solution is $a_1 = \frac{1}{2}$ and $a_2 = -\frac{1}{2}$ (verify). Hence

$$\mathbf{coord}_B(T(t^2)) = \mathbf{coord}_B(2t) = \begin{bmatrix} \frac{1}{2} \\ -\frac{1}{2} \end{bmatrix}_B.$$

Hence the matrix representing $T$ relative to the ordered bases $S$ and $B$ is

$$\mathbf{A} = \begin{bmatrix} 1 & 0 & \frac{1}{2} \\ 1 & 0 & -\frac{1}{2} \end{bmatrix}.$$

Let $\mathbf{v} = 5t^2 - 3t + 2$. Then to compute the image $T(\mathbf{v})$ using the computational path $\downarrow\rightarrow\uparrow$ as in Figure 3, we proceed as follows.

Step $\downarrow$:    $\mathbf{coord}_S(5t^2 - 3t + 2) = \begin{bmatrix} 5 \\ 2 \\ -3 \end{bmatrix}_S.$

Step $\rightarrow$:    $\mathbf{A} * \begin{bmatrix} 5 \\ 2 \\ -3 \end{bmatrix}_S = \begin{bmatrix} 1 & 0 & \frac{1}{2} \\ 1 & 0 & -\frac{1}{2} \end{bmatrix} \begin{bmatrix} 5 \\ 2 \\ -3 \end{bmatrix}_S = \begin{bmatrix} \frac{7}{2} \\ \frac{13}{2} \end{bmatrix}_B = \mathbf{coord}_B(T(\mathbf{v})).$

Step $\uparrow$:    $T(\mathbf{v}) = \frac{7}{2}(t+1) + \frac{13}{2}(t-1) = 10t - 3.$

Naturally, the image is the same as in Example 4. However, the matrix representing $T$ here is different from the representation developed in Example 4.    ■

If $T$ is a matrix transformation from $R^n$ to $R^m$ with associated matrix $\mathbf{A}$, then the matrix representation relative to the standard bases in both $R^n$ and $R^m$ is just the matrix $\mathbf{A}$. This follows immediately from the fact that the coordinates of a vector relative to the standard basis in $R^k$ is the vector itself. However, if bases other than the standard bases are used in $R^n$, $R^m$, or both, then we follow the same procedure as used in Examples 4 and 6 to determine the matrix representation of $T$ relative to the bases specified. Example 7 illustrates such a case.

EXAMPLE 7   Let $T$ be the matrix transformation from $R^3$ to $R^2$ whose associated matrix is

$$\mathbf{C} = \begin{bmatrix} 1 & 1 & 1 \\ 1 & 2 & 3 \end{bmatrix}.$$

Let

$$S = \left\{ \begin{bmatrix} 1 \\ 1 \\ 0 \end{bmatrix}, \begin{bmatrix} 0 \\ 1 \\ 1 \end{bmatrix}, \begin{bmatrix} 0 \\ 0 \\ 1 \end{bmatrix} \right\} \quad \text{and} \quad \mathbf{B} = \left\{ \begin{bmatrix} 1 \\ 2 \end{bmatrix}, \begin{bmatrix} 1 \\ 3 \end{bmatrix} \right\}$$

be ordered bases for $R^3$ and $R^2$ respectively. To determine the matrix representation of $T$ relative to these bases, we begin by computing the images of the vectors in the $S$-basis. We obtain

$$T\left( \begin{bmatrix} 1 \\ 1 \\ 0 \end{bmatrix} \right) = \mathbf{C} * \begin{bmatrix} 1 \\ 1 \\ 0 \end{bmatrix} = \begin{bmatrix} 2 \\ 3 \end{bmatrix},$$

$$T\left( \begin{bmatrix} 0 \\ 1 \\ 1 \end{bmatrix} \right) = \mathbf{C} * \begin{bmatrix} 0 \\ 1 \\ 1 \end{bmatrix} = \begin{bmatrix} 2 \\ 5 \end{bmatrix},$$

$$T\left( \begin{bmatrix} 0 \\ 0 \\ 1 \end{bmatrix} \right) = \mathbf{C} * \begin{bmatrix} 0 \\ 0 \\ 1 \end{bmatrix} = \begin{bmatrix} 1 \\ 3 \end{bmatrix}.$$

Next we compute the coordinates of these images relative to the $B$-basis. We must solve the linear systems

$$a_1 \begin{bmatrix} 1 \\ 2 \end{bmatrix} + a_2 \begin{bmatrix} 1 \\ 3 \end{bmatrix} = \begin{bmatrix} 2 \\ 3 \end{bmatrix},$$

$$a_1 \begin{bmatrix} 1 \\ 2 \end{bmatrix} + a_2 \begin{bmatrix} 1 \\ 3 \end{bmatrix} = \begin{bmatrix} 2 \\ 5 \end{bmatrix},$$

$$a_1 \begin{bmatrix} 1 \\ 2 \end{bmatrix} + a_2 \begin{bmatrix} 1 \\ 3 \end{bmatrix} = \begin{bmatrix} 1 \\ 3 \end{bmatrix}.$$

This can be done simultaneously by constructing the partitioned matrix with three augmented columns, as shown next:

$$\begin{bmatrix} 1 & 1 & | & 2 & | & 2 & | & 1 \\ 2 & 3 & | & 3 & | & 5 & | & 3 \end{bmatrix}.$$

Computing the reduced row echelon form of this matrix, we get (verify)

$$\begin{bmatrix} 1 & 0 & | & 3 & | & 1 & | & 0 \\ 0 & 1 & | & -1 & | & 1 & | & 1 \end{bmatrix}.$$

The last three columns of this matrix are the desired coordinate vectors of the $S$-basis relative to the $B$-basis. Hence the matrix $\mathbf{A}$ representing $T$ relative to the $S$ and $B$ bases is

$$\mathbf{A} = \begin{bmatrix} 3 & 1 & 0 \\ -1 & 1 & 1 \end{bmatrix}. \qquad \blacksquare$$

From Example 7 we see that if $T : R^n \to R^m$ is a matrix transformation with associated matrix $\mathbf{C}$, then a computationally efficient way to compute a matrix representation $\mathbf{A}$ of $T$ relative to ordered bases $S = \{\mathbf{v}_1, \mathbf{v}_2, \ldots, \mathbf{v}_n\}$ for $R^n$ and $B = \{\mathbf{w}_1, \mathbf{w}_2, \ldots, \mathbf{w}_m\}$ for $R^m$ is as follows:

Form the matrix

$$\begin{bmatrix} \mathbf{w}_1 & \mathbf{w}_2 & \cdots & \mathbf{w}_m \mid T(\mathbf{v}_1) \mid T(\mathbf{v}_2) \mid \cdots \mid T(\mathbf{v}_n) \end{bmatrix}$$
$$= \begin{bmatrix} \mathbf{w}_1 & \mathbf{w}_2 & \cdots & \mathbf{w}_m \mid \mathbf{Cv}_1 \mid \mathbf{Cv}_2 \mid \cdots \mid \mathbf{Cv}_n \end{bmatrix}$$

and then compute its reduced row echelon form. The matrix $\mathbf{A}$ consists of the last $n$ columns of the reduced row echelon form.

Suppose that $T: V \to W$ is a linear transformation and that $\mathbf{A}$ is the matrix representing $T$ relative to ordered bases $S$ and $B$ for $V$ and $W$, respectively. Information about transformation $T$ can be obtained from matrix $\mathbf{A}$ in the following sense:

- $\ker(T)$ is found by determining $\mathbf{ns(A)}$, which provides the coordinates of the vectors in $V$ relative to the $S$-basis whose images are $\mathbf{0}_W$.
- $\mathbf{R}(T)$ is found by determining a basis for $\mathbf{col(A)}$. Then the coordinates of all vectors in $\mathbf{R}(T)$ relative to the $B$-basis are linear combinations of the basis for $\mathbf{col(A)}$.
- $T$ is one-to-one if and only if $\mathbf{ns(A)} = \{\mathbf{0}\}$.
- $T$ is onto if and only if $\dim(\mathbf{col(A)}) = \dim(W)$.
- $T$ is invertible if and only if $\mathbf{A}$ is invertible.

Each of these statements requires verification; however, we will just accept them as following naturally from the representation process.

## Linear Operators

If $V = W$, then the linear transformation $T: V \to W$ is called a **linear operator**. For linear operators we introduce another concept, eigenvalues and eigenvectors. A nonzero vector $\mathbf{v}$ in $V$ is called an **eigenvector of linear operator** $T$ provided that there exists a scalar $\lambda$ such that $T(\mathbf{v}) = \lambda\mathbf{v}$. The scalar $\lambda$ is called the **eigenvalue corresponding to eigenvector v**. We illustrate this in Example 8.

EXAMPLE 8    Let $T: P_1 \to P_1$ be the linear operator given by

$$T(at + b) = (-8a + 5b)t + (-10a + 7b).$$

A nonzero vector $\mathbf{v} = at + b$ is an eigenvector of $T$ provided that there is a scalar $\lambda$ such that $T(\mathbf{v}) = \lambda\mathbf{v}$. Proceeding directly from the definition of $T$ leads to the equation

$$(-8a + 5b)t + (-10a + 7b) = \lambda(at + b).$$

Equating coefficients of like powers of $t$, we obtain the linear system

$$-8a + 5b = \lambda a$$
$$-10a + 7b = \lambda b.$$

Converting this system to matrix form, we have

$$\begin{bmatrix} -8 & 5 \\ -10 & 7 \end{bmatrix} \begin{bmatrix} a \\ b \end{bmatrix} = \lambda \begin{bmatrix} a \\ b \end{bmatrix}$$

which is a matrix eigen problem for matrix $\begin{bmatrix} -8 & 5 \\ -10 & 7 \end{bmatrix}$. The eigenpairs of this matrix are

$$\left(2, \begin{bmatrix} 1 \\ 2 \end{bmatrix}\right) \quad \text{and} \quad \left(-3, \begin{bmatrix} 1 \\ 1 \end{bmatrix}\right). \quad \text{(Verify.)}$$

It follows that the entries of the eigenvectors provide values for $a$ and $b$; hence we have that $t + 2$ and $t + 1$ are eigenvectors of the linear operator $T$ with eigenvalues 2 and $-3$, respectively. ■

The linear operator $T$ in Example 8 has a matrix representation $\mathbf{A}$ relative to any pair of ordered bases $S$ and $B$, and that matrix is square. Note that if $S = B = \{t, 1\}$, then $T(t) = -8t - 10$ and $T(1) = 5t + 7$. Hence it follows that

$$\mathbf{coord}_B(T(t)) = \begin{bmatrix} -8 \\ -10 \end{bmatrix}_B \quad \text{and} \quad \mathbf{coord}_B(T(1)) = \begin{bmatrix} 5 \\ 7 \end{bmatrix}_B$$

and so the matrix representing $T$ relative to $S$ and $B$ is

$$\mathbf{A} = \begin{bmatrix} -8 & 5 \\ -10 & 7 \end{bmatrix}.$$

This is the same matrix that appeared in Example 8 from equating the coefficients of like powers of $t$. The eigen equation for matrix $\mathbf{A}$ is $\mathbf{Ax} = \lambda\mathbf{x}$ or, equivalently, $(\mathbf{A} - \lambda\mathbf{I}_n)\mathbf{x} = \mathbf{0}$. Thus an eigenvector of $\mathbf{A}$ corresponding to eigenvalue $\lambda$ is a nonzero vector in $\mathbf{ker}(\mathbf{A} - \lambda\mathbf{I}_n)$. The eigenvectors of the matrix representation $\mathbf{A}$ provide coordinates of the eigenvectors of linear transformation $T$ relative to natural basis $B$. *The eigenvalues of the matrix representation are also the eigenvalues of the linear operator $T$.* We are led to the following general statement which we make without verification.

*If $T$ is a linear operator, an eigenvector of a matrix representation of $T$ gives the coordinates of an eigenvector of $T$.*

It also follows that there is a diagonal matrix that represents the linear operator $T$ if and only if a matrix representing $T$ is similar to a diagonal matrix. Thus we have another instance of parallel behavior for matrices and linear transformations.

Examples 9 and 10 use calculus. In calculus it is shown that the derivative operator satisfies the criteria of a linear transformation. Here we investigate eigenpairs of the derivative operator for different function spaces. These examples touch briefly on the topic of differential equations, which is an area that can use eigen information to solve a variety of problems and provide a characterization of a number of important classes of functions.

EXAMPLE 9    Let $V$ be the class of real-valued functions that have derivatives of all orders and are defined for all real values. It can be shown that $V$ is a vector space. Define $T : V \to V$ by $T(\mathbf{f}(t)) = \mathbf{f}'(t)$, the derivative of $\mathbf{f}$. Then $(\lambda, \mathbf{f}(t))$ is an eigenpair of $T$ provided that $T(\mathbf{f}(t)) = \lambda\mathbf{f}(t)$. That is, $\mathbf{f}'(t) = \lambda\mathbf{f}(t)$. We can determine eigenpairs as follows.

$$\mathbf{f}'(t) = \frac{d\mathbf{f}(t)}{dt} = \lambda\mathbf{f}(t)$$

is equivalent to

$$\frac{d\mathbf{f}(t)}{\mathbf{f}(t)} = \lambda\, dt.$$

Integrating both sides of this last expression gives $\ln(\mathbf{f}(t)) = \lambda t + C$, where $C$ is an arbitrary constant. Solving for $\mathbf{f}(t)$, we get $\mathbf{f}(t) = e^{\lambda t + C}$. Using properties of the exponential function, we can express this as $\mathbf{f}(t) = K e^{\lambda t}$, where $K = e^C$. This says that eigenvectors (they are called eigenfunctions when $V$ is a function space) of $T$ are scalar multiples of exponential functions, $e^{\lambda t}$ and the corresponding eigenvalue is $\lambda$. Thus we see that $T$ has infinitely many eigenvectors, one for each real value $\lambda$.    ■

**EXAMPLE 10**    Let $W$ be the real vector space span$\{\sin(t), \cos(t)\}$, which is a subspace of the vector space $V$ from Example 9. Let $T : W \rightarrow W$ be defined by $T(\mathbf{f}(t)) = \mathbf{f}'(t)$, the derivative of $\mathbf{f}$. As in Example 9, $T$ is a linear transformation. A nonzero vector in $V$ is a eigenvector (eigenfunction) of $T$ provided that $T(\mathbf{f}(t)) = \lambda \mathbf{f}(t)$ for some real scalar $\lambda$. To determine an eigenpair, let $\mathbf{f}(t)$ be a vector in span$\{\sin(t), \cos(t)\}$; that is, $\mathbf{f}(t) - c_1 \sin(t) + c_2 \cos(t)$. Then

$$T(\mathbf{f}(t)) = T(c_1 \sin(t) + c_2 \cos(t)) = c_1 \cos(t) - c_2 \sin(t)$$

and $T(\mathbf{f}(t)) = \lambda \mathbf{f}(t)$ implies that

$$c_1 \cos(t) - c_2 \sin(t) = \lambda(c_1 \sin(t) + c_2 \cos(t)).$$

Rearranging this equation, we obtain

$$(c_1 - \lambda c_2) \cos(t) - (\lambda c_1 + c_2) \sin(t) = 0.$$

It follows that

$$c_1 - \lambda c_2 = 0 \tag{2}$$

and

$$\lambda c_1 + c_2 = 0. \tag{3}$$

This pair of equations can be expressed in matrix form as

$$\begin{bmatrix} 1 & -\lambda \\ \lambda & 1 \end{bmatrix} \begin{bmatrix} c_1 \\ c_2 \end{bmatrix} = \begin{bmatrix} 0 \\ 0 \end{bmatrix}.$$

There will be a nontrivial solution only if

$$\det\left( \begin{bmatrix} 1 & -\lambda \\ \lambda & 1 \end{bmatrix} \right) = \lambda^2 + 1 = 0.$$

However, there is no real value $\lambda$ that makes this true. Hence $T$ has no real eigenpairs. If we consider $W$ a complex vector space, then we have eigenvalues $\lambda = i$ and $\lambda = -i$, where $i$ is the complex unit $\sqrt{-1}$. To determine the corresponding complex eigenvectors, we proceed as follows.

**Case $\lambda = i$:**    From (2) we have that $c_1 = i c_2$. Hence $c_2$ can be any nonzero complex number. Let $c_2 = 1$, then $c_1 = i$ so $\mathbf{f}(t) = i \sin(t) + \cos(t)$ is an eigenfunction of $T$ with corresponding eigenvalue $\lambda = i$. [The expression $\cos(t) + i \sin(t)$ is one of the most famous in mathematics and appears in Euler's formula, $e^{it} = \cos(t) + i \sin(t)$, which is valid for any real number t.]

**Case $\lambda = -i$:**    Using steps similar to those in the previous case, we can show that $f(t) = -i \sin(t) + \cos(t)$ is an eigenfunction of $T$ with corresponding eigenvalue $\lambda = -i$. (See Exercise 22.)

Thus when $W$ is considered as a complex vector space, $T$ has exactly two eigenpairs.    ■

Exercises 23 through 27 contain more eigenproblems for linear transformations.

## A Special Family of Linear Transformations (Optional)

We have seen that $R^n$ can model an $n$-dimensional vector space $V$. The linear transformation $\mathbf{coord}_S \colon V \to R^n$, where $S$ is an ordered basis for $V$, provides a correspondence that lets us manipulate coordinate vectors in the same manner we would manipulate vectors in $V$. This suggests that from an algebraic point of view, $V$ and $R^n$ behave rather similarly. The foundation for this behavior stems from the properties of **coord**. Here we clarify such ideas and define a family of linear transformations that share the same characteristics in terms of their algebraic impact.

Let $V$ be an $n$-dimensional vector space containing vectors $\mathbf{u}$ and $\mathbf{v}$. Let $S = \{\mathbf{v}_1, \mathbf{v}_2, \ldots, \mathbf{v}_n\}$ be an ordered basis for $V$; then we can write $\mathbf{u}$ and $\mathbf{v}$ uniquely in terms of the $S$-basis:

$$\mathbf{u} = \sum_{j=1}^{n} c_j \mathbf{v}_j \quad \text{and} \quad \mathbf{v} = \sum_{j=1}^{n} k_j \mathbf{v}_j.$$

The linear transformation $\mathbf{coord}_S$ tells us that

$$\mathbf{coord}_S(\mathbf{u} + \mathbf{v}) = \mathbf{coord}_S(\mathbf{u}) + \mathbf{coord}_S(\mathbf{v})$$

which implies that we can find the coordinates of the sum $\mathbf{u} + \mathbf{v}$ by adding the coordinates of vectors $\mathbf{u}$ and $\mathbf{v}$. Hence addition in $V$ is equivalent to addition of coordinate vectors. Similarly, for any scalar $t$,

$$\mathbf{coord}_S(t\mathbf{u}) = t\,\mathbf{coord}_S(\mathbf{u})$$

which implies that we can find the coordinates of the scalar multiple $t\mathbf{u}$ by multiplying the coordinates of vector $\mathbf{u}$ by the scalar $t$. Hence scalar multiplication in $V$ is equivalent to scalar multiplication of coordinate vectors. Naturally we can combine these properties for linear combinations of vectors.

The glue that holds together the algebraic modeling of $n$-dimensional vector space $V$ by the algebra of $n$-vectors is that **coord** is not just a linear transformation, but that it is onto and one-to-one. In fact, any such linear transformation can provide a means to model one vector space by another. Thus we are led to the following definition.

---

**Definition**    Let $V$ and $W$ be real vector spaces and $T$ a linear transformation mapping $V$ to $W$ that is one-to-one and onto. $T$ is called an **isomorphism**[1] and we say that $V$ **is isomorphic to** $W$.

---

To say that vector spaces $V$ and $W$ are isomorphic means that there exists an isomorphism between the spaces.

It follows that $\mathbf{coord}_S$ is an isomorphism and that the $n$-dimensional vector space $V$ is isomorphic to $R^n$. Another example of an isomorphism is given by $R^2$, the vector space of directed line segments emanating from the origin and the vector space of all ordered pairs of real numbers. There is a corresponding isomorphism for $R^3$. (See the discussion in Section 1.4 about the interchangeability of the concepts of point, vector, and matrix.)

Next we list important properties of isomorphisms and isomorphic spaces.

---

[1] Isomorphism is from the Greek *isos*, meaning "the same;" and *morphos*, meaning "structure."

> **Properties of Isomorphisms**
> (a) Every vector space $V$ is isomorphic to itself.
> (b) If $V$ is isomorphic to $W$, then $W$ is isomorphic to V.
> (c) If $U$ is isomorphic to $V$ and $V$ is isomorphic to $W$, then $U$ is isomorphic to $W$.
> (d) Two finite-dimensional vector spaces are isomorphic if and only if their dimensions are equal.

We leave the verification of these properties to Exercisess 28 through 31.

We have shown in this section that the idea of a $n$-dimensional vector space, which at first seems fairly abstract, is not so mysterious. In fact, such a vector space does not differ much from $R^n$ in its algebraic behavior.

## EXERCISES 7.3

*In Exercises 1 through 4, compute the coordinates of the vector* **v** *with respect to the ordered basis S for V.*

**1.** $V = R^2$, $S = \left\{ \begin{bmatrix} 1 \\ 2 \end{bmatrix}, \begin{bmatrix} -3 \\ 1 \end{bmatrix} \right\}$, $\mathbf{v} = \begin{bmatrix} -1 \\ 12 \end{bmatrix}$.

**2.** $V = R^3$, $S = \left\{ \begin{bmatrix} 1 \\ -1 \\ 0 \end{bmatrix}, \begin{bmatrix} 0 \\ 1 \\ 0 \end{bmatrix}, \begin{bmatrix} 1 \\ 0 \\ 2 \end{bmatrix} \right\}$, $\mathbf{v} = \begin{bmatrix} 2 \\ -1 \\ -2 \end{bmatrix}$.

**3.** $V = P_2$, $S = \{t^2 - t + 1, t + 1, t^2 + 1\}$, $\mathbf{v} = 4t^2 - 2t + 3$.

**4.** $V = M_{2\times 2}$,

$S = \left\{ \begin{bmatrix} 1 & -1 \\ 0 & 0 \end{bmatrix}, \begin{bmatrix} 0 & 1 \\ 1 & 0 \end{bmatrix}, \begin{bmatrix} 1 & 0 \\ 0 & -1 \end{bmatrix}, \begin{bmatrix} 1 & 0 \\ -1 & 0 \end{bmatrix} \right\}$,

$\mathbf{v} = \begin{bmatrix} 1 & 3 \\ -2 & 2 \end{bmatrix}$.

*In Exercises 5 through 8, compute the vector* **v** *whose coordinate vector* **coord**$_S$(**v**) *is given with respect to the ordered basis S for V.*

**5.** $V = R^2$, $S = \left\{ \begin{bmatrix} 2 \\ 1 \end{bmatrix}, \begin{bmatrix} -1 \\ 1 \end{bmatrix} \right\}$, $\mathbf{coord}_S(\mathbf{v}) = \begin{bmatrix} -1 \\ 2 \end{bmatrix}_S$.

**6.** $V = P_1$, $S = \{t, 2t - 1\}$, $\mathbf{coord}_S(\mathbf{v}) = \begin{bmatrix} -2 \\ 3 \end{bmatrix}_S$.

**7.** $V = P_2$, $S = \{t^2 + 1, t + 1, t^2 + t\}$, $\mathbf{coord}_S(\mathbf{v}) = \begin{bmatrix} 3 \\ -1 \\ -2 \end{bmatrix}_S$.

**8.** $V = M_{2\times 2}$,

$S = \left\{ \begin{bmatrix} -1 & 0 \\ 1 & 0 \end{bmatrix}, \begin{bmatrix} 2 & 2 \\ 0 & 1 \end{bmatrix}, \begin{bmatrix} 1 & 2 \\ -1 & 3 \end{bmatrix}, \begin{bmatrix} 0 & 0 \\ 2 & 3 \end{bmatrix} \right\}$,

$\mathbf{coord}_S(\mathbf{v}) = \begin{bmatrix} 2 \\ 1 \\ -1 \\ 3 \end{bmatrix}_S$.

**9.** Let $V$ be an $n$-dimensional vector space with ordered basis $S$. Show that transformation $\mathbf{coord}_S : V \to R^n$ is a linear transformation that is one-to-one and onto.

**10.** Explain why $\mathbf{coord}_S : V \to R^n$ from Exercise 9 is invertible.

**11.** Let $T : R^2 \to R^2$ be a linear transformation given by

$$T\left( \begin{bmatrix} v_1 \\ v_2 \end{bmatrix} \right) = \begin{bmatrix} v_1 + 2v_2 \\ 2v_1 - v_2 \end{bmatrix}.$$

(a) Let $S = B = $ the natural basis for $R^2$. Find the matrix representation of $T$ relative to bases $S$ and $B$.

(b) Let $S = $ the natural basis for $R^2$ and $B = \left\{ \begin{bmatrix} -1 \\ 2 \end{bmatrix}, \begin{bmatrix} 2 \\ 0 \end{bmatrix} \right\}$. Find the matrix representation of $T$ relative to bases $S$ and $B$.

**12.** Let $T : R^4 \to R^3$ be a linear transformation given by

$$T\left( \begin{bmatrix} v_1 \\ v_2 \\ v_3 \\ v_4 \end{bmatrix} \right) = \begin{bmatrix} v_1 \\ v_2 + v_3 \\ v_3 + v_4 \end{bmatrix}.$$

Let

$$S = \left\{ \begin{bmatrix} 1 \\ 0 \\ 0 \\ 1 \end{bmatrix}, \begin{bmatrix} 0 \\ 0 \\ 0 \\ 1 \end{bmatrix}, \begin{bmatrix} 1 \\ 1 \\ 0 \\ 0 \end{bmatrix}, \begin{bmatrix} 0 \\ 1 \\ 1 \\ 0 \end{bmatrix} \right\}$$

be an ordered basis for $R^4$ and

$$B = \left\{ \begin{bmatrix} 1 \\ 1 \\ 0 \end{bmatrix}, \begin{bmatrix} 0 \\ 1 \\ 0 \end{bmatrix}, \begin{bmatrix} 1 \\ 0 \\ 1 \end{bmatrix} \right\}$$

an ordered basis for $R^3$. Find the matrix representation of $T$ relative to bases $S$ and $B$.

**13.** Let $T : R_{2\times 2} \to R_{2\times 2}$ be a linear transformation given by

$$T(A) = \begin{bmatrix} 1 & 2 \\ 3 & 4 \end{bmatrix} A$$

for $\mathbf{A}$ in $R_{2 \times 2}$. Let

$$S = \left\{ \begin{bmatrix} 1 & 0 \\ 0 & 0 \end{bmatrix}, \begin{bmatrix} 0 & 1 \\ 0 & 0 \end{bmatrix}, \begin{bmatrix} 0 & 0 \\ 1 & 0 \end{bmatrix}, \begin{bmatrix} 0 & 0 \\ 0 & 1 \end{bmatrix} \right\}$$

and

$$B = \left\{ \begin{bmatrix} 1 & 0 \\ 0 & 1 \end{bmatrix}, \begin{bmatrix} 1 & 1 \\ 0 & 0 \end{bmatrix}, \begin{bmatrix} 1 & 0 \\ 1 & 0 \end{bmatrix}, \begin{bmatrix} 0 & 1 \\ 0 & 0 \end{bmatrix} \right\}$$

be ordered bases for $R_{2 \times 2}$. Find the matrix representation of $T$ relative to bases $S$ and $B$.

**14.** Let $T : R_{2 \times 2} \to R_{2 \times 2}$ be a linear transformation given by

$$T(\mathbf{A}) = \begin{bmatrix} 1 & 2 \\ 3 & 4 \end{bmatrix} \mathbf{A}$$

for $\mathbf{A}$ in $R_{2 \times 2}$. Let

$$S = \left\{ \begin{bmatrix} -1 & 0 \\ 1 & 0 \end{bmatrix}, \begin{bmatrix} 2 & 2 \\ 0 & 1 \end{bmatrix}, \begin{bmatrix} 1 & 2 \\ -1 & 3 \end{bmatrix}, \begin{bmatrix} 0 & 0 \\ 2 & 3 \end{bmatrix} \right\}$$

and

$$B = \left\{ \begin{bmatrix} 1 & 0 \\ 0 & 1 \end{bmatrix}, \begin{bmatrix} 1 & 1 \\ 0 & 0 \end{bmatrix}, \begin{bmatrix} 1 & 0 \\ 1 & 0 \end{bmatrix}, \begin{bmatrix} 0 & 1 \\ 0 & 0 \end{bmatrix} \right\}$$

be ordered bases for $R_{2 \times 2}$. Find the matrix representation of $T$ relative to bases $S$ and $B$.

**15.** Let $T : P_2 \to P_3$ be a linear transformation given by

$$T(at^2 + bt + c) = at^3 + ct^2 + b.$$

Let $S = \{t^2, t, 1\}$ and $B = \{t^3, t^2, t, 1\}$ be ordered bases for $P_2$ and $P_3$, respectively. Find the matrix representation of $T$ relative to bases $S$ and $B$.

**16.** Let $T : P_3 \to P_3$ be a linear transformation given by

$$T(at^3 + bt^2 + ct + d) = at^3 + ct^2 + b.$$

Let $S = \{t^3, t^2, t, 1\}$ and $B = \{t^3, t^2, t, 1\}$ be ordered bases for $P_3$.

(a) Find the matrix representation of $T$ relative to bases $S$ and $B$.

(b) Is $T$ one-to-one?

(c) Is $T$ onto?

**17.** Let $T : P_2 \to R_{2 \times 2}$ be a linear transformation given by

$$T(at^2 + bt + c) = \begin{bmatrix} a - 2b & 0 \\ b + c & a + c \end{bmatrix}.$$

Let $S = \{t^2, t, 1\}$ and

$$B = \left\{ \begin{bmatrix} 1 & 0 \\ 0 & 1 \end{bmatrix}, \begin{bmatrix} 1 & 1 \\ 0 & 0 \end{bmatrix}, \begin{bmatrix} 1 & 0 \\ 1 & 0 \end{bmatrix}, \begin{bmatrix} 0 & 1 \\ 0 & 0 \end{bmatrix} \right\}$$

be ordered bases for $P_2$ and $R_{2 \times 2}$, respectively.

(a) Find the matrix representation of $T$ relative to bases $S$ and $B$.

(b) Is $T$ one-to-one?

(c) Is $T$ onto?

**18.** Let $T : R_{2 \times 2} \to R_{2 \times 2}$ be a linear transformation given by

$$T(\mathbf{A}) = \begin{bmatrix} 1 & 2 \\ 3 & 4 \end{bmatrix} \mathbf{A} - \mathbf{A} \begin{bmatrix} 1 & 2 \\ 3 & 4 \end{bmatrix}$$

for $\mathbf{A}$ in $R_{2 \times 2}$. Let

$$S = \left\{ \begin{bmatrix} -1 & 0 \\ 1 & 0 \end{bmatrix}, \begin{bmatrix} 2 & 2 \\ 0 & 1 \end{bmatrix}, \begin{bmatrix} 1 & 2 \\ -1 & 3 \end{bmatrix}, \begin{bmatrix} 0 & 0 \\ 2 & 3 \end{bmatrix} \right\}$$

and

$$B = \left\{ \begin{bmatrix} 1 & 0 \\ 0 & 1 \end{bmatrix}, \begin{bmatrix} 1 & 1 \\ 0 & 0 \end{bmatrix}, \begin{bmatrix} 1 & 0 \\ 1 & 0 \end{bmatrix}, \begin{bmatrix} 0 & 1 \\ 0 & 0 \end{bmatrix} \right\}$$

be ordered bases for $R_{2 \times 2}$.

(a) Find the matrix representation of $T$ relative to bases $S$ and $B$.

(b) Is $T$ one-to-one?

(c) Is $T$ onto?

**19.** Let $V$ be the subspace of differentiable functions that has basis $S = \{\sin(t), \cos(t)\}$ and let $T : V \to V$ be the linear operator defined by $T(\mathbf{f}(t)) = \mathbf{f}'(t)$. Find the matrix representation of $T$ relative to the $S$-basis. (The $S$-basis is to be used in both the domain and range.)

**20.** Let $V$ be the subspace of differentiable functions that has basis $S = \{1, t, \sin(t), \cos(t), t \sin(t), t \cos(t)\}$ and let $T : V \to V$ be the linear operator defined by $T(\mathbf{f}(t)) = \mathbf{f}'(t)$. Find the matrix representation of $T$ relative to the $S$-basis. (The $S$-basis is to be used in both the domain and range.)

**21.** Let $V$ be the subspace of differentiable functions that has basis $S = \{1, t, e^t, te^t\}$ and let $T : V \to V$ be the linear operator defined by $T(\mathbf{f}(t)) = \mathbf{f}'(t)$. Find the matrix representation of $T$ relative to the $S$-basis. (The $S$-basis is to be used in both the domain and range.)

**22.** In Example 10, for the case of the eigenvalue $\lambda = -i$ determine the corresponding eigenfunction. Use arguments similar to those given for the case $\lambda = i$.

**23.** Let $W$ be the real vector space span$\{\sin(t), \cos(t)\}$. $T : W \to W$ is given by $T(\mathbf{f}(t)) = \mathbf{f}''(t)$, the second derivative of $\mathbf{f}$. It follows from calculus that $T$ is a linear transformation. A nonzero vector in $W$ is a eigenvector (eigenfunction) of $T$ provided $T(\mathbf{f}(t)) = \lambda \mathbf{f}(t)$ for some real scalar $\lambda$. Find all the eigenpairs of $T$.

**24.** Let $V$ be the vector space of Example 9. $T : V \to V$ is given by $T(\mathbf{f}(t)) = \mathbf{f}'(t)$. Determine all the eigenpairs of $T$. (*Hint*: $\mathbf{f}(t)$ is an eigenvector of $T$ provided $\mathbf{f}'(t)$ is a scalar multiple of $\mathbf{f}(t)$; that is, $\mathbf{f}'(t) = k\mathbf{f}(t)$, $k$ real. Use calculus to solve for $\mathbf{f}(t)$ in terms of $k$.)

**25.** Let $V$ be the real vector space span$\{1, e^{2t}, e^{-2t}\}$. $T: V \to V$ is given by $T(\mathbf{f}(t)) = \mathbf{f}'(t)$. Determine all the eigenpairs of $T$.

**26.** Let $V$ be the real vector space of all $2 \times 2$ lower triangular matrices and $T: V \to V$ be given by

$$T\left(\begin{bmatrix} a & 0 \\ b & c \end{bmatrix}\right) = \begin{bmatrix} a+c & 0 \\ a+b & a+b \end{bmatrix}.$$

(a) Show that $T$ is a linear transformation.

(b) Find all eigenpairs of $T$ using the definition of $T$ directly.

(c) Find a matrix representation of $T$ relative to the basis

$$S = \left\{\begin{bmatrix} 1 & 0 \\ 0 & 0 \end{bmatrix}, \begin{bmatrix} 0 & 0 \\ 1 & 0 \end{bmatrix}, \begin{bmatrix} 0 & 0 \\ 0 & 1 \end{bmatrix}\right\}.$$

(d) Find the eigenvalues and eigenvectors of the matrix representation from part (c) and show how they relate to the eigenpairs found in part (b).

**27.** Let $V$ be the real vector space of all $2 \times 2$ matrices and $T: V \to V$ be given by

$$T\left(\begin{bmatrix} a & b \\ c & d \end{bmatrix}\right) = \begin{bmatrix} 0 & a+b \\ c+d & 0 \end{bmatrix}.$$

(a) Show that $T$ is a linear transformation.

(b) Find all eigenpairs of $T$ using the definition of $T$ directly.

(c) Find a matrix representation of $T$ relative to the basis

$$\left\{\begin{bmatrix} 1 & 0 \\ 0 & 0 \end{bmatrix}, \begin{bmatrix} 0 & 1 \\ 0 & 0 \end{bmatrix}, \begin{bmatrix} 0 & 0 \\ 1 & 0 \end{bmatrix}, \begin{bmatrix} 0 & 0 \\ 0 & 1 \end{bmatrix}\right\}.$$

*In Exercises* 28 *through* 31, *assume that the vector spaces are finite dimensional.*

**28.** Show that every vector space $V$ is isomorphic to itself.

**29.** Show that if $V$ is isomorphic to $W$, then $W$ is isomorphic to $V$.

**30.** Show that if $U$ is isomorphic to $V$ and $V$ is isomorphic to $W$, then $U$ is isomorphic to $W$.

**31.** Show that two finite-dimensional vector spaces are isomorphic if and only if their dimensions are equal.

## True/False Review Questions

*Determine whether each of the following statements is true or false.*

**1.** In $R^n$, the coordinates of a vector relative to the natural basis are its entries.

**2.** If $V$ is a real vector space with $\dim(V) = k$ and $S$ is a basis for $V$, then $\mathbf{coord}_S$ is a linear transformation from $V$ to $R^k$.

**3.** Let $V = R^2$ and $S = \left\{\begin{bmatrix} 0 \\ 1 \end{bmatrix}, \begin{bmatrix} 1 \\ 1 \end{bmatrix}\right\}$ be a basis for $V$. Then $\mathbf{coord}_S\left(\begin{bmatrix} 5 \\ 4 \end{bmatrix}\right) = \begin{bmatrix} 5 \\ -1 \end{bmatrix}$.

**4.** If $V$ is a vector space with basis $S$ and $\mathbf{v}_1 \neq \mathbf{v}_2$, it is possible that $\mathbf{coord}_S(\mathbf{v}_1) = \mathbf{coord}_S(\mathbf{v}_2)$.

**5.** If $\mathbf{A}$ is the matrix representing linear transformation $T: V \to W$ relative to bases $S$ and $B$ respectively, then the columns of $\mathbf{A}$ are the coordinates of the images of the vectors in $S$.

## Terminology

| | |
|---|---|
| Ordered basis | Coordinates |
| Coordinate transformation | Matrix representation of a linear transformation |
| Linear operator | Eigenpairs of a linear operator |
| An isomorphism | Isomorphic vector spaces |

The main idea of this section is that a linear transformation $T$ between $n$-dimensional real vector space $V$ and $m$-dimensional real vector space $W$ can be represented by a matrix transformation between $R^n$ and $R^m$. In effect, this means that the action of $T$ on vectors from $V$ can be modeled (or reproduced) by a matrix multiplication. When we have enough information to describe completely $V$ and $W$, then we can construct the corresponding matrix transformation. This matrix model can provide information about the kernel and range of $T$ by using familiar matrix tools rather than the original definition of $T$. Thus we are led to an important instance of parallel behavior for matrices and linear transformations. It also suggests that models that mimic other mathematical concepts may provide indirect routes to determine properties of the concept. Respond to the following questions and statements to review the material in this section.

- Why do we use ordered bases in this section?
- Explain how to find the coordinates of a vector.
- If vector space $V$ is $n$-dimensional, how does the coordinate transformation on $V$ link it to $R^n$? Explain.
- What special properties does the coordinate transformation possess?
- If $T: V \rightarrow W$ is a linear transformation from $n$-dimensional space $V$ to $m$-dimensional space $W$, explain how we construct a matrix representation of $T$. Give details.
- Once we have a matrix representation of a linear transformation $T: V \rightarrow W$, how can we use the matrix to determine information about $T$? Be specific.
- If $T: V \rightarrow V$ is a linear operator, explain what we mean by an eigenpair of $T$.
- If matrix $\mathbf{A}$ is a matrix representation of a linear operator $T$, why must $\mathbf{A}$ be square? Explain how the eigenpairs of $\mathbf{A}$ are related to the eigenpairs of $T$.
- When is a linear transformation an isomorphism?
- Name a vector space $W$ that is isomorphic to $n$-dimensional vector space $V$ ($W \neq V$). Provide a detailed description of a correspondence that shows that they are isomorphic.
- Explain in everyday language what it means to say that two vector spaces are isomorphic.

## 7.4 ■ CHANGE OF BASIS (OPTIONAL)

In Section 7.3 we saw that a linear transformation $T$ of an $n$-dimensional vector space $V$ into an $m$-dimensional vector space $W$ could be "represented" by a matrix. For ordered bases $S$ and $B$ in $V$ and $W$, respectively, we found a matrix transformation from $R^n$ to $R^m$ that transformed coordinates of vectors in $V$ relative to the $S$-basis into coordinates of image vectors in $W$ relative to the $B$-basis. In this section we see how the matrix representing $T$ changes when the bases for $V$ and $W$ change. (We rely heavily on the concepts developed in Sections 7.2 and 7.3.)

Before beginning the development of the ideas in this section, let's elaborate on the situations we will encounter. Suppose we have computed the matrix representation of $T$ relative to ordered bases $S$ and $B$ of $V$ and $W$, respectively. To emphasize the dependence of the matrix representation on the bases involved we will denote the matrix by $\mathbf{A}_{B \leftarrow S}$. (In Section 7.3 we just used the notation $\mathbf{A}$.) As we saw in Section 7.3, there is significant work involved to obtain the matrix representation, and hence if we now want to change bases in $V$ and/or $W$ it seems reasonable to modify $\mathbf{A}_{B \leftarrow S}$ rather than start computations all over again. In addition, we will show that there is information to be gained about matrix representations of linear transformations when we use the approach developed in this section.

From Section 7.3 we know that for $\mathbf{v}$ in $V$ the matrix product

$$\mathbf{A}_{B \leftarrow S} * \mathbf{coord}_S(\mathbf{v}) = \mathbf{coord}_B(T(\mathbf{v})) \qquad (1)$$

provides the means to transform coordinates of $\mathbf{v}$ into coordinates of $T(\mathbf{v})$. Suppose that instead of having coordinates of $\mathbf{v}$ relative to the $S$-basis we have coordinates of **v** relative to a different basis for $V$, call it $S'$. Before computing the coordinates of the image $T(\mathbf{v})$, we will need to convert $\mathbf{coord}_{S'}(\mathbf{v})$ into $\mathbf{coord}_S(\mathbf{v})$. Ideally we would like to determine a matrix $\mathbf{P}_{S \leftarrow S'}$ such that

$$\mathbf{P}_{S \leftarrow S'} * \mathbf{coord}_{S'}(\mathbf{v}) = \mathbf{coord}_S(\mathbf{v}). \tag{2}$$

We say that $\mathbf{P}_{S \leftarrow S'}$ causes a "transition" from coordinates in the $S'$-basis to coordinates in the $S$-basis. The matrix $\mathbf{P}_{S \leftarrow S'}$ is called the **transition matrix from coordinates in the $S'$-basis to coordinates in the $S$-basis**. (We shorten this and just say that $\mathbf{P}_{S \leftarrow S'}$ is a **transition matrix**.)

We now show how to determine a transition matrix. Let $S = \{\mathbf{v}_1, \mathbf{v}_2, \ldots, \mathbf{v}_n\}$ and $S' = \{\mathbf{u}_1, \mathbf{u}_2, \ldots, \mathbf{u}_n\}$ be ordered bases for $V$. Since $S'$ is a basis for $V$, for any vector **v** in $V$ there exist unique scalars $c_1, c_2, \ldots, c_n$ such that $\mathbf{v} = \sum_{j=1}^{n} c_j \mathbf{u}_j$; that is, $\mathbf{c} = \mathbf{coord}_{S'}(\mathbf{v})$. It follows that in terms of the $S$-basis

$$\mathbf{coord}_S(\mathbf{v}) = \mathbf{coord}_S \left( \sum_{j=1}^{n} c_j \mathbf{u}_j \right)$$

$$= \sum_{j=1}^{n} c_j \, \mathbf{coord}_S(\mathbf{u}_j) \quad \text{(since } \mathbf{coord} \text{ is a linear transformation)}$$

$$= \begin{bmatrix} \mathbf{coord}_S(\mathbf{u}_1) & \mathbf{coord}_S(\mathbf{u}_2) & \cdots & \mathbf{coord}_S(\mathbf{u}_n) \end{bmatrix} * \mathbf{c}$$

(recognizing the previous summation as a linear combination of columns, we can rewrite it as a matrix-vector product)

$$= \begin{bmatrix} \mathbf{coord}_S(\mathbf{u}_1) & \mathbf{coord}_S(\mathbf{u}_2) & \cdots & \mathbf{coord}_S(\mathbf{u}_n) \end{bmatrix} * \mathbf{coord}_{S'}(\mathbf{v}).$$

The matrix

$$\begin{bmatrix} \mathbf{coord}_S(\mathbf{u}_1) & \mathbf{coord}_S(\mathbf{u}_2) & \cdots & \mathbf{coord}_S(\mathbf{u}_n) \end{bmatrix}$$

**is the transition matrix from the $S'$-basis to the $S$-basis**. We write

$$\mathbf{P}_{S \leftarrow S'} = \begin{bmatrix} \mathbf{coord}_S(\mathbf{u}_1) & \mathbf{coord}_S(\mathbf{u}_2) & \cdots & \mathbf{coord}_S(\mathbf{u}_n) \end{bmatrix}.$$

We make the following observations about transition matrices:

- Transition matrices are square.
- A transition matrix defines a matrix transformation from $R^n$ to $R^n$, when $\dim(V) = n$.
- Recall that the linear transformation **coord** is one-to-one and onto, hence invertible. (See Section 7.3.) Since $S$ is a basis for $V$ the set $\{\mathbf{coord}_S(\mathbf{u}_j), j = 1, 2, \ldots, n\}$ is a basis for $R^n$. (See Section 7.3.) It follows then that the columns of $\mathbf{P}_{S \leftarrow S'}$ are linearly independent so $\mathbf{P}_{S \leftarrow S'}$ is a nonsingular (or invertible) matrix.
- Since $\mathbf{P}_{S \leftarrow S'}$ is invertible, the transition matrix from the $S$-basis to the $S'$-basis, denoted $\mathbf{P}_{S' \leftarrow S}$, is $(\mathbf{P}_{S' \leftarrow S})^{-1}$. We see this from the matrix equation in (2).

EXAMPLE 1    Let $V = R^3$ and let $S = \{\mathbf{v}_1, \mathbf{v}_2, \mathbf{v}_3\}$ and $S' = \{\mathbf{u}_1, \mathbf{u}_2, \mathbf{u}_3\}$ be ordered bases for $V$ where

$$\mathbf{v}_1 = \begin{bmatrix} 2 \\ 0 \\ 1 \end{bmatrix}, \quad \mathbf{v}_2 = \begin{bmatrix} 1 \\ 2 \\ 0 \end{bmatrix}, \quad \mathbf{v}_3 = \begin{bmatrix} 1 \\ 1 \\ 1 \end{bmatrix}$$

and

$$\mathbf{u}_1 = \begin{bmatrix} 6 \\ 3 \\ 3 \end{bmatrix}, \quad \mathbf{u}_2 = \begin{bmatrix} 4 \\ -1 \\ 3 \end{bmatrix}, \quad \mathbf{u}_3 = \begin{bmatrix} 5 \\ 5 \\ 2 \end{bmatrix}.$$

To determine the transition matrix from the $S'$-basis to the $S$-basis we compute

$$\mathbf{P}_{S \leftarrow S'} = \begin{bmatrix} \mathbf{coord}_S(\mathbf{u}_1) & \mathbf{coord}_S(\mathbf{u}_2) & \mathbf{coord}_S(\mathbf{u}_3) \end{bmatrix}.$$

We need to determine scalars $k_1$, $k_2$, $k_3$ such that $k_1 \mathbf{v}_1 + k_2 \mathbf{v}_2 + k_3 \mathbf{v}_3 = \mathbf{u}_1$ and similarly for $\mathbf{u}_2$ and then $\mathbf{u}_3$. As noted in Section 7.3, we can combine the solution of this set of three linear systems into the following computation:

$$\mathbf{rref}\left(\begin{bmatrix} \mathbf{v}_1 & \mathbf{v}_2 & \mathbf{v}_3 \mid \mathbf{u}_1 \mid \mathbf{u}_2 \mid \mathbf{u}_3 \end{bmatrix}\right).$$

We have

$$\mathbf{rref}\left(\begin{bmatrix} 2 & 1 & 1 & 6 & 4 & 5 \\ 0 & 2 & 1 & 3 & -1 & 5 \\ 1 & 0 & 1 & 3 & 3 & 2 \end{bmatrix}\right) = \begin{bmatrix} 1 & 0 & 0 & 2 & 2 & 1 \\ 0 & 1 & 0 & 1 & -1 & 2 \\ 0 & 0 & 1 & 1 & 1 & 1 \end{bmatrix}.$$

Then the transition matrix $\mathbf{P}_{S \leftarrow S'}$ consists of the last three columns of the reduced row echelon form:

$$\mathbf{P}_{S \leftarrow S'} = \begin{bmatrix} 2 & 2 & 1 \\ 1 & -1 & 2 \\ 1 & 1 & 1 \end{bmatrix}.$$

    ■

To illustrate the use of a transition matrix, let

$$\mathbf{v} = \begin{bmatrix} 4 \\ -9 \\ 5 \end{bmatrix}.$$

The coordinates of $\mathbf{v}$ relative to the $S'$-basis in Example 1 are

$$\mathbf{coord}_{S'}(\mathbf{v}) = \begin{bmatrix} 1 \\ 2 \\ -2 \end{bmatrix}_{S'} \quad \text{(verify)}.$$

The coordinates of $\mathbf{v}$ relative to the $S$-basis in Example 1 are given by (verify)

$$\mathbf{P}_{S \leftarrow S'} * \mathbf{coord}_{S'}(\mathbf{v}) = \begin{bmatrix} 2 & 2 & 1 \\ 1 & -1 & 2 \\ 1 & 1 & 1 \end{bmatrix} \begin{bmatrix} 1 \\ 2 \\ -2 \end{bmatrix}_{S'} = \begin{bmatrix} 4 \\ -5 \\ 1 \end{bmatrix}_S = \mathbf{coord}_S(\mathbf{v}).$$

To check this result we form the linear combination $4\mathbf{v}_1 - 5\mathbf{v}_2 + \mathbf{v}_3$ and show that the result is $\mathbf{v}$. (Verify.)

Next we depict situations involving the matrix representation of a linear transformation $T$ combined with change of basis scenarios by using arrow diagrams like those in Figures 1 and 2 in Section 7.3. Our basic assumption is that $T : V \rightarrow W$ is a linear transformation, which is represented by the matrix $\mathbf{A}_{B \leftarrow S}$ where $S$ and $B$ are ordered bases for $V$ and $W$, respectively. We represent this in Figure 1.

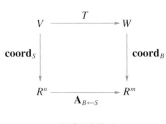

FIGURE 1

**Case 1:** The vectors in $V$ are specified in coordinates relative to basis $S'$. Hence we need a change of basis from $S'$ to $S$ in order to compute coordinates of $T(\mathbf{v})$ relative to the $B$-basis. We construct matrix $\mathbf{P}_{S \leftarrow S'}$ and combine it with matrix $\mathbf{A}_{B \leftarrow S}$ to determine the matrix $\mathbf{A}_{B \leftarrow S'}$ . We have

$$\mathbf{coord}_B(T(\mathbf{v})) = \mathbf{A}_{B \leftarrow S} * \mathbf{coord}_S(\mathbf{v}) = \mathbf{A}_{B \leftarrow S} * \mathbf{P}_{S \leftarrow S'} * \mathbf{coord}_{S'}(\mathbf{v}).$$

Hence the matrix representing $T$ relative to the $S'$-basis in $V$ and the $B$-basis in $W$ is

$$\mathbf{A}_{B \leftarrow S'} = \mathbf{A}_{B \leftarrow S} * \mathbf{P}_{S \leftarrow S'}.$$

Note that the transition matrix is used first and is then followed by the matrix representation of $T$. We represent this by the arrow diagram shown in Figure 2.

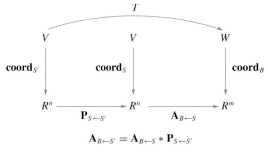

FIGURE 2

**Case 2:** The vectors in $V$ are specified in coordinates relative to the $S$-basis. We use the matrix $\mathbf{A}_{B \leftarrow S}$ to obtain coordinates of images relative to the $B$-basis, but we want these coordinates converted to coordinates relative to a $B'$-basis for $W$. Hence we need a change of basis from $B$ to $B'$. We construct transition matrix $\mathbf{Q}_{B' \leftarrow B}$ and combine it with matrix $\mathbf{A}_{B \leftarrow S}$ to determine the matrix $\mathbf{A}_{B' \leftarrow S}$ . (We have used $\mathbf{Q}$ instead of $\mathbf{P}$ merely to distinguish between change of bases in $V$ versus change of basis in $W$.) We have

$$\mathbf{coord}_{B'}(T(\mathbf{v})) = \mathbf{Q}_{B' \leftarrow B} * \mathbf{coord}_B(T(\mathbf{v})) = \mathbf{Q}_{B' \leftarrow B} * \mathbf{A}_{B \leftarrow S} * \mathbf{coord}_S(\mathbf{v}).$$

Hence the matrix representing $T$ relative to the $S$-basis in $V$ and the $B'$-basis in $W$ is

$$\mathbf{A}_{B' \leftarrow S} = \mathbf{Q}_{B' \leftarrow B} * \mathbf{A}_{B \leftarrow S}.$$

Note that the matrix representation of $T$ is used first and is then followed by the transition matrix. We represent this by the arrow diagram shown in Figure 3.

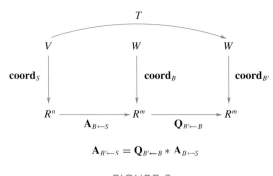

FIGURE 3

**Case 3:**　Now we consider a combination of cases 1 and 2. The vectors in $V$ are specified in coordinates relative to basis $S'$. Hence we need a change of basis from $S$ to $S'$ in order to compute coordinates of $T(\mathbf{v})$ relative to the $B$-basis. However, we also want these coordinates converted to coordinates relative to a $B'$-basis for $W$. Hence we need a change of basis from $B$ to $B'$. We construct transition matrix $\mathbf{P}_{S \leftarrow S'}$ as in Case 1 and transition matrix $\mathbf{Q}_{B' \leftarrow B}$ as in Case 2. Then the matrix representing $T$ relative to the $S'$-basis for $V$ and the $B'$-basis for $W$ is $\mathbf{Q}_{B' \leftarrow B} * \mathbf{A}_{B \leftarrow S} * \mathbf{P}_{S \leftarrow S'}$. We represent this by the arrow diagram shown in Figure 4.

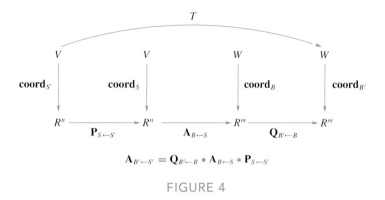

$$\mathbf{A}_{B' \leftarrow S'} = \mathbf{Q}_{B' \leftarrow B} * \mathbf{A}_{B \leftarrow S} * \mathbf{P}_{S \leftarrow S'}$$

FIGURE 4

EXAMPLE 2　Let $T: R^3 \rightarrow R^2$ be defined by

$$T\left(\begin{bmatrix} v_1 \\ v_2 \\ v_3 \end{bmatrix}\right) = \begin{bmatrix} v_1 + v_3 \\ v_2 - v_3 \end{bmatrix}.$$

Let $S$ be the natural basis for $R^3$ and $B$ the natural basis for $R^2$:

$$S = \left\{ \begin{bmatrix} 1 \\ 0 \\ 0 \end{bmatrix}, \begin{bmatrix} 0 \\ 1 \\ 0 \end{bmatrix}, \begin{bmatrix} 0 \\ 0 \\ 1 \end{bmatrix} \right\} \quad \text{and} \quad B = \left\{ \begin{bmatrix} 1 \\ 0 \end{bmatrix}, \begin{bmatrix} 0 \\ 1 \end{bmatrix} \right\}.$$

The matrix representing $T$ with respect to the $S$- and $B$-bases is easily determined. We recognize $T$ as a matrix transformation, and since $S$ and $B$ are the natural bases it follows that (verify)

$$\mathbf{A}_{B \leftarrow S} = \begin{bmatrix} 1 & 0 & 1 \\ 0 & 1 & -1 \end{bmatrix}.$$

(a)　Suppose that the vectors in $V$ are specified in coordinates relative to basis $S'$, where

$$S' = \left\{ \begin{bmatrix} 1 \\ 1 \\ 0 \end{bmatrix}, \begin{bmatrix} 0 \\ 1 \\ 1 \end{bmatrix}, \begin{bmatrix} 0 \\ 0 \\ 1 \end{bmatrix} \right\}.$$

Then to compute the coordinates of images relative to the $B$-basis we need to first construct the transition matrix $\mathbf{P}_{S \leftarrow S'}$. We have

$$\mathbf{P}_{S \leftarrow S'} = \begin{bmatrix} \mathbf{coord}_S \left( \begin{bmatrix} 1 \\ 1 \\ 0 \end{bmatrix} \right) & \mathbf{coord}_S \left( \begin{bmatrix} 0 \\ 1 \\ 1 \end{bmatrix} \right) & \mathbf{coord}_S \left( \begin{bmatrix} 0 \\ 0 \\ 1 \end{bmatrix} \right) \end{bmatrix}$$

but since $S$ is the natural basis the coordinates of the vectors in $S'$ are the vectors themselves. Hence

$$\mathbf{P}_{S\leftarrow S'} = \begin{bmatrix} 1 & 0 & 0 \\ 1 & 1 & 0 \\ 0 & 1 & 1 \end{bmatrix}.$$

It follows that the matrix representing $T$ relative to the $S'$- and $B$-bases is $\mathbf{A}_{B\leftarrow S} * \mathbf{P}_{S\leftarrow S'}$ and so

$$\mathbf{A}_{B\leftarrow S'} = \mathbf{A}_{B\leftarrow S} * \mathbf{P}_{S\leftarrow S'} = \begin{bmatrix} 1 & 0 & 1 \\ 0 & 1 & -1 \end{bmatrix}\begin{bmatrix} 1 & 0 & 0 \\ 1 & 1 & 0 \\ 0 & 1 & 1 \end{bmatrix} = \begin{bmatrix} 1 & 1 & 1 \\ 1 & 0 & -1 \end{bmatrix}.$$

(b) To find $\mathbf{A}_{B'\leftarrow S}$, the matrix representation of $T$ with respect to the $S$-basis for $R^3$ and the $B'$-basis for $R^2$ where

$$B' = \left\{ \begin{bmatrix} 1 \\ 1 \end{bmatrix}, \begin{bmatrix} 1 \\ 3 \end{bmatrix} \right\}$$

we determine the transition matrix from the $B$-basis to the $B'$-basis, $\mathbf{Q}_{B'\leftarrow B}$. We have

$$\mathbf{Q}_{B'\leftarrow B} = \left[ \mathbf{coord}_{B'}\left( \begin{bmatrix} 1 \\ 0 \end{bmatrix} \right) \quad \mathbf{coord}_{B'}\left( \begin{bmatrix} 0 \\ 1 \end{bmatrix} \right) \right]$$

which we can obtain from the results of

$$\mathbf{rref}\left( \begin{bmatrix} 1 & 1 & 1 & 0 \\ 1 & 3 & 0 & 1 \end{bmatrix} \right) = \begin{bmatrix} 1 & 0 & \frac{3}{2} & -\frac{1}{2} \\ 0 & 1 & -\frac{1}{2} & \frac{1}{2} \end{bmatrix}.$$

Hence $\mathbf{Q}_{B'\leftarrow B} = \begin{bmatrix} \frac{3}{2} & -\frac{1}{2} \\ -\frac{1}{2} & \frac{1}{2} \end{bmatrix}$ and

$$\mathbf{A}_{B'\leftarrow S} = \mathbf{Q}_{B'\leftarrow B} * \mathbf{A}_{B\leftarrow S} = \begin{bmatrix} \frac{3}{2} & -\frac{1}{2} \\ -\frac{1}{2} & \frac{1}{2} \end{bmatrix}\begin{bmatrix} 1 & 0 & 1 \\ 0 & 1 & -1 \end{bmatrix}$$

$$= \begin{bmatrix} \frac{3}{2} & -\frac{1}{2} & 2 \\ -\frac{1}{2} & \frac{1}{2} & -1 \end{bmatrix}.$$

(c) The matrix representation of $T$ relative to $S'$ and $B'$ is given by

$$\mathbf{A}_{B'\leftarrow S'} = \mathbf{Q}_{B'\leftarrow B} * \mathbf{A}_{B\leftarrow S} * \mathbf{P}_{S\leftarrow S'}$$

$$= \begin{bmatrix} \frac{3}{2} & -\frac{1}{2} \\ -\frac{1}{2} & \frac{1}{2} \end{bmatrix}\begin{bmatrix} 1 & 0 & 1 \\ 0 & 1 & -1 \end{bmatrix}\begin{bmatrix} 1 & 0 & 0 \\ 1 & 1 & 0 \\ 0 & 1 & 1 \end{bmatrix}$$

$$= \begin{bmatrix} 1 & \frac{3}{2} & 2 \\ 0 & -\frac{1}{2} & -1 \end{bmatrix}.$$

Portions of the computations needed in Example 2 were made easier by the fact that $S$ and $B$ were the natural bases. When $S$ and $B$ are not the natural bases, the same procedure is followed but there are more linear systems of equations that must be solved. Following the procedures specified by Cases 1 through 3 and using the associated arrow diagrams provides a step-by-step approach for determining the matrix representation of $T$ when a change of basis is involved. The key to these calculations is the modularity provided by the analysis we have developed.

A special case of the preceding development occurs when $V = W$, that is, $T$ is a linear operator. We have the following situation:

> $T$ is a linear operator from $n$-dimensional vector space $V$ into itself. $S$ is an ordered basis for $V$ and we have the matrix representation $\mathbf{A}_{S \leftarrow S}$. If we are given coordinates for vectors in $V$ in terms of another ordered basis $S'$ and want the coordinates of images in terms of $S'$ also, then what is the matrix representation of $T$?

To answer this we first look at the arrow diagram in Figure 5, which depicts the situation.

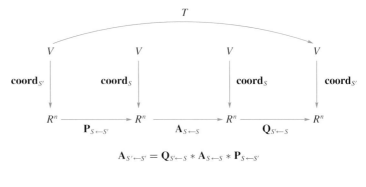

$$\mathbf{A}_{S' \leftarrow S'} = \mathbf{Q}_{S' \leftarrow S} * \mathbf{A}_{S \leftarrow S} * \mathbf{P}_{S \leftarrow S'}$$

FIGURE 5

We see that $\mathbf{P}_{S \leftarrow S'}$ is the transition matrix from $S'$ to $S$ and that $\mathbf{Q}_{S' \leftarrow S}$ is the transition matrix from $S$ to $S'$. It follows by a previous observation that $\mathbf{Q}_{S' \leftarrow S} = (\mathbf{P}_{S \leftarrow S'})^{-1}$ and hence

$$\mathbf{A}_{S' \leftarrow S'} = (\mathbf{P}_{S \leftarrow S'})^{-1} * \mathbf{A}_{S \leftarrow S} * \mathbf{P}_{S \leftarrow S'}.$$

In this case it is customary to drop the subscripts and say that the matrix representation of linear operator $T$ relative to the $S'$-basis is $\mathbf{P}^{-1}\mathbf{A}\mathbf{P}$. From Section 4.3 we recognize that this relationship between matrices $\mathbf{A}_{S' \leftarrow S'}$ and $\mathbf{A}_{S \leftarrow S}$ implies that they are similar matrices. Thus we have

> Any two matrix representations of the linear operator $T : V \rightarrow V$ are similar.

It is also true that if two matrices are similar, then they represent the same linear transformation relative to different ordered bases. (We will not pursue the verification of this statement.) These two statements tell us what matrices can represent a particular linear operator and how such matrices are related. This further emphasizes a result we noted in Section 7.3, and we can restate it as follows:

> $T : V \rightarrow V$ can be represented by a diagonal matrix if and only if there exists a basis $S$ for $V$ such that $\mathbf{A}_{S \leftarrow S}$ is diagonal.

## EXERCISES 7.4

**1.** Let $T: P_2 \rightarrow P_1$ be the linear transformation defined in Example 4 of Section 7.3. For ordered bases $S = \{t^2, t, 1\}$ and $B = \{t, 1\}$ we showed that the matrix representing $T$ was

$$\mathbf{A}_{B \leftarrow S} = \begin{bmatrix} 2 & 0 & 0 \\ 0 & 1 & 0 \end{bmatrix}.$$

(a) Let $S' = \{t^2 + t, t + 2, t - 1\}$ be another ordered basis for $P_2$. Determine the matrix representation $\mathbf{A}_{B \leftarrow S'}$.

(b) Let $B' = \{2t + 1, t + 2\}$ be another ordered basis for $P_1$. Determine the matrix representation $\mathbf{A}_{B' \leftarrow S}$.

(c) Determine the matrix representation $\mathbf{A}_{B' \leftarrow S'}$ where $S'$ and $B'$ are the bases in parts (a) and (b) respectively.

**2.** Let $T: R^3 \rightarrow R^2$ be the linear transformation defined in Example 7 of Section 7.3. For ordered bases

$$S = \left\{ \begin{bmatrix} 1 \\ 1 \\ 0 \end{bmatrix}, \begin{bmatrix} 0 \\ 1 \\ 1 \end{bmatrix}, \begin{bmatrix} 0 \\ 0 \\ 1 \end{bmatrix} \right\}$$

and

$$B = \left\{ \begin{bmatrix} 1 \\ 2 \end{bmatrix}, \begin{bmatrix} 1 \\ 3 \end{bmatrix} \right\}$$

we showed that the matrix representing $T$ was

$$\mathbf{A}_{B \leftarrow S} = \begin{bmatrix} 3 & 1 & 0 \\ -1 & 1 & 1 \end{bmatrix}.$$

(a) Let

$$S' = \left\{ \begin{bmatrix} 1 \\ 0 \\ 0 \end{bmatrix}, \begin{bmatrix} 0 \\ 1 \\ 0 \end{bmatrix}, \begin{bmatrix} 0 \\ 0 \\ 1 \end{bmatrix} \right\}$$

be another ordered basis for $R^3$. Determine the matrix representation $\mathbf{A}_{B \leftarrow S'}$.

(b) Let

$$B' = \left\{ \begin{bmatrix} 1 \\ 0 \end{bmatrix}, \begin{bmatrix} 0 \\ 1 \end{bmatrix} \right\}$$

be another ordered basis for $R^2$. Determine the matrix representation $\mathbf{A}_{B' \leftarrow S}$.

(c) Determine the matrix representation $\mathbf{A}_{B' \leftarrow S'}$ where $S'$ and $B'$ are the bases in parts (a) and (b), respectively.

**3.** Let $T: R^3 \rightarrow R^4$ be the linear transformation defined by

$$T \left( \begin{bmatrix} v_1 \\ v_2 \\ v_3 \end{bmatrix} \right) = \begin{bmatrix} v_1 \\ v_1 + v_2 \\ v_1 + v_2 + v_3 \\ v_2 + v_3 \end{bmatrix}.$$

(a) Let $S$ be the natural basis for $R^3$ and $B$ the natural basis for $R^4$. Find $\mathbf{A}_{B \leftarrow S}$.

(b) Let

$$S' = \left\{ \begin{bmatrix} 1 \\ 0 \\ 1 \end{bmatrix}, \begin{bmatrix} 0 \\ 1 \\ 1 \end{bmatrix}, \begin{bmatrix} 1 \\ 1 \\ 0 \end{bmatrix} \right\}$$

be another ordered basis for $R^3$. Determine the matrix representation $\mathbf{A}_{B \leftarrow S'}$.

(c) Let

$$B' = \left\{ \begin{bmatrix} 1 \\ 1 \\ 0 \\ 1 \end{bmatrix}, \begin{bmatrix} 2 \\ 0 \\ 1 \\ 1 \end{bmatrix}, \begin{bmatrix} 0 \\ 1 \\ 1 \\ 0 \end{bmatrix}, \begin{bmatrix} -1 \\ 0 \\ 0 \\ 2 \end{bmatrix} \right\}$$

be another ordered basis for $R^4$. Determine the matrix representation $\mathbf{A}_{B' \leftarrow S}$.

(d) Determine the matrix representation $\mathbf{A}_{B' \leftarrow S'}$ where $S'$ and $B'$ are the bases in parts (b) and (c), respectively.

**4.** Let $T: P_1 \rightarrow P_2$ be the linear transformation defined by $T(at + b) = (a + b)t^2 + (a - b)t$.

(a) Let $S$ be the natural basis for $P_1$ and $B$ the natural basis for $P_2$. Find $\mathbf{A}_{B \leftarrow S}$.

(b) Let $S' = \{t + 2, t - 3\}$ be another ordered basis for $P_1$. Determine the matrix representation $\mathbf{A}_{B \leftarrow S'}$.

(c) Let $B' = \{2t - 1, t + 1, t^2 + 2t\}$ be another ordered basis for $P_2$. Determine the matrix representation $\mathbf{A}_{B' \leftarrow S}$.

(d) Determine the matrix representation $\mathbf{A}_{B' \leftarrow S'}$ where $S'$ and $B'$ are the bases in parts (b) and (c), respectively.

**5.** Let $\mathbf{C} = \begin{bmatrix} 3 & 1 \\ 0 & 2 \end{bmatrix}$. Define linear operator $T: R^2 \rightarrow R^2$ by $T(\mathbf{v}) = \mathbf{C}\mathbf{v}$ for $\mathbf{v}$ in $R^2$.

(a) Let $S = B =$ natural basis for $R^2$. Find $\mathbf{A}_{B \leftarrow S}$.

(b) Let $S' = B' = \left\{ \begin{bmatrix} 1 \\ -1 \end{bmatrix}, \begin{bmatrix} 1 \\ 0 \end{bmatrix} \right\}$ be another ordered basis for $R^2$. Determine the matrix representation $\mathbf{A}_{B' \leftarrow S'}$.

(c) What is the relationship between $\mathbf{A}_{B \leftarrow S}$ and $\mathbf{A}_{B' \leftarrow S'}$?

**6.** Let

$$\mathbf{C} = \begin{bmatrix} 1 & 0 & 0 \\ 0 & -1 & 0 \\ 0 & 0 & 2 \end{bmatrix}.$$

Define linear operator $T: R^3 \rightarrow R^3$ by $T(\mathbf{v}) = \mathbf{C}\mathbf{v}$ for $\mathbf{v}$ in $R^3$.

(a) Let $S = B =$ natural basis for $R^3$. Find $\mathbf{A}_{B \leftarrow S}$.

(b) Let

$$S' = B' = \left\{ \begin{bmatrix} 1 \\ 0 \\ 0 \end{bmatrix}, \begin{bmatrix} 1 \\ 1 \\ 1 \end{bmatrix}, \begin{bmatrix} 1 \\ 0 \\ 1 \end{bmatrix} \right\}$$

be another ordered basis for $R^3$. Determine the matrix representation $\mathbf{A}_{B' \leftarrow S'}$.

(c) What is the relationship between $\mathbf{A}_{B \leftarrow S}$ and $\mathbf{A}_{B' \leftarrow S'}$?

7. Let $T: P_2 \to P_2$ be the linear operator defined by

$$T(at^2 + bt + c) = (2a + b)t^2 + (3b + c)t.$$

(a) Let $S = B =$ natural basis for $P_2$. Find $\mathbf{A}_{B \leftarrow S}$.

(b) Use the matrix from part (a) (and other computations) to show why there is a diagonal matrix that represents $T$.

8. Any two matrix representations of a linear operator $T: V \to V$ are similar matrices. Algebraically this means that for $n \times n$ matrices $\mathbf{E}$ and $\mathbf{H}$ representing $T$ with respect to different bases there exists a nonsingular matrix $\mathbf{P}$ such that $\mathbf{E} = \mathbf{PHP}^{-1}$. (See Section 4.3.) Thus show each of the following.

(a) $\mathbf{E}$ is nonsingular if and only if $\mathbf{H}$ is nonsingular. (This implies that every matrix representing an invertible linear operator is nonsingular.)

(b) $\mathbf{x}$ in $R^n$ is in $\mathbf{ns}(\mathbf{E})$ if and only if $\mathbf{P}^{-1}\mathbf{x}$ is in $\mathbf{ns}(\mathbf{H})$.

(c) $\mathbf{y}$ in $R^n$ is in $\mathbf{col}(\mathbf{E})$ if and only if $\mathbf{P}^{-1}\mathbf{y}$ is in $\mathbf{col}(\mathbf{H})$.

## True/False Review Questions

*Determine whether each of the following statements is true or false.*

1. A transition matrix is used to change coordinates relative to one basis into coordinates relative to another basis.

2. A transition matrix can be singular.

3. For a linear operator $T: V \to V$, if $\mathbf{A}$ and $\mathbf{B}$ are both matrix representations of $T$, then $\mathbf{A}$ and $\mathbf{B}$ are similar matrices.

4. The columns of $\mathbf{P}_{S \leftarrow S'}$ are the coordinates of vectors in basis $S'$ relative to basis $S$.

5. There exist linear transformations between finite-dimensional vector spaces that have no matrix representation.

## Terminology

| Change of basis | Transition matrix |
| --- | --- |

This section continues the study of the matrix representation of a linear transformation. Here we assumed that we had a matrix $\mathbf{A}$ representing linear transformation $T: V \to W$ with respect to a basis $S$ for $V$ and $B$ for $W$. We want to determine the matrix representing $T$ when we change bases by "transforming" matrix $\mathbf{A}$. The case in which $T$ is a linear operator is also considered and related to previous ideas. Respond to the following questions and statements to review the material from this section.

- Why do we use ordered bases when we determine a matrix representing a linear transformation?

- Suppose we have computed the matrix representation of $T$ relative to ordered bases $S$ and $B$ for $V$ and $W$, respectively, and it is denoted $\mathbf{A}_{B \leftarrow S}$. Explain how this matrix is used to transform information about vectors in $V$ to information about their images in $W$.

- If $S'$ is another basis for $V$, explain the role of the transition matrix $\mathbf{P}_{S \leftarrow S'}$.

- Given that we have $\mathbf{A}_{B \leftarrow S}$ and $\mathbf{P}_{S \leftarrow S'}$, how do we find $\mathbf{A}_{B \leftarrow S'}$?

- Given that we have $\mathbf{A}_{B \leftarrow S}$ and we want to use a different basis $B'$ for $W$, explain how we proceed to compute $\mathbf{A}_{B' \leftarrow S}$.

- Given that we have $\mathbf{A}_{B \leftarrow S}$ and we want to use a different basis $S'$ for $V$ and a different basis $B'$ for $W$, explain how we proceed to compute $\mathbf{A}_{B' \leftarrow S'}$.

- If $T$ is a linear operator, then what is the relationship between matrix representations $\mathbf{A}_{B \leftarrow S}$ and $\mathbf{A}_{B' \leftarrow S'}$ for $T$?

## 7.5 ■ FRACTALS: AN APPLICATION OF TRANSFORMATIONS (OPTIONAL)

The previous sections of this chapter have dealt with properties of linear transformations and their representation as matrix transformations. In this section we investigate a particular class of nonlinear transformations that have become exceedingly important in recent years. We will restrict our attention to mappings from $R^2 \to R^2$, but the general theory goes well beyond $R^2$. The ideas in this section have been applied to such diverse areas as the generation of backgrounds and landscapes for films, the compression of digital images, modeling of system dynamics, and describing aspects of meteorology, ecology, biology, economics, and astronomy. From a more esthetic point of view, when combined with color graphics, the mathematics in this section can generate surprisingly beautiful and complex patterns that are considered art.

In Section 1.4 we saw that vector $\mathbf{v}$ in $R^2$ can be viewed as a directed line segment between the origin $(0, 0)$ and the point $\mathbf{v} = (v_1, v_2)$. A line $L$ through the origin in the direction of $\mathbf{v}$ is the set of all scalar multiples of $\mathbf{v}$: $L = k\mathbf{v}$. We showed that any line $L'$ in $R^2$ not through the origin is a translation of some line $k\mathbf{v}$:

$$L' = k\mathbf{v} + \mathbf{b}, \text{ where } \mathbf{b} = \begin{bmatrix} b_1 \\ b_2 \end{bmatrix}.$$

In Exercise ML.3 in Section 1.4 we introduced the convex linear combination of a pair of vectors $\mathbf{u}$ and $\mathbf{v}$ in $R^2$ as the expression

$$t\mathbf{u} + (1 - t)\mathbf{v} \tag{1}$$

where $t$ is a scalar in $[0, 1]$. For $t$ in $[0, 1]$, expression (1) defines the line segment between points $(v_1, v_2)$ and $(u_1, u_2)$. For instance, with $t = \frac{1}{2}$ the expression in (1) gives $\frac{1}{2}\mathbf{u} + \frac{1}{2}\mathbf{v}$, which corresponds to the midpoint

$$\left( \frac{u_1 + v_1}{2}, \frac{u_2 + v_2}{2} \right)$$

of the line segment from $\mathbf{v}$ to $\mathbf{u}$. (Note that we have switched between vectors and points when convenient for the idea we want to focus on.)

Let $T \colon R^2 \to R^2$ be a linear transformation. Then we know that there exists a $2 \times 2$ matrix $\mathbf{A}$ such that $T(\mathbf{v}) = \mathbf{A}\mathbf{v}$; that is, $T$ is a matrix transformation. From the following expressions

$$T(k\mathbf{v}) = k(\mathbf{A}\mathbf{v})$$
$$T(k\mathbf{v} + \mathbf{b}) = k(\mathbf{A}\mathbf{v}) + \mathbf{A}\mathbf{b}$$
$$T(t\mathbf{v} + (1 - t)\mathbf{b}) = t(\mathbf{A}\mathbf{v}) + (1 - t)(\mathbf{A}\mathbf{b}),$$

we can say that matrix transformations map lines to lines and line segments to line segments.

In Section 1.5 we saw that matrix transformations can be used to alter the shape of a set of points in $R^2$ and to perform elementary graphic manipulations like reflections, scalings, rotations, and shears. Since $\mathbf{A0} = \mathbf{0}$, a matrix transformation cannot select a new origin for a coordinate system or figure. However, a translation can.

---

**Definition** Let $T \colon R^2 \to R^2$ be defined by $T(\mathbf{v}) = \mathbf{v} + \mathbf{b}$ where $\mathbf{b}$ is a fixed vector in $R^2$. We call this a **translation by vector b** (or just a **translation**) and use the notation $\mathbf{tran_b}$; $\mathbf{tran_b}(\mathbf{v}) = \mathbf{v} + \mathbf{b}$.

---

A translation by vector **b**, **b** $\neq$ **0**, is a nonlinear transformation. This follows from noting that $\mathbf{tran_b(0)} \neq \mathbf{0}$. (See Section 7.1 .)

The composition of a matrix transformation with a translation defines an important class of nonlinear transformations, which we define next.

---

**Definition**   The transformation $T: R^2 \rightarrow R^2$ defined by $T(\mathbf{v}) = \mathbf{Av} + \mathbf{b}$, where **A** is a specified $2 \times 2$ matrix and **b** is a fixed vector in $R^2$, is called an **affine transformation**.

---

By our previous discussion, for $\mathbf{b} \neq \mathbf{0}$, an affine transformation is nonlinear. If we let $T_1: R^2 \rightarrow R^2$ be defined by $T_1(\mathbf{v}) = \mathbf{Av}$ and $T_2: R^2 \rightarrow R^2$ be defined by $T_2(\mathbf{v}) = \mathbf{tran_b(v)} = \mathbf{v} + \mathbf{b}$, then the composition of $T_1$ with $T_2$, $T_2 \circ T_1$, is equivalent to the affine transformation $T(\mathbf{v}) = \mathbf{Av} + \mathbf{b}$; $(T_2 \circ T_1)(\mathbf{v}) = T_2(T_1(\mathbf{v})) = \mathbf{tran_b(Av)} = \mathbf{Av} + \mathbf{b}$. As we saw in Section 1.5, the order in which we compose transformations is important since

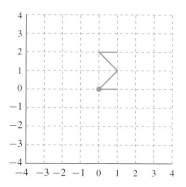

$$T_2(T_1(\mathbf{v})) = \mathbf{tran_b(Av)} = \mathbf{Av} + \mathbf{b}$$

but

$$T_1(T_2(\mathbf{v})) = T_1(\mathbf{tran_b(v)}) = T_1(\mathbf{v} + \mathbf{b}) = \mathbf{A(v + b)} = \mathbf{Av} + \mathbf{Ab}.$$

Both $T_2 \circ T_1$ and $T_1 \circ T_2$ are affine transformations, but the translations involved are, in general, different since **Ab** need not equal **b**. We illustrate this geometrically in Example 1.

FIGURE 1

EXAMPLE 1   Figure 1 displays a summation symbol, a sigma, with a large dot located at the origin. For our purposes we will consider the sigma as the set of ordered pairs $\{(1, 0), (0, 0), (1, 1), (0, 2), (1, 2)\}$ connected in the order listed by straight line segments. The dot is included so that we can see the image of the origin when transformations are applied to sigma. It is convenient to represent sigma as the matrix

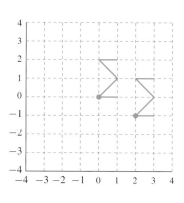

$$\mathbf{S} = \begin{bmatrix} 1 & 0 & 1 & 0 & 1 \\ 0 & 0 & 1 & 2 & 2 \end{bmatrix}$$

where the first row is the set of $x$-coordinates of the points defining sigma and the second row is the corresponding set of $y$ coordinates. We will assume that the line segments connecting successive points will be drawn as needed.

(a)   For ease of manipulation with vector $\mathbf{b} = \begin{bmatrix} b_1 \\ b_2 \end{bmatrix}$ we assume that a translation $\mathbf{tran_b(S)}$ is computed so that $b_1$ is added to each of the entries in $\mathbf{row}_1(\mathbf{S})$ and $b_2$ is added to each of the entries in $\mathbf{row}_2(\mathbf{S})$. Thus

FIGURE 2

$$\mathbf{tran_b(S)} = \begin{bmatrix} \mathbf{col}_1(\mathbf{S}) + \mathbf{b} & \mathbf{col}_2(\mathbf{S}) + \mathbf{b} & \mathbf{col}_3(\mathbf{S}) + \mathbf{b} & \mathbf{col}_4(\mathbf{S}) + \mathbf{b} & \mathbf{col}_5(\mathbf{S}) + \mathbf{b} \end{bmatrix}$$

$$= \begin{bmatrix} \mathbf{tran_b(col_1(S))} & \mathbf{tran_b(col_2(S))} & \mathbf{tran_b(col_3(S))} & \mathbf{tran_b(col_4(S))} & \mathbf{tran_b(col_5(S))} \end{bmatrix}.$$

For $\mathbf{b} = \begin{bmatrix} 2 \\ -1 \end{bmatrix}$, $\mathbf{tran_b(S)} = \begin{bmatrix} 3 & 2 & 3 & 2 & 3 \\ -1 & -1 & 0 & 1 & 1 \end{bmatrix}$. (Verify.) This image is shown in Figure 2 along with the original sigma. We see that the origin has been translated to $(2, -1)$.

(b) Let $T_1(\mathbf{v}) = \mathbf{A}\mathbf{v}$ where $\mathbf{A} = \begin{bmatrix} 0 & 1 \\ -1 & 0 \end{bmatrix}$; $T_1$ is a clockwise rotation of 90°. Let $T_2(\mathbf{v}) = \mathbf{tran_b}(\mathbf{v})$ as in part (a). Then Figure 3 shows the composition $(T_2 \circ T_1)(\mathbf{S})$ and Figure 4 the composition $(T_1 \circ T_2)(\mathbf{S})$. ■

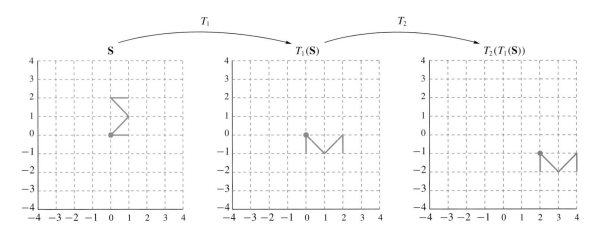

FIGURE 3

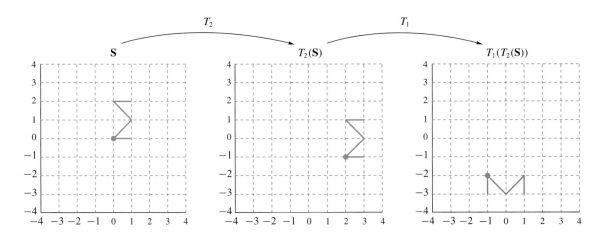

FIGURE 4

If we know the set of ordered pairs $\{(x_1, y_1), (x_2, y_2), \ldots, (x_n, y_n)\}$ that describe the original figure and the corresponding ordered pairs $\{(u_1, w_1), (u_2, w_2), \ldots, (u_n, w_n)\}$ that result from an affine transformation $T$, then we can determine the affine transformation. Let $T(\mathbf{v}) = \mathbf{A}\mathbf{v} + \mathbf{b}$ where $\mathbf{A} = \begin{bmatrix} p & r \\ s & t \end{bmatrix}$ and $\mathbf{b} = \begin{bmatrix} b_1 \\ b_2 \end{bmatrix}$. There are six unknowns $p, r, s, t, b_1$, and $b_2$ to be computed. Using the fact that

$$T\left(\begin{bmatrix} x_j \\ y_j \end{bmatrix}\right) = \begin{bmatrix} p & r \\ s & t \end{bmatrix}\begin{bmatrix} x_j \\ y_j \end{bmatrix} + \begin{bmatrix} b_1 \\ b_2 \end{bmatrix} = \begin{bmatrix} u_j \\ w_j \end{bmatrix}$$

for $j = 1, 2, \ldots, n$, we form the sum on the left side and then equate corresponding entries to obtain equations

$$px_j + ry_j + b_1 = u_j$$
$$sx_j + ty_j + b_2 = w_j. \tag{2}$$

For any three values of $j$ we will obtain a $3 \times 3$ linear system to determine $p$, $r$, and $b_1$ and another $3 \times 3$ linear system to determine $s$, $t$, and $b_2$. We illustrate this in Example 2.

EXAMPLE 2    Use the sigma defined in Example 1 and shown in Figure 1. Some affine transformation $T$ is applied to sigma and produces the image shown in Figure 5. It is known that

$$T(\mathbf{S}) = \begin{bmatrix} 1 & -1 & 2 & 1 & 3 \\ -\frac{1}{2} & \frac{1}{2} & \frac{1}{2} & \frac{5}{2} & \frac{3}{2} \end{bmatrix}.$$

It follows that $T(\mathbf{col}_j(\mathbf{S})) = \mathbf{col}_j(T(\mathbf{S}))$ using the procedure for determining $\mathbf{tran_b}(\mathbf{S})$ as given in Example 1a and properties of matrix multiplication. (Verify.) Selecting values 1, 2, and 3 for $j$ and using the equations in (2), we get the following:

$$T\left(\begin{bmatrix} 1 \\ 0 \end{bmatrix}\right) = \begin{bmatrix} 1 \\ -\frac{1}{2} \end{bmatrix} \quad \Rightarrow \quad \begin{matrix} p + 0r + b_1 = 1 \\ s + 0t + b_2 = -\frac{1}{2}, \end{matrix}$$

$$T\left(\begin{bmatrix} 0 \\ 0 \end{bmatrix}\right) = \begin{bmatrix} -1 \\ \frac{1}{2} \end{bmatrix} \quad \Rightarrow \quad \begin{matrix} b_1 = -1 \\ b_2 = \frac{1}{2}, \end{matrix}$$

$$T\left(\begin{bmatrix} 1 \\ 1 \end{bmatrix}\right) = \begin{bmatrix} 2 \\ \frac{1}{2} \end{bmatrix} \quad \Rightarrow \quad \begin{matrix} p + r + b_1 = 2 \\ s + t + b_2 = \frac{1}{2}. \end{matrix}$$

Grouping the equations with like variables to form two linear systems, we get

$$\begin{matrix} p \quad\quad + b_1 = 1 \\ b_1 = -1 \\ p + r + b_1 = 2 \end{matrix} \quad \Rightarrow \quad \begin{bmatrix} 1 & 0 & 1 & | & 1 \\ 0 & 0 & 1 & | & -1 \\ 1 & 1 & 1 & | & 2 \end{bmatrix}$$

and

$$\begin{matrix} s \quad\quad + b_2 = -\frac{1}{2} \\ b_2 = \frac{1}{2} \\ s + t + b_2 = \frac{1}{2} \end{matrix} \quad \Rightarrow \quad \begin{bmatrix} 1 & 0 & 1 & | & -\frac{1}{2} \\ 0 & 0 & 1 & | & \frac{1}{2} \\ 1 & 1 & 1 & | & \frac{1}{2} \end{bmatrix}.$$

It follows that the coefficient matrices of these systems are identical so we can form a partitioned matrix with two augmented columns and compute the reduced row echelon form to find $p$, $r$, $s$, $t$, $b_1$, and $b_2$:

$$\mathbf{rref}\left(\begin{bmatrix} 1 & 0 & 0 & | & 1 & | & -\frac{1}{2} \\ 0 & 0 & 1 & | & -1 & | & \frac{1}{2} \\ 1 & 1 & 1 & | & 2 & | & \frac{1}{2} \end{bmatrix}\right) = \begin{bmatrix} 1 & 0 & 0 & | & 2 & | & -1 \\ 0 & 1 & 0 & | & 1 & | & 1 \\ 0 & 0 & 1 & | & -1 & | & \frac{1}{2} \end{bmatrix}.$$

Hence $p = 2$, $r = 1$, $b_1 = -1$, $s = -1$, $t = 1$, and $b_2 = \frac{1}{2}$. Next we must check to see if these computed values for $\mathbf{A}$ and $\mathbf{b}$ do indeed define an affine transformation

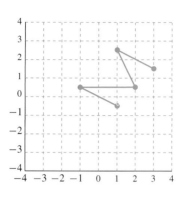

FIGURE 5

that maps the other vertices to the images specified previously. (There is no guarantee, because we have only used a subset of the mapping information to determine $\mathbf{A}$ and $\mathbf{b}$.) Checking, we see that

$$\begin{bmatrix} 2 & 1 \\ -1 & 1 \end{bmatrix} \begin{bmatrix} 0 \\ 2 \end{bmatrix} + \begin{bmatrix} -1 \\ \frac{1}{2} \end{bmatrix} = \begin{bmatrix} 1 \\ 2.5 \end{bmatrix}$$

and

$$\begin{bmatrix} 2 & 1 \\ -1 & 1 \end{bmatrix} \begin{bmatrix} 1 \\ 2 \end{bmatrix} + \begin{bmatrix} -1 \\ \frac{1}{2} \end{bmatrix} = \begin{bmatrix} 3 \\ 1.5 \end{bmatrix}$$

so we have the affine transformation such that

$$T(\mathbf{S}) = \begin{bmatrix} 1 & -1 & 2 & 1 & 3 \\ -\frac{1}{2} & \frac{1}{2} & \frac{1}{2} & \frac{5}{2} & \frac{3}{2} \end{bmatrix}.$$

If we had selected values 1, 3, and 5 for $j$ and used the equations in (2), we would have found that the coefficient matrix with two augmented matrix columns that results was

$$\begin{bmatrix} 1 & 0 & 1 & \bigm| & 1 & \bigm| & -\frac{1}{2} \\ 1 & 1 & 1 & \bigm| & 2 & \bigm| & \frac{1}{2} \\ 1 & 2 & 1 & \bigm| & 3 & \bigm| & \frac{3}{2} \end{bmatrix}$$

which is consistent with infinitely many solutions. (Verify.) We then need to determine the free variable in order to satisfy the other image properties. ■

Affine transformations seem simple enough; they map line segments to line segments. However, the simplicity of the process does not mean that repetitions of it will lead to simple patterns. In fact, it is surprising that a wide variety of complex patterns are the result of repeated applications of affine transformations. It was this type of observation that laid the foundation for the new areas of mathematics called **fractal geometry** and **chaos theory**. Here we focus on using affine transformations to illustrate fractal geometry.

To begin, what is a fractal? The pioneering work of Benoit Mandelbrot[1] used the term *fractal* to describe **figures that are self-similar** (that is, figures composed of infinite repetitions of the same shape). Hence such figures are made of smaller objects that are similar to the whole. For example, a tree is made up of branches, limbs, and twigs. Viewed at different scales, each of these has a similar shape or appearance. Examples 3 and 4 describe two fractals that are easy to visualize.

EXAMPLE 3    One of the most familiar fractals is the **Koch curve**. It is constructed by removing the middle third of a line segment, replacing it with an equilateral triangle, and then removing the base. We start with an initial line segment

_____

Following the preceding directions, we obtain Step 1 as shown in Figure 6. On each of the four line segments in Step 1 we repeat the process. The result is Step 2 in Figure 6.

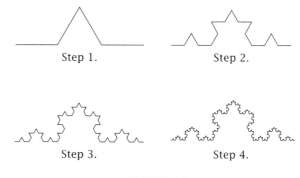

FIGURE 6

The directions are repeated on succeeding line segments to obtain Step 3 and then Step 4. The self-similarity is built into the construction process. If the initial line segment was 1 unit in length, then the four segments in Step 1 are $\frac{1}{3}$ of a unit, the sixteen segments in Step 2 are $\frac{1}{9}$ of a unit, and so on. At each step a line segment is scaled down by a factor of 3. Repeating the construction process infinitely often generates a curve that does not have a tangent line at any point.   ∎

EXAMPLE 4   Another simple fractal is obtained by repeatedly folding a strip of paper in half and opening it up. Take the original strip of paper and fold it in half; open it up keeping the angle between the halves a right angle; this is Step 1 in Figure 7.

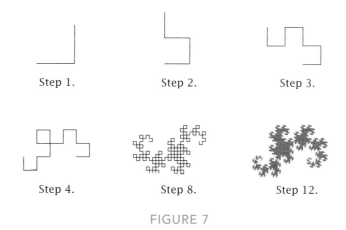

FIGURE 7

Next fold the original strip in half as before and then fold it in half again in the same direction. Open it up keeping the angles between the pieces a right angle; this is Step 2 in Figure 7. Repeat the folding in half successively in the same direction and opening, maintaining a right angle between the pieces. Figure 7 shows a number of steps. After a large number of steps the resulting figure resembles a dragon. Hence this fractal is called a **dragon curve**. (In Step 12, the resolution of the graphics screen is such that openings between successive line segments are quite small hence the solid appearance.) At the $k$-th step the resulting figure has $2^k$ segments. The folding in half is effectively replacing a segment by an isosceles right triangle. (See Example 5.)   ∎

Next we make the concept of self-similarity more mathematically precise. The key idea is that we use a succession of affine transformations on a figure. The affine transformations can be selected from a set that includes **translations**; **dilations** or **scalings**, which change the size of a figure; **reflections**; and **rotations**. (See Section 1.5.) The compositions from a set of affine transformations can be applied in a specified order or applied via random selection. We adopt the following terminology.

- Two geometric figures are called **similar** if one can be transformed into the other by a translation, dilation, reflection, or rotation or a combination of these transformations.
- A self-similar figure is the union of figures similar to itself.

We illustrate this mathematical formulation on the dragon curve in Example 5.

EXAMPLE 5    Let the original line segment of the dragon curve connect point $A(0, 0)$ and $B(\frac{\sqrt{2}}{2}, \frac{\sqrt{2}}{2})$. This segment is replaced by the legs of an isosceles right triangle that has $AB$ as its hypotenuse. Thus we have the transformation depicted in Figure 8.

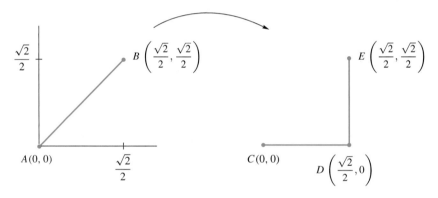

FIGURE 8

We view the image of segment $AB$ (Step 1 in Figure 7) as a union of two line segments each similar to segment $AB$. There is some affine transformation $T_1$ that maps $AB$ to $CD$ and another $T_2$ that maps $AB$ to $DE$. Transformation $T_1$ scales the length of $AB$ by a factor of $\frac{\sqrt{2}}{2}$ and then rotates it $45°$ clockwise. (See Example 7 in Section 1.5 for the matrix of a rotation; the angle of rotation is $-45°$.) Hence the matrix representing $T_1$ is the ($-45°$ rotation matrix) times (the dilation by $\frac{\sqrt{2}}{2}$ matrix); that is, the matrix representing $T_1$ is

$$\begin{bmatrix} \frac{\sqrt{2}}{2} & \frac{\sqrt{2}}{2} \\ -\frac{\sqrt{2}}{2} & \frac{\sqrt{2}}{2} \end{bmatrix} \begin{bmatrix} \frac{\sqrt{2}}{2} & 0 \\ 0 & \frac{\sqrt{2}}{2} \end{bmatrix} = \begin{bmatrix} \frac{1}{2} & \frac{1}{2} \\ -\frac{1}{2} & \frac{1}{2} \end{bmatrix}.$$

Transformation $T_2$ also scales $AB$ by a factor of $\frac{\sqrt{2}}{2}$, but also rotates it $45°$ counterclockwise, and then translates it by vector $\mathbf{b} = \begin{bmatrix} \frac{\sqrt{2}}{2} \\ 0 \end{bmatrix}$. Hence $T_2$ is the affine transformation that maps $(x, y)$ on segment $AB$ as follows:

$$T_2\left(\begin{bmatrix} x \\ y \end{bmatrix}\right) = \begin{bmatrix} \text{rotation} \\ \text{matrix} \end{bmatrix} * \begin{bmatrix} \text{dilation} \\ \text{matrix} \end{bmatrix} + \mathbf{b}$$

$$= \begin{bmatrix} \cos\left(\frac{\pi}{4}\right) & -\sin\left(\frac{\pi}{4}\right) \\ \sin\left(\frac{\pi}{4}\right) & \cos\left(\frac{\pi}{4}\right) \end{bmatrix} * \begin{bmatrix} \frac{\sqrt{2}}{2} & 0 \\ 0 & \frac{\sqrt{2}}{2} \end{bmatrix} \begin{bmatrix} x \\ y \end{bmatrix} + \begin{bmatrix} \frac{\sqrt{2}}{2} \\ 0 \end{bmatrix}$$

$$= \begin{bmatrix} \frac{1}{2} & -\frac{1}{2} \\ \frac{1}{2} & \frac{1}{2} \end{bmatrix} \begin{bmatrix} x \\ y \end{bmatrix} + \begin{bmatrix} \frac{\sqrt{2}}{2} \\ 0 \end{bmatrix}.$$

Thus Step 1 in Figure 7 is the union of $T_1(AB)$ and $T_2(AB)$. In a similar fashion, Step 2 in Figure 7 is the union of $T_1(CD)$, $T_2(CD)$, $T_1(DE)$, and $T_2(DE)$. Both transformations $T_1$ and $T_2$ are applied to each segment of a step of the dragon curve to obtain the next step as a union of the images. (The diagrams representing the steps of the dragon curve can vary depending upon the construction procedure employed.[2] We chose the form in Figure 7 because it was easy to construct. You can use the paper folding directions to construct the first few steps.)   ■

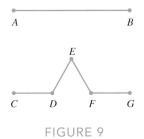

FIGURE 9

To determine the affine transformations used to obtain the Koch curve, we first note that one step generates four line segments from each line segment of the current figure (see Figure 6). Hence four affine transformations, $T_1$ through $T_4$, will be used to generate this fractal. Referring to Figure 9, $T_1(AB) = CD$, $T_2(AB) = DE$, $T_3(AB) = EF$, and $T_4(AB) = FG$. In each case there will be a scaling by a factor of $\frac{1}{3}$. We summarize these as follows (see Exercise 9):

• $T_1$ is a dilation by a factor of $\frac{1}{3}$.

• $T_2$ is a dilation by a factor of $\frac{1}{3}$ followed by a rotation of 60° counterclockwise and then a translation by vector $\mathbf{b} = \begin{bmatrix} \frac{1}{3} \\ 0 \end{bmatrix}$.

• $T_3$ is a dilation by a factor of $\frac{1}{3}$ followed by a rotation of 60° clockwise and then a translation by vector $\mathbf{b} = \begin{bmatrix} \frac{1}{2} \\ \frac{\sqrt{3}}{6} \end{bmatrix}$.

• $T_4$ is a dilation by a factor of $\frac{1}{3}$ followed by a translation by vector $\mathbf{b} = \begin{bmatrix} \frac{2}{3} \\ 0 \end{bmatrix}$.

Since every affine transformation can be written in the form $T(\mathbf{v}) = \mathbf{Av} + \mathbf{b}$, where $\mathbf{A} = \begin{bmatrix} p & r \\ s & t \end{bmatrix}$ and $\mathbf{b} = \begin{bmatrix} b_1 \\ b_2 \end{bmatrix}$, we can specify the transformations that generate a fractal as a table of coefficients for $p$, $r$, $s$, $t$, $b_1$, and $b_2$. The dragon fractal is specified in this manner in Table 1 and the Koch fractal in Table 2 that follow.

For both the Koch and dragon fractals we started with a straight line segment $AB$, computed images of the end points, and then connected them with a straight line. What happens if we take one of the affine transformations, say $T_2$ for the

[2]See "Fractals and Transformations," by T. Bannon, *Mathematics Teacher* vol. 84, March 1991, pp. 178–185.

TABLE 1  **The Dragon Fractal's Affine Transformations**

|       | $p$           | $r$            | $s$            | $t$           | $b_1$                  | $b_2$ |
|-------|---------------|----------------|----------------|---------------|------------------------|-------|
| $T_1$ | $\frac{1}{2}$ | $\frac{1}{2}$  | $-\frac{1}{2}$ | $\frac{1}{2}$ | 0                      | 0     |
| $T_2$ | $\frac{1}{2}$ | $-\frac{1}{2}$ | $\frac{1}{2}$  | $\frac{1}{2}$ | $\frac{\sqrt{2}}{2}$   | 0     |

TABLE 2  **The Koch Fractal's Affine Transformations**

|       | $p$           | $r$                     | $s$                     | $t$           | $b_1$         | $b_2$                  |
|-------|---------------|-------------------------|-------------------------|---------------|---------------|------------------------|
| $T_1$ | $\frac{1}{3}$ | 0                       | 0                       | $\frac{1}{3}$ | 0             | 0                      |
| $T_2$ | $\frac{1}{6}$ | $-\frac{\sqrt{3}}{6}$   | $\frac{\sqrt{3}}{6}$    | $\frac{1}{6}$ | $\frac{1}{3}$ | 0                      |
| $T_3$ | $\frac{1}{6}$ | $\frac{\sqrt{3}}{6}$    | $-\frac{\sqrt{3}}{6}$   | $\frac{1}{6}$ | $\frac{1}{2}$ | $\frac{\sqrt{3}}{6}$   |
| $T_4$ | $\frac{1}{3}$ | 0                       | 0                       | $\frac{1}{3}$ | $\frac{2}{3}$ | 0                      |

dragon, and use only it repeatedly on a point? That is, for any vector $\mathbf{v}$ in $R^2$, plot $\mathbf{v}$, compute and plot $T_2(\mathbf{v})$, $T_2(T_2(\mathbf{v}))$, $T_2(T_2(T_2(\mathbf{v})))$, .... The result is shown in Figure 10 for several choices of initial vector $\mathbf{v}$.

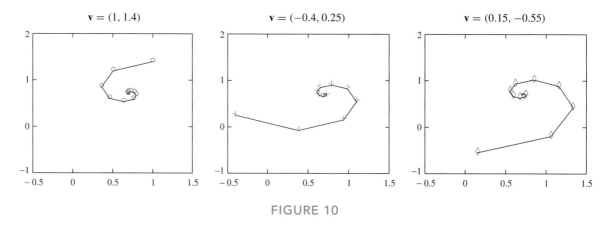

FIGURE 10

It appears that successive images get close to each other and stay close to a single point. In this case the points get close to $(0.7, 0.7)$. We call such a point a **point of attraction** or an **attractor** of the transformation $T_2$. To better determine the point of attraction, we algebraically seek a vector $\mathbf{v}$ such that $T_2(\mathbf{v}) = \mathbf{v}$. In this case we must solve

$$T_2\left(\begin{bmatrix} v_1 \\ v_2 \end{bmatrix}\right) = \begin{bmatrix} \frac{1}{2} & -\frac{1}{2} \\ \frac{1}{2} & \frac{1}{2} \end{bmatrix}\begin{bmatrix} v_1 \\ v_2 \end{bmatrix} + \begin{bmatrix} \frac{\sqrt{2}}{2} \\ 0 \end{bmatrix} = \begin{bmatrix} v_1 \\ v_2 \end{bmatrix}.$$

Expanding the matrix expression and equating corresponding components gives the linear system of equations

$$\frac{1}{2}v_1 - \frac{1}{2}v_2 + \frac{\sqrt{2}}{2} = v_1$$

$$\frac{1}{2}v_1 + \frac{1}{2}v_2 \qquad = v_2$$

whose solution is $v_1 = \frac{\sqrt{2}}{2}$ and $v_2 = \frac{\sqrt{2}}{2}$. (Verify.) Since

$$T_2\left(\begin{bmatrix} \frac{\sqrt{2}}{2} \\ \frac{\sqrt{2}}{2} \end{bmatrix}\right) = \begin{bmatrix} \frac{\sqrt{2}}{2} \\ \frac{\sqrt{2}}{2} \end{bmatrix},$$

this point is called a **fixed point** of transformation $T_2$.

Figure 10 is rather dull. No matter where we choose $\mathbf{v}$, successive images $T_2(\mathbf{v})$, $T_2(T_2(\mathbf{v}))$, $T_2(T_2(T_2(\mathbf{v})))$, ... approach the fixed point as though it were a target. There are two affine transformations that generate the dragon curve and three that generate the Koch curve. We call such a set of affine transformations an **iterated function system**, often abbreviated **IFS**. Instead of applying just one transformation, suppose we randomly select a member of the iterated function system to apply at each stage and plot the images as we did in Figure 10. Now something quite different happens. We illustrate this for the dragon iterated function system in Figure 11 for two different initial vectors $\mathbf{v}$ which are designated by the circles. The first three iterations are designated by $+$'s.

FIGURE 11

Regardless of the starting point, the dragon IFS pulls images toward the attractor of the system. (The attractor of an IFS is a set of points that is analogous to a fixed point of an affine transformation.) As the positions of the $+$ signs show, some of the initial images may not be in the attractor, but eventually images enter and do not leave the attractor. This is the case since once an image enters the attractor, here the dragon, its image is another point in the attractor. Thus more and more points begin to fill the attractor; the more points we plot, the more detail shown for the attractor. In Figure 11 we used 5000 points. You see that the images are not identical since the affine transformations of the dragon IFS are chosen randomly, but if we could plot an infinite number of points we would always get the same attractor for a specified IFS. The individual points of the attractor may be generated in a somewhat random fashion, but the result, the attractor, is certainly not random.

Michael Barnsley[3] referred to the random selection of affine transformations from an iterated function system as the **chaos game**. While the attractor of an IFS will be drawn differently every time, there is nothing chaotic about the outcome. However, what is interesting about the chaos game is that the attractor, some particular infinite set of points in $R^2$, can be generated by a small set of affine transformations, the IFS. This prompted Barnsley and his colleagues to ask the reverse question: Given an image, a set of points, can we find an iterated function system whose attractor is the specified image? This approach has led to a variety of interesting results, one of which is image compression and reconstruction using IFSs. (Earlier we mentioned wavelets and singular value decomposition in this context.)

Multimedia today requires that thousands of images be quickly available. Since graphic images are notoriously large files of digitized information, business and

---

[3]M. Barnsley, *Fractals Everywhere* (San Diego, CA: Academic Press, 1988).

industry are always on the lookout for new ways to reduce storage requirements and increase access speed to such information. Fractals and, in particular, iterated function systems have proven to be a commercially successful technique for image compression. In fact, the thousands of images that appear on a number of popular CD-ROMs have been encoded using a fractal image compression process. Fractal image compression starts with a digitized image and uses image-processing techniques to break it into segments. Each segment is then analyzed to remove as much redundancy as possible and then searched for self-similarity patterns. The final step of the process determines an iterated function system whose attractor is the self-similar pattern. Hence instead of saving detailed information about each pixel of the segment, only the information about the IFS need be retained to represent the image of the segment. The information about the IFS is just a table of coefficients for the affine transformations like those for the dragon and Koch fractals as given in Tables 1 and 2. To display the image, the chaos game is played with the IFS. It can be quite time consuming to develop the iterated function systems for the segments of an image, but once completed the IFS table of coefficients takes up significantly less memory and can generate the image quickly. The entire process has a firm theoretical foundation, which is based on the work of Barnsley and a result known as the "collage theorem."[4]

In the exercises we provide an opportunity to explore several other fractals and MATLAB routines that can be used to generate their images. We also provide an opportunity for you to experiment with fractals by choosing an iterated function system and generating its attractor. Keep in mind that the chaos game is easy to play:

1. Start with a set $S$ of affine transformations for the IFS and an initial point $(x, y)$.
2. Randomly choose a transformation in $S$, call it $T_k$.
3. Compute the image $T_k(x, y)$, plot it, and then set $x$ equal to the $x$-coordinate of the image and $y$ equal to the $y$-coordinate of the image.
4. Return to Step 2. Continue this loop until you are satisfied with the image generated.

## EXERCISES 7.5

**1.** Let $T : R^2 \to R^2$ be a translation by a vector $\mathbf{b} \neq \mathbf{0}$. Show that $T(\mathbf{0}) \neq \mathbf{0}$ and explain why this implies that $T$ is nonlinear.

**2.** Let $\mathbf{b} \neq \mathbf{0}$. Describe the behavior of compositions of $\mathbf{tran_b}$ with itself.

**3.** The "house" depicted in Figure 12 is made by connecting the set of ordered pairs $\{(0, 0), (0, 1), (1, 1), (2, 3), (3, 1), (3, 0), (0, 0)\}$ with straight line segments. Use a matrix $\mathbf{S}$, as in Example 1, to represent this figure. Compute the affine transformation of this image, $T(\mathbf{S}) = \mathbf{AS} + \mathbf{b}$, for each of the following pairs of $\mathbf{A}$ and $\mathbf{b}$ and then sketch the image.

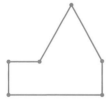

FIGURE 12

(a) $\mathbf{A} = \begin{bmatrix} 2 & -2 \\ 2 & 1 \end{bmatrix}$, $\mathbf{b} = \begin{bmatrix} -2 \\ 1 \end{bmatrix}$.

(b) $\mathbf{A} = \begin{bmatrix} 2 & -2 \\ 2 & -2 \end{bmatrix}$, $\mathbf{b} = \begin{bmatrix} -2 \\ 1 \end{bmatrix}$.

[4]For a brief overview of the collage theorem, see "Chaos and Fractals" by R. Burton, *Mathematics Teacher* 83, October 1990, pp. 524–529. For a detailed treatment, see "A Better Way to Compress Images" by M. Barnsley and A. Sloan, *BYTE* 13, January 1988, pp. 215–223, or "Fractal Image Compression" by M. Barnsley, *Notices of the American Mathematical Society* 43, No. 6, 1996, pp. 657–662.

(c) $\mathbf{A} = \begin{bmatrix} 2 & 2 \\ -2 & 1 \end{bmatrix}$, $\mathbf{b} = \begin{bmatrix} -2 \\ 1 \end{bmatrix}$.

4. The "v-wedge" depicted in Figure 13 is made by connecting the set of ordered pairs $\{(0, 0), (1.5, 1.5), (3, 1), (2.5, 2.5), (4, 4), (4, 0), (0, 0)\}$ with straight line segments. Use a matrix $\mathbf{S}$, as in Example 1, to represent this figure. Compute the affine transformation of this image, $T(\mathbf{S}) = \mathbf{AS} + \mathbf{b}$, for each of the following pairs of $\mathbf{A}$ and $\mathbf{b}$ and then sketch the image.

FIGURE 13

(a) $\mathbf{A} = \begin{bmatrix} -1 & 1 \\ -2 & 1 \end{bmatrix}$, $\mathbf{b} = \begin{bmatrix} -2 \\ 1 \end{bmatrix}$.

(b) $\mathbf{A} = \begin{bmatrix} 2 & 0 \\ -2 & 1 \end{bmatrix}$, $\mathbf{b} = \begin{bmatrix} -1 \\ 0 \end{bmatrix}$.

(c) $\mathbf{A} = \begin{bmatrix} 1 & -1 \\ -1 & 1 \end{bmatrix}$, $\mathbf{b} = \begin{bmatrix} 1 \\ -2 \end{bmatrix}$.

5. A polygon is constructed with vertices $\{(2, 0), (2, 2), (3, 3), (5, 4), (3, 1), (2, 0)\}$ connected by straight line segments in the order listed. The image of the polygon by affine transformation $T$ with matrix $\mathbf{A} = \begin{bmatrix} p & r \\ s & t \end{bmatrix}$ and vector $\mathbf{b} = \begin{bmatrix} b_1 \\ b_2 \end{bmatrix}$ is given by the corresponding set of vertices $\{(3, -4), (3, -2), (5, -3), (9, -6), (5, -5), (3, -4)\}$. Find $\mathbf{A}$ and $\mathbf{b}$.

6. A polygon is constructed with vertices $\{(1, 2), (1, 4), (2, 3), (3, 5), (3, 3), (1, 2)\}$ connected by straight line segments in the order listed. The image of the polygon by affine transformation $T$ with matrix $\mathbf{A} = \begin{bmatrix} p & r \\ s & t \end{bmatrix}$ and vector $\mathbf{b} = \begin{bmatrix} b_1 \\ b_2 \end{bmatrix}$ is given by the corresponding set of vertices $\{(1, 0), (-1, 0), (1, 1), (0, 2), (2, 2), (1, 0)\}$. Find $\mathbf{A}$ and $\mathbf{b}$.

7. Start with the line segment along the $x$-axis from $x = -1$ to $x = 1$. (There are two end points.) Draw a line segment one-half as long having its midpoint at an end point and perpendicular to the previous segment. (Now there are four end points.) Draw a line segment one-half as long as the previously drawn segment having its midpoint at an endpoint and perpendicular to the previous segment. Continue this for two more steps. The result is a portion of a curve known as the **H-fractal**.

(a) How many end points are there in the figure you have drawn?

(b) If you were to perform one more step, what is the length of the line segments that would be drawn?

8. Start at the origin. Draw the line segment from $(0, 0)$ to $(1, 0)$; call it $L$. To this figure adjoin two additional line segments of the same length. The first is that obtained from $L$ by rotating it about the origin $120°$ counterclockwise and the second is obtained from $L$ by rotating it about the origin $240°$ counterclockwise. Call the resulting figure a tripod. Scale the tripod down by a factor of $0.4$ and then translate it so that the origin of the scaled-down figure is at an end of the original tripod. Repeat the scaling and translating two more times, drawing the figure at each stage. (Each new tripod is drawn at an end of the previously drawn tripod.) The result is a portion of a curve known as the **ternary tree fractal**.

(a) In the figure you have drawn, how many "ends" are there that have no tripod attached?

(b) Starting at the origin, what is the longest distance you can travel before you reach an "end" with no tripod attached?

9. Let $T_1$, $T_2$, $T_3$ and $T_4$ be the affine transformations for the Koch fractal as given in Table 2. Let $A$ and $B$ be $(0, 0)$ and $(1, 0)$, respectively, in Figure 9. Determine the coordinates of $C$, $D$, $E$, $F$, and $G$ in Figure 9 from the relations $T_1(AB) = CD$, $T_2(AB) = DE$, $T_3(AB) = EF$, and $T_4(AB) = FG$.

10. Let $T: R^2 \rightarrow R^2$ be an affine transformation defined by $T(\mathbf{v}) = \mathbf{Av} + \mathbf{b}$ where

$$\mathbf{A} = \begin{bmatrix} p & r \\ s & t \end{bmatrix} \quad \text{and} \quad \mathbf{b} = \begin{bmatrix} b_1 \\ b_2 \end{bmatrix}.$$

A vector $\mathbf{v}$ in $R^2$ is a fixed point of $T$ provided $T(\mathbf{v}) = \mathbf{v}$. Thus it follows that $\mathbf{Av} + \mathbf{b} = \mathbf{v}$, which corresponds to the linear system $(\mathbf{A} - \mathbf{I}_2)\mathbf{v} = -\mathbf{b}$. Show that $T$ has a single fixed point whenever $(p - 1)(t - 1) - rs \neq 0$.

11. Determine the fixed point of each of the following affine transformations $T(\mathbf{v}) = \mathbf{Av} + \mathbf{b}$.

(a) $\mathbf{A} = \begin{bmatrix} \frac{1}{6} & \frac{\sqrt{3}}{6} \\ -\frac{\sqrt{3}}{6} & \frac{1}{6} \end{bmatrix}$, $\mathbf{b} = \begin{bmatrix} \frac{1}{2} \\ \frac{\sqrt{3}}{6} \end{bmatrix}$.

(b) $\mathbf{A} = \begin{bmatrix} -\frac{1}{2} & \frac{1}{2} \\ -\frac{1}{2} & -\frac{1}{2} \end{bmatrix}$, $\mathbf{b} = \begin{bmatrix} 2 \\ 0 \end{bmatrix}$.

12. Let $T: R^2 \rightarrow R^2$ be a linear transformation defined by $T(\mathbf{v}) = \mathbf{Av}$. When does $T$ have a fixed point and how many are there?

13. Rotations by an angle $\theta$ about the origin in $R^2$ are performed by the matrix transformation $T_\theta: R^2 \rightarrow R^2$ defined by $T_\theta(\mathbf{v}) = \mathbf{R}_\theta \mathbf{v}$ where

$$\mathbf{R}_\theta = \begin{bmatrix} \cos(\theta) & -\sin(\theta) \\ \sin(\theta) & \cos(\theta) \end{bmatrix}.$$

If we want to rotate **v** about a point **c** $= (c_1, c_2)$ different from the origin, then we use the composite transformation

$$T(\mathbf{v}) = \mathbf{tran_c}(T_\theta(\mathbf{tran_{-c}}(\mathbf{v}))).$$

Explain the three steps involved in this composite transformation and write a matrix expression for $T(\mathbf{v})$.

**14.** Let $S$ be the closed plane figure with vertices $\{(-1, 0), (0, 1), (2, 1), (2, 2), (0, 2), (-1, 0)\}$ that is formed by connecting the vertices by straight line segments in the order listed. (Refer to Exercise 13 for the notation involving rotation transformations.)

(a) Determine the vertices of the image $T_{90°}(S)$.

(b) Determine the vertices of the image when the figure is rotated $90°$ counterclockwise about point $(-1, 0)$.

(c) Draw the original figure and the images from parts (a) and (b) on the same set of axes.

## In MATLAB

*Each of the fractals and iterated function systems used in the examples or exercises in this section can be viewed using a MATLAB routine. The following is a list of the command names and a brief description. For more information and directions for using these routines in MATLAB type **help**, a space, and then the name of the routine.*

| | |
|---|---|
| **dragon** | Displays the dragon curve |
| **koch** | Displays the Koch curve |
| **hfractal** | Displays the H-fractal curve |
| **tree3** | Displays the ternary tree curve |
| **dragonifs** | Dragon iterated function system; attractor displayed |
| **kochifs** | Koch iterated function system; attractor displayed |

**ML. 1.** In Exercise 7 you were asked to construct the first few steps of the H-fractal.

(a) Execute MATLAB command **hfractal(5,0.6)** to see a portion of this fractal.

(b) Execute MATLAB command **hfractal(8,0.6)**. If the housing complex you lived in had its street network the same as the graph displayed, why would there be traffic flow problems? Explain.

(c) Execute command **hfractal(11,0.6)**. Do the following: At a MATLAB prompt $>>$ enter command **zoom on**[5]. Return to the picture generated by the command **hfractal(11,0.6)**. Position your mouse on one of the branches in the lower right portion of the figure. Click the left mouse button once. Click the left mouse button a second time. Besides enlarging a portion of the figure, how does this action relate to self-similarity? (For other options with **zoom** read the display from **help zoom**.)

**ML. 2.** In Exercise 8 you were asked to construct the first few steps of the ternary tree fractal.

(a) Execute MATLAB command **tree3(4,0.4)** to see a portion of this fractal.

(b) Execute MATLAB command **tree3(5,0.4)**. If the housing complex you lived in had its street network the same as the graph displayed, why would there be traffic flow problems? Explain.

(c) Execute command **tree3(7,0.4)**. Do the following: At a MATLAB prompt $>>$ enter command **zoom on**.[5] Return to the picture generated by the command **tree3(7,0.4)**. Position your mouse on one of the branches in the lower right portion of the figure. Click the left mouse button once. Click the left mouse button a second time. Besides enlarging a portion of the figure, how does this action relate to self-similarity?

**ML. 3.** Experiment with the iterated function system for the dragon fractal. In MATLAB type **help dragonifs** for directions.

**ML. 4.** Experiment with the iterated function system for the Koch fractal. In MATLAB type **help kochifs** for directions.

**ML. 5.** Another famous construction that leads to a fractal is shown when you execute the following MATLAB commands.

(a) In MATLAB execute command **sierptri(1)**. You will see a 'level zero' figure displayed. Follow the on screen directions to see 'level one'.

  (i) In terms of triangles, describe how level one was obtained from level zero.

  (ii) In terms of points and line segments connecting the points, describe how level one was obtained from level zero.

(b) In MATLAB execute command **sierptri(2)**. Go through levels zero and one as before. Now display level two. In terms of triangles, describe how level two was obtained from level one.

(c) Based upon your observations and responses to parts (a) and (b), describe the self-similarity that occurs in this demonstration.

(d) In MATLAB execute command **sierptri(3)**. Follow the on-screen directions. Explain how the figure should be completed to fully see level three.

---

[5]This command lets you zoom in on a portion of a figure. To use it, position your mouse over a point that you want to zoom in on. Click the left mouse button to zoom in on the point under the mouse; click the right mouse button to zoom out (shift-click on the Macintosh). Each time you click, the axes limits will be changed by a factor of 2, in or out.

(e) In MATLAB execute command **sierptri(6)**. Follow the on-screen directions until the routine ends. When you are back at a MATLAB prompt $>>$ enter command **zoom on**. (See footnote 5.) Return to the picture generated by the command **sierptri(6)**. Position your mouse in a black region near the second smallest white triangle. Click the left mouse button once. Click the left mouse button a second time.

  (i) Besides enlarging a portion of the figure, how does this action relate to self-similarity?

  (ii) Imagine that we executed command **sierptri** to get 100 levels and then repeatedly zoomed in as described previously. How would the original figure compare with the enlarged figure?

f) Execute command **zoom off** before using MATLAB further.

**ML. 6.** The attractor of the iterated function system that corresponds to the Sierpinski triangle discussed in Exercise ML.5 can be viewed using MATLAB routine **sierpifs**. Experiment with this routine. The attractor is often called the Sierpinski gasket.

**ML. 7.** Another well-known fractal is often referred to as the Barnsley fern. It differs from the other iterated system functions in that different probabilities are assigned for the use of the four affine transformations involved. In MATLAB type **help fernifs** for directions. Experiment with routine **fernifs** to generate a fern.

**ML. 8.** To experiment with your own iterated function system you can use the MATLAB routine **chaosgame**. In MATLAB type **help chaosgame** for directions. You can enter up to five affine transformations and assign probabilities for the use of the transformations. We suggest that you use entries between $-1$ and 1 for the matrix **A** and entries between $-2$ and 2 for the entries of vector **b**. If you assign unequal probabilities to the affine transformations, make sure the sum of the probabilities is 1.

(a) Use the chaos game with the following iterated function system.

| | $p$ | $r$ | $s$ | $t$ | $b_1$ | $b_2$ |
|---|---|---|---|---|---|---|
| $T_1$ | 0.6 | 0 | 0 | 0.6 | 0 | $-0.5$ |
| $T_2$ | 0.6 | 0 | 0 | 0.6 | 0 | 0.5 |
| $T_3$ | $\frac{\sqrt{2}}{4}$ | $-\frac{\sqrt{2}}{4}$ | $\frac{\sqrt{2}}{4}$ | $\frac{\sqrt{2}}{4}$ | $-0.5$ | $-0.25$ |
| $T_4$ | $\frac{\sqrt{2}}{4}$ | $\frac{\sqrt{2}}{4}$ | $-\frac{\sqrt{2}}{4}$ | $\frac{\sqrt{2}}{4}$ | 0.5 | $-0.25$ |

Describe the attractor that is displayed.

(b) Experiment with several iterated function systems of your own.

## True/False Review Questions

*Determine whether each of the following statements is true or false.*

**1.** The translation transformation by vector **b**, $T : R^2 \rightarrow R^2$ where $T(\mathbf{v}) = \mathbf{v} + \mathbf{b}$, is a linear transformation.

**2.** An affine transformation is a linear transformation.

**3.** Fractals are figures composed of infinite repetitions of the same shape.

**4.** A fixed point of transformation $T$ is the same as its image.

**5.** If $S$ represents a figure in $R^2$ to which we apply a rotation and then a dilation, the resulting figure is said to be similar to $S$.

## Terminology

| | |
|---|---|
| Translation | Affine transformation |
| Fractal | Similar geometric figures |
| Iterated function system | Chaos game |

This section focused on mappings from $R^2$ to $R^2$. We investigated how to use a combination of linear and special nonlinear transformations in compositions to generate fractals. Such figures can be quite simple or very complex depending upon the mappings used. This area of mathematics is quite new, but it has been applied to a variety of scientific areas. Use the following to review the ideas of this section.

- Explain why a translation by a nonzero vector is a nonlinear transformation.
- How is an affine transformation related to a matrix transformation?
- How was an affine transformation applied to a planar figure described by a set of points connected by line segments?
- What did we mean by saying that two figures are similar?
- What is a self-similar figure?
- What do we mean by a fractal?
- What is an iterated function system?
- How is a fractal image generated from an iterated function system?
- How is the chaos game different from using an iterated function system to determine an image?
- Why can we say that an iterated function system acts as a compressor of information?
- How can you construct your own personal fractal? How would you communicate your personal fractal to a friend?

## CHAPTER TEST

1. Let $V = W = R_{2 \times 2}$ with $T(\mathbf{A}) = \mathbf{A}^2$. Determine whether $T$ is linear or nonlinear.

2. Let $V = W = R^2$ with $T(\mathbf{v}) = r\mathbf{v} + \mathbf{s}$, where $r$ is any real number and $\mathbf{s}$ is in $R^2$. Determine all the values of $r$ and vectors $\mathbf{s}$ so that $T$ is a linear transformation.

3. Let $T: R^3 \rightarrow R^3$ be the linear transformation $T(\mathbf{v}) = \mathbf{Av}$ where $\mathbf{A} = \begin{bmatrix} 0 & -1 & 2 \\ -2 & 1 & 3 \\ -2 & 0 & 5 \end{bmatrix}$. Find $\mathbf{ker}(T)$.

4. Let $T: P_2 \rightarrow R$ be the linear transformation given by

$$T(at^2 + bt + c) = \int_0^1 (at^2 + bt + c)\, dt.$$

   (a) Show that $T$ is not one-to-one.
   (b) Show that $T$ is onto.

5. Let $T: P_2 \rightarrow P_2$ be the linear transformation given by $T(at^2 + bt + c) = (c + 2a)t + (b - c)$.
   (a) Find a basis for $\mathbf{ker}(T)$.
   (b) Find a basis for $\mathbf{R}(T)$.

6. Let $T: R^4 \rightarrow R^3$ be the linear transformation given by

$$T\left(\begin{bmatrix} v_1 \\ v_2 \\ v_3 \\ v_4 \end{bmatrix}\right) = \begin{bmatrix} v_2 + v_3 \\ v_1 \\ v_1 + v_2 \end{bmatrix}.$$

   (a) Find the matrix representation of $T$ relative to the natural bases for $R^4$ and $R^3$.
   (b) Find the matrix representation of $T$ relative to the bases

$$S = \left\{ \begin{bmatrix} 1 \\ 0 \\ 0 \\ 1 \end{bmatrix}, \begin{bmatrix} 0 \\ 0 \\ 0 \\ 1 \end{bmatrix}, \begin{bmatrix} 1 \\ 1 \\ 0 \\ 0 \end{bmatrix}, \begin{bmatrix} 0 \\ 1 \\ 1 \\ 0 \end{bmatrix} \right\} \quad \text{and} \quad B = \left\{ \begin{bmatrix} 1 \\ 1 \\ 0 \end{bmatrix}, \begin{bmatrix} 0 \\ 1 \\ 0 \end{bmatrix}, \begin{bmatrix} 1 \\ 0 \\ 1 \end{bmatrix} \right\}.$$

**7.** Let $S = \left\{ \begin{bmatrix} 2 \\ 0 \\ 1 \end{bmatrix}, \begin{bmatrix} 1 \\ 2 \\ 0 \end{bmatrix}, \begin{bmatrix} 1 \\ 1 \\ 1 \end{bmatrix} \right\}$ and $B = \left\{ \begin{bmatrix} 6 \\ 3 \\ 3 \end{bmatrix}, \begin{bmatrix} 4 \\ -1 \\ 3 \end{bmatrix}, \begin{bmatrix} 5 \\ 5 \\ 2 \end{bmatrix} \right\}$ be ordered bases for $R^3$.

(a) Find $\mathbf{coord}_S \left( \begin{bmatrix} 3 \\ -2 \\ 2 \end{bmatrix} \right)$.

(b) Find the transition matrix $P_{S \leftarrow B}$.

For Exercises 8 through 10, let linear transformation $T : V \rightarrow W$ be represented by matrix $\mathbf{A}_{B \leftarrow S}$, where $S$ is an ordered basis for $V$ and $B$ is an ordered basis for $W$.

**8.** If we have a vector $\mathbf{v}$ in $V$, explain how to compute $\mathbf{coord}_B(T(\mathbf{v}))$.

**9.** If $S'$ is another ordered basis for $V$ and we have $\mathbf{coord}_{S'}(\mathbf{v})$, then explain how to compute $\mathbf{coord}_B(T(\mathbf{v}))$.

**10.** If $S'$ is another ordered basis for $V$ and $B'$ is another ordered basis for $W$ and we have $\mathbf{coord}_{S'}(\mathbf{v})$, then explain how to compute $\mathbf{coord}_{B'}(T(\mathbf{v}))$.

C  H  A  P  T  E  R

# 8

# *MATLAB*[1]

This chapter provides a brief overview of MATLAB and is not intended as a substitute for a user manual. We focus on features that are needed to use the instructional files that accompany this book.

MATLAB is a versatile piece of computer software with linear algebra capabilities as its core. MATLAB stands for MATrix LABoratory. It incorporates portions of professionally developed projects of quality computer routines for linear algebra computation. The code employed by MATLAB is written in the C language and is upgraded as new versions of MATLAB are released.

MATLAB has a wide range of capabilities. In this book we use only a small portion of its features. We will find that MATLAB's command structure is very close to the way we write algebraic expressions and linear algebra operations. The names of many MATLAB commands closely parallel those of the operations and concepts of linear algebra. We give descriptions of commands and features of MATLAB that relate directly to this course. A more detailed discussion of MATLAB commands can be found in the *MATLAB User's Guide* that accompanies the software and in the books *Experiments in Computational Matrix Algebra*, by David R. Hill (New York: Random House, 1988) and *Linear Algebra LABS with MATLAB*, 2nd ed., by David R. Hill and David E. Zitarelli (Upper Saddle River, NJ.: Prentice Hall, 1996).

---

[1]This material on MATLAB refers to the Windows version for MATLAB 5. There may be minor variations in other versions.

Alternatively, the MATLAB software provides immediate on-screen descriptions using the **help** command. Typing

**help**

displays a list of MATLAB subdirectories and alternate directories containing files corresponding to commands and data sets. Typing **help name**, where **name** is the name of a command, accesses information on the specific command named. In some cases the description displayed goes much further than we need for this course. Hence you may not fully understand all of the description displayed by **help**. We provide a list of the majority of MATLAB commands we use in this book in Section 8.2.

Once you initiate the MATLAB software, you will see the MATLAB logo appear and the MATLAB prompt ››. The prompt ›› indicates that MATLAB is awaiting a command. In Section 8.1 we describe how to enter matrices into MATLAB and give explanations of several commands. However, there are certain MATLAB features you should be aware of before you begin the material in Section 8.1.

- *Starting execution of a command*
  After you have typed a command name and any arguments or data required, you must press ENTER before it will begin to execute.
- *The command stack*
  As you enter commands, MATLAB saves a number of the most recent commands in the stack. Previous commands saved on the stack can be recalled using the **up arrow** key. The number of commands saved on the stack varies depending on the length of the commands and other factors.
- *Editing commands*
  If you make an error or mistype something in a command, you can use the **left arrow** and **right arrow** keys to position the cursor for corrections. The **home** key moves the cursor to the beginning of a command, and the **end** key moves the cursor to the end. The **backspace** and **delete** keys can be used to remove characters from a command line. The **insert** key is used to initiate the insertion of characters. Pressing the insert key a second time exits the insert mode. If MATLAB recognizes an error after you have pressed ENTER, then MATLAB responds with a beep and a message that helps define the error. You can recall the command line using the up arrow key in order to edit the line.
- *Continuing commands*
  MATLAB commands that do not fit on a single line can be continued to the next line using an ellipsis, which is three consecutive periods, followed by ENTER.
- *Stopping a command*
  To stop execution of a MATLAB command, press **Ctrl** and **C** simultaneously, then press ENTER. Sometimes this sequence must be repeated.
- *Quitting*
  To quit MATLAB, type **exit** or **quit** followed by ENTER.

## 8.1 ■ MATRIX INPUT

To enter a matrix into MATLAB just type the entries enclosed in square brackets [...], with entries separated by a space and rows terminated with a semicolon. Thus, matrix

$$\begin{bmatrix} 9 & -8 & 7 \\ -6 & 5 & -4 \\ 11 & -12 & 0 \end{bmatrix}$$

is entered by typing

**[9  −8  7; −6  5  −4;11  −12  0]**

and the accompanying display is

**ans =**

| 9 | −8 | 7 |
|---|---|---|
| −6 | 5 | −4 |
| 11 | −12 | 0 |

Notice that no brackets are displayed and that MATLAB has assigned this matrix the name **ans**. Every matrix in MATLAB must have a name. If you do not assign a matrix a name, then MATLAB assigns it **ans**, which is called the **default variable name**. To assign a matrix name we use the assignment operator =. For example,

**A = [4  5  8;0  −1  6]**

is displayed as

**A =**

| 4 | 5 | 8 |
|---|---|---|
| 0 | −1 | 6 |

---

**Warning**

1. All rows must have the same number of entries.
2. MATLAB distinguishes between uppercase and lowercase letters. So matrix **B** is not the same as matrix **b**.
3. A matrix name can be reused. In such a case the "old" contents are lost.

---

To assign a matrix but *suppress the display of its entries*, follow the closing square bracket, ], with a semicolon.

**A = [4  5  8;0  −1  6];**

assigns the same matrix to name **A** as above, but no display appears. To assign a currently defined matrix a new name, use the assignment operator =. Command **Z=A** assigns the contents of **A** to **Z**. Matrix **A** is still defined.

To determine the matrix names that are in use, use the **who** command. To delete a matrix, use the **clear** command followed by a space and then the matrix name. For example, the command

**clear A**

deletes name **A** and its contents from MATLAB. The command **clear** by itself deletes all currently defined matrices.

To determine the number of rows and columns in a matrix, use the **size** command, as in

**size(A)**

which, assuming that **A** has not been cleared, displays

$$\textbf{ans} =$$

$$\textbf{2} \quad \textbf{3}$$

meaning that there are two rows and three columns in matrix **A**.

## Seeing a Matrix

*To see all of the components of a matrix, type its name.* If the matrix is large, the display may be broken into subsets of columns that are shown successively. For example, use the command

$$\textbf{hilb(9)}$$

which displays the first seven columns followed by columns 8 and 9. (For information on command **hilb**, use **help hilb**.) If the matrix is quite large, the screen display will scroll too fast for you to see the matrix. To see a portion of a matrix, type command **more on** followed by ENTER; then type the matrix name or a command to generate it. Press the Space Bar to reveal more of the matrix. Continue pressing the Space Bar until the "--more--" no longer appears near the bottom of the screen. Try this with **hilb(20)**. To disable this paging feature, type command **more off**. If a scroll bar is available, you can use your mouse to move the scroll bar to reveal previous portions of displays.

We have the following conventions to see a portion of a matrix in MATLAB. For purposes of illustration, type the following commands which will generate a $5 \times 5$ matrix:

$$\textbf{A=1:25;A=reshape(A,5,5)}$$

The display should be

$$\textbf{A} =$$

| | | | | |
|---|---|---|---|---|
| 1 | 6 | 11 | 16 | 21 |
| 2 | 7 | 12 | 17 | 22 |
| 3 | 8 | 13 | 18 | 23 |
| 4 | 9 | 14 | 19 | 24 |
| 5 | 10 | 15 | 20 | 25 |

To see the $(2, 3)$ entry of **A**, type

$$\textbf{A(2,3)}$$

To see the fourth row of **A**, type

$$\textbf{A(4,:)}$$

To see the first column of **A**, type

$$\textbf{A(:,1)}$$

In the above situations, the : is interpreted to mean "all." The colon can also be used to represent a range of rows or columns. For example, typing

$$\textbf{2:8}$$

displays the integers from 2 through 8 as shown next.

$$\textbf{ans} =$$

$$\textbf{2} \quad \textbf{3} \quad \textbf{4} \quad \textbf{5} \quad \textbf{6} \quad \textbf{7} \quad \textbf{8}$$

We can use this feature to display a subset of rows or columns of a matrix. As an illustration, to display rows 3 through 5 of matrix **A**, type

**A(3:5,:)**

Similarly, columns 1 through 3 are displayed by typing

**A(:,1:3)**

For more information on the use of the colon operator, type **help colon**. The colon operator in MATLAB is very versatile, but we will not need to use all of its features.

## Display Formats

MATLAB stores matrices in decimal form and does its arithmetic computations using a decimal-type arithmetic. This decimal form retains about 16 digits, but not all digits must be shown. Between what goes on in the machine and what is shown on the screen are routines that convert or format the numbers into displays. Here we give an overview of the display formats that we will use. (For more information, see the *MATLAB User's Guide* or type **help format**.)

- If the matrix contains *all* integers, then the entire matrix is displayed with integer entries; that is, no decimal points appear.
- If any entry in the matrix is not exactly represented as an integer, then the entire matrix is displayed in what is known as **format short**. Such a display shows four places behind the decimal point, and the last place may have been rounded. The exception to this is zero. If an entry is exactly zero, then it is displayed as an integer zero. Enter the matrix

$$Q = [5 \quad 0 \quad 1/3 \quad 2/3 \quad 7.123456]$$

into MATLAB. The display is

**Q =**

    **5.0000      0      0.3333      0.6667      7.1235**

**WARNING:**    If a value is displayed as **0.0000**, then it is not identically zero. You should change to **format long**, discussed later and display the matrix again.

To see more than four places, change the display format. One way to proceed is to use the command

**format long**

which shows 15 places. The matrix **Q** in format long is

**Q =**
    **Columns 1 through 4**
    **5.00000000000000    0    0.33333333333333    0.66666666666667**
    **Column 5**
    **7.12345600000000**

Other display formats use an exponent of 10. They are **format short e** and **format long e**. The "e-formats" are often used in numerical analysis. Try these formats with matrix **Q**.

MATLAB can display values in rational form. The command **format rat**, short for rational display, is used. Inspect the output from the following sequence of MATLAB commands:

$$\textbf{format short}$$
$$\textbf{V = [1   1/2   1/6   1/12]}$$

displays

$$\textbf{V =}$$

| **1.0000** | **0.5000** | **0.1667** | **0.0833** |
|---|---|---|---|

and

$$\textbf{format rat}$$
$$\textbf{V}$$

displays

$$\textbf{V =}$$

| **1** | **1/2** | **1/6** | **1/12** |
|---|---|---|---|

Finally, type **format short** to return to a decimal display form.

**WARNING:**   Rational output is displayed in what is called "string" form. Strings are not numeric data and hence cannot be used with arithmetic operators. Thus rational output is for "looks" only.

When MATLAB starts, the format in effect is **format short**. If you change the format, it remains in effect until another format command is executed. Some MATLAB routines change the format within the routine.

## 8.2 ■ MATLAB COMMANDS

The following is a list of MATLAB commands, designated by M, and supplemental instructional m-files, designated by S, that appear in the sections of this text.

1.1 Matrices
    M: **size**, **figure**, **plot**, **help**, **eye**, **ones**, **zeros**,
    **'**(prime), **:** (colon), **i** (complex unit)
    S: **igraph**

1.2 Matrix Operations
    M: **+**, **−**, **∗** (scalar multiplication), **polyval**
    S: **rgbexamp**

1.3 Matrix Products
    M: dot product for columns (**x'*y**), **^**, **∗** (matrix multiply), **reshape**
    S: **weather**

1.4 Geometry of Linear Combinations
    S: **vecdemo**, **lincombo**, **convex**

1.5 Matrix Transformations
    M: **figure(gcf)**, **print**, **grid**
    S: **mapit**, **planelt**, **project**

2.1 Linear Systems of Equations
    M: **[A b]** (forming an augmented matrix)
    S: **utriquik, utriview, utristep, bksubpr, reduce, rowop, paracub**

2.2 Echelon Forms
    M: **rref**
    S: **reduce, rowop, rrefquik, rrefview, rrefstep, vizrowop, vizplane**

2.3 Properties of Solution Sets
    M: **rank**
    S: **circ**

2.4 Nonsingular Linear Systems
    M:**rref, inv, flops**
    S: **compinv, elemmat, pinterp, pintdemo**

2.5 LU-Factorization
    S: **lupr**

2.6 The (Least Squares) Line of Best Fit
    S: **lsqgame**

3.1 The Magnification Factor
    S:**mapit**

3.2 The Determinant of a Matrix
    (No MATLAB exercises.)

3.3 Computing det(**A**) using Row Operations
    M: **det**
    S: **detpr, areapara**

3.4 Recursive Computation of the Determinant (Optional)
    S: **detrecur, cofactor, adjoint**

3.5 Cramer's Rule (Optional)
    (No MATLAB exercises.)

3.6 The Cross Product (Optional)
    (No MATLAB exercises.)

4.1 Eigenvalues and Eigenvectors
    S: **matvec, evecsrch, mapcirc**

4.2 Computing Eigen Information for Small Matrices
    M: **poly, roots, rref, eig, solve**
    S: **pmdemo**

4.3 Diagonalization and Similar Matrices
    M: **eig**

4.4 Orthogonal Matrices and Spectral Representation
    S: **scan, specgen**

4.5 The Gram-Schmidt Process
    S: **gsprac2, gsprac3**

4.6 The QR-Factorization (Optional)
    M: **qr**
    S: **basicqr**

4.7 Singular Value Decomposition (Optional)
    M: **svd**

No MATLAB Exercises in Chapter 5.

No MATLAB Exercises in Sections 6.1–6.3 and 6.5–6.6.

6.4 Least Squares
    S: **trigpoly**

No MATLAB Exercises in Sections 7.1–7.4.

7.5 Fractals: An Application of Transformations (Optional)
    S: **dragon, koch, hfractal, trec3, dragonifs, kochifs, sierptri, sierifs, chaosgame, fernifs**

# APPENDIX 1

## COMPLEX NUMBERS

Complex numbers are usually introduced in an algebra course to "complete" the solution to the quadratic equation

$$ax^2 + bx + c = 0.$$

In using the quadratic formula

$$x = \frac{-b \pm \sqrt{b^2 - 4ac}}{2a}$$

the case in which $b^2 - 4ac < 0$ is not resolved unless we can cope with the square roots of negative numbers. In the sixteenth century mathematicans and scientists justified this "completion" of the solution of the quadratic equations by intuition. Naturally, a controversy arose, with some mathematicians denying the existence of these numbers and others using them along with real numbers. The use of complex numbers did not lead to any contradictions, and the idea proved to be an important milestone in the development of mathematics.

A **complex number** $c$ is of the form $c = a + bi$, where $a$ and $b$ are real numbers and where $i = \sqrt{-1}$; $a$ is called the **real part** of $c$, and $b$ is called the **imaginary part** of $c$. The term *imaginary part* arose from the mysticism surrounding the beginnings of complex numbers; however, these numbers are as "real" as the real numbers.

EXAMPLE 1

(a) $5 - 3i$ has real part 5 and imaginary part $-3$.

(b) $-6 + \sqrt{2}\,i$ has real part $-6$ and imaginary part $\sqrt{2}$. ∎

The symbol $i = \sqrt{-1}$ has the property that $i^2 = -1$, and we can deduce the following relationships:

$$i^3 = -i, \quad i^4 = 1, \quad i^5 = i, \quad i^6 = -1, \quad i^7 = -i, \dots$$

These results will be useful for simplifying expressions involving complex numbers.

We say that two complex numbers $c_1 = a_1 + b_1 i$ and $c_2 = a_2 + b_2 i$ are **equal** if their real and imaginary parts are equal, that is, if $a_1 = a_2$ and $b_1 = b_2$. Of course, every real number $a$ is a complex number with its imaginary part zero: $a = a + 0i$.

We can operate on complex numbers in much that same way that we operate or manipulate real numbers. If $c_1 = a_1 + b_1 i$ and $c_2 = a_2 + b_2 i$ are complex numbers, then their **sum** is

$$c_1 + c_2 = (a_1 + a_2) + (b_1 + b_2)i,$$

and their **difference** is

$$c_1 - c_2 = (a_1 - a_2) + (b_1 - b_2)i.$$

In words, to form the sum of two complex numbers, add the real parts and add the imaginary parts; to form the difference of two complex numbers, form the difference of the real parts and the difference of the imaginary parts. The **product** of $c_1$ and $c_2$ is

$$c_1 c_2 = (a_1 + b_1 i) \cdot (a_2 + b_2 i) = a_1 a_2 + (a_1 b_2 + b_1 a_2)i + b_1 b_2 i^2$$
$$= (a_1 a_2 - b_1 b_2) + (a_1 b_2 + b_1 a_2)i.$$

A special case of multiplication of complex numbers occurs when $c_1$ is real. In this case we obtain the simple result

$$c_1 c_2 = c_1 \cdot (a_2 + b_2 i) = c_1 a_2 + c_1 b_2 i.$$

If $c = a + bi$ is a complex number, then the **conjugate** of $c$ is the complex number $\bar{c} = a - bi$. It is not difficult to show that if $c$ and $d$ are complex numbers, then the following basic properties of complex arithmetic hold:

1. $\bar{\bar{c}} = c.$
2. $\overline{c + d} = \bar{c} + \bar{d}.$
3. $\overline{cd} = \bar{c}\,\bar{d}.$
4. $c$ is a real number if and only if $c = \bar{c}.$
5. $c\bar{c}$ is a nonnegative real number and $c\bar{c} = 0$ if and only if $c = 0.$

We verify Property 4 here and leave the others as exercises. Let $c = a + bi$ so that $\bar{c} = a - bi$. If $c = \bar{c}$, then $a + bi = a - bi$, so $b = 0$ and $c$ is real. On the other hand, if $c$ is real, then $c = a$ and $\bar{c} = a$, so $c = \bar{c}$.

EXAMPLE 2    Let $c_1 = 5 - 3i$, $c_2 = 4 + 2i$, and $c_3 = -3 + i$.

(a) $c_1 + c_2 = (5 - 3i) + (4 + 2i) = 9 - i.$
(b) $c_2 - c_3 = (4 + 2i) - (-3 + i) = (4 - (-3)) + (2 - 1)i = 7 + i.$
(c) $c_1 c_2 = (5 - 3i) \cdot (4 + 2i) = 20 + 10i - 12i - 6i^2 = 26 - 2i.$
(d) $c_1 \bar{c}_3 = (5 - 3i) \cdot \overline{(-3 + i)} = (5 - 3i) \cdot (-3 - i)$
$\qquad = -15 - 5i + 9i + 3i^2$
$\qquad = -18 + 4i.$
(e) $3c_1 + 2\bar{c}_2 = 3(5 - 3i) + 2\overline{(4 + 2i)} = (15 - 9i) + 2(4 - 2i)$
$\qquad = (15 - 9i) + (8 - 4i) = 23 - 13i.$
(f) $c_1 \bar{c}_1 = (5 - 3i)\overline{(5 - 3i)} = (5 - 3i)(5 + 3i) = 34.$    ■

When we consider systems of linear equations with complex coefficients, we will need to divide complex numbers to complete the solution process and obtain a reasonable form for the solution. Let $c_1 = a_1 + b_1 i$ and $c_2 = a_2 + b_2 i$. If $c_2 \neq 0$, that is, if $a_2 \neq 0$ or $b_2 \neq 0$, then we can **divide** $c_1$ by $c_2$:

$$\frac{c_1}{c_2} = \frac{a_1 + b_1 i}{a_2 + b_2 i}.$$

To conform to our practice of expressing a complex number in the form real part $+$ imaginary part $\cdot i$, we must simplify the foregoing expression for $c_1/c_2$. To simplify this complex fraction, we multiply the numerator and the denominator by the conjugate of the denominator. Thus, dividing $c_1$ by $c_2$ gives the complex number

$$\frac{c_1}{c_2} = \frac{a_1 + b_1 i}{a_2 + b_2 i} = \frac{(a_1 + b_1 i)(a_2 - b_2 i)}{(a_2 + b_2 i)(a_2 - b_2 i)} = \frac{a_1 a_2 + b_1 b_2}{a_2^2 + b_2^2} - \frac{a_1 b_2 + a_2 b_1}{a_2^2 + b_2^2} i.$$

EXAMPLE 3    Let $c_1 = 2 - 5i$ and $c_2 = -3 + 4i$. Then

$$\frac{c_1}{c_2} = \frac{2 - 5i}{-3 + 4i} = \frac{(2 - 5i)(-3 - 4i)}{(-3 + 4i)(-3 - 4i)} = \frac{-26 + 7i}{(-3)^2 + (4)^2} = -\frac{26}{25} + \frac{7}{25}i.$$  ■

Finding the **reciprocal** of a complex number is a special case of division of complex numbers. If $c = a + bi$, $c \neq 0$, then

$$\frac{1}{c} = \frac{1}{a + bi} = \frac{a - bi}{(a + bi)(a - bi)} = \frac{a - bi}{a^2 + b^2}$$

$$= \frac{a}{a^2 + b^2} - \frac{b}{a^2 + b^2}i.$$

EXAMPLE 4

(a)  $\dfrac{1}{2 + 3i} = \dfrac{2 - 3i}{(2 + 3i)(2 - 3i)} = \dfrac{2 - 3i}{2^2 + 3^2} = \dfrac{2}{13} - \dfrac{3}{13}i.$

(b)  $\dfrac{1}{i} = \dfrac{-i}{i(-i)} = \dfrac{-i}{-i^2} = \dfrac{-i}{-(-1)} = -i.$  ■

Summarizing, we can say that complex numbers are mathematical objects for which addition, subtraction, multiplication, and division are defined in such a way that these operations on real numbers can be derived as special cases. In fact, it can be shown that complex numbers form a mathematical system that is called a field.

Complex numbers have a geometric representation. A complex number $c = a + bi$ may be regarded as an ordered pair $(a, b)$ of real numbers. This ordered pair of real numbers corresponds to a point in the plane. Such a correspondence naturally suggests that we represent $a + bi$ as a point in the **complex plane**, where the horizontal axis is used to represent the real part of $c$ and the vertical axis is used to represent the imaginary part of $c$. To simplify matters, we call these the **real axis** and the **imaginary axis**, respectively (see Figure 1).

EXAMPLE 5    Plot the complex numbers $c = 2 - 3i$, $d = 1 + 4i$, $e = -3$, and $f = 2i$ in the complex plane. See Figure 2.  ■

The rules concerning inequality of real numbers, such as less than and greater than, do not apply to complex numbers. There is no way to arrange the complex

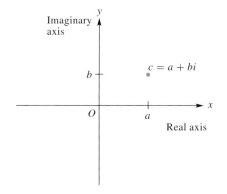

FIGURE 1

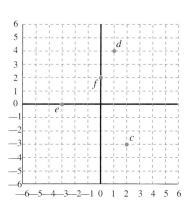

FIGURE 2

numbers according to size. However, using the geometric representation from the complex plane, we can attach a notion of size to a complex number by measuring its distance from the origin. The distance from the origin to $c = a + bi$ is called the **absolute value**, or **modulus**, of the complex number and is denoted by $|c| = |a + bi|$. Using the formula for the distance between ordered pairs of real numbers $(0, 0)$ and $(a, b)$, we obtain

$$|c| = |a + bi| = \sqrt{a^2 + b^2}.$$

It follows that $c\bar{c} = |c|^2$ (verify).

EXAMPLE 6    Referring to Example 5, $|c| = \sqrt{13}$; $|d| = \sqrt{17}$; $|e| = 3$; $|f| = 2$.

■

# EXERCISES

*In Exercises 1–8, let $c_1 = 3 + 4i$, $c_2 = 1 - 2i$, and $c_3 = -1 + i$. Compute each given expression and simplify as much as possible.*

**1.** $c_1 + c_2$.

**2.** $c_3 - c_1$.

**3.** $c_1 c_2$.

**4.** $c_2 \bar{c_3}$.

**5.** $4c_3 + \bar{c_2}$.

**6.** $(-i) \cdot c_2$.

**7.** $\overline{3c_1 - ic_2}$.

**8.** $c_1 c_2 c_3$.

*In Exercises 9–12, write each given expression in the form $a + bi$.*

**9.** $\dfrac{1 + 2i}{3 - 4i}$.

**10.** $\dfrac{1}{3 - i}$.

**11.** $\dfrac{(2 + i)^2}{i}$.

**12.** $\dfrac{1}{(3 + 2i)(1 + i)}$.

*In Exercises 13–16, represent each complex number as a point and as a vector in the complex plane.*

**13.** $4 + 2i$.

**14.** $-3 + i$.

**15.** $3 - 2i$.

**16.** $i(4 + i)$.

**17.** Compute the absolute value of each of the complex numbers in Exercises 13–16.

**18.** Verify properties 1, 2, 3, and 5 of the conjugate of a complex number.

# APPENDIX 2

## SUMMATION NOTATION

We will use the **summation notation**, and we now review this useful and compact notation. By $\sum_{j=1}^{n} c_j$ we mean

$$c_1 + c_2 + \cdots + c_n$$

and refer to $\sum_{j=1}^{n} c_j$ as a **summation** or **sum**. The letter $j$ is called the **index of summation**; it is an indicator that takes on the integer values 1 through $n$. The index of summation can be any letter, but we often use $i$, $j$, or $k$. We have

$$\sum_{j=1}^{n} c_j = \sum_{k=1}^{n} c_k = \sum_{i=1}^{n} c_i.$$

The following diagram shows the names commonly used for parts of the summation notation.

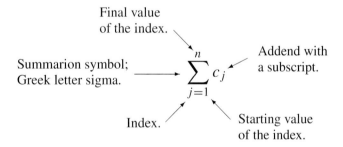

In this notation it is understood that the index, sometimes called the dummy variable, begins with the starting value (which may be something other than 1) and changes by 1 until it reaches the final value. We assume that the final value of the index is greater than or equal to the starting value; otherwise, the value of the summation is taken to be zero. The starting value of the index to the final value is called the **index range**.

EXAMPLE 1    Let $c_1 = 5$, $c_2 = 3$, $c_3 = 7$, $c_4 = -2$, $c_5 = 0$, and $c_6 = 10$.

(a) $\sum_{j=1}^{6} c_j = c_1 + c_2 + c_3 + c_4 + c_5 + c_6 = 5 + 3 + 7 + (-2) + 0 + 10 = 23$.

(b) $\sum_{j=2}^{4} c_k = c_2 + c_3 + c_4 = 3 + 7 + (-2) = 8$.    ∎

Summations with the same index range can be added together or subtracted from another:

$$\sum_{j=1}^{n} c_j + \sum_{j=1}^{n} b_j = \sum_{j=1}^{n} (c_j + b_j),$$

$$\sum_{j=1}^{n} c_j - \sum_{j=1}^{n} b_j = \sum_{j=1}^{n} (c_j - b_j).$$

If the index range is the same, but the index names are different we can still perform these operations by changing the names of the indices to be the same. For example,

$$\sum_{j=1}^{n} c_j + \sum_{k=1}^{n} b_k = \sum_{j=1}^{n} c_j + \sum_{j=1}^{n} b_j = \sum_{j=1}^{n} (c_j + b_j).$$

We can also multiply a summation by a constant:

$$t \sum_{j=1}^{n} c_j = \sum_{j=1}^{n} tc_j = tc_1 + tc_2 + \cdots + tc_n.$$

Example 2 illustrates various summation expressions.

EXAMPLE 2    Let

$$c_1 = 5, \quad c_2 = 3, \quad c_3 = 7, \quad c_4 = -2, \quad c_5 = 0, \quad c_6 = 10$$

and

$$b_1 = 2, \quad b_2 = -1, \quad b_3 = 6, \quad b_4 = 3, \quad b_5 = 7, \quad b_6 = 0.$$

(a) $\displaystyle\sum_{j=1}^{6} 3c_j = 3 \sum_{j=1}^{6} c_j = 3(23) = 69.$    (See Example 1(a).)

(b) $\displaystyle\sum_{k=1}^{6} 2c_k + \sum_{k=1}^{6} 4b_k = 2 \sum_{k=1}^{6} c_k + 4 \sum_{k=1}^{6} b_k$

$$= 2(23) + 4(17) = 46 + 68 = 114.    \text{(Verify.)}$$

(c) $\displaystyle\sum_{i=3}^{5} c_i b_i = c_3 b_3 + c_4 b_4 + c_5 b_5 = (7)(6) + (-2)(3) + (0)(7) = 36.$    ■

We will have occasion to use doubly indexed expressions like $c_{ij}$ where $i = 1, 2, 3$ and $j = 1, 2, 3, 4$. We can represent such expressions in a table as follows:

$$
\begin{array}{cccc}
c_{11} & c_{12} & c_{13} & c_{14} \\
c_{21} & c_{22} & c_{23} & c_{24} \\
c_{31} & c_{32} & c_{33} & c_{34}
\end{array}
$$

Each of the following is a summation involving doubly indexed expressions:

$$\sum_{i=1}^{3} c_{i2} = c_{12} + c_{22} + c_{32},$$

$$\sum_{i=2}^{3} \sum_{j=1}^{2} c_{ij} = \sum_{i=2}^{3} (c_{i1} + c_{i2}) = \sum_{i=2}^{3} c_{i1} + \sum_{i=2}^{3} c_{i2} = (c_{21} + c_{31}) + (c_{22} + c_{32}).$$

In an expression of the form $\sum_{i=1}^{n} \sum_{j=1}^{m} c_{ij}$ we first sum on the index for the inside summation, here it is $j$, and then sum the resulting expression on the outside index, here $i$. Since sums can be reordered without changing their total, we have

$$\sum_{i=1}^{n} \sum_{j=1}^{m} c_{ij} = \sum_{j=1}^{m} \sum_{i=1}^{n} c_{ij}.$$

That is, the order of summation can be interchanged.

## EXERCISES

Let

$$c_1 = 2, \quad c_2 = -3, \quad c_3 = 6, \quad c_4 = -4, \quad c_5 = 7$$

and

$$b_1 = 5, \quad b_2 = -1, \quad b_3 = -3, \quad b_4 = 3, \quad b_5 = 0.$$

*In Exercises 1–5, compute each given summation expression.*

**1.** $\displaystyle\sum_{k=1}^{4} c_k.$

**2.** $\displaystyle\sum_{j=3}^{5} (2b_j - c_j).$

**3.** $\displaystyle\sum_{j=2}^{4} b_{j-1}.$

**4.** $\displaystyle\sum_{k=1}^{5} (-1)^k c_k.$

**5.** $\displaystyle\sum_{i=1}^{5} c_i b_i.$

*In Exercises 6–8, verify each of the given expressions.*

**6.** $\displaystyle\sum_{j=1}^{n} (c_j + 1) = \sum_{j=1}^{n} c_j + n.$

**7.** $\displaystyle\sum_{j=1}^{m} \sum_{i=1}^{n} 1 = mn.$

**8.** $\displaystyle\sum_{j=1}^{n} \sum_{k=1}^{m} c_j b_k = \sum_{j=1}^{n} c_j \sum_{k=1}^{m} b_k.$

# ANSWERS/SOLUTIONS TO SELECTED EXERCISES

## CHAPTER 1

### Section 1.1, Page 9

**1.** $\begin{bmatrix} -2 & -\frac{3}{2} & -1 & -\frac{1}{2} & 0 & \frac{1}{2} & 1 & \frac{3}{2} \\ -8 & -\frac{27}{8} & -1 & -\frac{1}{8} & 0 & \frac{1}{8} & 1 & \frac{27}{8} \end{bmatrix}$.

**3.** $\begin{bmatrix} 0 & -5 & -10 \\ -25 & -31 & -38 \end{bmatrix}$.

**5.** Equation: $y = \frac{5}{9}(x - 32)$. Matrix:
$\begin{bmatrix} 40 & 50 & 60 & 70 & 80 \\ 4.4444 & 10 & 15.5556 & 21.1111 & 26.6667 \end{bmatrix}$.

**7.** Name the three parts $x$, $y$, and $z$ respectively. The equations are $x + y + z = 24{,}000$, $y = 2x$, $0.09x + 0.10y + 0.06z = 2210$. The matrix representing this system is
$\begin{bmatrix} 1 & 1 & 1 & 24{,}000 \\ -2 & 1 & 0 & 0 \\ 0.09 & 0.10 & 0.06 & 2210 \end{bmatrix}$.

**9.**

|       | $P_1$ | $P_2$ | $P_3$ | $P_4$ | $P_5$ |
|-------|-------|-------|-------|-------|-------|
| $P_1$ | 0 | 1 | 0 | 0 | 1 |
| $P_2$ | 1 | 0 | 1 | 1 | 1 |
| $P_3$ | 0 | 1 | 0 | 0 | 0 |
| $P_4$ | 0 | 1 | 0 | 0 | 0 |
| $P_5$ | 1 | 1 | 0 | 0 | 0 |

**12.**

|       | $P_1$ | $P_2$ | $P_3$ | $P_4$ | $P_5$ | $P_6$ |
|-------|-------|-------|-------|-------|-------|-------|
| $P_1$ | 0 | 1 | 0 | 0 | 0 | 1 |
| $P_2$ | 1 | 0 | 0 | 0 | 1 | 0 |
| $P_3$ | 0 | 0 | 0 | 1 | 1 | 0 |
| $P_4$ | 0 | 0 | 1 | 0 | 1 | 0 |
| $P_5$ | 0 | 1 | 1 | 1 | 0 | 1 |
| $P_6$ | 1 | 0 | 0 | 0 | 1 | 0 |

**13.** (Your graph may look different.)

**16.** (a) $\mathbf{A}^T = \begin{bmatrix} 5 & 4 \\ -3 & 9 \\ 1 & 6 \\ 2 & 7 \end{bmatrix}$. (b) $(\mathbf{A}^T)^T = \mathbf{A}$.

(c) The transpose changes rows to columns; applying it twice returns a row to its original position.

(d) The entries with values 5 and 9 do not change position. They are the diagonal entries of matrix $\mathbf{A}$.

**18.** No. **21.** $\begin{bmatrix} 0.2 & 0.5 & 0.4 \\ 0.3 & 0.3 & 0.1 \\ 0.5 & 0.2 & 0.5 \end{bmatrix}$. **25.** No.

**28.** (a) There are lots of answers; one is $\begin{bmatrix} 50 & 1 & -5 \\ 6 & 12 & -2 \\ 8 & -9 & 20 \end{bmatrix}$.

(b) No such matrix exists.

(c) No; the absolute value of the diagonal entry must be strictly larger than the sum of all the other entries in the row.

(d) No; a diagonal matrix can have a row of all zeros.

**31.** $\begin{bmatrix} 5 & 8 & -4 & 2 \\ 0 & -4 & 6 & 0 \\ 0 & 0 & 2 & 10 \end{bmatrix}$; the part corresponding to the coefficients is upper triangular.

**33.** $\mathbf{B}^* = \begin{bmatrix} -i \\ 0 \\ 2 + 4i \end{bmatrix}$.

**35.** (a) Yes. (b) No. (c) Yes.

(d) Take any $4 \times 4$ symmetric matrix.

(e) They must be real.

(f) Symmetric.

### Section 1.1 True/False, Page 12

**1.** T  **2.** T  **3.** T  **4.** T  **5.** T

**6.** T  **7.** T  **8.** T  **9.** T  **10.** F

### Section 1.2, Page 26

**1.** $p = 2, q = 4$; $\mathbf{B} = \begin{bmatrix} 2 & 6 \\ -1 & 4 \end{bmatrix}$.

**4.** $p = \pm 3$; $q = \pm\sqrt{2}$.

**6.** $\begin{bmatrix} 3 & 4 & -2 & | & 3 \\ -1 & 0 & 1 & | & -1 \\ 1 & -3 & 2 & | & 5 \end{bmatrix}$.

**7.** $\begin{aligned} 4x_1 - 2x_2 + x_3 &= 3 \\ 2x_1 \qquad - 5x_3 &= 7. \end{aligned}$

**9.** $\begin{bmatrix} 1 & 1 & | & 3 \\ 0 & 1 & | & 3 \end{bmatrix}$.

**11.** $\begin{bmatrix} -1 & 1 & | & 0 \\ -1 & 1 & | & 0 \end{bmatrix}$.

**12.** (a) $4 \times 4$. (b) 64%. (c) Six.

**13. (a)** There are 5050 entries in the upper triangular part. Hence we would need to send 5050 triples, that is, 15,150 pieces of information. But the entire matrix has $100^2 = 10{,}000$ entries. Thus there is no saving in this instance.

**(b)** The tridiagonal matrix has $99 + 100 + 99 = 298$ nonzero entries. To use triples we need to send $3(298) = 894$ values. The saving is $10{,}000 - 894 = 9106$ values; the percent of saving is 91.06%.

**15. (a)** One way is to send just the individual diagonal blocks together with the sizes of the blocks.

**(b)** The entire matrix has $100^2 = 10{,}000$ entries. The diagonal blocks account for
$8^2 + 15^2 + 10^2 + 17^2 + 20^2 + 10^2 + 10^2 + 10^2 = 1378$
entries. The percent of saving is 86.22%.

**17.** $\begin{bmatrix} 15 & -4 \\ -4 & 7 \\ 11 & 8 \end{bmatrix}$.    **19.** $\begin{bmatrix} -8 & 0 \\ 4 & 7 \end{bmatrix}$.

**21.** Not possible; matrices of different sizes.

**24.** $\begin{bmatrix} -9 + 5i & -2 - 2i \\ 7 + 7i & 9 - i \end{bmatrix}$.

**27.** Not possible; matrices of different sizes.

**29. (a)** $\mathbf{A} + \mathbf{B} = (p + q)\mathbf{I}_n$.

**(b)** $r\mathbf{A} = r(p\mathbf{I}_n) = (rp)\mathbf{I}_n$.

**(c)** a scalar matrix.

**31. (a)** Zero.    **(b)** One answer is $\begin{bmatrix} 0 & 1 & 2 \\ -1 & 0 & 3 \\ -2 & -3 & 0 \end{bmatrix}$.

**33.** Adding a pair of $2 \times 4$ matrices gives a $2 \times 4$ matrix and any scalar multiple of a $2 \times 4$ matrix is a $2 \times 4$ matrix.

**35.** Adding a pair of $3 \times 3$ diagonal matrices gives a $3 \times 3$ diagonal matrix and any scalar multiple of a $3 \times 3$ diagonal matrix is a $3 \times 3$ diagonal matrix.

**37.** Adding a pair of $m \times n$ matrices gives an $m \times n$ matrix and any scalar multiple of an $m \times n$ matrix is an $m \times n$ matrix.

**39.** The set is not closed.

**42. (a)**

**(b)** $\begin{bmatrix} M & M & M & M & M & M \\ M & M & M & M & M & M \\ M & M & M & M & M & M \\ M & M & M & M & M & M \\ M & M & M & M & M & M \\ M & M & M & M & M & M \end{bmatrix}$

**(c)**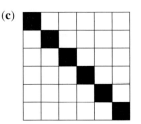

**44. (a)** Since the members of $P_2$ correspond to 3-vectors and the set of 3-vectors is closed, it follows that $P_2$ is closed.

**(b)** Since the members of $P_k$ correspond to $(k + 1)$-vectors and the set of $(k + 1)$-vectors is closed, it follows that $P_k$ is closed.

**(c)** The set of all constants.

**45. (a)** The set of all 2-vectors.

**(c)** The set of all $2 \times 2$ diagonal matrices.

**47.** The matrices in span$(T)$ are symmetric; this matrix is not symmetric.

**49.** $p = 3, q = 1$.    **51.** $p = 2, q = 3$.

**53. (a)** $\text{ent}_{ij}(r(s\mathbf{A})) = (r)\text{ent}_{ij}(s\mathbf{A}) = (r)(s)\text{ent}_{ij}(\mathbf{A}) = (rs)\text{ent}_{ij}(\mathbf{A})$.

**(b)** $\text{ent}_{ij}((r + s)\mathbf{A}) = (r + s)\text{ent}_{ij}(\mathbf{A}) = (r)\text{ent}_{ij}(\mathbf{A}) + (s)\text{ent}_{ij}(\mathbf{A}) = \text{ent}_{ij}(r\mathbf{A}) + \text{ent}_{ij}(s\mathbf{A})$.

**55.** We show that $\left(\frac{1}{2}(\mathbf{B} + \mathbf{B}^T)\right)^T = \frac{1}{2}(\mathbf{B} + \mathbf{B}^T)$ using properties of transposes.

$$\left(\tfrac{1}{2}(\mathbf{B} + \mathbf{B}^T)\right)^T = \tfrac{1}{2}\left(\mathbf{B}^T + (\mathbf{B}^T)^T\right)$$
$$= \tfrac{1}{2}(\mathbf{B}^T + \mathbf{B}) = \tfrac{1}{2}(\mathbf{B} + \mathbf{B}^T).$$

**59. (a)** They must be purely imaginary; that is, of the form $ki$, $k$ a real scalar.

**(b)** One such matrix is $\begin{bmatrix} 5i & 3 + i & 9 - 2i \\ -3 + i & 0 & 8 \\ -9 - 2i & -8 & -4i \end{bmatrix}$.

**61. (a)** $y = -7x + 11$.

**(b)** $\begin{bmatrix} x \\ y \end{bmatrix} = (1 - t)\begin{bmatrix} 2 \\ -3 \end{bmatrix} + t\begin{bmatrix} 1 \\ 4 \end{bmatrix}$.

**(d)** For $t = \frac{1}{2}$, the point is $\left(\frac{3}{2}, \frac{1}{2}\right)$. For $t = \frac{2}{3}$, the point is $\left(\frac{4}{3}, \frac{5}{3}\right)$.

**63.** $x = (1 - t)a + td$, $y = b$.

**65.** $\begin{bmatrix} x \\ y \\ z \end{bmatrix} = t^3\begin{bmatrix} a_1 \\ a_2 \\ a_3 \end{bmatrix} + t^2\begin{bmatrix} b_1 \\ b_2 \\ b_3 \end{bmatrix} + t\begin{bmatrix} c_1 \\ c_2 \\ c_3 \end{bmatrix} + \begin{bmatrix} d_1 \\ d_2 \\ d_3 \end{bmatrix}$.

## Section 1.2 True/False, Page 30

**1.** T    **2.** F    **3.** F    **4.** T    **5.** T

**6.** F    **7.** T    **8.** F    **9.** T    **10.** F

## Section 1.3, Page 43

**2.** $0$.    **4.** $x = -2$.    **5.** $-2$.    **7.** $0$.

**9.** $x = 4, y = -6$.

**11.** All the entries must be zero.    **13.** $1$.

**16.** (a) $\mathbf{v}\cdot(\mathbf{w} + 5\mathbf{u}) = \mathbf{v}\cdot\mathbf{w} + 5(\mathbf{v}\cdot\mathbf{u}) = 2 + 5(-3) = -13$.

   (c) $\mathbf{v}\cdot(2\mathbf{w} + 3\mathbf{u} - \mathbf{x}) = 2(\mathbf{v}\cdot\mathbf{w}) + 3(\mathbf{v}\cdot\mathbf{u}) - (\mathbf{v}\cdot\mathbf{x}) = 2(2) + 3(-3) - 4 = -9$.

**17.** $\mathbf{w} = \begin{bmatrix} 1 & 1 & 1 & 1 \end{bmatrix}$.

**19.** There is more than one way to perform the computations.

   (a) $(\mathbf{v1} + \mathbf{v2})\cdot\begin{bmatrix} 1 & 1 & 1 \end{bmatrix} =$
   $\begin{bmatrix} 47 & 51 & 43 \end{bmatrix}\cdot\begin{bmatrix} 1 & 1 & 1 \end{bmatrix} = 141$.

   (c) $\frac{1}{3}(\mathbf{v1} + \mathbf{v2} + \mathbf{v3}) = \frac{1}{3}\begin{bmatrix} 66 & 84 & 69 \end{bmatrix} =$
   $\begin{bmatrix} 22 & 28 & 23 \end{bmatrix}$.

   (e) $(\mathbf{s1} + \mathbf{s2}) - (\mathbf{v1} + \mathbf{v2} + \mathbf{v3}) = \begin{bmatrix} 165 & 135 & 117 \end{bmatrix} - \begin{bmatrix} 66 & 84 & 69 \end{bmatrix} = \begin{bmatrix} 99 & 51 & 48 \end{bmatrix}$.

**20.** (a) Polynomial: $2x^2 + x - 3$; value at $x = 4$ is
   $\begin{bmatrix} 2 & 1 & -3 \end{bmatrix}\cdot\begin{bmatrix} 16 & 4 & 1 \end{bmatrix} = 33$.

**22.** (a) $5 + 17i$.

**25.** $-3\begin{bmatrix} 4 \\ 2 \\ 1 \\ -1 \end{bmatrix} + 5\begin{bmatrix} 2 \\ 0 \\ 1 \\ 0 \end{bmatrix}$.

**27.** $\begin{bmatrix} 4 & 1 & 2 \\ 1 & -1 & 2 \end{bmatrix}\begin{bmatrix} c_1 \\ c_2 \\ c_3 \end{bmatrix}$.

**31.** $\begin{bmatrix} 5 & 3 & -2 \\ 1 & -6 & 1 \end{bmatrix}\begin{bmatrix} c_1 \\ c_2 \\ c_3 \end{bmatrix}$.

**33.** $\begin{bmatrix} 1 & 1 \\ 2 & -1 \\ 3 & 2 \\ -1 & 4 \end{bmatrix}\begin{bmatrix} c_1 \\ c_2 \end{bmatrix}$.

**35.** $\begin{bmatrix} 0 \\ 2 \\ -7 \\ 7 \end{bmatrix}$.    **37.** $\begin{bmatrix} 0 \\ 0 \\ 0 \end{bmatrix}$.

**39.** $\begin{bmatrix} 1 & 2 \\ -2 & 3 \\ 4 & -1 \end{bmatrix}\begin{bmatrix} x_1 \\ x_2 \end{bmatrix} = \begin{bmatrix} 5 \\ 4 \\ 2 \end{bmatrix}$.

**41.** $\begin{bmatrix} 1 & 1 \\ 2 & -1 \\ 3 & 2 \\ -1 & 4 \end{bmatrix}\begin{bmatrix} x_1 \\ x_2 \end{bmatrix} = \begin{bmatrix} 0 \\ 0 \\ 0 \\ 0 \end{bmatrix}$.

**43.** $\begin{bmatrix} 3 & 1 \\ 5 & 4 \\ -6 & 5 \end{bmatrix}$.    **45.** $\begin{bmatrix} 4 & 1 & 2 \\ 8 & 1 & 7 \\ -4 & -4 & 7 \end{bmatrix}$.

**47.** $\begin{bmatrix} 2 & -2 \\ 4 & 10 \end{bmatrix}$.    **49.** $\begin{bmatrix} 12 & 4 & 3 \\ 4 & -5 & 20 \end{bmatrix}$.

**52.** $\begin{bmatrix} 15 & 5 \\ 9 & 3 \\ -3 & 13 \end{bmatrix}$.    **54.** $\begin{bmatrix} -10 \\ 12 \end{bmatrix}$.

**56.** Not possible.

**57.** (a) $\mathbf{W1} + \mathbf{W2} = \begin{bmatrix} 39 & 32 & 25 & 25 \\ 27 & 44 & 51 & 65 \end{bmatrix}$.

   (c) $\text{row}_1(\mathbf{W3}) * \mathbf{I} = \$378.46$.

**63.** Let $\text{col}_j(\mathbf{B})$ be all zeros. Then
   $\text{col}_j(\mathbf{AB}) = \mathbf{A} * \text{col}_j(\mathbf{B}) = \mathbf{A}*$ column of all zeros; the
   result is a column of all zeros.

**64.** (a) $\begin{bmatrix} 5 & 12 \\ 0 & 5 \end{bmatrix}$.

**65.** (a) $\begin{bmatrix} 6 & 8 & 10 \\ -3 & -4 & -5 \end{bmatrix}$.

**68.** (a) $\mathbf{A}$ is $m \times n$ and $\mathbf{A}^T$ is $n \times m$, so $\mathbf{AA}^T$ is $m \times m$.
   Note: $(\mathbf{AA}^T)^T = (\mathbf{A}^T)^T\mathbf{A}^T = \mathbf{AA}^T$, hence $\mathbf{AA}^T$ is
   symmetric.

**70.** The steady state appears to be $\begin{bmatrix} 0.31 \\ 0.69 \end{bmatrix}$.

**71.** Forming a matrix we see the following:

$$\begin{bmatrix} a_1 \\ d_1 \end{bmatrix} = \begin{bmatrix} \frac{1}{2}\begin{bmatrix} 1 & 1 \end{bmatrix}\begin{bmatrix} x_1 \\ x_2 \end{bmatrix} \\ \frac{1}{2}\begin{bmatrix} 1 & -1 \end{bmatrix}\begin{bmatrix} x_1 \\ x_2 \end{bmatrix} \end{bmatrix}$$

$$= \frac{1}{2}\begin{bmatrix} 1 & 1 \\ 1 & -1 \end{bmatrix}\begin{bmatrix} x_1 \\ x_2 \end{bmatrix}.$$

Thus $\mathbf{M} = \frac{1}{2}\begin{bmatrix} 1 & 1 \\ 1 & -1 \end{bmatrix}$.

**72.** (a) $\begin{bmatrix} 4 \\ -2 \end{bmatrix}$.

**77.** (a) There are two paths of length 2 from $P_2$ to $P_1$. The
   list of paths of length 2 from $P_2$ to $P_3$ is $P_2 P_1 P_3$.

## Section 1.3 True/False, Page 49

**1.** T    **2.** T    **3.** F    **4.** T    **5.** F
**6.** F    **7.** F    **8.** T    **9.** T    **10.** T

## Section 1.4, Page 56

**1.** Let $\mathbf{z}$ be the $n$-vector with each entry zero. Then
   $\text{ent}_i(\mathbf{a} + \mathbf{z}) = a_i + z_i = a_i + 0 = a_i = \text{ent}_i(\mathbf{a})$. Hence
   $\mathbf{a} + \mathbf{z} = \mathbf{a}$.

**3.** For $\mathbf{b}$ in $R^n$, define $\mathbf{x} = \begin{bmatrix} -b_1 & -b_2 & \cdots & -b_n \end{bmatrix}$.
   Then $\text{ent}_i(\mathbf{b} + \mathbf{x}) = b_i + (-b_i) = 0$, so $\mathbf{b} + \mathbf{x} = \mathbf{z}$.
   Assume there is another vector $\mathbf{y}$ such that $\mathbf{b} + \mathbf{y} = \mathbf{z}$; we
   show that $\mathbf{y} = \mathbf{z}$ as follows:
   $\mathbf{y} = \mathbf{z} - \mathbf{b} = (\mathbf{b} + \mathbf{x}) - \mathbf{b} = \mathbf{x}$.

**5.**

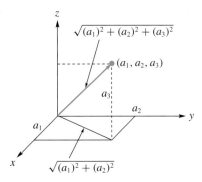

**7.** $\sqrt{41}$.              **9.** 1.

**11.** $t = \pm\sqrt{5}$.              **13.** $t = \pm 3$.

**15.** (a) $\dfrac{1}{\sqrt{10}}\begin{bmatrix} 1 \\ 3 \\ 0 \end{bmatrix}$   (b) $\dfrac{1}{\sqrt{27}}\begin{bmatrix} 1 & 4 & -3 & 1 \end{bmatrix}$.

   (c) **c**.

**21.** The following pairs are orthogonal: **a** and **b**, **a** and **c**, **a** and **d**, **a** and **e**, **b** and **d**, **b** and **e**.

**23.** $t = 0$ or $t = -1$.        **25.** $t = 4, s = 2$.

**27.** Following the hint, we get equations

$$2c_1 + c_2 = 0$$
$$-4c_1 - 2c_2 = 0.$$

Solving these we get $c_2 = -2c_1$. Hence, choose $c_1$ to be any nonzero value, say 1, then $c_1 = 1, c_2 = -2$ gives a vector **c** such that $A\mathbf{c} = \mathbf{0}$.

**30.** **c** is the zero vector in $R^4$. Answer: the zero vector.

**32.** $\cos(\theta) = 0.4983$.        **34.** $\cos(\theta) = 0$.

**36.** (a) $\dfrac{1}{\sqrt{34}}\begin{bmatrix} 2 - 3i \\ 1 + 2i \\ 4 \end{bmatrix} \approx \dfrac{1}{5.8310}\begin{bmatrix} 2 - 3i \\ 1 + 2i \\ 4 \end{bmatrix}$.

**38.** If $\overline{\mathbf{v}}\cdot\mathbf{w} = 0$, then taking the conjugate of both sides gives $0 = \overline{\mathbf{v}\cdot\mathbf{w}} = \overline{\mathbf{w}}\cdot\mathbf{v}$, so **w** is orthogonal to **v**.

## Section 1.4 True/False, Page 59

**1.** F        **2.** T        **3.** T        **4.** F        **5.** F

**6.** F        **7.** T        **8.** T        **9.** F        **10.** F

## Section 1.5, Page 70

**1.** Just use the columns of matrix **A**:

$$\mathbf{u} = \begin{bmatrix} 1 \\ 2 \end{bmatrix}, \mathbf{v} = \begin{bmatrix} -3 \\ 4 \end{bmatrix}, \mathbf{w} = \begin{bmatrix} -1 \\ 0 \end{bmatrix}.$$

**3.** $\begin{bmatrix} -1 \\ -5 \end{bmatrix}$.        **5.** $\begin{bmatrix} 10 \\ 4 \end{bmatrix}$.        **7.** $\begin{bmatrix} 0 \\ 0 \\ 0 \end{bmatrix}$.

**9.** $A\mathbf{c} = \begin{bmatrix} 1 & 2 \\ 1 & 1 \end{bmatrix}\begin{bmatrix} c_1 \\ c_2 \end{bmatrix} = \begin{bmatrix} c_1 + 2c_2 \\ c_1 + c_2 \end{bmatrix} = \mathbf{b} = \begin{bmatrix} 0 \\ 1 \end{bmatrix}$ gives two equations to solve for $c_1$ and $c_2$:

$$c_1 + 2c_2 = 0$$
$$c_1 + c_2 = 1.$$

Solving these gives $c_1 = 2$ and $c_2 = -1$. Thus
$$\mathbf{c} = \begin{bmatrix} 2 \\ -1 \end{bmatrix}.$$

**12.** Yes; $\mathbf{c} = \begin{bmatrix} 3 \\ -1 \end{bmatrix}$.        **14.** Yes; $\mathbf{c} = \begin{bmatrix} 0 \\ 0 \end{bmatrix}$.

**16.** No.

**17.** The range vector is $\begin{bmatrix} c_1 \\ 0 \end{bmatrix}$. It is the projection of the domain vector onto the $x$-axis.

**19.** The range vector is identical to the domain vector; note that the matrix $A = I_2$.

**21.** The range vector is $\begin{bmatrix} 0 \\ c_2 \\ c_3 \end{bmatrix}$. It is the projection of the domain vector onto the $yz$-plane.

**28.** (a) Each image is a rhombus with side of length 2. The individual images are in different positions. They have just been rotated.

   (b) The area of each image is the same as the area of the original rhombus. The original rhombus has its diagonals of length 2 and $2\sqrt{3}$, so Area $= \frac{1}{2}$ (product of the length of the diagonals) $= \frac{1}{2}(2)(2\sqrt{3}) = 2\sqrt{3}$.

**30.** (a) 6.              (c) 6.

**32.** (a) The matrix associated with $g(f(\mathbf{c}))$ is **BA**. We have

$$\mathbf{BA} = \begin{bmatrix} \cos(\phi) & -\sin(\phi) \\ \sin(\phi) & \cos(\phi) \end{bmatrix}\begin{bmatrix} \cos(\theta) & -\sin(\theta) \\ \sin(\theta) & \cos(\theta) \end{bmatrix}$$

$$= \begin{bmatrix} \cos(\phi)\cos(\theta) & -\cos(\phi)\sin(\theta) \\ -\sin(\phi)\sin(\theta) & -\sin(\phi)\cos(\theta) \\ \cos(\phi)\sin(\theta) & \cos(\phi)\cos(\theta) \\ +\sin(\phi)\cos(\theta) & -\sin(\phi)\sin(\theta) \end{bmatrix}$$

$$= \begin{bmatrix} \cos(\theta + \phi) & -\sin(\theta + \phi) \\ -\sin(\theta + \phi) & \cos(\theta + \phi) \end{bmatrix}.$$

   (b) The matrix associated with $f(g(\mathbf{c}))$ is **AB**. We have

$$\mathbf{AB} = \begin{bmatrix} \cos(\theta) & -\sin(\theta) \\ \sin(\theta) & \cos(\theta) \end{bmatrix}\begin{bmatrix} \cos(\phi) & -\sin(\phi) \\ \sin(\phi) & \cos(\phi) \end{bmatrix}$$

$$= \begin{bmatrix} \cos(\phi)\cos(\theta) & -\cos(\phi)\sin(\theta) \\ -\sin(\phi)\sin(\theta) & -\sin(\phi)\cos(\theta) \\ \cos(\phi)\sin(\theta) & \cos(\phi)\cos(\theta) \\ +\sin(\phi)\cos(\theta) & -\sin(\phi)\sin(\theta) \end{bmatrix}$$

$$= \begin{bmatrix} \cos(\theta + \phi) & -\sin(\theta + \phi) \\ -\sin(\theta + \phi) & \cos(\theta + \phi) \end{bmatrix}.$$

   (c) The order in which the rotations are performed is not relevant.

   (d) $\mathbf{AB} = \mathbf{BA}$.

**34. (a)** The matrix associated with $g(f(\mathbf{c}))$ is $\mathbf{W} = \mathbf{BA}$. We have

$$\mathbf{W} = \mathbf{BA} = \begin{bmatrix} 0 & 0 \\ 0 & 1 \end{bmatrix} \begin{bmatrix} 0 & 1 \\ 1 & 0 \end{bmatrix} = \begin{bmatrix} 0 & 0 \\ 1 & 0 \end{bmatrix}.$$

**(b)** The matrix associated with $f(g(\mathbf{c}))$ is $\mathbf{Q} = \mathbf{AB}$. We have

$$\mathbf{Q} = \mathbf{AB} = \begin{bmatrix} 0 & 1 \\ 1 & 0 \end{bmatrix} \begin{bmatrix} 0 & 0 \\ 0 & 1 \end{bmatrix} = \begin{bmatrix} 0 & 1 \\ 0 & 0 \end{bmatrix}.$$

**(c)** The order in which the two transformations are performed cannot be switched.

**(d)** $\mathbf{Q} = \mathbf{AB} \neq \mathbf{BA} = \mathbf{W}$.

**35. (a)** The matrix associated with $g(f(\mathbf{c}))$ is $\mathbf{W} = \mathbf{BA}$. We have

$$\mathbf{W} = \mathbf{BA} = \begin{bmatrix} 1 & 0 \\ 0 & 0 \end{bmatrix} \begin{bmatrix} \cos(30^\circ) & -\sin(30^\circ) \\ \sin(30^\circ) & \cos(30^\circ) \end{bmatrix}$$
$$= \begin{bmatrix} \cos(30^\circ) & -\sin(30^\circ) \\ 0 & 0 \end{bmatrix}.$$

**(b)** The matrix associated with $f(g(\mathbf{c}))$ is $\mathbf{Q} = \mathbf{AB}$. We have

$$\mathbf{Q} = \mathbf{AB}$$
$$= \begin{bmatrix} \cos(30^\circ) & -\sin(30^\circ) \\ \sin(30^\circ) & \cos(30^\circ) \end{bmatrix} \begin{bmatrix} 1 & 0 \\ 0 & 0 \end{bmatrix}$$
$$= \begin{bmatrix} \cos(30^\circ) & 0 \\ \sin(30^\circ) & 0 \end{bmatrix}.$$

**(c)** The order in which the two transformations are performed cannot be switched.

**(d)** $\mathbf{Q} = \mathbf{AB} \neq \mathbf{BA} = \mathbf{W}$.

**39.** $f(\mathbf{c}) = \text{proj}_{\mathbf{x}}\mathbf{c} + \text{proj}_{\mathbf{y}}\mathbf{c}$.

**43.** $\mathbf{p} = \frac{-2}{29}\mathbf{b}$.     **43.** $\mathbf{p} = \frac{11}{13}\mathbf{w}$.     **45.** $\mathbf{p} = \frac{19}{45}\mathbf{r}$.

**47.** False: $\text{proj}_{\mathbf{b}}\mathbf{c} = \dfrac{\mathbf{b} \cdot \mathbf{c}}{\mathbf{b} \cdot \mathbf{b}}\mathbf{b}$ while $\text{proj}_{\mathbf{c}}\mathbf{b} = \dfrac{\mathbf{b} \cdot \mathbf{c}}{\mathbf{c} \cdot \mathbf{c}}\mathbf{c}$. Note that $\text{proj}_{\mathbf{b}}\mathbf{c} = \dfrac{\mathbf{b} \cdot \mathbf{c}}{\mathbf{b} \cdot \mathbf{b}}\mathbf{b}$ is parallel to $\mathbf{b}$, while $\text{proj}_{\mathbf{c}}\mathbf{b} = \dfrac{\mathbf{b} \cdot \mathbf{c}}{\mathbf{c} \cdot \mathbf{c}}\mathbf{c}$ is parallel to $\mathbf{c}$.

## Section 1.5 True/False, Page 70

| | | | | |
|---|---|---|---|---|
| **1.** T | **2.** T | **3.** T | **4.** F | **5.** T |
| **6.** T | **7.** T | **8.** T | **9.** F | **10.** F |

## Chapter 1 Chapter Test, Page 77

**1.** $\begin{bmatrix} 8 & 4 \\ 4 & 11 \end{bmatrix}$; symmetric.    **2.** $x_1 = 3.5$, $x_2 = -1$.

**3.** $0.2\mathbf{b} = \begin{bmatrix} 0.6 \\ -0.2 \end{bmatrix}$.    **4.** $\begin{bmatrix} 8 \\ 32 \end{bmatrix}$.

**5.** $\dfrac{3}{\sqrt{10}} \approx 0.9487$.

**6. (a)** $\begin{bmatrix} -12 \\ -10 \end{bmatrix}$.

**(b)** $\mathbf{DA} = \begin{bmatrix} 6 & 12 \\ 0 & 5 \end{bmatrix} \neq \mathbf{AD} = \begin{bmatrix} 6 & 20 \\ 0 & 5 \end{bmatrix}$.

**7. (a)** $3 \times n$.     **(b)** No; $n$ may not equal $m$.

**(c)** One answer is $\begin{bmatrix} 3 \\ 6 \end{bmatrix}$.

**8. (a)** Parallel.          **(b)** Orthogonal.

**(c)** Linear combination.   **(d)** $-31\mathbf{x}$.

**9. (a)** We have that $\mathbf{u} \cdot \mathbf{v} = \mathbf{w} \cdot \mathbf{v} = 0$. Then $\mathbf{v} \cdot (\mathbf{u} + \mathbf{w}) = \mathbf{v} \cdot \mathbf{u} + \mathbf{v} \cdot \mathbf{w} = 0 + 0 = 0$.

**(b)** We have that $\mathbf{u} \cdot \mathbf{v} = 0$. Then $(k\mathbf{u}) \cdot \mathbf{v} = k(\mathbf{u} \cdot \mathbf{v}) = k0 = 0$.

**(c)** From parts (a) and (b), $S$ is closed under addition and scalar multiplication.

**10. (a)** $c_1\text{col}_1(\mathbf{A}) + c_2\text{col}_2(\mathbf{A}) + c_3\text{col}_3(\mathbf{A})$.

**(b)** $\begin{bmatrix} \text{row}_1(\mathbf{A}) \cdot \mathbf{c} \\ \text{row}_2(\mathbf{A}) \cdot \mathbf{c} \\ \text{row}_3(\mathbf{A}) \cdot \mathbf{c} \\ \text{row}_4(\mathbf{A}) \cdot \mathbf{c} \end{bmatrix}$.

**11.** One way to proceed is to show $\mathbf{A} + \mathbf{A}^T$ is equal to its transpose;

$$(\mathbf{A} + \mathbf{A}^T)^T = \mathbf{A}^T + (\mathbf{A}^T)^T = \mathbf{A}^T + \mathbf{A} = \mathbf{A} + \mathbf{A}^T.$$

**12.** $\text{row}_7(\mathbf{A}) * \text{col}_3(\mathbf{B})$.

**13.** All possible linear combinations of $\mathbf{v}_1$, $\mathbf{v}_2$, $\mathbf{v}_3$, and $\mathbf{v}_4$.

**14.** $\text{span}\left\{ \begin{bmatrix} 1 \\ 0 \\ 1 \\ 0 \end{bmatrix}, \begin{bmatrix} 0 \\ 1 \\ 0 \\ 1 \end{bmatrix} \right\} = S$.

## CHAPTER 2

## Section 2.1, Page 98

**1.** Consistent.                    **3.** Consistent.

**5.** Infinitely many solutions.

**7.** Infinitely many solutions.

**9.** One.                         **11.** Two.

**14.** Interchange rows to get it to upper triangular form

$$\begin{bmatrix} 4 & 3 & 2 & | & 1 \\ 0 & 1 & 0 & | & 1 \\ 0 & 0 & 2 & | & 6 \end{bmatrix}.$$

Back substitution gives $x_3 = 3$, $x_2 = 1$, $x_1 = (1 - 2x_3 - 3x_2)/4 = -2$.

**16.** The system is inconsistent, hence has no solutions.

**18.** $x_1 = -3.2$, $x_2 = 1.6$, $x_3 = 2.4$, $x_4 = -2.8$.

**20.** $x_1 = -2x_4$, $x_2 = 0$, $x_3 = -6x_4$, $x_4 =$ arbitrary constant.

**23. (a)** $x_n = \dfrac{b_n}{a_{nn}}$.

**(b)** $x_{n-1} = \dfrac{b_{n-1} - a_{n-1 \, n} x_n}{a_{n-1 \, n-1}}$.

**(c)** $x_{n-2} = \dfrac{b_{n-2} - a_{n-2 \, n} x_n - a_{n-2 \, n-1} x_{n-1}}{a_{n-2 \, n-2}}$.

**(d)** $x_i = \dfrac{b_i - a_{in} x_n - a_{i \, n-1} x_{n-1} - \cdots - a_{i \, i+1} x_{i+1}}{a_{ii}}$.

**(e)** Just expand the summation symbol.

**25.** $x_1 = \frac{3}{5}, x_2 = -\frac{4}{5}$.

**29.** $\mathbf{x} = \begin{bmatrix} 2 \\ 3 \\ 0 \end{bmatrix} + t \begin{bmatrix} 1 \\ 0 \\ 1 \end{bmatrix}$; thus $\mathbf{x}$ is the translation of $\begin{bmatrix} 1 \\ 0 \\ 1 \end{bmatrix}$

by $\begin{bmatrix} 2 \\ 3 \\ 0 \end{bmatrix}$.

**31.** $\mathbf{x} = \begin{bmatrix} -1 \\ 0 \\ 3 \\ 0 \end{bmatrix} + t \begin{bmatrix} -1 \\ 2 \\ 0 \\ 1 \end{bmatrix}$; thus $\mathbf{x}$ is the translation of

$\begin{bmatrix} -1 \\ 2 \\ 0 \\ 1 \end{bmatrix}$ by $\begin{bmatrix} -1 \\ 0 \\ 3 \\ 0 \end{bmatrix}$.

**33.** $\mathbf{x}$ belongs to span $\left\{ \begin{bmatrix} 2 \\ -3 \\ 1 \end{bmatrix} \right\}$.

**35.** **(a)** $\mathbf{b} = \begin{bmatrix} 0 \\ 0 \end{bmatrix}$ is in the range since $\mathbf{A} \begin{bmatrix} 0 \\ 0 \end{bmatrix} = \mathbf{b}$.

**(b)** There is no vector $\mathbf{c}$ such that $\mathbf{Ac} = \mathbf{b} = \begin{bmatrix} 1 \\ 2 \end{bmatrix}$, so $\mathbf{b}$

is not in the range.

**(c)** There is no vector $\mathbf{c}$ such that $\mathbf{Ac} = \mathbf{b} = \begin{bmatrix} 1 \\ 2 \end{bmatrix}$, so $\mathbf{b}$

is not in the range.

**37.** $t = 2$.

**39.** $r = 3, s = 1, t = 3, p = -2$.

**41.** $\mathbf{z} = \begin{bmatrix} 9 \\ 5 \\ 2 \end{bmatrix}, \mathbf{x} = \begin{bmatrix} 1 \\ 2 \\ 1 \end{bmatrix}$.

**43.** If any of the diagonal entries of $\mathbf{L}$ or $\mathbf{U}$ is zero, there will not be a unique solution.

**Strategy for Exercise 45:** We construct and solve the linear system from the equations $p(x_i) = ax_i^2 + bx_i + c = y_i$ for $i = 1, 2, 3$ where the pairs $(x_i, y_i)$ are from the data specified.

**45.** $2x^2 - x - 1$.

**47.** We have the following equations:
$p(1) = a + b + c = f(1) = 1$,
$p'(1) = 2a + b = f'(1) = 2, \ p''(1) = 2a = f''(1) = 3$.
It follows that we have linear system

$$\begin{bmatrix} 1 & 1 & 1 & 1 \\ 2 & 1 & 0 & 2 \\ 2 & 0 & 0 & 3 \end{bmatrix}.$$

The solution is

$$\begin{bmatrix} \frac{3}{2} \\ -1 \\ \frac{1}{2} \end{bmatrix},$$

hence the quadratic polynomial is $\frac{3}{2}x^2 - x + \frac{1}{2}$.

**49.** $\begin{bmatrix} 0.2 + 1.6i \\ 0.2 + 0.2i \end{bmatrix}$.

**51.** If we let $x_2 = k$, any complex number, then
$$x_1 = \frac{1 - (2 + 2i)k}{1 - i}.$$

## Section 2.1 True/False, Page 103

**1.** F      **2.** T      **3.** T      **4.** T      **5.** T

**6.** F      **7.** T      **8.** T      **9.** F      **10.** F

## Section 2.2, Page 118

**1.** REF.

**3.** RREF and REF.

**6.** RREF and REF.

**7.** $\begin{bmatrix} 1 & 3 & 2 & 1 \\ 0 & 1 & \frac{1}{24} & \frac{1}{12} \\ 0 & 0 & 0 & 1 \end{bmatrix}$. A matrix can have many REFs, so

your answer may not be the same.

**9.** $\mathbf{I}_3$; $\begin{bmatrix} 1 & 0 & \frac{15}{8} & 0 \\ 0 & 1 & \frac{1}{24} & 0 \\ 0 & 0 & 0 & 1 \end{bmatrix}$; $\begin{bmatrix} 1 & 4 & 0 \\ 0 & 0 & 1 \\ 0 & 0 & 0 \end{bmatrix}$.

**11.** $\left[\begin{array}{cc|c} 1 & 0 & c_1 \\ 0 & 1 & c_2 \end{array}\right]$, $\left[\begin{array}{cc|c} 1 & p & c_1 \\ 0 & 0 & c_2 \end{array}\right]$.

**13.** $\left[\begin{array}{ccc|c} 1 & 0 & 0 & c_1 \\ 0 & 1 & 0 & c_2 \\ 0 & 0 & 1 & c_3 \end{array}\right]$ where $\begin{array}{c} p \\ q \\ r \end{array}$, $\left[\begin{array}{cccc|c} 1 & 0 & p & 0 & c_1 \\ 0 & 1 & q & 0 & c_2 \\ 0 & 0 & 0 & 1 & c_3 \end{array}\right]$,

$\left[\begin{array}{cccc|c} 1 & p & 0 & 0 & c_1 \\ 0 & 0 & 1 & 0 & c_2 \\ 0 & 0 & 0 & 1 & c_3 \end{array}\right]$, $\left[\begin{array}{cccc|c} 1 & 0 & p & q & c_1 \\ 0 & 1 & r & s & c_2 \\ 0 & 0 & 0 & 0 & c_3 \end{array}\right]$,

$\left[\begin{array}{cccc|c} 1 & p & 0 & q & c_1 \\ 0 & 0 & 1 & r & c_2 \\ 0 & 0 & 0 & 0 & c_3 \end{array}\right]$, $\left[\begin{array}{cccc|c} 1 & p & q & 0 & c_1 \\ 0 & 0 & 0 & 1 & c_2 \\ 0 & 0 & 0 & 0 & c_3 \end{array}\right]$,

$\left[\begin{array}{cccc|c} 1 & p & q & r & c_1 \\ 0 & 0 & 0 & 0 & c_2 \\ 0 & 0 & 0 & 0 & c_3 \end{array}\right]$.

**15.** $\mathbf{x} = \begin{bmatrix} -\frac{7}{3} \\ \frac{4}{3} \\ 0 \end{bmatrix} + t \begin{bmatrix} -\frac{5}{3} \\ -\frac{1}{3} \\ 1 \end{bmatrix} = \begin{bmatrix} -\frac{7}{3} \\ \frac{4}{3} \\ 0 \end{bmatrix} + \text{span} \left\{ \begin{bmatrix} -\frac{5}{3} \\ -\frac{1}{3} \\ 1 \end{bmatrix} \right\}$.

**17.** $\mathbf{x} = \begin{bmatrix} 1 \\ 3 \\ 2 \\ 0 \end{bmatrix} + t \begin{bmatrix} -1 \\ 1 \\ -1 \\ 1 \end{bmatrix} = \begin{bmatrix} 1 \\ 3 \\ 2 \\ 0 \end{bmatrix} + \text{span} \left\{ \begin{bmatrix} -1 \\ 1 \\ -1 \\ 1 \end{bmatrix} \right\}$.

**19.** $\begin{bmatrix} 1 & 1 & -1 & | & 2 \\ 1 & 2 & 1 & | & 3 \\ 1 & 1 & a^2-5 & | & a \end{bmatrix} \begin{array}{c} \\ \\ {\scriptstyle -1R_1+R_2} \\ {\scriptstyle -1R_2+R_3} \end{array} \Rightarrow$

$\begin{bmatrix} 1 & 1 & -1 & | & 2 \\ 0 & 1 & 2 & | & 1 \\ 0 & 0 & a^2-4 & | & a-2 \end{bmatrix}.$

No solution, provided $a^2-4=0$, but $a-2 \neq 0$; thus $a=-2$. Unique solution, provided $a^2-4 \neq 0$; thus $a \neq \pm 2$. Infinitely many solutions, provided $a^2-4=0$ and $a-2=0$; thus $a=2$.

**21.** $\begin{bmatrix} 1 & 2 & -3 & | & p \\ 2 & 3 & 3 & | & q \\ 5 & 9 & -6 & | & r \end{bmatrix} \begin{array}{c} \\ \\ {\scriptstyle -2R_1+R_2} \\ {\scriptstyle -5R_1+R_3} \end{array} \Rightarrow$

$\begin{bmatrix} 1 & 2 & -3 & | & p \\ 0 & -1 & 9 & | & q-2p \\ 0 & -1 & 9 & | & r-5p \end{bmatrix} \begin{array}{c} \\ \\ {\scriptstyle -1R_2+R_3} \end{array} \Rightarrow$

$\begin{bmatrix} 1 & 2 & -3 & | & p \\ 0 & -1 & 9 & | & q-2p \\ 0 & 0 & 0 & | & r-3p-q \end{bmatrix}.$

The system will be consistent provided $r-3p-q=0$. Hence there are infinitely many values for which the system is consistent.

**23.** 
$$\begin{array}{llll} \text{At } A: & -x_1 + x_2 + x_3 & & = 300 \\ \text{At } B: & -x_3 - x_4 & & = -800 \\ \text{At } C: & x_4 + x_5 & = 900 \\ \text{At } D: & x_1 - x_2 & -x_5 & = -400 \end{array}$$

The corresponding linear system has RREF

$$\begin{bmatrix} 1 & -1 & 0 & 0 & -1 & | & -400 \\ 0 & 0 & 1 & 0 & -1 & | & -100 \\ 0 & 0 & 0 & 1 & 1 & | & 900 \\ 0 & 0 & 0 & 0 & 0 & | & 0 \end{bmatrix}.$$

**25.** (There are lots of possible answers here.) Another delivery plan is obtained by setting $x_4=25$ and $x_6=10$ in (9). It is convenient to display the plan as a (schedule) matrix as in the table below. We see that warehouse 1 will ship 5 units to store 1, 0 units to store 2, and 25 units to store 3, while warehouse number 2 will ship 15 units to store 1, 25 units to store 2, and 10 units to store 3.

| Warehouse $\Rightarrow$ Store $\Downarrow$ | #1 | #2 |
|---|---|---|
| #1 | 5 | 15 |
| #2 | 0 | 25 |
| #3 | 25 | 10 |

**27.** (a) $260. (b) $267.50.

**29.** $S(x)=$
$$\begin{cases} S_1(x) = 4 + 2(x-2) - \frac{1}{4}(x-2)^3 & \text{on } [2,4] \\ S_2(x) = 6 - (x-4) - \frac{3}{2}(x-4)^2 \\ \qquad + \frac{1}{2}(x-4)^3 & \text{on } [4,5] \end{cases}.$$

**31.** The RREF is $\begin{bmatrix} 1 & 0 & 0 & | & -3 \\ 0 & 1 & 0 & | & 1 \\ 0 & 0 & 1 & | & 1 \end{bmatrix}$; solution is $\begin{bmatrix} -3 \\ 1 \\ 1 \end{bmatrix}.$

## Section 2.2 True/False, Page 122

**1.** T  **2.** F  **3.** T  **4.** F  **5.** F

**6.** T  **7.** T  **8.** T  **9.** T  **10.** T

## Section 2.3, Page 136

**Strategy for Exercises 1–6:** Form a linear combination with unknown coefficients, set it equal to the zero vector, and determine if the only solution of the associated homogeneous linear system is the zero solution, that is, the trivial solution.

**1.** $c_1 \begin{bmatrix} 1 \\ 2 \end{bmatrix} + c_2 \begin{bmatrix} -1 \\ 3 \end{bmatrix} + c_3 \begin{bmatrix} 3 \\ 1 \end{bmatrix} = \begin{bmatrix} 0 \\ 0 \end{bmatrix}$ leads to a homogeneous system with augmented matrix

$$\begin{bmatrix} 1 & -1 & 3 & | & 0 \\ 2 & 3 & 1 & | & 0 \end{bmatrix}.$$

Its RREF is $\begin{bmatrix} 1 & 0 & 2 & | & 0 \\ 0 & 1 & -1 & | & 0 \end{bmatrix}$, which indicates that there is a nontrivial solution. Hence $T$ is a linearly dependent set of vectors.

**3.** $T$ is a linearly independent set of vectors.

**5.** $T$ is a linearly dependent set of vectors.

**7.** $T$ could be linearly dependent. To check linear independence we must show that the *only* linear combination that produces the zero vector is the one with all coefficients equal to zero.

**Strategy for Exercises 9–16:** Form the associated homogeneous linear system, compute its RREF, determine the general solution, and express it as a span of a set of vectors.

**9.** $\mathrm{rref}\left( [\, A \mid 0\, ] \right) = \begin{bmatrix} 1 & 2 & | & 0 \\ 0 & 0 & | & 0 \end{bmatrix}$, hence the general solution of $A\mathbf{x} = \mathbf{0}$ is $\mathbf{x} = k \begin{bmatrix} -2 \\ 1 \end{bmatrix}$ and

$$\mathbf{ns}(A) = \mathrm{span}\left\{ \begin{bmatrix} -2 \\ 1 \end{bmatrix} \right\}.$$

**11.** $\mathbf{ns}(A) = \mathrm{span}\left\{ \begin{bmatrix} -2 \\ 1 \\ 1 \end{bmatrix} \right\}.$

**13.** $\mathbf{ns}(A) = \mathrm{span}\left\{ \begin{bmatrix} -3 \\ -1 \\ 1 \end{bmatrix} \right\}.$

**15.** $\mathbf{ns}(A) = \mathrm{span}\left\{ \begin{bmatrix} -1 \\ 1 \\ 0 \\ 1 \\ 0 \end{bmatrix}, \begin{bmatrix} -2 \\ \frac{1}{2} \\ 0 \\ 0 \\ 1 \end{bmatrix} \right\}.$

**Strategy for Exercise 17:** Form the augmented matrix associated with the homogeneous system, determine its RREF, find the general solution, and write it as a span of a set of vectors.

**17.** The augmented matrix associated with the homogeneous system $(1\mathbf{I}_2 - \mathbf{A})\mathbf{x} = \mathbf{0}$ is $\left[\begin{array}{cc|c} -2 & -2 & 0 \\ -1 & -1 & 0 \end{array}\right]$. Its RREF is $\left[\begin{array}{cc|c} 1 & 1 & 0 \\ 0 & 0 & 0 \end{array}\right]$, so we have $x_1 = -x_2$, with $x_2$ arbitrary; the general solution is

$$\mathbf{x} = \begin{bmatrix} -k \\ k \end{bmatrix} = k\begin{bmatrix} -1 \\ 1 \end{bmatrix}.$$

Thus $\mathbf{ns}(1\mathbf{I}_2 - \mathbf{A})$ is span $\left\{\begin{bmatrix} -1 \\ 1 \end{bmatrix}\right\}$.

**21. (a)** $\mathbf{v} = 1\begin{bmatrix} 1 \\ -1 \\ 0 \end{bmatrix} + 1\begin{bmatrix} 2 \\ 1 \\ 1 \end{bmatrix} - 1\begin{bmatrix} 3 \\ 0 \\ 2 \end{bmatrix}$.

**(b)** $\mathbf{v} = 2\begin{bmatrix} 1 \\ -1 \\ 0 \end{bmatrix} - 1\begin{bmatrix} 2 \\ 1 \\ 1 \end{bmatrix} + 0\begin{bmatrix} 3 \\ 0 \\ 2 \end{bmatrix}$.

**Strategy for Exercise 23:** From the matrix equation $\mathbf{Ax} = t\mathbf{x}$, form the homogeneous linear system $(\mathbf{A} - t\mathbf{I}_n)\mathbf{x} = \mathbf{0}$. Next, form the augmented matrix associated with the homogeneous system, determine its RREF, find the general solution, and determine vector $\mathbf{x}$.

**23.** $(\mathbf{A} - 0\mathbf{I}_2)\mathbf{x} = \begin{bmatrix} 1 & 1 \\ 1 & 1 \end{bmatrix}\mathbf{x} = \mathbf{0}$ has augmented matrix $\left[\begin{array}{cc|c} 1 & 1 & 0 \\ 1 & 1 & 0 \end{array}\right]$ and RREF $\left[\begin{array}{cc|c} 1 & 1 & 0 \\ 0 & 0 & 0 \end{array}\right]$. It follows that $x_1 = -x_2$, with $x_2$ arbitrary. Hence the general solution is

$$\mathbf{x} = \begin{bmatrix} -k \\ k \end{bmatrix} = k\begin{bmatrix} -1 \\ 1 \end{bmatrix}.$$

One nonzero vector $\mathbf{x}$ such that $\mathbf{Ax} = 0\mathbf{x}$ is $\mathbf{x} = \begin{bmatrix} -1 \\ 1 \end{bmatrix}$.

**Strategy for Exercises 25–32:** Determine the RREF of $\mathbf{A}$, then the nonzero rows form a basis for $\mathbf{row}(\mathbf{A})$.

**25.** $\mathbf{rref}(\mathbf{A}) = \begin{bmatrix} 1 & 2 \\ 0 & 0 \end{bmatrix}$, thus a basis for $\mathbf{row}(\mathbf{A})$ is $\{[\begin{array}{cc} 1 & 2 \end{array}]\}$.

**27.** $\mathbf{rref}(\mathbf{A}) = \begin{bmatrix} 1 & 0 & 2 \\ 0 & 1 & -1 \\ 0 & 0 & 0 \end{bmatrix}$, thus a basis for $\mathbf{row}(\mathbf{A})$ is $\{[\begin{array}{ccc} 1 & 0 & 2 \end{array}], [\begin{array}{ccc} 0 & 1 & -1 \end{array}]\}$.

**29.** $\mathbf{rref}(\mathbf{A}) = \begin{bmatrix} 1 & -2 & 0 \\ 0 & 0 & 1 \\ 0 & 0 & 0 \end{bmatrix}$, thus a basis for $\mathbf{row}(\mathbf{A})$ is $\{[\begin{array}{ccc} 1 & -2 & 0 \end{array}], [\begin{array}{ccc} 0 & 0 & 1 \end{array}]\}$.

**31.** $\mathbf{rref}(\mathbf{A}) = \begin{bmatrix} 1 & 0 & 5 \\ 0 & 1 & -4 \\ 0 & 0 & 0 \end{bmatrix}$, thus a basis for $\mathbf{row}(\mathbf{A})$ is $\{[\begin{array}{ccc} 1 & 0 & 5 \end{array}], [\begin{array}{ccc} 0 & 1 & -4 \end{array}]\}$.

**33. Strategy:** Determine the RREF of $\mathbf{A}^T$. Then the nonzero rows, when written as columns, form a basis for $\mathbf{col}(\mathbf{A})$.

For Exercise 25: $\mathbf{rref}(\mathbf{A}^T) = \begin{bmatrix} 1 & -\frac{1}{2} \\ 0 & 0 \end{bmatrix}$, thus a basis for $\mathbf{col}(\mathbf{A})$ is $\left\{\begin{bmatrix} 1 \\ -\frac{1}{2} \end{bmatrix}\right\}$.

**Strategy for Exercises 34–39:** To determine a basis for $\mathbf{row}(\mathbf{A})$, use $\mathbf{rref}(\mathbf{A})$. To determine a basis for $\mathbf{col}(\mathbf{A})$, use $\mathbf{rref}(\mathbf{A}^T)$. To determine a basis for $\mathbf{ns}(\mathbf{A})$, form the general solution using $\mathbf{rref}([\begin{array}{c|c} \mathbf{A} & \mathbf{0} \end{array}])$.

**35.** $\mathbf{rref}(\mathbf{A}) = \begin{bmatrix} 1 & 0 & 0 \\ 0 & 1 & 0 \\ 0 & 0 & 1 \\ 0 & 0 & 0 \end{bmatrix}$, thus a basis for $\mathbf{row}(\mathbf{A})$ is $\{[\begin{array}{ccc} 1 & 0 & 0 \end{array}], [\begin{array}{ccc} 0 & 1 & 0 \end{array}], [\begin{array}{ccc} 0 & 0 & 1 \end{array}]\}$.

$\mathbf{rref}(\mathbf{A}^T) = \begin{bmatrix} 1 & 0 & 0 & -\frac{2}{9} \\ 0 & 1 & 0 & \frac{4}{9} \\ 0 & 0 & 1 & \frac{1}{3} \end{bmatrix}$, thus a basis for $\mathbf{col}(\mathbf{A})$ is

$$\left\{\begin{bmatrix} 1 \\ 0 \\ 0 \\ -\frac{2}{9} \end{bmatrix}, \begin{bmatrix} 0 \\ 1 \\ 0 \\ \frac{4}{9} \end{bmatrix}, \begin{bmatrix} 0 \\ 0 \\ 1 \\ \frac{1}{3} \end{bmatrix}\right\}.$$

$\mathbf{rref}([\begin{array}{c|c} \mathbf{A} & \mathbf{0} \end{array}]) = \begin{bmatrix} 1 & 0 & 0 & 0 \\ 0 & 1 & 0 & 0 \\ 0 & 0 & 1 & 0 \\ 0 & 0 & 0 & 0 \end{bmatrix}$, thus the general solution is $x_1 = x_2 = x_3 = 0$. In vector form the general solution is

$$\mathbf{x} = \begin{bmatrix} 0 \\ 0 \\ 0 \end{bmatrix}.$$

$\mathbf{ns}(\mathbf{A})$ is the zero subspace, which has no basis.

**37.** A basis for $\mathbf{ns}(\mathbf{A})$ is $\left\{\begin{bmatrix} -1 \\ -1 \\ 1 \end{bmatrix}\right\}$.

**39.** A basis for $\mathbf{ns}(\mathbf{A})$ is $\left\{\begin{bmatrix} 0 \\ -1 \\ 1 \\ 0 \end{bmatrix}\right\}$.

**41.** They are identical since $\mathbf{rref}(\mathbf{A})$ is the same as $\mathbf{rref}(k\mathbf{A})$.

**43.** LD, since one vector is a multiple of the other.

**45.** LD, since one vector is a multiple of the other.

**47.** If **A** and **B** are row equivalent, then **rref**(**A**) = **rref**(**B**). Hence the row spaces would be the same.

**50.** Since **A** has a row of all zeros, **rref** ([ **A** | **b** ]) has a row of the form [ 0  0  ⋯  0 | $t$ ]. If $t = 0$ and the system is consistent, there will be at least one free variable. If $t \neq 0$, then the system is inconsistent. In either case, there is no unique solution.

**52.** (a) We can multiply both sides by $\dfrac{1}{c_1}$ and then solve the expression for $\mathbf{v}_1$ to get

$$\mathbf{v}_1 = \frac{1}{c_1}(-c_2\mathbf{v}_2 - c_3\mathbf{v}_3 - \cdots - c_k\mathbf{v}_k),$$

which implies that $\mathbf{v}_1$ is a linear combination of the other vectors.

(b) For ease, assume $\mathbf{v}_1$ is a linear combination of vectors $\mathbf{v}_2, \mathbf{v}_3, \ldots, \mathbf{v}_k$ such that $\mathbf{v}_1 = a_2\mathbf{v}_2 + a_3\mathbf{v}_3 + \cdots + a_k\mathbf{v}_k$. Then to show that $T$ is a linearly dependent set, we display a linear combination of the vectors that gives the zero vector in which not all the coefficients are zero. We have $\mathbf{v}_1 - a_2\mathbf{v}_2 - a_3\mathbf{v}_3 - \cdots - a_k\mathbf{v}_k = \mathbf{0}$.

**56.** $\left\{ \begin{bmatrix} 1 \\ 0 \end{bmatrix}, \begin{bmatrix} 0 \\ 1 \end{bmatrix} \right\}$; dim($R^2$) = 2.

**58.** The columns of the $k$ by $k$ identity matrix; dim($R^k$) = $k$.

**60.** dim($V$) = 1.

**62.** The rows with leading 1's in **rref**(**A**) form a basis for **row**(**A**) and dim(**row**(**A**)) is the number of vectors in a basis for **row**(**A**).

**64.** Conjecture: dim(**row**(**A**)) = dim(**col**(**A**)).

**65.** (a) dim(**row**(**A**)) = $r$ since there are $r$ leading 1's in **rref**(**A**).

(b) Conjecture: dim(**col**(**A**)) = $r$.

**69.** (a) A basis for **ns**(**B**) is $\left\{ \begin{bmatrix} 2 \\ 1 \\ 0 \\ 0 \end{bmatrix}, \begin{bmatrix} -1 \\ 0 \\ -2 \\ 1 \end{bmatrix} \right\}$.

(b) dim(**ns**(**B**)) = 2.

(c) dim($R^4$) = 4.

**71.** (a) We consider the expression $k_1\mathbf{v} + k_2\mathbf{w} + k_3\mathbf{e}_1 + k_4\mathbf{e}_2 = \mathbf{0}$. The corresponding linear system has augmented matrix

$$\begin{bmatrix} 2 & -1 & 1 & 0 & 0 \\ 1 & 0 & 0 & 1 & 0 \\ 0 & -2 & 0 & 0 & 0 \\ 0 & 1 & 0 & 0 & 0 \end{bmatrix},$$

which has RREF

$$\begin{bmatrix} 1 & 0 & 0 & 1 & 0 \\ 0 & 1 & 0 & 0 & 0 \\ 0 & 0 & 1 & -2 & 0 \\ 0 & 0 & 0 & 0 & 0 \end{bmatrix}.$$

It follows that there exists a nontrivial solution, hence {$\mathbf{v}, \mathbf{w}, \mathbf{e}_1, \mathbf{e}_2$} is linearly dependent.

(b) {$\mathbf{v}, \mathbf{w}, \mathbf{e}_1, \mathbf{e}_2$} is not a basis since it is a linearly dependent set.

**73.** **rref** ([ **A** | **0** ]) = $\begin{bmatrix} 1 & 2 & 1 & 0 \\ 0 & 0 & 0 & 0 \\ 0 & 0 & 0 & 0 \end{bmatrix}$, hence the general solution is $x_1 = -2x_2 - x_3, x_2 = k$ and $x_3 = r$. It follows that

$$\{\mathbf{v}_1, \mathbf{v}_2\} = \left\{ \begin{bmatrix} -2 \\ 1 \\ 0 \end{bmatrix}, \begin{bmatrix} -1 \\ 0 \\ 1 \end{bmatrix} \right\}$$

is a basis for **ns**(**A**). To extend this to a basis for $R^3$, we form the matrix

$$\mathbf{C} = [ \mathbf{v}_1 \quad \mathbf{v}_2 \quad \mathbf{I}_3 ] = \begin{bmatrix} -2 & -1 & 1 & 0 & 0 \\ 1 & 0 & 0 & 1 & 0 \\ 0 & 1 & 0 & 0 & 1 \end{bmatrix}$$

and compute its RREF. We obtain

$$\begin{bmatrix} 1 & 0 & 0 & 1 & 0 \\ 0 & 1 & 0 & 0 & 1 \\ 0 & 0 & 1 & 2 & 1 \end{bmatrix},$$

which implies that

$$\left\{ \begin{bmatrix} -2 \\ 1 \\ 0 \end{bmatrix}, \begin{bmatrix} -1 \\ 0 \\ 1 \end{bmatrix}, \begin{bmatrix} 1 \\ 0 \\ 0 \end{bmatrix} \right\}$$

is a basis for $R^3$.

**75.** One answer: $\left\{ \begin{bmatrix} 1 \\ 0 \\ 0 \\ 2 \end{bmatrix}, \begin{bmatrix} 0 \\ 1 \\ 0 \\ -1 \end{bmatrix}, \begin{bmatrix} 0 \\ 0 \\ 1 \\ 0 \end{bmatrix}, \begin{bmatrix} 1 \\ 0 \\ 0 \\ 0 \end{bmatrix} \right\}$ is a basis for $R^4$.

## Section 2.3 True/False, Page 141

**1.** F  **2.** T  **3.** T  **4.** T  **5.** T

**6.** T  **7.** T  **8.** T  **9.** T  **10.** T

**11.** T  **12.** T  **13.** T  **14.** T  **15.** T

**16.** T  **17.** T  **18.** T  **19.** T  **20.** T

## Section 2.4, Page 158

**Strategy for Exercises 1–5:** We determine the RREF of the augmented matrix representing the linear system.

1. Note that the coefficient matrix is upper triangular with nonzero diagonal entries, thus the system has a unique solution;

$$\mathbf{rref}\left(\begin{bmatrix} 1 & -2 & 4 & 4 \\ 0 & 3 & 5 & 8 \\ 0 & 0 & -2 & -2 \end{bmatrix}\right) = \begin{bmatrix} 1 & 0 & 0 & 2 \\ 0 & 1 & 0 & 1 \\ 0 & 0 & 1 & 1 \end{bmatrix},$$

so the solution is $\mathbf{x} = \begin{bmatrix} 2 \\ 1 \\ 1 \end{bmatrix}$.

3. $\mathbf{x} = \begin{bmatrix} 0 \\ 1 \\ 2 \end{bmatrix}$.          5. $\mathbf{x} = \begin{bmatrix} -i \\ 1 \end{bmatrix}$.

**Strategy for Exercises 6–10:** Show that the RREF of each matrix is an identity matrix.

7. $\mathbf{rref}\left(\begin{bmatrix} 1 & 1 & 1 \\ 2 & 3 & 1 \\ 1 & 2 & 2 \end{bmatrix}\right) = \mathbf{I}_3$.

9. $\mathbf{rref}\left(\begin{bmatrix} 3 & 4 & -1 & 0 \\ 1 & -2 & 2 & 0 \\ -1 & 0 & 1 & 0 \\ 0 & 0 & 0 & 8 \end{bmatrix}\right) = \mathbf{I}_4$.

11. **Strategy:** Compute the reduced row echelon form of the matrix with the same size identity appended; the inverse is the matrix that results in the same position as the appended identity matrix.

For Exercise 6:

$$\mathbf{rref}\left(\begin{bmatrix} 2 & 0 & 0 & 1 & 0 & 0 \\ 3 & -1 & 0 & 0 & 1 & 0 \\ 4 & 0 & 2 & 0 & 0 & 1 \end{bmatrix}\right)$$

$$= \begin{bmatrix} 1 & 0 & 0 & \frac{1}{2} & 0 & 0 \\ 0 & 1 & 0 & \frac{3}{2} & -1 & 0 \\ 0 & 0 & 1 & -1 & 0 & \frac{1}{2} \end{bmatrix};$$

the inverse is $\begin{bmatrix} \frac{1}{2} & 0 & 0 \\ \frac{3}{2} & -1 & 0 \\ -1 & 0 & \frac{1}{2} \end{bmatrix}$.

**Strategy for Exercises 12–15:** Compute $\mathbf{x} = \mathbf{A}^{-1}\mathbf{b}$.

13. $\mathbf{x} = \begin{bmatrix} 0 \\ 1 \\ 2 \end{bmatrix}$.          15. $\mathbf{x} = \begin{bmatrix} -3 \\ 2 \\ -1 \end{bmatrix}$.

**Strategy for Exercise 22:** Rearrange the expression for $\mathbf{w}$ into a system of equations that can be solved for $\mathbf{w}$.

22. Multiply both sides by $\mathbf{A}$ to get $\mathbf{Aw} = (\mathbf{C} + \mathbf{F})\mathbf{v}$. Compute the expression $(\mathbf{C} + \mathbf{F})\mathbf{v}$ to obtain the right side of the linear system. Now solve the linear system.

$$\mathbf{Aw} = (\mathbf{C} + \mathbf{F})\mathbf{v} = \begin{bmatrix} 29 \\ 6 \\ 12 \end{bmatrix};$$

$$\mathbf{rref}\left(\begin{bmatrix} 1 & 0 & -2 & 29 \\ 1 & 1 & 0 & 6 \\ 0 & 1 & 1 & 12 \end{bmatrix}\right)$$

$$= \begin{bmatrix} 1 & 0 & 0 & -41 \\ 0 & 1 & 0 & 47 \\ 0 & 0 & 1 & -35 \end{bmatrix}.$$

Thus $\mathbf{w} = \begin{bmatrix} -41 \\ 47 \\ -35 \end{bmatrix}$.

**Strategy for Exercise 26:** Compute $\mathbf{rref}(\mathbf{A})$ and determine the values of $s$ so that the result will be an identity matrix.

26. $\mathbf{A}$ will be nonsingular provided $s \neq 0, \pm\sqrt{2}$.

**Strategy for Exercise 28:** To show that one matrix is the inverse of another, we verify that their product is the identity matrix.

28. Here we want to show that the inverse of $\mathbf{A}^T$ is $(\mathbf{A}^{-1})^T$, so we compute the product $\mathbf{A}^T(\mathbf{A}^{-1})^T$ as follows:

$$\begin{aligned} \mathbf{A}^T(\mathbf{A}^{-1})^T &= (\mathbf{A}^{-1}\mathbf{A})^T && \text{since } (\mathbf{AB})^T = \mathbf{B}^T\mathbf{A}^T \\ &= (\mathbf{I}_n)^T && \text{since } \mathbf{AA}^{-1} = \mathbf{I}_n \\ &= \mathbf{I}_n && \text{since } \mathbf{I}_n \text{ is symmetric} \end{aligned}$$

33. Here we have two propositions to verify. First, we show that if $\mathbf{D}$ is nonsingular then all of its diagonal entries are nonzero. Secondly, we must show that if all the diagonal entries of $\mathbf{D}$ are nonzero then $\mathbf{D}$ is nonsingular. We proceed as follows:

For the first statement: If $\mathbf{D}$ is nonsingular then its RREF is $\mathbf{I}_n$. Since $\mathbf{D}$ is a diagonal matrix, the only way this can be true is if each of the row operations $(1/d_{jj})\mathbf{R}_j$ is valid. Hence each diagonal entry of $\mathbf{D}$ must be nonzero.

For the second statement: If each of the diagonal entries of $\mathbf{D}$ is nonzero, then all the row operations $(1/d_{jj})\mathbf{R}_j$ are valid. The resulting matrix is $\mathbf{I}_n$. This is the RREF of $\mathbf{D}$, hence $\mathbf{D}$ is nonsingular.

37. Nonsingular. the inverse is $\begin{bmatrix} \frac{1}{5} & 0 & 0 \\ 0 & -\frac{1}{6} & 0 \\ 0 & 0 & \frac{1}{3} \end{bmatrix}$.

**39.** Nonsingular. The inverse is $\begin{bmatrix} \frac{1}{3} & 0 & 0 \\ -\frac{2}{3} & 1 & 0 \\ \frac{4}{3} & -2 & 1 \end{bmatrix}$.

**41.** Nonsingular. The inverse is $\begin{bmatrix} 1 & 0 & 0 & 0 \\ 0 & 2 & 0 & 0 \\ 0 & 0 & 3 & 0 \\ 0 & 0 & 0 & 4 \end{bmatrix}$.

**For Exercises 43–47:** Refer to Exercises 6–10 in which we showed that each of these matrices is nonsingular. Then from the list of properties of a nonsingular matrix given in this section we have the following:

**43.** (i) The rows of $\mathbf{I}_3$ are a basis for the row space.
    (ii) The columns of $\mathbf{I}_3$ are a basis for the column space.
    (iii) The null space is $\{\mathbf{0}\}$ and hence there is no basis.
    (iv) The rank is 3.

**45.** (i) The rows of $\mathbf{I}_2$ are a basis for the row space.
    (ii) The columns of $\mathbf{I}_2$ are a basis for the column space.
    (iii) The null space is $\{\mathbf{0}\}$ and hence there is no basis.
    (iv) The rank is 2.

**50.** There is a column of all zeros.

**52.** Row 3 is a scalar multiple of row 1.

**54.** (a) $\mathbf{E}_1\mathbf{A} = \begin{bmatrix} 1 & 0 & 0 \\ 0 & 0 & 1 \\ 0 & 1 & 0 \end{bmatrix} \begin{bmatrix} a & b & c \\ d & e & f \\ g & h & i \end{bmatrix} =$
$\begin{bmatrix} a & b & c \\ g & h & i \\ d & e & f \end{bmatrix} = \begin{bmatrix} a & b & c \\ d & e & f \\ g & h & i \end{bmatrix}_{R_2 \leftrightarrow R_3}$.

**55.** (a) $\begin{bmatrix} 2 & 4 & 0 \\ 2 & 3 & 2 \\ -3 & 8 & 1 \end{bmatrix}_{\frac{1}{2}R_1} \Rightarrow$

$\begin{bmatrix} 1 & 2 & 0 \\ 2 & 3 & 2 \\ -3 & 8 & 1 \end{bmatrix}_{-2R_1 + R_2} \Rightarrow$

$\begin{bmatrix} 1 & 2 & 0 \\ 0 & -1 & 2 \\ -3 & 8 & 1 \end{bmatrix}_{3R_1 + R_3} \Rightarrow$

$\begin{bmatrix} 1 & 2 & 0 \\ 0 & -1 & 2 \\ 0 & 14 & 1 \end{bmatrix}$.

**57.** Let $\mathbf{E}_1 = \begin{bmatrix} 1 & 0 & 0 \\ 0 & 1 & 0 \\ 0 & 0 & 1 \end{bmatrix}_{\frac{1}{2}R_1} = \begin{bmatrix} \frac{1}{2} & 0 & 0 \\ 0 & 1 & 0 \\ 0 & 0 & 1 \end{bmatrix}$,

then $\mathbf{E}_1\mathbf{A} = \begin{bmatrix} 1 & -2 & 0 \\ 0 & 1 & 3 \\ 0 & 0 & 6 \end{bmatrix}$.

Let $\mathbf{E}_2 = \begin{bmatrix} 1 & 0 & 0 \\ 0 & 1 & 0 \\ 0 & 0 & 1 \end{bmatrix}_{\frac{1}{6}R_3} = \begin{bmatrix} 1 & 0 & 0 \\ 0 & 1 & 0 \\ 0 & 0 & \frac{1}{6} \end{bmatrix}$,

then $\mathbf{E}_2(\mathbf{E}_1\mathbf{A}) = \begin{bmatrix} 1 & -2 & 0 \\ 0 & 1 & 3 \\ 0 & 0 & 1 \end{bmatrix}$.

Let $\mathbf{E}_3 = \begin{bmatrix} 1 & 0 & 0 \\ 0 & 1 & 0 \\ 0 & 0 & 1 \end{bmatrix}_{-3R_3 + R_2} = \begin{bmatrix} 1 & 0 & 0 \\ 0 & 1 & -3 \\ 0 & 0 & 1 \end{bmatrix}$,

then $\mathbf{E}_3(\mathbf{E}_2(\mathbf{E}_1\mathbf{A})) = \begin{bmatrix} 1 & -2 & 0 \\ 0 & 1 & 0 \\ 0 & 0 & 1 \end{bmatrix}$.

Let $\mathbf{E}_4 = \begin{bmatrix} 1 & 0 & 0 \\ 0 & 1 & 0 \\ 0 & 0 & 1 \end{bmatrix}_{2R_2 + R_1} = \begin{bmatrix} 1 & 2 & 0 \\ 0 & 1 & 0 \\ 0 & 0 & 1 \end{bmatrix}$,

then $\mathbf{E}_4(\mathbf{E}_3(\mathbf{E}_2(\mathbf{E}_1\mathbf{A}))) = \begin{bmatrix} 1 & 0 & 0 \\ 0 & 1 & 0 \\ 0 & 0 & 1 \end{bmatrix}$.

**59.** (a) RREF of $\mathbf{E}_1$ is an identity matrix.
    (b) RREF of $\mathbf{E}_2$ is an identity matrix.
    (c) RREF of $\mathbf{E}_3$ is an identity matrix.

**61.** The computation of the RREF via row operations is equivalent to the construction of a set of elementary matrices whose product with the original matrix gives the RREF. Since the product of elementary matrices is nonsingular, we have that there exists a nonsingular matrix $\mathbf{F}$ such that $\mathbf{F}\mathbf{A} = \mathbf{rref}(\mathbf{A})$.

**63.** We show that $\mathbf{A}$ times the proposed solution gives $\mathbf{b}$;

$\begin{bmatrix} 1 & 1 & 1 & 1 \\ 0 & 1 & 3 & 7 \\ 0 & 0 & 1 & 0 \\ 0 & 0 & 0 & 1 \end{bmatrix} \begin{bmatrix} 0 \\ -2 \\ 6+4\sqrt{2} \\ -4-4\sqrt{2} \end{bmatrix} = \begin{bmatrix} 0 \\ 4\sqrt{2} \\ -2 \\ 0 \end{bmatrix}$.

**65.** Arrange the message into a 3 by 5 matrix

$$\mathbf{M} = \begin{bmatrix} -40 & -13 & -25 & -37 & -37 \\ 4 & 24 & 29 & 12 & 10 \\ 55 & 31 & 46 & 57 & 51 \end{bmatrix}.$$

Next compute

$$\mathbf{D} = \mathbf{C}^{-1}\mathbf{M} = \begin{bmatrix} 1 & -1 & -2 \\ 1 & 1 & -1 \\ -1 & 2 & 2 \end{bmatrix}^{-1} \begin{bmatrix} -40 & -13 & -25 & -37 & -37 \\ 4 & 24 & 29 & 12 & 10 \\ 55 & 31 & 46 & 57 & 51 \end{bmatrix}$$

$$= \begin{bmatrix} -4 & 2 & -3 \\ 1 & 0 & 1 \\ -3 & 1 & -2 \end{bmatrix} \begin{bmatrix} -40 & -13 & -25 & -37 & -37 \\ 4 & 24 & 29 & 12 & 10 \\ 55 & 31 & 46 & 57 & 51 \end{bmatrix}$$

$$= \begin{bmatrix} 3 & 7 & 20 & 1 & 15 \\ 15 & 18 & 21 & 20 & 14 \\ 14 & 1 & 12 & 9 & 19 \end{bmatrix}.$$

Rearrange matrix **D** by stringing together its columns; thus

$$\mathbf{D} = \begin{bmatrix} 3 & 15 & 14 & 7 & 18 & 1 & 20 & 21 & 12 & 1 & 20 & 9 & 15 & 14 & 19 \end{bmatrix}.$$

Now decode the string of numbers in **D** using the following correspondence between numbers and letters.

| 1 | 2 | 3 | 4 | 5 | 6 | 7 | 8 | 9 | 10 |
|---|---|---|---|---|---|---|---|---|----|
| ⇕ | ⇕ | ⇕ | ⇕ | ⇕ | ⇕ | ⇕ | ⇕ | ⇕ | ⇕ |
| A | B | C | D | E | F | G | H | I | J |

| 11 | 12 | 13 | 14 | 15 | 16 | 17 | 18 | 19 | 20 |
|----|----|----|----|----|----|----|----|----|----|
| ⇕ | ⇕ | ⇕ | ⇕ | ⇕ | ⇕ | ⇕ | ⇕ | ⇕ | ⇕ |
| K | L | M | N | O | P | Q | R | S | T |

| 21 | 22 | 23 | 24 | 25 | 26 |
|----|----|----|----|----|----|
| ⇕ | ⇕ | ⇕ | ⇕ | ⇕ | ⇕ |
| U | V | W | X | Y | Z |

We have

| 3 | 15 | 14 | 7 | 18 | 1 | 20 | 21 | 12 | 1 | 20 | 9 | 15 | 14 | 19 |
|---|----|----|---|----|---|----|----|----|---|----|---|----|----|----|
| C | O | N | G | R | A | T | U | L | A | T | I | O | N | S |

**67.** The encoded message will consist of integer values, which are easy to transmit.

**69.** $\mathbf{M}^{-1} = \begin{bmatrix} 1 & 1 \\ 1 & -1 \end{bmatrix}.$

**72.** (a) $x_3 = 21, \mathbf{d}_1 = \begin{bmatrix} 2 & -5 & -1 & 7 \end{bmatrix},$
$\mathbf{d}_2 = \begin{bmatrix} 12 & 8 \end{bmatrix}, \mathbf{d}_3 = \begin{bmatrix} 2 \end{bmatrix}.$

(b) $x_3 = 21, \tilde{\mathbf{d}}_1 = \begin{bmatrix} 0 & -5 & 0 & 7 \end{bmatrix}, \tilde{\mathbf{d}}_2 = \begin{bmatrix} 12 & 8 \end{bmatrix},$
$\tilde{\mathbf{d}}_3 = \begin{bmatrix} 0 \end{bmatrix}.$

(c) $\begin{bmatrix} a \\ b \end{bmatrix} = \mathbf{M}^{-1} \begin{bmatrix} 21 \\ 0 \end{bmatrix} = \begin{bmatrix} 21 \\ 21 \end{bmatrix},$

$\begin{bmatrix} c & d \\ e & f \end{bmatrix} = \mathbf{M}^{-1} \begin{bmatrix} 21 & 21 \\ 12 & 8 \end{bmatrix} = \begin{bmatrix} 33 & 29 \\ 9 & 13 \end{bmatrix},$

$\begin{bmatrix} g & h & i & j \\ k & l & m & n \end{bmatrix} = \mathbf{M}^{-1} \begin{bmatrix} 33 & 9 & 29 & 13 \\ 0 & -5 & 0 & 7 \end{bmatrix} = $
$\begin{bmatrix} 33 & 4 & 29 & 20 \\ 33 & 14 & 29 & 6 \end{bmatrix}.$

**Section 2.4 True/False, Page 164**

| 1. F | 2. F | 3. T | 4. F | 5. F |
|------|------|------|------|------|
| 6. T | 7. F | 8. T | 9. T | 10. F |
| 11. F | 12. T | 13. T | 14. T | 15. T |

**Section 2.5, Page 171**

**1.** $\mathbf{x} = \begin{bmatrix} 1 \\ 2 \\ 1 \end{bmatrix}.$

**3.** $\mathbf{x} = \begin{bmatrix} 1 \\ 0 \\ 2 \\ -4 \end{bmatrix}.$

For Exercises 5–9 we follow the procedure in Example 1 to find the LU-factorization.

**5.** $\mathbf{A} = \begin{bmatrix} 2 & 3 & 4 \\ 4 & 5 & 10 \\ 4 & 8 & 2 \end{bmatrix};$

$$\mathbf{U}_1 = \mathbf{A}_{\substack{-2R_1 + R_2 \\ -2R_1 + R_3}} = \begin{bmatrix} 2 & 3 & 4 \\ 0 & -1 & 2 \\ 0 & 2 & -6 \end{bmatrix}$$

$$\mathbf{L}_1 = \begin{bmatrix} 1 & 0 & 0 \\ 2 & 1 & 0 \\ 2 & * & 1 \end{bmatrix}$$

$$\mathbf{U}_2 = \mathbf{A}_{2R_2 + R_3} = \begin{bmatrix} 2 & 3 & 4 \\ 0 & -1 & 2 \\ 0 & 0 & -2 \end{bmatrix}$$

$$\mathbf{L}_2 = \begin{bmatrix} 1 & 0 & 0 \\ 2 & 1 & 0 \\ 2 & -2 & 1 \end{bmatrix}.$$

Define $\mathbf{L} = \mathbf{L}_2$ and $\mathbf{U} = \mathbf{U}_2$, then linear system $\mathbf{Ax} = \mathbf{b}$ is equivalent to $\mathbf{LUx} = \mathbf{b}$. Let $\mathbf{z} = \mathbf{Ux}$, then we solve $\mathbf{Lz} = \mathbf{b}$ for $\mathbf{z}$ by forward substitution. We get

$$\mathbf{z} = \begin{bmatrix} 6 \\ 4 \\ -2 \end{bmatrix}.$$ Next we solve $\mathbf{Ux} = \mathbf{z}$ by back substitution

and get $\mathbf{x} = \begin{bmatrix} 4 \\ -2 \\ 1 \end{bmatrix}.$

**7.** $\mathbf{x} = \begin{bmatrix} 1 \\ -2 \\ 5 \\ -4 \end{bmatrix}.$    **9.** $\mathbf{x} = \begin{bmatrix} i \\ 1 \\ 1 \end{bmatrix}.$

## Section 2.5 True/False, Page 172
**1.** F    **2.** T    **3.** F

## Section 2.6, Page 178
**1.** $y = mx + b = 0.3x + 1.3.$

**3.** **(a)** Using Equation 2, we obtain the linear system

$$\begin{bmatrix} 49555 & 545 \\ 545 & 6 \end{bmatrix} \begin{bmatrix} m \\ b \end{bmatrix} = \begin{bmatrix} 368237.1 \\ 4025.3 \end{bmatrix}$$

and

$$\mathbf{rref}\left( \begin{bmatrix} 49555 & 545 & | & 368237.1 \\ 545 & 6 & | & 4025.3 \end{bmatrix} \right)$$
$$= \begin{bmatrix} 1 & 0 & | & 51.2593 \\ 0 & 1 & | & -3985.1738 \end{bmatrix}.$$

(The results of the RREF are displayed to 4 decimal digits.) Hence $m \approx 51.2593$ and $b \approx -3985.1738$ so the least squares line is (approximately) $y = mx + b = 51.2593x - 3985.1738.$

**(b)** In the equation developed in part (a), set $x = 99$. It follows that the expenditure estimate is $y = 51.2593(99) - 3985.1738 = 1089.50131$, or about 1089.5 billion.

**Strategy for Exercises 5–7:** We form the associated normal system and solve it to get the least squares solution.

**5.** $\mathbf{x} = \begin{bmatrix} -\frac{1}{11} \\ \frac{39}{11} \end{bmatrix} \approx \begin{bmatrix} -0.0909 \\ 3.5455 \end{bmatrix}$ is the least squares solution.

**7.** $\mathbf{x} = \begin{bmatrix} -\frac{2}{11} \\ \frac{16}{11} \\ -2 \end{bmatrix} \approx \begin{bmatrix} -0.1818 \\ 1.4545 \\ -2 \end{bmatrix}$ is the least squares solution.

**9.** **(a)** Since the $x_i$ are distinct, the columns are not multiples of one another; hence they are linearly independent.

**(b)** Since the columns of $\mathbf{A}$ are linearly independent and $\mathbf{Ak}$ is a linear combination of the columns, the only way $\mathbf{Ak} = \mathbf{0}$ is if $\mathbf{k} = \mathbf{0}$.

**(c)** This follows from properties of the transpose of a product of matrices.

**(d)** If $\mathbf{Ck} = \mathbf{0}$, then we have $\mathbf{A}^T \mathbf{Ak} = \mathbf{0}$ and multiplying both sides on the left by $\mathbf{k}^T$ gives $\mathbf{k}^T \mathbf{A}^T \mathbf{Ak} = \mathbf{k}^T \mathbf{0} = 0$. Thus by part (b), we have $(\mathbf{Ak})^T (\mathbf{Ak}) = 0.$

**(e)** This follows because $\mathbf{Ak}$ is a column, so $(\mathbf{Ak})^T (\mathbf{Ak})$ is a row-by-column product and is equivalent to a dot product.

**(f)** The only time the dot product of a vector with itself is zero is when the vector is the zero vector.

**(g)** Suppose there is a vector $\mathbf{x}$ such that $\mathbf{Cx} = \mathbf{0}$. Then it follows that $\mathbf{x}^T \mathbf{Cx} = \mathbf{x}^T \mathbf{0} = 0$. But it follows that $\mathbf{x}$ must be the zero vector, hence the only solution of $\mathbf{Cx} = \mathbf{0}$ is the trivial solution so $\mathbf{C}$ is nonsingular.

## Section 2.6 True/False, Page 180
**1.** T    **2.** T    **3.** T

## Chapter 2 Chapter Test, Page 181
**1.** Form a linear combination of $\mathbf{w}_1$, $\mathbf{w}_2$, and $\mathbf{w}_3$.

**2.** No. The sum of two matrices in $V$ will have its $(2, 2)$-entry equal to 10, not 5.

**3.** There exists a linear combination of $\mathbf{v}_1$, $\mathbf{v}_2$, and $\mathbf{v}_3$ using coefficients that are not all zero, so that the result is the zero vector of $R^5$.

**4.** $\mathbf{ns}(\mathbf{A})$ is the set of all vectors $\mathbf{x}$ such that $\mathbf{Ax} = \mathbf{0}$.

**5.** The general solution will be $\mathbf{x}_p + \mathbf{x}_h$, where $\mathbf{x}_p$ is a particular solution and $\mathbf{x}_h$ is a solution of $\mathbf{Ax} = \mathbf{0}$ containing 2 free variables.

**6.** $\begin{bmatrix} 1 & 0 & 0 & -6 \\ 0 & 1 & 0 & -13 \\ 0 & 0 & 1 & 4 \end{bmatrix}.$

**7.** Linearly dependent.

**8.** $\mathbf{x} = \mathbf{x}_p + \mathbf{x}_h = \begin{bmatrix} 1 \\ 2 \\ -1 \\ 0 \end{bmatrix} + t \begin{bmatrix} -1 \\ 1 \\ 1 \\ 1 \end{bmatrix}.$

**9.** Vector **v** is not in span($T$). This follows since system

$$\begin{bmatrix} 1 & 2 & 4 \\ 2 & 0 & 4 \\ 1 & 1 & 3 \end{bmatrix} \mathbf{x} = \mathbf{v}$$

is inconsistent.

**10. (a)** $\left\{ \begin{bmatrix} 1 & 0 & 1.8 \end{bmatrix}, \begin{bmatrix} 0 & 1 & -0.6 \end{bmatrix} \right\}$

**(b)** $\left\{ \begin{bmatrix} 1 \\ 0 \\ 1 \\ 2 \end{bmatrix}, \begin{bmatrix} 0 \\ 1 \\ -2 \\ -1 \end{bmatrix} \right\}$    **(c)** $\left\{ \begin{bmatrix} -1.8 \\ 0.6 \\ 1 \end{bmatrix} \right\}$.

**11.** $\mathbf{A}^{-1} = \begin{bmatrix} 1 & -5 & 14 \\ 1 & 0 & -2 \\ -1 & 1 & -1 \end{bmatrix}$.

**12. (a)** 0.      **(b)** 1.      **(c)** $\mathbf{x} = \mathbf{A}^{-1}\mathbf{b}$.

**(d)** 4.      **(e)** 0.

**(f)** Linearly independent since **rref**($\mathbf{A}$) = $\mathbf{I}_4$.

**13. (a)** The zero vector, $\begin{bmatrix} 0 \\ 0 \end{bmatrix}$.

**(b)** $\mathbf{x} = \mathbf{A}^{-1}\mathbf{y}$.      **(c)** 0.

## Chapter 3

### Section 3.1, Page 190

**1.** The vertices are: $(0, 1), (2, 2), (3, -1)$. Thus let

$$\mathbf{x} = \begin{bmatrix} 0 \\ 2 \\ 3 \end{bmatrix} \quad \text{and} \quad \mathbf{y} = \begin{bmatrix} 1 \\ 2 \\ -1 \end{bmatrix}.$$

From Equation (4) we have that the area of the triangle is given by

$$\frac{1}{2} |\mathbf{x}^T \mathbf{C} \mathbf{y}| = \frac{1}{2} \left| \begin{bmatrix} 0 & 2 & 3 \end{bmatrix} \begin{bmatrix} 0 & -1 & 1 \\ 1 & 0 & -1 \\ -1 & 1 & 0 \end{bmatrix} \begin{bmatrix} 1 \\ 2 \\ -1 \end{bmatrix} \right|$$

$$= 3.5 \text{ square units.}$$

**3.** 2 square units.

**For Exercise 5:** There are a variety of ways to triangulate a polygonal region. The computational details depend upon the scheme you choose.

**5.** One triangulation is shown in the accompanying figure where we have labeled the triangular regions A, B, C, D, and E. We now compute the area of each of the five triangles and add the results to find the area of the polygonal region.

Triangle A: vertices $(-3, 1), (-2, 3), (-1, 1)$;
      area 2 sq. units.

Triangle B: vertices $(-1, 1), (2, 5), (3, 2)$;
      area 6.5 sq. units.

Triangle C: vertices $(3, 2), (4, 3), (4, -2)$;
      area 2.5 sq. units.

Triangle D: vertices $(-3, 1), (-1, 1), (4, -2)$;
      area 3 sq. units.

Triangle E: vertices $(-1, 1), (3, 2), (4, -2)$;
      area 8.5 sq. units.

Area of region = 22.5 sq. units.

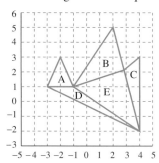

**Strategy for Exercise 7:** Construct a 2 by 3 matrix **V** containing the vertices of the triangle and then compute **AV** to determine the vertices of the image.

$$\mathbf{V} = \begin{bmatrix} -2 & 3 & 0 \\ 2 & 4 & 0 \end{bmatrix}; \quad \mathbf{AV} = \begin{bmatrix} -6 & 2 & 0 \\ 0 & 7 & 0 \end{bmatrix}.$$

**9.** $\mathbf{C}^T = \begin{bmatrix} 0 & -1 & 1 \\ 1 & 0 & -1 \\ -1 & 1 & 0 \end{bmatrix}^T = \begin{bmatrix} 0 & 1 & -1 \\ -1 & 0 & 1 \\ 1 & -1 & 0 \end{bmatrix} =$

$(-1)\begin{bmatrix} 0 & -1 & 1 \\ 1 & 0 & -1 \\ -1 & 1 & 0 \end{bmatrix} = -\mathbf{C}.$

**11.** $\mathbf{y}^T \mathbf{C}\mathbf{x}$ is a scalar, hence it equals its transpose. Thus we have
$\mathbf{y}^T \mathbf{C}\mathbf{x} = (\mathbf{y}^T \mathbf{C}\mathbf{x})^T = \mathbf{x}^T \mathbf{C}^T \mathbf{y} = \mathbf{x}^T (-\mathbf{C})\mathbf{y} = -\mathbf{x}^T \mathbf{C}\mathbf{y}.$

### Section 3.1 True/False, Page 191

**1.** T      **2.** T      **3.** F      **4.** T      **5.** T

### Section 3.2, Page 198

**1.** By definition, for $\mathbf{A} = \begin{bmatrix} a & b \\ c & d \end{bmatrix}$ we have that

$$\det(\mathbf{A}) = \det\left( \begin{bmatrix} a & b \\ c & d \end{bmatrix} \right) = ad - bc.$$

**(a)** $(-1)^{1+1}a * \det(\begin{bmatrix} d \end{bmatrix}) + (-1)^{1+2}c * \det(\begin{bmatrix} b \end{bmatrix}) =$
$(1)ad + (-1)bc = ad - bc = \det(\mathbf{A}).$

**(b)** $(-1)^{1+2}c * \det(\begin{bmatrix} b \end{bmatrix}) + (-1)^{2+2}d * \det(\begin{bmatrix} a \end{bmatrix}) =$
$(-1)bc + (1)ad = ad - bc = \det(\mathbf{A}).$

**5.** $-2$.      **7.** 0.      **9.** 1.      **11.** $-1$.

**13.** We have that $\mathbf{A}\mathbf{A}^{-1} = \mathbf{I}$. Then from Exercise 12 we have
$\det(\mathbf{A}\mathbf{A}^{-1}) = \det(\mathbf{I}_n) = 1$. From Equation 3 we have
$\det(\mathbf{A}\mathbf{A}^{-1}) = \det(\mathbf{A})\det(\mathbf{A}^{-1})$. Since $\mathbf{A}$ is nonsingular
we know $\det(\mathbf{A})$ is not zero, hence combining the
preceding equations we get $\det(\mathbf{A})\det(\mathbf{A}^{-1}) = 1$ and
hence $\det(\mathbf{A}^{-1}) = 1/\det(\mathbf{A})$.

**15. (a)** diagonal.      **(b)** $\det(\mathbf{E}) = k$.

**(b)** $\det(\mathbf{EA}) = \det(\mathbf{E})\det(\mathbf{A}) = k\det(\mathbf{A})$.

**17. (a)** The determinants of the respective matrices are 10, $-12$, $-30$, and 0.

**(b)** the diagonal entries.

**19. (a)** upper

**(b)** lower

**(c)** $\det(\mathbf{E}) = 1$.

**(d)** $\det(\mathbf{A}_{kR_i + R_j}) = 1$.

**21. (a)** $\det(\mathbf{A}^T) = \det\left(\begin{bmatrix} a & c \\ b & d \end{bmatrix}\right) = ad - bc = \det(\mathbf{A})$.

**(b)** $\det(\mathbf{A}^T) = \det\left(\begin{bmatrix} a & d & g \\ b & e & h \\ c & f & i \end{bmatrix}\right) =$
$aei + cdh + bfg - ceg - afh - bdi = \det(\mathbf{A})$.

**(c)** $\det(\mathbf{A})$.

**23.** $\left| \det\left(\begin{bmatrix} x_1 & y_1 & 1 \\ x_2 & y_2 & 1 \\ x_3 & y_3 & 1 \end{bmatrix}\right) \right|$.

**25.** Using the result in Exercise 24, if

$$\det\left(\begin{bmatrix} x_1 & y_1 & 1 \\ x_2 & y_2 & 1 \\ x_3 & y_3 & 1 \end{bmatrix}\right) = 0,$$

then the points are collinear.

## Section 3.2 True/False, Page 200

**1.** T    **2.** T    **3.** T    **4.** T    **5.** T

**6.** T    **7.** T    **8.** T    **9.** T    **10.** F

## Section 3.3, Page 204

**Strategy for Exercises 1–4:** There are many sets of row operations that will lead us to an upper triangular form. The operations we choose may not be the most efficient, but we try to avoid fractions during the first few steps.

**1.** Let $\mathbf{A} - \begin{bmatrix} 4 & 1 & 3 \\ 2 & 3 & 0 \\ 1 & 3 & 2 \end{bmatrix}$.

$\det(\mathbf{A}) = -\det(\mathbf{A}_{R_1 \leftrightarrow R_3})$

$= -\det\left(\begin{bmatrix} 1 & 3 & 2 \\ 2 & 3 & 0 \\ 4 & 1 & 3 \end{bmatrix}\right)$

$= -\det\left(\begin{bmatrix} 1 & 3 & 2 \\ 2 & 3 & 0 \\ 4 & 1 & 3 \end{bmatrix}\right)_{\substack{-2R_1 + R_2 \\ -4R_1 + R_3}}$

$= -\det\left(\begin{bmatrix} 1 & 3 & 2 \\ 0 & -3 & -4 \\ 0 & -11 & -5 \end{bmatrix}\right)$

$= -(-3)\det\left(\begin{bmatrix} 1 & 3 & 2 \\ 0 & -3 & -4 \\ 0 & -11 & -5 \end{bmatrix}\right)_{\frac{-1}{3}R_2}$

$= 3\det\left(\begin{bmatrix} 1 & 3 & 2 \\ 0 & 1 & \frac{4}{3} \\ 0 & -11 & -5 \end{bmatrix}\right)$

$= 3\det\left(\begin{bmatrix} 1 & 3 & 2 \\ 0 & 1 & \frac{4}{3} \\ 0 & -11 & -5 \end{bmatrix}\right)_{11R_2 + R_3}$

$= 3\det\left(\begin{bmatrix} 1 & 3 & 2 \\ 0 & 1 & \frac{4}{3} \\ 0 & 0 & \frac{29}{3} \end{bmatrix}\right)$

$= (3)\left(\frac{29}{3}\right) = 29.$

**3.** 72.

**Strategy for Exercises 5–8:** There are many sets of row operations that will lead us to an upper triangular form. The operations we choose may not be the most efficient, but we try to avoid fractions during the first few steps.

**5.** Let $\mathbf{A} = \begin{bmatrix} 2 & 0 & 1 & 4 \\ 1 & 2 & -6 & -10 \\ 2 & 3 & -1 & 0 \\ 11 & 8 & -4 & 6 \end{bmatrix}$.

$\det(\mathbf{A}) = -\det(\mathbf{A}_{R_1 \leftrightarrow R_2})$

$= -\det\left(\begin{bmatrix} 1 & 2 & -6 & -10 \\ 2 & 0 & 1 & 4 \\ 2 & 3 & -1 & 0 \\ 11 & 8 & -4 & 6 \end{bmatrix}\right)$

$= -\det\left(\begin{bmatrix} 1 & 2 & -6 & -10 \\ 2 & 0 & 1 & 4 \\ 2 & 3 & -1 & 0 \\ 11 & 8 & -4 & 6 \end{bmatrix}\right)_{\substack{-2R_1 + R_2 \\ -2R_1 + R_3 \\ -11R_1 + R_4}}$

$= -\det\left(\begin{bmatrix} 1 & 2 & -6 & -10 \\ 0 & -4 & 13 & 24 \\ 0 & -1 & 11 & 20 \\ 0 & -14 & 62 & 116 \end{bmatrix}\right)$

$= \det\left(\begin{bmatrix} 1 & 2 & -6 & -10 \\ 0 & -4 & 13 & 24 \\ 0 & -1 & 11 & 20 \\ 0 & -14 & 62 & 116 \end{bmatrix}\right)_{R_2 \leftrightarrow R_3}$

$= \det\left(\begin{bmatrix} 1 & 2 & -6 & -10 \\ 0 & -1 & 11 & 20 \\ 0 & -4 & 13 & 24 \\ 0 & -14 & 62 & 116 \end{bmatrix}\right)$

$$= \det \begin{pmatrix} \begin{bmatrix} 1 & 2 & -6 & -10 \\ 0 & -1 & 11 & 20 \\ 0 & -4 & 13 & 24 \\ 0 & -14 & 62 & 116 \end{bmatrix} \end{pmatrix}_{\substack{-4R_2 + R_3 \\ -14R_2 + R_4}}$$

$$= \det \begin{pmatrix} \begin{bmatrix} 1 & 2 & -6 & -10 \\ 0 & -1 & 11 & 20 \\ 0 & 0 & -31 & -56 \\ 0 & 0 & -92 & -164 \end{bmatrix} \end{pmatrix}$$

$$= \det \begin{pmatrix} \begin{bmatrix} 1 & 2 & -6 & -10 \\ 0 & -1 & 11 & 20 \\ 0 & 0 & -31 & -56 \\ 0 & 0 & -92 & -164 \end{bmatrix} \end{pmatrix}_{\frac{-92}{-31}R_3 + R_4}$$

$$= \det \begin{pmatrix} \begin{bmatrix} 1 & 2 & -6 & -10 \\ 0 & -1 & 11 & 20 \\ 0 & 0 & -31 & -56 \\ 0 & 0 & 0 & \frac{68}{31} \end{bmatrix} \end{pmatrix} = 68.$$

**7.** 80.

**9. (a)** $\det \begin{pmatrix} \begin{bmatrix} \lambda - 1 & 2 \\ 3 & \lambda - 2 \end{bmatrix} \end{pmatrix} = (\lambda - 1)(\lambda - 2) - 6 =$
$\lambda^2 - 3\lambda + 2 - 6 = \lambda^2 - 3\lambda - 4.$

**(b)** $\det(\lambda \mathbf{I}_2 - \mathbf{A})$, where $\mathbf{A} = \begin{bmatrix} 4 & 2 \\ -1 & 1 \end{bmatrix}$ is computed as

$$\det \begin{pmatrix} \begin{bmatrix} \lambda & 0 \\ 0 & \lambda \end{bmatrix} - \begin{bmatrix} 4 & 2 \\ -1 & 1 \end{bmatrix} \end{pmatrix}$$

$$= \det \begin{pmatrix} \begin{bmatrix} \lambda - 4 & -2 \\ 1 & \lambda - 1 \end{bmatrix} \end{pmatrix}$$

$$= (\lambda - 4)(\lambda - 1) + 2$$

$$= \lambda^2 - 5\lambda + 4 + 2$$

$$= \lambda^2 - 5\lambda + 6.$$

**11. (a)** $(\lambda - 1)(\lambda^2 - 3\lambda + 3)$.

**(b)** $(\lambda + 1)(\lambda)(\lambda - 1)$.

**12. (a)** $(\lambda - 1)(\lambda^2 - 3\lambda + 3) = 0$ provided $\lambda = 1$ or
$\lambda = \frac{3}{2} \pm \frac{\sqrt{3}}{2}$.

**(b)** $(\lambda + 1)(\lambda)(\lambda - 1) = 0$ provided $\lambda = 0$, $\lambda = 1$, or
$\lambda = -1$.

**14. (a)** $\det(\mathbf{A}^2) = \det(\mathbf{A})\det(\mathbf{A}) = (\det(\mathbf{A}))^2 =$
$(-3)(-3) = 9$.

**(b)** $\det(\mathbf{A}^5) = \det(\mathbf{A})\det(\mathbf{A})\det(\mathbf{A})\det(\mathbf{A})\det(\mathbf{A}) =$
$(\det(\mathbf{A}))^5 = (-3)^5 = -243$.

**(c)** $(\det(\mathbf{A})^{-1}) = 1/\det(\mathbf{A}) = -\frac{1}{3}$.

**19.** We have that $\det(\mathbf{A}^T) = \det(\mathbf{A})$ and
$\det(\mathbf{A}^{-1}) = 1/\det(\mathbf{A})$. Since $\mathbf{A}^T = \mathbf{A}^{-1}$ we have
$\det(\mathbf{A}^T) = \det(\mathbf{A}^{-1})$, or equivalently that
$\det(\mathbf{A}) = 1/\det(\mathbf{A})$. This implies that $\det(\mathbf{A})^2 = 1$,
which gives $\det(\mathbf{A}) = 1$ or $-1$.

**21.** $\mathbf{A}^T$ will have two identical rows, so from Exercise 20,
$\det(\mathbf{A}^T) = 0$. But $\det(\mathbf{A}) = \det(\mathbf{A}^T)$, so $\det(\mathbf{A}) = 0$.

**25.** From Exercise 24, $\det(\mathbf{A} + \mathbf{B})$ is not always
$\det(\mathbf{A}) + \det(\mathbf{B})$. Hence det cannot be a linear
transformation.

## Section 3.3 True/False, Page 206

| 1. T | 2. F | 3. F | 4. F | 5. T |
| 6. F | 7. F | 8. T | 9. T | 10. T |

## Section 3.4, Page 212

**1. (a)** $\mathbf{A}_{31} = \begin{bmatrix} 0 & 1 \\ 3 & 4 \end{bmatrix}$, $\mathbf{A}_{22} = \begin{bmatrix} 2 & 1 \\ 5 & -2 \end{bmatrix}$,
$\mathbf{A}_{23} = \begin{bmatrix} 2 & 0 \\ 5 & 1 \end{bmatrix}$.

**(b)** $C_{31} = (-1)^{3+1}\det(\mathbf{A}_{31}) = \det\begin{pmatrix}\begin{bmatrix} 0 & 1 \\ 3 & 4 \end{bmatrix}\end{pmatrix} = -3;$
$C_{22} = (-1)^{2+2}\det(\mathbf{A}_{22}) = \det\begin{pmatrix}\begin{bmatrix} 2 & 1 \\ 5 & -2 \end{bmatrix}\end{pmatrix} = -9;$
$C_{23} = (-1)^{2+3}\det(\mathbf{A}_{23}) = \det\begin{pmatrix}\begin{bmatrix} 2 & 0 \\ 5 & 1 \end{bmatrix}\end{pmatrix} = -2.$

**3.** Use expansion about the first row since $a_{13} = 0$. We get

$$\det(\mathbf{A}) = (-1)C_{11} + (-2)C_{12}$$
$$= (-1)(-1)^{1+1}\det\begin{pmatrix}\begin{bmatrix} 5 & 1 \\ 4 & -3 \end{bmatrix}\end{pmatrix}$$
$$+ (-2)(-1)^{1+2}\det\begin{pmatrix}\begin{bmatrix} 2 & 1 \\ 3 & -3 \end{bmatrix}\end{pmatrix}$$
$$= (-1)(-19) + (2)(-9) = 1.$$

**5.** Use expansion about the second column since $a_{12} = 0$
and $a_{32} = 0$. We get

$$\det(\mathbf{A}) = (1 + i)C_{22}$$
$$= (1 + i)(-1)^{2+2}\det\begin{pmatrix}\begin{bmatrix} i & 1 \\ 2 & i \end{bmatrix}\end{pmatrix}$$
$$= (1 + i)(i^2 - 2)$$
$$= (1 + i)(-3) = -3 - 3i.$$

**7.** Use expansion about the first row since $a_{13} = 0$ and
$a_{14} = 0$. We get

$$\det(\mathbf{A}) = (2)C_{11} + (-1)C_{12}$$
$$= (2)(-1)^{1+1}\det\begin{pmatrix}\begin{bmatrix} 2 & -1 & 0 \\ -1 & 2 & -1 \\ 0 & -1 & 2 \end{bmatrix}\end{pmatrix}$$
$$+ (-1)(-1)^{1+2}\det\begin{pmatrix}\begin{bmatrix} -1 & -1 & 0 \\ 0 & 2 & -1 \\ 0 & -1 & 2 \end{bmatrix}\end{pmatrix}$$
$$= (2)(4) + (1)(-3) = 5.$$

9. $\text{adj}(\mathbf{A}) = \begin{bmatrix} C_{11} & C_{12} & C_{13} \\ C_{21} & C_{22} & C_{23} \\ C_{31} & C_{32} & C_{33} \end{bmatrix}^T = \begin{bmatrix} 0 & 6 & 2 \\ 19 & -9 & -3 \\ 19 & -1 & -13 \end{bmatrix}.$

11. $\text{adj}(\mathbf{A}) = \begin{bmatrix} 2 & -1+i & -1-i \\ 2i & -1+i & 1-i \\ 0 & 1 & i \end{bmatrix}.$

15. (a) Since $\mathbf{A}$ has all integer entries the submatrices of $\mathbf{A}$ will have integer entries. Hence each cofactor of $\mathbf{A}$ will be an integer times the determinant of a submatrix with integer entries. But the determinant of a matrix with integer entries is an integer. (See part (b).) Therefore, each entry of $\text{adj}(\mathbf{A})$ will be a product of integers, hence an integer.

   (b) If $\mathbf{A}$ were $2 \times 2$ or $3 \times 3$, then our special devices for this size matrix give us the determinant as a linear combination of a product of entries. If $\mathbf{A}$ is a matrix of larger size, the expansion by cofactors can give us an expression for $\det(\mathbf{A})$ which is a linear combination of determinants of submatrices of size $2 \times 2$ or $3 \times 3$. Thus it follows that $\det(\mathbf{A})$ is an integer.

   (c) From equation (8), we must have $\det(\mathbf{A}) = 1$ or $-1$ in order for $\mathbf{A}^{-1}$ to be guaranteed to have integer entries.

17. $\text{row}_2(\mathbf{A}) = \begin{bmatrix} 4 & 8 & -7 \end{bmatrix}$; $\mathbf{A}_{21}$ implies

   $\mathbf{A} = \begin{bmatrix} 6 & 1 \\ 4 & 8 & -7 \\ 0 & 3 \end{bmatrix}$; $\mathbf{A}_{22}$ implies $\mathbf{A} = \begin{bmatrix} 2 & 6 & 1 \\ 4 & 8 & -7 \\ 9 & 0 & 3 \end{bmatrix}.$

19. (a) $\det(\mathbf{B}) = 2(21) - 3(44) - 5(-18) = 0.$

   (b) $\mathbf{B} = \begin{bmatrix} 2 & 3 & -5 \\ 2 & 3 & -5 \\ 6 & 0 & 7 \end{bmatrix}.$

21. (a) $\begin{matrix} + & - & + \\ - & + & - \\ + & - & + \end{matrix}$.   (b) $\begin{matrix} + & - & + & - \\ - & + & - & + \\ + & - & + & - \\ - & + & - & + \end{matrix}$.

## Section 3.4 True/False, Page 214

1. T      2. T      3. T      4. T      5. T

6. F      7. T      8. T      9. T      10. F

## Section 3.5, Page 216

1. $\mathbf{B}_1 = \begin{bmatrix} 2 & 4 & 6 \\ 0 & 0 & 2 \\ -5 & 3 & -1 \end{bmatrix}$, $\mathbf{B}_2 = \begin{bmatrix} 2 & 2 & 6 \\ 1 & 0 & 2 \\ 2 & -5 & -1 \end{bmatrix}$,

   $\mathbf{B}_3 = \begin{bmatrix} 2 & 4 & 2 \\ 1 & 0 & 0 \\ 2 & 3 & -5 \end{bmatrix}.$

   $$x_1 = \frac{\det(\mathbf{B}_1)}{\det(\mathbf{A})} = \frac{-52}{26} = -2,$$

   $$x_2 = \frac{\det(\mathbf{B}_2)}{\det(\mathbf{A})} = \frac{0}{26} = 0,$$

   $$x_3 = \frac{\det(\mathbf{B}_3)}{\det(\mathbf{A})} = \frac{26}{26} = 1.$$

3. The system is homogeneous with $\det(\mathbf{A}) = 11$, so $\mathbf{A}$ is nonsingular. Hence the solution is the trivial solution.

5. $y = \dfrac{\det\left( \begin{bmatrix} 1 & 100{,}000 \\ 0.05 & 7{,}800 \end{bmatrix} \right)}{\det\left( \begin{bmatrix} 1 & 1 \\ 0.05 & 0.09 \end{bmatrix} \right)} = \dfrac{2800}{0.04} = 70{,}000.$

7. $x_2 = \$14{,}000.$

9. Let the cost of the items be denoted as $x_1, x_2, x_3$, and $x_4$, respectively.

   (a) The linear system consists of the following equations:

   $$x_1 + x_2 + x_3 + x_4 = T$$
   $$x_1 = x_2$$
   $$x_3 = x_1 + 2$$
   $$x_4 = x_2 + x_3 - 5.$$

   (b) We use the expression for Cramer's Rule for $x_4$, which we know, is $\$10$, in order to solve for $T$.

   $$x_4 = \frac{\det\left( \begin{bmatrix} 1 & 1 & 1 & T \\ 1 & -1 & 0 & 0 \\ -1 & 0 & 1 & 2 \\ 0 & -1 & -1 & -5 \end{bmatrix} \right)}{\det\left( \begin{bmatrix} 1 & 1 & 1 & 1 \\ 1 & -1 & 0 & 0 \\ -1 & 0 & 1 & 0 \\ 0 & -1 & -1 & 1 \end{bmatrix} \right)}$$

   $$= \frac{13 - 2T}{-5} = 10$$

   hence $T = \$31.50.$

11. Denote the coefficient matrix $\mathbf{A}$. Then there is a unique solution provided $\det(\mathbf{A})$ is not zero. We find that $\det(\mathbf{A}) = 3b - cab = b(3 - ac)$. Thus we see that the solution will be unique provided that $b \neq 0$ and $ac \neq 3$.

## Section 3.5 True/False, Page 217

1. F      2. T      3. F      4. T      5. T

## Section 3.6, Page 223

**1.** $\mathbf{u} \times \mathbf{v} = \det \left( \begin{bmatrix} \mathbf{i} & \mathbf{j} & \mathbf{k} \\ 3 & 2 & -1 \\ -1 & 3 & 2 \end{bmatrix} \right) =$
$4\mathbf{i} + \mathbf{j} + 9\mathbf{k} + 2\mathbf{k} + 3\mathbf{i} - 6\mathbf{j} = 7\mathbf{i} - 5\mathbf{j} + 11\mathbf{k}.$

**Strategy for Exercise 5:** Use the expression in Equation (7).

**5.** $(\mathbf{u} \times \mathbf{v}) \cdot \mathbf{w} = \det \left( \begin{bmatrix} 3 & 0 & 2 \\ 2 & -1 & 3 \\ 0 & 2 & -1 \end{bmatrix} \right) = -7.$

**9.** Let $\mathbf{p} = c\mathbf{u} + d\mathbf{v}$. then

$$\mathbf{n} \cdot \mathbf{p} = (\mathbf{u} \times \mathbf{v}) \cdot (c\mathbf{u} + d\mathbf{v})$$
$$= (\mathbf{u} \times \mathbf{v}) \cdot (c\mathbf{u}) + (\mathbf{u} \times \mathbf{v}) \cdot (d\mathbf{v})$$
$$= c(\mathbf{u} \times \mathbf{v}) \cdot \mathbf{u} + d(\mathbf{u} \times \mathbf{v}) \cdot \mathbf{v}$$
$$= c0 + d0 = 0.$$

The preceding development used the results from (8).

**Strategy for Exercises 10–13:** Compute the cross product of the pairs of vectors.

**11.** $\mathbf{n} = -\mathbf{j}$.

**13.** $\mathbf{n} = \mathbf{i} - 5\mathbf{j} - 4\mathbf{k}$.

**Strategy for Exercise 15:** Compute $\|\mathbf{u} \times \mathbf{v}\|$.

**15.** $\|\mathbf{u} \times \mathbf{v}\| = \|3\mathbf{i} - 6\mathbf{j} - 4\mathbf{k}\| = \sqrt{9 + 36 + 16} = \sqrt{61}.$

**17.** From (7) we have

$$(\mathbf{u} \times \mathbf{v}) \cdot \mathbf{w} = \det \left( \begin{bmatrix} u_1 & u_2 & u_3 \\ v_1 & v_2 & v_3 \\ w_1 & w_2 & w_3 \end{bmatrix} \right).$$

Using (7) again we have that

$$\mathbf{u} \cdot (\mathbf{v} \times \mathbf{w}) = (\mathbf{v} \times \mathbf{w}) \cdot \mathbf{u}$$
$$= \det \left( \begin{bmatrix} v_1 & v_2 & v_3 \\ w_1 & w_2 & w_3 \\ u_1 & u_2 & u_3 \end{bmatrix} \right)$$
$$= \det \left( \begin{bmatrix} v_1 & v_2 & v_3 \\ w_1 & w_2 & w_3 \\ u_1 & u_2 & u_3 \end{bmatrix} \begin{matrix} \\ \\ R_1 \leftrightarrow R_3 \\ R_2 \leftrightarrow R_3 \end{matrix} \right)$$
$$= \det \left( \begin{bmatrix} u_1 & u_2 & u_3 \\ v_1 & v_2 & v_3 \\ w_1 & w_2 & w_3 \end{bmatrix} \right)$$
$$= (\mathbf{u} \times \mathbf{v}) \cdot \mathbf{w}$$

**20. Strategy:** Use Equation (11).

**(a)** volume $= \left| \det \left( \begin{bmatrix} 5 & 1 & -2 \\ 1 & 2 & -1 \\ 0 & 1 & 3 \end{bmatrix} \right) \right| = |30| = 30.$

**23.** $t\mathbf{u} \times \mathbf{v} = (tu_2 v_3 - tu_3 v_2)\mathbf{i} + (tu_3 v_1 - tu_1 v_3)\mathbf{j}$
$+ (tu_1 v_2 - tu_2 v_1)\mathbf{k}$
$= t(u_2 v_3 - u_3 v_2)\mathbf{i} + t(u_3 v_1 - u_1 v_3)\mathbf{j}$
$+ t(u_1 v_2 - u_2 v_1)\mathbf{k}$
$= t(\mathbf{u} \times \mathbf{v})$

**24.** Approximately 17.2192.

**26.** $\mathbf{v} = \begin{bmatrix} 2 & 6 & 2 \end{bmatrix}$.

**28.** The vector corresponding to line segment $PQ$ is $\mathbf{u} = \begin{bmatrix} 2 & -2 & 3 \end{bmatrix}$ and that corresponding to $PR$ is $\mathbf{v} = \begin{bmatrix} 1 & 4 & 6 \end{bmatrix}$. A normal to the plane containing $P$, $Q$, and $R$ is $\mathbf{n} = \mathbf{u} \times \mathbf{v} = \begin{bmatrix} -24 & -9 & 10 \end{bmatrix}$. Thus an equation of the plane containing $P$, $Q$, and $R$ is $-24(x - 3) - 9(y - 1) + 10(z - 0) = 0$.

**30.** The normals to the respective planes are: $\mathbf{n}_1 = \begin{bmatrix} 4 & -3 & 6 \end{bmatrix}$ and $\mathbf{n}_2 = \begin{bmatrix} 2 & -4 & 1 \end{bmatrix}$. The angle between the normals is the angle between the planes. We have

$$\cos(\theta) = \frac{\mathbf{n}_1 \cdot \mathbf{n}_2}{\|\mathbf{n}_1\| \, \|\mathbf{n}_2\|} = \frac{26}{\sqrt{61} \sqrt{21}} \approx 0.7264$$

and it follows that the angle is about 43.4114 degrees.

**33.** Let $\mathbf{u} = \begin{bmatrix} -4 & -4 & 3 \end{bmatrix}$ and $\mathbf{v} = \begin{bmatrix} 4 & -3 & 1 \end{bmatrix}$.

$$\text{area}(R) = \tfrac{1}{2}\|\mathbf{n}\|$$
$$= \tfrac{1}{2}\|\mathbf{u} \times \mathbf{v}\|$$
$$= \tfrac{1}{2} \|\begin{bmatrix} 5 & 16 & 28 \end{bmatrix}\| \approx 16.3172.$$

**(a)** $\cos(\gamma) = \dfrac{\begin{bmatrix} 5 & 16 & 28 \end{bmatrix} \cdot \begin{bmatrix} 0 & 0 & 1 \end{bmatrix}}{\|\begin{bmatrix} 5 & 16 & 28 \end{bmatrix}\|} = \dfrac{28}{\sqrt{1065}},$

$$\left|\text{area}(\text{proj}_{xy} R)\right| = \left|\tfrac{1}{2}\|\mathbf{n}\| \, \cos(\gamma)\right|$$
$$= \left|\tfrac{1}{2}\|\mathbf{n}\| \frac{28}{\|\mathbf{n}\|}\right|$$
$$= 14 \text{ square units.}$$

**(b)** The projections of the points onto the $xy$-plane are respectively $(1, 2, 0)$, $(-3, -2, 0)$, and $(5, -1, 0)$. Let $\mathbf{u} = \begin{bmatrix} -4 & -4 & 0 \end{bmatrix}$ and $\mathbf{v} = \begin{bmatrix} 4 & -3 & 0 \end{bmatrix}$, then compute

$$\tfrac{1}{2}\|\mathbf{u} \times \mathbf{v}\| = \tfrac{1}{2} \|\begin{bmatrix} 0 & 0 & 28 \end{bmatrix}\| = 14.$$

## Section 3.6 True/False, Page 224

**1.** T    **2.** F    **3.** T    **4.** T    **5.** T

## Chapter 3 Chapter Test, Page 225

**1.** There are many answers. For example, $\mathbf{A} = \begin{bmatrix} 5 & 1 \\ 2 & 1 \end{bmatrix}$ or $\mathbf{A} = \begin{bmatrix} -7 & -2 \\ -2 & -1 \end{bmatrix}$.

**2.** 7.5.

**3.** Use triangulation to subdivide the region into triangles with no overlaps. Then compute the sum of the areas of the individual triangles.

**4.** $\det(\mathbf{A}) = 3k - 15$; thus $k = 5$ is the only value so that $\mathbf{A}$ is singular.

**5.**  −68.

**6.**  $5(-1)^{2+3} \det \left( \begin{bmatrix} 2 & -1 & 2 \\ 1 & 2 & 7 \\ 3 & 4 & -3 \end{bmatrix} \right) +$

$1(-1)^{3+3} \det \left( \begin{bmatrix} 2 & -1 & 2 \\ 0 & 1 & 6 \\ 3 & 4 & -3 \end{bmatrix} \right) =$

$-5(-96) + 1(-78) = 402.$

**7.**  $\mathbf{A}^{-1} = \frac{1}{2} \begin{bmatrix} 2 & -2 & 0 \\ -1 & 2 & 1 \\ -1 & 0 & 1 \end{bmatrix}.$

**8.**  $x_2 = \dfrac{\det \left( \begin{bmatrix} 4 & 9 & 1 \\ 3 & 10 & 1 \\ 1 & -2 & -1 \end{bmatrix} \right)}{\det \left( \begin{bmatrix} 4 & 2 & 1 \\ 3 & 0 & 1 \\ 1 & 2 & -1 \end{bmatrix} \right)} = \dfrac{-12}{6} = -2.$

**9.**  A normal to the plane is $\mathbf{n} = \begin{bmatrix} 8 & 1 & -10 \end{bmatrix}$. Thus we can write the equation of the plane as $\mathbf{n} \cdot ((x-1)\mathbf{i} + (y-2)\mathbf{j} + (z-1)\mathbf{k}) = 0$, which simplifies to $8(x-1) + (y-2) - 10(z-1) = 0$ or $8x + y - 10z = 0.$

**10.**  $\dfrac{-9}{\sqrt{38}\,\sqrt{6}} \approx -0.5690.$

**11.**  **(a)** True.     **(b)** True.     **(c)** True.

    **(d)** True.     **(e)** False.

## Chapter 4

## Section 4.1, Page 236

**1.**  $\mathbf{Ap} = t\mathbf{p} \Leftrightarrow \begin{bmatrix} y \\ x \end{bmatrix} = \begin{bmatrix} tx \\ ty \end{bmatrix} \Leftrightarrow y = tx$ and $x = ty$. If either $x$ or $y$ is zero, it follows that so is the other one. Since $\mathbf{p} \neq \mathbf{0}$, it must be the case that neither $x$ nor $y$ is zero. Then we have $y = tx = t(ty) = t^2 y$. Since $y \neq 0$, we have that $t^2 = 1$ and hence $t = 1$ or $-1$. We consider these two cases to determine the vector $\mathbf{p} \neq \mathbf{0}$.

Case $t = 1$: $\begin{bmatrix} y \\ x \end{bmatrix} = \begin{bmatrix} x \\ y \end{bmatrix} \Leftrightarrow y = x$.

Thus $\mathbf{p} = \begin{bmatrix} x \\ y \end{bmatrix} = \begin{bmatrix} x \\ x \end{bmatrix} = x\begin{bmatrix} 1 \\ 1 \end{bmatrix}$ for $x \neq 0$. Hence we have that $t = 1$ and $\mathbf{p} = x\begin{bmatrix} 1 \\ 1 \end{bmatrix}$, $x \neq 0$, is an eigenpair of $\mathbf{A}$. (Note that there are infinitely many eigenvectors corresponding to the eigenvalue $t = 1$.)

Case $t = -1$: $\begin{bmatrix} y \\ x \end{bmatrix} = \begin{bmatrix} -x \\ -y \end{bmatrix} \Leftrightarrow t = -x$.

Thus $\mathbf{p} = \begin{bmatrix} x \\ y \end{bmatrix} = \begin{bmatrix} x \\ -x \end{bmatrix} = x\begin{bmatrix} 1 \\ -1 \end{bmatrix}$ for $x \neq 0$. Hence we have that $t = -1$ and $\mathbf{p} = x\begin{bmatrix} 1 \\ -1 \end{bmatrix}$, $x \neq 0$, is an eigenpair of $\mathbf{A}$. (Note that there are infinitely many eigenvectors corresponding to the eigenvalue $t = -1$.)

**5.**  **(a)**  $\mathbf{p} = \begin{bmatrix} -1 \\ 1 \end{bmatrix}$ is an eigenvector corresponding to eigenvalue $\lambda = 0$.

    **(b)**  $\mathbf{q} = \begin{bmatrix} 1 \\ 1 \end{bmatrix}$ is an eigenvector corresponding to eigenvalue $\mu = 2$.

**13.**  **(a)**  We must show that $\left\{ \begin{bmatrix} 1 \\ 1 \end{bmatrix}, \begin{bmatrix} 1 \\ 2 \end{bmatrix} \right\}$ spans $R^2$ and is a linearly independent set. This set is linearly independent since there are only two vectors and they are not scalar multiples of one another. To show they span $R^2$, we let $\mathbf{b} = \begin{bmatrix} b_1 \\ b_2 \end{bmatrix}$ and show that there exist scalars $x_1$ and $x_2$ so that

$$x_1 \begin{bmatrix} 1 \\ 1 \end{bmatrix} + x_2 \begin{bmatrix} 1 \\ 2 \end{bmatrix} = \mathbf{b}.$$

This follows since the preceding linear combination can be expressed as the linear system

$$\begin{bmatrix} 1 & 1 \\ 1 & 2 \end{bmatrix} \begin{bmatrix} x_1 \\ x_2 \end{bmatrix} = \mathbf{b}$$

and the coefficient matrix is nonsingular. (Note: $\det \left( \begin{bmatrix} 1 & 1 \\ 1 & 2 \end{bmatrix} \right) = 1$.)

**17.**  **(b)**  The beetle population will grow, in the long run, about 2.85% per year.

## Section 4.1 True/False, Page 239

**1.** F     **2.** T     **3.** T     **4.** T     **5.** T

## Section 4.2, Page 249

**1.**  $\det \left( \begin{bmatrix} 4 & 2 \\ 3 & 3 \end{bmatrix} - \lambda \mathbf{I}_2 \right) = \det \left( \begin{bmatrix} 4-\lambda & 2 \\ 3 & 3-\lambda \end{bmatrix} \right) = \lambda^2 - 7\lambda + 6.$

**3.**  $\lambda^2 - 7\lambda + 12.$

**5.**  $\lambda^3 - 4\lambda^2 + 7.$

**7.**  $\det \left( \begin{bmatrix} 3 & 2 \\ 6 & 2 \end{bmatrix} - \lambda \mathbf{I}_2 \right) = \det \left( \begin{bmatrix} 3-\lambda & 2 \\ 6 & 2-\lambda \end{bmatrix} \right) = \lambda^2 - 5\lambda - 6 = (\lambda - 6)(\lambda + 1);$ eigenvalues $\lambda = 6$ and $\lambda = -1.$

**9.**  $\lambda = 4 \pm i.$

**11.**  $\lambda = -1, 4, 2.$

**13.**  $\lambda = \pm i.$

**15.**  $\lambda = 0, -i, 2.$

**17.**  $\mathbf{A} = \begin{bmatrix} 5 & 2 \\ -1 & 2 \end{bmatrix}$, $\lambda = 4$. Form the system $(\mathbf{A} - 4\mathbf{I}_2)\mathbf{x} = \mathbf{0}$. Compute $\mathbf{rref}\left( \begin{bmatrix} \mathbf{A} - 4\mathbf{I}_2 & | & \mathbf{0} \end{bmatrix} \right)$. We get $\begin{bmatrix} 1 & 2 & | & 0 \\ 0 & 0 & | & 0 \end{bmatrix}$. It follows that $\mathbf{x} = t \begin{bmatrix} -2 \\ 1 \end{bmatrix}$, hence we find that $\begin{bmatrix} -2 \\ 1 \end{bmatrix}$ is a corresponding eigenvector.

**19.** $\begin{bmatrix} -2 \\ 1 \\ 0 \end{bmatrix}$ is a corresponding eigenvector.

**21.** We have a pair of linearly independent eigenvectors,

$\begin{bmatrix} 0 \\ 1 \\ 0 \end{bmatrix}$ and $\begin{bmatrix} 0 \\ 0 \\ 1 \end{bmatrix}$.

**23.** $\mathbf{A} = \begin{bmatrix} 0 & -9 \\ 1 & 0 \end{bmatrix}$; characteristic polynomial: $\lambda^2 + 9$;

eigenvalues: $\pm 3i$; corresponding eigenvectors:

$\begin{bmatrix} 3i \\ 1 \end{bmatrix}$ and $\begin{bmatrix} -3i \\ 1 \end{bmatrix}$.

**25.** Eigenvalues: $i$, $-i$ and 3; corresponding eigenvectors:

$\begin{bmatrix} -3i \\ -3+i \\ 1 \end{bmatrix}$, $\begin{bmatrix} 3i \\ -3-i \\ 1 \end{bmatrix}$ and $\begin{bmatrix} 1 \\ 0 \\ 1 \end{bmatrix}$.

**27.** Eigenvalues: $\pm i$; corresponding eigenvectors:

$\begin{bmatrix} -i \\ 1 \end{bmatrix}$ and $\begin{bmatrix} i \\ 1 \end{bmatrix}$.

**Strategy for Exercises 29–33:** Find a basis for $\mathbf{ns}(\mathbf{A} - \lambda \mathbf{I})$.

**29.** $\mathbf{A} - \lambda \mathbf{I}_3 = \begin{bmatrix} -1 & 0 & 1 \\ 0 & 0 & 0 \\ 1 & 0 & -1 \end{bmatrix}$;

$\mathbf{rref}\left(\left[\, \mathbf{A} - \lambda \mathbf{I}_3 \mid \mathbf{0} \,\right]\right) = \begin{bmatrix} 1 & 0 & -1 & 0 \\ 0 & 0 & 0 & 0 \\ 0 & 0 & 0 & 0 \end{bmatrix}$.

It follows that the eigenvectors are $\begin{bmatrix} t \\ s \\ t \end{bmatrix}$, where $s$ and $t$ are arbitrary constants. We have

$$\begin{bmatrix} t \\ s \\ t \end{bmatrix} = t\begin{bmatrix} 1 \\ 0 \\ 1 \end{bmatrix} + s\begin{bmatrix} 0 \\ 1 \\ 0 \end{bmatrix}$$

and thus $\left\{ \begin{bmatrix} 1 \\ 0 \\ 1 \end{bmatrix}, \begin{bmatrix} 0 \\ 1 \\ 0 \end{bmatrix} \right\}$ is a basis for the eigenspace.

**31.** $\left\{ \begin{bmatrix} -1 \\ 0 \\ 1 \end{bmatrix}, \begin{bmatrix} 0 \\ 1 \\ 0 \end{bmatrix} \right\}$ is a basis for the eigenspace.

**33.** $\left\{ \begin{bmatrix} 0 \\ 0 \\ 1 \\ 0 \end{bmatrix} \right\}$ is a basis for the eigenspace.

**Strategy for Exercises 34–38:** Find a basis for the eigenspace corresponding to any eigenvalue of multiplicity greater than 1.

**35.** $\mathbf{A} - 3\mathbf{I}_3 = \begin{bmatrix} 0 & 0 & 0 \\ -2 & 0 & -2 \\ 2 & 0 & 2 \end{bmatrix}$

$\mathbf{rref}\left(\left[\, \mathbf{A} - 3\mathbf{I}_3 \mid \mathbf{0} \,\right]\right) = \begin{bmatrix} 1 & 0 & 1 & 0 \\ 0 & 0 & 0 & 0 \\ 0 & 0 & 0 & 0 \end{bmatrix}$.

It follows that the eigenvectors are $\begin{bmatrix} -t \\ s \\ t \end{bmatrix}$, where $s$ and $t$ are arbitrary constants. We have

$$\begin{bmatrix} -t \\ s \\ t \end{bmatrix} = t\begin{bmatrix} -1 \\ 0 \\ 1 \end{bmatrix} + s\begin{bmatrix} 0 \\ 1 \\ 0 \end{bmatrix}$$

and thus $\left\{ \begin{bmatrix} -1 \\ 0 \\ 1 \end{bmatrix}, \begin{bmatrix} 0 \\ 1 \\ 0 \end{bmatrix} \right\}$ is a basis for the eigenspace. Since the number of basis vectors equals the multiplicity, $\mathbf{A}$ is not defective.

**37.** This matrix is not defective.

**43.** If $\mathbf{U}$ is upper triangular, then the characteristic polynomial is
$\det(\mathbf{U} - \lambda \mathbf{I}_n) = (u_{11} - \lambda)(u_{22} - \lambda) \cdots (u_{nn} - \lambda)$. The eigenvalues are the roots of the characteristic polynomial and in this case are just the diagonal entries of $\mathbf{U}$. (Recall that the determinant of an upper triangular matrix is the product of the diagonal entries and $\mathbf{U} - \lambda \mathbf{I}_n$ is upper triangular.)

**47.** $\mathbf{A}\,\mathbf{col}_j(\mathbf{I}_n) = 0\,\mathbf{col}_1(\mathbf{A}) + 0\,\mathbf{col}_2(\mathbf{A}) + \cdots + 0\,\mathbf{col}_{j-1}(\mathbf{A})$
$+ 1\,\mathbf{col}_j(\mathbf{A}) + 0\,\mathbf{col}_{j+1}(\mathbf{A}) + \cdots$
$+ 0\,\mathbf{col}_n(\mathbf{A})$
$= \mathbf{col}_j(\mathbf{A}) = k\,\mathbf{col}_j(\mathbf{I}_n)$.

**49. (a)**

| $j$   $k \to$ $\downarrow$ | 0 | 1 | 2 | 3 | 4 | 5 | 6 | 7 |
|---|---|---|---|---|---|---|---|---|
| 1 | $-1$ | 5 | 2.6 | 2.2308 | 2.1034 | 2.0492 | 2.0240 | 2.0119 |
| 2 | 0.5 | $-1$ | $-5$ | $-2.6$ | 2.2308 | 2.1034 | 2.0492 | 2.0240 |

**(b)** Conjecture: 2.

**(c)** $\mathbf{A}^{20}\mathbf{u}_0 = \begin{bmatrix} -2097149 \\ -1048573 \end{bmatrix}$, $\mathbf{A}^{21}\mathbf{u}_0 = \begin{bmatrix} -4194301 \\ -2097149 \end{bmatrix}$;

the ratio of corresponding components is 2.

## Section 4.2 True/False, Page 253

1. F    2. T    3. F    4. F    5. F

6. T    7. F    8. T    9. T    10. T

## Section 4.3, Page 259

**Strategy for Exercises 1–4:** Determine a nonsingular matrix **P** and compute $\mathbf{P}^{-1}$ (the given matrix) times **P**. Hence there are many possible answers.

1. Let $\mathbf{P} = \begin{bmatrix} 1 & 2 \\ 0 & 8 \end{bmatrix}$ and $\mathbf{Q} = \begin{bmatrix} 0 & 1 \\ -1 & 2 \end{bmatrix}$. Then

$$\mathbf{P}^{-1} \begin{bmatrix} 2 & 3 \\ 1 & 0 \end{bmatrix} \mathbf{P} = \begin{bmatrix} 1.75 & 27.5 \\ 0.125 & 0.25 \end{bmatrix}$$

is similar to $\begin{bmatrix} 2 & 3 \\ 1 & 0 \end{bmatrix}$. Also,

$$\mathbf{Q}^{-1} \begin{bmatrix} 2 & 3 \\ 1 & 0 \end{bmatrix} \mathbf{Q} = \begin{bmatrix} -6 & 15 \\ -3 & 8 \end{bmatrix}$$

is similar to $\begin{bmatrix} 2 & 3 \\ 1 & 0 \end{bmatrix}$.

5. We know that **A** and **B** are similar, thus there exists a nonsingular matrix **P** such that $\mathbf{P}^{-1}\mathbf{A}\mathbf{P} = \mathbf{B}$. It follows that $\mathbf{P}\mathbf{B}\mathbf{P}^{-1} = \mathbf{A}$, hence if we let $\mathbf{Q} = \mathbf{P}^{-1}$, we have a nonsingular matrix **Q** such that $\mathbf{Q}^{-1}\mathbf{B}\mathbf{Q} = \mathbf{A}$. Thus **B** is similar to **A**.

9. (a) $\mathbf{B} = \mathbf{P}^{-1}\mathbf{A}\mathbf{P} = \begin{bmatrix} 2 & 0 \\ 0 & 5 \end{bmatrix}$.

   (b) Eigenpairs of **A** are $\left(2, \begin{bmatrix} 1 \\ 1 \end{bmatrix}\right)$ and $\left(5, \begin{bmatrix} 2 \\ 3 \end{bmatrix}\right)$.
   (Note: There are many possible answers for the eigenvectors.)

   (c) Eigenpairs of **B** are $\left(2, \begin{bmatrix} 1 \\ 0 \end{bmatrix}\right)$ and $\left(5, \begin{bmatrix} 0 \\ 1 \end{bmatrix}\right)$.
   (Note: There are many possible answers for the eigenvectors.)

13. This matrix has distinct eigenvalues 6 and 1, hence it is diagonalizable.

15. This matrix has distinct eigenvalues 1, 2, and 4, hence it is diagonalizable.

17. The eigenvalues are 1 and 1. However, to compute eigenvectors we form the linear system

$$\left( \begin{bmatrix} 1 & 1 \\ 0 & 1 \end{bmatrix} - 1 \begin{bmatrix} 1 & 0 \\ 0 & 1 \end{bmatrix} \right) \begin{bmatrix} x_1 \\ x_2 \end{bmatrix} = \begin{bmatrix} 0 \\ 0 \end{bmatrix}$$

and we find that there is only one linearly independent eigenvector since the RREF of the augmented matrix is $\begin{bmatrix} 0 & 1 & | & 0 \\ 0 & 0 & | & 0 \end{bmatrix}$.

20. Define $\mathbf{P} = \begin{bmatrix} -1 & 1 \\ 2 & 1 \end{bmatrix}$ and $\mathbf{D} = \begin{bmatrix} 2 & 0 \\ 0 & -3 \end{bmatrix}$. The matrix

$$\mathbf{A} = \mathbf{PDP}^{-1} = \begin{bmatrix} -\frac{4}{3} & -\frac{5}{3} \\ -\frac{10}{3} & \frac{1}{3} \end{bmatrix}$$

has the given eigenpairs.

## Section 4.3 True/False, Page 261

1. T    2. F    3. T    4. T    5. T

6. T    7. F    8. T    9. F    10. F

## Section 4.4, Page 271

1. $\begin{bmatrix} 1 \\ 2 \end{bmatrix} \cdot \begin{bmatrix} -2 \\ 1 \end{bmatrix} = 1(-2) + 2(1) = 0.$

5. An orthogonal set is linearly independent. The largest possible set of linearly independent vectors in $R^2$ contains two vectors.

7. $\begin{bmatrix} 1 \\ 2 \\ 0 \\ -1 \end{bmatrix} \cdot \begin{bmatrix} x_1 \\ x_2 \\ x_3 \\ x_4 \end{bmatrix} = x_1 + 2x_2 - x_4 = 0,$

$\begin{bmatrix} -1 \\ 0 \\ 3 \\ -1 \end{bmatrix} \cdot \begin{bmatrix} x_1 \\ x_2 \\ x_3 \\ x_4 \end{bmatrix} = -x_1 + 3x_3 - x_4 = 0,$

$\begin{bmatrix} -2 \\ 2 \\ 0 \\ 2 \end{bmatrix} \cdot \begin{bmatrix} x_1 \\ x_2 \\ x_3 \\ x_4 \end{bmatrix} = -2x_1 + 2x_2 + 2x_4 = 0.$

We solve the system consisting of these three equations;

$$\mathbf{rref}\left( \begin{bmatrix} 1 & 2 & 0 & -1 & | & 0 \\ -1 & 0 & 3 & -1 & | & 0 \\ -2 & 2 & 0 & 2 & | & 0 \end{bmatrix} \right)$$

$$= \begin{bmatrix} 1 & 0 & 0 & -1 & | & 0 \\ 0 & 1 & 0 & 0 & | & 0 \\ 0 & 0 & 1 & -\frac{2}{3} & | & 0 \end{bmatrix},$$

thus we have $x_1 = x_4$, $x_2 = 0$, and $x_3 = \frac{2}{3}x_4$. Setting $x_4 = 3$, we find that one vector orthogonal to all of the vectors in this set is $\begin{bmatrix} 3 \\ 0 \\ 2 \\ 3 \end{bmatrix}$.

8. **Strategy:** We convert each vector to a unit vector.
   For Exercise 1: $\left\{ \frac{1}{\sqrt{5}} \begin{bmatrix} 1 \\ 2 \end{bmatrix}, \frac{1}{\sqrt{5}} \begin{bmatrix} -2 \\ 1 \end{bmatrix} \right\}.$

17. The inverse of a unitary matrix is just its conjugate transpose, which requires no computation other than taking conjugates and interchanging rows with columns.

**27.** **(a)** Since $\mathbf{A}$ is symmetric and the eigenvalues are distinct, the corresponding eigenvectors are orthogonal. In addition, we know that a scalar multiple of an eigenvector is another eigenvector, hence we can multiply each eigenvector by the reciprocal of its length. Thus we have

$$\mathbf{p}_1 = \frac{1}{\sqrt{2}}\begin{bmatrix} 1 \\ 1 \end{bmatrix} \text{ and } \mathbf{p}_2 = \frac{1}{\sqrt{2}}\begin{bmatrix} 1 \\ -1 \end{bmatrix} \text{ are}$$

orthonormal eigenvectors.

**(b)** $\mathbf{S}_1 = 10\mathbf{p}_1\mathbf{p}_1^* = 10 \begin{bmatrix} \frac{1}{\sqrt{2}} \\ \frac{1}{\sqrt{2}} \end{bmatrix} \begin{bmatrix} \frac{1}{\sqrt{2}} & \frac{1}{\sqrt{2}} \end{bmatrix}$

$$= 10 \begin{bmatrix} \frac{1}{2} & \frac{1}{2} \\ \frac{1}{2} & \frac{1}{2} \end{bmatrix}$$

and

$\mathbf{S}_2 = 1\mathbf{p}_2\mathbf{p}_2^* = 1 \begin{bmatrix} \frac{1}{\sqrt{2}} \\ -\frac{1}{\sqrt{2}} \end{bmatrix} \begin{bmatrix} \frac{1}{\sqrt{2}} & -\frac{1}{\sqrt{2}} \end{bmatrix}$

$$= 1 \begin{bmatrix} \frac{1}{2} & -\frac{1}{2} \\ -\frac{1}{2} & \frac{1}{2} \end{bmatrix}.$$

**(c)** $\mathbf{S}_1 + \mathbf{S}_2 = 10 \begin{bmatrix} \frac{1}{2} & \frac{1}{2} \\ \frac{1}{2} & \frac{1}{2} \end{bmatrix} + 1 \begin{bmatrix} \frac{1}{2} & -\frac{1}{2} \\ -\frac{1}{2} & \frac{1}{2} \end{bmatrix}$

$$= \begin{bmatrix} 5.5 & 4.5 \\ 4.5 & 5.5 \end{bmatrix} = \mathbf{A}.$$

**(d)** Note that

$$\mathbf{A} - \mathbf{S}_1 = \mathbf{S}_2 = \begin{bmatrix} \frac{1}{2} & -\frac{1}{2} \\ -\frac{1}{2} & \frac{1}{2} \end{bmatrix};$$

the square root of the sum of the squares of the entries is

$$\left(\tfrac{1}{4} + \tfrac{1}{4} + \tfrac{1}{4} + \tfrac{1}{4}\right)^{\frac{1}{2}} = 1.$$

## Section 4.4 True/False, Page 274
**1.** T     **2.** T     **3.** F     **4.** T     **5.** T
**6.** T     **7.** T     **8.** T     **9.** T     **10.** F

## Section 4.5, Page 280
**Strategy for Exercises 1–3:** For an orthonormal basis, any vector $\mathbf{w}$ in the same space can be expressed as

$$\mathbf{w} = (\mathbf{u}_1^*\mathbf{w})\mathbf{u}_1 + (\mathbf{u}_2^*\mathbf{w})\mathbf{u}_2 + \cdots + (\mathbf{u}_n^*\mathbf{w})\mathbf{u}_n.$$

**1.** $\mathbf{w} = \dfrac{1}{\sqrt{2}}\mathbf{v}_1 + \dfrac{1}{\sqrt{3}}\mathbf{v}_2 + \dfrac{1}{\sqrt{6}}\mathbf{v}_3.$

**3.** $\mathbf{w} = \dfrac{1}{\sqrt{2}}\mathbf{v}_1 + \dfrac{2}{\sqrt{3}}\mathbf{v}_2 + \dfrac{-1}{\sqrt{6}}\mathbf{v}_3.$

**4.** **(b)** We get orthonormal basis

$$T = \left\{ \frac{1}{\sqrt{2}}\begin{bmatrix} 1 \\ 1 \end{bmatrix}, \frac{1}{\sqrt{\frac{1}{2}}}\begin{bmatrix} -\frac{1}{2} \\ \frac{1}{2} \end{bmatrix} \right\}.$$

**5.** Let $\mathbf{q}_1 = \begin{bmatrix} 2 \\ 1 \\ 0 \end{bmatrix}$ and $\mathbf{q}_2 = \begin{bmatrix} 1 \\ 0 \\ 1 \end{bmatrix}$. Define

$$\mathbf{v}_1 = \mathbf{q}_1$$

and

$$\mathbf{v}_2 = \mathbf{q}_2 - \left(\frac{\mathbf{v}_1^*\mathbf{q}_2}{\mathbf{v}_1^*\mathbf{v}_1}\right)\mathbf{v}_1$$

$$= \mathbf{q}_2 - \text{proj}_{\mathbf{v}_1}\mathbf{q}_2$$

$$= \begin{bmatrix} 1 \\ 0 \\ 1 \end{bmatrix} - \left(\frac{2}{5}\right)\begin{bmatrix} 2 \\ 1 \\ 0 \end{bmatrix}$$

$$= \begin{bmatrix} \frac{1}{5} \\ -\frac{2}{5} \\ 1 \end{bmatrix}.$$

Now convert both $\mathbf{v}_1$ and $\mathbf{v}_2$ to unit vectors. We get orthonormal basis

$$T = \left\{ \frac{1}{\sqrt{5}}\begin{bmatrix} 2 \\ 1 \\ 0 \end{bmatrix}, \frac{1}{\sqrt{\frac{30}{25}}}\begin{bmatrix} \frac{1}{5} \\ -\frac{2}{5} \\ 1 \end{bmatrix} \right\}.$$

**9.** No. The vectors are linearly dependent.

**11.** $\left\{ \dfrac{1}{\sqrt{\frac{21}{4}}}\begin{bmatrix} -\frac{1}{2} \\ -2 \\ 1 \\ 0 \end{bmatrix}, \dfrac{1}{\sqrt{\frac{9}{7}}}\begin{bmatrix} -\frac{2}{7} \\ \frac{1}{7} \\ \frac{3}{7} \\ 1 \end{bmatrix} \right\}$ is an orthonormal basis for the null space.

## Section 4.5 True/False, Page 282
**1.** T     **2.** T     **3.** F     **4.** F     **5.** T

## Section 4.6, Page 291
**1.** **(a)** $\{\mathbf{q}_1, \mathbf{q}_2, \mathbf{q}_3\} =$

$$\left\{ \frac{1}{\sqrt{5}}\begin{bmatrix} 1 \\ 0 \\ 2 \end{bmatrix}, \frac{1}{\sqrt{1.8}}\begin{bmatrix} 1.2 \\ 0 \\ -0.6 \end{bmatrix}, \begin{bmatrix} 0 \\ 1 \\ 0 \end{bmatrix} \right\}.$$

**(b)** $\mathbf{R} = \begin{bmatrix} \|\mathbf{v}_1\| & \dfrac{(\mathbf{v}_1^*\mathbf{a}_2)}{\|\mathbf{v}_1\|} & \dfrac{(\mathbf{v}_1^*\mathbf{a}_3)}{\|\mathbf{v}_1\|} \\ & \|\mathbf{v}_2\| & \dfrac{(\mathbf{v}_2^*\mathbf{a}_3)}{\|\mathbf{v}_2\|} \\ & & \|\mathbf{v}_3\| \end{bmatrix}$

$= \begin{bmatrix} \sqrt{5} & \dfrac{4}{\sqrt{5}} & \dfrac{1}{\sqrt{5}} \\ 0 & \sqrt{1.8} & \dfrac{1.2}{\sqrt{1.8}} \\ 0 & 0 & 1 \end{bmatrix}$

$\approx \begin{bmatrix} 2.2361 & 1.7889 & 0.4472 \\ 0 & 1.3416 & 0.8944 \\ 0 & 0 & 1 \end{bmatrix}$

**(c)** $\mathbf{A} = \mathbf{QR} = \begin{bmatrix} \dfrac{1}{\sqrt{5}} & \dfrac{1.2}{\sqrt{1.8}} & 0 \\ 0 & 0 & 1 \\ \dfrac{2}{\sqrt{5}} & -\dfrac{0.6}{\sqrt{1.8}} & 0 \end{bmatrix}$

$= \begin{bmatrix} \sqrt{5} & \dfrac{4}{\sqrt{5}} & \dfrac{1}{\sqrt{5}} \\ 0 & \sqrt{1.8} & \dfrac{1.2}{\sqrt{1.8}} \\ 0 & 0 & 1 \end{bmatrix}.$

**(d)** Just form the product $\mathbf{QR}$ and we will get $\mathbf{A}$.

**7.** $\mathbf{x} = \begin{bmatrix} 2 \\ 0.75 \\ 1.5 \end{bmatrix}.$

## Section 4.6 True/False, Page 292

**1.** F       **2.** T      **3.** T      **4.** T      **5.** T

## Section 4.7, Page 300

**Strategy for Exercises 1–4:** The singular values of a matrix $\mathbf{A}$ are the square roots of the eigenvalues of $\mathbf{A} * \mathbf{A}$.

**1.** The singular values of $\mathbf{A}$ are 5 and 1.

**3.** The singular values of $\mathbf{A}$ are $\sqrt{6}$ and $\sqrt{5}$.

**5. Strategy:** Follow the steps used in Example 1.
$\mathbf{A}^*\mathbf{A} = \begin{bmatrix} 9 & 0 \\ 0 & 36 \end{bmatrix}$, hence its eigenvalues are 9 and 36. A pair of corresponding eigenvectors is $\begin{bmatrix} 1 \\ 0 \end{bmatrix}$ and $\begin{bmatrix} 0 \\ 1 \end{bmatrix}$, respectively. Label the eigenvalues in decreasing magnitude as $\lambda_1 = 36$ and $\lambda_2 = 9$ with corresponding

eigenvectors $\mathbf{v}_1 = \begin{bmatrix} 0 \\ 1 \end{bmatrix}$ and $\mathbf{v}_2 = \begin{bmatrix} 1 \\ 0 \end{bmatrix}$. Hence

$$\mathbf{V} = \begin{bmatrix} \mathbf{v}_1 & \mathbf{v}_2 \end{bmatrix} = \begin{bmatrix} 0 & 1 \\ 1 & 0 \end{bmatrix}$$

and $s_{11} = 6$ and $s_{22} = 3$. It follows that

$$\mathbf{S} = \begin{bmatrix} 6 & 0 \\ 0 & 3 \\ 0 & 0 \end{bmatrix}.$$

Next we determine the matrix $\mathbf{U}$ starting with the first two columns

$$\mathbf{u}_1 = \frac{1}{s_{11}}\mathbf{A}\mathbf{v}_1 = \frac{1}{6}\begin{bmatrix} -4 \\ 2 \\ 4 \end{bmatrix}$$

and

$$\mathbf{u}_2 = \frac{1}{s_{22}}\mathbf{A}\mathbf{v}_2 = \frac{1}{3}\begin{bmatrix} 1 \\ -2 \\ 2 \end{bmatrix}.$$

The remaining column is found by extending the linearly independent vectors in $\{\mathbf{u}_1, \mathbf{u}_2\}$ to a basis for $R^3$ and then applying the Gram–Schmidt process. We proceed as follows. We first compute

$$\mathbf{rref}\left(\begin{bmatrix} \mathbf{u}_1 & \mathbf{u}_2 & \mathbf{I}_3 \end{bmatrix}\right)$$
$$= \begin{bmatrix} 1 & 0 & 0 & 1 & 1 \\ 0 & 1 & 0 & -1 & 0.5 \\ 0 & 0 & 1 & 1 & 0.5 \end{bmatrix}.$$

This tells us that the set $\{\mathbf{u}_1, \mathbf{u}_2, \mathbf{e}_1\}$ is a basis for $R^3$. Next we apply the Gram–Schmidt process to this set to find the unitary matrix $\mathbf{U}$. We get

$$\mathbf{U} = \begin{bmatrix} -\frac{2}{3} & \frac{1}{3} & \frac{2}{3} \\ \frac{1}{3} & -\frac{2}{3} & \frac{2}{3} \\ \frac{2}{3} & \frac{2}{3} & \frac{1}{3} \end{bmatrix}.$$

Hence we conclude that the singular value decomposition of $\mathbf{A}$ is

$$\mathbf{A} = \mathbf{USV}^* = \begin{bmatrix} -\frac{2}{3} & \frac{1}{3} & \frac{2}{3} \\ \frac{1}{3} & -\frac{2}{3} & \frac{2}{3} \\ \frac{2}{3} & \frac{2}{3} & \frac{1}{3} \end{bmatrix}\begin{bmatrix} 6 & 0 \\ 0 & 3 \\ 0 & 0 \end{bmatrix}\begin{bmatrix} 0 & 1 \\ 1 & 0 \end{bmatrix}.$$

In Exercise 9, we record the output to four decimal digits.

**9.** $\mathbf{U} = \begin{bmatrix} 0.8404 & -0.0201 & 0.5415 \\ -0.1970 & -0.9423 & 0.2708 \\ -0.5048 & 0.3343 & 0.7959 \end{bmatrix}$,

$\mathbf{S} = \begin{bmatrix} 18.9245 & 0 & 0 \\ 0 & 3.8400 & 0 \\ 0 & 0 & 0.3440 \end{bmatrix}$,

$\mathbf{V} = \begin{bmatrix} 0.1762 & -0.4352 & 0.8829 \\ 0.7567 & -0.5138 & -0.4042 \\ -0.6295 & -0.7394 & -0.2388 \end{bmatrix}$

## Section 4.7 True/False, Page 301

**1.** T    **2.** T    **3.** F    **4.** T    **5.** T

## Chapter 4 Chapter Test, Page 302

**1.** $\lambda^3 - 6\lambda^2 + 11\lambda - 6$.

**2.** $\left( 0, \begin{bmatrix} 1 \\ 0 \\ 0 \end{bmatrix} \right) \left( 1, \begin{bmatrix} 0 \\ 1 \\ 0 \end{bmatrix} \right), \left( 1, \begin{bmatrix} 1 \\ 0 \\ 1 \end{bmatrix} \right)$.

**3.** $\lambda = 3$ is a twice-repeated eigenvalue with only one linearly independent eigenvector $\begin{bmatrix} 1 \\ 0 \\ 0 \end{bmatrix}$.

**4.**

| Yes | No |
|-----|-----|
| IITD | Yes |
| Yes | No |

**5.** $\begin{bmatrix} -\frac{1}{\sqrt{2}} \\ \frac{1}{\sqrt{2}} \\ 0 \end{bmatrix}, \begin{bmatrix} -\frac{1}{\sqrt{6}} \\ -\frac{1}{\sqrt{6}} \\ \frac{2}{\sqrt{6}} \end{bmatrix}, \begin{bmatrix} \frac{1}{\sqrt{3}} \\ \frac{1}{\sqrt{3}} \\ \frac{1}{\sqrt{3}} \end{bmatrix}$.

**6.** $\left\{ \begin{bmatrix} 1 \\ 0 \\ 0 \\ 0 \end{bmatrix}, \begin{bmatrix} 0 \\ 1 \\ 0 \\ 0 \end{bmatrix} \right\}$.

**7.** There exists a matrix $\mathbf{P}$ so that $\mathbf{P}^{-1}\mathbf{AP} = \mathbf{D}$, a diagonal matrix. Since $\mathbf{A}$ is nonsingular, so is $\mathbf{D}$. Thus we have $(\mathbf{P}^{-1}\mathbf{AP})^{-1} = \mathbf{D}^{-1}$ or $\mathbf{P}^{-1}\mathbf{A}^{-1}\mathbf{P} = \mathbf{D}^{-1}$. So $\mathbf{A}^{-1}$ is diagonalizable since $\mathbf{D}^{-1}$ is diagonal.

**8.** $1\mathbf{vv}^T + 3\mathbf{uu}^T - 2\mathbf{ww}^T$.

**9.** No. $\begin{bmatrix} \frac{3}{\sqrt{10}} & 0 & \frac{1}{\sqrt{10}} \\ 0 & 1 & 0 \\ -\frac{1}{\sqrt{10}} & 0 & \frac{3}{\sqrt{10}} \end{bmatrix}$.

**10.** (a) True.    (b) False.    (c) True.

(d) True.    (e) False.

## Chapter 5

## Section 5.1, Page 311

**1.** Corresponding angles are equal. The ratios of the lengths of corresponding sides are equal.

**3.** Their determinants are not zero. Their rows (columns) are linearly independent. Their RREFs are $\mathbf{I}_2$. Their null space is the zero vector. A basis for their row space (column space) consists of the rows (columns) of $\mathbf{I}_2$. The dimension of the row space (column space) is 2. They each have an inverse.

**5.** Each is a subspace of the plane. They have equations of the form $y = kx$ or $x = 0$.

**7.** If both $(0, x, y)$ and $(0, r, s)$ are in $V$, then $(0, x, y) \oplus (0, r, s) = (0, x + r, y + s)$ is in $V$ since it is an ordered triple with first entry zero. Thus $V$ is closed under addition. For scalar $k$, we have $k \odot (0, x, y) = (0, 0, ky)$ is an ordered triple with first entry zero, thus it is in $V$. It follows that $V$ is closed under scalar multiplication.

**11.** Let $V = P_n$ be the set of all polynomials of degree $n$ or less with real coefficients. Let

$$\mathbf{p} = a_n x^n + a_{n-1} x^{n-1} + \cdots + a_1 x + a_0,$$
$$\mathbf{q} = b_n x^n + b_{n-1} x^{n-1} + \cdots + b_1 x + b_0,$$

and

$$\mathbf{r} = c_n x^n + c_{n-1} x^{n-1} + \cdots + c_1 x + c_0$$

be vectors in $V$.

Property (a)
$$\mathbf{p} \oplus \mathbf{q} = (a_n + b_n)x^n + (a_{n-1} + b_{n-1})x^{n-1} + \cdots$$
$$+(a_1 + b_1)x + (a_0 + b_0)$$
$$= (b_n + a_n)x^n + (b_{n-1} + a_{n-1})x^{n-1} + \cdots$$
$$+(b_1 + a_1)x + (b_0 + a_0) = \mathbf{q} \oplus \mathbf{p}.$$

Property (b)
$$(\mathbf{p} \oplus \mathbf{q}) \oplus \mathbf{r} = ((a_n + b_n)x^n + (a_{n-1} + b_{n-1})x^{n-1} + \cdots$$
$$+(a_1 + b_1)x + (a_0 + b_0))$$
$$+c_n x^n + c_{n-1} x^{n-1} + \cdots + c_1 x + c_0$$
$$= ((a_n + b_n) + c_n)x^n +$$
$$((a_{n-1} + b_{n-1}) + c_{n-1})x^{n-1} + \cdots$$
$$+((a_1 + b_1) + c_1)x + ((a_0 + b_0) + c_0)$$
$$= (a_n + (b_n + c_n))x^n$$
$$+(a_{n-1} + (b_{n-1} + c_{n-1}))x^{n-1} + \cdots$$
$$+(a_1 + (b_1 + c_1))x + (a_0 + (b_0 + c_0))$$
$$= \mathbf{p} \oplus (\mathbf{q} \oplus \mathbf{r}).$$

Property (c) Let $\mathbf{z} = 0x^n + 0x^{n-1} + \cdots + 0x + 0$, then by the definition of addition here, we have $\mathbf{p} \oplus \mathbf{z} = \mathbf{z} \oplus \mathbf{p} = \mathbf{p}$. To show that there is only one vector $\mathbf{z}$ for which this is true, assume that there is another vector $\mathbf{w}$ such that $\mathbf{p} \oplus \mathbf{w} = \mathbf{w} \oplus \mathbf{p} = \mathbf{p}$ for every vector $\mathbf{p}$ in $V$. Then it must be that $\mathbf{z} \oplus \mathbf{w} = \mathbf{w} \oplus \mathbf{z} = \mathbf{z}$, but also $\mathbf{w} \oplus \mathbf{z} = \mathbf{z} \oplus \mathbf{w} = \mathbf{w}$ and thus $\mathbf{z} = \mathbf{w}$.

Property (d) Let
$\mathbf{w} = -a_n x^n - a_{n-1} x^{n-1} - \cdots - a_1 x - a_0$; then
$\mathbf{p} \oplus \mathbf{w} = \mathbf{w} \oplus \mathbf{p} = \mathbf{z}$, the zero vector. Thus for every vector $\mathbf{p}$ in $V$ there exists a negative of $\mathbf{p}$. To show that $\mathbf{w}$ is unique, assume that there is another vector $\mathbf{s}$ in $V$ such that $\mathbf{p} \oplus \mathbf{s} = \mathbf{s} \oplus \mathbf{p} = \mathbf{z}$. Add $\mathbf{w}$ to each of the preceding terms and get $\mathbf{w} \oplus \mathbf{p} \oplus \mathbf{s} = \mathbf{w} \oplus \mathbf{s} \oplus \mathbf{p} = \mathbf{w} \oplus \mathbf{z}$. Using properties verified so far we have
$(\mathbf{w} \oplus \mathbf{p}) \oplus \mathbf{s} = \mathbf{s} \oplus (\mathbf{w} \oplus \mathbf{p}) = \mathbf{w} \oplus \mathbf{z} = \mathbf{w}$, which simplifies to give $\mathbf{z} \oplus \mathbf{s} = \mathbf{s} \oplus \mathbf{z} = \mathbf{w}$, and further to $\mathbf{s} = \mathbf{w}$.

Property (e)
$$\begin{aligned} k \odot (\mathbf{p} \oplus \mathbf{q}) &= k((a_n + b_n)x^n + (a_{n-1} + b_{n-1})x^{n-1} + \cdots \\ &\quad + (a_1 + b_1)x + (a_0 + b_0)) \\ &= (ka_n + kb_n)x^n + (ka_{n-1} + kb_{n-1})x^{n-1} \\ &\quad + \cdots + (ka_1 + kb_1)x + (ka_0 + kb_0) \\ &= (ka_n x^n + ka_{n-1}x^{n-1} + \cdots + ka_1 x + ka_0) \\ &\quad + (kb_n x^n + kb_{n-1}x^{n-1} + \cdots \\ &\quad + kb_1 x + kb_0) \\ &= k\mathbf{p} \oplus k\mathbf{q}. \end{aligned}$$

Property (f)
$$\begin{aligned} (k + j) \odot \mathbf{p} &= (k + j)a_n x^n + (k + j)a_{n-1}x^{n-1} + \cdots \\ &\quad + (k + j)a_1 x + (k + j)a_0 \\ &= (ka_n x^n + ja_n x^n) \\ &\quad + (ka_{n-1}x^{n-1} + ja_{n-1}x^{n-1}) + \cdots \\ &\quad + (ka_1 x + ja_1 x) + (ka_0 + ja_0) \\ &= (ka_n x^n + ka_{n-1}x^{n-1} + \cdots + ka_1 x + ka_0) \\ &\quad + (ja_n x^n + ja_{n-1}x^{n-1} + \cdots \\ &\quad + ja_1 x + ja_0) \\ &= (k \odot \mathbf{p}) \oplus (j \odot \mathbf{p}). \end{aligned}$$

Property (g)
$$\begin{aligned} k \odot (j \odot \mathbf{p}) &= k(ja_n x^n + ja_{n-1}x^{n-1} + \cdots \\ &\quad + ja_1 x + ja_0) \\ &= kja_n x^n + kja_{n-1}x^{n-1} + \cdots \\ &\quad + kja_1 x + kja_0 \\ &= (kj) \odot \mathbf{p}. \end{aligned}$$

Property (h)
$1 \odot \mathbf{p} = 1a_n x^n + 1a_{n-1}x^{n-1} + \cdots + 1a_1 x + 1a_0 = \mathbf{p}$.

15. $k \odot \mathbf{z} = k \odot (\mathbf{z} \oplus \mathbf{z})$; expanding the right side we get $k \odot \mathbf{z} = (k \odot \mathbf{z}) \oplus (k \odot \mathbf{z})$; now add $-(k \odot \mathbf{z})$ to each side and simplifying we get $\mathbf{z} = k \odot \mathbf{z}$.

17. Since a negative vector is unique, to show that a vector is the negative of another vector we show that their sum is the zero vector. We have

$$\begin{aligned} \mathbf{u} \oplus ((-1) \odot \mathbf{u}) &= (1 \oplus \mathbf{u}) \oplus ((-1) \odot \mathbf{u}) \\ &= (1 - 1) \odot \mathbf{u} \\ &= 0 \odot \mathbf{u} \\ &= \mathbf{z}. \end{aligned}$$

22. This is not a vector space since
$\mathbf{v} \oplus \mathbf{w} = \mathbf{u} - \mathbf{w} \neq \mathbf{w} - \mathbf{v} = \mathbf{w} \oplus \mathbf{v}$.

## Section 5.1 True/False, Page 312

1. F    2. T    3. F    4. T    5. T
6. T    7. T    8. F    9. F    10. F

## Section 5.2, Page 322

1. This follows because a subspace is a vector space in its own right. That is, it satisfies the ten properties listed in Section 5.1.

3. No. (Consider the case $n = 2$.) Note that $\mathbf{A} = \begin{bmatrix} 1 & 0 \\ 0 & 0 \end{bmatrix}$ and $\mathbf{B} = \begin{bmatrix} 0 & 0 \\ 0 & 1 \end{bmatrix}$ are both singular, yet $\mathbf{A} + \mathbf{B} = \mathbf{I}_2$, which is nonsingular. (This situation can be generalized to $n \times n$ matrices.)

6. Yes. The sum of diagonal matrices is another diagonal matrix and a scalar multiple of a diagonal matrix is also a diagonal matrix. Thus $W$ is closed under addition and scalar multiplication.

7. No. Note that $0(ax^3 + bx + 1) = 0$, which is not in $W$.

14. Yes. Let $\mathbf{p} = (a, b, c)$ and $\mathbf{u} = (d, e, f)$ be in $W$. Then $\mathbf{v} \cdot \mathbf{p} = 0$ and $\mathbf{v} \cdot \mathbf{u} = 0$. From the properties of dot products, it follows that
$\mathbf{v} \cdot (\mathbf{p} + \mathbf{u}) = \mathbf{v} \cdot \mathbf{p} + \mathbf{v} \cdot \mathbf{u} = 0 + 0 = 0$. So $\mathbf{p} + \mathbf{u}$ is in $W$. Also from properties of dot products we have that $\mathbf{v} \cdot k\mathbf{p} = k(\mathbf{v} \cdot \mathbf{p}) = k0 = 0$ and hence $k\mathbf{p}$ is in $W$.

17. No. Let $\mathbf{A} = \begin{bmatrix} a & b & c \\ 0 & 0 & d \end{bmatrix}$ be in $W$. It follows that $0\mathbf{A}$ does not have the $(1, 2)$-entry positive, hence $W$ is not closed under scalar multiplication.

**Strategy for Exercises 21–24:** Determine if $\mathbf{u}$ is a linear combination of the vectors $\mathbf{v}_1, \mathbf{v}_2, \mathbf{v}_3$.

21. Compute
$$\mathbf{rref}\left(\begin{bmatrix} 1 & 1 & 1 & 2 \\ 0 & -1 & 1 & 1 \\ 1 & 0 & 2 & 3 \end{bmatrix}\right) = \begin{bmatrix} 1 & 0 & 2 & 3 \\ 0 & 1 & -1 & -1 \\ 0 & 0 & 0 & 0 \end{bmatrix}.$$

Since the system is consistent, $\mathbf{u}$ is in span$\{\mathbf{v}_1, \mathbf{v}_2, \mathbf{v}_3\}$.

**Strategy for Exercises 25–28:** Determine if $\mathbf{u}$ is a linear combination of the vectors $\mathbf{v}_1, \mathbf{v}_2, \mathbf{v}_3$. Forming a linear combination of the matrices $\mathbf{v}_1, \mathbf{v}_2, \mathbf{v}_3$ with unknown coefficients, we set it equal to the matrix $\mathbf{u}$ and then equate corresponding entries to get a linear system of 4 equations in 3 unknowns to solve.

25. Compute
$$\mathbf{rref}\left(\begin{bmatrix} 1 & 1 & 2 & 2 \\ 0 & 0 & -1 & -1 \\ -1 & 1 & 2 & 1 \\ 3 & 2 & 1 & 9 \end{bmatrix}\right) = \begin{bmatrix} 1 & 0 & 0 & 0 \\ 0 & 1 & 0 & 0 \\ 0 & 0 & 1 & 0 \\ 0 & 0 & 0 & 1 \end{bmatrix}.$$

Since the system is inconsistent, $\mathbf{u}$ is not in span$\{\mathbf{v}_1, \mathbf{v}_2, \mathbf{v}_3\}$.

**Strategy for Exercises 29–31:** Follow the steps used in Example 7.

**29.** Let $\mathbf{w} = \begin{bmatrix} r \\ s \end{bmatrix}$, where $r > 0$ and $s > 0$, then we have

$$c_1 \begin{bmatrix} 4 \\ 1 \end{bmatrix} + c_2 \begin{bmatrix} 6 \\ 1 \end{bmatrix} = \begin{bmatrix} 4^{c_1} 6^{c_2} \\ 1^{c_1} 1^{c_2} \end{bmatrix} = \begin{bmatrix} r \\ s \end{bmatrix}.$$

Equating corresponding matrix entries leads to the (nonlinear) system of equations

$$4^{c_1} 6^{c_2} = r$$
$$1^{c_1} 1^{c_2} = s.$$

Taking the natural logarithm of both sides of each equation and using properties of logarithms leads to the linear system of equations

$$c_1 \ln(4) + c_2 \ln(6) = \ln(r)$$
$$c_1 \ln(1) + c_2 \ln(1) = \ln(s).$$

The second equation implies that $0 = \ln(s)$, hence we must have $s = 1$. It follows that this linear system is not consistent for all positive values of $r$ and $s$ so $\text{span}(S) \neq V$.

**32. Strategy:** We form a linear combination of the vectors in $S$ and set it equal to $\mathbf{w}$. We then determine the choices for the entries of $\mathbf{w}$ that make the system consistent. To do this, we row reduce the corresponding augmented matrix until we get a row of zeros in the coefficient matrix part and an expression in the same row in the augmented column.
A sequence of row operations applied to

$$\begin{bmatrix} 2 & -4 & 0 & 10 & | & r \\ 3 & 1 & 7 & 1 & | & s \\ -1 & 0 & -2 & -1 & | & t \end{bmatrix}$$

gives

$$\begin{bmatrix} 1 & 0 & 2 & 1 & | & -t \\ 0 & 1 & 1 & -2 & | & 3t + s \\ 0 & 0 & 0 & 0 & | & 14t + 4s + r \end{bmatrix}.$$

Thus we see that only those values of $r$, $s$, and $t$ that make the expression $14t + 4s + r = 0$ will yield a vector $\mathbf{w}$ that is in $\text{span}(S)$.

## Section 5.2 True/False, Page 323

**1.** T     **2.** F     **3.** T     **4.** T     **5.** F
**6.** T     **7.** T     **8.** T     **9.** T     **10.** T

## Section 5.3, Page 336

**1.** This set is linearly independent since there are only two vectors and they are not multiples of one another.

**3.** Here we have four 3-vectors. Since the dimension of $R^3$ is 3, the largest number of vectors in a linearly independent set is 3, so this set must be linearly dependent.

**Strategy for Exercises 4–8:** Since $\dim(P_2) = 3$, the largest number of linearly independent vectors in a set is 3. Hence any set with more than 3 vectors must be linearly dependent. If the set contains three or fewer vectors, then we must check it directly by using the definition of linearly independent or make appropriate observations utilizing the theory established in this section.

**5.** Form the linear combination

$$c_1(2x^2 + x + 2) + c_2(x^2 - 3x + 1) + c_3(x^2 + 11x + 1) = 0.$$

Expand, collect terms of like powers of $x$, and we obtain

$$(2c_1 + c_2 + c_3)x^2 + (c_1 - 3c_2 + 11c_3)x + (2c_1 + c_2 + c_3) = 0.$$

Equating coefficients of like power terms on either side of the equals sign gives the linear system

$$\begin{aligned} 2c_1 + c_2 + c_3 &= 0 \\ c_1 - 3c_2 + 11c_3 &= 0 \\ 2c_1 + c_2 + c_3 &= 0. \end{aligned}$$

Solving this system, we find that one solution is $c_1 = -2, c_2 = 3, c_3 = 1$. Hence the vectors are linearly dependent.

**Strategy for Exercises 9–11:** A basis must be a linearly independent set that spans the space. It must also have the same number of vectors as the dimension of the space. If a set has fewer, then it cannot span the space and if it has more it is linearly dependent. In this case if we have exactly 3 vectors in the set, we must check that they are linearly independent and span. However, from the theory in this section, since $\dim(R^3) = 3$, any set of 3 linearly independent vectors will also span the space. Hence for sets with exactly 3 vectors we need only to check that they are linearly independent.

**11.** Checking to see if this set of 3 vectors is linearly independent, we construct a homogeneous linear system with coefficient matrix

$$\begin{bmatrix} 1 & 2 & 3 \\ 2 & 1 & 0 \\ 1 & 2 & 3 \end{bmatrix}.$$

Since its determinant is zero (note the two identical rows), there is a nontrivial solution of the homogeneous system. Thus the set is linearly dependent and is not a basis.

**15. (a)** Looking at the set $S$, we see that it contains the zero vector. Certainly we can have no hope of having a basis. Let's try the three nonzero vectors in $S$. Since we are in $R^3$, which has dimension 3, we need only check that these vectors are linearly independent. Constructing the corresponding homogeneous linear system, we find that the coefficient matrix is

$$\begin{bmatrix} 1 & -2 & 1 \\ -2 & 4 & 1 \\ 1 & -2 & 0 \end{bmatrix}.$$

We also find that the determinant of this matrix is 0, hence the homogeneous linear system has a nonzero solution indicating this set of 3 vectors is linearly dependent. Look at these 3 vectors again. Note that

$$\begin{bmatrix} -2 \\ 4 \\ -2 \end{bmatrix} = -2 \begin{bmatrix} 1 \\ -2 \\ 1 \end{bmatrix}.$$

Thus we can omit vector $\begin{bmatrix} -2 \\ 4 \\ -2 \end{bmatrix}$, so we now have the set

$$\left\{ \begin{bmatrix} 1 \\ -2 \\ 1 \end{bmatrix}, \begin{bmatrix} 1 \\ 1 \\ 0 \end{bmatrix} \right\}.$$

Note these vectors are not multiples of one another, hence are linearly independent. Thus we can choose

$$T = \left\{ \begin{bmatrix} 1 \\ -2 \\ 1 \end{bmatrix}, \begin{bmatrix} 1 \\ 1 \\ 0 \end{bmatrix} \right\}.$$

(This choice of $T$ is not unique; we could have also chosen $\left\{ \begin{bmatrix} -2 \\ 4 \\ -2 \end{bmatrix}, \begin{bmatrix} 1 \\ 1 \\ 0 \end{bmatrix} \right\}$.)

**(b)** No, since span$(T)$ would only have dimension 2.

**19.** Let $S = \{\mathbf{v}_1, \mathbf{v}_2, \ldots, \mathbf{v}_k\}$ where $\mathbf{v}_1 = \mathbf{0}$, the zero vector. Then we have that $99\mathbf{v}_1 + 0\mathbf{v}_2 + 0\mathbf{v}_3 + \cdots + 0\mathbf{v}_k = \mathbf{0}$, which is a linear combination that produces the zero vector where not all the scalar coefficients are zero. Thus set $S$ is linearly dependent. (Note, any nonzero scalar could be used as the coefficient of $\mathbf{v}_1$. In addition, we can reorder such a set of vectors so that the first one is the zero vector without any loss of generality.)

**21.** Let $S = \{\mathbf{v}_1, \mathbf{v}_2, \ldots, \mathbf{v}_k\}$ be a linearly dependent set of vectors. Then there exists a nontrivial linear combination of these vectors which gives the zero vector. Suppose we have $c_1\mathbf{v}_1 + c_2\mathbf{v}_2 + c_3\mathbf{v}_3 + \cdots + c_k\mathbf{v}_k = \mathbf{0}$, where $c_1 \neq 0$. Then we can solve the expression for $\mathbf{v}_1$ by multiplying both sides by $1/c_1$. This gives $\mathbf{v}_1$ as a linear combination of the other vectors. (Note, we can reorder the vectors so that a vector with a nonzero coefficient is named $\mathbf{v}_1$ without any loss of generality.)

**27.** $S = \{1, x, x^2, x^3, \ldots, x^k, \ldots\}$ is a basis for $P$. $S$ contains infinitely many vectors. (There are many other such bases.)

## Section 5.3 True/False, Page 337
1. F    2. T    3. T    4. T    5. F
6. T    7. T    8. F    9. T    10. T

## Chapter 5 Chapter Test, Page 338
1. No. Operation $\oplus$ is not commutative.
2. Yes. It is closed under addition and scalar multiplication.
3. Any value of $m$, but $b$ must be zero.
4. They must contain the zero vector.
5. $\left\{ \begin{bmatrix} 1 & 1 \\ 0 & 0 \end{bmatrix}, \begin{bmatrix} 0 & 1 \\ 0 & 1 \end{bmatrix} \right\}.$
6. Let $f(t)$ and $g(t)$ be in $W$. Then

$$\int_0^1 f(t)\,dt = \int_0^1 g(t)\,dt.$$

It follows that

$$\int_0^1 (f(t) + g(t))\,dt = 0$$

and

$$\int_0^1 kf(t)\,dt = 0,$$

so $W$ is closed under addition and scalar multiplication, hence a subspace.

7. $\left\{ \begin{bmatrix} 1 & 0 \\ 0 & 0 \end{bmatrix}, \begin{bmatrix} 1 & 0 \\ 0 & 1 \end{bmatrix}, \begin{bmatrix} 0 & 0 \\ 1 & 0 \end{bmatrix}, \begin{bmatrix} 0 & 1 \\ 0 & 0 \end{bmatrix} \right\}.$

8. ns$(\mathbf{A})$ contains $\begin{bmatrix} 1 \\ 1 \end{bmatrix}$ if and only if $\mathbf{A}\begin{bmatrix} 1 \\ 1 \end{bmatrix} = \begin{bmatrix} 0 \\ 0 \end{bmatrix}$.
   If $\mathbf{A} = \begin{bmatrix} a & b \\ c & d \end{bmatrix}$, then $a + b = 0$ and $c + d = 0$. Hence $a = -b$ and $c = -d$, so $\mathbf{A} = \begin{bmatrix} -b & b \\ -d & d \end{bmatrix}$. It follows that

$$\left\{ \begin{bmatrix} -1 & 1 \\ 0 & 0 \end{bmatrix}, \begin{bmatrix} 0 & 0 \\ -1 & 1 \end{bmatrix} \right\}$$

is a basis for $W$.

9. Since $\dim(V) = k$, every basis for $V$ contains exactly $k$ vectors and any set of $k$ linearly independent vectors is a basis. If a set had $k + 1$ or more linearly independent vectors, it would contain a basis with at least one vector that would not be a linear combination of the basis it contains. Hence $\dim(V) > k$. A contradiction.

10. **(a)** True.    **(b)** False.    **(c)** False.
    **(d)** True.    **(e)** True.    **(f)** True.
    **(g)** False.    **(h)** True.    **(i)** True.
    **(j)** True.

# Chapter 6

## Section 6.1, Page 343

1. (a) $\dfrac{\mathbf{x}}{\|\mathbf{x}\|} = \dfrac{1}{\sqrt{5}}\begin{bmatrix} 2 \\ 1 \\ 0 \end{bmatrix}.$

(b) $(\mathbf{x}, \mathbf{y}) = 8$, $(\mathbf{x}, \mathbf{s}) = 2$,
$(\mathbf{x}, 2\mathbf{y} - 3\mathbf{s}) = 2(\mathbf{x}, \mathbf{y}) - 3(\mathbf{x}, \mathbf{s}) = 2(8) - 3(2) = 10.$

(c) $D(\mathbf{x}, \mathbf{y}) = \|\mathbf{x} - \mathbf{y}\| = \left\| \begin{bmatrix} -1 \\ -1 \\ 1 \end{bmatrix} \right\| = \sqrt{3};$

$D(\mathbf{x}, \mathbf{s}) = \|\mathbf{x} - \mathbf{s}\| = \left\| \begin{bmatrix} 3 \\ -3 \\ -2 \end{bmatrix} \right\| = \sqrt{22}.$

Vector $\mathbf{y}$ is closer to $\mathbf{x}$ than vector $\mathbf{s}$, since $D(\mathbf{x}, \mathbf{y}) < D(\mathbf{x}, \mathbf{s})$.

(d) $\mathbf{v} = \begin{bmatrix} 2 \\ 0 \\ 1 \end{bmatrix}$ is one such vector. There are infinitely many such vectors.

(e) Here we compute $(\mathbf{x}, \mathbf{w})$ and $(\mathbf{y}, \mathbf{w})$ and set each expression equal to zero. This leads to the set of homogeneous equations

$$2w_1 + w_2 \qquad = 0$$
$$3w_1 + 2w_2 - w_3 = 0.$$

Solving this system we find there is one free variable and the general solution can be represented as

$$\mathbf{w} = t\begin{bmatrix} -1 \\ 2 \\ 1 \end{bmatrix},$$

where $t$ is any scalar. Choosing $t$ to be nonzero, we can find many vectors $\mathbf{w}$ that are each orthogonal to both $\mathbf{x}$ and $\mathbf{y}$.

3. (a) $(\mathbf{x}, \mathbf{y}) = 0$ and $(\mathbf{y}, \mathbf{s}) = 0$.

(b) $(a\mathbf{x} + b\mathbf{y}, \mathbf{s}) = a(\mathbf{x}, \mathbf{s}) + b(\mathbf{y}, \mathbf{s}) = 0 + 0 = 0.$

(c) Consider the equation $a\mathbf{x} + b\mathbf{y} = \mathbf{s}$ and determine if there is a pair of scalars $a$ and $b$ which satisfy the equation. This leads to the linear system with augmented matrix

$$\begin{bmatrix} 1 & 1 & 1 \\ 0 & 1 & -3 \\ -1 & 2 & 1 \end{bmatrix}$$

whose RREF is $\begin{bmatrix} 1 & 0 & 0 \\ 0 & 1 & 0 \\ 0 & 0 & 1 \end{bmatrix}$. It follows that the system is inconsistent, hence $\mathbf{s}$ is not in span$\{\mathbf{x}, \mathbf{y}\}$.

5. (a) We show $W$ is closed under addition and scalar multiplication.

Let $\mathbf{p}$ and $\mathbf{q}$ be vectors in $W$, then
$(\mathbf{p}, \mathbf{w}) = (\mathbf{q}, \mathbf{w}) = 0$. We show $(a\mathbf{p} + b\mathbf{q}, \mathbf{w}) = 0$;

$$(a\mathbf{p} + b\mathbf{q}, \mathbf{w}) = a(\mathbf{p}, \mathbf{w}) + b(\mathbf{q}, \mathbf{w}) = 0 + 0 = 0.$$

(b) Let $\mathbf{p} = \begin{bmatrix} p_1 \\ p_2 \\ p_3 \end{bmatrix}$ be in $W$. Then we must have

$(\mathbf{p}, \mathbf{w}) = p_1 + p_2 - 2p_3 = 0$. This is one equation in 3 unknowns. Solving for $p_1$ we obtain $p_1 = -p_2 + 2p_3$. Letting $p_2 = r$ and $p_3 = s$, we see that

$$\mathbf{p} = \begin{bmatrix} -r + 2s \\ r \\ s \end{bmatrix} = r\begin{bmatrix} -1 \\ 1 \\ 0 \end{bmatrix} + s\begin{bmatrix} 2 \\ 0 \\ 1 \end{bmatrix}.$$

Hence it follows that $\left\{ \begin{bmatrix} -1 \\ 1 \\ 0 \end{bmatrix}, \begin{bmatrix} 2 \\ 0 \\ 1 \end{bmatrix} \right\}$ is a basis for $W$.

(c) $\dim(W) = 2$.

(d) We need only show that set

$$\left\{ \mathbf{w}, \begin{bmatrix} -1 \\ 1 \\ 0 \end{bmatrix}, \begin{bmatrix} 2 \\ 0 \\ 1 \end{bmatrix} \right\}$$

$$= \left\{ \begin{bmatrix} 1 \\ 1 \\ -2 \end{bmatrix}, \begin{bmatrix} -1 \\ 1 \\ 0 \end{bmatrix}, \begin{bmatrix} 2 \\ 0 \\ 1 \end{bmatrix} \right\}$$

is linearly independent. But this follows immediately since $\mathbf{w}$ is orthogonal to the vectors in

$$\left\{ \begin{bmatrix} -1 \\ 1 \\ 0 \end{bmatrix}, \begin{bmatrix} 2 \\ 0 \\ 1 \end{bmatrix} \right\}$$

and these vectors are linearly independent by the construction in part (b).

7. $D(\mathbf{x}, \mathbf{y}) = \|\mathbf{x} - \mathbf{y}\| = \left\| \begin{bmatrix} 1 \\ t + 2 \end{bmatrix} \right\| = \sqrt{1 + (t + 2)^2} = \sqrt{t^2 + 4t + 5}$. Set $\sqrt{t^2 + 4t + 5} = 2$ and solve for $t$. We find $t = -2 \pm \sqrt{3}$.

## Section 6.1 True/False, Page 344

1. F    2. F    3. T    4. T    5. F

## Section 6.2, Page 352

1. By Property (b) in the Definition of an inner product we have that $(\mathbf{v} + \mathbf{w}, \mathbf{u}) = (\mathbf{u}, \mathbf{v} + \mathbf{w})$. Now use Property (c) and the result follows.

5. 12.      7. $\dfrac{1}{3}$.      9. $\dfrac{2\sqrt{2}}{3} - \dfrac{1}{3}$.

11. $(\mathbf{u}, \mathbf{u}) = u_1^2 + 4u_2^2 > 0$ if $\mathbf{u} \neq \mathbf{0}$. $(\mathbf{u}, \mathbf{u}) = 0$ if and only if $\mathbf{u} = \mathbf{0}$.

$$(\mathbf{u}, \mathbf{v}) = u_1 v_1 + 4u_2 v_2 = v_1 u_1 + 4v_2 u_2 = (\mathbf{v}, \mathbf{u})$$
$$(\mathbf{u}, \mathbf{v} + \mathbf{w}) = u_1(v_1 + w_1) + 4u_2(v_2 + w_2)$$
$$= (u_1 v_1 + 4u_2 v_2) + (u_1 w_1 + 4u_2 w_2)$$
$$= (\mathbf{u}, \mathbf{v}) + (\mathbf{u}, \mathbf{w})$$
$$(k\mathbf{u}, \mathbf{v}) = ku_1 v_1 + 4ku_2 v_2$$
$$= k(u_1 v_1 + 4u_2 v_2)$$
$$= k(\mathbf{u}, \mathbf{v}).$$

This is an inner product since all four properties are satisfied.

19. If $(\mathbf{u}, \mathbf{v}) = 0$ for all vectors $\mathbf{v}$ in $V$, then $(\mathbf{u}, \mathbf{u}) = 0$, which implies $\mathbf{u} = \mathbf{z}$.

21. $\|\mathbf{u}\| = \sqrt{6}$, $\|\mathbf{v}\| = 5$, $\|\mathbf{w}\| = \sqrt{30}$.

23. $\|e^t\| = \sqrt{\int_0^1 e^{2t}\, dt} = \sqrt{\dfrac{e^2}{2} - \dfrac{1}{2}}$.

27. $D(\mathbf{u}, \mathbf{v}) = \|\mathbf{u} - \mathbf{v}\| = \sqrt{(\mathbf{u} - \mathbf{v}, \mathbf{u} - \mathbf{v})}$
$$= \sqrt{(\mathbf{u}, \mathbf{u}) - 2(\mathbf{u}, \mathbf{v}) + (\mathbf{v}, \mathbf{v})}$$
$$= \sqrt{(\mathbf{v}, \mathbf{v}) - 2(\mathbf{u}, \mathbf{v}) + (\mathbf{u}, \mathbf{u})}$$
$$= \sqrt{(\mathbf{v} - \mathbf{u}, \mathbf{v} - \mathbf{u})} = \|\mathbf{v} - \mathbf{u}\|.$$

29. $\sqrt{2}$.      31. $\sqrt{\dfrac{4}{3}}$.

33. By definition, $N(\mathbf{v}) = \|\mathbf{v}\|_\infty = \max\{|v_1|, |v_2|, \dots, |v_n|\}$. Then we have for $\mathbf{v}$ in $R^n$, that $N(\mathbf{v}) > 0$ provided any component of $\mathbf{v}$ is not zero and $N(\mathbf{v}) = 0$ if and only if $\mathbf{v} = \mathbf{0}$. Thus Property 1 is verified. Next we see that

$$N(k\mathbf{v}) = \|k\mathbf{v}\|_\infty = \max\{|kv_1|, |kv_2|, \dots, |kv_n|\}$$
$$= \max\{|k|\,|v_1|, |k|\,|v_2|, \dots, |k|\,|v_n|\}$$
$$= |k|\max\{|v_1|, |v_2|, \dots, |v_n|\}$$
$$= |k| N(\mathbf{v})$$

so that Property 2 is valid. Finally

$$N(\mathbf{u} + \mathbf{v}) = \|\mathbf{u} + \mathbf{v}\|_\infty$$
$$= \max\{|u_1 + v_1|, |u_2 + v_2|, \dots, |u_n + v_n|\}$$
$$\leq \max\{(|u_1| + |v_1|), (|u_2| + |v_2|), \dots,$$
$$(|u_n| + |v_n|)\}$$
$$\leq \max\{|u_1|, |u_2|, \dots, |u_n|\}$$
$$+ \max\{|v_1|, |v_2|, \dots, |v_n|\}$$
$$= \|\mathbf{u}\|_\infty + \|\mathbf{v}\|_\infty$$
$$= N(\mathbf{u}) + N(\mathbf{v}).$$

which verifies Property 3.

35. (a) $\|\mathbf{u}\|_1 = 4$, $\|\mathbf{u}\|_\infty = 2$.

37. $\|\mathbf{w}\|_1 = 8$, $\|\mathbf{w}\|_\infty = 5$.

41. This set of vectors will have end points at ordered pairs where the component of maximum absolute value is 1. This leads to the accompanying picture.

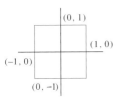

## Section 6.2 True/False, Page 354

1. T      2. T      3. T      4. T      5. T

## Section 6.3, Page 368

Note: Integrals were computed using MATLAB's symbolic toolbox.

1. $\cos(\theta) = \dfrac{4}{\sqrt{2}\,\sqrt{10}} = \dfrac{4}{\sqrt{20}} = \dfrac{4}{2\sqrt{5}} = \dfrac{2}{\sqrt{5}}$.

3. $\cos(\theta) = 0$.

5. $\cos(\theta) = \dfrac{(t, t^2)}{\|t\|\,\|t^2\|} = \dfrac{\int_0^1 t^3\, dt}{\sqrt{\int_0^1 t^2\, dt}\,\sqrt{\int_0^1 t^4\, dt}}$
$$= \dfrac{\frac{1}{4}}{\sqrt{\frac{1}{3}}\,\sqrt{\frac{1}{5}}} = \dfrac{\sqrt{15}}{4}.$$

7. $\cos(\theta) = \dfrac{(\sin(t), \cos(t))}{\|\sin(t)\|\,\|\cos(t)\|}$

$$= \dfrac{\int_0^1 \sin(t)\cos(t)\, dt}{\sqrt{\int_0^1 \sin(t)^2\, dt}\,\sqrt{\int_0^1 \cos(t)^2\, dt}}$$

$$= \dfrac{-0.5\cos(1)^2 + 0.5}{\sqrt{-0.5\sin(1)\cos(1) + 0.5}\,\sqrt{0.5\sin(1)\cos(1) + 0.5}}$$

$$\approx 0.7950.$$

9. (a) Let

$$5t - t^2 = a_1 \mathbf{u}_1 + a_2 \mathbf{u}_2$$
$$= a_1 t\sqrt{3} + a_2 \sqrt{80}\left(t^2 - \tfrac{3}{4}t\right).$$

Expanding the right side, collecting terms and equating coefficients of like power terms of $t$ we get the linear system of equations

$$a_1\sqrt{3} - a_2\dfrac{3\sqrt{80}}{4} = 5$$

$$a_2\sqrt{80} = -1$$

whose solution is

$$a_2 = \frac{-1}{\sqrt{80}} = \frac{-\sqrt{5}}{20}$$

$$a_1 = \frac{17}{4\sqrt{3}} = \frac{17\sqrt{3}}{12}.$$

However, we can do this in a simpler way using the fact that $\{\mathbf{u}_1, \mathbf{u}_2\}$ is an orthonormal basis. From our discussion about Linear Combinations and Orthonormal Bases we find that

$$a_1 = (5t - t^2, \mathbf{u}_1)$$

$$= \int_0^1 (5t - t^2)(t\sqrt{3})\, dt = \frac{17\sqrt{3}}{12}$$

and

$$a_2 = (5t - t^2, \mathbf{u}_2)$$

$$= \int_0^1 (5t - t^2)(\sqrt{80})\left(t^2 - \frac{3}{4}t\right) dt = \frac{-\sqrt{5}}{20}.$$

Using properties of orthonormal sets we avoid the need to solve a linear system.

**Strategy for Exercises 12–14:** We apply the Gram–Schmidt process but do not scale the orthogonal set that is generated to be unit vectors.

**13.** Let $\mathbf{q}_1 = 1$, $\mathbf{q}_2 = t$, and $\mathbf{q}_3 = e^t$. Define $\mathbf{v}_1 = \mathbf{q}_1$ and

$$\mathbf{v}_2 = \mathbf{q}_2 - \frac{(\mathbf{v}_1, \mathbf{q}_2)}{(\mathbf{v}_1, \mathbf{v}_1)}\mathbf{v}_1 = t - \frac{\int_0^1 t\, dt}{\int_0^1 1\, dt} 1$$

$$= t - \frac{\frac{1}{2}}{1} 1 = t - 0.5.$$

Next we define

$$\mathbf{v}_3 = \mathbf{q}_3 - \frac{(\mathbf{v}_1, \mathbf{q}_3)}{(\mathbf{v}_1, \mathbf{v}_1)}\mathbf{v}_1 - \frac{(\mathbf{v}_2, \mathbf{q}_3)}{(\mathbf{v}_2, \mathbf{v}_2)}\mathbf{v}_2$$

$$= e^t - \frac{\int_0^1 e^t\, dt}{\int_0^1 1\, dt} 1 - \frac{\int_0^1 (t - 0.5)e^t\, dt}{\int_0^1 (t - 0.5)^2\, dt}(t - 0.5)$$

$$= e^t - \frac{e - 1}{1} 1 - \frac{1 - 0.5(e - 1)}{\frac{1}{12}}(t - 0.5)$$

$$= e^t + (-18 + 6e)t + (10 - 4e).$$

Thus

$$\{1, t - 0.5, e^t + (-18 + 6e)t + (10 - 4e)\}$$

is an orthogonal basis for the subspace spanned by $\{1, t, e^t\}$.

**16.** We first normalize the vectors $\mathbf{w}_1$ and $\mathbf{w}_2$ to obtain an orthonormal basis. Then we use the fact that the projection is given by

$$\text{proj}_W \mathbf{v} = (\mathbf{v}, \mathbf{w}_1)\mathbf{w}_1 + (\mathbf{v}, \mathbf{w}_2)\mathbf{w}_2.$$

Thus redefine

$$\mathbf{w}_1 = \frac{1}{\sqrt{14}}\begin{bmatrix} 3 \\ -1 \\ -2 \end{bmatrix} \quad \text{and} \quad \mathbf{w}_2 = \frac{1}{\sqrt{13}}\begin{bmatrix} 2 \\ 0 \\ 3 \end{bmatrix}.$$

**(a)** $\text{proj}_W \mathbf{v} = (\mathbf{v}, \mathbf{w}_1)\mathbf{w}_1 + (\mathbf{v}, \mathbf{w}_2)\mathbf{w}_2$

$$= 0\frac{1}{\sqrt{14}}\begin{bmatrix} 3 \\ -1 \\ -2 \end{bmatrix} + \frac{8}{\sqrt{13}}\frac{1}{\sqrt{13}}\begin{bmatrix} 2 \\ 0 \\ 3 \end{bmatrix}$$

$$= \frac{8}{13}\begin{bmatrix} 2 \\ 0 \\ 3 \end{bmatrix}.$$

**21.** **(a)** $a_0 = \left(t, \frac{1}{\sqrt{2}}\right) = 0$, $a_1 = (t, \cos(t)) = 0$,

$b_1 = (t, \sin(t)) = 2$.

Thus the first order Fourier approximation is $2\sin(t)$.

## Section 6.3 True/False, Page 369

**1.** T    **2.** F    **3.** T    **4.** T    **5.** T

## Section 6.4, Page 380

**1.** The expenditures in 1999 are $y = m(1999) + b \approx (51.25)(1999) - 101377.92 \approx 1089.5$ billion dollars.

**6.** The normal system is

$$\begin{bmatrix} \int_0^1 1\, dt & \int_0^1 t\, dt \\ \int_0^1 t\, dt & \int_0^1 t^2\, dt \end{bmatrix}\begin{bmatrix} c_1 \\ c_2 \end{bmatrix} = \begin{bmatrix} \int_0^1 e^t\, dt \\ \int_0^1 te^t\, dt \end{bmatrix}.$$

Computing the integrals we have

$$\begin{bmatrix} 1 & \frac{1}{2} \\ \frac{1}{2} & \frac{1}{3} \end{bmatrix}\begin{bmatrix} c_1 \\ c_2 \end{bmatrix} = \begin{bmatrix} e - 1 \\ 1 \end{bmatrix}.$$

The solution of this system gives

$$\begin{bmatrix} c_1 \\ c_2 \end{bmatrix} = \begin{bmatrix} 4e - 10 \\ 18 - 6e \end{bmatrix} \approx \begin{bmatrix} 0.8731 \\ 1.6903 \end{bmatrix},$$

hence the least squares approximation to $e^t$ from $P_1$ is $0.8731 + 1.6903t$.

## Section 6.4 True/False, Page 382

**1.** T    **2.** T    **3.** T    **4.** F    **5.** T

## Section 6.5, Page 386

**1.** By definition, $W^\perp$ is the set of all vectors in $V$ that are orthogonal to every vector in subspace $W$. Let $\mathbf{p}$ and $\mathbf{q}$ be in $W^\perp$, then for any vector $\mathbf{w}$ in $W$ we have $(\mathbf{q}, \mathbf{w}) = (\mathbf{p}, \mathbf{w}) = 0$. We show that $W^\perp$ is closed:

$$(\mathbf{p} + \mathbf{q}, \mathbf{w}) = (\mathbf{p}, \mathbf{w}) + (\mathbf{q}, \mathbf{w}) = 0 + 0 = 0$$

and for any scalar $k$,

$$(k\mathbf{p}, \mathbf{w}) = k(\mathbf{p}, \mathbf{w}) = k0 = 0.$$

It follows that $W^\perp$ is a subspace of $V$.

**7.** Let $W = \operatorname{span}\{1, t^2\}$. Then to compute $W^\perp$ we proceed as follows. We determine all the vectors $\mathbf{v} = at^2 + bt + c$ in $V$ that are orthogonal to all the vectors in $W$. It suffices to determine all the vectors $\mathbf{v}$ that are orthogonal to the basis vectors $1$ and $t^2$. Hence we require that

$$(\mathbf{v}, 1) = \int_0^1 (at^2 + bt + c)\, dt = \frac{a}{3} + \frac{b}{2} + c = 0$$

$$(\mathbf{v}, t^2) = \int_0^1 (at^2 + bt + c)t^2\, dt = \frac{a}{5} + \frac{b}{4} + \frac{c}{3} = 0.$$

This gives us the homogeneous linear system

$$\left[\begin{array}{ccc|c} \frac{1}{3} & \frac{1}{2} & 1 & 0 \\ \frac{1}{5} & \frac{1}{4} & \frac{1}{3} & 0 \end{array}\right]$$

whose reduced row echelon form is

$$\left[\begin{array}{ccc|c} 1 & 0 & -5 & 0 \\ 0 & 1 & \frac{16}{3} & 0 \end{array}\right].$$

Then we have $a = 5c$ and $b = -\frac{16}{3}c$ and so

$$W^\perp = \left\{ c(5t^2 - (\tfrac{16}{3})t + 1),\ c \text{ any scalar} \right\}.$$

**13. (a)** A basis for $\mathbf{ns}(\mathbf{A})$ is

$$\left\{ \begin{bmatrix} -1 \\ \frac{1}{2} \\ 1 \\ 0 \end{bmatrix}, \begin{bmatrix} -1 \\ \frac{1}{2} \\ 0 \\ 1 \end{bmatrix} \right\}.$$

**(b)** $\mathbf{ns}(\mathbf{A})^\perp = \mathbf{col}(\mathbf{A}^T)$. A basis for

$$\mathbf{col}(\mathbf{A}^T) = \left\{ \begin{bmatrix} 1 \\ 0 \\ 1 \\ 1 \end{bmatrix}, \begin{bmatrix} 0 \\ 1 \\ -\frac{1}{2} \\ -\frac{1}{2} \end{bmatrix} \right\}.$$

**(c)** We solve the linear system

$$c_1 \begin{bmatrix} -1 \\ \frac{1}{2} \\ 1 \\ 0 \end{bmatrix} + c_2 \begin{bmatrix} -1 \\ \frac{1}{2} \\ 0 \\ 1 \end{bmatrix} + c_3 \begin{bmatrix} 1 \\ 0 \\ 1 \\ 1 \end{bmatrix}$$

$$+ c_4 \begin{bmatrix} 0 \\ 1 \\ -\frac{1}{2} \\ -\frac{1}{2} \end{bmatrix} = \begin{bmatrix} 3 \\ 2 \\ 4 \\ 0 \end{bmatrix}.$$

We get $c_1 = 2$, $c_2 = -2$, $c_3 = 3$, $c_4 = 2$.

## Section 6.5 True/False, Page 387

**1.** T  **2.** T  **3.** T  **4.** F  **5.** T

## Chapter 6 Chapter Test, Page 390

**1.** $\dfrac{3}{2}$.

**2.** $\|\mathbf{f} - \mathbf{g}\| = \sqrt{\dfrac{4}{5}}$.

**3.** $\dfrac{\frac{3}{2}}{\sqrt{\frac{4}{5}}\sqrt{3}} \approx 0.9682$.

**4.** $\dfrac{(\mathbf{f}, \mathbf{g})}{(\mathbf{f}, \mathbf{f})}\mathbf{f} = \dfrac{\frac{3}{2}}{\frac{4}{5}}\mathbf{f} = \frac{15}{8}(2t^2) = \frac{15}{4}t^2$.

**5.** $\{\mathbf{f}, \mathbf{g} - \operatorname{proj}_{\mathbf{f}}\mathbf{g}\} = \{2t^2, 3t - \frac{15}{4}t^2\}$.

**6.** Vector $\mathbf{w} = at^3 + bt^2 + ct + d$ is in $W^\perp$ provided $(\mathbf{w}, 1) = 0$ and $(\mathbf{w}, t^3) = 0$. This gives us the pair of equations

$$\frac{a}{4} + \frac{b}{3} + \frac{c}{2} + d = 0$$

$$\frac{a}{7} + \frac{b}{6} + \frac{c}{5} + \frac{d}{4} = 0.$$

Solving this pair of equations, we find that the general solution is

$$r \begin{bmatrix} \frac{14}{5} \\ -\frac{8}{5} \\ 1 \\ 0 \end{bmatrix} + s \begin{bmatrix} 14 \\ -\frac{27}{2} \\ 0 \\ 1 \end{bmatrix}.$$

Hence a basis for $W^\perp$ is

$$\left\{ \tfrac{14}{5}t^3 - \tfrac{18}{5}t^2 + t,\ 14t^3 - \tfrac{27}{2}t^2 + 1 \right\}.$$

**7.** $5t^3 + 4t^2 + 3t = a + bt^3 + c(\frac{14}{5}t^3 - \frac{18}{5}t^2 + t) + d(14t^3 - \frac{27}{2}t^2 + 1)$. Solving for $a, b, c$, and $d$, we find that $a = \frac{148}{135}$, $b = \frac{1613}{135}$, $c = 3$, $d = -\frac{148}{135}$.

**8.** $\{1, t^3 - \operatorname{proj}_1 t^3\} = \{1, t^4 - \frac{1}{4}\}$.

**9.** $\dfrac{(t^2, 1)}{(1, 1)} 1 + \dfrac{(t^2, t^3 - \frac{1}{4})}{(t^3 - \frac{1}{4}, t^3 - \frac{1}{4})}(t^3 - \frac{1}{4}) =$

$\frac{1}{3} + \dfrac{\frac{1}{12}}{(\frac{1}{7} - \frac{1}{16})}(t^3 - \frac{1}{4}).$

**10.** $9.6553t^2 - 26.5805t + 19.4258.$

## Chapter 7

### Section 7.1, Page 395

**1.** Let $\mathbf{u}$ and $\mathbf{v}$ be vectors in $V$. Then $T(\mathbf{u}) = T(\mathbf{v}) = \mathbf{0}$.
Since $\mathbf{u} + \mathbf{v}$ is in $V$, $T(\mathbf{u} + \mathbf{v}) = \mathbf{0}$. We note that
$T(\mathbf{u}) + T(\mathbf{v}) = \mathbf{0} + \mathbf{0} = \mathbf{0}$ also, so
$T(\mathbf{u} + \mathbf{v}) = T(\mathbf{u}) + T(\mathbf{v})$. For any scalar $k$, $k\mathbf{u}$ is in $V$.
Hence $T(k\mathbf{u}) = \mathbf{0}$. But also we have that
$kT(\mathbf{u}) = k\mathbf{0} = \mathbf{0}$. Thus $T(k\mathbf{u}) = kT(\mathbf{u})$. It follows that
$T$ is a linear transformation.

**3.** Let $\mathbf{A}$ and $\mathbf{B}$ be $2 \times 2$ matrices. To show that both
properties of a linear transformation are violated, we will
use particular choices of matrices $\mathbf{A}$ and $\mathbf{B}$.
Let $\mathbf{A} = \begin{bmatrix} 1 & 2 \\ 0 & 0 \end{bmatrix}$ and $\mathbf{B} = \begin{bmatrix} 0 & 0 \\ 1 & 2 \end{bmatrix}$;

$$T(\mathbf{A} + \mathbf{B}) = \mathbf{rref}\left(\begin{bmatrix} 1 & 2 \\ 1 & 2 \end{bmatrix}\right) = \begin{bmatrix} 1 & 2 \\ 0 & 0 \end{bmatrix}$$

while

$$T(\mathbf{A}) + T(\mathbf{B}) = \begin{bmatrix} 1 & 2 \\ 0 & 0 \end{bmatrix} + \begin{bmatrix} 1 & 2 \\ 0 & 0 \end{bmatrix} = \begin{bmatrix} 2 & 4 \\ 0 & 0 \end{bmatrix}.$$

also we have

$$T(3\mathbf{A}) = \mathbf{rref}\left(\begin{bmatrix} 3 & 6 \\ 0 & 0 \end{bmatrix}\right) = \begin{bmatrix} 1 & 2 \\ 0 & 0 \end{bmatrix}$$

$$\neq 3T(A) = \begin{bmatrix} 3 & 6 \\ 0 & 0 \end{bmatrix}.$$

Hence $T$ is nonlinear.

**7.** $\mathbf{A} = \begin{bmatrix} 0 & 1 \\ 1 & 0 \end{bmatrix}$. This transformation is a reflection about
the line $y = x$.

**9.** $\mathbf{A} = \begin{bmatrix} 1 & 0 \\ 0 & 0 \end{bmatrix}$. This transformation is a projection onto
the $x$-axis.

**13.** $T\left(\begin{bmatrix} v_1 \\ v_2 \end{bmatrix}\right) = \begin{bmatrix} v_1 + v_2 \\ v_2 \\ v_1 - v_2 \end{bmatrix} = \begin{bmatrix} 1 & 1 \\ 0 & 1 \\ 1 & -1 \end{bmatrix}\begin{bmatrix} v_1 \\ v_2 \end{bmatrix}$, so
this is a matrix transformation, hence linear.

**15.** Nonlinear since

$$T\left(k\begin{bmatrix} v_1 \\ v_2 \\ v_3 \end{bmatrix}\right) = T\left(\begin{bmatrix} kv_1 \\ kv_2 \\ kv_3 \end{bmatrix}\right)$$

$$= \begin{bmatrix} kv_1 - kv_2 + kv_3 \\ (kv_2)^2 \end{bmatrix}$$

$$\neq k\begin{bmatrix} v_1 - v_2 + v_3 \\ (v_2)^2 \end{bmatrix}$$

$$= kT\left(\begin{bmatrix} v_1 \\ v_2 \\ v_3 \end{bmatrix}\right).$$

**21.** Linear.

$$T(p(t) + q(t)) = t^3(p(0) + q(0)) + (p(t) + q(t))$$
$$= t^3 p(0) + p(t) + t^3 q(0) + q(t)$$
$$= T(p(t)) + T(q(t))$$

$$T(kp(t)) = t^3(kp(0)) + kp(t)$$
$$= k(t^3 p(0) + p(t))$$
$$= kT(p(t)).$$

### Section 7.1 True/False, Page 396

**1.** T    **2.** F    **3.** T    **4.** T    **5.** F
**6.** F    **7.** F    **8.** T    **9.** F    **10.** T

### Section 7.2, Page 405

**1.** (a) No.    (b) Yes.

**3.** (a) If $\mathbf{w}_1$ and $\mathbf{w}_2$ are in $\mathbf{R}(T)$, then there exist vectors $\mathbf{v}_1$
and $\mathbf{v}_2$ in $V$ such that $T(\mathbf{v}_1) = \mathbf{w}_1$ and $T(\mathbf{v}_2) = \mathbf{w}_2$.
We note that since $T$ is a linear transformation then

$$T(\mathbf{v}_1 + \mathbf{v}_2) = T(\mathbf{v}_1) + T(\mathbf{v}_2) = \mathbf{w}_1 + \mathbf{w}_2.$$

Hence vector $\mathbf{v}_1 + \mathbf{v}_2$ of $V$ is such that its image is
$\mathbf{w}_1 + \mathbf{w}_2$, so $\mathbf{w}_1 + \mathbf{w}_2$ is in $\mathbf{R}(T)$. Thus $\mathbf{R}(T)$ is
closed under addition.

(b) If $\mathbf{w}$ is in $\mathbf{R}(T)$, then there exists a vector $\mathbf{v}$ in $V$
such that $T(\mathbf{v}) = \mathbf{w}$. Since $T$ is a linear
transformation, we have $T(k\mathbf{v}) = kT(\mathbf{v}) = k\mathbf{w}$.
Hence vector $k\mathbf{v}$ of $V$ has image $k\mathbf{w}$, so $\mathbf{R}(T)$ is
closed under scalar multiplication.

**9.** (a) We find the general solution of the homogeneous
linear system $\mathbf{Ax} = \mathbf{0}$. The RREF of the
corresponding augmented matrix is

$$\begin{bmatrix} 1 & 0 & 0 & 0 & | & 0 \\ 0 & 1 & 0 & 1 & | & 0 \\ 0 & 0 & 1 & -1 & | & 0 \end{bmatrix},$$

hence the general solution is

$$\mathbf{x} = t \begin{bmatrix} 0 \\ -1 \\ 1 \\ 1 \end{bmatrix}$$

where $t$ is an arbitrary constant. Thus we have the basis stated for $\ker(T)$.

(b) We form the matrix $\begin{bmatrix} \mathbf{v} & \mathbf{I}_4 \end{bmatrix}$ and compute its RREF. The leading ones of the RREF point to the columns of this matrix that are linearly independent. We get

$$\mathbf{rref}\left(\begin{bmatrix} \mathbf{v} & \mathbf{I}_4 \end{bmatrix}\right) = \begin{bmatrix} 1 & 0 & 0 & 0 & 1 \\ 0 & 1 & 0 & 0 & 0 \\ 0 & 0 & 1 & 0 & 1 \\ 0 & 0 & 0 & 1 & -1 \end{bmatrix}$$

so the basis for $R^4$ that includes $\mathbf{v}$ is

$$\mathbf{Q} = \left\{ \begin{bmatrix} 0 \\ -1 \\ 1 \\ 1 \end{bmatrix}, \begin{bmatrix} 1 \\ 0 \\ 0 \\ 0 \end{bmatrix}, \begin{bmatrix} 0 \\ 1 \\ 0 \\ 0 \end{bmatrix}, \begin{bmatrix} 0 \\ 0 \\ 1 \\ 0 \end{bmatrix} \right\}.$$

11. We use the fact that $T$ is one-to-one if and only if $\ker(T) = \{\mathbf{0}\}$. In this case, the kernel of $T$ is the solution set of $\mathbf{Ax} = \mathbf{0}$. Next we show that $T$ one-to-one implies that $\mathbf{A}$ is nonsingular and conversely that if $\mathbf{A}$ is nonsingular, then $T$ is one-to-one.

   • If $T$ is one-to-one, then the linear system $\mathbf{Ax} = \mathbf{0}$ has only the zero solution, hence $\mathbf{A}$ is nonsingular.

   • If $\mathbf{A}$ is nonsingular, then the linear system $\mathbf{Ax} = \mathbf{0}$ has a unique solution (which is the zero vector), thus $\ker(T) = \{\mathbf{0}\}$.

19. We use the fact that for a linear transformation $T : V \to W$, $\dim(\ker(T)) + \dim(\mathbf{R}(T)) = \dim(V)$. Here $T$ is onto so $\dim(V) = \dim(R^5) = 5$.

   (a) Here $T$ is onto so $\dim(\mathbf{R}(T)) = \dim(R^5) = 5$, thus $\dim(\ker(T)) + \dim(\mathbf{R}(T)) = \dim(V)$ implies $2 + 5 = \dim(V)$, so $\dim(V) = 7$.

   (b) Here $T$ is one-to-one and onto so $\dim(\mathbf{R}(T)) = \dim(R^5) = 5$ and $\dim(\ker(T)) = 0$, thus $\dim(\ker(T)) + \dim(\mathbf{R}(T)) = \dim(V)$ implies $0 + 5 = \dim(V)$, so $\dim(V) = 5$.

## Section 7.2 True/False, Page 407

| | | | | |
|---|---|---|---|---|
| 1. F | 2. T | 3. T | 4. T | 5. T |
| 1. T | 7. F | 8. F | 9. T | 10. F |

## Section 7.3, Page 422

**Strategy:** For Exercises 1–4, construct a linear system to express vector $\mathbf{v}$ in terms of the specified basis. The coordinates of $\mathbf{v}$ are the coefficients that are obtained from the solution of the linear system.

1. $\mathbf{coord}_S(\mathbf{v}) = \begin{bmatrix} 5 \\ 2 \end{bmatrix}$.

3. $\mathbf{coord}_S(\mathbf{v}) = \begin{bmatrix} 1 \\ -1 \\ 3 \end{bmatrix}$.

**Strategy:** For Exercises 5–8, use the entries of the coordinate vector as coefficients of the corresponding basis vectors to determine $\mathbf{v}$.

5. $\mathbf{v} = (-1)\begin{bmatrix} 2 \\ 1 \end{bmatrix} + (2)\begin{bmatrix} -1 \\ 1 \end{bmatrix} = \begin{bmatrix} -4 \\ 1 \end{bmatrix}$.

7. $\mathbf{v} = (3)(t^2 + 1) + (-1)(t + 1) + (-2)(t^2 + t) = t^2 - 3t + 2$.

11. We show the steps of the development of the matrix representation and use the arrow notation as indicated in this section; that is, $\downarrow \to \uparrow$. For this problem, we need only the $\to$ step.

   (a) $S = B = \left\{ \begin{bmatrix} 1 \\ 0 \end{bmatrix}, \begin{bmatrix} 0 \\ 1 \end{bmatrix} \right\}$ and $T(\mathbf{v}) = \begin{bmatrix} v_1 + 2v_2 \\ 2v_1 - v_2 \end{bmatrix}$.

   In this case, we will not need to solve linear systems in order to determine coordinates since the bases involved are the natural bases.

   $\to$ step.

   $$T\left(\begin{bmatrix} 1 \\ 0 \end{bmatrix}\right) = \begin{bmatrix} 1 \\ 2 \end{bmatrix} = 1\begin{bmatrix} 1 \\ 0 \end{bmatrix} + 2\begin{bmatrix} 0 \\ 1 \end{bmatrix}$$

   and

   $$T\left(\begin{bmatrix} 0 \\ 1 \end{bmatrix}\right) = \begin{bmatrix} 2 \\ -1 \end{bmatrix} = 2\begin{bmatrix} 1 \\ 0 \end{bmatrix} - 1\begin{bmatrix} 0 \\ 1 \end{bmatrix}.$$

   Thus

   $$\mathbf{coord}_B\left(T\left(\begin{bmatrix} 1 \\ 0 \end{bmatrix}\right)\right) = \begin{bmatrix} 1 \\ 2 \end{bmatrix}$$

   and

   $$\mathbf{coord}_B\left(T\left(\begin{bmatrix} 0 \\ 1 \end{bmatrix}\right)\right) = \begin{bmatrix} 2 \\ -1 \end{bmatrix};$$

   then the matrix containing the images of the $S$-basis is

   $$\mathbf{A} = \begin{bmatrix} 1 & 2 \\ 2 & -1 \end{bmatrix}.$$

   This is the matrix representation of $T$ with respect to the bases $S$ and $B$.

**(b)** $S = \left\{ \begin{bmatrix} 1 \\ 0 \end{bmatrix}, \begin{bmatrix} 0 \\ 1 \end{bmatrix} \right\}$, $B = \left\{ \begin{bmatrix} -1 \\ 2 \end{bmatrix}, \begin{bmatrix} 2 \\ 0 \end{bmatrix} \right\}$,

and $T(\mathbf{v}) = \begin{bmatrix} v_1 + 2v_2 \\ 2v_1 - v_2 \end{bmatrix}$.

In this case, we will need to solve linear systems in order to determine coordinates relative to the $B$-basis.

$\to$ step.

$$T\left( \begin{bmatrix} 1 \\ 0 \end{bmatrix} \right) = \begin{bmatrix} 1 \\ 2 \end{bmatrix} = a_{11} \begin{bmatrix} -1 \\ 2 \end{bmatrix} + a_{21} \begin{bmatrix} 2 \\ 0 \end{bmatrix},$$

which leads to a linear system with augmented matrix

$$\begin{bmatrix} -1 & 2 & | & 1 \\ 2 & 0 & | & 2 \end{bmatrix}$$

whose RREF is $\begin{bmatrix} 1 & 0 & | & 1 \\ 0 & 1 & | & 1 \end{bmatrix}$, hence

$$\mathbf{coord}_B\left( T\left( \begin{bmatrix} 1 \\ 0 \end{bmatrix} \right) \right) = \begin{bmatrix} 1 \\ 1 \end{bmatrix}.$$

$$T\left( \begin{bmatrix} 0 \\ 1 \end{bmatrix} \right) = \begin{bmatrix} 2 \\ -1 \end{bmatrix} = a_{12} \begin{bmatrix} -1 \\ 2 \end{bmatrix} + a_{22} \begin{bmatrix} 2 \\ 0 \end{bmatrix},$$

which leads to a linear system with augmented matrix

$$\begin{bmatrix} -1 & 2 & | & 2 \\ 2 & 0 & | & -1 \end{bmatrix}$$

whose RREF is $\begin{bmatrix} 1 & 0 & | & -\frac{1}{2} \\ 0 & 1 & | & \frac{3}{4} \end{bmatrix}$, hence

$$\mathbf{coord}_B\left( T\left( \begin{bmatrix} 0 \\ 1 \end{bmatrix} \right) \right) = \begin{bmatrix} -\frac{1}{2} \\ \frac{3}{4} \end{bmatrix}.$$

It follows that the matrix containing the coordinates of the images of the $S$-basis is

$$\mathbf{A} = \begin{bmatrix} 1 & -\frac{1}{2} \\ 1 & \frac{3}{4} \end{bmatrix}.$$

This is the matrix representation of $T$ with respect to the bases $S$ and $B$.

**15.** We show the steps of the development of the matrix representation and use the arrow notation as indicated in this section; that is, $\downarrow \to \uparrow$. For this problem we need only the $\to$ step.

$\to$ step.

$$S = \{t^2, t, 1\}, \; B = \{t^3, t^2, t, 1\},$$

and

$$T(at^2 + bt + c) = at^3 + ct^2 + b.$$

We compute images of the $S$-basis vectors:

$$T(t^2) = t^3, \quad T(t) = 1, \quad T(1) = t^2.$$

Next we express each of the images as a linear combination of the $B$-basis functions. In this case, it is easy to perform this step and we can record the coordinates of the images:

$$\mathbf{coord}_B(T(t^2)) = \mathbf{coord}_B(t^3) = \begin{bmatrix} 1 \\ 0 \\ 0 \\ 0 \end{bmatrix},$$

$$\mathbf{coord}_B(T(t)) = \mathbf{coord}_B(1) = \begin{bmatrix} 0 \\ 0 \\ 0 \\ 1 \end{bmatrix},$$

$$\mathbf{coord}_B(T(1)) = \mathbf{coord}_B(t^2) = \begin{bmatrix} 0 \\ 1 \\ 0 \\ 0 \end{bmatrix}.$$

It follows that the matrix representing $T$ relative to bases $S$ and $B$ is

$$\mathbf{A} = \begin{bmatrix} 1 & 0 & 0 \\ 0 & 0 & 1 \\ 0 & 0 & 0 \\ 0 & 1 & 0 \end{bmatrix}.$$

**19.** We show the steps of the development of the matrix representation and use the arrow notation as indicated in this section; that is, $\downarrow \to \uparrow$. For this problem, we need only the $\to$ step.

$\to$ step.

$$S = \{\sin(t), \cos(t)\} \quad \text{and} \quad T(\mathbf{f}(t)) = \mathbf{f}'(t).$$

We compute the images of the $S$-basis vectors and then express them as linear combinations of the $S$-basis vectors. We have $T(\sin(t)) = \cos(t)$ and so

$$\mathbf{coord}_S(T(\sin(t))) = \begin{bmatrix} 0 \\ 1 \end{bmatrix};$$

$T(\cos(t)) = -\sin(t)$ and so

$$\mathbf{coord}_S(T(\cos(t))) = \begin{bmatrix} -1 \\ 0 \end{bmatrix}.$$

Thus the matrix representing $T$ relative to the $S$-basis is

$$\mathbf{A} = \begin{bmatrix} 0 & -1 \\ 1 & 0 \end{bmatrix}.$$

**23.** Let $\mathbf{f}(t) = c_1 \sin(t) + c_2 \cos(t)$. The eigen equation is

$$(c_1 \sin(t) + c_2 \cos(t))'' = \lambda(c_1 \sin(t) + c_2 \cos(t)).$$

Computing the left side, we have

$$-c_1 \sin(t) - c_2 \cos(t) = \lambda(c_1 \sin(t) + c_2 \cos(t))$$

and then equating coefficients of like terms we have $-c_1 = \lambda c_1$ and $-c_2 = \lambda c_2$. It follows that the eigenvalues are $\lambda = -1, -1$. Hence any linear combination $c_1 \sin(t) + c_2 \cos(t)$ is a corresponding eigenfunction to $\lambda = -1$, provided one of $c_1$ and $c_2$ is not zero.

**27. (a)** $T\left( \begin{bmatrix} a & b \\ c & d \end{bmatrix} + \begin{bmatrix} p & q \\ r & s \end{bmatrix} \right)$

$$= T\left( \begin{bmatrix} a+p & b+q \\ c+r & d+s \end{bmatrix} \right)$$

$$= \begin{bmatrix} 0 & a+p+b+q \\ c+r+d+s & 0 \end{bmatrix}$$

$$= \begin{bmatrix} 0 & a+b \\ c+d & 0 \end{bmatrix} + \begin{bmatrix} 0 & p+q \\ r+s & 0 \end{bmatrix}$$

$$= T\left( \begin{bmatrix} a & b \\ c & d \end{bmatrix} \right) + T\left( \begin{bmatrix} p & q \\ r & s \end{bmatrix} \right)$$

$$T\left( k \begin{bmatrix} a & b \\ c & d \end{bmatrix} \right) = T\left( \begin{bmatrix} ka & kb \\ kc & kd \end{bmatrix} \right)$$

$$= \begin{bmatrix} 0 & ka+kb \\ kc+kd & 0 \end{bmatrix}$$

$$= k \begin{bmatrix} 0 & a+b \\ c+d & 0 \end{bmatrix}$$

$$= kT\left( \begin{bmatrix} a & b \\ c & d \end{bmatrix} \right).$$

**(b)** We seek matrices $\begin{bmatrix} a & b \\ c & d \end{bmatrix}$ such that

$$T\left( \begin{bmatrix} a & b \\ c & d \end{bmatrix} \right) = \lambda \begin{bmatrix} a & b \\ c & d \end{bmatrix}.$$

This leads to the equation

$$\begin{bmatrix} 0 & a+b \\ c+d & 0 \end{bmatrix} = \lambda \begin{bmatrix} a & b \\ c & d \end{bmatrix} = \begin{bmatrix} \lambda a & \lambda b \\ \lambda c & \lambda d \end{bmatrix}.$$

Equating corresponding entries we have $0 = \lambda a$, $a + b = \lambda b$, $c + d = \lambda c$, and $0 = d\lambda$. Let's consider cases here for $\lambda$.

Case $\lambda = 0$: $a + b = 0$ and $c + d = 0$. Solving this system of two equations in 4 unknowns gives $a = -b$ and $c = -d$. Thus we get matrix

$$\begin{bmatrix} -b & b \\ -d & d \end{bmatrix}$$

is an eigenvector corresponding to eigenvalue $\lambda = 0$. Since $b$ and $d$ are arbitrary, it follows that there will be two linearly independent eigenvectors;

$$\begin{bmatrix} -b & b \\ 0 & 0 \end{bmatrix} \quad \text{for } b \neq 0$$

and

$$\begin{bmatrix} 0 & 0 \\ -d & d \end{bmatrix} \quad \text{for } d \neq 0.$$

If we set $b$ and $d$ to 1 (for convenience), it follows that we have eigenpairs

$$\left\{ \left( 0, \begin{bmatrix} -1 & 1 \\ 0 & 0 \end{bmatrix} \right), \left( 0, \begin{bmatrix} 0 & 0 \\ -1 & 1 \end{bmatrix} \right) \right\}.$$

Case $\lambda \neq 0$: We have that $a = d = 0$. But then $b = \lambda b$ and $c = \lambda c$. Hence $\lambda = 1$. Thus

$$\begin{bmatrix} 0 & b \\ c & 0 \end{bmatrix}$$

is an eigenvector. Since $b$ and $c$ are arbitrary, it follows that there will be two linearly independent eigenvectors;

$$\begin{bmatrix} 0 & b \\ 0 & 0 \end{bmatrix} \quad \text{for } b \neq 0$$

and

$$\begin{bmatrix} 0 & 0 \\ c & 0 \end{bmatrix} \quad \text{for } c \neq 0.$$

If we set $b$ and $c$ to 1 (for convenience), it follows that we have eigenpairs

$$\left\{ \left( 1, \begin{bmatrix} 0 & 1 \\ 0 & 0 \end{bmatrix} \right), \left( 1, \begin{bmatrix} 0 & 0 \\ 1 & 0 \end{bmatrix} \right) \right\}.$$

There are no other eigenpairs since we have considered all possible cases (either $\lambda = 0$ or $\lambda \neq 0$).

**(c)** We have

$$T\left( \begin{bmatrix} 1 & 0 \\ 0 & 0 \end{bmatrix} \right) = \begin{bmatrix} 0 & 1 \\ 0 & 0 \end{bmatrix},$$

$$T\left( \begin{bmatrix} 0 & 1 \\ 0 & 0 \end{bmatrix} \right) = \begin{bmatrix} 0 & 1 \\ 0 & 0 \end{bmatrix},$$

$$T\left( \begin{bmatrix} 0 & 0 \\ 1 & 0 \end{bmatrix} \right) = \begin{bmatrix} 0 & 0 \\ 1 & 0 \end{bmatrix},$$

$$T\left( \begin{bmatrix} 0 & 0 \\ 0 & 1 \end{bmatrix} \right) = \begin{bmatrix} 0 & 0 \\ 1 & 0 \end{bmatrix}.$$

Thus the coordinates of these images relative to the specified basis are respectively,

$$\left\{ \begin{bmatrix} 0 \\ 1 \\ 0 \\ 0 \end{bmatrix}, \begin{bmatrix} 0 \\ 1 \\ 0 \\ 0 \end{bmatrix}, \begin{bmatrix} 0 \\ 0 \\ 1 \\ 0 \end{bmatrix}, \begin{bmatrix} 0 \\ 0 \\ 1 \\ 0 \end{bmatrix} \right\}.$$

Hence the matrix representing $T$ relative to the specified bases is

$$\mathbf{A} = \begin{bmatrix} 0 & 0 & 0 & 0 \\ 1 & 1 & 0 & 0 \\ 0 & 0 & 1 & 1 \\ 0 & 0 & 0 & 0 \end{bmatrix}.$$

## Section 7.3 True/False, Page 424

1. T    2. T    3. F    4. F    5. T

## Section 7.4, Page 432

For these exercises, it is important to plan the steps for the completion of the exercises. In this regard, the three cases detailed in the section will be extremely useful, as will the accompanying diagrams in Figures 2, 3, and 4. We suggest you first decide which of the three cases apply to the exercise and proceed accordingly.

1. (a) **Strategy:** Here we must find the transition matrix from coordinates in the $S'$-basis to coordinates in the $S$-basis. See Case 1.

If we let $\mathbf{u}_1 = t^2 + t$, $\mathbf{u}_2 = t + 2$, and $\mathbf{u}_3 = t - 1$, then the transition matrix we seek is

$$\mathbf{P}_{S \leftarrow S'} = \begin{bmatrix} \mathbf{coord}_S(\mathbf{u}_1) & \mathbf{coord}_S(\mathbf{u}_2) & \mathbf{coord}_S(\mathbf{u}_3) \end{bmatrix}.$$

Following the ideas in Example 1, we determine the coordinates of the $S'$-basis vectors in terms of the $S$-basis. Since $S$ is the natural basis for $P_2$, these computations are easy. We get

$$\mathbf{P}_{S \leftarrow S'} = \begin{bmatrix} 1 & 0 & 0 \\ 1 & 1 & 1 \\ 0 & 2 & -1 \end{bmatrix}.$$

It follows that

$$\mathbf{A}_{B \leftarrow S'} = \mathbf{A}_{B \leftarrow S} * \mathbf{P}_{S \leftarrow S'}$$

$$= \begin{bmatrix} 2 & 0 & 0 \\ 0 & 1 & 0 \end{bmatrix} \begin{bmatrix} 1 & 0 & 0 \\ 1 & 1 & 1 \\ 0 & 2 & -1 \end{bmatrix}$$

$$= \begin{bmatrix} 2 & 0 & 0 \\ 1 & 1 & 1 \end{bmatrix}.$$

(b) **Strategy:** Here we must find the transition matrix from coordinates in the $B$-basis to coordinates in the $B'$-basis. See Case 2.

If we let $\mathbf{u}_1 = t$ and $\mathbf{u}_2 = 1$, then the transition matrix we seek is

$$\mathbf{Q}_{B' \leftarrow B} = \begin{bmatrix} \mathbf{coord}_{B'}(\mathbf{u}_1) & \mathbf{coord}_{B'}(\mathbf{u}_2) \end{bmatrix}.$$

Following the ideas in Example 2 part (b), we determine the coordinates of the $B$-basis vectors in terms of the $B'$-basis. We must find scalars $a$ and $b$ so that $a(2t + 1) + b(t + 2) = t$ and scalars $c$ and $d$ so that $c(2t + 1) + d(t + 2) = 1$. This leads to a pair of linear systems with the same coefficient matrix but different right sides that we can represent as a partitioned augmented matrix as follows:

$$\begin{bmatrix} 2 & 1 & 1 & 0 \\ 1 & 2 & 0 & 1 \end{bmatrix}.$$

The RREF of this matrix is

$$\begin{bmatrix} 1 & 0 & \frac{2}{3} & -\frac{1}{3} \\ 0 & 1 & -\frac{1}{3} & \frac{2}{3} \end{bmatrix},$$

thus

$$\mathbf{Q}_{B' \leftarrow B} = \begin{bmatrix} \frac{2}{3} & -\frac{1}{3} \\ -\frac{1}{3} & \frac{2}{3} \end{bmatrix}.$$

It follows that

$$\mathbf{A}_{B' \leftarrow S} = \mathbf{Q}_{B' \leftarrow B} * \mathbf{A}_{B \leftarrow S}$$

$$= \begin{bmatrix} \frac{2}{3} & -\frac{1}{3} \\ -\frac{1}{3} & \frac{2}{3} \end{bmatrix} \begin{bmatrix} 2 & 0 & 0 \\ 0 & 1 & 0 \end{bmatrix}$$

$$= \begin{bmatrix} \frac{4}{3} & -\frac{1}{3} & 0 \\ -\frac{2}{3} & \frac{2}{3} & 0 \end{bmatrix}.$$

(c) **Strategy:** Here we follow the ideas for Case 3.

Using the results from parts (a) and (b) we have

$$\mathbf{A}_{B' \leftarrow S'} = \mathbf{Q}_{B' \leftarrow B} * \mathbf{A}_{B \leftarrow S} * \mathbf{P}_{S \leftarrow S'}$$

$$= \begin{bmatrix} \frac{2}{3} & -\frac{1}{3} \\ -\frac{1}{3} & \frac{2}{3} \end{bmatrix} \begin{bmatrix} 2 & 0 & 0 \\ 0 & 1 & 0 \end{bmatrix} \begin{bmatrix} 1 & 0 & 0 \\ 1 & 1 & 1 \\ 0 & 2 & -1 \end{bmatrix}$$

$$= \begin{bmatrix} 1 & -\frac{1}{3} & -\frac{1}{3} \\ 0 & \frac{2}{3} & \frac{2}{3} \end{bmatrix}.$$

**3. (a)** From Section 7.3, we have that

$$\mathbf{A}_{B \leftarrow S} = \left[ \ \mathbf{coord}_B \left( T \left( \begin{bmatrix} 1 \\ 0 \\ 0 \end{bmatrix} \right) \right) \right.$$

$$\mathbf{coord}_B \left( T \left( \begin{bmatrix} 0 \\ 1 \\ 0 \end{bmatrix} \right) \right)$$

$$\left. \mathbf{coord}_B \left( T \left( \begin{bmatrix} 0 \\ 0 \\ 1 \end{bmatrix} \right) \right) \right]$$

$$= \begin{bmatrix} 1 & 0 & 0 \\ 1 & 1 & 0 \\ 1 & 1 & 1 \\ 0 & 1 & 1 \end{bmatrix}.$$

This was easy to do since $B$ was the natural basis.

**(b) Strategy:** Here we must find the transition matrix from coordinates in the $S'$-basis to coordinates in the $S$-basis. See Case 1.

If we let $\mathbf{u}_1 = \begin{bmatrix} 1 \\ 0 \\ 1 \end{bmatrix}$, $\mathbf{u}_2 = \begin{bmatrix} 0 \\ 1 \\ 1 \end{bmatrix}$, and $\mathbf{u}_3 = \begin{bmatrix} 1 \\ 1 \\ 0 \end{bmatrix}$,

then the transition matrix we seek is

$$\mathbf{P}_{S \leftarrow S'} = \left[ \ \mathbf{coord}_S(\mathbf{u}_1) \quad \mathbf{coord}_S(\mathbf{u}_2) \quad \mathbf{coord}_S(\mathbf{u}_3) \ \right].$$

Following the ideas in Example 1, we determine the coordinates of the $S'$-basis vectors in terms of the $S$-basis. But since $S$ is the natural basis we have

$$\mathbf{P}_{S \leftarrow S'} = \left[ \ \mathbf{u}_1 \quad \mathbf{u}_2 \quad \mathbf{u}_3 \ \right] = \begin{bmatrix} 1 & 0 & 1 \\ 0 & 1 & 1 \\ 1 & 1 & 0 \end{bmatrix}.$$

Hence it follows that

$$\mathbf{A}_{B \leftarrow S'} = \mathbf{A}_{B \leftarrow S} * \mathbf{P}_{S \leftarrow S'}$$

$$= \begin{bmatrix} 1 & 0 & 0 \\ 1 & 1 & 0 \\ 1 & 1 & 1 \\ 0 & 1 & 1 \end{bmatrix} \begin{bmatrix} 1 & 0 & 1 \\ 0 & 1 & 1 \\ 1 & 1 & 0 \end{bmatrix}$$

$$= \begin{bmatrix} 1 & 0 & 1 \\ 1 & 1 & 2 \\ 2 & 2 & 2 \\ 1 & 2 & 1 \end{bmatrix}.$$

**(c) Strategy:** Here we must find the transition matrix from coordinates in the $B$-basis to coordinates in the $B'$-basis. See Case 2.

If we let $\mathbf{u}_1 = \begin{bmatrix} 1 \\ 0 \\ 0 \\ 0 \end{bmatrix}$, $\mathbf{u}_2 = \begin{bmatrix} 0 \\ 1 \\ 0 \\ 0 \end{bmatrix}$, $\mathbf{u}_3 = \begin{bmatrix} 0 \\ 0 \\ 1 \\ 0 \end{bmatrix}$, and

$\mathbf{u}_4 = \begin{bmatrix} 0 \\ 0 \\ 0 \\ 1 \end{bmatrix}$, then the transition matrix we seek is

$$\mathbf{Q}_{B' \leftarrow B} = [\mathbf{coord}_{B'}(\mathbf{u}_1) \quad \mathbf{coord}_{B'}(\mathbf{u}_2)$$
$$\mathbf{coord}_{B'}(\mathbf{u}_3) \quad \mathbf{coord}_{B'}(\mathbf{u}_4)].$$

Following the ideas in Example 2 part (b), we determine the coordinates of the $B$-basis vectors in terms of the $B'$-basis. It follows that we must solve four linear systems with the same coefficient matrix, but different right sides. We can represent this situation by the partitioned augmented matrix

$$\left[ \begin{array}{cccc|cccc} 1 & 2 & 0 & -1 & 1 & 0 & 0 & 0 \\ 1 & 0 & 1 & 0 & 0 & 1 & 0 & 0 \\ 0 & 1 & 1 & 0 & 0 & 0 & 1 & 0 \\ 1 & 1 & 0 & 2 & 0 & 0 & 0 & 1 \end{array} \right]$$

whose RREF is

$$\left[ \begin{array}{cccc|cccc} 1 & 0 & 0 & 0 & \frac{1}{4} & \frac{5}{8} & -\frac{5}{8} & \frac{1}{8} \\ 0 & 1 & 0 & 0 & \frac{1}{4} & -\frac{3}{8} & \frac{3}{8} & \frac{1}{8} \\ 0 & 0 & 1 & 0 & -\frac{1}{4} & \frac{3}{8} & \frac{5}{8} & -\frac{1}{8} \\ 0 & 0 & 0 & 1 & -\frac{1}{4} & -\frac{1}{8} & \frac{1}{8} & \frac{3}{8} \end{array} \right].$$

It follows that

$$\mathbf{Q}_{B' \leftarrow B} = \begin{bmatrix} \frac{1}{4} & \frac{5}{8} & -\frac{5}{8} & \frac{1}{8} \\ \frac{1}{4} & -\frac{3}{8} & \frac{3}{8} & \frac{1}{8} \\ -\frac{1}{4} & \frac{3}{8} & \frac{5}{8} & -\frac{1}{8} \\ -\frac{1}{4} & -\frac{1}{8} & \frac{1}{8} & \frac{3}{8} \end{bmatrix}$$

and that

$$\mathbf{A}_{B' \leftarrow S} = \mathbf{Q}_{B' \leftarrow B} * \mathbf{A}_{B \leftarrow S}$$

$$= \begin{bmatrix} \frac{1}{4} & \frac{5}{8} & -\frac{5}{8} & \frac{1}{8} \\ \frac{1}{4} & -\frac{3}{8} & \frac{3}{8} & \frac{1}{8} \\ -\frac{1}{4} & \frac{3}{8} & \frac{5}{8} & -\frac{1}{8} \\ -\frac{1}{4} & -\frac{1}{8} & \frac{1}{8} & \frac{3}{8} \end{bmatrix} \begin{bmatrix} 1 & 0 & 0 \\ 1 & 1 & 0 \\ 1 & 1 & 1 \\ 0 & 1 & 1 \end{bmatrix}$$

$$= \begin{bmatrix} \frac{1}{4} & \frac{1}{8} & -\frac{1}{2} \\ \frac{1}{4} & \frac{1}{8} & \frac{1}{2} \\ \frac{3}{4} & \frac{7}{8} & \frac{1}{2} \\ -\frac{1}{4} & \frac{3}{8} & \frac{1}{2} \end{bmatrix}.$$

**(d)** Here we follow the ideas for Case 3.

Using the results from parts (a), (b), and (c) we have

$$\mathbf{A}_{B' \leftarrow S'} = \mathbf{Q}_{B' \leftarrow B} * \mathbf{A}_{B \leftarrow S} * \mathbf{P}_{S \leftarrow S'}$$

$$= \begin{bmatrix} \frac{1}{4} & \frac{5}{8} & -\frac{5}{8} & \frac{1}{8} \\ \frac{1}{4} & -\frac{3}{8} & \frac{3}{8} & \frac{1}{8} \\ -\frac{1}{4} & \frac{3}{8} & \frac{5}{8} & -\frac{1}{8} \\ -\frac{1}{4} & -\frac{1}{8} & \frac{1}{8} & \frac{3}{8} \end{bmatrix}$$

$$\begin{bmatrix} 1 & 0 & 0 \\ 1 & 1 & 0 \\ 1 & 1 & 1 \\ 0 & 1 & 1 \end{bmatrix}$$

$$\begin{bmatrix} 1 & 0 & 1 \\ 0 & 1 & 1 \\ 1 & 1 & 0 \end{bmatrix}$$

$$= \begin{bmatrix} -\frac{1}{4} & -\frac{3}{8} & \frac{3}{8} \\ \frac{3}{4} & \frac{5}{8} & \frac{3}{8} \\ \frac{5}{4} & \frac{11}{8} & \frac{13}{8} \\ \frac{1}{4} & \frac{7}{8} & \frac{1}{8} \end{bmatrix}.$$

**5. (a)** $\mathbf{A}_{B \leftarrow S} = \begin{bmatrix} 3 & 1 \\ 0 & 2 \end{bmatrix}$.

**(b)** $\mathbf{Q}_{B' \leftarrow B} = \begin{bmatrix} 0 & -1 \\ 1 & 1 \end{bmatrix}$. $\mathbf{A}_{B' \leftarrow S'} = \begin{bmatrix} 2 & 0 \\ 0 & 3 \end{bmatrix}$.

**(c)** The matrices are similar.

## Section 7.4 True/False, Page 433

**1.** T    **2.** F    **3.** T    **4.** T    **5.** F

## Section 7.5, Page 444

**1.** $\mathbf{tran}_b(\mathbf{0}) = \mathbf{0} + \mathbf{b} = \mathbf{b} \neq \mathbf{0}$, hence
$\mathbf{tran}_b(k\mathbf{0}) = k\mathbf{0} + \mathbf{b} = \mathbf{b} \neq k\,\mathbf{tran}_b(\mathbf{0}) = k(\mathbf{0}+\mathbf{b}) = k\mathbf{b}$.
Thus $\mathbf{tran}_b$ is nonlinear.

**3.** $\mathbf{S} = \begin{bmatrix} 0 & 0 & 1 & 2 & 3 & 3 & 0 \\ 0 & 1 & 1 & 3 & 1 & 0 & 0 \end{bmatrix}$. Recall from Example
1 that to compute $T(\mathbf{S})$ we compute $\mathbf{AS}$ then add vector
$\mathbf{b}$ to each column of the result of $\mathbf{AS}$.

**(a)** $T(\mathbf{S}) = \mathbf{AS} + \mathbf{b}$

$$= \begin{bmatrix} 2 & -2 \\ 1 & 2 \end{bmatrix} \begin{bmatrix} 0 & 0 & 1 & 2 & 3 & 3 & 0 \\ 0 & 1 & 1 & 3 & 1 & 0 & 0 \end{bmatrix}$$

$$+ \begin{bmatrix} -2 & -2 & -2 & -2 & -2 & -2 & -2 \\ 1 & 1 & 1 & 1 & 1 & 1 & 1 \end{bmatrix}$$

$$= \begin{bmatrix} -2 & -4 & -2 & -4 & 2 & 4 & -2 \\ 1 & 2 & 4 & 8 & 8 & 7 & 1 \end{bmatrix}.$$

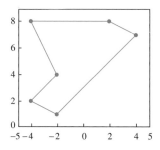

**5.** We have the following information:

$$T\left(\begin{bmatrix} 2 \\ 0 \end{bmatrix}\right) = \begin{bmatrix} p & r \\ s & t \end{bmatrix} \begin{bmatrix} 2 \\ 0 \end{bmatrix} + \begin{bmatrix} b_1 \\ b_2 \end{bmatrix}$$
$$= \begin{bmatrix} 2p + b_1 \\ 2s + b_2 \end{bmatrix} = \begin{bmatrix} 3 \\ -4 \end{bmatrix}.$$

$$T\left(\begin{bmatrix} 2 \\ 2 \end{bmatrix}\right) = \begin{bmatrix} p & r \\ s & t \end{bmatrix} \begin{bmatrix} 2 \\ 2 \end{bmatrix} + \begin{bmatrix} b_1 \\ b_2 \end{bmatrix}$$
$$= \begin{bmatrix} 2p + 2r + b_1 \\ 2s + 2t + b_2 \end{bmatrix} = \begin{bmatrix} 3 \\ -2 \end{bmatrix}.$$

$$T\left(\begin{bmatrix} 3 \\ 3 \end{bmatrix}\right) = \begin{bmatrix} p & r \\ s & t \end{bmatrix} \begin{bmatrix} 3 \\ 3 \end{bmatrix} + \begin{bmatrix} b_1 \\ b_2 \end{bmatrix}$$
$$= \begin{bmatrix} 3p + 3r + b_1 \\ 3s + 3t + b_2 \end{bmatrix} = \begin{bmatrix} 5 \\ -3 \end{bmatrix}.$$

Equating corresponding entries in the relations above
leads to two systems of equations with three equations in
each system. We get systems

$$2p + 0r + b_1 = 3$$
$$2p + 2r + b_1 = 3$$
$$3p + 3r + b_1 = 5$$

and

$$2s + 0t + b_2 = -4$$
$$2s + 2t + b_2 = -2$$
$$3s + 3t + b_2 = -3.$$

We form the corresponding augmented matrices and
compute the RREF to solve the systems. (Note that the
coefficient matrices are identical.) We obtain $p = 2$,
$r = 0$, $b_1 = -1$ and $s = -2$, $t = 1$, $b_2 = 0$. Thus

$$\mathbf{A} = \begin{bmatrix} 2 & 0 \\ -2 & 1 \end{bmatrix} \quad \text{and} \quad \mathbf{b} = \begin{bmatrix} -1 \\ 0 \end{bmatrix}.$$

**11.** We compute the solution(s) to $(\mathbf{A} - \mathbf{I}_2)\mathbf{v} = -\mathbf{b}$ in each
case.

**(a)** rref $\left(\begin{bmatrix} -\frac{5}{6} & \frac{\sqrt{3}}{6} & -\frac{1}{2} \\ -\frac{\sqrt{3}}{6} & -\frac{5}{6} & -\frac{\sqrt{3}}{6} \end{bmatrix}\right)$

$$= \begin{bmatrix} 1 & 0 & 0.6429 \\ 0 & 1 & 0.1237 \end{bmatrix}$$

hence $\mathbf{v} \approx \begin{bmatrix} 0.6429 \\ 0.1237 \end{bmatrix}$.

## Section 7.5 True/False, Page 447

**1.** F      **2.** F      **3.** T      **4.** T      **5.** T

## Chapter 7 Chapter Test, Page 448

**1.** Nonlinear.

**2.** $r$ = any real number; $s = \begin{bmatrix} 0 \\ 0 \end{bmatrix}$.

**3.** span $\left\{ \begin{bmatrix} \frac{5}{2} \\ 2 \\ 1 \end{bmatrix} \right\}$

**4.** (a) $T(t^2) = \frac{1}{3}$ and $T(\frac{1}{3}) = \frac{1}{3}$.

    (b) Let $r$ be any real number, then $T(at^2 + bt + c) = r$
provided

$$\frac{a}{3} + \frac{b}{2} + c = r.$$

This is one equation in 3 unknowns so it has infinitely many solutions.

**5.** (a) $\left\{ -\frac{1}{2}t^2 + t + 1 \right\}$.

    (b) A basis for $P_2$ containing $-\frac{1}{2}t^2 + t + 1$ is

$$\left\{ -\frac{1}{2}t^2 + t + 1, t^2, t \right\}.$$

Hence $T(t^2)$ and $T(t)$ is a basis for $\mathbf{R}(T)$; that is, $\{2t, 1\}$ is a basis for $\mathbf{R}(T)$.

**6.** (a) $\begin{bmatrix} 0 & 1 & 1 & 0 \\ 1 & 0 & 0 & 0 \\ 1 & 1 & 0 & 0 \end{bmatrix}$.

    (b) $\begin{bmatrix} -1 & 0 & -1 & 0 \\ 2 & 0 & 2 & 0 \\ 1 & 0 & 2 & 1 \end{bmatrix}$.

**7.** (a) $\begin{bmatrix} 2 \\ -1 \\ 0 \end{bmatrix}$.

    (b) $\begin{bmatrix} 2 & 2 & 1 \\ 1 & -1 & 2 \\ 1 & 1 & 1 \end{bmatrix}$.

**8.** $\mathbf{A}_{B \leftarrow S} \, \text{coord}_S(\mathbf{v})$.

**9.** $\mathbf{A}_{B \leftarrow S} \mathbf{A}_{S \leftarrow S'} \, \text{coord}_S(\mathbf{v})$.

**10.** $\mathbf{A}_{B \leftarrow B'} \mathbf{A}_{B \leftarrow S} \mathbf{A}_{S \leftarrow S'} \, \text{coord}_S(\mathbf{v})$.

## Appendix 1, Page 462

**1.** $c_1 + c_2 = 4 + 2i$.

**3.** $c_1 c_2 = 11 - 2i$.

**5.** $4c_3 + \overline{c_2} = -3 + 6i$.

**7.** $\overline{3c_1 - ic_2} = 7 - 11i$.

**9.** $\dfrac{1 + 2i}{3 - 4i} = -0.2 + 0.4i$.

**11.** $\dfrac{(2 + i)^2}{i} = 4 - 3i$.

**13.** $(4, 2)$.

**15.** $(3, -2)$.

**17.** For Exercise 13: $\sqrt{20}$.
For Exercise 14: $\sqrt{10}$.
For Exercise 15: $\sqrt{13}$.
For Exercise 16: $\sqrt{17}$.

## Appendix 2, Page 465

**1.** $\displaystyle\sum_{k=1}^{4} c_k = 1$.

**3.** $\displaystyle\sum_{j=2}^{4} b_{j-1} = 1$.

**5.** $\displaystyle\sum_{i=1}^{5} c_i b_i = -17$.

**7.** $\displaystyle\sum_{j=1}^{m}\sum_{i=1}^{n} 1 - \sum_{j=1}^{m}\left(\sum_{i=1}^{n} 1\right) = \sum_{j=1}^{m} n = n \sum_{j=1}^{m} 1 = mn$.

# I N D E X

MATLAB commands are indicated in **boldface**.

507